Lecture Notes on Data Engineering and Communications Technologies

Volume 98

Series Editor

Fatos Xhafa, Technical University of Catalonia, Barcelona, Spain

The aim of the book series is to present cutting edge engineering approaches to data technologies and communications. It will publish latest advances on the engineering task of building and deploying distributed, scalable and reliable data infrastructures and communication systems.

The series will have a prominent applied focus on data technologies and communications with aim to promote the bridging from fundamental research on data science and networking to data engineering and communications that lead to industry products, business knowledge and standardisation.

Indexed by SCOPUS, INSPEC, EI Compendex.

All books published in the series are submitted for consideration in Web of Science.

More information about this series at http://www.springer.com/series/15362

John Macintyre · Jinghua Zhao ·
Xiaomeng Ma
Editors

The 2021 International Conference on Machine Learning and Big Data Analytics for IoT Security and Privacy

SPIoT-2021 Volume 2

 Springer

Editors
John Macintyre
University of Sunderland
Sunderland, UK

Jinghua Zhao
University of Shanghai for Science
and Technology
Shanghai, China

Xiaomeng Ma
Shenzhen University
Shenzen, China

ISSN 2367-4512   ISSN 2367-4520   (electronic)
Lecture Notes on Data Engineering and Communications Technologies
ISBN 978-3-030-89510-5   ISBN 978-3-030-89511-2   (eBook)
https://doi.org/10.1007/978-3-030-89511-2

Foreword

SPIOT 2021 is the 2nd international conference dedicated to promoting novel theoretical and applied research advances in the interdisciplinary agenda of Internet of things. The "Internet of things" heralds the connections of a nearly countless number of devices to the Internet, thus promising accessibility, boundless scalability, amplified productivity and a surplus of additional paybacks. The hype surrounding the IoT and its applications is already forcing companies to quickly upgrade their current processes, tools, and technology to accommodate massive data volumes and take advantage of insights. Since there is a vast amount of data generated by the IoT, a well-analyzed data is extremely valuable. However, the large-scale deployment of IoT will bring new challenges and IoT security is one of them.

The philosophy behind machine learning is to automate the creation of analytical models in order to enable algorithms to learn continuously with the help of available data. Continuously evolving models produce increasingly positive results, reducing the need for human interaction. These evolved models can be used to automatically produce reliable and repeatable decisions. Today's machine learning algorithms comb through data sets that no human could feasibly get through in a year or even a lifetime's worth of work. As the IoT continues to grow, more algorithms will be needed to keep up with the rising sums of data that accompany this growth.

One of the main challenges of the IoT security is the integration with communication, computing, control, and physical environment parameters to analyze, detect and defend cyber-attacks in the distributed IoT systems. The IoT security includes: (i) the information security of the cyber-space, and (ii) the device and environmental security of the physical space. These challenges call for novel approaches to consider the parameters and elements from both spaces and get enough knowledge for ensuring the IoT's security. As the data has been collecting in the IoT, and the data analytics has been becoming mature, it is possible to conquer this challenge with novel machine learning or deep learning methods to analyze the data which synthesize the information from both spaces.

We would like to express our thanks to Professor John Macintyre, University of Sunderland, Professor Junchi Yan, Shanghai Jiaotong University, for being the keynote speakers at the conference. We thank the general chairs, program committee chairs, organizing chairs, and workshop chairs for their hard work. The local organizers' and the students' help are also highly appreciated.

Our special thanks are also to editors Dr. Thomas Ditzinger and Prof. Xhafa, Fatos, for publishing the proceedings in Lecture Notes on Data Engineering and Communications Technologies.

Organization

General Chairs

Bo Fei (President)
Shanghai University of Medicine and Health Sciences, China

Program Committee Chairs

John Macintyre (Pro Vice Chancellor)
University of Sunderland, UK

Jinghua Zhao
University of Shanghai for Science and Technology, China

Xiaomeng Ma
Shenzhen University, China

Publicity Chairs

Shunxiang Zhang
Anhui University Science and Technology, China

Dandan Jiang
DiDi Research Center, DiDi Global Inc., China

Xianchao Wang
Fuyang Normal University, China

Publication Chairs

Jun Ye
Hainan University, China

Ranran Liu
The University of Manchester, UK

Qingyuan Zhou
Changzhou Institute of Mechatronic Technology, China

Local Organizing Chairs

Xiao Wei
Shanghai University, China

Shaorong Sun
University of Shanghai for Science and Technology, China

Program Committee Members

Paramjit Sehdev	Coppin State University, USA
Khusboo Pachauri	Dayanand Sagar University, Hyderabad, India
Khusboo Jain	Oriental University of Engineering and Technology, Indore, India
Akshi Kumar	Delhi Technological University, New Delhi, India
Sumit Kumar	Indian Institute of Technology (IIT), India
Anand Jee	Indian Institute of Technology (IIT), New Delhi, India
Arum Kumar Nachiappan	Sastra Deemed University, Chennai, India
Afshar Alam	Jamia Hamdard University, New Delhi, India
Adil Khan	Institute of Technology and Management, Gwalior, India
Amrita Srivastava	Amity University, Gwalior, India
Abhisekh Awasthi	Tshingua University, Beijing, China
Dhiraj Sangwan	CSIR-CEERI, Rajasthan, India
Jitendra Kumar Chaabra	National Institute of Technology, Kurkshetra, India
Muhammad Zain	University of Louisville, USA
Amrit Mukherjee	Jiangsu University, China
Nidhi Gupta	Institute of Automation, Chinese Academy of Sciences, Beijing, China
Neil Yen	University of Aizu, Japan
Guangli Zhu	Anhui Univ. of Sci. and Tech., China
Xiaobo Yin	Anhui Univ. of Sci. and Tech., China
Xiao Wei	Shanghai Univ., China
Huan Du	Shanghai Univ., China
Zhiguo Yan	Fudan University, China
Jianhui Li	Computer Network Information Center, Chinese Academy of Sciences, China
Yi Liu	Tsinghua University, China
Kuien Liu	Pivotal Inc, USA
Feng Lu	Institute of Geographic Science and Natural Resources Research, Chinese Academy of Sciences, China
Wei Xu	Renmin University of China, China
Ming Hu	Shanghai University, China

Contents

Data-Driven Co-design
of Communication, Computing
and Control for IoT Security

Analysis of Sentiment Tendency of Tourists' Comments Based on Text Mining

Shaona Feng[1], Cui Wen[1], and Lingjian Kong[2(✉)]

[1] Hainan College of Software Technology, Qionghai, Hainan, China
[2] Changzhou Vocational Institute of Mechatronic Technology, Changzhou, Jiangsu, China
zqy2131@czimt.edu.cn

Abstract. Based on text mining technology, it analyzes the sentiment tendency of the tourist comment data. Three major tourism websites are selected as the sample data source platforms, and through the construction of emotional dictionaries suitable for the study of this article and the design of sentiment analysis algorithms, the sentiment color and sentiment tendency of each tourist review data are finally obtained, and the tourists' evaluation of tourist destinations can be understood. According to the results of text analysis and the actual situation, optimization suggestions are put forward at the level of the relevant tourism website and the management department of the tourism destination.

Keywords: Text mining · Tourist comments · Sentiment analysis

1 Preface

In the Internet age, tourists often share their travel feelings on the Internet, and their comments are highly subjective and reliable. Many potential tourists can learn about the general situation of the tourism destination by viewing the data of tourists' comments, and choose whether to travel or not. There are too many new tourist comments on the Internet every day, so it is difficult to view and analyze such data in a timely and efficient way only by human.

Throughout the domestic and foreign research on the application of text mining technology to visitor web comment data, it can be found that text mining has accumulated fruitful results in both theoretical and technical methods and applied research in various fields. But so far, most of the research on tourist online reviews is limited to a single travel website, and each platform has its own user group. If only a single travel website is researched, the results are often limited.

Therefore, this study selects three typical domestic tourism websites as the sample data source platform, and obtains the evaluation content of tourists at any time period on the tourist destination based on text mining technology, analyzes the sentiment tendency of the comment content, and calculates the sentiment value of the tourist comments. It is convenient for tourism-related administrative departments to monitor the online public opinion at any time and provide appropriate guidance, promptly improve the destination tourism image, software and hardware management, and improve the management level and make scientific decisions in a targeted manner.

J. Macintyre et al. (Eds.): SPIoT 2021, LNDECT 98, pp. 3–9, 2022.
https://doi.org/10.1007/978-3-030-89511-2_1

2 Research Objects and Research Methods

2.1 Research Status at Home and Abroad

As an emerging field of tourism research, tourist network comment data is sought after by many experts and scholars. For example, Stella Kladou et al. evaluated the cognitive and emotional components of the tourist destination image from the perspective of tourists through the comment data of tourists on TripAdvisor [1]. M.A. Rodríguez-Molina and others analyzed the effect of the design of tourist destination websites. When visitors use the website and do not feel overloaded, they show positive emotions towards the image of the tourist destination [2]. Költringer and Dickinger collected 5,719 related documents on the Internet, conducted text mining and analysis on them, and obtained the brand image content of a certain destination [3]. Xu Yayuan and others used the content mining software ROSTCM6.0 to analyze the online comments of tourists, and made a more accurate image positioning for the scenic spot [4]. Zhang Hongying took the Lingshan Giant Buddha scenic spot in Wuxi as an example, and used content mining software to analyze the tourist review text, and found that tourists are relatively satisfied with the overall image perception of the Lingshan Giant Buddha [5]. Feng Xiaobing took Mount Emei as a tourist destination, and studied the tourism image of Mount Emei from the aspects of space, landscape, services and facilities, and tourist experience through text mining methods [6]. Xu Jinya and others used Ctrip.com's online reviews of Lijiang scenic spots as textual materials to evaluate the factors affecting the satisfaction of tourists in Lijiang scenic spots [7]. Zhang Yan and others took Jiuzhaigou as a tourist destination, collected data from major online forums, and summarized and extracted 8 factors that affect the satisfaction of tourists on self-driving tours: transportation, planning, economy, nature, personal, region, service and Safety, it is concluded that safety, nature and region have a greater impact on self-driving tour satisfaction [8]. Based on the analysis of more than 3000 travel notes of representative tourism websites in recent three years, Gao Xiang and others put forward a conceptual model of tourist satisfaction, which is composed of eight elements: Travel expectation, travel behavior, experience of expectation elements, experience of other elements, emotion regulation, tourist satisfaction, tourist loyalty and tourist complaint [9]. Lang Zhijun takes haxian island in Dalian City as an example, using the research method of combining descriptive statistical analysis and network text analysis to analyze the island tourists' satisfaction [10].

Through literature research, it is found that the research on tourist online comment data has also fallen into an upsurge in recent years. But so far, most of the research on online reviews of tourists is limited to a single travel website, but there are many travel platforms, and each platform has its own user group. If only a single travel website is studied, the results are often limited. Therefore, this study selects three typical domestic tourism websites as the sample data source platform to carry out a comparative study of tourist review data of typical tourism websites. Through text content mining, it analyzes sentiment tendency and researches the comment data of tourists visiting Hainan, aiming to provide decision-making basis for tourism website and other tourism service providers and tourism-related management departments.

2.2 Research Object

Before comparing tourist review data of typical travel websites, it is necessary to clarify the research object and the selected data source platform. The choice of the research object is Hainan, and the data source platform selects three typical travel websites: Ctrip, Tuniu and Lvmama. This article crawled and captured some tourist comment data about Hainan tourism by Ctrip, Tuniu, and Lvmama through the octopus collector. In order to make the structure of the data collected by the three platforms consistent, three content including user name, comment content, and comment time were captured and collected from the three platforms. After data cleaning and standardization processing, the comment data composition table of the three platforms obtained is shown in Table 1.

Table 1. The comment data composition table of the three platforms after preprocessing

Platform	Ctrip	Tuniu	Lvmama
Quantity (items)	1659	1548	1483

2.3 Sentiment Analysis

1) Dictionary-based sentiment analysis algorithm. Dictionary-based sentiment analysis is mainly to match the word segmentation results of the text with the entries in the pre-built sentiment dictionary, obtain the corresponding weights, and perform simple calculations to finally get the text sentiment score. However, this method has certain flaws, that is, if there are negative words and modifying adverbs in the sentence, the negative words and modifying adverbs have different sentiment values in different positions. Therefore, when performing sentiment analysis, it is also necessary to pay attention to situations similar to this that lead to unreasonable results.

2) The preparation of emotional words dictionary. To analyze the sentiment tendency of the tourist review text based on the sentiment dictionary, it is necessary to prepare the related sentiment dictionary, including the basic sentiment dictionary, the negative word dictionary, the degree adverb dictionary, and the stop word dictionary.

(1) Basic sentiment dictionary. Choose the open source BosonNLP sentiment dictionary, each line has an emotional word and its corresponding emotional score, separated by spaces, including 114767 words in total.

(2) Negative word dictionary. The existence of negative words will reverse the sentiment of the sentence. Negative words are assigned the value "−1".

(3) Degree adverb dictionary. The introduction of adverbs of degree can express the strength of emotion. For example, "beautiful hotel" and "super beautiful hotel", or "poor service attitude" and "very bad service attitude", the emotions expressed are different, and the tone of the latter is obviously with a strong emotional tendency. This article assigns different weights to adverbs of different degrees. The weight assignment of Chinese degree adverbs is shown in Table 2 below.

(4) Stop word dictionary. This paper adopts the Chinese stop word list of 1208 stop words published by the Chinese Natural Language Processing Open Platform of the Institute of Computing Technology of the Chinese Academy of Sciences.

3) Sentiment analysis algorithm design. The first step is to read the tourist comment data.

Table 2. Chinese degree adverbs graded weight assignment table

Degree level	Degree words	Assignment
"Extreme/most"	One hundred percent, extreme, absolute…	6
"Over"	Super, over, overpowering…	5
"Very"	Very, quite, special, especially, a lot…	4
"More"	More, so, big deal…	3
"A littlel-ish"	Slightly, a little, not exempt…	2
"Insufficiently"	Not a little bit, mild, not so…	1

The second step is to preprocess the tourist comment data, including word segmentation and stop words.

The third step is to load the sentiment dictionary and determine the position of the word segmentation results in the sentiment dictionary, negative dictionary, and degree adverb dictionary.

The fourth step is to traverse the result of word segmentation, find the emotional word, and record the position of the current emotional word.

The fifth step is to obtain the position of the next emotional word, and determine whether there is a negative word or degree adverb between the current emotional word and the next emotional word.

The sixth step, if there is a negative word, update the weight and take the opposite.

The seventh step, if there is an adverb of degree, update the weight, and multiply the score by the degree score of the degree adverb.

The eighth step is to calculate the sentiment value of a comment.

The ninth step, iteratively, calculate the sentiment value of all comments.

According to the above algorithm design, use python language for programming.

3   Results of Sentiment Analysis

Combining the above algorithm design, using python to analyze the sentiment tendency of the tourist comment text data, and get the sentiment value of each tourist comment data. From the magnitude of the emotional value and the positive and negative directions, it can be judged whether the tourist's emotional tendency is positive, neutral, or negative, and the tourist experience of the tourist destination can be obtained from this. Through the calculation of sentiment value and the sorting of sentiment tendency on the tourist comment text data, combined with the original data of the tourist comment text, the sentiment distribution table of the tourist comment text is obtained as shown in Table 3.

Table 3. Emotional distribution of tourist review texts on typical tourism websites

Overall			Classified websites			
Emotion type	Quantity (items)	Proportion (%)	Site name	Emotion type	Quantity (items)	Proportion (%)
Positive	4113	87.70%	Ctrip	Positive	1536	92.59%
				Neutral	103	6.21%
				Negative	20	1.20%
Neutral	316	6.74%	Tuniu	Positive	1297	83.79%
				Neutral	133	8.59%
				Negative	118	7.62%
Negative	261	5.56%	Lvmama	Positive	1280	86.31%
				Neutral	80	5.40%
				Negative	123	8.29%

As can be seen from the above table, on the whole, there are a total of 4113 tourist comment data reflecting the positive emotional tendency of tourists, accounting for more than 80%. The proportion of negative affective tendencies is relatively small, 5.56%. That is, the emotional attitudes of tourists who travel to Hainan are mainly positive, indicating that Hainan as an international tourist island is still very attractive to tourists, and the overall tourist experience of tourists is relatively good.

Judging from the visitor sentiment type data presented by various websites, Ctrip.com has 1536 comments reflecting the positive sentiments of tourists, accounting for 92.59%. It is the website with the largest proportion of positive sentiments among the three typical travel websites. Secondly, there are 1280 comments on Lvmama.com, which reflect the positive feelings of tourists, accounting for 86.31%.Tuniu.com has 1,297 visitor reviews that reflect positive sentiment, accounting for 83.79%. It is the website with the least positive sentiment among the three websites. Through the data in the table, we know that positive emotions have obvious advantages in the three websites. But negative emotions also exist. Ctrip.com shows the least negative emotions, accounting for only 1.20%. Tuniu.com and Lvmama.com respectively accounted for 7.62% and 8.29% of their negative emotions.

4 Tourism Service Improvement Strategy Based on Tourist Comments

4.1 Enlightenment on the Level of Related Travel Websites

Travel websites should combine their own positioning to improve travel services horizontally and vertically. For example, from the sentimental analysis of the tourist review data, it is known that Tuniu.com tourists are more praising for group meals and play items. Then Tuniu.com can connect with travel suppliers to optimize and improve the board and lodging and play items for group tours. For another example, Lvmama.com focuses on free travel. When tourists enter a scenic spot without a guide to explain, the website can provide pictures, audio and video information of the scenic spot. It even locates the specific play locations of tourists and provides voice explanations, so that tourists can get a high-quality travel experience. In addition to the vertical and refined extension of tourism products, specialized tourism websites such as Tuniu and Lvmama should also broaden their business content horizontally. For example, it can cooperate with local tourism suppliers in Hainan to carry out tourism promotion activities, and cooperate with car rental companies, photography companies, major scenic spots and various hotels.

4.2 Enlightenment on the Level of Tourism Destination Management Department

From the semantic mining of tourist emotional tendency analysis, it is found that tour guides, hotels and scenic spots are the key content that affects the positive or negative direction of tourists' emotional tendency.

In terms of tour guides, first of all, the government should formulate various policies to protect tour guides. Secondly, tourism bureaus of tourist destinations can establish professional tour guide training institutions. Third, government departments should take the lead in establishing a tour guide information database to facilitate supervision of tour guide services.

In terms of hotels, firstly, while building high-end hotels, it is necessary to reasonably plan the proportion of low-end hotels. The second is that the hotel must integrate the functions of accommodation, food, appreciation, and play to become an independent tourist destination.

In terms of scenic spots, one is that tourist spots can predict and monitor the passenger flow, and timely divert and alleviate tourists in the scenic spot. Second, by recording tourists' personal information, such as fingerprints, facial features, etc., help scenic spots to timely monitor tourists' bad behaviors in scenic spots and quickly stop them. The third is to deploy virtual mixed reality technology in the scenic area to allow tourists to experience the new era of tourism in depth. The fourth is to record tourists' behavior information in scenic spots through big data, so as to accurately grasp the situation and needs of tourists in the future.

Acknowledgments. This work was financially supported by Hainan Provincial Natural Science Foundation of China (Project approval number: 719MS071).

References

1. Kladou, S., Mavragani, E.: Assessing destination image: An online marketing approach and the case of TripAdvisor. J. Destination Mark. Manage. **4**(3), 187–193 (2015)
2. Rodríguez-Molina, M.A., Frías-Jamilena, D.M., Castañeda, J.A.: The contribution of website design to the generation of tourist destination image: the moderating effect of involvement. Tour. Manage. **47**, 303–317 (2015)

3. Költringer, C., Dickinger, A.: Analyzing destination branding and image from online sources: a web content mining approach. J. Bus. Res. **68**(9), 1836–1843 (2015)
4. Xu, Y., Yao, G.: Research on Huangshan Scenic area tourism image perception based on online reviews. World Geogr. Res. **25**(2), 158–168 (2016)
5. Zhang, H.: Research on the image perception of tourist attractions based on online reviews—taking the Lingshan Giant Buddha Scenic area in Wuxi as an example. J. Huaibei Normal Univ. (Philos. Soc. Sci. Ed.) **6**, 82–85 (2017)
6. Feng, X.: Research on tourist image perception based on internet comments—taking Mount Emei as an example. West. Econ. Manage. Forum **28**(1), 70–76 (2017)
7. Xu, J., Duan, W.: Analysis of tourist satisfaction in Lijiang Scenic area based on internet public opinion. J. Central South Univ. For. Technol. (Soc. Sci. Ed.) **11**(5), 89–95 (2017)
8. Zhang, Y., Guo, X., Zhang, Y., Zhang, T.: Research on the influencing factors of self-driving tourist satisfaction based on grounded theory—a case study of Jiuzhaigou Tourism in Sichuan province. Bus. Econ. **12**, 63–65 (2017)
9. Gao, X., Chen, W.: A new interpretation of tourist satisfaction model in tourist destinations-based on the rooted analysis of Guilin internet travel notes. Tour. Res. **9**(5), 78–86 (2017)
10. Lang, Z.: Research on the satisfaction of island tourists based on Internet travel notes–taking Dalian Haxian Island as an example. Prod. Res. **9**, 112–115 (2017)

Analysis of Smart City Construction Based on 5G Data Technology

Jibing Meng(⊠)

Dalian Art College, Liaoning 116600, China

Abstract. The advent of the 5G era has a very important influence and effect on the development of cities in our country, and it can also play a certain role in promoting the development of urban sculpture in our country in the future. In a modern society, 5G technology can continue to develop. Full use of this technology in the process of urban sculpture business can accurately and comprehensively analyze the current development of urban sculpture business, meet the needs of stable development of the city, provide innovations in new sculpture business, and promote the sustainable development of cities in our country. Promote the continuous progress of the city towards the direction of intelligence. This paper takes the development status and innovation transformation of urban sculpture under the background of 5G as the theoretical basis of the research, and integrates its important content to analyze and research the improvement of urban sculpture. This article takes the development of sculpture as the research object, and optimizes and improves the decision-making method of urban sculpture under the background of 5G. Through the combination of the new data analysis technology in the 5G era and the urban sculpture career, based on this, this article outlines the urban sculpture career and proposes the application of the urban sculpture career in the construction of smart cities in the 5G era. The application of 5G technology in the urban sculpture business can create many opportunities for the development of cities and sculptures by improving the development and innovation of urban sculpture business, provide many conveniences, and also help comprehensively enhance the innovation and transformation of urban sculptures. The experimental results show that this research has a better effect on improving the development and innovation of urban sculpture undertakings using the technology in the era of 5G background.

Keywords: 5G era · City · Sculpture career · Innovation

1 Introduction

With the continuous progress and rapid development of science and technology, more and more new technologies are widely used in various fields, especially in the era of 5G background, more advanced technologies are developed and utilized [1]. With the introduction of the concept of "smart earth", a "smart city" has emerged in our country, and it is regarded as a very important development strategy in the future [2]. The advent

J. Macintyre et al. (Eds.): SPIoT 2021, LNDECT 98, pp. 10–16, 2022.
https://doi.org/10.1007/978-3-030-89511-2_2

of the 5G era, with faster transmission speed, lower time limit, and less power consumption, has opened the era of the Internet of Everything, promoted the transformation and upgrading of all walks of life, and provided great opportunities for the diversification and enrichment of urban sculpture. Technical support has transformed the traditional development model of sculpture business and promoted the development of the city [3]. Moreover, it can meet the needs of sculpture enthusiasts in all aspects and multiple fields, and bring high-quality sculpture experience to sculpture enthusiasts. Through the calculation of big data, it has a very important influence and effect on our country's economic development. "5G technology is in our country" "Smart city" plays a very important role in the construction, and it can also play a certain role in promoting the development of urban sculpture in our country in the future [4].

Sculptures in the city reflect the image of the city, a historical imprint, and a witness to urban development. The urban sculpture design is located in the center of the city, fully demonstrating its regional characteristics, containing the sustenance of humanistic feelings, and increasing invisibly the cohesion of the city [5]. Sculpture art is the general term for the three creative methods of carving, engraving and plastic. The status of urban sculpture in the city, small is an appreciation, and large is the soul of a city. It conveys the beautiful yearning of the people in this city, reflects the ruler's ideal planning of the city, etc., reflects all the culture of the city The epitome is an out-and-out urban witness [6]. Urban sculpture can reflect the history of this city, the aesthetics of this city, and from the perspective of social status, it can lead the people's aesthetics of this city. Every sculpture work reflects the characteristics of the times, and there are not all works displayed in different historical stages [7]. Deeply studied by Xiang, the sculptures presented in the economy, culture, religion, military, and the pursuit of the people at that time are inseparable, reflecting the value choice of this city [8].

This paper believes that the future development of urban sculpture should be based on the systematic study of the 5G background era, and its essence should be used in contemporary urban construction [9]. Through research and research on the development background and development momentum of our country's urban design and urban sculpture for decades, identify the relationship between the two, explore and summarize their development rules more rationally and deeply, and look forward to the background of the 5G era [10]. And learn from the successful experience of urban sculpture to create a sculpture business that is truly suitable for the development and construction of modern cities in the era of 5G background.

2 Method

2.1 Urban Sculpture Management Design Under 5G Technology

As the premise of this research, the 5G background era plays a primary role in the 5G technology in the 5G background era, and it is more important for urban sculpture management and design requirements. As an important part of the locality, the city is an extremely important focus of local economic and social development. Its role in promoting local economic development, so it has a certain analogy, microcosm and immutability. The statistics of urban sculpture management design sample data can be expressed as follows:

$$x = (w_1, w_2, \ldots, w_k) \tag{1}$$

$$w_k = \left\{ x^{(k,1)}, x^{(k,1)}, \ldots, x^{(k,N)} \right\} \tag{2}$$

$$X^{(k,i)} = (x_1^{(k,i)}, x_2^{(k,i)}, \ldots, x_d^{(k,i)})^T \tag{3}$$

2.2 Define the Inter-class Divergence Matrix

As the basis of this research, data plays an important role in the development of urban sculpture by 5G technology. It is more important for the use and analysis of big data. In the above formula, according to the total number of samples, m represents the mean value of the k-th sample, then:

$$m = \frac{1}{N} \sum_{K=1}^{K} \sum_{x=\partial_k} x \tag{4}$$

Each sample should be separated as far as possible, while keeping the sample variance as much as possible while keeping the mean square error of the original sample as small as possible, which corresponds to the eigenvector of the first d largest eigenvalues of the matrix. m is also the overall mean. The definition of the inter-class divergence matrix is:

$$S_b = \sum_{k=1}^{k} \frac{N_k}{N} (m_k - m)(m_k - m)^T = \Phi_b \Phi_b^T \tag{5}$$

By defining the inter-class divergence matrix, when the criterion reaches the maximum, after orthogonal transformation, the product of the overall defined inter-class divergence matrix also reaches the maximum value. Using 5G technology to calculate the minimum value of the error function in the data analysis of urban sculpture samples, and iteratively adjust the weights and thresholds to improve the calculation accuracy.

3 Methods and Experimental Research Design

3.1 People-Oriented Modern Innovation

The so-called people-oriented is to consider the aesthetic level of the public in the city, and to guide the public to improve their aesthetic ability and guide the public to correct values. This requires the designer to be detailed, serious and perfect in the research work before the design. Urban sculpture presents a good environment for the public and can please the audience. The value of urban sculpture is not only to inherit the traditional culture, but also to delight the citizens. It is not just talking about it, it runs through the concept, shape and material of sculpture creation. Even the fun of sculpture landing, etc., the people-oriented of urban sculpture is not just sculpture itself. If today's sculpture practitioners want to meet people's spiritual needs and the increasing aesthetic needs of sculpture art, they must not be trapped in something. Within the rules, but to get out of these rules, to truly achieve the urban sculpture that the people like to see, let the people understand, aim at the people's life, and improve the people's artistic aesthetics. This kind of urban sculpture can be regarded as a person-oriented innovative design of sculpture.

3.2 Strengthen the Planning of Urban Sculpture Construction

Special plans for urban sculpture or public environmental art can be prepared in various places, so urban sculpture construction must be implemented in a planned way under the premise of meeting the requirements of urban planning, so as to ensure the quality of urban sculpture works. 5G has always been regarded as a leading field in the formulation and specific application of digital strategy, and systematically promote the development of urban sculpture and urban public environmental art under the guidance of planning. It is necessary to strictly check the basic data research, subject selection, spatial layout, etc., and even recreate the specific location, spatial volume, material and structure, architectural spatial relationship, site environment coordination, and title of the work. With noble aesthetic orientation, good art works, exquisite project planning, exquisite production technology, and innovative expression techniques, promote the healthy development of urban sculpture and urban public environmental art. The governments of various countries are in the daily operation and development process. Digital reform and innovation are usually realized through 5G, which can provide a new development opportunity for economic growth and development.

3.3 Experimental Investigation Object

First of all, the investigation and research method is used to conduct detailed and in-depth investigation and research on the combined application of 5G technology in the urban sculpture industry in the 5G background era, research data, research rules, and refine and summarize the first-hand information. This paper selects 80 sculptors in multiple cities, and conducts a practical investigation and research on the application of 5G technology from factors such as the work model of the sculpture career.

Table 1. Questionnaire survey report

Questionnaire issuance and recovery	Corporate city sculptor	Personal urban sculptor	Total
Issue	35	45	80
Recycle	34	44	78
Effective	34	43	77
Efficient	97.1%	95.5%	96.3%

As shown in Table 1, in the investigation of the research report of the experiment, a total of 80 urban sculptors were reasonably selected and given an experimental research questionnaire, and a questionnaire survey was conducted on the 80 urban sculptures to evaluate the work of the sculpture career. The model explores the application degree of 5G technology. For this, we first need to study the various working modes of urban sculpture in detail and thoroughly, and at the same time study the direct assimilation of data information in order to apply research and research methods, as well as to concretize and summarize specific cases. Secondly, it analyzes the use of case analysis method. For this reason, it is necessary to investigate the sculpture work mode of urban sculptors in order to analyze the current situation of urban sculpture career work mode through case study.

4   Results

4.1   Urban Sculpture is Affected by Various Factors

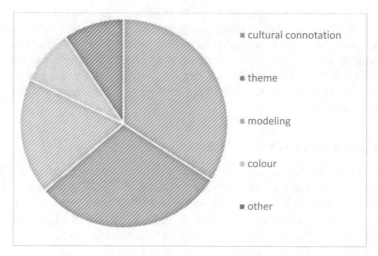

Fig. 1. Statistics of the degree of influence of various factors on urban sculpture

As shown in Fig. 1, urban sculptures are affected to varying degrees by cultural connotations, themes, shapes, colors and other aspects of the construction of sculptures. The topic of urban sculpture must pay attention to the cultural connotation of the city and the artistic influence of sculpture, which is the spiritual power guided by the public. Cultural connotation and spiritual power are an indispensable part of urban sculpture. Urban sculptures without chemical connotation and spiritual power are just like vases. Cultural connotation and spiritual power are an essential part of urban sculpture. Urban sculptures that have transformed their connotation and spiritual power are just like vases. The theme of urban sculpture must complement the environment, the cultural connotation of the city, the history of the city, and the city, and the values conveyed by the city. Starting from the students of sculpture, the theme of sculpture must be properly matched with the city where the sculpture is placed. The concept, leading them to apply this concept in the future work design. Modeling can be said to be the most important element in sculpture design. There is a kind of expressive technique in Chinese sculpture called "freehand" freehand which does not take shape as the main design consideration, but emphasizes the expression of the author's inner world. The shape design is a comparative test of the ability of the sculpture designer. Therefore, it is particularly important to use these elements to determine the theme and content of urban sculptures, to achieve a perfect combination of content and modeling. Color is the most important point except for the shape. Let the color and the sculpture shape match. The mood conveyed by the color matches the meaning contained in the sculpture. It will make people shine, so that it conforms to the connotation of the city and can be integrated with the city. Symbiosis of urban sculpture, this will also be a big trend in the development of urban sculpture in the future.

4.2 Current Urban Sculptures Are Affected by the 5G Era

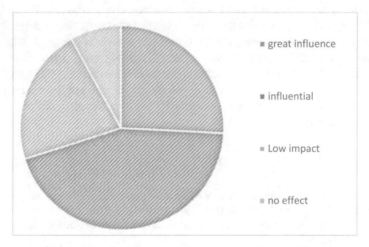

Fig. 2. Statistics of urban sculptures affected by the 5G era

As shown in Fig. 2, through a questionnaire survey of urban sculpture designers, statistics are made on the degree of influence of urban sculptures in the 5G era. With the progress of the times and social development, the 5G era is coming soon. Under such circumstances, in the process of smart city construction, it is necessary to make full use of 5G technology to realize the new era of urban sculpture. When building urban sculptures, it is necessary to use corresponding measurement and sensing operations on the basis of two-way and integrated networks, make full use of advanced control methods and facilities and equipment, and use a scientific and reasonable decision-making system to create a good 5G. The environment ensures the efficiency and reliability of the normal operation of intelligence. our country's urban sculpture needs a long way to go if it wants to make progress, and the relevant research is also the same, especially the Chinese society is in a period of uneven economic development and cultural development, which sets the corresponding urban sculpture research. Therefore, we should dare to seize opportunities and opportunities, such as the arrival of the 5G era, which has brought a variety of new technologies, and the development of urban sculpture should be fully integrated with 5G technology for development and innovation.

5 Conclusion

In this new environment, our country pays more and more attention to the construction of smart cities. Therefore, it is necessary to make full use of advanced 5G technology. Urban sculpture is the business card of the city, which embodies the humanistic characteristics of the entire city, improves the overall image of the city and has a positive role in promoting the development of the city. Whether it is political economy or cultural thoughts, it will play a positive role in the city. Therefore, protecting urban sculptures is also protecting urban culture, highlighting urban connotation, inheriting urban culture, and promoting urban construction. Urban sculpture not only bears the value of showing the city's individual signs, but also reflects the innovative characteristics of the city's region. Its construction process is not only a material shape shaping process, but also a shaping process of urban spiritual civilization. Only when development is in harmony with the spirit of the city can a sculpture business that truly belongs to the Chinese nation and suitable for the development of modern cities can be created. The background of the 5G era is the feasible and innovative development direction of urban sculpture, and it is of great significance to fully integrate the 5G era to make urban sculpture carry out innovative development.

References

1. Zheng, J.: Producing Chinese urban landscapes of public art: the urban sculpture scene in Shanghai. China Q. **239**, 1–29 (2019)
2. Culig, D., Kenney, J.: The four decade career of sculptor Candyce Garrett. Sculpt. Rev. **68**(4), 36–39 (2019)
3. Hochman, A.: RACSO art gallery presents "'ContraFuerte": conceiving an urban sculpture'. Sculpt. Rev. **68**(1), 26–31 (2019)
4. Smith, C.W.: Yakuglas' Legacy: The Art and Times of Charlie James. Ronald W. Hawker. University of Toronto Press, Toronto, Buffalo (2016). Museum Anthropology, 42(1):51–52
5. Moen, K.: Expressive motion in the early films of Mary Ellen Bute. Animation **14**(2), 102–116 (2019)
6. Ye, Q., Zhuang, W., Li, X., et al.: End-to-end delay modeling for embedded VNF chains in 5G core networks. IoT J. IEEE **6**(1), 692–704 (2019)
7. Fernández-Maimó, L., Huertas-Celdrán, A., Gil Pérez, M., García Clemente, F.J., Martínez Pérez, G.: Dynamic management of a deep learning-based anomaly detection system for 5G networks. J. Ambient Intell. Humaniz. Comput. **10**(8), 3083–3097 (2018). https://doi.org/10. 1007/s12652-018-0813-4
8. Nonaka, K., Watanabe, R., et al.: Real-time free-viewpoint video streaming technology by using 5G wireless network. J Inst. Image Inf Telev. Eng. **74**(1), 180–186 (2020)
9. Cao, B., Sun, Z., Zhang, J., et al.: Resource allocation in 5G IoV architecture based on SDN and fog-cloud computing. IEEE Trans. Intell. Transp. Syst. **22**(6), 3832–3840 (2021)
10. Alsharif, M.H., Yahya, K., Chaudhry, S.A.: Analyzing and evaluating the energy efficiency based on multi-5G small cells with a mm-waves in the next generation cellular networks. Int. J. Electr. Comput. Eng. **10**(4), 3492–3500 (2020)

Prediction of Stock Price Based on Artificial Intelligence Algorithm

Rui Xia[✉] and Kaili Wang

School of Business, Nantong Institute of Technology, Nantong 226001, Jiangsu, China

Abstract. The study of stock value estimation is an important process of stock investment. The formation of research methods marks the maturity and perfection of stock value estimation models. It can not only make stock investment deeply rooted in the hearts of the people, but also provide investors with more accurate stock investment methods, which has important theoretical and practical significance. This article first studies the development status of the stock market, and then points out the key problems faced by the stock transaction price prediction, and gives the method of predicting the fluctuation law of the stock transaction price. Next, this paper proposes the fusion of the support vector machine algorithm based on the AdaBoost boosting algorithm and the BP artificial neural network algorithm. Finally, the method is verified experimentally in this paper, and the results show that the algorithm model can improve the prediction accuracy of the final stock trading price fluctuation law.

Keywords: Stock price · Support vector machines · Neural networks · AdaBoost

1 Introduction

Since the establishment of the stock market, it has gradually become one of the important means of investment and financial management in people's daily life because of its flexible investment methods compared with funds and bonds, and less initial capital demand compared with investment industries such as real estate [1, 2]. However, the coexistence of high risk and high return is the main feature of the stock trading market. This main feature makes the stock investment trading market a very attractive capital trading market, which deeply attracts a group of investment enthusiasts. However, it is difficult for ordinary investors to grasp the changing law of stock trading prices in the stock market, and it is difficult to sell high and buy low [3, 4]. Therefore, what kind of method can be used to accurately predict the changing law of stock trading price fluctuations has always been a question of great concern to many investors. A prediction model that can accurately predict the law of stock transaction price fluctuations can not only provide important useful information for the national government to stabilize the financial investment market, but also help the majority of stock investors to increase the rate of return and reduce investment risks [5, 6].

In order to eliminate the multicollinearity of stock indicators, Chinese scholar Ma Juan uses AdaptiveL asso algorithm to screen indicator variables. After achieving data

© The Author(s), under exclusive license to Springer Nature Switzerland AG 2022
J. Macintyre et al. (Eds.): SPIoT 2021, LNDECT 98, pp. 17–24, 2022.
https://doi.org/10.1007/978-3-030-89511-2_3

dimensionality reduction, grey prediction is used to predict stock price influence indicators [7]. Hu Di first uses the nearest neighbor propagation algorithm to select other stocks that are similar to the stock price change to be predicted, and then uses the stock to be predicted and other stocks with similar price changes as input data to train a support vector machine to realize the prediction of the rise and fall of stock prices [8]. On the basis of the BP neural network method, Qifangzhong introduced the principal component analysis method and the improved fruit fly algorithm, and proposed a stock price prediction model based on the BP neural network. The improved fruit fly algorithm is used to optimize the initial weights and thresholds of the BP neural network, and a stock price prediction model based on PCA and IFOA-BP neural network is established [9].

Value investment in the securities market is a process of value discovery, input and value presentation, and everything is carried out around its value. Therefore, the correct estimation of the stock value is conducive to promoting the maturity and perfection of the stock value estimation model, helping to curb excessive speculation, and making the stock value investment theory popular [10, 11]. At the same time, it has a positive influence on the functions of the stock market, promoting the healthy development of the stock market, and forming a rational investment market atmosphere. Combining intelligent algorithms with stock pricing models to estimate the intrinsic value of stocks is conducive to promoting the application of the intrinsic value of stocks in the investment field [12]. At the same time, the value estimation based on intelligent algorithms also provides investors with a more accurate and scientific stock selection method for stock investment, which enriches the research content of stock value estimation. Deepening the study of stock value estimation methods can guide investors to invest in scientific and effective investment strategies and methods.

2　Prediction of Stock Price Based on Artificial Intelligence Algorithm

2.1　Key Issues Facing the Prediction of Stock Exchange Prices

For relevant researchers and stock investors, if you want to accurately predict the law of stock trading price fluctuations, you must first conduct a comprehensive and detailed investment analysis of the intricate trading environment in the stock trading market. In this high-risk and high-yield stock trading market, if you blindly invest in the stock market without systematically analyzing the stock market in advance, you will lose your money. Therefore, in order to avoid blind and risky stock trading market investments, before investing in any stock, the majority of investors need to analyze all the possible factors that affect the price fluctuation of the stock with patience. Only in this way can the majority of investors be able to reduce risks and increase yields as much as possible. Highly complex nonlinear systems are often difficult to be well understood and mastered by humans. The stock exchange market is such a typical nonlinear system. In this nonlinear system, there are many kinds of factors that affect the fluctuation law of stock trading prices, and there are often a large number of intricate connections between these factors. How to avoid risks and increase returns under such extremely complex adverse conditions is a hot academic issue jointly studied by investors and related researchers.

2.2   Prediction Method of Stock Exchange Price Fluctuation Law

Although the traditional analysis methods are simple and practical, they can't solve the problem of predicting the fluctuation law of stock trading prices well—the problem of nonlinear system learning. Fortunately, the artificial neural network can gradually approach the target value through repeated calculations of existing samples, and finally generate an ideal prediction model. The artificial neural network algorithm can simulate the human thinking mode, human intelligence and the biological structure of the human brain, so that it can learn a predictive model with strong ability to deal with the learning problem of the nonlinear mapping system, so that the trained predictive model has strong adaptability ability. In this article, firstly, the research problem is changed from the traditional regression learning problem of predicting the rise and fall of stock trading prices on a single trading day to the classification learning problem of predicting whether the highest rise of stock trading prices in a period of time exceeds the expected value. In order to solve this two-class classification problem, it is necessary to extract feature vectors with good discriminating ability based on the historical information of stock transaction price fluctuations. Then, deep learning algorithms will be used to extract potential information from the original information in order to improve the prediction accuracy of the final model.

2.3   Feature Representation and Feature Processing

For the research questions on the prediction of stock trading price fluctuations, there are a lot of data information that can be used for reference, including the turnover rate, amplitude, opening price rise and fall of each normal trading day, and other technical indicators. Although these technical indicators can reflect the specific situation of the stock trading price trend of the day, the overall fluctuation of the stock trading price cannot be accurately reflected. It is still necessary to use all the technical indicators of a single trading day in a certain period of time and the period of time. The overall technical index serves as the characteristic information of the fluctuation law of stock trading prices. In this article, for judging whether a stock has investment value on a certain reference trading day, the relevant information of the stock in the 100 trading days before the reference trading day is used as the characteristic information, from the trading information of the one hundred trading days, extract useful features as the original feature vector. Within 10 trading days after the reference trading day of the stock, if the closing price of a trading day relative to the closing price of the reference trading day has increased by more than investors' expectations, the stock is considered to have investment on the reference trading day value. In order to obtain the inherent feature information implicit in the original features, a deep learning network with a Boltzmann machine as the basic structure is used for feature processing, in order to obtain a feature numerical representation with strong expressive ability. And in the end, a prediction model with higher prediction accuracy can be obtained.

3 The Fusion of Support Vector Machine Algorithm and BP Artificial Neural Network Algorithm Based on AdaBoost Lifting Algorithm

3.1 Support Vector Machines

When support vector machines were first proposed, they were mainly aimed at linearly separable two-class classification and prediction problems. Compared with linear classification problems, non-linear problems are more common in practical applications, and it is difficult to separate the data in the non-linear feature space. In other words, it is definitely not easy to directly solve the nonlinear classification problem perfectly. Therefore, relevant scholars hope that they can borrow mature methods for solving the linear classification problem to solve the nonlinear classification problem. The solution they proposed is mainly to use the kernel function technique for nonlinear transformation, and then use a linear learning algorithm to learn a classification model with higher prediction accuracy from the training data set. Most scholars believe that the projection of nonlinear data in a low-dimensional feature space to a high-dimensional feature space will be linearized to a large extent. Based on this idea, they applied the kernel function technique that can map low-dimensional to high-dimensional to the linear support vector machine algorithm.

3.2 Build BP Artificial Neural Network Algorithm Model

In a general sense, the final output layer of the BP artificial neural network algorithm will use ordinary linear functions for integration and conversion by default. But in this article, for the convenience of representation and calculation, a unified mathematical function is used as the excitation function. To build a good BP artificial neural network algorithm model, it is inevitable to determine all the parameter variables in the model. The selection and initialization of these parameters directly affect the performance of the BP artificial neural network algorithm model. According to specific learning problems, such as the problem of predicting the law of stock trading price fluctuations, the number of elements contained in the feature vector often determines the number of nodes in the input layer of the network, and the attributes of the predicted target value also directly determine the number of output nodes of the model. However, there is no complete theoretical system on how to determine the hidden nodes so far. Fortunately, an empirical formula can be adjusted to obtain the optimal number of hidden nodes. The empirical formula is:

$$M = \sqrt{N + K} + a \tag{1}$$

3.3 Ada Boost Boosting Algorithm

The main guiding ideology of the promotion method is: For any learning problem with high complexity, use multiple comprehensive learning analysis and judgments, and finally reach a consistent conclusion, which is often better than any one of them. The learning system alone is good for learning judgment. In the learning framework where the probability is approximately correct, it is a sufficient condition for judging that any learnable problem is strong and learnable. It is that the problem is weak and learnable, and the necessary condition is also that the learning problem is weak and learnable. For a specific classification problem, such as the prediction of whether the increase in stock trading prices over a period of time exceeds the expected learning problem in this article. When a set of training samples in a specific feature space is given, it is much easier to learn a weak classifier model than to learn a strong classifier model. The AdaBoost algorithm is such an effective method for learning many weak classifier models and then further fusing each weak classifier model. The final strong classification model obtained in this paper is:

$$f(x) = \text{sign}(\sum\nolimits_{m=1}^{M} \alpha_m f_m(x)) \qquad (2)$$

4 Experimental Verification

In order to verify the effectiveness of the method, this article will focus on the 10-fold cross-validation and Leave-One-Out cross-validation on the Shanghai Stock Exchange Index and Industrial and Commercial Bank of China stock data from January 1, 2006 to the present. 10-fold cross-validation means that all the instance samples in the training data set are randomly divided into 10 equal parts, and then 9 equal parts of the experimental sample data are used to train a corresponding prediction model, and the remaining 1 equal parts are used to train a corresponding prediction model. The experimental test, up to 10 equal parts were used for testing without exception, and the 10-fold cross-validation was terminated. Leave-One-Out cross-validation can be understood as an extension of 10-fold cross-validation. It takes a sample point from all sample points in the training data set as a test sample each time, and the sample points in the rest of the training data set are used To train the predictive model. Such steps will be repeated until each sample point in the training data set is tested, and only tested once.

4.1 10-Fold Cross-Validation Experiment Results Before and After the Feature Processing of the Deep Learning Algorithm

According to Table 1 and Fig. 1, we can know that when investors expect a 10% increase, MCC can achieve the maximum value. In addition, for different investor expectations, the features processed by the deep learning algorithm are all higher than the MCC values obtained by the unprocessed features. This shows that the deep learning algorithm has a good learning ability for the features used in the prediction of stock trading price fluctuations, and can discover its inherent essential features and help improve the prediction accuracy. In addition, it can be seen that the higher the investors' expectation, the lower the prediction accuracy of stock trading price fluctuations.

Table 1. The prediction results of deep learning algorithms on stock data

Expected increase	MCC of Shanghai stock index stocks processed by deep learning algorithm	MCC of Shanghai stock index stocks that have not been processed by deep learning algorithms	ICBC stock MCC processed by deep learning algorithm	ICBC stock MCC that has not been processed by deep learning algorithms
5	0.18	0.12	0.33	0.34
10	0.39	0.36	0.48	0.46
15	0.32	0.31	0.45	0.44
20	0.12	0.11	0.43	0.42
25	0.05	0.06	0.26	0.25

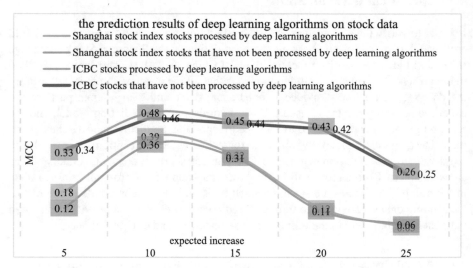

Fig. 1. The prediction results of deep learning algorithms on stock data

4.2 Leave-One-Out Cross-Validation Test Results

According to Table 2 and Fig. 2, it can be known that the AdaBoost boosting algorithm model fusing the support vector machine and the BP neural network algorithm can improve the prediction accuracy of the final stock transaction price fluctuation law. Especially in the stock data of Industrial and Commercial Bank of China, the MCC index of the AdaBoost boosting algorithm is 0.532, which is 0.9% higher than the MCC of the support vector machine and 0.8% higher than the BP neural network algorithm. It is worth mentioning that under the condition of large stock investment, even if the forecast accuracy is increased by one ten thousandth, considerable gains may be obtained.

Table 2. Leave-one-out cross-validation experiment results of Shanghai stock index stocks and ICBC stocks

Data	Algorithm	Sen (%)	Spe (%)	Acc (%)	MCC
Shanghai Stock Exchange Index Stocks	Support vector machine algorithm	71.67	76.12	75.03	0.429
	BP neural network algorithm	71.68	76.09	75.01	0.429
	Ada boost boosting algorithm	71.94	76.15	75.12	0.432
Industrial and Commercial Bank of China	Support vector machine algorithm	76.44	80.12	79.15	0.523
	BP neural network algorithm	76.50	80.14	79.18	0.524
	Ada boost boosting algorithm	77.36	80.24	79.48	0.532

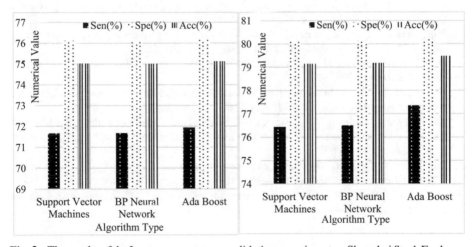

Fig. 2. The results of the Leave-one-out cross-validation experiment on Shanghai Stock Exchange Index stocks and Industrial and Commercial Bank of China stocks

5 Conclusions

With the continuous development of the stock trading market, a large amount of stock trading information will be generated. How to learn and predict the law of stock trading price fluctuations from this large amount of stock information is of great practical help to the majority of stock investors to achieve high yields. Moreover, new technical indicators are constantly being created, and they can all help improve the prediction accuracy of stock trading price fluctuations to a certain extent. They are characteristic information worth considering in the future. In addition, the field of artificial intelligence is also rapidly developing, and new artificial intelligence algorithms can be used to predict the fluctuations of stock trading prices. All in all, although the prediction of stock transaction price fluctuation law has made certain achievements, its prediction accuracy is far from the expectations of investors. Predicting the law of stock trading price fluctuation is still a very challenging research topic.

References

1. Zhou, X., Zhao, M.: Research on portfolio optimization under the background of artificial intelligence. J. Changzhou Inst. Technol. **32**(002), 31–35 (2019)
2. Zhang, P., Liu, H., Pei, D., et al.: Stock price prediction based on SVM-KNN. Stat. Appl. **8**(6), 13 (2019)
3. Zhao, C., Ye, Y., Yao, M.: Forecast of stock volatility based on financial text sentiment. Comput. Sci. **47**(05), 87–91 (2020)
4. Duan, M., Liu, M., Xue, B., et al.: Shanghai stock Exchange Index forecast based on long and short-term memory network. Rural Econ. Technol. **30, 470**(18), 84–86 (2019)
5. Chen, J., Yan, M., Yu, R.: Legal problems caused by artificial intelligence in the financial field and solutions. J. Hangzhou Inst. Finan. Res. **284**(11), 66–68 (2020)
6. He. C.: Intelligent service and risk management of securities companies driven by financial technology. Prog. Sci. Technol. **37,505**(21), 168 (2020)
7. Ma, J., Wang, L., Zuo, L.: Research on ChiNext stock price prediction based on grey system and neural network. Contemp. Financ. Res. **11**(02), 92–102 (2019)
8. Di, H., Wei, H.: Stock price prediction based on AP-SVM combination model. J. Wuhan Inst. Technol. **041**(003), 296–302 (2019)
9. Qi, F., Lin, S., Yu, T.: Stock price prediction model based on PCA and IFOA-BP neural network. Comput. Appl. Softw. **037**(001), 116–121, 156 (2020)
10. Zhu, C., Kang, L., Feng, W.: A stock market closing price prediction algorithm based on adaptive whale optimization algorithm combined with Elman neural network. J. Comput. Appl. **040**(005), 1501–1509 (2020)
11. Liu, Y., He, X.: BP neural network stock price prediction based on adaptive firefly algorithm. J. Weinan Normal Univ. **034**(002), 87–96 (2019)
12. Li, Z., Wu, Q.: Research on stock prediction algorithm based on LSTM neural network. Fujian Comput. **035**(007), 41–43 (2019)

Variation Translation Strategy System of Intangible Cultural Heritage Based on Data Mining

Cong Wang[⊠] and Lu Zhang

School of English Language and Literature, Xi'an Fanyi University, Xi'an 710105,
Shaanxi, China
wangcong@xafy.edu.cn

Abstract. With the development of human society, intangible cultural heritage
has gradually become an important information resource, and its value is constantly increasing. Intangible cultural heritage has gradually become one of the
comprehensive competitiveness standards of a country or region's economic, cultural and political strength. Therefore, it is very important to study intangible
cultural heritage translation strategies. To this end, this article systematically studies the strategies of intangible cultural heritage translation based on data mining
algorithms. This article mainly uses case analysis method, questionnaire survey
method and expert interview method to study the translation strategies of intangible cultural heritage in translation. The research results show that translation
strategies at different levels are different. 80.2% of intangible cultural heritage
translation focuses on literal translation, free translation and foreignization and
domestication. These methods all hide the shadow of translation.

Keywords: Data mining · Intangible cultural heritage · Translation strategy ·
Variant translation

1 Introduction

With the development of the times, the protection of intangible cultural heritage has
attracted attention, and international cooperation and exchanges have become increasingly close. In this context, foreign publicity translation plays a vital role in the research,
communication, promotion and protection of intangible cultural heritage [1, 2].

Zeng Yanwen pointed out that since 2006, my country has paid more and more
attention to the protection and inheritance of intangible cultural heritage [3]. Wang Yan
clarified and discussed the main factors affecting the translation of Wuxi intangible
cultural heritage, providing theoretical support for translation practice [4]. Qi Jiantao
said that the translation of intangible cultural heritage is an important means of deepening
national cultural outreach under the background of reform and opening up, and the quality
of translation is directly related to the effectiveness of outreach [5].

J. Macintyre et al. (Eds.): SPIoT 2021, LNDECT 98, pp. 25–33, 2022.
https://doi.org/10.1007/978-3-030-89511-2_4

This article mainly analyzes the characteristics and status quo of intangible cultural heritage, and then introduces the concept, essence, elements, and methods of variant translation. After that, data mining technology is used to search and research the application of variant translation.

2 Variation Translation Strategies for Intangible Cultural Heritage

2.1 Intangible Cultural Heritage

Intangible cultural heritage is not only an important foundation for inheriting national culture and enhancing national feelings, but also has important functions such as inheriting and maintaining cultural diversity [6, 7].

(1) Features of intangible cultural heritage text

The intangible cultural heritage text has its own distinctive features: First, there are a large number of proper nouns. Second, the texts related to intangible cultural heritage have a distinctive national character. In terms of sentences, related texts are mixed with long and short sentences [8, 9].

(2) There is a problem

Currently, intangible cultural heritage protection has problems such as insufficient attention, insufficient investment, imperfect laws, and lack of planning [10].

2.2 Variation Translation

(1) Concept

The theory of variation translation is the main theory of translation of the original text. It was first proposed by Professor Huang Zhonglian in 1997. He pointed out that variable translation is an activity aimed at editing the relevant content of the original text through appropriate methods. Translation variation is to edit, add, delete, state, condense, and adapt according to the specific needs of readers [11, 12].

(2) Essence

Variation translation, in popular terms, is to adapt and translate, highlighting the word "variation". The translation system is shown in Fig. 1:

(3) Three elements of translation system
1) The subject of the translation: The subject of the translation is divided into two parts: the reader and the translator.
2) Translation intermediary: flexible plus translation

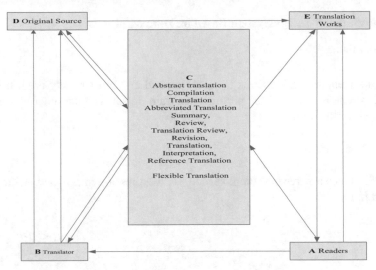

Fig. 1. Variation translation system

3) Variable translation object: The variable translation object mainly refers to the object accepted by the reader of the variable translation.
(4) Seven methods
1) Increase, increase in information based on the original work. Including elaboration, commenting and adding related content, etc.
2) Subtract and delete unnecessary things.
3) Edit, organize and order the content of the original work.
4) Narrate, write or say the original content in the target language to convey the meaning.
5) Shrink, compress the content of the original work, and compress the original text with condensed voice.
6) And, combine two or more parts of the same kind or sequential logic in the original work.
7) Revise, make obvious changes to the original work, such as content, form, etc.
(5) Features
1) A lot of information. Reduce translation media and increase the ratio of information capacity to media.
2) Quickly and quickly respond to external information.
3) Good, high information quality. The amount of information is sufficient and the information is accurate.
4) Save time and time.
5) Targeted and pertinent.

2.3   Translation Strategy System Based on Data Mining

(1) Data mining

Among them, classification is the most studied and most widely used. It has two main steps: one is to model the performance training set, and the other is to perform predictive classification.

(2) Decision tree

The final decision tree is a complete model and expression rules. The specific process is shown in Fig. 2:

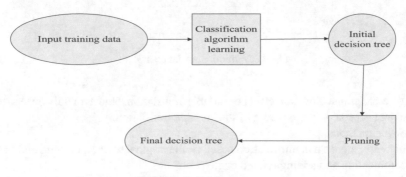

Fig. 2. Build flow chart

Entropy is a measure of the suddenness, uncertainty or randomness of a data set. Given the probability $S_1, S_2, \ldots, S_v$, among them, $\sum_{i=1}^{v} S_i = 1$, entropy is defined as

$$P(S_1, S_2, \ldots, S_v) = -\sum_{i=1}^{v} S_i \log_2 \frac{1}{S_i} \qquad (1)$$

The value of entropy is between 0 and 1, and the entropy is 0, which means that the data in the set has no uncertainty and all belong to the same category.

Suppose W is a set containing w data samples, and there are n different category attributes Di, where i = 1, 2, ..., n. Therefore, the number of samples in the category Di is Wi, then the sample set W is classified by the amount of information required is defined as:

$$K(v_1, v_2, \ldots, v_n) = -\sum_{i=1}^{n} D_i \log_2 D_i \qquad (2)$$

Based on the classification algorithm of data mining, we can know that the English translation strategy can be divided into the following parts:

(3) Translation strategy
1) In terms of vocabulary, most English translation vocabulary is ambiguous. There are three ways to choose the meaning of a word:

Supplementary words refer to specific meanings in the original text but not directly expressed through vocabulary. These meanings must be added to the translation. Omission means that certain words are omitted without affecting the completeness.

2) In terms of sentences, sentence structure translation techniques mainly include word order, combination and conversion.

The word order categories include sequential translation and reverse translation, preposition, induction, etc.
There are clause syntax and co-syntax in combination class.
The conversion categories include the conversion of sentence components, the conversion of passive voice, and so on.

2.4 Examples of Vocabulary Translation Based on the Translation Strategy

(1) Translation of proper nouns
How to make the translation acceptable to the target readers while retaining the cultural characteristics of the source language is a difficult problem for the translator in the translation process.
1) Literal translation

A translation method that not only maintains the original content, but also maintains the original form.
Tiannv SanHua—"The goddess scatters flowers."
Cao Shoulianer—thin-faced Cao.

2) Transliteration and annotation

The method of transliteration can bring the target language readers a kind of foreign cultural experience.
Yumin Haozi—Fisherme' shaozi (rhythmic work songs).
Tiao Ma Fu—Tiao Ma Fu (dance of men dressed like stablemen).

3) Paraphrase

A translation method that only keeps the original content and does not keep the original form.
"Shao Ma Xiangfu Zhe—the dancers.
Zhen Zhi—paperweight.

4) "Internet + Intangible Cultural Heritage" model

The translation of intangible cultural heritage text materials is compatible with the "Internet +" model.

Zhongkui—Zhongkui: a god in the Chinese folk lore who can ward off evil things (https://en.wikipedia.org/wiki/Zhong_Kui).

(2) Translation of culturally loaded words

These words reflect the unique types of activities of a particular ethnic group that are different from other ethnic groups.

1) Naturalization

The naturalization method naturalizes foreign values into the target language culture and invites the original author to come home.

Qi Pa—a unique part.

La Gua—chatting.

2) Alienation

The alienation method expresses the language and cultural differences in foreign texts and sends readers abroad.

Yifu—Yifu (foster/adoptive father).

Guifei Chu yu—bathing concubine.

3 Investigation on Translation Strategies of Intangible Cultural Heritage Translation

3.1 Investigation Background

Nowadays, the protection of intangible cultural heritage in our country is in full swing. Publicity translation focuses mainly on the political and economic fields, and there are very few publicity translations involving intangible cultural heritage. Based on this, this article aims to study the translation strategies of intangible cultural heritage, so as to promote the inheritance and development of intangible cultural heritage.

3.2 Investigation Process

This article mainly sampled 50 English translation students and a number of intangible cultural heritage experts. The content of the questionnaire includes: commonly used translation strategies, translation methods, translation strategies used in different situations, and frequency of translation methods. A total of 50 questionnaires were issued this time Two rounds of questionnaires were conducted, and the final effective recovery rate reached 100%.

4   Analysis of the Use of Intangible Cultural Heritage Translation Strategies

Based on the questionnaire survey results of 50 English translation majors, combined with the effective questions and solutions put forward by experts, this paper draws Table 1. The specific translation strategy is as follows:

Table 1. The application of translation strategies of vocabulary, sentence and rhetoric in variation translation

	Literal translation	Free translation	Annotation	Foreignization	Domestication
Vocabulary	20	23	12	18	20
Sentence	15	26	7	24	16
Rhetoric	6	21	19	16	18

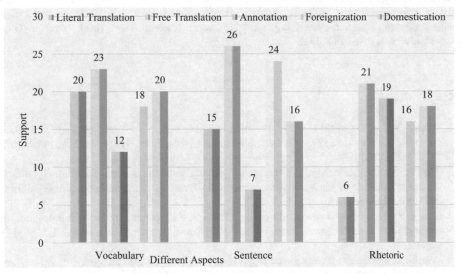

Fig. 3. The application of translation strategies of vocabulary, sentence and rhetoric in variation translation

As shown in Fig. 3, we can find that in terms of vocabulary, students often use literal translation and free translation methods to translate intangible cultural heritage. For sentence translation, they tend to use translation strategies of free translation and foreignization.

5 Conclusion

In view of the characteristics of intangible cultural heritage texts, the translation strategy is summarized as follows: At the lexical level, the author believes that the translator must carefully study the cultural connotation of the source language text. While conveying its cultural connotation, it is accepted by readers. At the syntactic level, there are big differences between Chinese and English sentence structure. In the translation process, the sentence structure must be carefully analyzed to clarify the hidden relationships between sentences. At the rhetoric level, due to the influence of social context and other factors, there are certain differences in rhetoric between Chinese and English.

Acknowledgements. This work was supported by the Scientific Research Major Project of Xi'an Fanyi University, PN: 2021Z02, 2021 Scientific Research General Project of Xi'an Fanyi University, PN: 21B47, the "Double First-class" Construction Project of Xi'an Fanyi University, non-number and Belt and Road Language and Culture Research Base (Think Tank) of Xi'an Fanyi University, PN: 20KYJD02.

References

1. Wang, Y.: Let the world listen to the voice of Wuxi: a study on Wuxi's intangible cultural heritage translation strategy. Chin. J. Multimedia Netw. Educ. (Electron. Edn.) **000**(006), 310–312 (2017)
2. Zhong, A., Hu, Y.: Analysis of the status quo and countermeasures of intangible cultural heritage translation research—taking Shaoguan City as an example. J. Suzhou Educ. Inst. **021**(005), 22–25 (2018)
3. Zeng, Y.: Analysis and discussion on the Status Quo of research on English translation of intangible cultural heritage—based on 2007–2016 data. Sichuan Drama **000**(001), 138–141 (2018)
4. Wang, Y.: Influencing factors and corresponding strategies of Wuxi intangible cultural heritage translation practice. Jiangsu Educ. Res. **408**(12), 34–37 (2019)
5. Qi, J.: A corpus-based analysis of intangible cultural heritage text translation research path. J. Mudanjiang Univ. **27**(226(06)), 84–87 (2018)
6. Ren, Z.: The interpretation and translation of culturally loaded words in intangible cultural heritage propaganda text from the perspective of relevance and adaptation. J. Bengbu Univ. **009**(001), 69–73 (2020)
7. Wang, Y.: Research on translation of intangible cultural heritage text based on text type theory. J. Jilin Province Educ. Inst. **35**(01), 151–154 (2019)
8. Zhou, N., Wu, L.: A study on the English translation of intangible cultural heritage from the perspective of ecological translation: taking Chongqing Wulong intangible cultural heritage as an example. J. Chongqing Radio Telev. Univ. **30**(1), 49–56 (2018)

9. Sun, Y.: Research on translation and dissemination path of "intangible cultural heritage" under the background of new media—taking the representative intangible cultural heritage of Jinan as an example. Hebei Agricultural Mach. **242**(08), 61–62 (2018)
10. Wang, Z.: Status Quo and thoughts on translation studies of my country's intangible cultural heritage—based on the analysis of related documents from 2006 to 2017. Popular Lit. Art **000**(020), 179–181 (2018)
11. Guo, L., Wu, B., Liu, Y.: Research on intangible cultural heritage tourism publicity from the perspective of relevance translation theory——taking the English translation of Manchu cultural heritage in Chengde as an Example. China Nationalities Expo. **000**(002), 96–97 (2019)
12. Qiu, M.: Research on the English translation of non-heritage tourism publicity under the guidance of the theory of translation adaptation and selection. J. Anhui Univ. Technol. Soc. Sci. Edn. **034**(001), 58–60 (2017)

A Computer-Aided Comparative Study on Grammatical Cohesion in Abstracts of Sci-Tech Journal Papers by Chinese and American Scholars

Xiuyun Xu[✉] and Dandan Yao

Xi'an Jiaotong University City College, Xi'an, Shaanxi, China

Abstract. Cohesion, as one of the important ways to achieve discourse coherence, is a crucial part of discourse analysis. Therefore, employing computer-aided Independent Sample T-Test in SPSS software, this paper explores the similarities and, mainly, differences of grammatical cohesive devices used by Chinese and American scholars in their English abstracts of Sci-tech journal Papers. The results show that: 1) there are significant differences in the use of reference between Chinese scholars and American scholars, with the former using a higher frequency than the latter; 2) both Chinese scholars and American scholars employ few cohesive devices of substitution, but the using frequency of Chinese scholars is still higher than that of American scholars, and the difference in the frequency is statistically significant; 3) neither Chinese nor American scholars use ellipsis; 4) Chinese scholars use conjunction more frequently than American scholars, and the difference is statistically significant.

Keywords: Computer-aided · Grammatical cohesion · Abstract · Chinese scholar · American scholar

1 Research Background

Halliday & Hasan holds that any discourse that can be a text must have texture, which is one of the characteristics that distinguish it from non-text [1]. Cohesion is one of the most important ways to achieve discourse coherence and it is a crucial part of discourse analysis. We can distinguish any part of a text by the number and type of cohesive devices [1]. Therefore, it can be seen that cohesion plays an important role in the construction of a coherent discourse.

Since the publication of the book Cohesion In English by Halliday & Hasan in 1976, cohesion theory has received extensive attention from scholars both at home and abroad, and many scholars have begun to study cohesion in discourse. Hoey redefined lexical cohesion on the basis of Halliday & Hasan's study of cohesion theory. Hu Zhuanglin, Zhang Delu and Miao Xingwei were the first Chinese scholars to conduct researches on cohesion theory [2–6]. They wrote a large number of papers and monographs from different angles, enriching and developing cohesion theory in China.

© The Author(s), under exclusive license to Springer Nature Switzerland AG 2022
J. Macintyre et al. (Eds.): SPIoT 2021, LNDECT 98, pp. 34–41, 2022.
https://doi.org/10.1007/978-3-030-89511-2_5

In addition, some scholars conducted applied researches on cohesion theory, such as researches on stylistics, cognitive linguistics, translation, and English teaching. Li Wen believes that some texts are marginal members in the category of coherence, so it is necessary to combine stylistic knowledge and consider how these texts can achieve specific stylistic effects under the interaction of "form" and "meaning" [7]. Wei Lin, starting from the semantic theory of cognitive linguistics, elaborated on the importance of discourse cohesion and coherence in cognitive semantics [8]. Wang Wenqian and Liu Hongqiang applied discourse cohesion theory to the translation of political documents, and discussed its specific application in the English version of the 2017 Government Work Report [9]. Xu Wenyu reviewed and discussed the importance of the theory of discourse cohesion and coherence, analyzed the errors and causes in college students' English writing, and proposed measures to improve the ability of achieving discourse cohesion and coherence [10].

Abstract is an extremely important part of a paper and its quality may directly affect whether readers will choose to read the whole paper and the paper will be accepted to be published or not. In view of these, many scholars have studied the cohesive devices in abstracts. Li Shipin and Hu Qinqin conducted a comparative study of the cohesive devices between Chinese and English abstracts and pointed out the similarities and differences between Chinese and English [11]. Based on Halliday's cohesion theory and taking the Chinese version The Little Prince translated by Zhou Kexi as the research object, Lao Tongwei and Wang Xin made a comparative study between English and Chinese cohesive devices in fairy tale discourses in order to find out the similarities and differences [12]. Some scholars also studied the similarities and differences in the employment of English cohesive devices between Chinese scholars and English native speakers in different majors. For example, Lu Yumei and Ren Peihong conducted a corpus based comparative study of grammatical cohesive devices in English abstracts of 30 medical and political papers published in domestic and foreign journals and the results show that there are certain differences [13]. For instance, the frequency of Chinese scholars using grammatical cohesive devices is higher than that of native speakers, and colloquial conjunction appears. Furthermore, Li Shuangyan conducted a comparative study on lexical cohesive devices in 45 news reports published in Times from 2011 to 2013 and 45 abstracts of doctoral dissertations published in PQDT database from 2011 to 2013 [14].

To sum up, many scholars have done researches on cohesion theory and its application in stylistics, cognitive linguistics, translation, and English teaching. Other scholars have conducted comparative studies on English and Chinese cohesive devices in abstracts; Some scholars have studied the use of cohesive devices in different fields. However, few scholars have studied the similarities and differences in the use of English cohesive devices in sci-tech journal papers between Chinese scholars and English native scholars. Therefore, this article, taking the English abstracts of sci-tech journal papers as the research corpus and employing the Independent Sample T-test in the SPSS software, analyzes the similarities and differences in the use of grammatical cohesive devices in English abstracts by Chinese and American scholars, with a view to providing some enlightenment for Chinese scholars to write English abstracts.

2 Research Design

2.1 Theoretical Basis

Halliday & Hasan divides cohesive devices into two categories: grammatical cohesion and lexical cohesion. The former includes reference, substitution, ellipsis and conjunction and the latter contains lexical restatement and collocation [1]. Based on Halliday & Hasan's classification of cohesive devices, this paper mainly focuses on the similarities and differences of grammatical cohesive devices used by Chinese and American scholars, namely, reference, substitution, ellipsis and conjunction. This paper takes sentence as a basic unit of analysis, and studies how many cohesive devices are used between sentences, because Halliday & Hasan hold that "sentences are the basic unit of discourse, and only cohesion between sentences is meaningful" [1].

2.2 Corpus Collection

The corpus of this paper includes randomly chosen 30 English abstracts written by Chinese scholars from a core journal of China's science and technology from January 2019 to December 2019 and 30 English abstracts written by American scholars from one core journal of America's science and technology from 2015 to 2019. The reason for choosing papers in American journal in the past five years lies in that the number of papers published in the journal every year is too small. For example, there are only 21 articles in 2019 and 27 articles in 2018, so the number of papers available for analysis is too small. From 2015 to 2019, there are a total number of 193 papers, but 77 papers do not meet the requirements and are excluded (the authors of 18 papers are not American, the authors of 10 papers could not be confirmed American, 27 papers have no author introduction, 8 are front covers, 8 back covers, 6 announcements). As a result, there are 116 papers left and the author chooses 30 abstracts among them according to the same standard as the author chooses abstracts of Chinese scholars. Finally, the corpus of Chinese scholars and American scholars respectively contain 14 abstracts with 8 sentences, 13 abstracts with 9 sentences, and 3 abstracts with 10 sentences, a total of 30 abstracts, 259 Sentences. The fact that the 30 abstracts of Chinese scholars and American scholars have the same total number of sentences and similar number of words, ensures the reliability and validity of the comparison results.

2.3 Corpus Analysis

For the 60 English abstracts, the 259 sentences in 30 abstracts by Chinese scholars are set as the first group (numbered from 1 to 259) and input into the SPSS software. Similarly, the 259 sentences in 30 abstracts by American scholars are set as the second group (numbered as 260 to 518) and input into the same SPSS software in succession to the 259 sentences by Chinese scholars. Two experienced college English teachers analyze the cohesive devices of reference, substitution, ellipsis, conjunction, and then input the total number of the four in each sentence into the corresponding sentence number. Then, open SPSS - Analyze - Compare the Mean - Independent Sample T Test in sequence, to analyze whether the difference in the number of cohesive devices in each sentence between the two groups of scholars is statistically significant.

3   Results and Discussion

Through SPSS Independent Sample T Test, we know the usage of four cohesive devices altogether in each sentence in the abstracts of Chinese scholars and American scholars (see Table 1 & Table 2).

Table 1. Mean of grammatical cohesive device

Type	Group	Total number of sentence	Mean	Standard deviation	Standard error of mean
G	1	259	.79	.951	.059
	2	259	.57	.848	.053

Note: G-Grammatical Cohesive Device, 1-Chinese Scholars, 2-American Scholars

Table 2. Sig. value of grammatical cohesive device

		Levene test of variance equation		Mean equation t test		
		F	Sig	T	df	Sig. (two-tailed)
G	Assuming equal variances	3.658	.056	2.731	516	.007
	Assuming inequal variances			2.731	509.338	.007

Note: G-Grammatical Cohesive Device

From Table 1, we can know that the average number of grammatical cohesive devices (including reference, substitution, ellipsis and conjunction) of Chinese scholars and American scholars are 0.79 and 0.57 respectively. That is, Chinese scholars, on average, use 0.79 grammatical cohesive devices per sentence, while American scholars use 0.57 grammatical cohesive devices per sentence, and Chinese scholars use 0.22 more grammatical cohesive devices per sentence than American scholars. From Table 2, we know that the difference in the number of grammatical cohesive devices used in each sentence is statistically significant, because the Sig (two-tailed) value is 0.007, which is between 0.000 and 0.05 and indicates a significant difference.

Next, we analyze the usage of each type of cohesive devices by Chinese and American scholars.

3.1   Reference

Table 3 shows that Chinese scholars use 0.62 cohesive devices of reference per sentence on average, while American scholars use 0.47 per sentence. In other words, the former uses 0.15 more references than the latter in each sentence. Table 4 indicates that the difference in the number of reference used by two groups is statistically significant,

Table 3. Mean of reference, substitution, ellipsis and conjunction

Type	Group	Total number of sentence	Mean	Standard deviation	Standard error of mean
R	1	259	.62	.865	.054
	2	259	.47	.799	.050
S	1	259	.02	.124	.008
	2	259	.00	.000	.000
E	1	259	.00	.000	.000
	2	259	.00	.062	.004
C	1	259	.15	.358	.022
	2	259	.10	.296	.018

Note: R- Reference, S-Substitution, E-Ellipsis, C-Conjunction, 1-Chinese Scholars, 2-American Scholars

Table 4. Sig. value of reference, substitution, ellipsis and conjunction

		Levene test of variance equation		Mean equation t test		
		F	Sig	T	df	Sig. (two-tailed)
R	Assuming equal variances	4.115	.043	2.006	516	.045
	Assuming inequal variances			2.006	512.780	.045
S	Assuming equal variances	16.708	.000	2.012	516	.045
	Assuming inequal variances			2.012	258.000	.045
E	Assuming equal variances	4.031	.045	1.000	516	.318
	Assuming inequal variances			1.000	258.000	.318
C	Assuming equal variances	14.333	.000	1.872	516	.062
	Assuming inequal variances			1.872	498.169	.062

Note: R- Reference, S-Substitution, E-Ellipsis, C-Conjunction

because the Sig (two-tailed) value is 0.045, which belongs to 0.000–0.05, and means significant difference. Through further analysis, we find that among the 162 references used by Chinese scholars, the is the most frequently used, accounting for 66.67% of all the references, followed by the comparative degree of adjectives, which is used nine times. Next is this, which is used 6 times, other 5 times, and it 4 times. There are other references used in Chinese scholars' abstracts, such as those, same, similar, different, but they are not frequently used. After analyzing the references used by American scholars, we can find that of the 122 references used by American scholars, the is used 47 times, accounting for 38.52% of all the references, followed by these, this, - er, their, other, further, they, additional, respectively used 17, 10, 12, 7, 7, 5, 3 and 3 times. There are other references used, such as it, those, same, different, similar, but they are not used with high frequency. Therefore, we can draw the conclusion that when Chinese scholars and American scholars use reference, there are not only differences in the number, but also differences in the specific words chosen.

3.2 Substitution and Ellipsis

From Table 3, we can see that both Chinese scholars and American scholars rarely use substitution and ellipsis. On average, Chinese scholars use 0.02 cohesive devices of substitution per sentence, while American scholars do not use it at all. The four substitutions used by Chinese scholars are: stay the same, is the same, is the same, and was so. Table 4 tells us that the difference in the number of substitution used by the two groups is statistically significant, because the Sig (two-tailed) value is 0.045, which is between 0.000 and 0.05, and indicates significant difference. From Table 3, we can see that Chinese scholars and American scholars do not employ ellipsis, and Table 4 also gives the corresponding result, because the Sig (two-tailed) value is 0.318, which belongs to 0.1–1, and indicates there is no significant difference. We might give the reason that because substitution and ellipsis often appear in spoken English, both Chinese scholars and American scholars seldom use them. However, the frequency of substitution used by Chinese scholars is still higher than that of American scholars, and the difference in frequency is statistically significant.

3.3 Conjunction

Table 3 shows that Chinese scholars and American scholars use 0.15 and 0.10 cohesive devices of conjunction per sentence on average. That is, the former uses 0.05 more conjunctions per sentence than the latter. Table 4 tells us that the difference in the number of conjunction by the two groups is statistically significant, because the Sig (two-tailed) value is 0.062, which is between 0.05 and 0.1, and indicates statistic difference. Through further analysis, we know that among the 39 conjunctions used by Chinese scholars, the most frequently used is however, firstly, finally, each used 4 times. Next is secondly, thirdly, then, each used 3 times and Therefore, and, so, in the first place are used twice respectively. Additionally, at last, at the same time, but, conversely, especially, fourthly, furthermore, on this basis, to that end are used once. By contrast, among the 26 conjunctions used by American scholars, however is the most frequently used, altogether 5 times, followed by additional, in general, to this end, each used 3 times. Also, thus, specifically are used twice each, and as a result, consequently, for example, furthermore, in conclusion 1 time. Thus, it can be seen from the above that the most frequently used conjunction no matter by Chinese scholars or American scholars is however. But, generally speaking, the conjunctions that Chinese scholars use the second and third most frequently are relatively simple and colloquial, while the conjunctions that American scholars use the second most frequently are relatively complex and written.

In conclusion, generally speaking, there are significant differences in the frequency of grammatical cohesive devices in the abstracts by Chinese and American scholars, and the frequency of using grammatical cohesive devices by Chinese scholars is higher than that of American scholars. Specifically: 1) There are significant difference in the frequency of reference between Chinese and American scholars. The frequency of the former is higher than that of the latter. In addition, the frequency of reference the in Chinese scholars' abstracts is much higher than that of American scholars', 66.67%, and 38.52% respectively. 2) Chinese scholars and American scholars use few substitutions, but the frequency in the abstracts by Chinese scholars is still higher than that by American scholars, and the difference in the frequency is statistically significant. 3) Neither Chinese nor American scholars use ellipsis. 4) Chinese scholars use conjunctions more frequently than American scholars, and the difference is statistically significant. Besides, the conjunctions used by Chinese scholars are relatively simple and colloquial, while those by American scholars are relatively complex and written.

4 Conclusion

Cohesion is one of the important means to achieve textual coherence. However, quantitative and qualitative analyses reveal that there are certain differences in the use of grammatical cohesive devices both Chinese scholars and American scholars, not only in the frequency, but also in the choice of specific grammatical cohesive words. In response to these differences, the author puts forward some suggestions in order to help Chinese scholars to write English abstracts at a level close to that of native speakers. They include: 1) to get to know the differences in the use of cohesive devices between Chinese language and English language; 2) to read more English academic papers written by native English speakers, and pay attention to the use of English cohesive devices.

References

1. Halliday, M.A.K, Hasan, R.: Cohesion in English. Longman, London (1976)
2. Hoey, M.: Patterns of Lexis in Text. Oxford University Press, Oxford (1991)
3. Hu, Z.: Discourse Cohesion and Coherence. Shanghai Foreign Language Education Press, Shanghai (1994)
4. Zhang, D.: Discourse coherence and discourse information structure- on the conditions of discourse coherence. Foreign Lang. Res. **3**, 7–12 (1992)
5. Zhang, D.: On the study of text coherence. J. Foreign Lang. **6**, 24–32 (1999)
6. Miao, X.: On the relationship between cohesion and coherence. J. Foreign Lang. **4**, 44–49 (1998)
7. Li, W.: A study of the relationship between cohesion and coherence from the perspective of stylistics. J. English Stud. **3**, 23–27 (2014)
8. Wei, L.: Semantic representation and discourse cohesion from the perspective of cognitive linguistics. Lang. Plann. **14**, 79–80 (2016)
9. Wang, W., Liu, H.: The application of discourse cohesion theory in the translation of political documents-taking the 2017 government work report as an example. High. Educ. Online **7**, 157–158 (2018)
10. Xu, W.: Error analysis of discourse cohesion and coherence in college English writing. Data Cult. Educ. **21**, 201–203 (2018)
11. Li, S., Hu, Q.: Quantitative analysis on cohesive words in Chinese and English patent abstracts based on parallel corpus. English Teacherss (20), 53–55 (2019)
12. Lao, T., Xin, W.: A comparative study of grammatical cohesion in English and Chinese fairy tales-taking the little prince as an example. J. Kaifeng Inst. Educ. **2**, 76–77 (2019)

13. Lu, Y., Peihong, R.: A corpus-based contrastive study on grammatical cohesive devices in English abstracts of academic journals in English and Chinese. J. Changchun Univ. Sci. Technol. (Soc. Sci. Edn.) **7**, 147–149 (2012)
14. Li, S.: A comparative study on lexical cohesion between news reports and academic abstracts. Xi'an. Chang'an University (2015)

Computer Graphics and Image Software in Advertising Design

Jichao Huai[✉]

Nanchang Vocational University, Nanchang, Jiangxi, China

Abstract. Graphic image is an indispensable element in advertising design, which has a decisive impact on the expression of advertising content. This paper mainly analyzes the various application methods of computer graphics and image software in advertising design, and discusses the advantages and disadvantages of computer-aided design software in advertising design through questionnaire survey and practical operation methods. A total of 120 questionnaires were issued and 110 valid questionnaires were recovered. The results show that the advantages of computer graphics and image processing software in advertising design are mainly to reduce the workload (95%), improve efficiency (90%), and enhance creativity (80%); the disadvantages are mainly reflected in the use of a single computer graphics and image software (80%), poor image presentation effect (78%), and insufficient layout and connotation (75%). Thus, the relevant personnel should constantly strengthen the analysis of the advantages and disadvantages of computer graphics and image software in the use of advertising, while meeting the market demand, give full play to the advantages of graphics and image software, make up for the shortcomings, and promote the sustainable development of the advertising industry.

Keywords: Advertising design · Computer aided design · Graphics software · Innovative use

1 Introduction

With the development and progress of society, people's ideas are also constantly improving [1]. Advertising design is no longer a new thing, more abundant and creative advertising design is welcomed and recognized by more and more people, which requires more and more advertising design [2, 3]. The theme, style, color and content of advertising design need to be constantly improved and innovated. Therefore, the application of computer graphics software becomes indispensable. It can greatly improve the influence of advertising and give a rich visual experience [4, 5].

The application of graphic image in advertising design is very important for realizing the ideal advertising effect [6, 7]. The appearance and use of computer graphics and image software broke through the traditional pattern of graphic image application in advertising, and greatly reduced the proportion of manual input [8, 9]. At the same time, the effect of the graphics produced by advertisements is better than that of traditional graphics.

© The Author(s), under exclusive license to Springer Nature Switzerland AG 2022
J. Macintyre et al. (Eds.): SPIoT 2021, LNDECT 98, pp. 42–49, 2022.
https://doi.org/10.1007/978-3-030-89511-2_6

The application of computer graphics and image software described here is essentially computer-aided design. Computer aided design can not only make the audience more intuitive understand and understand the specific content of the advertisement, but also enhance the overall effect of advertising [10].

In the process of advertising design, scientific and reasonable use of computer graphics and image software can enrich the innovation of advertising design to a large extent, and bring people a variety of design sense. This paper mainly analyzes the mainstream design in advertising, and analyzes the computer graphics and image software. Through questionnaire survey and practical operation, the advantages and disadvantages of CAD software in advertising design are further discussed, and some suggestions are put forward on how to promote its application in advertising design.

2 Overview of Computer Graphics and Image Software

2.1 Introduction and Application of Computer Graphics and Image Software

(1) Introduction of computer graphics and image software. The main application of computer graphics and image software is plane graphics and 3D graphics. For example, we often use software such as Photo shop, CorelDRAW, index, etc., but at present, it is still more used than Photo shop, CorelDRAW, and Auto CAD, pro/e, ug.3dmax and other software used in 3D graphics.

(2) The application of computer graphics and image software. Adobe photo shop software application. Adobe photo shop software runs on Macintosh (MAC Apple) and windows based computers (PC), which has powerful functions, such as image processing, special role effects, vector rendering, and many filters (such as KPT) and shortcut keys.

Application of 3D software. Maya, soft image/XSI and lightwave3d are very popular 3D software. It is more competitive than 3dsmax in many ways, such as Maya with particle system, dynamics and character animation system; soft image/XSI with nonlinear animation system and mental ray super render system.

2.2 Application Mode of Computer Aided Design in Graphic Advertisement Design

(1) Increase the use of color elements in advertisement design. The application of color elements in advertising design is a very common phenomenon. By using computer graphics and image software, we can greatly improve the use of color elements in advertisements, especially professional graphics and image software. It has many different color enhancement functions and related color processing tools, so it is very convenient to use in the group. The application of professional graphics and image software in advertising design can quickly adjust the color, and ensure the expression effect of advertising design can be displayed accurately, and the effect of inaccurate expression can be reduced.

(2) Increase the use of various graphic elements in advertising design. Another element widely used in graphic advertising design is graphics. The rational use of graphics can fully display the content of advertisement to a large extent, deepen the audience's memory of the advertising content, so as to achieve the ultimate purpose of advertising. Professional graphics and image software can meet the requirements of various performance effects in graphics processing, thus reducing the difficulty of graphic processing and making the application of graphics more convenient.

(3) Convenient adjustment of the whole and effect in advertising design. In the design of advertisement, the control between the whole is not to be ignored. When the scale of advertising design is large, the traditional design method is not convenient for the overall adjustment, but it becomes more convenient after the application of computer graphics software. Professional computer graphics and image software can easily adjust the whole advertisement in advertisement design, especially the vector image processing software can ensure the authenticity of the image to a great extent in the amplification and shrinking hours, and avoid over amplification or reduction without knowing. In the design of advertisement, the scientific and reasonable use of graphics and image software can adjust the overall effect in time, clearly display the content to be expressed in the advertisement, and facilitate the overall adjustment.

2.3 Bresenham Algorithm

Bresenham algorithm is the most frequently used linear scan conversion algorithm in computer graphics. Let a straight line from the starting point $(x1, Y1)$ to the end point $(X2, Y2)$, which can be expressed as the equation: $y = ms + b$, where $b = y_1 - ms_1$, $m = y_2 - y_1/(x_2 - x_1) = d_y/d_x$.

The calculation formula is as follows:

$$y = m(x_i + 1) + b \tag{1}$$

$$d_1 = y - y_i \tag{2}$$

$$d_2 = y_i + 1 - Y \tag{3}$$

3 Ideas and Methods

3.1 Research Ideas

This paper first provides a solid theoretical basis for the research through literature analysis. Secondly, the paper uses the method of questionnaire survey and practical operation to further study, analyzes the application of computer graphics and image processing software in advertising design, analyzes the advantages and disadvantages of CAD software in the specific design of advertising process, at the same time, provides more text information related to the research, and further analyzes the application of computer graphics and image software in advertising design In order to improve the overall design quality and efficiency of advertising to a large extent.

3.2 Experimental Methods

(1) Questionnaire survey and practical operation. The questionnaire is divided into two parts, including the advantages and disadvantages of CAD software in the process of advertising design; among them, 120 questionnaires are distributed and 110 effective questionnaires are collected; at the same time, some suggestions and suggestions are put forward on how to promote the application of computer graphics and image software in advertising design.
(2) Data analysis and experimental evaluation. In this study, spss20.0 was used to process the data, and the incomplete induction method was used to analyze the questionnaire, observe and record the data collected in the notes, and evaluate the CAD software index in the process of advertising design.

4 Computer Graphics and Image Software in Advertising Design

4.1 Analysis of Advantages and Disadvantages of Computer Graphics and Image Processing Software in Advertising Design

(1) Investigation and analysis of the advantages of computer graphics and image processing software in advertising design. At present, in advertising design, computer graphics and image processing software have been effectively applied. Compared with the previous methods used in advertising design for graphics and image processing, computer graphics and image processing methods have very significant advantages. The results are shown in Table 1 and Fig. 1.

Table 1. Advantages of computer graphics and image processing software in advertising design

Advantage	Percentage	Reliability test
Reduce workload	95%	0.57
Increase of efficiency	92%	0.55
Improve creativity	90%	0.51

From Table 1 and Fig. 1, the main advantages of computer graphics and image processing software in advertising design are as follows: reduce workload (95%), improve efficiency (90%), and improve creativity (80%). In the traditional advertising design, the main design is manual design, which takes a long time in the process of design and drawing, and the design efficiency is relatively low. In the picture format, the expression of color and content will be limited to a great extent, and it is not easy to modify. At present, the computer graphics and image software can fully solve this problem. With the powerful platform of computer application system, more update information can be provided for advertising design, and the selection range of graphic design can be increased, and the design content is greatly enriched. In the process of advertising design

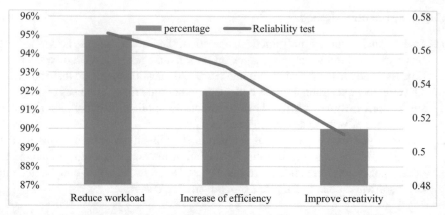

Fig. 1. Investigation and analysis of advantages of computer graphics and image processing software in advertising design

using computer graphics and image software, many links in manual design are saved, design tools are simplified and materials to be used in traditional advertising design are saved. In traditional advertising design, manual operation is affected by many factors, such as style and color, which is difficult to meet the requirements of creation. Creative advertising design can attract people's attention to a large extent, and effectively reflect the value of face-to-face advertising design, which can be realized by computer graphics to a large extent.

(2) Analysis of the deficiency of computer graphics and image processing software in advertisement design. Although computer graphics and image software have many advantages in advertising design, this does not mean that graphics and image software are universal, and the software itself has obvious defects. The results of the inferior investigation are shown in Table 2 and Fig. 2.

Table 2. Disadvantages of computer graphics and image processing software in advertising design

Inferiority	Percentage	Standard deviation
Single software	80%	4.429
Poor image rendering	78%	2.436
Lack of design format and connotation	75%	3.425

From Table 2 and Fig. 2, the disadvantages of computer graphics and image processing software in advertising design are mainly shown as: using single computer graphics and image processing software (80%), poor image rendering effect (78%), design layout and connotation (75%). When using computer graphics and image processing software to design, other computer graphics and image software should be used for auxiliary

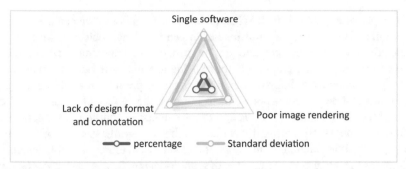

Fig. 2. Investigation and analysis of the disadvantages of computer graphics and image processing software in advertising design

cooperation. Otherwise, the rendering effect of the image will be ambiguous, and the horizontal contrast of the image will be reduced, which will lead to the unsatisfactory effect of the picture design, which makes the plane report difficult to meet the actual expectations. At the same time, when using computer graphics and image processing software in graphic design, some technicians did not carefully observe the details of the design process, which makes the layout and connotation of graphic design unsatisfactory.

4.2 Measures for Better Application of Computer Graphics and Image Processing Software in Advertising Design

(1) Enrich the color of advertisement design. In order to make computer graphics and image processing software better applied in advertising design, we should enrich the color display in advertising design. Color display is a link that attracts people in the process of graphic design. Designers can enrich the graphics or images in the design by using computer graphics and image processing software. Color is the first impression in advertising design, which plays an indispensable role in improving the quality and aesthetic feeling of advertising design.

(2) Reasonable innovation and utilization of advertising design. Through the computer image processing software, it is very important to carry on the reasonable innovation to the advertisement design. By enhancing the creativity between graphics or images, the charm of graphic design can be fully displayed to a great extent, making graphic design more valuable. Designers can use computer graphics and image processing software to design graphics, such as graphics backbone, contrast, size and other parameters, which need to be designed according to the actual design requirements. On the one hand, it is necessary to deeply apply graphics and image processing software, understand and be familiar with the operation specifications of graphics and image processing software; on the other hand, it can also give full play to the advantages of image processing software, greatly improve the creativity and expressiveness of graphics design, so as to increase the color of graphics design. Technicians can carry out reasonable innovation in graphic design, such as reasonable graphic design, that is, in the case of retaining the original characteristics of the graphics, all-round processing and design of the graphics, which is also the key

to display the graphic effect. When technicians use photos, we can use the graphic design tools in the software during the software running, such as making full use of the advantages of color pen and graphic pen, creating and changing graphics, and selecting and moving graphics in the existing graphics software. It can be seen that designers can further improve the graphics on the premise of retaining the graphic features and design concepts, so as to improve the quality of graphics and create graphics that meet the actual design requirements; in addition, designers can also filter the images. By adding filters outside, they can synthesize or match with other images to show the images to the public Good visual effect of shape design.

(3) Improve the layout and connotation of graphic design. Besides rich color expression and superb creativity, improving the layout and content of advertising design is also one of the important measures to promote the application of computer graphics and image processing software. Good design layout and valuable connotation is an indispensable part of improving the quality of advertising design. Designers need to make full use of computer graphics and image processing software to arrange and match the text or graphic style scientifically and reasonably, so as to ensure a good graphic design layout to a great extent, and give full play to its advantages.

5 Conclusions

The application of computer graphics and image software in advertising design has an important impact on the development of advertising industry. We should make full use of graphics and images as an important element to realize the innovation of advertising design. The excellent performance of computer graphics and image software can effectively expand the space of advertising design and enrich the concept of advertising design. Through the continuous analysis of the advantages and disadvantages of computer graphics and image software in advertising design, give full play to the advantages of graphics and image software, make up for its shortcomings, improve the design function of advertising, enhance the expression and creativity, and actively promote the rapid development of advertising industry.

References

1. Zhang, B., Rui, Z.: Application analysis of computer graphics and image aided design in art design teaching. Comput. Aided Des. Appl. **18**(S4), 13–24 (2021)
2. Weber, G.H., Carpendale, S., Ebert, D., et al.: Apply or die: on the role and assessment of application papers in visualization. IEEE Comput. Graph. Appl. **38**(3), 96–104 (2017)
3. Zhao, Y., Kong, X., Taubman, D. (eds.): ICIG 2017. LNCS, vol. 10666. Springer, Cham (2017). https://doi.org/10.1007/978-3-319-71607-7
4. Zhang, N.: Application of computer graphics and image software in marine graphic design. J. Coastal Res. **106**(sp1), 600 (2020)
5. Yuhang, C.: Research on application of computer aided design in environmental engineering. Agro Food Ind. Hi Tech **28**(1), 2769–2773 (2017)
6. Liu, G.: Research on the application of 3D digital technology in traditional arts and crafts design. Paper Asia **2**(1), 180–185 (2019)

7. Fan, M., Li, Y.: The application of computer graphics processing in visual communication design. J. Intell. Fuzzy Syst. **39**(8), 1–9 (2020)
8. Zhu, Y.: Research on the application of multimedia computer in news technology. Revista de la Facultad de Ingenieria **32**(14), 183–187 (2017)
9. Hu, X.: Research on the application of computer-aided art image generation based fractal method. Revista de la Facultad de Ingenieria **32**(3), 1–9 (2017)
10. Majeed, J.G., Ibrahim, Y.K., Al-Aziz, G.A.A.: Influence of heat treatment on corrosion of low carbon steel in sulfuric acid solutions. Eur. J. Eng. Res. Sci. **4**(4), 64–69 (2019)

Design and Research of Production Information Management System for Project Based Mechanical Manufacturing Enterprises

Di Wu[✉]

Network CEN, HuLunBuir University, Hulunbuir 021008, Inner Mongolia Autonomous Region, China

Abstract. With the development of the times, many enterprises' production information management system design has begun to adopt project management. Because of their customer-oriented production concept, project-based machinery manufacturing enterprises have stronger competitiveness and market advantages in the market, so this type of enterprises began to become the mainstream. In order to study and analyze the difference between the design of production information management system and the original traditional system design of project-based machinery manufacturing enterprises, we specially selected two mechanical manufacturing enterprises, but e enterprise is a project manufacturing enterprise, and enterprise f is a traditional machinery manufacturing enterprise. Then the relevant experimental research shows that the production efficiency of enterprise E is higher, the minimum efficiency is 80% in this experiment, 97% in the highest production efficiency, while the minimum production efficiency of enterprise f is 70%, and the highest efficiency is only 80%.

Keywords: Project type machinery manufacturing · Enterprise production · Information management · Design research

1 Introduction

Since the reform and opening up, China's economy has developed rapidly, and the development speed of science and technology is also faster and faster. Due to the influence of the changes of the times, many work and management methods have been improved compared with before [1, 2]. Especially for machinery manufacturing enterprises, many machinery manufacturing enterprises have improved and upgraded the methods of production information management system. Due to many advantages of project-based, such as clear and centralized objectives, fast response speed and simple hierarchy structure, many machinery manufacturing enterprises now focus on Project-based machinery manufacturing, because the manufacturing industry is an economic development in society. As a pillar of economic development, its future development trend and direction have attracted the attention of the society [3, 4]. Based on this, this paper focuses on the design and research of production information management system of project-based machinery manufacturing enterprises [5].

J. Macintyre et al. (Eds.): SPIoT 2021, LNDECT 98, pp. 50–56, 2022.
https://doi.org/10.1007/978-3-030-89511-2_7

With the continuous development of machinery manufacturing industry, the whole machinery manufacturing industry has made great progress. The original traditional production information management method is no longer applied to the current manufacturing industry, because the traditional information management system is old, inefficient and has a high error rate, which can be said to be very backward [6]. Therefore, we need to find a good way to solve this problem, so as to upgrade and optimize the existing production information management system, and promote the rapid development of machinery manufacturing industry. Because of its intelligent, scientific and standardized advantages, project-based production information management system has become the best choice for many machinery manufacturing industries. In recent years, modern machinery manufacturing enterprises have taken many measures in the development process to break the unfavorable development situation by increasing market competition and slowing down the economic situation. Moreover, they continue to study new technologies in order to improve the competitiveness of enterprises in the market, including improving technical efficiency and in-depth study of product technology. At the same time, they also introduce and deepen project management innovation on this basis. Many practices show that the effective application of project-based production information management has become the key to accelerate the development of modern machinery. In the production process of machinery manufacturing enterprises, many enterprises actively improve the production site and production process of top production technicians, stimulate innovative talents, create a relaxed and pleasant working atmosphere for all employees, train employees, and then design a standardized working mechanism. So as to avoid unnecessary waste of time, improve work efficiency, provide paperless management and one-stop service for project management, this practice will greatly reduce the workload of production information management staff. The system can provide a service platform for employees to submit proposals and projects at any time, so as to follow up the whole work process [7, 8].

Project based machinery manufacturing enterprises have become the mainstream of the industry, and the development prospect of this industry is also very broad. The enterprise production information management is the key part of the whole production process, so it is necessary to upgrade and optimize the management system to improve the work efficiency of the production management information system, which has become the consensus of many professionals in the industry. Then, how much impact this new system mode will bring to the machinery manufacturing industry, and whether the impact is positive or negative, are the objectives of this study [9, 10].

2 Method

2.1 Project Concept

The research and production period from signing the project contract to the completion of the delivery product is long. There are many uncertain factors in the process, and the implementation process is complicated, and there is no inventory product basically. In addition, because the production plan is usually made according to the specific project conditions, there is no fixed product specification and standard. In this way, although the production efficiency is not high, it can meet the needs of users generally, thus improving the product profit.

(1) Project type organizational structure. Project organization structure is the main work of the project independently, and the project team is responsible for the specific work of the project. In short, it is to establish a project organization outside the functional departments of the company.

(2) Advantages of project organization structure. (1) The real project manager is the project manager;

(3) The project manager mainly issues project management instructions, so the instructions are consistent; (3) The work goal is relatively simple and consistent with the team members; (4) Facilitate communication within the project team;(5) The project management level is relatively simple, which improves the speed of decision-making and response; (6) Project management is relatively simple, which is convenient to control the cost, quality and progress of the project.

(4) The disadvantages of project organization structure. (1) The problems such as configuration repetition and resource waste are easy to occur; (2) Because the project organization is independent of the functional departments of the company, it is relatively closed, which may hinder the implementation of the company's management and countermeasures, and the resource support of each functional department will also be retained; (3) As only project manager is responsible for communicating with the company on behalf of the project team, it is easy to have insufficient communication or insufficient communication, which leads to some problems; (4) Because the project team is built for a specific project, it is easy to generate the idea of small group and the idea of "hoarding" resources, which leads to waste of resources, and the members of the project team are easy to have no sense of belonging in the later stage of the project;

2.2 Definition of Machinery

The relative movement of the parts of the machine is determined and is a kind of artificial combination of physical components. Besides the characteristics of mechanism, machines need to have the function of replacing human labor, so as to realize mechanical energy conversion or complete useful mechanical work. Machines and mechanisms can be referred to as machinery from the perspective of structure and function. From the perspective of service industry, machinery can be divided into agricultural machinery, textile machinery, packaging machinery and mining machinery; from the perspective of working principle, it can be divided into bionic machinery, fluid machinery and thermal machinery; from the perspective of function, it can be divided into pulverizer, material handling machinery, power machinery, etc.

2.3 Project Type Manufacturing Enterprises

Bidding is the main way for the project manufacturing enterprises to obtain most orders. Price and delivery time are the key factors to success of bidding. Of course, besides, bidding will be affected by many other factors. It is very difficult to give accurate quotation during the competitive bidding period, because the products of the project manufacturing enterprises are very complex, among which tens of thousands of parts will participate in the quotation. The list of materials can reach more than ten layers, and it is difficult to accurately estimate the processing cost. In addition, another difficult problem is that enterprises need to give a reasonable delivery time within the scope of production capacity under the condition of production capacity constraints, which requires enterprises to consider the impact of production capacity on delivery time. With the rapid development of market economy and more and more fierce competition, price and delivery time become the focus of the customer's most attention, and their requirements are also higher and higher, which also leads to the price and delivery time becoming the most important issue whether the enterprise can bid successfully on the premise of profit.

2.4 Some Algorithm Formulas Involved in the Experimental Steps

When we design experiments, we often use some mathematical algorithm formulas to help us deal with experimental data and analysis problems. The formulas involved in this experiment are as follows:

$$\sup \text{port}(X) = P(X) = \frac{\text{num}(X)}{\text{num}(ALL)} \tag{1}$$

$$\sup \text{port}(X, T) = P(X, Y) = \frac{\text{num}(X, Y)}{\text{num}(ALL)} \tag{2}$$

$$\text{Lift}(X \Leftarrow Y) = \frac{P(X, Y)}{P(X) * P(Y)} = \frac{P(X|Y)}{P(X)} \tag{3}$$

3 Experiment

3.1 Research Objects

In order to explore the difference between the design of production information management system of project-based mechanical manufacturing enterprises and that of traditional mechanical manufacturing enterprises, we select two mechanical manufacturing enterprises E and F, The mechanical manufacturing enterprise E adopts the project thinking method in the design of its production information management system, while the f enterprise still adopts the original design idea. Then we conducted a random survey of 150 employees in the two factories, in order to find out people's attitudes towards the two methods.

3.2 Experimental Steps

We take the work efficiency of two mechanical manufacturing enterprises as the experimental research object, in order to investigate whether the new system design method has brought influence and change to the production efficiency of enterprises, and then we record and analyze the survey results of employees interviewed by the two enterprises, so as to explore people's attitude towards this new method.

4 Discussion

4.1 Investigation on Production Efficiency of Two Machinery Manufacturing Enterprises

In this experiment, we conducted a follow-up survey on the production efficiency of the two enterprises, and then recorded five groups of relevant data. Each group of experimental data represents the production efficiency of the two enterprises in one month. The results are shown in the following chart:

Table 1. Production efficiency of two machinery manufacturing enterprises

	Production efficiency	
	Enterprise E	Enterprise F
Group one	80%	72%
Group two	83%	75%
Group three	88%	70%
Group four	93%	80%
Group five	97%	76%

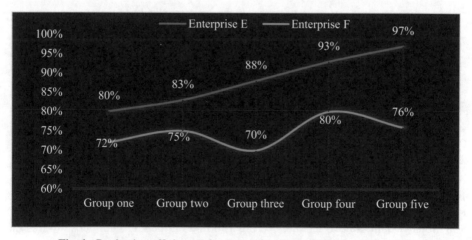

Fig. 1. Production efficiency of two machinery manufacturing enterprises

From the above Table 1 and Fig. 1, we can see that the production efficiency of enterprise E is significantly higher than that of enterprise, and the production efficiency of enterprise E has been maintaining a steady growth. From Fig. 2, we can see that the five groups of efficiency of enterprise E are almost in a straight line. Enterprise E starts from the first group, its work efficiency has been increasing. The first group efficiency is 80%, the second group efficiency is 83%, the third group efficiency is 88%, the fourth group efficiency is 93%, and the fifth group efficiency is 97%. As we can see from Fig. 1, the production efficiency curve of enterprise f is a wavy line. This shows that the production efficiency of enterprise f is very unstable, and its curve position is always lower than that of enterprise E.

4.2 Opinion Survey of Employees in Two Enterprises

We interviewed and investigated the employees of two machinery manufacturing enterprises by random sampling, so as to ask their attitude towards the two production methods. We conducted four groups of surveys according to different preferences:

Table 2. A survey of the views of employees in two enterprises

	Enterprise E	Enterprise F
Very satisfied	33%	13%
Satisfied	40%	20%
Unconcerned	15%	31%
Dissatisfied	12%	36%

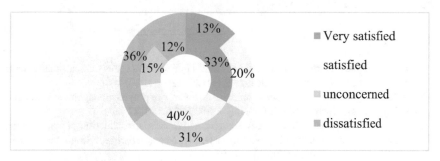

Fig. 2. A survey of the views of employees in two enterprises

According to the chart above, 40% of employees in company E are satisfied with the use of the new method, and 33% of them are very satisfied. The proportion of the two units accounts for 73% of the total number of employees surveyed. The number of employees who are indifferent to the new technology in enterprise E is 15%, and the number of dissatisfied employees accounts for 12% of the total employees. These data can show to some extent that the employees of enterprise E are still positive about the application of project-based mechanical manufacturing to the design of enterprise production information management system. When observing the relevant data of enterprise F, we can find that only 13% of the 150 employees surveyed and studied in the enterprise are very satisfied with this, 20% of them are satisfied with this, and the total number of both is less than half of the total number. The Table 2 shows that the number of employees who are not concerned about this is 31%, indicating the number of employees dissatisfied with the traditional production information management method 36%, which shows that most employees of enterprise F have no great affection for the original traditional methods, and the overall impression is biased.

5 Conclusions

With the change of the times and the improvement of living standards, people's various needs are more and more diverse and complex. In order to meet people's preferences as much as possible, many manufacturing industries now adopt project-based production information management, especially mechanical manufacturing. Nowadays, project-based machinery manufacturing enterprises occupy a large proportion in the whole machinery manufacturing enterprises. Their production information management system can analyze and study the specific situation of the enterprise, and then quickly find the problems in the production process, and then solve these problems one by one, so as to improve the existing production mode. As we can see in the experiment in this paper, the production information management system of project-based machinery manufacturing enterprises can improve the production efficiency of enterprises, and this method has been affirmed by most employees. This shows that the information management system of project-based machinery manufacturing enterprises is due to the traditional production information management system, and we also expect this new system to be able to be in the mechanical system in the future The manufacturing industry plays a greater role and creates more value.

References

1. Zheng, L.: Analysis of project management application in mechanical manufacturing enterprises. Eng. Cconstruction Ddes. **433**(11), 296–297 (2020)
2. Yang, R., Wang, L., Cui, D.: Modern application research of project management in machinery manufacturing enterprises. Sci. Inf. Technol. **000**(031), 180182 (2017)
3. Gao, Y., Wang, Y., Zhao, B., et al.: Research and practice of production planning management method based on network planning principle. Manag. Technol. Small Medium-Sized Enterprises (zhongxunyuan) **578**(06), 44–45 (2019)
4. Zhang N.: The application of project management in modern machinery manufacturing enterprises. Econ. Manag. Abstract **729**(15), 71–72 (2019)
5. Wang, Y., Liu, Z., Daiqiang, et al.: Design and application of a management system for staff site improvement and innovation and innovation. Commun. World **026**(010), 89–90
6. Lihuaze: Application of project management mode in teaching of secondary machinery manufacturing specialty. Internal Combust. Engine Accessories **000**(004), 261–262 (2020)
7. Wang, J., Wu, F.H.: Exploration of CDIO project teaching mode guided by engineering capability - taking the direction of mechanical manufacturing and automation as an example. Teach. Res. **39**(002), 98–102 (2016)
8. Xu, W.: Exploration and practice of new project based teaching mode of mechanical manufacturing process course. Sino Foreign Eexchange **000**(042), 60–61 (2017)
9. Qiang, S.: Research on the reform of practical teaching of mechanical design and manufacturing and automation specialty based on scientific research projects. Forum Ind. Sci. Technol. **18**(08), 167–168 (2019)
10. Zhang, J.: Project-based teaching design of machinery manufacturing technology course in Higher Vocational Colleges. Southern Agricultural Mach. **50**(333(17)), 212–212 (2019)

Impact of Computer Network Technology on Regional Economic Development

Tingting Li[✉]

Department of Economics and Management, North China Electric Power University, Baoding, Heibei, China

Abstract. China is in a critical period of continuous adjustment of industrial structure and continuous upgrading of economic growth model. With the rapid development of information technology, the original resource-consuming economic growth model is gradually transformed into a knowledge-based, technology-based and innovative economic growth model. This paper mainly studies the influence of computer network technology on regional economic development. This paper firstly expounds the relationship between informationization and economic development, and discusses the influence path of information and communication technology on economic output. Then, on the basis of spatial exploratory analysis, the output elasticity of information elements is estimated by using spatial econometric model, and the results are compared with the traditional growth accounting results. Finally, the paper calculates the influence of provincial informatization on economic output, and then puts forward some relevant policy suggestions for China's informatization construction.

Keywords: Computer network technology · Information technology · Regional economy · Spatial econometric model

1 Introduction

After the third industrial revolution, the development of science and technology has rapidly become the focus of the development of all countries. Especially since the beginning of the last century, when entering the era of knowledge economy, countries around the world have begun to develop knowledge-intensive high-tech industries one after another. The development of science and technology promotes the application of information technology in various industries and the optimization and upgrading of industrial structure. For example, in the aspect of industrial production, the application of information technology transforms the industrial production structure into information production and promotes the development of industrial production. With the promoting role of informatization in economy, politics, society and culture, countries all over the world pay more and more attention to the construction of informatization infrastructure, so as to promote economic growth through the development of informatization [1]. At present, regional economy tends to unbalanced development, so how to make regional economy coordinated development is the main research topic of economists. It can be

J. Macintyre et al. (Eds.): SPIoT 2021, LNDECT 98, pp. 57–64, 2022.
https://doi.org/10.1007/978-3-030-89511-2_8

seen from the course of economic growth in developed countries that the development of computer network technology industry plays a more and more important role in economic growth and industrial structure upgrading of various countries. At present, the influence of computer network technology industry on regional economic growth in China is facing the problem of unbalanced development between the eastern and western regions, that is, the eastern coastal areas are developing rapidly, while the central and western regions are relatively backward. How to narrow this gap is an economic problem to be solved in China.

In recent years, many scholars have conducted empirical studies on the correlation between informatization level and economic growth, and most of them have found that informatization has a driving effect on economic growth. Rajan analyzed the impact of information technology on the economic growth of a certain region, and the research results showed that the development of information technology contributed significantly to the growth of the overall economy and manufacturing industry, and found that the use of information technology and the growth of total factor productivity in the information technology industry had indirect effects on the development of the production sector [2]. Mahadevan used the GMM model analysis method to select the mobile phone penetration rate index to study the impact of informatization development on economic growth. The results show that the growth of mobile phone penetration rate has a significant promoting effect on economic growth, so the development level of informatization has a significant promoting effect on economic growth [3].

Based on spatial econometrics, this paper uses R and Matlab software to conduct data processing, model building and result visualization to explore the impact of informatization on economic development in China, and carries out comparative analysis among regions.

2   Impact of Informatization on Regional Economic Development

2.1   Main Impact Mechanisms

(1) The development of informatization is an important node to promote economic growth. From the macro point of view, informationization constantly creates new informationization products and expands the market scale. With the popularization of social networking, the scale effect and expansion effect of economic development are gradually improved to promote economic growth. From microscopic perspective, the information such as the development of information industry to drive the development of information technology, software operating service market development prospect is good, the e-commerce platform is more widely used, forming a complete set of electronic business service system, the system is used for economic transactions platform, therefore, the rapid development of the electricity industry to promote economic growth [4]. The development model of information industry is constantly innovated. The combination of information economy and real economy creates many new industrial models, such as logistics service industry, digital products, network culture industry, etc. These new industrial models serve as new economic growth points to promote economic growth.

(2) Informatization has optimized the economic and industrial structure. The rapid development of information technology and the wide application of information and communication technology to reform the traditional industry technology, optimize the structure of the traditional industry, changing the traditional mode of production and operation, increase the content of the traditional industry of information technology, and can extend other economic chain of industry, such as science and technology consulting, software and services, Internet of things industry. The structural characteristics of traditional industries in China are labor-intensive and capital-intensive. The development of information technology promotes the transformation of the structural characteristics of traditional industries into knowledge-intensive and technology-intensive. At the same time, the development of information technology has also promoted the transformation of labor structure from industrial and agricultural labor to high-tech industry and service industry, and the demand for knowledge and technical talents is increasingly strong [5].

(3) Development of information technology improves labor productivity and stimulates economic consumption. There are two ways of informationization to promote economic growth, namely, to promote the development of labor productivity and to stimulate economic consumption. With the rapid development of information technology, employment opportunities increase, especially the number of positions with higher requirements for knowledge and technology increases exponentially, and a large amount of human capital has been accumulated. The development of informatization has driven the development of other industries, and the number of jobs has increased significantly, so the labor productivity has increased on a large scale [6]. Information led to the development of Internet payment on behalf of information consumption pattern, the network shopping gradually become people's main shopping way, thus greatly promoted the growth of online sales of daily consumer goods, therefore, the development of information technology not only promote the increase of labor productivity, at the same time also can drive economic consumption, promote economic growth.

(4) Development of information technology improves efficiency and reduces investment cost. The operation of networked economy reduces the cost of the economic behavior of the economic subject and improves the efficiency of the economic behavior of the economic subject. The structure of economic organization is gradually transformed into network organization structure, thus improving management efficiency [7]. Through the network management mode, the decision makers can effectively carry on the resource allocation, improve the management efficiency and economic benefits.

(5) Popularization of information technology is conducive to narrowing the gap between urban and rural areas. The popularization of informationization in rural areas is conducive to improving the efficiency of agricultural production, upgrading and optimizing the industrial structure, and realizing the integration of urban and rural economy. In terms of public social services, rural population can join the social service system through informatization and enjoy basic public security services. At the same time, it can enrich individual network technology knowledge and improve their skills, which helps to narrow the gap between urban and rural areas.

2.2 Growth Accounting Framework

The use of information technology refers to the marginal contribution of the increase in output value and production efficiency of traditional industries brought about by the use of information and communication technology. The calculation idea is to separate the contribution of information technology in the output of different traditional industries and calculate the impact of the input of information and communication technology on economic growth. According to the transmission mechanism of the production and utilization of information technology products to economic development, the method adopted in this paper is the economic growth accounting framework of neoclassical economics. Informatization is regarded as a special capital goods, and its contribution can be measured on the basis of measuring the capital stock. This method assumes that an economy has A total production function with constant returns to scale, and includes information and communication technology capital (KICT), non-information capital (KN), labor (L) and technological progress (A) together into the input-output equation [8, 9].

$$Y = Af(K_{ICT}, K_N, L) = AK_{ICT}^{\alpha} K_N^{\beta} L^{\gamma} \tag{1}$$

If the scale effect is unchanged, then the sum of elasticity of each factor of production should be 1, that is, $\alpha + \beta + \gamma = 1$; If returns to scale are diminishing (increasing), $\alpha + \beta + \gamma < 1(>1)$. The production equation divides the capital stock into two parts on the whole, in which the calculation of information technology capital stock includes three parts: hardware, software and communication equipment, and is determined by the components of the narrow sense information economy. Non-information technology capital stock is other capital, such as plant, land and other fixed assets. Y is the total economic output GDP; L is the number of labor force or labor remuneration; A stands for Hicks Neutral Technology Progress.

In order to calculate the contribution of each factor to the output, the above production function needs to be differentiated. Take the logarithm of both sides of Eq. (1) and take the derivative with respect to time, then the above expression can be converted into:

$$d(\ln Y) = \alpha d(\ln K_{ICT}) + \beta d(\ln K_N) + \gamma d(\ln L) \tag{2}$$

d is the derivative of the variable with respect to time, representing the growth rate of the variable.

2.3 Spatial Measurement Model

The economic development of a region is not only related to the resource input within the region, but also affected by neighboring regions. Therefore, to study the contribution of information technology capital stock to economic growth, it is necessary to consider the existing spatial dependence. Therefore, the research of this paper needs to use spatial econometrics model to capture the possible spatial influence and bring the spatial correlation of economic variables into the original model. According to different influence modes of spatial correlation, spatial econometric models are usually divided into two classical models: spatial lag model (SLM) and spatial error model (SEM) [10].

Spatial lag model is usually used to study whether there is spatial diffusion or spatial overflow of various variables in a certain area:

$$y = \rho W y + X \beta + \varepsilon \tag{3}$$

For the spatial error model, the spatial interrelationship of the studied object is represented by the spatial error term, so the influence degree of the error impact of the dependent variable in the neighboring area on the observed value in the region is mainly investigated.

$$\begin{aligned} y &= X\beta + \varepsilon \\ \varepsilon &= \lambda W \varepsilon + \mu \\ \mu &- N(0, \sigma_z^2) \end{aligned} \tag{4}$$

Before using the spatial econometric model, it is necessary to conduct exploratory spatial analysis, investigate the spatial correlation among economic variables, and test the spatial correlation. If the real spatial correlation is found through the test, the spatial dependence effect should be taken into consideration, and the contribution of ICT capital stock to economic growth should be investigated by selecting an appropriate spatial econometric model.

3 Contribution of Informatization to Regional Economic Development

This paper will try to use the data of all the years from 2012 to 2016 to establish the spatial panel econometric model, and then select the spatial econometric model by using the Lagrange multiplier test method on the basis of the above description. Calculation related statistics, found that LM Error than LM Lag in statistically more significant, and Robust LM Error is more significant than Robust LM Lag, in measuring period, provincial spatial correlation between the level of economic development mainly space Error not Lag in the form of space, so the options panel Error model is analyzed.

Through calculation, the Hausmann test statistic is 63, and the p value is zero, indicating that the fixed effect is a better choice in the spatial error model. Therefore, the model selected in this paper is the error fixed effect model of spatial panel data. Due to the existence of spatial correlation, the parameters estimated by the least square method are biased or invalid. Here, the maximum likelihood estimation method is used for parameter estimation, and the final model selected is the fixed effect model of spatial panel error.

According to the spatial error panel model, the output elasticity of each input element is estimated. After normalization of the output elasticity of each element, the contribution degree of information and communication technology capital stock to economic growth can be calculated according to the above formula, and the specific contribution rate can be further obtained.

Table 1. Description statistics of the contribution rate of informatization over the years

Year	Average value	Min value	Max value
2012	7.31%	−1.69%	17.26%
2013	7.95%	−0.28%	15.89%
2014	14.23%	1.34%	28.37%
2015	14.62%	0.87%	33.41%
2016	15.21%	0.59%	30.05%

4 Regional Information Contribution

4.1 Statistics of Information Contribution Rate

As shown in Table 1 and Figure 1, before analyzing the contribution rate of regional informatization to economy, the descriptive statistics of the contribution rate of regional informatization from 2012 to 2016 and the descriptive statistics of the contribution rate of regional informatization over the years are first given. During the period of 2012–2016, the average annual contribution of informatization has gradually increased, especially in recent years, the contribution rate of informatization has increased significantly, from 7.95% in 2013 to 15.21% in 2016.

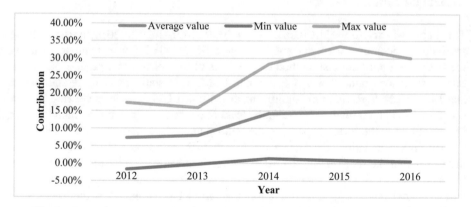

Fig. 1. Description statistics of the contribution rate of informatization over the years

4.2 Results of Regional Information Contribution Rate

For the comparison of different regions, according to the division of administrative regions in China, this paper divides all the regions investigated into three parts: eastern, central and western regions, conducts comparative analysis among regions, and gives the contribution rate and change trend chart of regional informatization during 2012–2016.

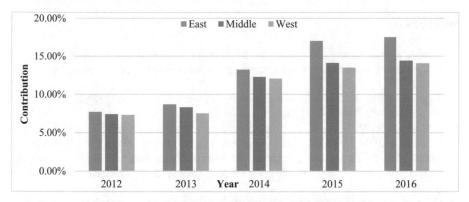

Fig. 2. Contribution rate of informationization between regions

As shown in Fig. 2, for different regions, the average contribution rate of informatization to economic growth is gradually increasing, and the eastern region is the highest, increasing from 7.71% in 2012 to 17.52% in 2016, indicating that regional informatization plays an increasingly significant role in promoting economic growth. In addition, regional differences still exist in the contribution of informatization. The contribution rate of the central and western regions is significantly lower than that of the eastern region, while the difference of the contribution rate of the central and western regions is not very obvious.

5 Conclusions

This paper, through the construction of informationization index system, measures the informationization level of our country, and obtains a series of data of informationization index, which represents the development level of informationization in our country. Through the results of informatization index, it can be found that the informatization level of eastern China is obviously higher than that of central and western regions. This paper constructs different spatial econometric models and discusses the relationship between inter-provincial informationization and economic output by using the model based on spatial correlation. The results show that information economy plays an important role in promoting regional economic development, but the regional differences are obvious. After considering the influence of spatial autocorrelation on regional economic activities, the elasticity coefficients of information technology capital, non-information technology capital and labor estimated by the spatial econometric model are all positive, indicating that the three factors jointly promote the economic development of a province. Although the elasticity coefficient of information technology capital is small compared with non-information capital and labor factors, the contribution of information capital is still considerable due to the rapid accumulation of information capital.

References

1. Rohozian, Y.S.: Strategic dominants for forming the organizational and economic potential of the regional economy development: innovative and entrepreneurial aspect. Econ. Law **4**, 44–50 (2020)
2. Rajan, S., Ratna, et al.: Regional comprehensive economic partnership (RCEP) FTA. Korea World Econ. **17**(2), 213–242 (2016)
3. Mahadevan, R., Nugroho, A.: Can the regional comprehensive economic partnership minimise the harm from the United States-China trade war? World Econ. **42**(11), 3148–3167 (2019)
4. Cui, X., Wu, X., He, X., et al.: Regional suitability of virtual water strategy: evaluating with an integrated water-ecosystem-economy index. J. Cleaner Prod. **199**(PT.1–1130), 659–667 (2018)
5. Ostapenko, E.A.: Assessment of the potential of the regional economy and its competitive advantages. Voprosy regionalnoj ekonomiki **39**(2), 65–72 (2019)
6. Myzrova, O., Serdyukova, L., Labaznova, E.: Assessment of innovative potential as a criterion for evolution of the mesoeconomic system. Regional Science Inquiry, xii (2020)
7. Mishchenko, K.N., Shevchenko, D.A., Abramyan, G.A.: Business financial support institutions in the context of enhancing the competitiveness of the regional economy. Econ. Manag. **7**, 89–96 (2019)
8. Kim, Y.J., Lee, C.I.: Sovereign debt crisis in a monetary union: accounting for excessive debt, housing bubbles, and the transmission of crises. Econ. Inq. **57**(2), 1098–1119 (2019)
9. Elgin, C., Birinci, S.: Growth and informality: a comprehensive panel data analysis. J. Appl. Econ. **19**(2), 271–292 (2016)
10. Tang, D., Li, L., Yang, Y.: Spatial econometric model analysis of foreign direct investment and Haze pollution in China. Pol. J. Environ. Stud. **25**(1), 317–324 (2016)

Chaos Algorithm of Electrical Control System Based on Neural Network Technology

Zhiwei Zhu[✉]

Wuhan Railway Vocational College of Technology, Wuhan 430073, Hubei, China

Abstract. Chaos is a common physical phenomenon in electrical control systems. The chaotic movement has high flexibility and sensitivity to the initial value, ergonomics and randomness of the movement trajectory. When maximizing and optimizing the design of the system, we can consider making full use of the chaotic ergonomics as the restraint of system maximization and optimization, to prevent the entire system from being trapped into a local minimum when it falls into search. The main reason for chaos optimization is to use chaotic variables to search for data within a certain range according to its availability and regularity, so that the search for chaotic variables exceeds the local minimum, and finally achieves the global maximum excellent. System optimization based on chaos theory is developing into a new global optimization method. In this paper, the chaotic algorithm of electrical control system based on neural network technology is researched. On the basis of related data, the chaotic phenomenon is generally understood, and then the characteristics of chaos are summarized on the basis, and then the chaos of electrical control system is explained. The discrimination method has laid the groundwork for the following experiments. Finally, the chaos algorithm of the electrical control system based on neural network technology is optimized. According to the experimental results, the optimized algorithm is nearly 10 s faster than the original algorithm. It can be seen that the optimized algorithm has better performance than the unoptimized algorithm.

Keywords: Neural network · Electrical control · Chaos algorithm · Control system

1 Introductions

China is a big power user. With the development of the west-to-east power transmission project, the interconnection of large power grids has become the development trend of future power grids. As the scale of the power grid expands, the problem of power grid stability will become more prominent [1, 2]. So far, there have been many incidents of voltage collapse and system disconnection due to network instability at home and abroad, causing serious economic losses [3, 4]. For example, when the United States was connected to the grid in 1966, there were 6 low-frequency oscillations per minute, which eventually led to a large-scale blackout [5, 6]. Similar accidents have occurred in Moscow. These serious accidents have aroused people's vigilance, and the stability of the power system has attracted more and more attention from researchers [4, 7].

© The Author(s), under exclusive license to Springer Nature Switzerland AG 2022
J. Macintyre et al. (Eds.): SPIoT 2021, LNDECT 98, pp. 65–72, 2022.
https://doi.org/10.1007/978-3-030-89511-2_9

In the research on chaos algorithm of electrical control system based on neural network technology, some researchers have carried out hybrid control on interconnected power system. Because the control objective is relatively simple, that is, the mathematical model is a first-order system, a single-state variable feedback system is used to suppress the initial chaotic phenomenon and make the system tend to a steady state [8]. The advantage is that this method only needs to receive system output information for control applications, can ignore the irregularities of system parameters, and the test results have strong durability. The disadvantage is that the controlled system can only be stabilized on a periodic orbit [9]. The researchers also verified the hybrid effect of the power system through simulation. The controller is designed with a variable structure control method of fuzzy sliding operation. The first is to design the switching control law through the approximation law, so that the system quickly reaches the switching surface, combined with fuzzy control, to suppress. Finally, the stability of the system is verified by simulation and its effectiveness is verified [10]. Some researchers pointed out for the chaotic algorithm that the method of calculating the lowest point of the hopfield energy function of the neural network is difficult to obtain the correct choice of the simulated annealing parameters due to the cumbersome process and the long calculation duration. How to directly transform the actual problems into some optimization problems that can be handled and easily solved in the genetic algorithm. Some control parameters in the genetic algorithm are difficult to accurately judge by relying on actual experience; these factors are very important in the optimization search. It may fall into a local minimum [11].

This paper studies the chaos algorithm of the electrical control system based on neural network technology. First, on the basis of the relevant literature, a simple understanding of the chaotic phenomenon of the electrical control system is made, and the relevant characteristics are summarized, and then the chaos of the electrical control system is judged. The method is explained, and the theoretical foundation is laid for the following experiments. Finally, the chaos algorithm of the electrical control system based on neural network technology is optimized experiment, and the relevant conclusions are drawn through the experimental results.

2   Research on Chaos of Electrical Control System

2.1   Chaos Characteristics

(1) Fractal characteristics
Dimensions are a quantitative description of the complexity of attractive geometric structures. In Euclidean space, space is considered to be three-dimensional, plane or spherical surface is considered to be two-dimensional, and straight line or curved surface is considered to be one-dimensional. Tractors such as balance points, limit circles, and two-dimensional turns have overall dimensions. Chaos has similar characteristics, expressed as the non-integer dimension in the dimension, that is, the fractal dimension.
(2) Randomness
The chaotic state is an instability phenomenon spontaneously produced by a deterministic system. The system exhibits various complex motions similar to random from its persistent dynamic properties. This property is also called internal randomness; the basis for the generation and formation of chaotic phenomena is rooted in the system, and has

no direct relationship with external factors. Generally speaking, chaotic systems have the integrity and stability of the system, and their parts are unstable. The local instability in the system is an important feature of internal randomization, and it is also the fundamental reason that affects the flexibility and sensitivity of its initial value.

(3) Long-term unpredictable

Since the initial conditions are strictly limited to a certain limited accuracy, and the small differences between these initial conditions will have a great impact on the future time and evolution mode, we cannot predict whether the chaotic system will behave outside a specific time in the future, has unpredictable dynamic characteristics.

(4) It has the nature of fractal

The word fractal is actually a proper term in fractal geometry theory. Fractal geometry theory is a geometric theory that takes irregular geometric figures as the main research object. The fractal dimension attribute refers to the various geometric attributes that need to be defined in the n-dimensional space. They can have an infinitely thin structure, and they can have the same scale or completely similar parts and characteristics, and they can also has a non-integer dimension that is smaller than the n dimension in the space in which it is located. The quantification of fractal dimension is the use of non-integer dimension quantification and the basic properties of fractal dimension quantification.

(5) Ergodicity

The "steady state" of chaos is not the three steady states of deterministic motion in the usual sense: static (balance), periodic motion and quasi-periodic motion; it is always limited to a limited area and the trajectory will never repeat. Therefore, as time goes by, the chaotic process will never stay in a certain state, passing through every point in the outer space.

2.2 Distinguishing Method of Chaotic Behavior of Electrical Control System

Chaos is an extremely complex dynamic phenomenon, but not all nonlinear systems have chaotic behavior. Therefore, it is necessary to study the method of judging whether there is chaos in the power system. In the early stages of chaotic development, most researchers believed that once it was determined that the system had more complex and irregular motion, chaotic behavior should be considered. It now appears that this idea is not entirely correct. After there are more ideas, some researchers suggest to create a specific mathematical model of the system and use simulation methods to observe the chaotic behavior. However, this system modeling is not a simple process. Most nonlinear systems are extremely complex, cannot accurately create high-end models, and are susceptible to external interference.

(1) Direct observation method

This method uses numerical simulation to draw the traversal graph of system variables, and then judges whether there is chaos phenomenon by analyzing the graph. If it exists, the strange attractor can definitely be found in the phase space.

(2) Lyapunov exponent analysis method

If chaos appears in the system, there are strange attractors in the phase space, and the adjacent orbits of the attractors are separated exponentially. The image based on the index can vividly reflect the behavior of the chaotic attractor and reflect the chaotic characteristics of sensitivity to the initial value. The dynamic behavior is: when the

Lyapunov exponent is less than zero, the trajectory on the exponential image is in a compact shape and is insensitive to the initial value, and the system orbit is in a fixed point state; when the Lyapunov exponent is greater than zero, the trajectory on the exponential image is in a separated shape, that is quite sensitive to the initial value, and the system orbit is in a chaotic state; when the Lyapunov exponent is equal to zero, the trajectory on the index image is in a straight line state, and the system orbit does not change.

(3) Kolmogorov entropy method

The abbreviation of K entropy was proposed by Kolmogorov. He suggested using information entropy to describe the operating state of the system, such as irregularities, low-frequency oscillations and chaotic oscillations. According to the different values of entropy K, the summary is as follows: if $K = 0$, the trajectory of the system is in a periodic orbit, showing a steady state; if $K \to$ infinity, the trajectory of the system is on the irregular orbit of the system, showing instability state; if $0 < K < \infty$, the trajectory of the system is in a chaotic trajectory, and the complexity of chaos increases as the value of K increases.

3   Chaos Algorithm Optimization Experiment of Electrical Control System

3.1   Problems in Algorithm Optimization

(1) The basic design idea of the chaotic optimization algorithm is to linearly map the chaotic variable within the range of the obtained value of the optimized variable, and then use the chaotic variable method to search. The chaotic search method is to use the characteristics and operability of chaos to traverse various situations within a certain time range according to its own rules, to conduct the entire global search and optimize it. The chaotic search algorithm is easy to exceed the local minimum, the structure is simple, the intermediate function is relatively small, and the execution speed is high. It is suitable for completing the optimization search of the entire global minimum. This method has strong applicability, and there is no special requirement for the difference between objective functions. It is only based on the repeated evaluation of the objective function. It can also use the complex objective function required by the specific application problem without paying too much extra programming and calculation time. It can be used for nonlinear, non-convex and undirected mechanical optimization problems.

(2) Neural network mapping is usually used as the chaotic variable in the chaotic search optimization algorithm. Here is a study and analysis of the problems of using neural networks as chaotic variables: the probability of chaotic variables based on neural network mapping distribution is not uniform, so no matter how the number of repetitions increases, many points are still close, and some points are difficult to reach. If the optimal point is close to these hard-to-reach points, the optimization result may not be very good. This is why the F function is not optimized in some cases.

3.2 Algorithm Optimization

(1) Suppose the optimization problem of a class of continuous objects is:

$$minf(x), x \in R \tag{1}$$

(2) The basic steps to solve this problem with the second carrier chaotic search method are:

1)) Final initialization of chaotic variables: for x in Eq. (1), n finalized values with slight differences in trajectories between other initialization variables are given, and n chaotic variables that are different from the initialization trajectory are obtained. Is x, set k = 1.

2) Linearly map the chaotic variable to the optimized variable interval according to formula (2):

$$x_{i_n n+1} = c_i + d_i x_{i,n+1} \tag{2}$$

In the formula, x is the chaotic change that occurs after the nth iteration of Eq. (2); c and d are gain coefficients. One of its main functions is to directly map the value range of the chaotic variable to the corresponding is equivalent to future optimization variables. $x'i$, n + 1 is a chaotic search variable for an optimization problem.

3) Search with chaotic variables. Let $x'(k) = x'i$, n + 1, calculate the corresponding performance index fl;(k). Let x = x(0) and f' = f(0).

4) Set k = k + 1. Turn to Step3, if f' remains unchanged after m-step search, perform the secondary carrier according to formula (3); otherwise, return to Step3.

$$x''_{i,n+1} = x^* + \alpha_i x'_{i,n+1} \tag{3}$$

5) Use the chaotic variables after the secondary carrier to continue the iterative fine search. Let $x'(k) = x'i$, n + 1, calculate the corresponding performance index fi(k).

6) If the termination criterion is satisfied, the search ends, and the optimal solution x*, f* is output; otherwise, it returns to Step5.

4 Analysis of Experimental Results

4.1 Using One-Dimensional Iterative Chaotic Self-mapping as the Simulation of Chaotic Variables

Through the optimized algorithm and the original algorithm, the results are shown in Table 1.

Table 1. Using one-dimensional iterative chaotic self-mapping as the simulation result of chaotic variables

	Based on one-dimensional iteration	Based on neural network
1000	41 s	51 s
1500	45 s	55 s
2000	75 s	85 s
2500	91 s	97 s
5000	106 s	120 s
5500	101 s	127 s

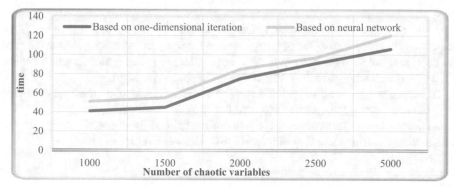

Fig. 1. Using one-dimensional iterative chaotic self-mapping as the simulation result of chaotic variables

It can be seen from the optimization results in Fig. 1 that the optimized algorithm is nearly 10 s faster than the original algorithm. From this, it can be seen that the optimized algorithm has better performance than the unoptimized algorithm.

4.2 Simulation of the New Chaotic Optimization Algorithm

In order to further verify the superiority, the optimization algorithm was optimized and analyzed for another parameter. The results are shown in Table 2:

It can be seen from Fig. 2 that the chaotic neural network mapping optimization algorithm fails due to the complexity of the target operation and the large intersection. A new type of chaotic optimization using one-dimensional repeated chaotic self-mapping, the algorithm can find the optimal value, and the optimization speed of the new chaotic optimization algorithm is faster. The number of chaotic variables has a specific relationship with the optimization results. The more chaotic variables selected, the better the optimization result and the longer it takes. Although the chaos optimization algorithm can be used to finally approximate the optimal value, it is time-consuming and the

Table 2. Simulation results of the new chaotic optimization algorithm

	Based on one-dimensional iteration	Based on neural network
3000	102 s	205 s
3500	218 s	420 s
5000	481 s	653 s
5500	534 s	774 s
6000	691 s	991 s
10000	1260 s	1960 s

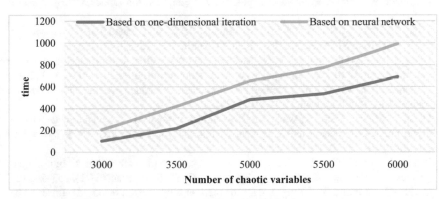

Fig. 2. Simulation results of the new chaotic optimization algorithm

performance is relatively low, but it proves that the one-dimensional repeated chaotic self-mapping algorithm is more superior.

5 Conclusions

This paper studies the chaotic algorithm of electrical control system based on neural network technology. Firstly, it puts forward the existing problems of the chaotic algorithm of electrical control system based on neural network technology. Then the algorithm is optimized according to the problem. According to experiments, it is verified that the chaotic neural network mapping optimization algorithm fails due to the complexity of the target operation and the large intersection. A new type of chaotic optimization using one-dimensional repeated chaotic self-mapping, the algorithm can find the optimal value, and the optimization speed of the new chaotic optimization algorithm is faster. The number of chaotic variables has a specific relationship with the optimization results. The more chaotic variables selected, the better the optimization result and the longer it takes. Although the chaos optimization algorithm can be used to finally approximate the optimal value, it is time-consuming and the performance is relatively low, but it proves that the one-dimensional repeated chaotic self-mapping algorithm is more superior.

References

1. Chai, X., Chen, Y., Broyde, L.: A novel chaos-based image encryption algorithm using DNA sequence operations. Optics Lasers Eng. **88**(Complete), 197–213 (2017)
2. Wang, X., Liu, C., Xu, D., Liu, C.: Image encryption scheme using chaos and simulated annealing algorithm. Nonlinear Dyn. **84**(3), 1417–1429 (2016). https://doi.org/10.1007/s11071-015-2579-y
3. Saha, S., Mukherjee, V.: Optimal placement and sizing of DGs in RDS using chaos embedded SOS algorithm. IET Gener. Transm. Distrib. **10**(14), 3671–3680 (2016)
4. Hongtao, L., Fengju, K.: Adaptive chaos parallel clonal selection algorithm for objective optimization in WTA application. Optik Int. J. Light Electron Optics **127**(6), 3459–3465 (2016)
5. Jiang, J., Chen, Y., Narayan, A.: A goal-oriented RBM-Accelerated generalized polynomial chaos algorithm. SIAM/ASA J. Uncertainty Quantif. **4**(1), 1398–1420 (2016)
6. Bian, J., Peng, X.G., Wang, Y., et al.: An efficient cost-sensitive feature selection using chaos genetic algorithm for class imbalance problem. Math. Probl. Eng. **2016**(pt.6), 1–9 (2016)
7. Doğan, Ş.: A new data hiding method based on chaos embedded genetic algorithm for color image. Artif. Intell. Rev. **46**(1), 129–143 (2016)
8. Chen, L., Ma, B., Zhao, X., Wang, S.: Differential cryptanalysis of a novel image encryption algorithm based on chaos and Line map. Nonlinear Dyn. **87**(3), 1797–1807 (2016). https://doi.org/10.1007/s11071-016-3153-y
9. Huang, X., Ye, C., Cao, L.: Chaos invasive weed optimization algorithm for multi-objective permutation flow shop scheduling problem. Syst. Eng. Theory Pract. **37**(1), 253–262 (2017)
10. Liu, X.-H., Shan, M.-Y., Zhang, L.-H.: Low-carbon supply chain resources allocation based on quantum chaos neural network algorithm and learning effect. Nat. Hazards **83**(1), 389–409 (2016). https://doi.org/10.1007/s11069-016-2320-2
11. Alhadawi, H.S., Lambi, D., Zolkipli, M.F., et al.: Globalized firefly algorithm and chaos for designing substitution box. J. Inf. Secur. Appl. **55**(102671), 1–13 (2020)

Pulse Signal Acquisition System Based on Match Pursuit Algorithm

Yinxing Li[✉] and Guokun Xie

Electrical Engineering Institute, Xi'an Traffic Engineering Institute, Xi'an, Shaanxi, China

Abstract. Pulse diagnosis is an important disease diagnosis method in ancient Chinese medicine. TCM physicians use their fingers to feel the strength and frequency of the patient's pulse to diagnose diseases. The reason why the pulse can be used as the basis for diagnosing diseases is determined by the cause and propagation characteristics of the pulse signal. Pulse signal is a kind of bioelectric signal that can reveal the activity state of human organs. The comprehensive information of many organs can be reflected by pulse signals, so pulse signals are widely used in medical research and diagnosis of clinical diseases. Pulse condition is an important physiological feature of the human body. The pulse signal contains a lot of physiological information. However, the subjectivity of traditional Chinese medicine is too strong. The objective and scientific pulse diagnosis is a problem that needs to be solved urgently. Therefore, pulse signals are widely used in medical research and diagnosis of clinical diseases. The purpose of this article is to study the pulse signal acquisition system based on matching pursuit algorithm. Tests show that the pulse signal acquisition system designed in this paper can collect human pulse signals objectively, and realize the real-time display and storage functions of the collected pulse signals.

Keywords: Matching pursuit algorithm · Pulse signal · Signal acquisition system · Pulse diagnosis

1 Introduction

Nowadays, the speed of social development is gradually accelerating, and people's life rhythm is also accelerating, and the pressure of life is increasing. In addition to the improvement of living standards, changes in work and rest time and diet, high blood pressure, heart disease and other cardiovascular and cerebrovascular diseases the morbidity is getting higher and higher [1, 2]. If it cannot be effectively prevented, it will seriously affect our future lives. Pulse condition is an important physiological feature of the human body. Pulse signal contains a lot of physiological information. However, the subjectivity of traditional Chinese medicine is too strong. The objective and scientific pulse diagnosis is a problem that needs to be solved urgently. Therefore, pulse signals are widely used in medical research and diagnosis of clinical diseases [3, 4].

The research of pulse wave, whether at home or abroad, starts with pulse diagnosis. "Pliving the pulse" appears in the traditional medicine of almost all ethnic groups.

J. Macintyre et al. (Eds.): SPIoT 2021, LNDECT 98, pp. 73–81, 2022.
https://doi.org/10.1007/978-3-030-89511-2_10

Western pulse studies have a very long history [5]. At present, many domestic companies have developed diagnostic instruments for cardiovascular and other related diseases, and most of them use single-chip microcomputers or data acquisition cards for the collection and analysis of pulse signals [3]. In a system with a single-chip microcomputer as the processor, the work flow is to amplify and filter the signals collected by the sensor. The acquired analog voltage signal is then processed by the analog-to-digital converter. Finally, the data is processed by the program in the single-chip microcomputer. Although the research history of pulse wave is very long, and a lot of achievements have been made in theoretical research and clinical practice [6]. However, it is still far away from the real modernization and objectification of "Chinese medicine pulse diagnosis", and there is still a lot of work to be done [7].

Based on the matching pursuit algorithm, this paper proposes a pulse signal acquisition system based on the matching pursuit algorithm. First, the matching pursuit algorithm is introduced, and then the software design of the pulse signal acquisition system based on the matching pursuit algorithm is introduced, and the system is simulated and tested.

2 Pulse Signal Acquisition System Based on Match Pursuit Algorithm

2.1 Matching Pursuit Algorithm

The process of decomposing the image by the matching pursuit (MP) method is as follows:

First, select the atom $g_{\gamma 0}$ that best matches the image to be decomposed from the over-complete dictionary, which satisfies the following conditions:

$$|\langle f, g_{\gamma 0}\rangle| = \text{Sup}_{\gamma \in \Gamma}|\langle f, g_{\gamma}\rangle| \tag{1}$$

After n-step decomposition, the image is decomposed into:

$$f = \sum_{k=0}^{n-1} < R^k f, g_{\gamma k} > g_{\gamma k} + R^n f \tag{2}$$

Under the condition that the image signal meets the limited length (for digital images, this is completely possible and must be satisfied), $\|R^n f\|$ will exponentially decay to 0 as n increases. Thus the image can be decomposed into:

$$f = \sum_{k=0}^{\infty} < R^k f, g_{\gamma k} > g_{\gamma k} \tag{3}$$

In fact, due to the attenuation characteristics of $\|R^n f\|$, generally speaking, a few atoms (compared to the size of the image) can represent the main components of the image, namely:

$$f \approx \sum_{k=0}^{m-1} < R^k f, g_{\gamma k} > g_{\gamma k} \tag{4}$$

Among them, $m \ll M_1 \times M_2$. Equation (4) and the condition $m \ll M_1 \times M_2$ reflect the sparsity expressed by the MP method.

2.2 Software Design of Pulse Signal Acquisition System Based on Matching Pursuit Algorithm

(1) Driver design

The pulse signal acquisition system is first initialized after power-on, including the initialization of LCD and CH376, and then the LCD displays the operation interface, including start, send, and storage options. By touching different options, the system performs corresponding operations [8, 9].

Qsys provides some commonly used NiosII peripherals, most of which are controllers for some off-chip devices, but there are thousands of off-chip components that may be used in SOPC development, each of which has a different model. It is impossible for Altera to provide IP cores for all component controllers. In order to enable NiosII to meet various needs in SOPC development, Altera provides two communication methods between NiosII processors and off-chip components or on-chip hardware logic. One is through custom peripherals, and the other is through the general-purpose input and output port (PIO) provided by NiosII.

1) LCD driver design

The driver is a set of programs that directly manipulate the registers of the device. The communication between NiosII and ILI9325 is carried out through PIO, including the address, commands and display data sent by NiosII to ILI9325. Therefore, the PIO is operated first in the LCD display driver design. It can be seen from the foregoing that the PIO core mainly includes 4 registers, namely Data, Direction, Interruptmask, and Edgecapture. The base address, interrupt information, data width, etc. of each PIO are given in system.h [10, 11].

2) Touch drive design

NiosII communicates with the touch panel controller XPT2046 through the SPI bus to read the touch point coordinates. The communication process between NiosII and XPT2046 is that NiosII first writes a control word to XPT2046 through the SPI bus, the length of the control word is 1byte: MSB S is the start bit, set high to start data communication; BIT6 ~ BIT4 A2 ~ A0 are channel selection Bit, when reading the X-axis coordinate, it should be set to 101, when reading the Y-axis, set to 001; SER/nDFR of BIT2 is the single-ended/differential reference selection bit, 1 is the single-ended input mode, and 0 is the differential Input mode; MODE of BIT3 is a 12bit/8bit conversion mode selection, the high is 8bit, and the low is 12bit, which determines the number of bits of the next read data on the SPI bus; PD1 ~ PD0 of BIT1 ~ BIT0 are power saving mode selections. The system uses a 12-bit differential mode. According to the structure of the control word, when reading X + coordinates, the control word that needs to be written is 0xD0 (11010000), and when reading Y + coordinates, the control word that needs to be written is 0x90 (10010000).

3) USB driver design.

The writing of the driver program is mainly based on the read and write timing of CH376 to control and read and write to the U disk. The connection between NiosII system and CH376 is realized by PIO components added in SOPC, and then the program controls these components. It should be noted that the bit width and direction of different signals in PIO components are different. Similarly, before writing the driver program, first define a structure according to the register file of the PIO core, and then define the pointer variables of the structure type so that these variables point to the base address of each PIO. Writing the driver program is actually Assign values to these variables according to the read and write timing of CH376 to control the data and command transmission between CH376 and NiosII. In addition, the FAT file system is integrated in CH376. If you want to read the U disk, you also need to use the processing file system program provided by the manufacturer, mainly including file creation, deletion and other functions. The USB driver written in this system mainly includes the following functions:

Set the working mode of CH376: UINT16Set_Usb_Mode (UINT8type).

Write command to CH376: voidWrite_CH376_Cmd (UINT8command).

Write data to CH376: voidWrite_CH376_Data(UINT8data).

Read data from CH376: unsignedcharRead_CH376_Data(void).

(2) Application program design of pulse signal acquisition system

1) Design of pulse signal monitoring program

In the pulse signal collection process, the collector first uses his fingers to find the position of the inch, off, and ruler, and then places the sensor in the corresponding position. However, in the process of placing the sensor, the position of the sensor needs to be fine-tuned to accurately correspond to the three points of inch, off and feet. The pulse signal detection program can display the signal collected by the sensor on the LCD in real time. By observing the amplitude of the displayed waveform and whether it has regularity, it can be judged whether the pulse sensor is accurately placed on the three points of Cun, Guan and Chi.

The sampling frequency of the pulse signal in the system is 1K, that is, one thousand points are collected per second, which is far greater than the frequency of the pulse signal. The sampling frequency can be appropriately reduced when the pulse signal is displayed in real time. The specific method is as follows: first set up a 1000-depth buffer, and display a point every 10 depths during display, which is equivalent to reducing the sampling frequency to 100. Since the pulse signal frequency is less than 30 Hz, 100 points can completely capture the pulse. The signal waveform is completely reconstructed. The LCD resolution used in this design is 320×240, and it can be considered that there are 320 points on the ordinate and 240 points on the abscissa. The specific implementation is, first, set the display area size, set the lower left corner (20, 50) as the origin, the abscissa range is (10–239), the ordinate (50–220). Secondly, set the color of the pulse signal waveform and the background color. Once again, the collected data is converted into coordinates. The vertical axis of the display area is set to 170 points, which can be converted into the relative vertical coordinate in the display area according to the number of pulse signal bits. Then you can directly call the drawing point function in the driver to display. It is worth noting that when the abscissa reaches the maximum value, the display area is partially cleared first, and then the next cycle display is started, and cyclically display in the display area. In addition, because the system clock frequency is too high, it is necessary to adjust the speed of the trace point through the delay function to make it easy to observe, so that the dynamic display of the pulse signal can be realized.

2) Design of pulse signal storage program

The storage of pulse signals is an important function of this system. Because the driver design of the USB management chip CH376 has been completed before, the application part mainly calls the driver to realize the creation of files in the U disk and the writing of pulse data. First set up a buffer in NiosII, the length is set to 5000, because the sampling rate of AD in the system is 1K, so 5 s can just fill the buffer. Since the FAT file system is integrated in the CH376, the collected pulse signal can be stored in a TXT file.

3 Introduction to Development and Debugging Tools

3.1 Development Environment

Keil was acquired by ARM in 2005, which can better provide a complete solution for the rapidly growing field of 32-bit microcontrollers, and it can also support the compilation of a variety of different processors. This article selects this piece of software as the prerequisite for program writing, compiling and debugging. It is that the author has mastered the method of writing processor programs under this software, which will help improve the efficiency of development.

This design uses the latest Keil5 development environment, which is used to write and compile STM32 programs. Its powerful functions are very convenient for us to debug programs and can help us write and debug programs such as single-chip microcomputers.

3.2 Filter Processing

In order to obtain the period of the pulse wave, the differential threshold method used in this article is used to calculate. From the analysis and research of the waveform characteristics of the pulse wave, it is found that the descending branch of the pulse wave is obviously steeper than the ascending branch, and there is a more significant differential signal. Increase the point, and the remaining points are relatively small. According to this principle, the maximum and minimum points can be found from the pulse wave. The starting point of the local pulse wave is regarded as the signal between adjacent minimum values, so that the local pulse waves are separated, and the pulse frequency is calculated.

4 System Test

4.1 Effect of Pulse Signal Image Sparsity on the Accuracy of Parameter Estimation

When the number of scattering points is constant, according to the definition of sparsity, the larger the image, the sparser the image.

We conduct the following experiments: take the imaging simulation parameters as: center frequency 10 GHz, bandwidth 400 MHz, then the range resolution Rx = 0.375 m, select the imaging accumulation angle 2.3°, so that the azimuth resolution Ry = 0.375

m. Set 4 ideal point targets at the middle distance in the simulated pulse signal image, the coordinates are (0.435 m, 0), (−0.435 m, 0), (0,0.45 m), (0, −0.45m), the scattering complex amplitude is 1. Choose M1 = N1 = 16, M2 = N2 = 32 to obtain 16 × 16 and 32 × 32 phase history domain data respectively. Add complex Gaussian white noise with peak signal-to-noise ratio to this phase history domain data to obtain noisy phase history domain observation data. After imaging, two noisy pulse signal images I and II are obtained. For these two images, the same dictionary is selected, and the matching pursuit algorithm is used to improve the resolution. The results are shown in Table 1:

Table 1. Point scattering center parameter estimation results

Parameter	Point target 1	Point target 2	Point target 3	Point target 4
Actual coordinate value	(0.435,0)	(−0.435,0)	(0,0.435)	(0, −0.435)
Figure 1 Estimated coordinate values	(0.45,0)	(−0.45,0)	(0,0.45)	(0, −0.45)
Figure 2 Estimated coordinate values	(0.45,0)	(−0.45,0)	(0,0.45)	(0, −0.45)
Actual complex amplitude value	1	1	1	1
Figure 1 Estimated complex amplitude value	1.012–0.002i	0.995 + 0.021i	1.015 + 0.012i	0.997 + 0.024i
Figure 1 Estimated amplitude modulus error	1.5%	0.5%	1.1%	0.3%
Figure 2 Estimated complex amplitude value	1.014–0.005i	0.986 + 0.031i	1.003 + 0.004i	0.994–0.003i
Figure 2 Estimated amplitude modulus error	1.2%	1.5%	0.5%	0.2%

Because the angle errors of the complex amplitude values are close to 1, only the amplitude mode errors are considered here. Similarly, in the following simulation experiments, only the amplitude mode errors are considered. It can be seen from the above table that when the scattering point is constant, the overall sparsity has an impact on the accuracy of parameter estimation, but the impact is very small and can be considered as having no impact.

4.2 Influence of Pulse Signal Image Noise on the Accuracy of Parameter Estimation

Noise is a common problem in signal processing. In the pulse signal image domain, it mainly appears as multiplicative noise, while in the phase history domain it appears as additive noise. Therefore, only the influence of additive noise on the accuracy of parameter estimation will be discussed in the following.

Taking imaging simulation parameters as: center frequency 10 GHz, bandwidth 400 MHz, the range resolution is R, = 0.375 m, and the imaging accumulation angle is selected to be 2.3°, so that the azimuth resolution Ry = 0.375 m. In the simulated pulse signal image, two ideal point targets with a scattering complex amplitude of 1 are set, and the initial coordinates are (0.75 m, 0) and (−0.75 m, 0) respectively. Choose M1 = N1 = 16 to obtain phase history domain data, add complex Gaussian white noise with different variances to this phase history domain data, obtain a series of noisy phase history domain observation data, and obtain noisy pulse signal images by imaging. The matching pursuit algorithm is used to improve the resolution, and the amplitude modulus estimation curve is as follows:

As can be seen from Figs. 1 and 2, when the noise variance is less than the threshold θ1 (this simulation experiment θ1≈5), a higher parameter estimation accuracy will be obtained: when it is greater than θ1, the parameter estimation accuracy will increase with the noise Increases and decreases; when greater than another threshold θ2 (the simulation experiment θ2≈6.2), the algorithm will appear false points. The signal-to-noise ratio is similar, except that the higher the signal-to-noise ratio, the higher the accuracy of parameter estimation; when the signal-to-noise ratio is higher than a certain threshold λ1 (this simulation experiment λ1 ≈ −24 dB), higher parameter estimation accuracy will be obtained. When the signal-to-noise ratio is lower than λ1, the parameter estimation accuracy is greatly reduced. When it is lower than another threshold λ2 (this simulation experiment λ2 ≈ −30 dB), the algorithm will have false points. And the signal-to-noise ratio has universal significance.

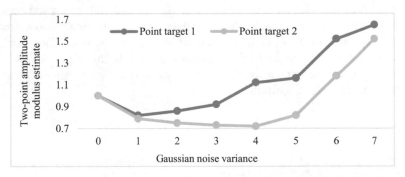

Fig. 1. The variation curve of the amplitude modulus estimate with the noise variance2

In summary, the smaller the noise, the higher the signal-to-noise ratio, and the better the performance of this algorithm to improve the resolution of the pulse signal image. When the signal-to-noise ratio is lower than the noise, the larger the signal-to-noise ratio is, and the worse the algorithm improves the resolution of the pulse signal image. When the signal-to-noise ratio is lower than −30 dB, the point target will be submerged. At this time, this algorithm will cause the appearance of false point targets.

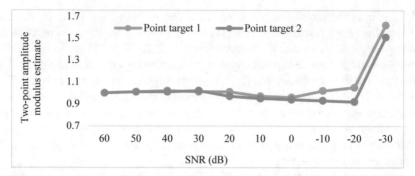

Fig. 2. The variation curve of the amplitude modulus estimate with the signal-to-noise ratio

5   Conclusions

Pulse is an important indicator of human health, and accurate measurement of pulse clinically is of great significance to the prevention and treatment of diseases. In Chinese history, most people used the pulse diagnosis method. Although the pulse diagnosis method is simple and easy to implement and does not require hardware equipment, we all know that the ancient traditional pulse diagnosis method is easily affected by external or internal interference, such as easy Influenced by the doctor's subjective emotions and medical level, or interference from the external natural environment, the use of pulse data acquisition systems can effectively avoid similar shortcomings, regardless of the difficulty or accuracy of the actual pulse measurement, it is greatly improved compared to the traditional pulse diagnosis. With the rapid development of science and technology and technological breakthroughs in the medical field, advanced pulse data acquisition devices may replace the original pulse diagnosis method. In this paper, a pulse signal acquisition system based on matching pursuit algorithm is designed. After experimental tests, the system has good performance.

References

1. Zhou, B., Wang, X., Li, Q., et al.: Human pulse signal acquisition system based on PVDF Piezoelectric film. J. Phys. Conf. Ser. **1924**(1), 012019 (9 pp.) (2021)
2. Chen, J., Sun, K., Zheng, R., et al.: Three-dimensional arterial pulse signal acquisition in time domain using flexible pressure-sensor dense arrays. Micromachines, **12**(5), 569 (2021)
3. Lee, S., Tsou, C., Huang, P.: Ultra-high-frequency radio-frequency-identification baseband processor design for bio-signal acquisition and wireless transmission in healthcare system. IEEE Trans. Consum. Electron. **66**(1), 77–86 (2019)
4. Cerina, L., Iozzia, L., Mainardi, L.: Influence of acquisition frame-rate and video compression techniques on pulse-rate variability estimation from vPPG signal. Biomed. Tech. **64**(1), 53–65 (2019)
5. Becchetti, F.D., Damron, N., Torres-Isea, R.O.: Applications of high-speed digital pulse acquisition and software-defined electronics (SDE) in advanced nuclear teaching laboratories. Am. J. Phys. **88**(1), 70–80 (2020)
6. Lebedev, S.G., Yants, V.E.: Electroluminescent fission chamber for neutron registration in counting mode with fiber optic signal acquisition. J. Instrum. **14**(6), P06002–P06002 (2019)

7. Kim, D.H., Lee, E., Kim, J., et al.: A sleep apnea monitoring IC for respiration, heart-rate, SpO2 and pulse-transit time measurement using thermistor, PPG and body-channel communication. IEEE Sens. J. **20**(4), 1997–2007 (2020)
8. Yan, W., Zhao, K., Li, S., et al.: Precise Loran-C signal acquisition based on envelope delay correlation method. Sensors **20**(8), 2329 (2020)
9. Kuncoro, C., Luo, W.J., Kuan, Y.D.: Wireless photoplethysmography sensor for continuous blood pressure biosignal shape acquisition. J. Sens. **2020**(60), 1–9 (2020)
10. Rui, J., Guan, R., Zhang, J., et al.: Design of information acquisition system for high voltage pulse power supply. J. Phys. Conf. Ser. **1894**(1), 012094 (7 pp.) (2021)
11. Xue, S., Hao, Z., An, Y., et al.: The research of three regions acquisition and analysis system of pulse based on flexible sensor. E3S Web Conf. **271**(10), 03056 (2021)

Data Analysis of Power System Engineering Construction Based on PPSO Algorithm

Zhuan Zhou[2], Chaoshan Xin[2(✉)], Shoutao Tian[2], Yan Zhang[1], and Xiumin Ran[1]

[1] State Grid Xinjiang Electric Power Co., Ltd., Urumqi, Xinjiang, China
[2] State Grid Xinjiang Electric Power Co., Ltd., Economic and Technical Research Institute, Urumqi, Xinjiang, China

Abstract. With the management reform of power grid enterprises and the construction of "big marketing" system, higher requirements are put forward for the data management and collaborative operation of power system engineering. Particle Swarm Optimization (PSO) is a new and advanced algorithm. Because of its simple, easy-to-operate, general-purpose, and parallel processing advantages, it can perform data analysis on power system engineering construction. In view of this, this paper designs an improved particle swarm algorithm (PPSO) based on the basic particle swarm algorithm, and conducts research on the analysis of power system engineering construction data. In this paper, the basic particle swarm algorithm is summarized first, and then the principle of the algorithm is researched and analyzed. Based on again, the improved particle swarm algorithm designed in this paper is proposed. And combined with the current status of power system engineering construction data analysis, based on its existing problems and deficiencies, data analysis of power system engineering construction. This article systematically expounds the system architecture, core module realization and forecasting model construction of the construction of power data analysis system. And using comparative method, field survey method and other research forms to carry out research on the theme of this article. Experimental research shows that compared with the traditional power construction data analysis system, the power construction data analysis system based on the PPSO algorithm designed in this paper is superior in many aspects, especially the status analysis is more than 15% higher, which fully reflects this article Research the feasibility of the theme.

Keywords: PPSO · Power system · Data analysis · Application research

1 Introduction

With the change and growth of energy demand, new and new energy sources such as electric energy promote the development of social demand [1, 2]. In the past ten years, our country has summed up international experience and decided to develop UHV power grids. In the national economic system, the power industry has gradually developed into a pillar industry. Electric power information has the characteristics of large volume, many types, low density value, and strong real-time performance [3, 4]. Traditional analysis methods are no longer used.

© The Author(s), under exclusive license to Springer Nature Switzerland AG 2022
J. Macintyre et al. (Eds.): SPIoT 2021, LNDECT 98, pp. 82–90, 2022.
https://doi.org/10.1007/978-3-030-89511-2_11

Internationally, there are "smart grid evaluation and investment optimization decision-making systems" that help power companies rationally optimize various construction and transformation investment plans [5, 6]; there are also "smart power outage management systems" that can manage and optimize corporate outage processing plans [7, 8]; etc. In addition, Oracle has proposed a public data model under smart grid big data [9, 10]; research institutions such as the American Electric Power Research Institute have launched several smart grid big data research projects [11, 12].

The purpose of this paper is to improve the accuracy of power system data analysis, and to analyze the power system engineering construction data based on the PPSO algorithm. By comparing the traditional power construction data analysis system with the PPSO algorithm-based power construction data analysis system studied in this article, the feasibility of the research content of this article can be judged.

2 Application Research of Power Construction Data Analysis System Based on PPSO Algorithm

2.1 Particle Swarm Algorithm

(1) Overview

The particle swarm algorithm is evolved from the swarm intelligence algorithm (Swarm Intelligence, SI). So, for example, ants can coordinate and cooperate to move things collectively and build a solid ant nest; swallows can unite and cooperate to create a strong and delicate bird's nest. Although these individuals have relatively weak abilities and simple behaviors, they can show unusually complex and orderly group behaviors when placed in a group. This intelligent phenomenon with the characteristics of self-organization, self-adaptation, and indirect communication has caused more and more. The more scholars pay attention and value.

Mathematical description of PSO: The search space is n-dimensional, the population size is m, and the population $X = \{x1,...,xi,...,xm\}$, where the position of the i-th particle is $Xi = (xi1,xi2,...,xiN)T$, The speed is $Vi = (Vi1,vi2,...,viN)T$. The individual extreme value of particle i is $Pi = (pi1,pi2,...,piN)T$, and the global extreme value of the population is $Pg = (Pg1,Pg2,...,PgN)T$. The particle Xi updates its speed and position according to (1) and (2).

$$v_{id}(k+1) = v_{id}(k) + c_1 \text{ rand} ((p_{ia}(k) - x_{id}(k)) + c_2 \text{ rand} ((p_{gd}(k) - x_{id}(k)) \quad (1)$$

$$v_{id}(k+1) = v_{id}(k) + v_{id}(k+1) \quad (2)$$

(2) Steps of basic particle swarm algorithm

1) Initialize the population, including c1, c2, and w values, the upper and lower limits of the search space Ld and Ud; the maximum number of iterations of the algorithm Tmax; the range of speed [Vmin, Vmax]; and set the current algebra to t = 0;

2) Evaluate the particles and calculate the fitness of each particle in the group;

3) Update the particle state, if Vi > vmax, use it as vmax, if vi < vmax, use it as vmin;

4) Update the current individual optimal and global optimal of the particle according to formulas (3) and (4);

5) Judge whether the condition is met, if it is met, it will jump out of the loop and get the optimal Pg value and the corresponding fitness value,

If not satisfied, skip back to the second step.

2.2 Feasibility Analysis

Feasibility analysis refers to the corresponding analysis of technology, economy and engineering, so that they can obtain a more comprehensive design plan more systematically, scientifically and correctly, analyze the economic benefits of the software to the society, and carry out the feasibility In the process of analysis, constructive opinions and suggestions should be put forward for the entire project, so as to avoid blind investment and make the designed software practical. The feasibility analysis of this paper mainly focuses on the following aspects:

(1) Economic feasibility

Mainly the cost of the operation and maintenance of the system, which also includes the equipment and personnel of the entire system. At present, computers have become popular, so you only need to install software on the computer in the office to use it. During this period, equipment costs will be incurred. And personnel costs, personnel costs mainly come from the early development costs and the later software maintenance costs, because the power equipment itself is a routine work, so the funds will be supported. Once the system is successfully developed, manpower and material resources can be greatly saved, and economic costs can be effectively saved.

(2) Operational feasibility

The interface settings are simple to operate, because this allows users to quickly become familiar with the operation of the system, and the administrator can use the system to analyze the operating data of power equipment after simple training, so the feasibility of operation. It is also certain.

2.3 System Functional Requirements

This system has multiple modules, so it is necessary to make a demand analysis for the specific functions of each module. The specific module business is: user management module. The main function of this module is to conduct unified management of personnel and departments. Set the password for simple operation; the data acquisition module is mainly to complete the data acquisition work of the operating status of the power equipment, and provide data support for the status monitoring of the equipment; test battery performance module, this module mainly provides battery reorganization plan based on various data indicators of the battery to extend battery life; detects substation equipment, provides real-time equipment status information, and provides a basis for operation, maintenance, and repair; detects power transmission equipment, monitor the equipment in the power transmission system in order to find problems in time; system parameter maintenance: This module can maintain the system parameters, and the data in the system can be entered, modified, deleted, viewed and other operations, and normal work can be carried out.

(1) Data acquisition module

The main responsibility of this module is to complete the data collection and pre-processing of the black box, and collect the power equipment operating data collected on the power site through the Internet data network to the server, and share the collected instantaneous data with the power that interacts with it. The other major type of power equipment involved is the power black box. Like the black box of the aircraft, the equipment records a wide variety of operating parameters. The intermediate parameters can be used to understand the historical operation of the system and equipment alarms.

(2) Test battery performance module

Battery performance data statistics mainly include battery AC charge and discharge capacity, battery charge power, battery discharge power, battery cumulative running time, battery cumulative charge and discharge data, etc. Through analysis and calculation of these data, a report of battery information can be obtained, and convey information to users. Calculating battery performance data can help users understand the most comprehensive battery performance, provide an effective reference for later data mining and analysis, and predict battery life.

(3) Power transmission equipment status analysis module

Therefore, a detailed demand analysis of its functions is carried out, and its main functions are subdivided into six aspects. The function of the information overview is to present the transmission line, high-voltage cable load information (35 kV and above) in the dispatching system, and the load data of prefecture and city companies to users in a simple and intuitive chart form. The equipment summary function is very important. In order to ensure the normal and stable operation of the power system, it is necessary to monitor the equipment in the power transmission system in order to discover problems in time.

2.4 System Architecture Design

The system adopts the B/S structure, the server is deployed on the open source cloud computing platform Hadoop, the web content is deployed and published on tomcat, and the browser and server directly use the RESTful architecture for communication. In the process of using RESTful technology to remotely call the service, the requested data is encapsulated into JSON format, and after data mining, the result obtained is transmitted to the business platform layer in the same format, and processed by the business logic of the business platform layer.

2.5 Reactive Power Optimization of Power System Based on Improved Particle Swarm Algorithm

Compared with the standard particle swarm algorithm, the quantum particle swarm algorithm introduces delta potential drop. Assuming that the particle is in a potential drop of δ, because of the movement of the particle, the speed and position are dynamically and randomly changed in the quantum space, then the wave function (Y) is used to represent the dynamic of the particle, and the Monte Carlo method. The position equation of the particle is:

$$X = P \pm \frac{L}{2}\ln(\frac{1}{u}) \tag{3}$$

Among them, u ~ u(0,1), because it changes with time t, the position equation of the particle can be expressed as:

$$X_{t+1} = P \pm \frac{L(t)}{2}\ln(\frac{1}{u(t)}) \tag{4}$$

Since t is a discrete time, X(t) is a random variable. When t approaches infinity, if L(t) approaches 0, the position X(t) converges to point p.

PPSO integrates quantum wave theory into the framework of PSO, so that population particles can be optimized in quantum space. Data proves that QPSO has better global convergence performance. The reactive power optimization problem of power system has high dimensionality, many constraints, and both continuous and discrete variables. Compared with the optimization results of standard particle swarm optimization and standard quantum particle swarm optimization, the simulation results show that the adaptive quantum particle swarm optimization is better efficient.

Reactive power optimization design steps:

Step1. Enter the corresponding information, namely system node information and branch information;

Step2. Randomly generate a group of populations that meet the constraints of Ug, Tt, and Qc, and set the maximum number of iterations n;

Step3. Substitute the control variables into the power flow calculation using the Newton-Rason method, find the voltage amplitude of each node bus of the power system, and calculate the active power loss of the system at the same time, substitute the objective function and calculate the result. This result is the particle's adaptability. In addition, calculate the optimal value and global optimal value of everyone in the population;

Step4. Calculate the average fitness of the particles, and judge the aggregation state of the population through the ratio of individual fitness to average fitness.

Step5. If the population is too concentrated, increase the shrinkage expansion factor to help the algorithm get rid of the local optimum; if it is too scattered, reduce the shrinkage expansion factor to improve the local search ability of the algorithm and enhance the accuracy of optimization;

Step6. Calculate the position of the particles;

Step7. Calculate the current fitness value f(x) of the particle and compare it with the values of pBest and gBest to re-determine the particle's pBest and gBest;

Step8. Judge whether the program has run to the given number of iterations n, the program will end when the conditions are met and the optimal solution will be output, otherwise skip to Step3 to continue running.

3 Data Analysis Experiment Research of Power System Engineering Construction Based on PPSO Algorithm

3.1 Experimental Protocol

In order to make this experiment more scientific and effective, this study compares and analyzes the traditional power data analysis system with the power construction data analysis system based on the PPOS algorithm studied in this paper. This experiment uses the same type of two systems. The data set is simulated and the data obtained is analyzed by tomographic analysis. On this basis, this paper conducts face-to-face interviews with professors of electric power majors in colleges and universities in a certain place. A total of 20 people was interviewed this time, among which the male to female ratio was three to one to ensure the scientificity of the experimental data.

3.2 Research Methods

(1) Comparative analysis method

This experiment compares the traditional power construction data analysis system with the PPSO algorithm-based power construction data analysis system studied in this article to judge the feasibility of the research topic in this article.

(2) Interview method

In this study, we conducted face-to-face interviews with professors of electric power-related majors in a university and recorded data. These data not only provided theoretical support for the topic selection of this article, but also provided data support for the final research results of this article.

(3) AHP

Use the analytic hierarchy process to analyze the final research results.

4 Data Analysis and Experiment Analysis of Power System Engineering Construction Based on PPSO Algorithm

4.1 Comparative Analysis of Data Analysis Systems

In order to make this experiment more scientific and effective, this experiment uses the same data set to run the two systems on the simulation software. The data obtained is shown in Table 1.

Table 1. Data analysis system comparative analysis

	User Management	Data collection	State analysis	Others
PPSO	61.2%	69.7%	71.6%	59.4%
Traditional	53.6%	55.1%	56.4%	51.9%

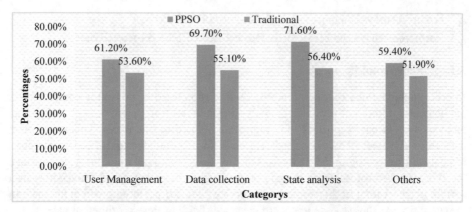

Fig. 1. Data analysis system comparative analysis

It can be seen from Fig. 1 that compared to the traditional power construction data analysis system, the power construction data analysis system based on the PPSO algorithm designed in this paper is superior in many aspects, especially in the state analysis aspect, which is more than 15% higher, which fully reflects the feasibility of the research topic in this article.

4.2   Performance Analysis of Power Construction Data Analysis System Based on PPSO Algorithm

In order to further research and analyze this experiment, this paper conducts face-to-face interviews with professors and records data, and organizes and analyzes the data obtained. The final result is shown in Table 2.

It can be seen from Fig. 2 that the performance evaluation of the power construction data analysis system based on the PPSO algorithm designed in this paper is above 5,

Table 2. Performance analysis of power construction data analysis system based on PPSO algorithm

	Convenience	Safety	Accuracy	Others
1	6.72	7.05	7.23	6.12
2	7.03	7.26	7.62	6.35
3	6.58	7.14	7.18	6.24
4	6.24	6.79	6.98	6.11
5	6.30	6.92	7.19	5.89
6	6.81	6.82	8.21	6.20
7	5.99	7.00	7.69	6.30
8	7.38	7.51	7.79	5.78
...				
20	6.82	6.87	7.69	6.08

which shows that the content of this research is feasible. Among them, the accuracy of the data analysis of the system is the highest, which fully reflects the excellent performance of the system designed in this paper.

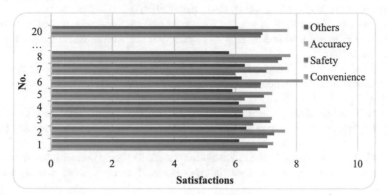

Fig. 2. Performance analysis of power construction data analysis system based on PPSO algorithm

5   Conclusion

Based on the current problems in data analysis, this paper proposes a PPSO-based power equipment operation data analysis system, which uses big data analysis methods to systematically analyze massive amounts of data, and combines various technical supports to make a complete information processing system realizes massive technical analysis work. It has realized the centralized management, real-time monitoring and visual management of big data resources of electric power enterprises, established stable, efficient and flexible system development, operation and maintenance and management support tools, and improved the practicability and ease of use of ecosystem management and development operations. Through the research on the project management process of multi-type data integration and correlation analysis, the direction of project management improvement under the background of new business and new technology is proposed, which has laid a theoretical and practical basis for supporting the application of power data analysis business, and has achieved certain research results.

References

1. Yang, L., Peng, J., Wang, T., et al.: Compliance verification and probabilistic analysis of state-wide power quality monitoring data. Global Energy Internet English Ed. **001**(003), 391–395 (2018)
2. Dekkiche, M., Tahri, T., Bettahar, A., et al.: Weather data analysis and optimal design of hybrid PV-wind-diesel power system for a village in Chlef Algeria. Desalination Water Treat. **79**(6), 125–134 (2017)
3. Cabrera, I.R., Barocio, E., Betancourt, R.J., et al.: A semi-distributed energy-based framework for the analysis and visualization of power system disturbances. Electr. Power Syst. Res. **143**(2), 339–346 (2017)

4. Jiang, J., Huang, J., Yang, Q., et al.: Research on data platform for power distribution and utilization based on CIM and OPC UA. Dianli Xitong Baohu yu Kongzhi/Power Syst. Protect. Control **47**(3), 160–167 (2019)
5. Jiang, X., Sheng, G.: Research and application of big data analysis of power equipment condition. High Voltage Eng. **44**(4), 1041–1050 (2018)
6. Yun-Hwan, L.: An analysis of delayed voltage recovery phenomenon according to the characteristics of motor load in Korean power system. Trans. Korean Inst. Electric. Eng. **65**(3), 178–182 (2016)
7. Wu, X.D., Xia, X.H., Chen, G.Q., et al.: Embodied energy analysis for coal-based power generation system-highlighting the role of indirect energy cost. Appl. Energy **184**(5), 936–950 (2016)
8. Simone, P., Alexandros, A., Ileana, B., et al.: Electrolyzer performance analysis of an integrated hydrogen power system for greenhouse heating. Case Study. Sustainability **8**(7), 629 (2016)
9. Liu, Y., Wang, X., Liu, Y., et al.: Resolution-enhanced harmonic and interharmonic measurement for power quality analysis in cyber-physical energy system. Sensors **16**(7), 946 (2016)
10. Sawicka-Chudy, P., Rybak-Wilusz, E., Cholewa, M.: Thermal efficiency of a solar power system in a collective residential structure based on performance tests. J. Renew. Sustain. Energy **8**(5), 533–582 (2016)
11. Hernando, G.P., Gonzalez, L., Escudero, F.: Analysis of the variable renewable energy in the Spanish power system based on Kernel probabilistic distributions. DYNA **96**(1), 179–185 (2021)
12. Novikov, S.O., Novikova, L.I., Titkov, V.Y.: Control modes for frequency tracking equipment in the power system. Syst. Anal. Appl. Inf. Sci. **2020**(3), 10–13 (2020)

Reactive Optimization of Power System Based on K-means Algorithm

Wei Sun[1], Shibo Jing[2(✉)], Zhiyong Yu[2], Ya Zuo[1], and Xiangping Li[1]

[1] State Grid Xinjiang Electric Power Co., Ltd., Urumqi, Xinjiang, China
[2] State Grid Xinjiang Electric Power Co., Ltd., Economic and Technical Research Institute, Urumqi, Xinjiang, China

Abstract. The RP optimization problem in power systems is a very large and complex problem. Based on the analysis of the k-means algorithm, a new improvement method is proposed. In this paper, we establish a reactive optimization mathematical model and a power system reactive optimization model with the minimum active loss as the objective function. Based on k-means algorithm, it is applied to power system reactive optimization, proposing the solution problem of power system and calculating the node system, which proves the effectiveness of K-means algorithm in solving the reactive optimization problem of power system.

Keywords: K-means algorithm · Power system · RP optimization · Power optimization

1 Introduction

With the rapid development of power system, power quality has become a very important factor to ensure the safe and stable operation of power system, and voltage, waveform and frequency are the three main reference bases to judge power quality. Therefore, ensuring that voltage quality is qualified plays an important role in ensuring the insulation performance of electrical equipment, the safe production of power users and product quality [1, 2].

In the long-term research and exploration, experts and scholars have done a series of research in the field of RP optimization. On the whole, the research topics can be divided into two categories: classical RP optimization algorithm and artificial intelligence algorithm. In recent years, researchers have also tried to combine or integrate different optimization algorithms [3, 4]. For example, combine the traditional RP optimization algorithm with artificial intelligence algorithm to synthesize their advantages, or combine different artificial intelligence algorithms to make use of the advantages and characteristics of different algorithms to learn from each other and achieve the purpose of optimization [5, 6].

However, with the more and more complexity of the power network, various voltage regulation methods and reactive compensation equipment also come, which makes it more and more difficult to reasonably optimize the RP. The traditional optimization methods gradually leak its disadvantages and increasingly fail to meet the needs of modern networks. In addition, reactive optimization itself is a complex non-linear problem. In view of its various characteristics, scholars at home and abroad have not found a truly reasonable and mature method. Therefore, this paper discusses the reactive optimization of the power system and proposes effective suggestions [7, 8].

2   Basic Theory of RP Optimization in Power System

2.1   Principle of RP Optimization in Power System

When the circuit is in normal operation, the resistance emits the electric energy in the form of heat energy, which needs to consume the active power. However, the inductor first stores the obtained electric energy in the form of magnetic energy in a cycle, and then converts the stored magnetic energy into the outward circuit release of electric energy. This cycle inductor does not consume any energy, but only the exchange and flow of energy [9, 10].

If the generator outputs excessive RP and then transmits it through a long distance transmission line, the excess RP will not only increase the active power network loss of the system, but also increase the voltage drop of the line. For a modern society that is increasingly considering economic transmission, this approach is clearly not in line with the requirements. The only way is to compensate for RP nearby. The most significant advantage of nearest compensation is to reduce the active power network loss caused by excessive RP transmission in transmission lines, in which shunt capacitors are widely used as a typical representative of RP compensation devices [5, 11].

System voltage loss AU represents:

$$AU = \frac{RP + XQ}{U} \tag{1}$$

In formula (1) It is concluded that the voltage loss consists of two parts: RP/U, XQ/U, in which XQ/U is mainly related to electrical resistance and RP. In the high voltage level, XQ/U occupies the main position in the low voltage level. Since the circuit resistance is far greater than the circuit resistance, RP/U occupies the main consumption position in AU. Therefore, reasonably reducing the transmission amount of RP in the transmission lines can effectively improve the voltage quality.

2.2   RP Compensation Equipment in Power System

The generator is the only equipment to convert the mechanical energy of the host machine into electricity. The PQ limit curve of the generator determines the output RP, and during normal operation, some induction elements on the engine and the line consume a lot of active power and induction RP, increasing the power and voltage output of the generator. It can reduce the power factor of the system, improve the power supply, and stabilize the voltage levels. The working point of the generator shall be within the working limit curve of the generator, otherwise the power supply will be limited, which will adversely affect the user and the performance system.

2.3   Common RP Control Methods

Change the terminal voltage of the generator

With the increase of generator terminal voltage, the power supply also increases. If the generator clamp voltage decreases, even if the generator set runs backward, the RP of the generator will also be reduced. Therefore, in the control process, the RP of the generator must be limited to a certain range, when the RP of the generator reaches this limit, the voltage control stops. This method can correctly make use of the unique ability of the generator to generate and absorb RP, which is simple to operate and low in cost. It is the first choice for all kinds of voltage control methods, but this method can only meet the requirements of simple control of power supply voltage in some power plants. If the power supply is large, such as multi-stage transformer, the adjustment of generator voltage is difficult to meet the requirements of RP of the system, so it must be coordinated with other measures to meet the requirements of blind power.

Change of variable ratio of adjustable transformer

The transformer adopts the principle of electromagnetic induction. In order to change the power transmission voltage, the transformer will produce blind power consumption in the process of operation. In the process of transformer operation, due to the existence of series impedance, RP and voltage will be interrupted, and only the RP distribution on the high and low voltage side of the transformer will be changed when generating power. That is to say, reduce (or increase) the voltage on the low voltage side, so as to reduce (or increase) the RP on the high voltage side, and the load step switch, also known as the load step switch, has the advantage of wide voltage control range. The control time is not limited, can be adjusted at any time, and is easy to meet the needs of users. Because the change of variable ratio of adjustable transformer will consume RP, only RP control and voltage control can be carried out when the RP of the system is insufficient, the voltage control by changing the variable ratio of adjustable transformer will lead to the decrease of primary side voltage. This will eventually destroy the stability of the whole system or even paralyze. When the blind power of the system is insufficient, RP compensation devices such as bypass capacitors must be installed to ensure the RP balance of the system.

Installation of RP compensation equipment

By installing the RP compensation device in the correct position of the power grid, the voltage loss and power loss can be reduced. The compensation device is widely used in the power grid to improve the system voltage; when the system load is reduced, the number of shunt capacitors will be reduced, only in special cases, can static compensators and synchronous capacitors be used as RP compensators. Compared with synchronous capacitors, the RP regulation of static compensator is fast and stable. If the load changes unevenly, it can compensate the dynamic requirements of phase. It is a kind of simple operation and RP compensation device. Because of its convenient maintenance and good performance, the RP compensation device adopts static compensation device, and the price is similar. All kinds of reactions of power system equipment will have an impact on the system. Therefore, blind tuning should first ensure the safety and stability of the system, and then choose advanced technology and simple operation, economic voltage control means, in order to obtain the best optimization effect.

2.4   Improve the Search Strategy Based on the K-means Algorithm

Improved algorithm for the k-means algorithm. We call the improvement of the k-means plus, k-means plus algorithm is precisely aimed that the traditional algorithm does not fully extract useful information in the iteration process in each iteration, it is simply observed after each iteration to analyze whether the convergence of each result cluster improves. For any distance from the element Q to the central point of each cluster, we can give this formula:

Assuming k clusters, for each element Qi during the m iteration, the distance to the center of the i cluster is calculated as follows:

$$D_{\text{plus}}(w, 0) = (1 - \frac{e/k}{k^2}) \times d(w, o) \tag{2}$$

Among them, w represents the i element, o means the center of mass of the j cluster, e means the number of times the i element is assigned to the m cluster during iteration, (w, o) represents the traditional Euclidean distance, the division by k here is the implemented data. We should ensure that the weights cannot be too large, and the smaller weights are uncertain, so it is selected for power reduction here.

3   Construction of a RP Optimization Model Based on the K-means Algorithm

3.1   RP Optimization Power Grid Design

In this paper, the RP control variables in the same power supply area are divided into one area. However, in the actual power supply system, because of the complexity of the power supply area, the complexity of the problem has not been substantially reduced, so it is necessary to divide these areas with large scale control variables.

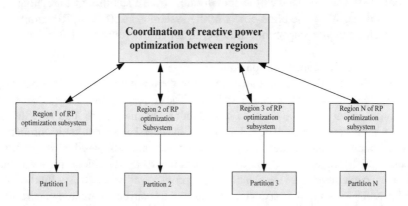

Fig. 1. Layered control RP optimization diagram

Figure 1 is a schematic diagram of the hierarchical control. It can be seen from the diagram, the first layer is the reactive optimization of each subsystem, according to the optimization of each subsystem, through information exchange with the second layer (coordination layer), iterative calculation until the optimality criterion is satisfied.

3.2 Characteristics of RP Optimization Problem

The RP optimization of power system has the following characteristics:

(1) Nonlinear

In order to meet the requirements of active power and RP balance in the system, the power flow calculation equation with nonlinear characteristics is brought into the equality constraint conditions. On this basis, the mathematical model of RP optimization has typical nonlinear characteristics.

(2) Large scale

The power grid is gradually developing in the direction of UHV AC/DC interconnection, large capacity units and long distance transmission, and the nodes and line multiples contained in the network structure are increasing. With the increase of the number of nodes, the corresponding electrical equipment is increased, and the network structure becomes more complex.

(3) Discreteness

Dealing with the discrete variables in the optimization problem is helpful to improve the efficiency of algorithm optimization and save time and cost.

(4) Convergence depends on the initial value

The constraint conditions of power system include power flow calculation equation, which is a transcendental equation with diversity. In the actual operation process, the RP optimization problem itself has more than one solution. Coupled with the nonlinear characteristics of constraints, it is more difficult to realize the differentiable and continuous requirements of RP optimization after introducing discrete variables into the equation. Therefore, how to select and determine the initial value has a very important impact on the convergence of RP optimization.

3.3 Construction of Mathematical Model

The RP optimization of power system can reduce the network loss of the system, reduce the occurrence of voltage overrun, improve the voltage quality, and make the power grid safe, stable and economical to supply power to users.

In this model, the power balance is used as the equality constraint. Under the power flow distribution of the power grid, the active power and RP consumed by the system should be equal to those generated by the system, as follows:

$$P - V \sum_{}^{M}(H\cos\alpha + J_{\sin}\alpha) = 1 \qquad (3)$$

In formula P is the active power of node i; V is the node i RP; H, J is the voltage of node i and j. Due to the complexity of the reactive optimization problem itself, the characteristics of its mathematical model should be considered in which algorithm is used to solve the reactive optimization problem. The reactive optimization mathematical model is high dimensional and nonlinear, summarized as follows:

(1) High dimension

RP optimization is a large-scale optimization problem in power system. The characteristics of high dimension are mainly reflected in the large number of nodes in the power grid itself, various control variables and state variables, as well as various equality and inequality constraints on variables.

(2) Nonlinear

The constraints of variables in RP optimization are nonlinear, and the phenomenon of non-convergence will occur when the traditional algorithm is used for iterative optimization. Therefore, this characteristic should be fully considered in RP optimization, and reasonable equality and inequality constraints should be adopted.

(3) Inseparability

Obviously, the RP optimization problem is inseparable. This characteristic makes it impossible to simply decompose RP optimization problems into various small system problems. Different objective functions are used to solve the corresponding separability. When the RP optimization of power system takes the system network loss as the objective function, the inseparability is especially obvious at this time.

(4) Sparsity

Due to the characteristics of the power grid itself, it is sparse to solve the time variable constraints of RP optimization. Reasonable use of this characteristic, set reasonable equality and inequality constraints, so that the algorithm calculation time will not increase exponential with the expansion of the scale of the system.

(5) Nonconvexity

The RP optimization problem of power system is no longer a large-scale, nonlinear, non-differentiable and other characteristics. Therefore, to be exact, solving RP optimization can only be solved.

4   Study on Reactive Optimization of Power System Based on K-means Algorithm

4.1   Rereactive Optimization Node Test Model Based on K-means Algorithm

The IEEE14 node test system includes 4 generators (corresponding node number 1,2,3,6) corresponding to G1, G2, G3, G6, three load voltage regulating transformers (corresponding branch is 1–6,4–7,4–9) corresponding to T1, T2, T3 and one RP compensation node (node 9) respectively.

Node 1 is selected as the balance node, node 2,3,6,8 as PV nodes, and nodes 4, 5, 7, 9, 10, 11, 12, 13, 14 as QP nodes for optimization processing. The voltage value range in the setting system is [0.95, 1.1], the upper and lower limits of the transformer transformer ratio are [0.9, 1.1] and the adjustment step length is 0.025.

4.2 Various Data of Node Test Model

The following Table 1 include IEEE14 The generator limit table, node data and other parameters of the node test system.

(1) The upper and lower limits of reactive and active power of generator

Table 1. Generator limit value table of IEEE14 node test system

Node number	Upper limit of RP	Lower limit of RP	Upper limit of active power	Lower limit of active power
Two	1.2	−0.21	1.6	0.22
Three	0.9	0.16	1.3	0.13
Six	0.7	−0.25	0.8	0.1
Eight	0.4	−0.2	0.7	0.2

(2) System node parameters

Table 2. IEEE14 node test system node data sheet

Node	Generator output		Load		Node voltage	
	Active merit	RP	Active merit	RP	Amplitude	Phase angle
One	Two hundred	−16.9	0	0	1.06	0
Two	Fifty	41.2	23.3	12.7	1.045	−4.98
Three	0	24.5	Ninety-five	Nineteen	1.01	−12.72
Four	Forty-eight	Seventeen	41.2	−3.9	1.019	−10.33
Five	Twenty-five	-	2.36	1.6	1.02	−8.78
Six	0	Thirteen	4.36	7.5	1.07	−14.22
Seven	Thirty-five	One	0	0	1.062	−13.37
Eight	0	17.6	0	0	1.09	−13.36
Nine	-	-	29.5	16.6	1.056	−14.94
Ten	Forty-five	-	Nine	5.8	1.051	−15.1

According to Table 2, the RP of the generator meets the requirements of the RP in Table 1, and the voltage amplitude of the QP node is also within the required range.

4.3 Comparative Analysis of Simulation Results

It is obvious from Fig. 2 that most of the system node voltage values optimized by the K-means algorithm are located in the middle of several curves and are more stable

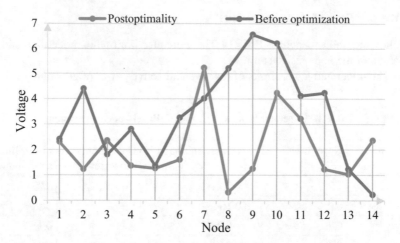

Fig. 2. Comparison of node voltage before and after the optimization

relative to the pre-optimization voltage values. When the system voltage fluctuates, the node voltage optimized by the K-means algorithm has more change space within the upper and lower voltage limit range, which increases the stability of the system to some extent.

5 Conclusion

With the gradual expansion of the scale of the power grid, the demand for electricity consumption of residents is also increasing. Power RP optimization is one of the important means to control the safety, stability and economic operation of power system, is the prerequisite for safe and reliable operation of power system, and is the scientific allocation of RP resources. This can effectively reduce the system loss and further improve the system performance is the key to ensure the stability of the network.

References

1. Xu, Y., Liu, Y., Xie, Q.: Research on RP planning of distribution network based on life cycle cost. Dianli Xitong Baohu yu Kongzhi/Power Syst. Protect. Control **46**(11), 30–36 (2018)
2. Lizhen, W.U., Jiang, L., Hao, X.: RP optimization of active distribution network based on optimal scenario generation algorithm. Power Syst. Protect. Control **45**(15), 152–159 (2017)
3. Zuo, T., Meng, K., Tong, Z., et al.: Offshore wind farm collector system layout optimization based on self-tracking minimum spanning tree. Int. Trans. Electr. Energy Syst. **29**(2), e2729.1e2729.16 (2019)
4. Su, J., Sun, J., Liu, Y., et al.: Research on AVC system optimization of power plant based on human-simulated intelligent control. Dianli Xitong Baohu yu Kongzhi/Power Syst. Protect. Control **46**(2), 157–162 (2018)
5. Zheng, F., Chiao, C.: Research on prediction and recommendation of financial stocks based on K-means clustering algorithm optimization. J. Comput. Meth. Sci. Eng. **2**, 1–9 (2021)

6. Fuchs, K., Kuiava, R., Fernandes, T.S.P., Block, P.A.B., de Souza Benedito, R.A.: Transient stability constrained optimal power flow based on multi-time scale power system models. J. Control Autom. Electr. Syst. **28**(3), 418–427 (2017). https://doi.org/10.1007/s40313-017-0304-4

7. Yu, L., Zhang, X., Liu, Y., et al.: An adjustment method of distribution network RP compensation partition balance with distributed power sources. Dianli Xitong Baohu yu Kongzhi/Power Syst. Protect. Control **45**(5), 58–64 (2017)

8. Ethmane, I.A., Maaroufi, M., Mahmoud, A.K., et al.: Optimization for electric power load forecast. Int. J. Electr. Comput. Eng. **8**(5), 3453–3462 (2018). (ISSN: 2088-8708)

9. Christakou, K., Paolone, M., Abur, A.: Voltage control in active distribution networks under uncertainty in the system model: a robust optimization approach. IEEE Trans. Smart Grid **9**(6), 5631–5642 (2017)

10. Chakrabarty, M., Sarkar, D., Basak, R.: An interactive partitioning algorithm-based electrical power crisis management for service restoration with existing black-start resources considering load priority. J. Inst. Eng. (India) Ser. B **102**(2), 169–178 (2020). https://doi.org/10.1007/s40031-020-00519-9

11. Du, Y.A.: Research on the route pricing optimization model of the car-free carrier platform based on the BP neural network algorithm. Complexity **2021**(4), 1–10 (2021)

Design and Structure Analysis of Manipulator Based on Acceleration Sensor

Shuiqin Zhu[1]([⊠]) and Jianliang Cai[2]

[1] Ningbo Polytechnic, Ningbo, Zhejiang, China
zsq6879@sina.com
[2] Hangzhou Anmaisheng Intelligent Technology Co., Ltd., Hangzhou, Zhejiang, China

Abstract. In the process of movement, the mechanical arm is easy to be affected by external impact factors, leading to its shock or unstable work. This present the design and structural analysis of mechanical arm based on acceleration sensor. Firstly, the dynamic equation and the bottom control algorithm. Measure the mechanical arm joint angle feedback control parameters to realize the dynamic analysis of the mechanical arm feedback control. The acceleration sensor is designed on the analysis basis and realizes the sensor signal fusion. Thus, we realize the mechanical arm design and structure analysis based on the acceleration sensor. Finally, the results show that the designed mechanical arm can realize the balance control of the main and underdriving arms, with good signal output stability and parameter adjustment ability, and low automatic control error.

Keywords: Acceleration · Flexible joint · Vibration suppression · Robot arm · Sensor

1 Introduction

With the rapid development of the industrial economies and the progress of science and technology in modern society, industrial robots, with their high sensitivity, large working space, simple and convenient structure and many other advantages, have made great contributions to the development of the industry economies. At the same time, as an important part of industrial robots, the rationality of their structure design and good performance are directly related to the operation efficiency of industrial robots to a great extent [1]. Robots play an important role in industrial control, and the demand for reconfigurable manipulators with strong versatility is also increasing. However, in practice, different manipulator configurations lead to different uncertainties, such as friction, physical parameter perturbation, load variation and external interference, which become uncertain or partially uncertain inputs as well as measurement errors. Under this imperfect knowledge, how to obtain high-precision performance for trajectory tracking problem is very important. Therefore, it is a challenging task to design a general controller with reliable performance for various configurations of reconfigurable manipulator [2–4]. At the same time, the analysis and exploration of the mechanical arm structure design and its related properties also have an important theoretical significance and practical value that cannot be ignored.

J. Macintyre et al. (Eds.): SPIoT 2021, LNDECT 98, pp. 100–115, 2022.
https://doi.org/10.1007/978-3-030-89511-2_13

There are two modes for robot to complete task operation, namely, position control in free space and force control when constrained by environment, which requires the manipulator to easily switch from one mode to another. Generally, there is no problem in the transition from constrained force control to unconstrained position control, but there will be a serious problem of impact force in the force control from free space to constrained space [5]. Because the normal contact velocity is not zero, there is discontinuity in dynamics, which will lead to excessive impact force. In the contact point/surface force control, there will be oscillation and even instability of the system. The classification of basic force control algorithms for robots is based on the following relationships: the relationship between position/velocity and force, or the use of direct force feedback, or a combination of them. A common characteristic of these control methods is that in the "transition process", the feedback information of speed and force is mainly used to stabilize the control system and improve the force control performance. However, due to the narrow bandwidth of the speed loop, the energy consumption of the contact force control is slow when the speed feedback is used as the damping, so the system will vibrate during the contact transition. In order to increase the robustness of force control system, robust force control algorithm, nonlinear and adaptive control and nonlinear torque control can be used. However, due to the complexity of the algorithm and the large amount of calculation, it is difficult to apply in the contact force control. It shows that although most current methods realize the effective control of the mechanical arms, they all need to be improved in some respect [6, 7].

In order to solve the poor feedback control and poor balance control ability, this paper proposes the design and structural analysis of the mechanical arm based on the acceleration sensor. Using the feedback control in the acceleration sensor to improve the damping for the contact force control of the mechanical arm, the proposed method can not only avoid the limitation of the simple speed feedback control band width, but also improve the accuracy of the control of the mechanical arm.

2 Structure Design of Manipulator

2.1 Mechanical Arm Structure Analysis

Looking at the whole development process of industrial robots manipulating, it can be seen that traditional mechanical equipment often needs to occupy a large space for use, and it is difficult to use in some narrow industrial places or workshops. However, with the development of modern society, some special industrial production places with narrow space for use need to liberate the labor force to a greater extent, so the degree of freedom is higher. The research on the structure design and function of the manipulator with more flexibility and smaller space use area is of great significance [8–10]. Mechanical structure is an important part of the manipulator. In order to better control the actual use process and work efficiency of the manipulator, it is necessary to study the mechanical structure of the manipulator and discuss its power transmission method and power source. Generally speaking, the mechanical arm which depends on the circuit transmission has a wider application range. The diversified power transmission methods such as gear type, connecting rod type and rope type make its application prospect more broad. In this process, as the main structure of the circuit drive manipulator, gear type has the advantages of compact structure, strong flexibility, high bearing capacity and good accuracy. But at the same time, gear manipulator often needs a reducer in its actual use process, so it also has some disadvantages, such as large space occupation and large quality. In addition, with the further development of power and electrical technology in modern society, a certain number of motors are installed in the special joint structure of some manipulators, which greatly improves the accuracy and safety of the manipulator.

2.2 Mechanical Arm Structure Design

Figure 1 shows the design of the manipulator, which is mainly a two degree of freedom motion mechanism composed of the first joint and the second joint.

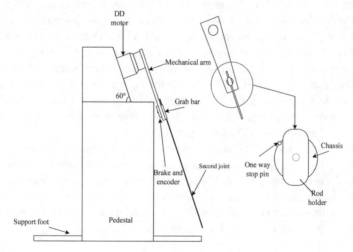

Fig. 1. Robot structure chart

Base and supporting feet: the base is the Coordinate Datum of the robot motion, so the stability and balance of the base must be maintained under the high-speed motion of the robot arm. As the reaction force of the arm against the base is quite large when the high-speed swing is in motion, a heavy block is added on the base to increase the stability (as shown in Fig. 1). At the same time, the unfolded support foot is designed to increase the support area and prevent the robot from overturning.

First joint: driven by a DD motor. In our research, the control method is to give the driving torque directly to control the torque of the mechanical arm. Direct drive motor to improve control accuracy and reaction speed. The actual maximum output torque of the motor is 110 nm. In our experiment, the maximum torque is limited to 100 nm.

The second joint is a non driving joint. In this joint, an encoder is used to measure the joint movement angle, and a brake device is used to provide the braking force after the action. A one-way stop pin is set on the joint of the robot to prevent the grab rod from rotating too much clockwise.

2.3 Kinetic Equation

The robot arm motion plane is used as the coordinate plane to establish $x - y$ coordinate, and the first joint axis is used as the coordinate origin. θ_1, θ_2 is defined as the motion angle of the first joint and the second joint. Then the forward kinematics solution of the grab bar is as follows:

$$\begin{cases} x = l_1 cos\theta_1 + l_2 cos(\theta_1 + \theta_2) \\ y = l_1 sin\theta_1 + l_2 sin(\theta_1 + \theta_2) \end{cases} \tag{1}$$

Among them, l_1 and l_2 are the length of two connecting rods respectively, and l_1 is the distance from the first joint axis to the second joint axis, which is 0.34M; l_2 is the distance from the axis of the second joint to the hitting point. The experimental value is 0.955 m (varying according to the club).

The inverse kinematics solution is:

$$\begin{cases} \theta_1 = arctan(\frac{(l_1+l_2cos\theta_2)y-l_2xsin\theta_2}{l_2ysin\theta_2+(l_1+l_2cos\theta_2)x}) \\ \theta_2 = arccos(\frac{x^2+y^2-l_1^2-l_2^2}{2l_1l_2}) \end{cases} \tag{2}$$

According to the different arm shape, there are multiple solutions, that is, there are positive and negative solutions when solving θ_2 in the above formula.

2.4 Control Structure of Robot Bottom

Because only one drive joint needs motor control, the bottom control hardware of the robot is relatively simple, but because the grab rod action is a high speed motion, the whole process has only about 0.8 s. There are complex torque changes in this process, so the performance of the underlying hardware is also quite high [4].

Figure 2 shows the hardware structure of the control system.

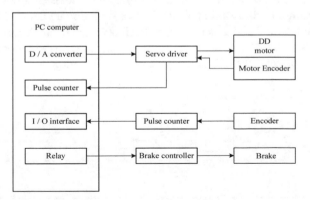

Fig. 2. Robot control hardware

The control computer is NEC PC-9801 microcomputer, which adopts DOS 6.0 operating system. The D/A converter converts the calculated control torque into analog quantity and outputs it to the motor driver to drive the motor. At the same time, a pulse counter is used to receive the signal from the motor code disk. Due to the high precision of DD motor's code disk (319 488 pulses per revolution), the frequency of code disk signal is very high under high-speed operation, so qpg45 high-speed pulse counter is used, and its counting frequency can reach 1.8 MHz. For the non driving joint, an encoder is used to feed back the angle of the joint, and the brake is used for braking. The signal of encoder is read through I/O interface of computer after counting, and the brake is controlled by computer through relay. The reaction time of braking is 50 ms, and the braking torque is 25 nm when the braking is released for 20 ms.

The sampling period of the whole control is 1 ms, and timing interrupt is provided by 8253 chip. In the bottom control, the current loop (torque) is controlled by the motor driver, and the speed loop and position loop are controlled by the computer software, that is, the PID controller of software servo. Since there is no control torque for the non driving joint, the PID controller has two inputs and one output, that is, to adjust the two joints at the same time, which is actually equivalent to the superposition effect of the outputs of the two PID controllers, as follows:

$$u = K_{p1}\Delta\theta_1 + K_{I1}\int \Delta\theta_1 + K_{d1}\Delta\dot{\theta}_1 + K_{p2}\Delta\theta_2 + K_{I2}\int \Delta\theta_2 + K_{d2}\Delta\dot{\theta}_2 \quad (3)$$

The coefficients are adjusted according to the experimental results. As explained in the introduction of the first chapter, this kind of PID feedback controller can not guarantee the uniform convergence of the error. In fact, in the process of motion, the effect of two joint errors on the controller will cancel each other [11]. But it can keep the balance between the errors of two joints and prevent one-sided increase of the error of a joint. Moreover, it can effectively reduce the error when the error effect of two joints is the same.

The control structure of the whole system is shown in Fig. 3, in which the track generation part is calculated offline, and the calculated track data is transferred to the lower computer for the underlying control as a file. The robot is controlled by the lower computer. Dynamic feedforward and PID feedback control are adopted in the control. The feedforward control signal is the torque track obtained by off-line calculation. The PID controller is realized by the lower computer with software. The motion track obtained by off-line calculation is the reference signal of PID regulation.

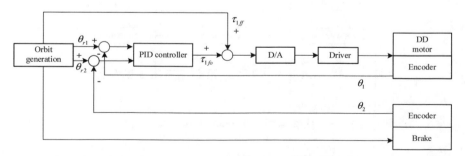

Fig. 3. Overall control structure of the robot

3   Design of Mechanical Arm Based on Acceleration Sensor

3.1   Robot Arm Control Algorithm

As we all know, there are many control methods used in the control system of robot arm. Fuzzy control method, PID control method and adaptive control method can be used as important control means of the control system of robot arm [12, 13]. In terms of the development of various control methods at this stage, adaptive control method and fuzzy control method are the methods with high accuracy and strong safety. The two control methods are not only widely used in the manipulator of industrial robots, but also widely used in aerospace engineering and military industry. PID control method as a common control method of manipulator, its control accuracy is relatively low, but in some industrial production areas with relatively low accuracy requirements, using PID control algorithm to design manipulator control system has good economic advantages. But with the further development of modern society, a variety of control technologies are widely used in many industries to achieve the comprehensive optimization of price, control performance and design.

3.2   Analysis of Joint Angle Feedback Control Parameters of Flexible Joint Manipulator

(1)   Joint angle feedback control parameter measurement of flexible joint manipulator

In order to realize the joint angle feedback control of flexible joint manipulator, the constraint parameter model is established. Combined with the sensor measurement method of the joint angle of the manipulator, the information fusion and feature extraction are carried out. Combined with the upper computer communication control technology, the operating system and user control system of the joint angle feedback control of the manipulator are constructed. The manipulator realizes the control state parameter fusion and control law design in the data processing module. The overall structure is shown in Fig. 4.

According to the overall structure of the manipulator joint angle feedback control shown in Fig. 4, the sensitive parameter identification model of the manipulator joint angle feedback control is established by using the method of sensitive elements and sensor information collection. The method of position and pose tracking and identification is realized by using paper filter. The system equation of the manipulator joint angle feedback control is obtained as follows:

$$A = r(1 - \frac{2x_1 + \sigma x_2}{N})$$ (4)

Among them, r is the movement distance of the manipulator joint, σ is the sensitive parameter of the manipulator joint angle feedback control, x_1, x_2 is the parameters of the manipulator joint angle before and after the change of the end pose, and N is the control coefficient. The linear parameter fusion model of manipulator joint angle feedback control is constructed by using dual joint differential drive mode. Combined with the pose parameter sensing measurement method of

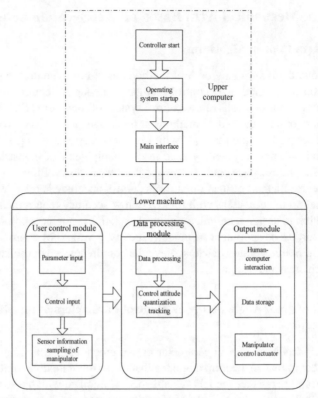

Fig. 4. General structure of mechanical arm joint angle feedback control

manipulator joint angle, the information fusion and feature extraction are carried out

$$B = A\{p(a_1), p(a_2), \cdots, p(a_m)\} \tag{5}$$

Among them, $0 \le p(a_i) \le 1(i = 0, 1, 2, \cdots, m)$ is the fusion feature of the structural parameters of the manipulator joint angle control, and $\sum_{i=1}^{m} p(a_i) = 1$, according to the fusion results of the manipulator joint angle feedback parameters, carries out dynamic tracking control and parameter identification.

(2) Dynamic analysis of feedback control of manipulator's joint angle

According to the reliability parameter distribution of the two sensors, the attitude positioning of the joint angle is carried out. According to the process of the manipulator joint angle feedback control, the linear equation of the manipulator joint angle is derived:

$$J = \varphi \sum_{i=1}^{m} p(a_i) - f(x) \tag{6}$$

Where, ϕ is the dynamic positioning variable of the manipulator joint angle, $f(x)$ is the iterative function of the output trajectory about x = WVD(n, k(n)). considering the scale coefficient of the state component, the measurement parameter between the characteristic sampling time intervals $n \in [n_1, n_2]$, n_1 and n_2 is $k(n)$, and the dynamic parameter adjustment equation of the manipulator joint angle is obtained

$$\delta p(a_1)\text{-}2p(a_2) + \rho A - J = 0 \tag{7}$$

Where δ is the adjusted value of kinetic parameters. In the process of pose parameter identification of manipulator joint angle feedback control [14], through the steady-state control and dynamic measurement of the input state term of the manipulator joint angle, the feedback gain control model u of the manipulator joint angle is established. The periodic spatial interconnected system is used to suppress the state oscillation. It can be seen that the stability convergence condition satisfies $|\rho| \leq 1$ under the influence of constrained sideslip angle, the stability convergence condition satisfies $|\rho| \leq 1$, The fitting state equation of the manipulator joint angle excitation control is obtained

$$F = (u - v\sin\alpha) + \sqrt{\frac{1 + i\tan\alpha}{1 + k\tan\alpha}} \tag{8}$$

Among them, v is the excitation control speed of the manipulator joint angle, α is the motion angle of the manipulator joint, and i, k is the joint angle position of the manipulator after convergence, and the hierarchical fusion model $D(C_i)$ is obtained by using the dynamic programming model.

$$D(C_i) = \begin{cases} \frac{\sum D(C_i)}{1-F} & C \neq \varnothing \\ 0 & C = \varnothing \end{cases} \tag{9}$$

According to the distribution of sensor credibility parameters, the joint angle is located, and the joint angle feedback parameter measurement and hierarchical fusion model are established. In the process of position and attitude tracking, the fuzzy parameter identification method is used to identify the dynamic parameters.

3.3 Analysis of Acceleration Sensor Feedback Control

The sensor is an important means for robots to interact with the external environment. In order to ensure that the control target does not fall off and the control force does not damage the controlled object in the control process, this paper attempts to apply acceleration sensor to the contact force control of flexible joint manipulator.

1) Acceleration sensor design

The acceleration sensor is installed at the end of the manipulator finger, which is mainly used to obtain the information and contact state when the manipulator finger contacts with the control target. In order to make the output signal of the sensor more accurate, it should have the following characteristics:

(1) The sensor material is wear-resistant and has strong compliance.

(2) The sensor should have good repeatability and stability without hysteresis.
(3) The sensor should have high sensitivity and fast frequency response.

Therefore, in this paper, PVDF is selected as the material of acceleration sensor. If PVDF is subjected to external pressure, it will generate electric charge.

$$\Delta q = \sum_{j=1}^{3} d_{3j} \Delta\sigma_j \tag{10}$$

Where Δq is the amount of charge per unit area, $\Delta\sigma_j$ is the change of stress in different directions, and d_{3j} is the piezoelectric coefficient in different directions. According to Eq. (10), the contact force is proportional to the strain of the piezoelectric material, and the voltage generated by the material under the influence of pressure is also proportional to the amount of charge. In other words, the relationship between the pressure generated by the manipulator and the maximum value of the feedback signal is approximately linear, so the contact pressure can be as follows:

$$|F| = K \cdot |U| \tag{11}$$

Where, F is the contact force between the controlled object and the sensor, and U is the output voltage of the sensor. According to Eq. (11), the peak value of the feedback signal of the sensor is proportional to the stress. When the signal strength reaches the maximum value, it will be attenuated according to a fixed law. Since the force of the robot finger includes two processes of touching and loosening, the force of the robot finger will be reduced, The feedback signal generated by the sensor also contains two peaks.

In order to make the sensor detect the feedback information better, assuming that there is a threshold of M_T, the rules of contact logic flag P are as follows:

$$P = \begin{cases} 0 & F < M_T \\ 1 & F \geq M_T \end{cases} \tag{12}$$

First, a fixed force value is set to make the object undamaged by the actual situation of the controlled object. When the threshold value is less than the output voltage, the pressure acting on the sensor is greater than this value, and it is judged to be in contact, otherwise it is non-contact. Figure 5 is the feedback response curve of the sensor. From Fig. 5, it can be seen that the feedback process of the manipulator is the loading process of the material, and the peak signal formed is the upper peak signal. When the manipulator releases the object, it is the unloading process, which produces the lower peak signal, and the last two peak values gradually decay to zero according to the exponential law.

By observing the feedback response curve, the sensor's feedback can be reflected by the amplitude of the output signal. If z_i, $i = 1, 2, \cdots , N$ is the discrete value of the feedback signal and N is the number of sampling points, the eigenvalue of the feedback signal can be:

$$T_z = \frac{1}{N} \left(\sum_{i=k}^{k+N-1} |z_i| + |z_{k+N-1}| + |z_k| \right) k = 1, 2, 3 \tag{13}$$

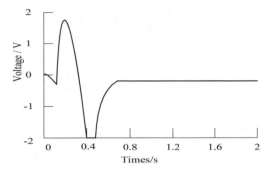

Fig. 5. Feedback response curve

In the process of data processing, a data queue is drawn up. When the system samples and fills the data queue, the data of team head is removed and new data is added to the end of the team. Through the above operations, the real-time update of data can be ensured and the calculation speed can be improved.

Considering that acceleration sensor also needs to be used to sense the distance between robot arm and object under test, prevent transition impact, realize flexible contact and avoid damage to controlled object. The main function of this system is to feedback the proximity between the robot and the control target in real time.

Assume that the output voltage V_{out} of the feedback signal is:

$$V_{out} = F\left(\alpha, d, s, P, A', A''\right) \tag{14}$$

Among them, α is the reflection coefficient, d is the duty cycle of the emission modulation of the sensor, s is the distance, P is the emission power of the infrared light of the sensor, and A', A'' is the amplification factor of the two amplifiers before and after the sensor. The size of α depends on the color, surface roughness and color depth of controlled object. Set to a fixed value according to the actual situation, the distance between the sensor and the controlled object is one-to-one corresponding to the output voltage of the sensor.

2) Sensor signal fusion

Sensor information fusion can reduce the loss of single or overall sensing information in detection, effectively improve the performance of the system composed of multi model sensors, and fuse the feedback sensing information of the manipulator. The overall structure of the fusion algorithm is shown in Fig. 6.

The measurement equations of the algorithm are as follows:

$$\begin{aligned} x_k &= f(x_{k-1}) + w_{k-1} \\ z_k &= h(x_k) + v_k \end{aligned} \tag{15}$$

Where w_{k-1} is the process noise; v_k is the observation noise, which is Gaussian white noise.

According to the calculation results, the optimal design of the manipulator structure is achieved, as shown in Fig. 7.

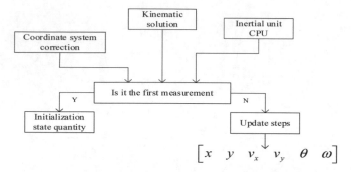

Fig. 6. Overall structure of the fusion algorithm

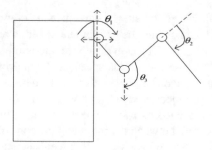

Fig. 7. Optimization structure diagram of mechanical arm

The position and attitude of each member of the manipulator in Fig. 7 can be described by the following formula.

$$T_i = T_{i-1}A_i (i = 1, 2, \cdots, n) \tag{16}$$

Among them, T_i is the matrix of controlling, and A_i is relationship between the connecting members of the manipulator. In the basic coordinate system, the position and posture of the fingers of the manipulator can be expressed as:

$$T_n = T_0 A_i A_2 \cdots A_n \tag{17}$$

The symbolic method is used to automatically derive the motion method. The basic method is as follows: A_i matrix is regarded as a symbolic matrix, and two recursive matrix multiplication forms are obtained by using symbolic matrix multiplication

$$T_n = T_0(A_1 C \cdots (A_{n-1}A_n)) \tag{18}$$

$$T_n = ((T_0 A_1) \cdots)A_n \tag{19}$$

The motion equation can be obtained by counting the number of calculations, and the structural optimization modeling of the manipulator can be completed.

4   Experimental Verification

To verify the effectiveness of the research content, simulation software. Matlab is used to carry out the simulation experiment, assuming that the initial angle of the manipulator is $\theta_{20} = 0.8[rad]$ and the initial angular velocity is $\dot{\theta}_{20} = 0[rad/s]$. the expected points to be achieved by the control are: angle $\theta_{2d} = 0.5[rad]$ and angular velocity $\dot{\theta}_{2d} = 0[rad/s]$. The setting of other relevant parameters is shown in Table 1.

Table 1. System parameter setting

Parameter item	Value
Quality	46 kg
Operation load of robot	27 KN.s
Space parameters of inverse kinematics	0.35
Stiffness of balance spring of robot	390 KN/s
Delay time	2.5 ms

According to the simulation environment and parameter setting shown in Table 1, the parameter characteristic information of manipulator joint control is tested, and the vertical fluctuation and horizontal velocity distribution are shown in Fig. 8(a). The acquisition results of roll angle, pitch angle and yaw angle are shown in Fig. 8(b).

According to the analysis of Fig. 8, the linear parameter fusion model of the manipulator joint angle feedback control is constructed by using the acceleration sensor driving

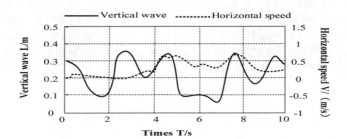

(a) Vertical fluctuations and horizontal velocity distribution

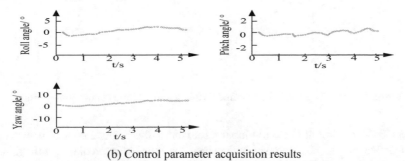

(b) Control parameter acquisition results

Fig. 8. Parameter measurement results

method in this paper, so the parameter fusion of the manipulator joint control method in this paper is high, and the output stability of the manipulator joint angle feedback control is tested, and the results are shown in Fig. 9.

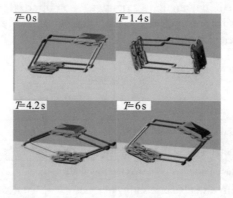

Fig. 9. Arbicular joint angle feedback control output

According to the analysis Fig. 9, the method of this paper realizes the integration and incremental control of the angle feedback parameters of the manipulator through the method of attitude compensation and feedback correction. When the angle control of the manipulator joint is carried out, the output is stable and the information fusion is good, which improves the angle control ability of the manipulator joint.

(1) Balance control test of manipulator

The initial angular velocity of the first joint and the second joint is 0, the maximum simulation step is 1ms, and the simulation time is 5S. The angle value of the first joint and the second joint freely swinging near the stable balance position vertically down is shown in Fig. 10.

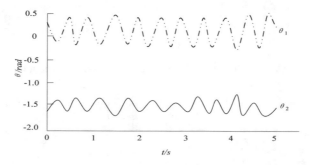

Fig. 10. The Angle value of the free swing near the stable equilibrium position

By changing the position input parameters, the balance control is carried out near the vertical unstable balance position, and the balance control of the first joint and the second joint is tested. The experimental results are shown in Fig. 11.

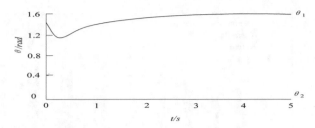

Fig. 11. The Angle value of the free swing near the unstable equilibrium position

As can be seen from Fig. 11, the angle of the driving arm (the first joint) θ 1 close to π/2. Angle of underactuated arm (second joint) θ It is proved that the designed method can be used for balance control.

(2) Control stability and error test.

Taking the Fig. 8(b) control parameter as the constraint object, the automatic control of the manipulator is carried out. The method in this paper is compared with the method in reference [2] and reference [3], and the stability result of the automatic control of the manipulator is obtained, as shown in Fig. 12.

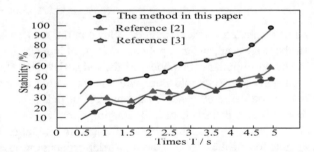

Fig. 12. Stability results of mechanical arm automatic control parameters

According to the analysis in Fig. 12, the output stability and parameter adjustment ability of the design and structure analysis method based on acceleration sensor are better, and the maximum can reach 98% over time, which is higher than the methods in reference [2] and reference [3]. The main reason is that the acceleration sensor is used to capture and determine the feedback signal, which further improves the performance of the manipulator.

Test the control error and get the comparison results as shown in Fig. 13.

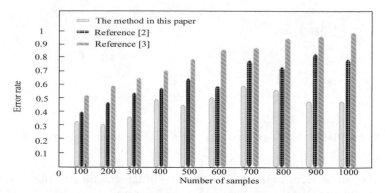

Fig. 13. Control error analysis

According to the analysis of Fig. 13, in different sample sequences, the automatic control error of the design method is lower than that of the contrast method. The first and second joints of the manipulator are further optimized to improve the control accuracy.

5   Conclusion

The automation development level and intelligence level in the modern social industry field have been further improved, and the application scope of industrial robots has been further expanded. As an important component of the industrial robot, the mechanical arm plays an increasingly important role along with the continuous development of the industrial automation and the intelligence. In the process of industrial production, the mechanical arm can further with the help of computer programming to complete all kinds of expected work objectives, more can strengthen through the imitation of the human arm some action function of the robot in the industrial production process of grasp, handling objects and even the operation of some production tools, the national economy in various industries has a broad application prospects. Therefore, this paper studies the design and structural analysis of the manipulator based on an acceleration sensor. In order to realize the smooth and stable control of the contact transition process from free motion state to contact force control, the experimental study of contact force control was carried out, and the results show that the theoretical analysis is correct. At the same time, the following conclusions can be drawn: in the process of optimal design of the manipulator, enterprises or relevant researchers should further strengthen the working sensitivity and efficiency of robots on the basis of fully understanding the manipulator structure design points and industrial production characteristics of enterprises, so as to reduce the production energy consumption of the manipulator and reduce its self weight and occupation of the production workshop area, such as to make the industrial production efficiency and economic benefits to maximize lay a solid foundation.

References

1. Jain, R.K., Khan, A., Inamuddin, Asiri, A.M.: Design and development of non-perfluorinated ionic polymer metal composite-based flexible link manipulator for robotics assembly. Polymer Compos. **40**(7), 2582–2593 (2019)
2. Yang, X., Zhu, W.L.: Design, analysis and control of a XY parallel nanomanipulator with multiple actuation modes. IEEE Trans. Ind. Electron. **PP**(99), 1 (2019)
3. Bastos, G., Brüls, O.: Analysis of open-loop control design and parallel computation for underactuated manipulators. Acta Mech. **231**(6), 2439–2456 (2020). https://doi.org/10.1007/s00707-020-02656-0
4. He, Y., Zhang, X., Zhu, L., Sun, G., Dong, M.: Curvature and force measurement of soft manipulator based on stretchable helical optic fibre. Opt. Fiber Technol. **53**, 102010 (2019)
5. Wang, J., Yang, X., Li, P., Song, S., Liu, L., Meng, M.-H.: Design of a multi-arm concentric-tube robot system for transnasal surgery. Med. Biol. Eng. Comput. **58**(3), 497–508 (2020). https://doi.org/10.1007/s11517-019-02093-9
6. Roy, S., Kosmatopoulos, E.B., Baldi, S.: On vanishing gains in robust adaptation of switched systems: a new leakage-based result for a class of Euler–Lagrange dynamics. Syst. Control Lett. **144**, 104773 (2020)
7. Sarojini, V., Cameron, A.J., Varnava, K.G., Denny, W.A., Sanjayan, G.: Cyclic Tetrapeptides from nature and design: a review of synthetic methodologies, structure, and function. Chem. Rev. **119**(17), 10318–10359 (2019)
8. Eichhorn, S., Hrschlger, A., Steiner, M., Laimer, J., Gadermaier, G.: Rational design, structure-activity relationship, and immunogenicity of hypoallergenic Pru p 3 Variants. Mol. Nutr. Food Res. **63**(18), 1900336 (2019)
9. Kim, B.J., Jang, Y., Ji, H.M., Baughman, R., Kim, S.J.: Self-powered carbon nanotube yarn for acceleration sensor application. IEEE Trans. Ind. Electron. **PP**(99), 1 (2020)
10. Alambeigi, F., Pedram, S.A., Speyer, J.L., Rosen, J., Armand, M.: SCADE: simultaneous sensor calibration and deformation estimation of FBG-equipped unmodeled continuum manipulators. IEEE Trans. Robot. **PP**(99), 1–18 (2019)
11. Ganji, B.A., Kheiry, S., Soleimani, S.: Design of small size and high sensitive less-invasive wireless blood pressure sensor using MEMS technology. IET Circ. Devices Syst. **13**(1), 39–44 (2019)
12. Hachicha, S., Zaoui, C., Dallagi, H., Nejim, S., Maalej, A.: Innovative design of an underwater cleaning robot with a two arm manipulator for hull cleaning. Ocean Eng. **181**(JUN.1), 303–313 (2019)
13. Mo, K., He, M., Cao, X., Chang, C.: Direct current electric field induced gradient hydrogel actuators with rapid thermo-responsive performance as soft manipulators. J. Mater. Chem. C **8**(8), 2756–2763 (2020)
14. Xu, J., Li, K., Yang, J., Zhang, Z.: Research on motion planning scheme for redundant manipulator with zero initial acceleration. Comput. Simul. **37**(1), 311–316 (2020)

Discussion on Decision Tree Algorithm in University Teaching Management System

Ying Zhao[1] and Kai Zhou[2(✉)]

[1] The Tourism College of Changchun University, Changchun, Jilin, China
[2] The Cyberspace Security College of Changchun University, Changchun, Jilin, China
zhouhuiyan0309@sina.com

Abstract. The Ministry of Education issued (February 4, 2020) the guiding opinions on organizing and managing the online teaching in ordinary colleges and universities during the period of epidemic prevention and control. To jointly implement and protect online teaching in the period of epidemic prevention and control, to achieve the goal of "ceasing teaching and stopping classes without stopping learning". How to effectively link up online and offline teaching, how to continuously consolidate the effectiveness of online teaching in the post epidemic era is a challenge for university teaching management.

Keywords: Epidemic period · University teaching management · Online teaching

1 Online Educations During the Outbreak

1.1 Courses Offered During the Outbreak

Taking the Z of a comprehensive university in Jilin province of higher education as an example, Z university plans to start 2342 sessions this semester, with 2314 courses actually opened, and the opening rate is 98.8%, as show in Fig. 1.

A variety of teaching methods are adopted by teachers, including 1254 by live broadcast Gate, accounting for 53.4%; 424 recording and broadcasting, 492 online courses, using other software, tutoring 144 lectures, as show in Fig. 2.

J. Macintyre et al. (Eds.): SPIoT 2021, LNDECT 98, pp. 116–123, 2022.
https://doi.org/10.1007/978-3-030-89511-2_14

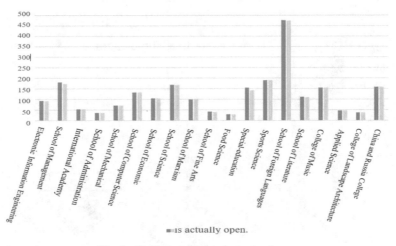

Fig. 1. Statistics of school opening hours

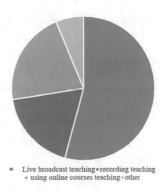

Fig. 2. Statistics of teachers' teaching methods

1.2 Distribution of Online Teaching Platforms During the Epidemic Period

Teachers use various teaching platforms to carry out online teaching. The utilization rate of rain class and intelligent tree is about 49%, which is the main teaching platform for teachers in our school. Platform functions can basically meet the basic requirements of daily teaching and learning for teachers and students, as show in Fig. 3.

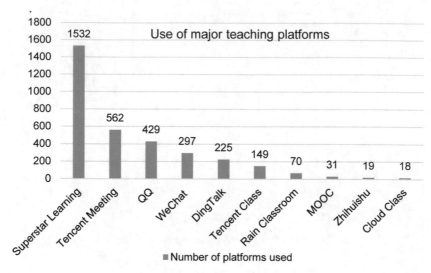

Fig. 3. Use of major teaching platforms

1.3 Online Course Review Summary

This week, a total of 17 school supervisors were involved in the online teaching activities of 34 courses in the whole school.

Spot check and spot check results are shown in the Table 1 below:

Table 1. Spot check results

Record content	Yes	No
Is the teacher on time or after class?	33	1
Is there any teaching activity organized by teachers during class?	34	0
Is teaching activity closely related to this course?	34	0
Is teaching activity suitable for students' learning?	34	0
Is the design of teaching activities reasonable?	34	0
Is there any teaching material related to this course for students' autonomous learning on the teaching platform?	34	0
Is there any obscene or insulting language, indecent video or image?	0	34
Is there any illegal or illegal speech and act?	0	34

2   Analyses of Problems in Teaching Management in Universities During the Epidemic Period

2.1   Teaching and Learning Aspect

According to the survey, 58.59% of the teachers' teaching resources come from self-construction. 56.06% of the teachers used the method of online interactive discussion to answer questions; 87.37% of the teachers had 1–2 assignments per week; 48.48% of the teachers set up the quiz; 80.3% of the teachers thought their courses were suitable for online teaching; 81.31% of the teachers were willing to continue online teaching after the school began. At the same time, 84.34% of the teachers indicated that the rate of students can reach 100%; 68.18% of the teachers were satisfied with the students' participation in online teaching interaction; 41.41% of the teachers believed that online teaching could achieve the effect of offline teaching; 49.32% of the students preferred the live online teaching method, while 30.36% preferred the online teaching method.70.65% of the students were basically satisfied with the effect of online teaching. 20.47% of students were willing to continue online learning after returning to school.

2.2   Teaching Quality Supervision

The two level supervision of the school supervision and teaching units persisted in the "online classroom" teaching, and conducted an improvement plan for the teachers' teaching problems. Individual schools supervised the difficulties of unfamiliar with the teaching platform. It was difficult to help teachers find out the problems and worth learning in the online teaching process. Students' grading of teachers' teaching resources and teaching conditions are as Table 2.

Table 2. Students' grading of teachers' teaching resources and teaching conditions

Title option	Very dissatisfied	Dissatisfied	Commonly	Satisfied	Very satisfied	Average
Teaching resources	177 (3.23%)	42 (0.77%)	575 (10.49%)	1769 (32.28%)	2918 (53.24%)	4.32
Classroom interaction	176 (3.21%)	33 (0.6%)	498 (9.09%)	1813 (33.08%)	2961 (54.02%)	4.34
Online answering questions	176 (3.21%)	28 (0.51%)	489 (8.92%)	1825 (33.3%)	2963 (54.06%)	4.34
Homework correction	173 (3.16%)	28 (0.51%)	480 (8.76%)	1817 (33.15%)	2983 (54.42%)	4.35
Engineering service	185 (3.38%)	80 (1.46%)	673 (12.28%)	1743 (31.8%)	2800 (51.09%)	4.26
Subtotal	887 (3.24%)	211 (0.77%)	2715 (9.91%)	8967 (32.72%)	14625 (53.37%)	4.32

Note: the average score of the matrix is 4.32

Therefore, students' attendance rate and classroom participation have been improved compared with offline teaching, while teachers' recognition of online teaching has improved significantly, but there is still much room for improvement.

2.3 Feature Analysis Based on Decision Tree Algorithm

Step 1: According to the different characteristics of teachers' teaching resources and teaching conditions, with the number column deleted, they are treated as follows Table 3:

Table 3. Feature annotation

Characteristic	Possible value	Feature annotation
Teaching resources	Abundant	1
	Sparse	0
Classroom interaction	Yes	1
	No	0
Online answering questions	Yes	1
	No	0
Homework correction	Yes	1
	No	0
Engineering service	Yes	1
	No	0

The satisfaction degree is the evaluation result, and the evaluation result is processed as follows Table 4:

Table 4. Evaluation results

Evaluation results	Possible value	Class label
Satisfaction degree	Very dissatisfied, Dissatisfied, commonly	Dissatisfied
	Satisfied, Very satisfied	Satisfied

Step 2: The above samples are called data sets. Let the data set be D, see in formula (1), and let the attribute set be B, see in formula (2).

$$D = \{(x_1, y_1), (x_2, y_2), \cdots, (x_N, Y_N)\} \tag{1}$$

$$B = (B, B_2, \cdots, B_n) \tag{2}$$

In formula (1), $x_i = (x_i^{(1)}, x_i^{(2)}, \cdots, x_i^{(n)})^T$ is the feature vector, n is the number of features, in this data set, $n = 5$, yi is the category, $1 \leq i \leq n$, $i \in Z$, N is the sample size, so N = 5481.

In formula (2), $B = (B_1, B_2, B_3, B_4, B_5) = \{$teaching resources, classroom interaction, online answering questions, homework correction, engineering service$\}$.

Step 3: Calculating information gain ratio.

(1) The empirical entropy $H(D)$ of calculative data set D, see in the particular formula (3):

$$H(D) = -\sum_{K=1}^{K} \frac{|C_k|}{|D|} \log_2 \frac{|C_k|}{|D|} \tag{3}$$

Among them, the K is the number of categories, so in this paper, K is equal to the largest 2, $|C_K|$ for the number of data sets belong to the category of C_K $|D|$ for the number of data sets, so $|D| = 5481$, according to the classification of evaluation results, namely $\sum_{K=1}^{K} |C_k| = C_1 + C_2 = |D|$.

(2) Calculating attribute sets B on the experience of the data set D conditional entropy $H(D \mid B)$, see in the particular formula (4):

$$H(D \mid B) = \sum_{i=1}^{n} \frac{|D_i|}{|D|} H(D_i) = -\sum_{i=1}^{n} \frac{|D_i|}{|D|} \sum_{k=1}^{k} \frac{|D_{ik}|}{|D_i|} \log_2 \frac{|D_{ik}|}{|D_i|} \tag{4}$$

Among them, according to the attribute set B the data set D is divided into n subsets, i.e., $\sum_{i=1}^{n} |D_i| = |D|$, $|D_i|$ for the subset of sample number, in this thesis, n = 5, and $D = \{D_1, D_2, D_3, D_4, D_5\}$. At the same time the subset Dibelong to C_K in the collection of samples forD_{ik}, $|D_{ik}|$ for the number of samples in the D_{ik}.

(3) Calculating information gain g(D,B), see in the particular formula (5):

$$g(D, B) = H(D) - H(D|B) \tag{5}$$

(4) Calculating information gain ratio gr(D,B), see in the formula (6), (7).

$$H_A(D) = -\sum_{i=1}^{n} \frac{|D_i|}{|D|} \log_2 \frac{|D_i|}{|D|} \tag{6}$$

$$g_r(D, A) = \frac{g(D, A)}{H_A(D)} \tag{7}$$

For the meaning of the letters in the formula, see the above explanation of the letters.

Step 4: C4.5 algorithm is used to generate decision tree T, as shown in flowchart 3:

(1) Input data set and attribute set;
(2) If all the categories in D belong to the same class CK, CK is taken as the category of this node [1, 2]. In this paper, it is satisfied or dissatisfied, and return the decision tree T.

(3) If B = Ø or samples in D have the same value on B, and take the category CK with the largest number in D as the category of this node, and return the decision tree T;

(4) Otherwise, calculate the information gain ratio of each feature in B to D according to Formula (7), select the feature Bg with the largest information gain ratio to construct node Bg.

(5) For each possible value bi in Bg, the data set D is divided into several subsets Di according to Bg = bi, and the node is constructed for each subset [3].

(6) For the i-th sub node, with Di as the data set and B-{Bg} as the feature set, recursively call steps (1) to (4) to get the sub tree Ti and return Ti.

Step 5: According to the decision tree T, the nodes are output from top to bottom, and the output order of the nodes corresponds to the ranking order of the features from high to low. Therefore, the first ranking feature is the most important factor affecting students' satisfaction with the teaching resources and teaching conditions of the school.

3   The Need for Continuous Online and Offline Learning Revolution During the Epidemic Period

First, teaching administrators and teachers must further emancipate their minds, renew their ideas, uphold the concept of "open" and "share", focus on the learning and development of students, pay attention to the new educational information and teaching methods in the Internet age, reform the traditional teaching mode, constantly condense consensus and gather advantages, improve the teaching level and the quality of personnel training [4].Second, many kinds of flipped classes, such as "under class study, class practice", "class study, class discussion" and other forms, have promoted students to transform from single learning to self-learning mutual learning. It also promoted the transformation of teachers' role from a single knowledge imparting role to the "guidance supervision" role. The traditional "cramming" teaching mode gradually changed to "practice speaking instead of speaking", "student speaking and student assessment", "student defense and student assessment" classroom transformation. Mixed teaching only applies line teaching to increase teacher-student interaction and interaction time, without losing the systematic teaching of teachers [5]. Third, online teaching requires teachers to make use of modern information technology to promote changes in course contents and teaching methods and methods, optimize teaching design, rebuild curriculum system, reform teaching contents, explore online and offline interactive integrated teaching, and mobilize students' enthusiasm, initiative and creativity in learning[6]. Continuous improvement of teaching quality. Teachers ask students to lower their cell phones to attend lectures than to help students use mobile phones to attend lectures [7].

4   Suggestions on Teaching Management in Colleges and Universities During the Epidemic Period

First, we should further clarify the relevant policies from the aspects of the construction and operation, application and practice, management and evaluation, support and guarantee of online open courses, encourage teachers to participate in online teaching, and effectively solve the practical problems that teachers are concerned about in the process of online open curriculum construction and operation [8]. Second, the school curriculum operation platform will be built through the combination of self-building and purchasing services, and the basic conditions for online and offline teaching will be well prepared [9]. Special funds will be set up to support online curriculum construction, and online course construction as the breakthrough point. Third, we should actively do well in line teaching plans, and promote the formation of positive interaction between online and offline teaching. Enhance the initiative and initiative of autonomous learning [10]. Pay close attention to student dynamics; collect information about students' specific needs and suggestions on online teaching through various channels.

Acknowledgements. Fund Project: Humanities and Social Sciences Research Project of Jilin provincial education department (JJKH202114508K).

References

1. Chen, W., Li, G.: Teaching management change in China's higher education institutions in the context of universalization of higher education. Univ. Educ. Sci. **2020**(06), 46–51 (2020)
2. Rao, H., et al.: Feature selection based on artificial bee colony and gradient boosting decision tree. Appl. Soft Comput. **74**, 634–642 (2019)
3. Li, M., Xu, H., Deng, Y.: Evidential decision tree based on belief entropy. Entropy **21**(9), 897 (2019)
4. Zeng, L.: Analysis on the implementation effect and influencing factors of offline teaching under the background of epidemic prevention and control – Based on the investigation of universities in Guangdong Province. High. Educ. Explor. **07**, 85–91 (2020)
5. Yang, J., Shi, G., et al.: 5G+Smart education: educational reform based on intelligent technology. China Educ. Technol. **2021**(04), 1–7 (2021)
6. Tortorella, G.L., Narayanamurthy, G., Cauchick-Miguel, P.A.: Operations management teaching practices and information technologies adoption in emerging economies during COVID-19 outbreak. Technol. Forecasting Soc. Change **171**, 120996 (2021)
7. Nghia, T.L.H., Phuong, P.T.N., Huong, T.L.K.: Implementing the student-centred teaching approach in Vietnamese universities: the influence of leadership and management practices on teacher engagement. Educ. Stud. **46**, 1–17 (2018)
8. Zhao, D., Jie, Z.: Research on the construction of cross regional teaching community under the perspective of "Internet+Education"—rethinking the application of "Three Classrooms." China Educ. Technol. **02**, 97–104 (2021)
9. Han, J., Wei, J.: Opportunities and challenges of novel coronavirus pneumonia network teaching under the background of new crown pneumonia epidemic situation. Party Build. Ideol. Educ. Schools **17**, 57–59 (2020)
10. Duan, C., Hong, J.: The relationship between internet teacher-student interaction and online learning performance: the mediating effect of internet learning self-efficacy and internet learning motivation. Psychol. Dev. Educ. **35**(02), 184–191 (2019)

Application of Road Vertical Algorithm Based on Directed Graph in Reconstruction of Villages and Towns

Rui Zhang[1](✉), Yonghui Zhang[2], Ying Li[3], and Lin Ma[1]

[1] Architecture Department School of Mechanics and Civil Engineering,
Northwestern Polytechnic University, Xi'an, Shaanxi, China
[2] China Telecom Shaanxi Branch, Xi'an, Shaanxi, China
[3] Xi'an University of Architecture and Technology, Xi'an, Shaanxi, China

Abstract. The road vertical design is one of the more complicated problems in the rural reconstruction of villages and towns. In order to make a construction drawing effectively and quickly, the algorithm and a calculation program of the directed graph were used to solve this problem. After analyzing the geometric model, slope direction and elevation limit of the road network, the algorithm is designed by depth first approach and elevation limit analysis. The algorithm was designed of gradually adjust elevation in step searches, and the design validity is verified by breadth first search. Compared with the traditional manual calculation design adjustment, it greatly improves the design efficiency.

Keywords: Road vertical design · Digraph · Depth first search · Breadth first search

1 Introduction

Since the 18th National Congress of the Communist Party of China, it is mandated for governments at all levels to strengthen the construction of new rural infrastructure. The reconstruction and construction of villages and towns across the country is a complex task with a large workload. For the architectural design industry, the redesign of each village and town involves a series of problems, such as the renovation of the original buildings, the improvement of the street landscape, and the renewing and planning of the roads.

In the work of village and town reconstruction, the vertical design of the road has become a time-consuming and laborious problem, and it is difficult to form an optimal solution [10]. Due to the original villages and towns formed naturally, the roads in the villages and towns are lack of planning, and there are no drainage and sewage pipeline facilities. After more than decades of long-term repair, the problem of road water accumulation in villages and towns has become a persistent problem. At the same time, the conflict between homestead and public facilities makes limits the road repair more. In the village and town reconstruction project of Chang'an District, Xi'an City, the author used methods of operation research and graph theory [2] to establish a model and designed an algorithm for the vertical road planning, which achieved good results and greatly improved the efficiency.

2 Road Network Topology Based on Directed Graph

2.1 Network Geometry Model of Village Road

In the northwest, due to historical reasons, most of the roads in villages and towns have no sewage pipes for rainwater. The rainwater flows along the road to the low-lying place. If there is no waterlogging pool, it will either drain to the farmland or form waterlogging. The rainwater and sewage will accumulate at the intersection of the low-lying road. The vertical design of the road is to adjust the elevation of the road, to solve the problem of waterlogging without affecting the original road and homestead of the villagers.

The road and intersection data include elevation, distance between intersections, and the resulting slope. In addition, whether the road intersect at a farmland or the planned location of waterlogging pool has become one of the attributes of the road intersection. Digraphs are composed of ordered triples, involving vertices, neighbors, and associated functions [2]. Each vertex forms a geometric network, and the direction is the connected edge (correlation function). The drainage of the road is directional, and the drainage direction is abstracted into the direction of the directed graph [5]. In this way, the intersection and slope of the road are abstracted into the geometric network model of the directed graph. Where vertex set n = {N1, N2, N3 NN} is a none empty set, where the element is the road intersection (hereinafter referred to as node), n > 2. The set s of each vertex n is composed of the road intersection connected with it and the corresponding slope (directed edge) element. Figure 1 is an excerpt from the rural road map, and Fig. 2 is an abstract digraph.

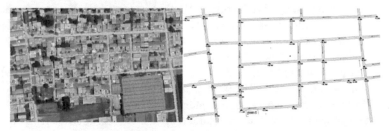

Fig. 1. An excerpt from a rural road map

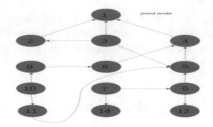

Fig. 2. The abstract digraph of a rural road map

2.2 Effectiveness Inspection of Road Vertical

The main purpose of road vertical adjustment is to solve the problem of waterlogging. According to the actual terrain and the actual layout of villages and towns, determine the location of waterlogging pool, then the road to the location of waterlogging pool for rainwater drainage, becomes the main requirement of vertical design. The height difference between two road intersections (nodes) determines the slope between them and the flow direction of rainwater. Including outflow (from the node to outflow) and inflow (to the node). Before the vertical adjustment, the road nodes of the whole village and town shall be inspected as a whole. The first thing is to check and judge whether the node has accumulated water or not. Then check whether each road conforms to the vertical design basic conditions [1].

(1) Judgment of water accumulation point. For the node of the road, as long as all the edges (roads) connected with it are the direction of the internal flow. In a directed graph, this node is an outlier. Then it can be judged that the node of this road is a water accumulation point. In order to facilitate the subsequent vertical adjustment, only the flow direction is judged here, but not whether the gradient of the road is appropriate.

The node to be judged is the original node, its elevation is H0, its adjacent node is the neighbor node, its elevation is H1, and its distance from the original node is L1. The corresponding slope between them is $s = ((h0-h1))/((0-l1))$. If the slope is less than 0, it is outflow, otherwise it is inflow.

(2) The validity judgment of road slope. According to GB 50352-2019, the gradient of road shall not be less than 3/1000. For roads with slope $s < 0.3\%$, in the second step of effectiveness inspection, it is judged as disconnected.

The following figure is a schematic diagram of local road nodes in a village. Ponding points and roads that do not conform to the slope are marked in the Fig. 3.

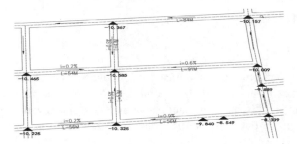

Fig. 3. A schematic diagram of local road nodes in a village

(3) Effectiveness analysis of road nodes. According to the structure of directed graph, the principle of judgment is that each road node is connected with the direction of outflow and the slope s ≥ 0.3%. Only one neighbor node has weak connectivity, and more than two are strong connectivity. As shown in Table 1:

Table 1. Analysis rules of node

Flow type	Original node	Slope	Effectiveness
Outflow	Yes	<abs(0.3%)	Ineffective
Inflow	Yes	Unlimited	Ineffective
Outflow	Yes	≥abs(0.3%)	Effective

2.3   Build Sequence Model of Traversal

In the vertical adjustment of road nodes, the depth first search algorithm (DFS) is used to judge whether the original node and neighbor node are connected to the waterlogging point. For the road nodes after vertical adjustment, the breadth first search algorithm (BFS) is used to simulate the flow direction of rainwater to get the shortest path of each node [3].

The basic idea is: (1) DFS traversal, starting from any node, access to the effective neighbor nodes, and continue to visit until the end point (waterlogging point). After traversing the tags, if you can't access them, go back [7]. (2) BFS traversal starts from any node, traverses all its neighbor nodes, and then traverses the next level of nodes. Mark nodes that have been traversed [6].

In this way, DFS traversal can judge whether this road node connects to waterlogging pool, while BFS traversal can get the shortest path [8]. Based on the digraph of Fig. 2, a node is searched to get the path of the node connecting the waterlogging point.

3 Algorithm Design

3.1 Data Structure

According to the definition of the directed graph, every node of the road relates to at least one edge except the isolated water point. That is to say, the node of the road has at least one road surface discharging water outwards. The isolated vertex is not considered in the directed graph, but in the vertical design of the road, the isolated vertex is the problem that must be solved.

Table 2. Data structure of road node

Road nodes	Node attributes
Name	Node name: int ID of node(non-repetitive)
Height	High: float measured elevation
Distance	Long: float measured distance to origin
Slope	Slip: float calculated slope
Waterlogging	Laochinode: int pool being 9, farm being 1, others 0

The data structure of the digraph can be expressed in the form of an adjacency matrix [9], but in addition to connectivity (slope), there are elevation, node distance, and whether this node is the attribute of nodes such as waterlogging pool point or farmland point (automatic drainage point) in the vertical design of the road. At the same time, the description of the road can be abstracted to the node attributes at both ends, that is, the connectivity (slope) of neighbor nodes. Therefore, the description of digraph in this paper constructs data structure in the way of a dictionary [4]. See Table 2 for details.

Because the attributes of the road are merged into nodes, each node is a dictionary of two to five key value pairs. The first key value pair is the source node, and the others are the nodes of the road to which it is connected. Key is the node number. The example is as follows:

{13: [−10.009, 0, 0, 0], 14: [−9.889, 19, 0.0063, 0], 12: [−10.585, 91, −0.0063, 0], 20: [−10.157, 48, −0.0031, 0]}.

This way, the dictionary of these nodes as the basic elements constitutes the list dictionary Road node of the whole village road geometry network. The node dictionary can also be converted to Dataframe data format by using Python's data package [4].

3.2 Road Vertical Design Algorithm

In the vertical design of roads in villages and towns, the most complex is that there are many restrictions, not only considering the requirements of design specifications, but also according to the actual construction situation, which cannot cause the complexity of construction. According to many factors, the mathematical expression of vertical adjustment is as follows:

$$S(i) = h0 - h(i)/l0 - l(i) \quad i = 1, \ldots, 4$$

$$|S(i)| > 0.003$$

$$h\_min \leq \theta(h(i)) \leq h\_max \; i = 1, \ldots, 4$$

$$h\_min = -400, h\_max = 300 \text{(Initial limit)}$$

The first formula is used to calculate and adjust the vertical of the road, followed by restrictions. However, the third constraint, in the process of adjustment, will also change with the actual elevation of the neighboring nodes, which will be less than the initial limit value of the above formula.

Based on the complexity of the adjustment, the vertical adjustment of the road is designed into two processes. First, adjust the ponding point, that is, connect the isolated nodes of the digraph. In order to ensure the connection, the slope condition will be relaxed here. Second, adjust the slope of the road to make the road conform to the standard specifications.

(1) the pseudo code of adjusting ponding point

```
input  roadnode
output  High Slip roadnode
While water_node is true
h = h(i)-(-0.003)*(l0-l(i))
θ = h -h0
h_max h_min
h_max,h_min=check(other node)
if θ<h_max
   h0 = h
else
   h0 = h0 + h_max
calculate Slip
dfs node to laochinode
if true
   break
else
   return
```

The Primary method is to raise the isolated points which are not connected according to the limited conditions, so that they have at least one connected node, that is, to form the edge of the digraph. In the initial condition, all the neighbor nodes are evaluated symmetrically to get the nodes that need to get through the edge of the directed graph. At the same time, the original node limit is not higher than the neighbor nodes except the connected nodes. There are two factors for evaluation. First, the adjusted elevation is the smallest, and second, the points which need to be connected can be connected to the waterlogging pool.

(2) the pseudo code for adjusting road slope

```
input  nodename roadnode
output High Slip roadnode
While exist abs(nodeslip)< 0.003
  For h in node  # In a set of inconsistent nodes
  Find neighbor_change #Find neighboring nodes that need to be changed
  h_min ,h_max = check(neighbor_change.node) #check the limiting range
  sign = graph(node) #adjust verticals by graph connectivity
  θ = (h+0.003*sign*(0-l(i)))-h#calculate the amount of vertical adjust
  if h_min<θ<h_max
    h = h + θ
  else
    h = h + h_max
  calculate Slip
  θ(i) = (h(i)+0.003*sign*(0-l(i)))-h(i) #calculate for neighboring node
  if h_min<θ(i)<h_max
    h(i) = h(i) + θ(i) #adjust for neighbors with inconsistent slope
  calculate Slip
  if exist abs(only nodeslip)< 0.001  #judge if the only paring node has slope less
than 0.001
  break # end iteration
```

The main steps are as follows: for the nodes with inconsistent slope, first adjust the node, then adjust the neighbor node under the condition that the connectivity direction of the directed graph is unchanged. The range of the adjusted vertical value is limited by the height and slope of the surrounding nodes. The iteration ends at the end of adjustment, or when the nodes of the whole digraph only have two pairs of nodes with a slope of less than 0.001 (almost flat).

(3) the pseudo code for adding nodes between nodes that cannot be adjusted

```
input:node1 node2 roadnode
output:newnode roadnode
New_h1 =h1-0.003*(0-l/2)
New_h2 =h2 -0.003*(0-l/2)
If New_h1 > New_h2
  New_h = New_h1
Else
  New_h = New_h2
calculate rodenode  #Recalculate node network, add new nodes
calculate Slip
```

This is mainly for establishing a new node in the middle of the two nodes, making the new node the highest, outflow to the two nodes.

3.3 Verification and Simulation of Road Rainwater Flow Direction

After the vertical adjustment design, all nodes of the digraph are simulated according to the BFS algorithm, and the shortest path from each node to the waterlogging point is obtained. At the same time, the results of vertical adjustment are verified.

The pseudo code is as follows:

```
input:starnode endnode graph
output:path
Search stack (starnode ,None) #Creates stack for search
While Search stack
Popleft (node) #FIFO
If end_node
  path
else
  add node in Search stack
```

The following Fig. 4 shows the results of breadth traversal node:

```
start search...
need traversal node:  deque([19])
current node:  19 | need traversal node:  deque([12, 18])
current node:  12 | need traversal node:  deque([18, 11])
current node:  18 | need traversal node:  deque([11])
current node:  11 | need traversal node:  deque([5])
current node:  5 | need traversal node:  deque([4])
current node:  4 | need traversal node:  deque([1])
current node:  1 | Search complete,The shortest path is: [19, 12, 11, 5, 4, 1] length: 5
```

Fig. 4. Path from searching a node to waterlogging point by breadth traversal

4 Algorithm Implementation and Analysis

4.1 Experimental Environment

The algorithm is implemented by calling numpy and pandas libraries in Python. The development environment is visual studio code and the running environment is windows 10. The data are the road network data of Zhangqu village, Sima village and Dazhao village in Chang'an District of Xi'an city. There are 78, 115 and 425 road nodes, corresponding elevation data, road distance data and relevant drawings.

4.2 Experimental Analysis

The algorithm of breadth traversal was used to verify the flow direction of all nodes adjusted by vertical design, which meets the design requirements of road reconstruction. Compared with manual adjustment, the efficiency was greatly improved. It used to take a designer a week to adjust the vertical of a village's road. The running time of this program was only a few minutes. In terms of accuracy. With annual adjustment, there will

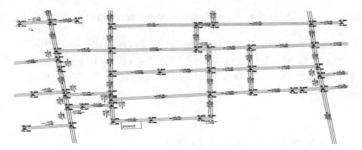

Fig. 5. The local drawings of Dazhao village

be some loose slopes slightly less than 0.003 which will not happen with the algorithm. The computer program completely meets the specification.

We will import the results of the final road vertical design into the drawings, and finally output the complete road adjustment construction drawings of villages and towns to guide the actual construction and transformation of villages and towns. The Fig. 5 shows the local drawings of Dazhao village.

5 Conclusion

In this paper, directed graph method is used to analyze the vertical design of village roads. Through depth traversal algorithm, considering the constraints of actual construction and other factors, all road nodes are extracted and instantiated. Through step-by-step adjustment, the results for vertical adjustment are obtained quickly and effectively, which reduces the complexity of time and improves efficiency and accuracy of artificial design.

The algorithm design of this paper has been applied in the transformation of villages and towns. Due to the limitation of reconstruction funds, only one waterlogging pool has been set up. For large villages and towns, in the future actual transformation, there are many waterlogging pools. Then the designer only needs to divide the waterlogging point of the road node and the road network into multiple digraphs. The algorithm in this paper can still be applied. In the face of complex reconstruction of villages and towns, the algorithm can replace the original manual calculation of designers, which improves the efficiency. It has practical applications in the coming large number of village and town reconstructions.

References

1. Wu, Z.: Operations Research. Beijing University of Technology Press, Beijing (2014)
2. Bu, Y., Wang, W., Lv, X.: Graph Theory and Its Application, 2nd edn. Southeast University Press, Beijing (2015)
3. Fu, J.: Operational Research Methods and Models. 2nd edn. Fudan University Press, Beijing (2014)
4. Wes, M.: Data Analysis Using Python. China Machine Press (2013)
5. Yi, C., Jianhui, W., Meng, Z.: Analysis of automatic flow direction of drainage network by using directed graph. J. Wuhan Univ. Inf. Sci. Ed. **44**(8), 62–67 (2019)

6. Martin, L.L., Wong, D.F., Hwu, W.-M.W.: An effective GPU implementation of breadth-first search. In: Proceedings of the 47th Design Automation Conference, DAC 2010, Anaheim, California, USA, 13–18 July 2010, pp. 52–55 (2010)
7. Hu, Y., Wei, O.: Research on route searching algorithm by DFS. Railway Comput. Appl. **16**(9), 9 (2007)
8. Bang, J., Gutin, G.: Theory, Algorithm and Application of Digraph. Science Press, Beijing (2009)
9. Jay, W.: A Common-Sense Guide to Data Structures and Algorithms. The People's Posts and Telecommunications Press, Beijing (2019)
10. Yang, M.: Vertical Design of Road and General Layout: China Architecture & Building Press, Beijing (2018)

An Improved Microphone Array Noise Reduction Algorithm for Speech Recognition

Li Qian$^{(\boxtimes)}$ and Qian Zhu

Department of Computer Science, Sichuan Vocational and Technical College,
Suining, Sichuan, China

Abstract. With the development of science and technology, the computing power of human electronic devices is increasing, which makes the application of array signal processing requiring large computing power in daily life possible. People have started to study and apply microphone array speech noise reduction technology. Microphone arrays can obtain spatial and temporal information of the signal while receiving speech signals, and the algorithm can use such information more flexibly to improve the noise reduction performance. In this paper, we introduce the adaptive noise reduction technology of microphone array for speech recognition, and optimize the RLS algorithm to design the QR-RLS algorithm without complex iterations. The simulation results show that the algorithm can effectively reduce noise for microphone array.

Keywords: Microphone array · Noise reduction · Speech recognition · Adaptive algorithm

1 Introduction

The advancement of science and technology has brought the advancement of speech recognition technology, and the current technology for recognizing pure speech signals has reached a fairly mature level [1, 2]. The original voice signal capture tool is a single microphone, such as a wearable microphone. In an environment without external interference, when the voice signal captured by the wearable microphone is recognized, the recognition effect of the system is good [3, 4]. However, it is inevitable that there will be environmental noise in the actual communication environment, so the signal captured by the microphone is a noisy speech signal composed of noise superimposed on the sound source [5, 6].

In the 1950s, research work on speech recognition began to sprout. In 1952, Davis and others of Bell Labs developed the first recognizer that can recognize isolated numbers for a specific speaker. In recent years, speech enhancement algorithms based on subspace have been greatly developed [7]. Ephrim and Van Tees systematically studied the subspace speech enhancement algorithm, using KL transform to decompose the signal into the subspace, and proposed two signal estimation methods: the algorithm they proposed must assume that the noise is white noise, and for colored noise only can perform pre-whitening processing [8]. The domestic work on the research of speech

© The Author(s), under exclusive license to Springer Nature Switzerland AG 2022
J. Macintyre et al. (Eds.): SPIoT 2021, LNDECT 98, pp. 134–140, 2022.
https://doi.org/10.1007/978-3-030-89511-2_16

recognition methods is a little late, but the progress is very fast. Since the mid-1990s, Chinese speech recognition technology has reached a new height, and it has been practiced and applied in some products. A typical product is the ViaVoice system for Chinese speech developed by IBM [9].

This paper introduces the least squares (RLS) algorithm, then optimizes the RLS algorithm, designs the QR-RLS algorithm without complex iterations, and finally performs simulation experiment to verify its feasibility.

2 A Study of Microphone Array Noise Reduction Algorithms for Speech Recognition

2.1 Improved Adaptive Noise Reduction Technology for Microphone Array

The main steps in the implementation of the RLS noise reduction algorithm for microphone array are: data acquisition and generation, taking the main input signal d(n), X(n); initialization of the parameters; adaptive filtering process; filter coefficient update. The least squares algorithm is a recursive algorithm that works by observing the new parameters, comparing them with the previous ones, and correcting the previous data to reduce the estimation error according to the recursive algorithm, thus updating the parameter estimates over and over again. Such a process is an iterative optimization search, where parameter estimation is performed as new data are updated until the parameter estimates meet the requirements of the experiment.

In this paper, the RLS algorithm is selected as the noise processing algorithm of the microphone array, but its algorithm is complex and slow in iteration, so its algorithm should be optimized [10].

(1) QR-RLS noise reduction algorithm

The RLS algorithm is very sensitive to broadband noise because of its fast convergence speed, and it can handle it very well. This algorithm not only inherits the advantages of the RLS algorithm, but also controls the computational effort effectively and improves the noise control ability of the algorithm.

The QR-RLS algorithm is an orthogonal diagonalization method in mathematics to diagonalize the input matrix X(n). What is useful to the algorithm is the new vector for each iteration. For orthogonal diagonalization, we use the longest-used Givens rotation-orthogonal diagonalization method. The QR-RLS algorithm of Givens rotation needs to perform M square root operations in the process of processing the signal.

(2) Improvement of QR-RLS algorithm

Since the RLS algorithm based on QR decomposition needs to perform M square root operations, and each operation must consider the value of the previous moment, the processing speed is very unremarkable when processing a large amount of data. The calculation process is unit by unit, so the calculation speed of the square root is much slower than the multiplication and division calculation, which makes the real-time calculation rate lower.

In this paper, the QR-RLS algorithm is improved to make it omit the open square operation, and the operation process is as follows.

$$R(n) = \sqrt{D(n)}\overline{R}(n) \tag{1}$$

where

$$
D(n) = \begin{bmatrix} d_{11}(n) & & & \\ & d_{11}(n) & & \\ & & \cdots & \\ & & & d_{MM}(n) \end{bmatrix} = \begin{bmatrix} r_{11}^2(n) & & & \\ & r_{22}^2(n) & & \\ & & \cdots & \\ & & & r_{MM}^2(n) \end{bmatrix}
$$

$$
\bar{R}(n) = \begin{bmatrix} 1 & \bar{r}_{12}(n) & \bar{r}_{13}(n) & \cdots & \bar{r}_{1M}(n) \\ 0 & & \bar{r}_{23}(n) & \cdots & \bar{r}_{2M}(n) \\ & \cdots & \cdots & \cdots & \\ 0 & 0 & 0 & \cdots & \bar{r}_{M-1,M}(n) \\ 0 & 0 & 0 & \cdots & 1 \end{bmatrix} \tag{2}
$$

At this point, R(n) is given by the following equation

$$
R(n-1) = \begin{bmatrix} \cdots & & & & & \\ 0 & \cdots 0 & \sqrt{d(n-1)} & \cdots & \sqrt{d(n-1)} r_k(n-1) & \cdots \\ \cdots & & & & & \\ 0 & \cdots 0 & \sqrt{\gamma} x_i(n) & \cdots & \sqrt{\gamma} x_k(n) & \cdots \end{bmatrix} \tag{3}
$$

After QR variation the following equation is obtained.

$$
R(n) = \begin{bmatrix} \cdots & & & & & \\ 0 & \cdots 0 & \sqrt{d(n)} & \cdots & \sqrt{d(n)} r_k(n) & \cdots \\ \cdots & & & & & \\ 0 & \cdots 0 & 0 & \cdots & \sqrt{\gamma} x_k^{(i)}(n) & \cdots \end{bmatrix} \tag{4}
$$

And because

$$
\sqrt{d(n)} = \sqrt{d(n-1) + \gamma x_i^2(n)} \tag{5}
$$

where

$$
\gamma = \frac{\lambda r_{ii}^2(n-1) - d_{ii}(n-1)}{|x_i^{(i)}(n)|^2} + 1 \tag{6}
$$

This algorithm improvement not only solves the shortcomings of the RLS algorithm and QR-RLS algorithm with more iterations and slower processing speed due to the large number of open square operations, but also merges the advantages of the RLS algorithm with fast convergence speed, which is a good solution algorithm for microphone array noise reduction system.

2.2 Evaluation of the Effect of Microphone Array Noise Reduction

(1) Signal-to-noise ratio

The output signal-to-noise ratio directly reflects the noise reduction effect, and the calculation formula is as follows:

$$SNR(s, n) = 10\log_{10}(\frac{\sum_{n=1}^{L} |h_{s,i}(n)^* s(n)|^2}{\sum_{n=1}^{L} |n_i(n)|^2}) \tag{7}$$

Where, s(n) denotes the speech signal collected by the microphone as, h_(s,j) (n) denotes the impulse response between the main source and the ith microphone, n_i (n) denotes the ambient noise, "*" denotes the linear convolution.

(2) Log area ratio

The lower the LAR value, the smaller the distortion, and the higher the noise reduction level of the microphone array. Let the lth area ratio function of the speech signal be g_s (l,m) and the lth area ratio function of the output signal be g_out (l,m), then the logarithmic area ratio is defined as

$$LAR(m) = \frac{1}{L} \sum_{l=1}^{L} |20\log_{10}| \frac{g_s(l, m)}{g_{uot}(l, m)} \| \tag{8}$$

$$g(l, m) = \frac{1 + k(l, m)}{1 - k(l, m)} \tag{9}$$

Where k(l, m) is the LAR coefficient.

3 Simulation Experiment

3.1 Microphone Array Hardware Experiment Platform

The microphone array improved noise reduction algorithm designed in this paper is mainly applied to speech recognition in a home environment. The experimental process involves proper placement of microphone positions - keeping the microphones away from mechanical and acoustic noise sources. In addition, the arrangement structure of the microphones directly affects the acoustic effect of the system, including the number of microphones, the type of microphones, etc.

When the ambient noise is low or the main sound source is close to the microphone, a two-element microphone array is used.

When the ambient noise is large or the distance between the main sound source and the microphone is far, we use a four-element microphone array is more appropriate, the distance of the two-element microphone array should not exceed 0.6 m in general, and the four-element microphone array can still obtain a good output even in the case of a distance of more than 2 m.

This experiment was chosen for the signal acquisition performed by the four-microphone arrangement.

3.2 Simulation Environment

In order to analyze the performance of the noise reduction method used for speech recognition microphone array speech signals, we simulate and analyze the actual speech array signals. The microphone array speech database used for the experiments is from Carnegie Mellon University (CMU), and the data in this database includes 10 male speakers, each of whom has spoken 14 speech data.

The recording environment was in a noisy laboratory, with interference noise coming from multiple computer cooling fans and disk drive noise. The data sampling rate was 16 kHz.

4   Simulation Experiment Analysis

4.1   SNR Comparison Analysis

In this paper, the signal-to-noise ratio (SNR) is used as an objective evaluation equation to analyze the SNR of different algorithms using noise-free environments. The data for the comparison of SNR performance of noise reduction algorithms are shown in Table 1 below.

Table 1. SNR results of different noise reduction algorithms

Algorithm	−5 dB	0 dB	5 dB	10 dB	15 dB
RLS algorithm	1.3	1.8	2.4	2.6	2.9
QR-RLS algorithm	2.1	2.3	2.7	2.9	3.2
Improved QR-RLS algorithm	2.5	2.7	3.4	3.8	4.3

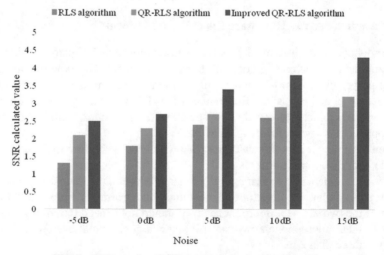

Fig. 1. SNR results of different noise reduction algorithms

As can be seen from Fig. 1, the performance of the RLS algorithm, QR-RLS algorithm, and improved QR-RLS algorithm gradually improves as the noise received by the microphone increases, which is reflected in the higher SNR scores; the RLS algorithm, QR-RLS algorithm, and improved QR-RLS algorithm increase sequentially in each of the same noise environments. From the results, it can be seen that the QR-RLS algorithm has a significantly higher SNR value than the RLS algorithm, and the improved QR-RLS algorithm also has a greater performance improvement compared to the traditional algorithm.

4.2 LAR Comparison Analysis

Next, the LAR of the microphone array is compared.

As in case Sect. 4.1, a noisy environment is simulated, and the simulated four microphones received signals are processed with the RLS algorithm, QR-RLS algorithm, and improved QR-RLS algorithm, respectively, and the LAR values are calculated. The results are shown in Fig. 2.

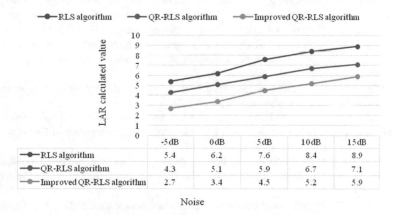

Fig. 2. LAR results of different noise reduction algorithms

It can be seen from Fig. 2 that with the increase of noise, although the LAR value of the RLS algorithm, QR-RLS algorithm and improved QR-RLS algorithm has increased, the increase is not very large. It can be seen that in the same noise environment, among the three algorithms, the improved QR-RLS algorithm has the smallest LAR value and the smallest distortion, that is, the microphone array of this algorithm has the highest noise reduction level. It can be seen through experiments that the improved QR-RLS algorithm still has obvious advantages based on other algorithms.

5 Conclusions

As an important branch of speech signal processing, the research of microphone array noise reduction algorithms for speech recognition has been conducted for decades, and both RLS algorithm and microphone array speech noise reduction algorithms have achieved remarkable research results. Microphone array is used in many fields such as speech recognition and video conferencing, but the signal received by microphones is generally a mixed signal consisting of ambient noise and multiple sound sources and their echoes. The research of microphone array noise reduction system for speech recognition is very challenging. In this paper, we study the structure of microphone array system and the basis of adaptive algorithm and optimization, and verify it through simulation experiments. The improved noise reduction algorithm proposed in this paper obviously improves the performance of all aspects of the algorithm, reduces the complexity of calculation, and achieves a better noise reduction effect.

References

1. Kumar, N., Singh, A., et al.: Study of microphone array characteristics and noise reduction. Int. J. Appl. Eng. Res. **13**(12 Pt.5), 10826–10830 (2018)
2. Chen, Y.Y.: Speech enhancement of mobile devices based on the integration of a dual microphone array and a background noise elimination algorithm. Sensors **18**(5), 1467 (2018)
3. Chen, W., Mao, L., Xiang, K., et al.: The application of a linear microphone array in the quantitative evaluation of the blade trailing-edge noise reduction. Appl. Sci. **11**(2), 572 (2021)
4. Schädler, M.R., Warzybok, A., Kollmeier, B.: Objective evaluation of binaural noise-reduction algorithms for the hearing-impaired in complex acoustic scenes. J. Acoust. Soc. Am. **141**(5), 3971–3971 (2017)
5. Zhao, S., Cheng, E., Qiu, X., et al.: Spatial decorrelation of wind noise with porous microphone windscreens. Acoust. Soc. Am. J. **143**(1), 330–339 (2018)
6. Yee, D., Kamkar-Parsi, H., Martin, R., et al.: A noise reduction postfilter for binaurally linked single-microphone hearing aids utilizing a nearby external microphone. IEEE/ACM Trans. Audio Speech Lang. Process. **26**(1), 5–18 (2017)
7. Li, M., et al.: A study on noise reduction for dual-energy CT material decomposition with autoencoder. Radiat. Detect. Technol. Meth. **3**(3), 1–13 (2019). https://doi.org/10.1007/s41 605-019-0122-2
8. Desjardins, J.L.: The effects of hearing aid directional microphone and noise reduction processing on listening effort in older adults with hearing loss. J. Am. Acad. Audiol. **27**(1), 29–41 (2016)
9. Ando, A., Yamauchi, K., Onishi, K.: Noise reduction for sound capturing by creating spatial null sensitivity point. Acoust. Soc. Am. J. **140**(4), 2966 (2016)
10. Hong, J., Park, S., Jeong, S., et al.: Dual-microphone noise reduction in car environments with determinant analysis of input correlation matrix. IEEE Sens. J. **16**(9), 3131–3140 (2016)
11. Id, S.C.L., Wang, J.F., Chen, M.H.: Threshold-based noise detection and reduction for automatic speech recognition system in human-robot interactions. Sensors **18**(7), 2068 (2018)
12. Djendi, M., Sayoud, A.: A new dual subband fast NLMS adaptive filtering algorithm for blind speech quality enhancement and acoustic noise reduction. Int. J. Speech Technol. **22**(2), 391–406 (2019). https://doi.org/10.1007/s10772-019-09614-9

Reactive Power Optimization of Power Project Management System Based on K-Means Algorithm

Chao Chai[1][✉], Tielei Liu[1], and Changhe Wang[2]

[1] Beijing Fibrlink Communications Co., Ltd., Beijing 100070, China
[2] Power Grid Technology Center, Jinxiandai Information Industry Co., Ltd., Shandong 250101, Jinan, China

Abstract. With the continuous improvement of our country's economic and technological level, the rapid expansion of the scale of the power system, and the continuous application of advanced power technology in the power system, which makes the complexity and urgency of reactive PO more prominent. Its connotation is also changing day by day, which increases the randomness, uncertainty, unpredictability and other factors in the reactive PO of the power system. It is precisely because of the emergence of these new challenges that it has always been one of the hot issues in the field of power system optimization. The reactive PO calculation method has been developed so far, and it is mainly divided into the bionic intelligent optimization algorithm and the traditional conventional mathematical algorithm. Conventional mathematical algorithms face the shortcomings of non-linear, discrete and continuous variable mixed, multi-constrained reactive PO inverse problems, it is easy to fall into the local optimal solution or even fail to find the optimal solution. When the bionic intelligent optimization method deals with the above problems, it is widely used in the field of power system reactive PO due to its high flexibility, high robustness, strong adaptability and other characteristics. This article explores the reactive PO of the power project management system based on the K-means algorithm. First, consult the relevant information to get a general understanding of the reactive PO of the power project management system, and then summarize the reactive PO methods of the power project management system based on the data. According to the method, the reactive PO steps of the power project management system based on the K-means algorithm are proposed, and then the experimental verification is carried out for the optimized power project management system. The experimental results show that the active power loss after optimization using the standard particle swarm algorithm is reduced from the initial 0.159 to 0.133, which is a 16.352% reduction; when the IPSO algorithm is used for optimization, the active power loss value is reduced to 0.131, which is a decrease of 17.610%; after adopting the K-means algorithm, the active power loss value drops to 0.125, and the reduction rate is 21.384%.

Keywords: K-means algorithm · Reactive PO · Power system · Particle swarm optimization

© The Author(s), under exclusive license to Springer Nature Switzerland AG 2022
J. Macintyre et al. (Eds.): SPIoT 2021, LNDECT 98, pp. 141–149, 2022.
https://doi.org/10.1007/978-3-030-89511-2_17

1 Introductions

As the scale of the power system continues to expand, the topology of the power grid is becoming more and more complex, and the problems of reactive PO and compensation have become more prominent [1, 2], and the requirements for reactive PO schemes and control methods have become more stringent, such as optimization algorithms. Can the optimal solution be approached quickly and reliably, can the infeasible solution problem be detected and solved, etc. [3, 4]. Moreover, the reactive PO problem is a very complex problem, coupled with its own characteristics including multiple extreme points [5, 6], continuous variables and discrete variables are mixed, and the system scale determines the complexity of the problem. It is very difficult to solve the problem. So far, for the solution of reactive PO problems, no practical and mature optimization methods have been found. Therefore, it is still necessary to explore reactive PO methods [7, 8].

Regarding the research on reactive PO of power project management systems, some studies have proposed that in recent years, bionic intelligent optimization methods such as ant colony algorithm, particle swarm algorithm, genetic algorithm, etc., are characterized by their high flexibility, high robustness, and strong adaptability. It has been widely used in many disciplines [9]. Compared with traditional optimization algorithms, the bionic intelligent optimization algorithm does not need to use any gradient information and has less dependence on the characteristics of the function itself. In dealing with non-linear, multi-constrained, multi-variable, discontinuous, non-convex and other optimization problems, the bionic intelligent optimization algorithm mainly uses group optimization technology to make it inherently parallel and can be easily applied to high-performance parallel computing [10]. The bionic intelligent optimization algorithm has a strong ability to jump out of local extreme points, and the theory converges to the global optimum. It can well adapt to the high-dimensional, non-convex, non-linear, multi-extreme and other characteristics of reactive PO problems, and overcomes the limitations of traditional conventional mathematical programming methods, and can handle continuous and discrete control variables at the same time [11]. Therefore, intelligent optimization algorithms are increasingly applied to power system optimization problems, such as power system reactive voltage control, grid planning and generation planning, capacitor optimal configuration, short-term load forecasting, power flow optimization, etc. On the basis of in-depth summary of the research on reactive PO by domestic and foreign researchers, some researchers believe that the use of algorithm to solve the reactive PO problem is a good choice, but due to the complexity of the reactive PO problem and the particle the limitations of the swarm optimization algorithm make the algorithm easy to fall into the local optimal and slow convergence speed in the later iterations. In order to further improve the performance of the algorithm, this paper analyzes the principles and characteristics of the algorithm in detail. And proposed its own improvement measures for the algorithm parameters. During the in-depth research, the paper found that the idea of co-evolutionary algorithm is of great help in dealing with reactive PO problems, so it introduced the idea of co-evolution, designed a collaborative algorithm and used it to solve reactive PO problems [12].

In this paper, the reactive PO of power project management system based on K-means algorithm is researched. On the basis of relevant literature data, the reactive PO method of power project management system is generally understood, and then based on K-means algorithm is proposed. The design of reactive PO steps of means algorithm is verified by experiments according to the optimized algorithm, and experimental conclusions are drawn.

2 Research on Reactive PO of Power Project Management System

2.1 Reactive PO Method of Power Project Management System

The main ways to realize the reactive PO of the power system are: changing the number of switching capacitor banks; controlling the position of the on-load tapping transformer; adjusting the terminal voltage of the generator. The optimization of reactive power flow is realized by controlling the above-mentioned control variables, so as to reduce the power system network loss and improve the voltage quality. Among the above-mentioned control variables, the number of switching capacitor banks and the position of the on-load tapping transformer are discrete variables, and the terminal voltage of the generator is a continuous variable. From the perspective of mathematical analysis, reactive PO problems include nonlinear equality and inequality constraints, that is, discrete variables need to be processed, and continuous variables need to be considered. It is not difficult to conclude that power system reactive PO is a non-convex, multi-extreme, multi-constraint, nonlinear, multi-constraint, discrete and continuous variable optimization problem. Therefore, it is quite difficult to solve it quickly and accurately.

2.2 Problems with Traditional Optimization Algorithms

(1) These traditional optimization algorithms can only converge quickly and obtain a global optimal solution only when a good initial solution is given. Therefore, these algorithms are relatively harsh on the selection of the initial solution. If the selected initial solution is poor, the iterative search space will be expanded, and the result obtained is likely to be a sub-optimal solution or even an infeasible solution, resulting in poor convergence accuracy of the algorithm.

(2) These traditional optimization algorithms have high requirements for the accuracy of the mathematical model, but the construction of the precise mathematical model will increase the amount of calculation, so the actual optimization problem is still very different from the construction of the theoretical mathematical model.

(3) The modern power system has a large number of control variables and a complex structure, resulting in a huge amount of calculation for many optimization algorithms and a substantial increase in calculation time.

(4) Most traditional optimization algorithms require the objective function and constraint conditions to be differentiable and differentiable as much as possible, and the control variables require continuity, but a large number of discrete variables will cause the difficulty and accuracy of the solution in practical problems.

2.3 Design of Reactive PO Steps Based on K-Means Algorithm

(1) Input system raw data, including power flow calculation data, reactive power control variable description, upper and lower limits of each variable, and particle swarm algorithm operation data.

(2) Divide the control variables of the entire power system into M groups by region, each group represents a subgroup of reactive PO, the control variable of the i-th group is Xi, then the reactive power control variable of the entire system is X = [X1, X2, X3, ..., Xn],

(3) Initialize each population, get X10, X20, X30, ..., Xn0 and set the current population number i = 1.

(4) IPSO optimization algorithm is used for independent update speed and position within each population.

(5) Choose from each of the remaining populations-a particle representative, and the individual X of the population i; together constitute the reactive power control variable of the entire system.

(6) Combine the objective function to calculate the fitness value in each partition and update the individual extreme value and global extreme value in the population according to the size of the fitness value.

(7) Judgment of a complete co-evolution. If i < M, go to the i + 1th population, otherwise go to the next step.

(8) Determine whether the algorithm has reached the maximum number of iterations, if it is, the algorithm will end the operation and output the result, if not, move to the first population and skip to step 4 for calculation.

2.4 System Related Calculations

The specific principle of reactive power compensation: where i and i are the grid current before and after reactive power compensation, respectively, which are decomposed into active and reactive components in turn, then i = 1μ + jlx, i2 = 12p + jl2x. When the compensation capacitor C is connected in parallel to the system, according to the nature of the capacitor, the phase difference between the current ic and the voltage U of the capacitor is 90°. It can be seen from that the active components of the system before and after compensation are still satisfied. The formula LR = L2p, that is, the capacitor C does not affect the active component of the system current. When the parameters in the power grid are known, the active power loss of the system can be expressed as:

$$P_{loss} = I^2R = \frac{(p^2 + Q^2)R}{U^2} \tag{1}$$

The line losses before and after compensation are respectively

$$P_{loss1} = I^2R = (\frac{I_{1R}}{\cos\varphi_1})^2 R \tag{2}$$

$$P_{ioss2} = I^2R = (\frac{I_{2R}}{\cos\varphi_2})^2 R \tag{3}$$

Since the active component of the system current is not affected by the parallel capacitor, that is, i = IR, the difference between the network loss before and after compensation is

$$\Delta P = [1 - (\frac{\cos\varphi_1}{\cos\varphi_2})^2] \times 100\% \tag{4}$$

3 Simulation Experiment Verification of Reactive PO Based on K-Means Algorithm

3.1 IEEE-14 Node Test System Model

For the IEEE-14 node system, before the reactive PO, the voltage of all generator nodes is set to 1.0, the transformer ratio is set to 1.0, and the reactive power compensation capacity is set to 0. The parameters and calculation results in this article are all set to standard unit values. The value is 100 MVA, and the structure diagram of the system can be checked. The node test system contains a total of 5 generators (corresponding node numbers are 1, 2, 3, 6, 8, and 1 is the balance node number). Three generators have Load regulating transformer (branch 5–6, 4–7, 4–9) and a reactive power compensation node (node 9).

3.2 Algorithm Parameter Setting

(1) In order to verify the idea of the CPSO algorithm, compare the calculated results with the results of the SPSO and IPSO algorithms. The mixed coding method is adopted, the reactive power compensation capacity and the transformer transformation ratio adopt integer coding, and the adjustment compensation is set separately. The generator node voltage adopts real number coding. The state variables of the system are generator reactive power output and load node voltage, set the penalty coefficient of load node voltage cross-border as 100, and generator node reactive power cross-border penalty coefficient as 100.

(2) In this paper, the population numbers of the three algorithms are set to 50, and the maximum number of iterations is $T = 100$. The parameter setting in the SPSO algorithm is: the inertia weight adopts the strategy of formula, and the learning factor is $c = c1 = 2$. In the IPSO algorithm, the inertia weight adopts the strategy of grouping and changing the inertia weight, and the learning factor adopts the strategy of formula. Since the simulation node system is not complicated, the parameter setting method for the collaborative algorithm is: the algorithm is divided into three populations, and no further division is performed. The three populations represent three sub-optimization problems, and each sub-optimization all problems adopt IPSO strategy. For the accuracy of the algorithm performance, each algorithm was run 100 times in the simulation, and the optimal optimization results were selected for comparison.

4 Comparison and Analysis of Simulation Results

4.1 Node Optimization Results

The state voltage value of each node is measured through the simulation experiment, and the result reflects the stability of the system. The relevant data results are shown in Table 1:

Table 1. Node optimization results

	SPSO	IPSO	CPSO
U4	1.02	1.03	1.04
U5	1.03	1.03	1.04
U7	1.04	0.98	1.03
U9	1.02	0.99	1.02
U10	1.02	0.99	1.03
U11	1.03	1.00	1.04
U12	1.03	1.01	1.03
U13	1.02	1.00	1.03
U14	1.00	0.98	1.01

It can be clearly seen from Fig. 1 that the voltage value of the system node optimized by the cooperative particle swarm algorithm is more stable than the voltage value optimized by the standard particle swarm algorithm and the IPSO algorithm. When the system voltage fluctuates, within the range of the upper and lower limits of the voltage, the node voltage optimized by the collaborative particle swarm algorithm has more room for change, which increases the stability of the system to a certain extent.

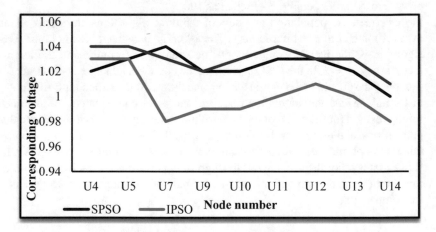

Fig. 1. Node optimization results

4.2 The Active Power Loss of the System

The active power loss of the system is measured through the simulation experiment, and the result reflects the stability of the system. The relevant data results are shown in Table 2:

Table 2. Active power loss of the system

	SPSO	IPSO	CPSO
1	0.1436	0.1428	0.1436
5	0.1427	0.1409	0.141
10	0.139	0.137	0.1372
15	0.137	0.135	0.13
20	0.137	0.134	0.126
25	0.135	0.134	0.126
30	0.1345	0.133	0.125
35	0.134	0.133	0.125
40	0.1338	0.132	0.125
45	0.1338	0.132	0.125
50	0.133	0.131	0.125
55	0.132	0.131	0.125
60	0.132	0.131	0.125
65	0.132	0.131	0.125
70	0.132	0.131	0.125
75	0.132	0.131	0.125
80	0.132	0.131	0.125
85	0.132	0.131	0.125
90	0.132	0.131	0.125
95	0.132	0.131	0.125
100	0.132	0.131	0.125

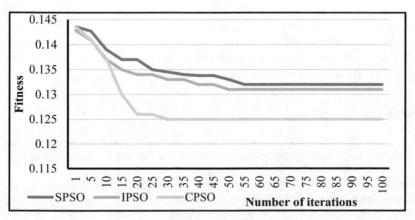

Fig. 2. Active power loss of the system

Figure 2 shows that the active network loss value after optimization using the standard particle swarm algorithm is reduced from the initial 0.159 to 0.133, which is a 16.352% reduction; when the IPSO algorithm is used for optimization, the active network loss value is reduced to 0.131, which is a 17.610% reduction; after adopting the K-means algorithm, the active power loss value drops to 0.125, and the reduction rate is 21.384%. From the perspective of the distribution of the solution, the SPSO algorithm is relatively divergent. Once the algorithm finds the active power loss value of 0.133, it is easy to fall into the local optimum, resulting in the algorithm not being able to converge better; the IPSO algorithm is better, but the obtained the result is also unstable, and the probability of seeking a better quality solution is also low. When using the CPSO algorithm, this problem will not occur, so the improved algorithm can better jump out of the local optimal value and perform global optimization. Moreover, the optimization efficiency of the CPSO algorithm is also high. Generally, the optimization will stabilize after the 20th generation, while the SPSO algorithm and the IPSO algorithm need to be optimized until the 40th generation before they gradually become stable. In general, the CPSO algorithm is obviously stronger than the SPSO algorithm and the IPSO algorithm in terms of active power loss value and iteration time. CPSO algorithm is ideal in dealing with reactive PO problems and can adapt to the actual situation of power system reactive PO.

5   Conclusions

In this paper, the reactive PO of the power project management system of the K-means algorithm is studied. Due to the limitation of time and personal ability, the K-means algorithm still has many limitations. Therefore, I still need to learn about the K-means algorithm in the future. The algorithm research and exploration mainly has the following two points: (1) Real-time problem. Electric power systems are increasingly developing in the direction of informatization and automation, and the network's requirements for real-time performance are getting higher and higher. Therefore, the algorithm still needs further research, how to solve the global optimal solution in a shorter time. (2) This article only takes active power loss as the optimization objective, but in the actual system, not only the active power loss must be considered, but also the total reactive power compensation consumption of the system and other factors. The next step will be combined with the actual regional power grid operation to design multi-objective power system reactive PO that truly conforms to the actual operation of our country.

References

1. Wang, Q., Chao, L., Yong, L., et al.: A dynamic reactive power optimization method of receiving power system containing infeasible buses with violated voltage constraint. Diangong Jishu Xuebao/Trans. China Electrotech. Soc. **33**(5), 1096–1105 (2018)
2. Zheng, F., Wei, W., Pu, T., et al.: Power system reactive power optimization based on fuzzy formulation and interior point filter algorithm. Energy Power Eng. **05**(4), 693–697 (2016)
3. Jangir, P., Parmar, S., et al.: A novel hybrid particle swarm optimizer with multi verse optimizer for global numerical optimization and optimal reactive power dispatch problem. Eng. Sci. Technol. Int. J. **20**(2), 570–586 (2017)

4. Abdelhady, S., Osama, A., Shaban, A., et al.: A real-time optimization of reactive power for an intelligent system using genetic algorithm. IEEE Access **8**(1), 11991–12000 (2020)
5. Hao, W., Liu, B., Yao, S., et al.: Reactive power optimization of distribution network with distributed generation based on genetic and immune algorithm. J. Eng. **2019**(16), 1280–1284 (2018)
6. Kassem, A.M., Abdelaziz, A.Y.: Firefly optimization algorithm for the reactive power control of an isolated wind-diesel system. Electr. Power Componen. Syst. **45**(13), 1413–1425 (2017)
7. Tabrizi, N., Babaei, E., Mehdinejad, M.: An interactive fuzzy satisfying method based on particle swarm optimization for multi-objective function in reactive power market. Iran. J. Electr. Electron. Eng. **12**(1), 65–72 (2016)
8. Zhu, Y., Zhuo, F., Wang, F., et al.: A virtual impedance optimization method for reactive power sharing in networked microgrid. IEEE Trans. Power Electron. **31**(4), 2890–2904 (2016)
9. Sheng, W., Liu, K.-y., Liu, Y., Ye, X., He, K.: A reactive power coordinated optimization method with renewable distributed generation based on improved harmony search. IET Gener. Transm. Distrib. **10**(13), 3152–3162 (2016)
10. Zhang, X., Yu, T., Yang, B., Cheng, L., et al.: Accelerating bio-inspired optimizer with transfer reinforcement learning for reactive power optimization. Knowl.-Based Syst. **116**, 26–38 (2017)
11. Lizhen, W.U., Jiang, L., Hao, X.: Reactive power optimization of active distribution network based on optimal scenario generation algorithm. Power Syst. Prot. Control **45**(15), 152–159 (2017)
12. Niu, T., Guo, Q., Sun, H., et al.: Dynamic reactive power reserve optimization in wind power integration areas. IET Gener. Transm. Distrib. **12**(2), 507–517 (2017)

Variant Translation Strategy of Dough Sculpture Art Based on Deep Neural Network Algorithm

Lu Zhang[(⊠)] and Cong Wang

School of English Language and Literature, Xi'an Fanyi University, Xi'an 710105, Shaanxi, China

Abstract. With the development of globalization and information diversification, translation plays a pivotal role in the exchanges of countries around the world; many translation theories have also emerged from this and become increasingly mature. As an efficient tool, machine translation can realize the equivalent conversion between different languages while retaining the original semantics, which has important practical significance. This article aims to study the translation strategy of dough art translation based on deep neural network algorithms. Based on the analysis of the characteristics of dough art, the main content of the translation and the neural machine translation model, it aims at the inability of neural machine translation technology knowledge in the existing language translation system. It makes good use of the current situation and existing problems, and proposes a new neural machine translation system model with a large amount of speech sequence information. Based on the theoretical basis of fully integrating people's attention and thinking mechanism gru two-way translation language model, the stanford parser is used to analyze the language and syntax, and after obtaining relevant data and information about the part of speech and the order of sentences in the language. The language and encoder components integrated into the language translation model in the form of two-way encoding, using vector splicing methods and forms to jointly construct a vector background. Experiments have proved that adding a word sequence of information can greatly improve the translation model.

Keywords: Deep neural network · Art of dough sculpture · Translation strategy

1 Introduction

Machine translation is an important research direction in the field of natural language processing and artificial intelligence. Through computer-related functions, while retaining its original meaning, one natural language is transformed into another natural language, thereby realizing two natural languages. Interlingual translation completes the equivalent conversion of information between different languages [1, 2]. It is particularly important to obtain translation results quickly and efficiently through computers, and it is even more important to replace human translation in the professional translation industry. At this stage, it is not a perfect translation, which also means that there is more research space in this field [3, 4].

© The Author(s), under exclusive license to Springer Nature Switzerland AG 2022
J. Macintyre et al. (Eds.): SPIoT 2021, LNDECT 98, pp. 150–157, 2022.
https://doi.org/10.1007/978-3-030-89511-2_18

The theory of variation translation was put forward at the 1997 International Translation Symposium. Although Chinese scholars proposed and practiced translation as early as 100 years ago, due to various reasons, this theory has not received enough attention from future generations [5, 6]. However, with the increasing exchanges and cooperation between countries, China's international status has been further improved, and the market has higher requirements for the practicality of theories, and the practice of local translation in China is the research object, and the market is oriented. The variation translation theory undoubtedly satisfies this requirement [7, 8].

Based on the analysis of the characteristics of dough sculpture art, the main content of translation and the neural machine translation model, this paper proposes a new type of belt based on the current situation and existing problems of neural machine translation technology in the existing language translation system. A neural machine translation system model with a large amount of speech sequence information.

2   Research on the Translation Strategy of Dough Sculpture Art Translation Based on Deep Neural Network Algorithm

2.1   Features of Dough Sculpture Art

(1) Luliang area

The main characteristics of Luliang dough sculptures are: mostly composed of plain noodles and red dates, and the shape is simple and general. It is different from the dough sculptures in other places, and they are mostly used as sacrificial tributes. For example, the chaotic bun and Zaoshan are mostly used to worship ancestors and worship gods in the first month. Every year in the Luliang area of the Lantern Festival, a face lamp is made to worship the god of the river in order to pray for a year of good weather. The image is mostly in the shape of the twelve zodiacs, especially the dragon and tiger, and a candlestick is placed on it. Every year on the 23rd of the twelfth lunar month, the local also steamed cold swallows for the children. It is different from the past dough sculptures. It looks like a bird but not a bird. After the dough sculpture is out of the pot, it will be inserted into the jujube branch to show people's worship of the gods. The liveliest event is the Lancheng Festival held on February 19 of the lunar calendar every year. Every family will prepare dough sculpture tributes on this day and gather on the main street of Lancheng County to pay homage to the "white-clothed priest" [9, 10].

(2) Jinbei area

Compared with the dough sculptures in other parts of Shanxi, the folk dough sculptures in northern Shanxi are relatively pure, which should not be underestimated in Shanxi. Due to its geographic location, the art of dough sculpture in northern Shanxi is less affected by the surrounding environment. Folk dough sculptures in many places still retain local customs and habits [11, 12].

Especially Dingxiang County and Dai County in northern Shanxi are the most distinctive. There are many types of dough sculptures in the northern part of Shanxi. According to the age of the year, different dough sculptures are represented, including Zaoshan, Hanyan, Climbing Baby, Unidentified, etc., with various shapes.

Every New Year's Day in Dingxiang County, the prepared Zaoshan will be taken to the fire kang to worship the stove king. The locals call the rice noodles a mountain, which means ample food and clothing. During the Spring Festival in Dai County, there is a custom of steaming offerings to the gods. Before the holiday, the fermented dough is shaped into sweet bergamot, ingot buns, lotus offerings, red jujube flowers and other shapes, which respectively represent good fortune and longevity, wealth fortune, many children, and sweet life. Every July fifteenth generation of counties also knead dough people. It is said that dough people are things that convey information. The shapes include crawling babies and sleeping babies, etc., and the shape is lovely and lovable.

2.2   The Main Content of the Translation

(1) Readers' needs catalyze "changes"

Readers are not only the starting point and driving force of translation activities, but also the foothold and destination of information transmission. In the target language environment, the use value of the source language information and the degree of realization of its potential value are reflected by the reader. The different needs of readers when reading articles, or thirst for knowledge, or pleasure of self, or cultivation of sentiment, or enjoyment of beauty, all promote the modification of the original work. The needs of the target language readers promote the transformation of the original work. When the target language readers receive foreign information, their mental state, psychological and cultural literacy determine the translator's choice and adaptation of the original translation; in addition, in the new era, readers' reading also presents new characteristics: fear of clichés and dull forms. The big tome, the vivid language, and the flexible, short and concise articles are full of interest. Therefore, how to make the translated work see the big from the small, the small paragraph contains the big world, improve the efficiency of text expression, and meet the reader's psychological needs of reading, has also become a major factor that hastened the birth of variable translation.

(2) The translator's operation leads to "change"

Translation is an activity that requires the translator to give full play to his subjectivity, and the theory of variation translation reflects this idea. In the subject-object theory of variation translation theory, the translator is the subject and implementer of variation translation and adaptation. First of all, when selecting the source language information, the attitude towards the original work is driven by the needs of the target language audience. The translator needs to be uncharacteristically, changing the previous "respectful and respectful" attitude towards the original work, etc. The attitude that the translator intervenes and the source language information is used by me, translators need to use their initiative to understand the richness and diversity of the original work, and to measure whether the original work is changeable or not. Secondly, in the process of translation, the translator needs to change the source language information according to the reader's special needs to produce a translation that meets the needs of the target language audience. What kind of workarounds and translation methods to adopt requires the consideration, judgment and choice of the translator. At the same time, it is also necessary to keep in mind the purpose of the translation activity and the reader's acceptance and understanding, and pay attention to the reader's reading experience in order to achieve the best Effective cross-cultural information exchange.

2.3 Neural Machine Translation Model

(1) Building an encoder-decoder framework

The structure of "encoder-decoder" is used for translation conversion work, which marks the emergence of neural machine translation modeling ideas. When the input and output are both variable-length sequences, a decoder encoder can be used to create a model frame, where the encoder corresponds to the input sequence and the decoder corresponds to the output sequence. The encoding step encodes the entire sequence source into a vector, and the decoding step decodes the entire target sequence through the largest conversion predicted sequence probability.

The actual meaning of the framework structure is that, for the sequence $X = \{x_1, x_2, \ldots, x_n\}$ along the language and the sequence $Y = \{y_1, y_2, \ldots, y_m\}$ of the target language, the probability of generating the target language can be constructed as shown in formula (1).

$$P(y_1, y_2, \ldots, y_m / x_1, x_2, \ldots, x_n) = \prod_{t-1}^{t-m} P(y_t / c, y_1, \ldots, y_{t-1}) \tag{1}$$

Among them, c represents the fixed-dimensional background vector generated by the encoder stage, which contains the relevant information of the source language sequence. The generation probability for each target language vocabulary in the second half of the announcement is calculated by softmax, as shown in formula (2).

$$P(y_t / x, y < t; 0) = \frac{\exp(\varphi(y_t, x, y < t, \theta))}{\sum_{y \in Y} \exp(\varphi(y_t, x, y < t, \theta))} = \frac{\exp(\varphi(V_{yt}, c_s, c_t, \theta))}{\sum_{y \in Y} \exp(\varphi(V_y, c_s, c_t, \theta))} \tag{2}$$

(2) Translation model based on two-way GRU

When performing translation tasks, the context information of the source language translation at the current moment should be considered comprehensively. The recurrent neural network can store time series information by sequentially scanning the input sequence, and has the ability to remember context information. However, the word vector obtained by this method only contains the left context information, and does not consider the right context information.

3 Experiment

The experimental data uses the machine translation task data set provided by the AI Challenger Global AI Challenge, which includes training data about 10 million Chinese-English parallel sentence pairs, and 8000 sentences each in the development set and the test set. The experiment process randomly selected 200,000 parallel sentence pairs in the training set as training data, while limiting the length of source and target language sentences to no more than 40 words. A neural scanner translation system is built on the deep learning framework TensorFlow, and the open source syntactic analysis tool Stanford Parser is used to perform word segmentation, part-of-speech tagging and dependency syntax analysis of Chinese sentences, and obtain the part-of-speech sequence information of source language sentences through this method. The relevant parameter settings of the neural network are shown in Table 1.

Table 1. Parameter settings

Parameter type	Value
Vocabulary size	30000
Word vector dimension	512
Hidden layer nodes	512
Number of network layers	2
Column search width	3
Learning rate	0.1
Dropout	0.5
Batch-size	128
Optimizer	SGD

The optimizer selects the stochastic gradient descent method for parameter optimization, and the column search algorithm is used in the decoding stage. Similarly, the baseline system uses Niutrans phrase-based statistical machine translation system to automatically evaluate the translation quality with the case-insensitive BLEU-4 value.

4   Discussion

By comparing the translation performance of the neural machine translation model based on the two-way GRU proposed in the previous chapter before and after the coding information of the part-of-speech sequence is added, the experimental results are shown in Fig. 1.

It can be seen from the experimental results that the baseline system still has the best performance. The translation model with part-of-speech sequence information has a performance improvement of 0.16 BLEU value compared to the previous one, and it is gradually approaching the translation performance of the baseline system. This result prove the effectiveness of the method proposed in this paper to add part-of-speech sequence information to the coding stage to improve the performance of neural machine translation.

In order to verify the sensitivity of the translation model to sentence length after adding part-of-speech sequence information, the test set data is classified according to

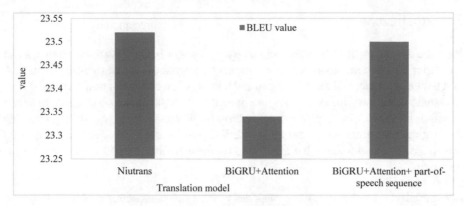

Fig. 1. Experimental results

sentence length for testing, and experiments are carried out for different sentence length distributions. The final BLEU value is used to measure the sentence length sensitivity of the translation model.

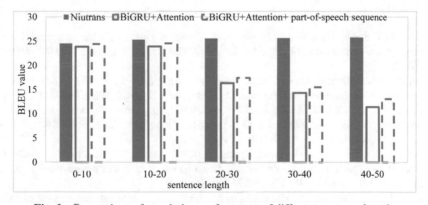

Fig. 2. Comparison of translation performance of different sentence lengths

By analyzing Fig. 2, it can be found that the performance of the Niutrans translation model is optimal regardless of the length of the interval, and as the sentence length gradually increases, the translation performance gradually increases. The experimental results here also verify that the current machine translation model is at it still has obvious advantages in the face of long sequences of sentences. For the neural machine translation model, in the process of increasing the sentence length from 10 to 20, the translation performance has an average improvement of 0.1 BLEU value, and the translation model with part-of-speech sequence information has better performance. When the sentence length exceeds 20, as the sentence length increases, the performance of the two neural machine translation models shows a downward trend, and the performance of the translation model with part-of-speech sequence information is always better than that of ordinary neural machine translation.

5   Conclusions

With the development of Internet technology, exchanges between various countries and industries have become more and more frequent. Language is the carrier of communication between people, and the conversion and transmission of information between different languages is very important. As an efficient tool for language conversion, machine translation can perform equivalent conversion of different languages while preserving the original semantics, which has great practical significance. The development of deep learning related technologies has also improved the method and performance of machine translation.

Acknowledgements. This work was supported by 2021 Scientific Research General Project of Xi'an Fanyi University, PN: 21B47, the Scientific Research Major Project of Xi'an Fanyi University, PN: 2021Z02, the "Double First-class" Construction Project of Xi'an Fanyi University, non-number and Belt And Road Language and Culture Research Base (Think Tank) of Xi'an Fanyi University, PN: 20KYJD02.

References

1. Cao, J., Cui, H., Hao, S., et al.: Big data: a parallel particle swarm optimization-back-propagation neural network algorithm based on MapReduce. PLoS ONE **11**(6), e0157551 (2016)
2. Guerrero-Enamorado, A., Ceballos-Gastell, D.: An experimental study of evolutionary product-unit neural network algorithm. Computacion Y Sistemas **20**(2), 205–218 (2016)
3. Liu, X.-H., Shan, M.-Y., Zhang, L.-H.: Low-carbon supply chain resources allocation based on quantum chaos neural network algorithm and learning effect. Nat. Hazards **83**(1), 389–409 (2016). https://doi.org/10.1007/s11069-016-2320-2
4. Jae-Hong, L., Do-Hyung, K., Seong-Nyum, J., et al.: Diagnosis and prediction of periodontally compromised teeth using a deep learning-based convolutional neural network algorithm. J. Periodontal Implant Sci. **48**(2), 114–123 (2018)
5. Bramslw, L., Naithani, G., Hafez, A., et al.: Improving competing voices segregation for hearing impaired listeners using a low-latency deep neural network algorithm. J. Acoust. Soc. Am. **144**(1), 172–185 (2018)
6. Gao, F., Yue, Z., Wang, J., et al.: A Novel active semisupervised convolutional neural network algorithm for SAR image recognition. Comput. Intell. Neurosci. **2017**(24), 3105053 (2017)
7. Katz, P.P., Barton, J., Trupin, L., et al.: Poverty, depression, or lost in translation? Ethnic and language variation in patient-reported outcomes in rheumatoid arthritis. Arthritis Care Res. **68**(5), 621–628 (2016)
8. Araúz, P.L., Cabezas-García, M.: Term and translation variation of multiword terms. MonTi Monografías de Traducción e Interpretación, **2020**(Special Issue 6), 210–247
9. Lytnev, V., Fujiwara, K., Kiyota, N., et al.: Postural control and contingent negative variation during transient floor translation while standing with the ankle fixed. J. Physiol. Anthropol. **36**(1), 7 (2017)
10. Cheesman, T., Flanagan, K., et al.: Multi-retranslation corpora: visibility, variation, value, and virtue. Digit. Sch. Hum. **32**(4), 739–760 (2017)

11. Jiménez-Crespo, M.A., Tercedor Sánchez, M.: Lexical variation, register and explicitation in medical translation: a comparable corpus study of medical terminology in US websites translated into Spanish. Translation and Interpreting Studies. J. Am. Transl. Interpreting Stud. Assoc. **12**(3), 405–426 (2017)
12. Khanh, D.D., Song, Y.S., Tamir, T.: The impact of ribosomal interference, codon usage, and exit tunnel interactions on translation elongation rate variation. PLoS Genet. **14**(1), e1007166 (2018)

AI-Based IWrite Assisted English Writing Teaching

Jinling Xia[1], Huafeng Liu[2], and Wenbin Liu[3,4](✉)

[1] Economics and Business Foreign Language School, Wuhan Technology and Business University, Wuhan, Hubei, China
[2] Information and Technology Center, Wuhan Institute of Shipbuilding Technology, Wuhan, Hubei, China
[3] Wuhan Technology and Business University, Wuhan, Hubei, China
[4] Wuchang Institute of Technology, Wuhan, Hubei, China

Abstract. With the rapid development of modern information technology, the application of artificial intelligence in the field of education has brought new opportunities and challenges to educators. Iwrite, based on artificial intelligence, is an English writing teaching and evaluation system that includes writing teaching, intelligent correction and automatic grading, which can play a great role in English writing teaching. In order to explore whether AI-based iWrite assisted English writing teaching mode can effectively improve students' English writing ability, the study adopts two different teaching methods to compare, in order to observe the effect of AI-based iWrite assisted English writing teaching. The results show that AI-based iWrite assisted English writing teaching mode can effectively improve students' English writing ability, and students are more receptive to it. Therefore, the study suggests that in the future teaching, more and more teachers should follow the trend of educational informatization and integrate AI-based iWrite into their teaching.

Keywords: AI-based IWrite · English writing teaching · An empirical study

1 Introduction

1.1 Application of AI in the Education Field

With the rapid development of Internet, big data, cloud computing, artificial intelligence and other information technologies, the application of artificial intelligence in the field of education has become a trend in the future. At present, from the national policy level, in 2019, *Chinese Education Modernization 2035* pointed out that education should use modern technology to accelerate the pace of talent training, to achieve the organic combination of large-scale education and personalized education, to accelerate the education reform in the information age. Therefore, it is an important arrangement policy to advocate the integration of artificial intelligence and education. At the technical level, the teaching platform with artificial intelligence has improved functions greatly, which can provide strategic learning scaffolding and teaching aids in the interaction and collaboration among teachers, machines and students [1], thus escorting the development of online and offline teaching.

© The Author(s), under exclusive license to Springer Nature Switzerland AG 2022
J. Macintyre et al. (Eds.): SPIoT 2021, LNDECT 98, pp. 158–165, 2022.
https://doi.org/10.1007/978-3-030-89511-2_19

1.2 Difficulties in Traditional English Writing Teaching

In the current English writing teaching, there are still many problems: teachers indoctrinate students in class and students absorb them passively. Mechanical knowledge instillation reduces students' motivation for independent learning, and students' absorption of what teachers say is very limited. In the large-class teaching, the number of students is too large, and teachers are unable to give targeted guidance to students [2]. Teachers are often unable to timely feedback students' problems in writing, such as wrong spelling, incorrect grammar, improper words, disordered sentence structure and so on. The score of English writing is often high in the test, but it is difficult for students to get high marks. Therefore, English writing has troubled not only students, but also teachers. Faced with such many problems as heavy teaching tasks, lack of time and large-class teaching, how to improve the effect of teaching has become a difficult problem [3, 4].

1.3 AI-Based IWrite Assisted English Writing Teaching

AI-based iWrite is an teaching and evaluation system of English writing combining theoretical support, technical application and teaching practice, which can play a great positive role in teaching. As the chief designer of iWrite, Liang Maocheng, a professor in *Beijing Foreign language University*, said that iWrite uses a comprehensive evaluation concept, which assesses students' writing from four dimensions: language, content, discourse structure and technical specifications. He especially emphasized the importance of content evaluation.

IWrite, a teaching and evaluation system in English writing, which uses big data and cloud computing, is simple and easy to operate. When Reviewing, it can accurately point out students' grammatical errors and other errors, which can provide great convenience for teaching [5]. Undoubtedly, if AI-based iWrite system collaborates with teachers, it can not only help teachers to complete teaching tasks more effectively, but also greatly improve the teaching quality and students' learning [6]. However, AI-based iWrite system cannot completely replace or neglect the role of teachers [7], and teachers still play a significant role in teaching. Machine evaluation is just a helper, and man-machine combination can improve English writing teaching better [8].

2 Theory

The constructivism theory holds that learning is the learning subject absorbs and digests the information obtained according to his own experience, so as to construct the understanding of the object. In other words, learners themselves have different experiences and backgrounds and observe things from different perspectives. Therefore, education should be taught in accordance with their aptitude, instead of adopting the same method for everyone. Different people establish their understanding of the object according to their own understanding. The process of knowledge construction is individualized and situational, but not unified. AI-based iWrite assisted English writing teaching is exactly to achieve this purpose, it helps students to learn more effectively, to achieve personalized learning. In addition, the constructivism theory also believes that teachers are students' assistants and learning partners, whose main task is to help students construct the understanding of the object by creating situations. When students encounter difficulties, teachers should guide them appropriately and make appropriate teaching plans to keep students' enthusiasm. In the AI-based iWrite assisted English writing teaching, teachers mainly provide basic materials and give guidance to students, which is consistent with the constructivism theory. This teacher-led teaching mode with AI-based iWrite as a tool is conducive to stimulating students' interest and potential in writing and improving the teaching quality.

3   Teaching Practice

In view of current teaching situation, this study attempts to use AI-based iWrite assisted English writing teaching mode, and compares it with traditional writing teaching mode, aiming to answer two questions.

Q1. Whether AI-based iWrite assisted English writing teaching can improve students' writing ability more effectively?

Q2. What is the attitudes of students towards the new teaching mode?

In the following part of the study, two teaching modes were compared to verify the effect of the new teaching.

3.1   Teaching Experimental Subjects

In order to ensure the effectiveness of the experiment, the study conducted a pre-test for non-English major sophomores by using iWrite, and conducted a survey of students' self-learning ability. The data was collected and then analyzed by SPSS software. According to the results analysis, the class of accounting major is finally determined as an experimental class (class 1 in the table), which uses AI-based iWrite assisted English writing teaching mode, while the class of law major is a control class (class 2 in the table), which uses traditional teaching mode. Independent sample t-test is used to compare the writing level of students in two classes, and data analysis is shown as follows in Table 1.

Table 1. Comprison of pre-test

	Class	N	Mean	Std. Deviation	F	Sig.	t	Sig. (2-tailed)
Pre-test	1.00	38	88.2500	17.98824	4.352	.040	1.647	0.104
	2.00	40	80.2262	25.63587				

Two classes are roughly similar in size, with 38 students in the experimental class and 40 in the control class. In order to ensure that there is no significant difference in writing proficiency between two classes, independent sample t-test is used for analysis. It can be found that the first sig. value is $0.040 < 0.05$, indicating that the variance is not homogeneity. Only the second sig. value can be referred to, and it can be seen that the second sig. value is 0.104 > 0.05, indicating that the writing level of two classes before experiment are balanced and they are comparable.

In addition, since self-learning ability plays a pivotal role, the study also conducted a questionnaire on self-learning ability of the students.

Table 2. Comprison of self-learning ablity

Dimensions	F	Sig	t	df	Sig. (2-tailed)
1st dimension	.111	.740	.776	80	.440
2nd dimension	.418	.520	−.373	80	.710
3rd dimension	.609	.438	.129	80	.898
4th dimension	.085	.771	.665	80	.508
5th dimension	3.331	.072	1.809	80	.074
Total score	.266	.607	.872	80	.386

In the Table 2 below, the questionnaire of students' self-learning ability is divided into five dimensions. The first dimension is to understand teaching objectives and requirements of teachers. The second dimension is to make learning goals and plans. The third dimension is effective use of learning strategies. The fourth dimension monitors the use of learning strategies. The fifth dimension monitors and evaluates the English learning process. The last one represents the overall level of self-learning. It can be found from the Table 2 that, under the condition of equal homogeneity of variance, the Sig. value of each dimension is lager than 0.05, indicating that there is no significant difference between two classes.

3.2 Teaching Process

Since the study is to explore whether AI-based iWrite assisted English writing teaching mode can effectively improve students' writing ability compared with traditional teaching mode, it is necessary to make a comparison between two kinds of teaching. Here it is necessary to describe two teaching briefly.

About traditional English writing teaching. The teacher uses it in the control class. Before writing, the teacher provides students with reading materials to increase their vocabulary by means of reading. In class, the teacher makes students clear the composition structure and writing method, and then assigns the writing task. In writing, students finish the composition in an exercise book after class, and then hand it in to the teacher next class. After writing, the teacher reviews it, and then comments on it in class next time. Because the teacher's energy and time are limited, the evaluation and scores are subjective and evaluation standard is sometimes not unified. The disadvantages of this teaching model are obvious. Teachers often pays too much attention to the results and neglect the process, which leads to less classroom participation of students, excessive reliance on teachers and lack of independent thinking. Besides, teachers' feedback can not be presented in time, so that students' problems in writing can not be solved in time.

About AI-based iWrite assisted English writing teaching. The teacher uses it in the experimental class. Before writing, the process is the same as the control class. In writing, teachers publish writing tasks in the iWrite and set requirements such as writing topics, time and word count. Students complete writing and submit it on the iWrite according to the teacher's requirements. Because iWrite carefully evaluates students' essays from four dimensions: language, content, discourse structure, and technical specifications, students can revise their writing repeatedly until they get a satisfactory score based on the feedback from iWrite. Because automatic evaluation system will unconsciously guide students to adjust their writing to meet the evaluation criteria of the system [9]. After writing, teachers do not need to correct the composition, so they can focus on the error correction and composition evaluation.

3.3 Result Analysis

The teaching experiment went on for a whole semester. Students were required to have at least one writing practice every two weeks. In order to ensure the validity of the experiment, the teaching content, teaching progress and teachers of the two classes were kept the same each time in experiment, so as to minimize the loss of irrelevant variables. After experiment, in order to verify the effect of AI-based iWrite assisted English writing teaching mode, two classes were conducted a post-test through iWrite and scores are given directly by iWrite. The data are analyzed as follows in Table 3.

Table 3. Comparison of post-test

	Class	N	Mean	Std. Deviation	t	P
Post-test	1.00	38	8.8000	2.00256	.139	.000
	2.00	40	6.9524	2.38851		

The total score is 10 points. The average score of the experimental class is 8.8000, and that of the control class is 6.9524. Experimental class is significantly higher than control class. In the case of variance heterogeneity ($P = 0.139 > 0\ 0.05$), there is a significant difference in the writing scores between two classes ($P = 0.000 < 0.05$).

In order to determine the reliability of the results, the scores of students' compositions in the final examination were also analyzed.

Table 4. Comprison of composition in the fnal exam

	Class	N	Mean	Std. Deviation	t	p
Composition	1.00	38	12.3750	3.79397	.019	.000
	2.00	40	7.8810	4.52205		

The total score is 15 points, all of which are handwritten on paper. The compositions are scored manually by the teachers. The Table 4 above shows that the average score of the experimental class is 12.3750, and that of the control class is 7.8810. Experimental class is much higher than control class. In the case of variance heterogeneity ($p = 0.139 > 0.05$), there is a significant difference in the writing scores between two classes ($P = 0.000 < 0.05$).

In conclusion, based on the analysis above, it is obvious that AI-based iWrite assisted English writing teaching mode can improve students' English writing scores more effectively, which indicates that it has better effect than traditional English writing teaching mode.

What is the attitudes of students towards AI-based iWrite assisted English writing teaching mode? At the end of experiment, the study conducted a questionnaire in experimental class.

Table 5. Students' attitudes to iWrite assisted English writing

Questions	Strongly disagree	Not quite agree	Not sure	Relatively agree	Strongly agree
It can reduce the grammer and vocabulary errors in the composition	3.80%	4.50%	2.20%	80.10%	9.40%
The idea and content of the composition are improved	6.80%	9.10%	9.50%	65.00%	9.60%
It can improve the structure and cohesion of the composition	2.68%	8.70%	13.60%	56.32%	18.70%
The interest in writing is enhanced	4.89%	10.50%	17.60%	57.61%	9.40%
It can enhance the self-learning ability	2.40%	5.78%	5.80%	68.62%	17.40%
This teaching method is preferred compared with traditional teaching	2.30%	8.19%	10.80%	68.31%	10.40%

From the above Table 5, the vast majority of students believe that AI-based iWrite assisted English writing teaching mode is helpful to reduce the vocabulary and grammar errors in writing (80.10% relatively agree, 9.40% strongly agree). The majority of students believe that new teaching mode helps to improve the idea and content of the composition (65.00% relatively agree, 9.60% strongly agree). Similarly, in terms of improving composition structure and content cohesion, 56.71% relatively agree and 18.70% strongly agree. In terms of increasing students' interest in writing and improving students' self-learning ability, the majority agreed. Compared with traditional teaching, the majority of students prefer AI-based iWrite assisted English writing teaching (68.31% relatively agree, 10.40% strongly agree). Therefore, students believe the new teaching can improve writing in words, syntax and discourse structure, and students accept this new teaching mode generally.

Based on above, the research can answer the two questions previously. Firstly, AI-based iWrite assisted English writing teaching mode can improve students' English writing ability more effectively. Secondly, the majority of students accept this new teaching mode.

4   Suggestions and Prospects

AI-based iWrite is the result of long-term research in the field of education, and its wide application in colleges and universities is of great significance to English writing teaching. The future education reform will definitely be an era of coexistence of teachers and artificial intelligence, both of which will give full play to their own advantages, especially the advantages of artificial intelligence in terms of convenience, speed and accuracy, cooperating to provide support for education and students [10]. Therefore, the study suggests that in the future teaching, more and more teachers should follow the trend of educational informatization and integrate AI-based iWrite into teaching, which can not only effectively save their own energy and time, but also improve the teaching quality.

Acknowledgements. This work was supported by Special Project for Scientific Planning and Research of Private Higher Education (Project No.: GB2020002).

References

1. Kizilcec, R.F., Reich, J., Yeomans, M., Dann, C., Brunskill, E., Lopez, G., et al.: Scaling up behavioral science interventions in online education. Proc. Natl. Acad. Sci. **117**(26), 14900–14905 (2020)
2. Li, X.: Teaching design of writing course for English majors based on output-oriented approach. Public Relat. World **18**, 32–33 (2020). (in Chinese)
3. Luo, N.: Deep learning based college English writing teaching in the intelligent age. Mod. Commun. **08**, 23–25 (2021). (in Chinese)
4. Wang, H.: Research on English writing teaching in the context of ESP in the age of artificial intelligence: opportunities and challenges. J. Univ. Shanghai Sci. Technol. (Soc. Sci. Ed.) **42**(03), 209–214 (2020). (in Chinese)
5. Zhang, Z.: A study on the integration of IWrite2.0 and college English writing teaching. J. Hubei Correspondence Univ. **31**(16), 179–180 + 183 (2020). (in Chinese)

6. Yu, H.: Research on the practice of teaching intelligent platform and college English writing teaching -taking iWrite2.0 English writing teaching and evaluation system as an example. J. Jilin Agric. Sci. Technol. Univ. **29**(06), 101–103 (2020). (in Chinese)
7. Sharkey, A.J.C.: Should we weclome robot teachers? Ethics Inf. Technol. **18**(4), 283–297 (2016)
8. Yu, S., Liu, Q.: Research on computer assisted English writing teaching in the background of artificial intelligence-analysis and thinking based on Awrite survey data. Inf. Technol. Educ. China **20**(04), 103–106 (2020). (in Chinese)
9. Ebyary, K.E., Windeatt, S.: The impact of computer- based feedback on students' written work. Int. J. English Stud. **10**, 121–142 (2020)
10. Chen, L.: The construction and implementation of digital teaching model for college English writing. J. Jilin Radio TV Univ. **3**(15), 132–134 (2021). (in Chinese)

Innovative Design of Immersion Teaching Aids for Children's Art Based on Augmented Reality Technology

Yu Sun[✉]

Changchun Normal University, Changchun 130031, Jilin, China

Abstract. With the continuous development of science and technology, augmented reality (AR) technology is emerging. It is a multi-disciplinary interdisciplinary research field, involving many fields of development research. It also has development and application in children's teaching. At present, more and more attention has been paid to children's education. Parents attach importance to children's physical and mental education and cultivate their artistic accomplishment. Traditional children's teaching products mainly use the operator or parents, and children are in the position of passive learning and acceptance. Children's teaching products based on AR technology use realistic three-dimensional models to interact with children, so that children can immerse themselves in happy learning. Through intelligent devices, children can immerse themselves in visual, auditory, tactile and other senses, realize interactive teaching. This paper mainly studies the innovative design of immersion teaching aids for children's art based on AR technology. This paper expounds the advantages of children's art teaching aids (CATA) based on AR technology, expounds the design principles of CATA based on AR technology, and still chooses children's reading books, AR children's picture books, cognitive cards and game teaching aids for innovation in the form of CATA design innovation. In this paper, through the collection of children's sensitivity to the color design of teaching aids, we also conducted a questionnaire survey to understand the importance of teaching aids to children's art education in the hearts of parents. The results show that 51.02% of parents think art teaching aids are important to children, 20.19% think art teaching aids are not very important to children, and 28.79% think art teaching aids are not important to children.

Keywords: Immersion teaching · Children's art · Teaching aids · Augmented reality technology

1 Introduction

Children's cognition of the world is mainly through the interactive understanding of vision, hearing, touch and other senses, relying on bright color images, some songs and music, as well as hand touch to learn. Therefore, grasping this feature is the key to realize the design of immersion children's teaching products [1, 2]. Traditional children's teaching products are lack of interaction, most of them are in the form of books and picture books, they rely on the guidance of parents to complete the teaching, children are passive receivers [3, 4]. AR children's teaching products can use the current information technology with audio and video to achieve interactive effects, display the teaching content in multiple ways, and guide children to immerse themselves in learning, increased the entertainment and interaction of learning [5, 6]. At present, people pay attention to comprehensive education and cultivate children's artistic accomplishment. As one of the main forms of art education, art is also chosen by many parents. Art teaching aids are indispensable tools for children's Art Education [7, 8]. Innovative design of CATA, the use of AR technology to achieve its interesting interactive use, all-round, multifaceted to improve the quality of children's art teaching effect is a major focus of the current children's Art Education [9, 10].

In the research on the innovative design of immersion teaching aids for children's art based on AR technology, many scholars at home and abroad have studied it and achieved certain research results. Some scholars pointed out that designers should consider their psychological changes from the perspective of children, start with children's five senses for interesting design, and infiltrate the development of intelligence and emotional interaction into the packaging design of teaching aids, so that packaging can not only protect teaching aids, but also inspire children's intelligence [11]. Other scholars pointed out that the domestic research on the application of AR technology in the field of children's education started late, so the form of children's education products is relatively single. At present, the forms of domestic AR children's education products are mainly cognitive card + app, cognitive card + intelligent hardware and book + app [12].

This paper mainly studies the innovative design of immersion teaching aids for children's art based on AR technology. This paper expounds the advantages of CATA based on AR technology, which can make children use both hands and brain, inspire children's interest, increase interaction, and develop their intelligence and cognition. This paper also expounds the design principles of CATA based on AR technology, with social sharing as the guidance, children's cognition as the center and interactive experience as the process. In the innovative form of CATA design, children's reading books, AR children's picture books, cognitive cards and game teaching aids are still selected. In this paper, through the collection of children's sensitivity to the color design of teaching aids, in the innovative design of art teaching aids, high sensitivity colors are selected for use, and a questionnaire survey is conducted to understand the importance of teaching aids to children's art education in parents' mind.

2   The Innovative Design of Immersion CATA Based on AR Technology

2.1   Advantages of AR in CATA

(1) Hand brain combination

When children are in close contact with teaching aids, they often need to use eyes, ears, hands and brain. In this way, the visual, auditory, tactile and other senses are mobilized at the same time, so that children's attention is focused on art teaching.

The use of art teaching aids plays an extremely important role in the development of children's right hemisphere and the cultivation of practical ability. At the same time, it also promotes the cultivation and development of children's emotion, emotion, cognition, expression and social skills.

(2) Inspire interest and increase interaction

At present, parents pay attention to the development of comprehensive education, and art education is also a part of it. As a part of art education, art education is also chosen by many parents. The use of art teaching aids can not only bring children the joy of learning, but also cultivate the aesthetic feeling and quality of art, shape a sound character and form a noble sense of morality. The most suitable art learning situation for children is to learn in play. Only when the learning content is rich and interesting, and the interactive experience is fascinating, can children's interest be fully aroused, and art teaching tools can turn static art into vivid teaching materials, adding glue to the cultivation and inspiration of children's art literacy.

(3) Developing intelligence and recognizing things

Every child has an artist's heart. Children's perception and cognitive ability are innate, but they need to be well developed with the help of practical activities. Art teaching tools can help them better understand the external things, obtain external information, and make positive feedback on the information, so as to establish a healthy and good interpersonal relationship.

2.2 Design Principles of Art Teaching Aids for Children with AR Technology

(1) Social Sharing Oriented

Children are eager to play with their peers. The quickest way for children to distance themselves is to share toys. In the design of CATA, we should pay attention to creating a platform for children to cooperate and share. For example, in the design of touch art teaching aids, we should pay attention to the configuration of sensitive hot touch. The action of pressing with fingers on the screen is a basic way for children to recognize things. Touch art teaching aids support many people to carry out color matching, accessory design and image discussion on the application screen, but when the application program has no feedback on multi touch, it will cause preschool children's interest to decline.

(2) Focus on children's cognition

The design of CATA mainly serves for children's art teaching. It needs to take children's cognition as the center, combine with children's senses to design, maximize the role of senses, and establish an interactive mode. When designing art teaching aids for children, in view of the cognitive psychological characteristics, we should pay attention to both pictures and text, and focus on pictures. For excellent children's picture books, although children can't recognize all the words, they can independently complete the cognition of the story through the image.

(3) Interactive experience as the process

When designing art teaching aids for children, it is not simply to reduce the interactive content or enlarge the text and image information, but we need to explore the characteristics of user behavior and interaction from their cognition and perspective. For the unfamiliar world, children are adventurous and curious explorers.

They will boldly slide and drag on the screen to explore and discover new things in the unknown field. In the design of CATA, timely and positive feedback should be given. Children's good cognition is formed in the positive feedback. In the design, it is very important to give children timely, positive and appropriate feedback. At the same time, the design should be convenient for children to use and operate, and the size of the icon on the interface should be designed obviously. In order to reflect the sense of security, the rounded shape should be used as far as possible in the modeling. In terms of color matching, the color with higher purity and lightness can cause children's visual perception, animation and sound effects are also essential. Children are very sensitive to external information, adults can continue to focus on the same thing. However, children are easily disturbed by external information. Therefore, it is necessary to attract children's attention through the combination of animation, sound effects and other forms of feedback, so that they can concentrate on the completion of art learning.

2.3   Innovative Design Form of AR CATA

(1) Children's reading books

The communication between children and traditional books is mainly realized by reading words or pictures, which requires children to have a higher understanding, but also have a strong logical reasoning ability and spatial judgment ability. It is not difficult for adults to understand book knowledge, but it is not easy for children. Books contain many abstract languages. We maintain the advantages of traditional art books, and use real, three-dimensional scenes, audio, video and other technologies to supplement incomprehensible knowledge. Electronic art books based on AR technology enhance the expressiveness of art learning content, improve children's learning interest, and make art learning easier.

(2) AR picture book for children

In the digital age, the single picture of traditional picture books can not meet the needs of contemporary children. The emergence of AR technology perfectly meets the needs of children's learning psychology. With AR technology, the creation of children's picture books can jump out of the two-dimensional space and present in the form of three-dimensional space on intelligent devices. In the virtual learning space, children can interact with a variety of senses and virtual environment, and can get feedback results in time. Based on the feedback effect, children can take the next step in the picture book and complete the learning content. In digital children's picture education, digital picture books present the content of traditional picture books, and use intelligent devices to spread knowledge to children in the form of animation and video. In the display of digital picture books, children's picture books stimulate children's participation and interaction by displaying three-dimensional virtual models, and pay attention to the diversification of interaction forms.

(3) Cognitive card

At present, most of the existing forms of children's teaching products in China are in the form of cognitive card + app, cognitive card + hardware device. The learning content on the cognitive card is mainly color cognition and picture knowledge. For example, the integration of three-dimensional model and voice elements, art pictures in the form of cards, so that children in the process of learning can be associated with memory. Another is the form of Book + app, the 4D bookstore, which makes the popular science knowledge in the book into a three-dimensional model, and then scans to produce a three-dimensional physical model.

(4) Game AIDS

In terms of game teaching aids, there are many applications abroad. While children continue to play, they also form the memory of art knowledge. Game teaching aids are both entertaining and educational. There are various forms of game teaching aids, such as plasticine, cards, colored toys and so on. In the form of games, they help children understand the shape of art color and cultivate children's sense of color.

2.4 Questionnaire Analysis Algorithm

The design of CATA needs to be based on children's sensory cognition, and analyze the needs of CATA, so that the existing art teaching aids can be innovated. In this paper, through the form of questionnaire survey, to understand the needs of innovative design of CATA under AR technology,the specific formula is as follows:

$$A = \sum_{i=1}^{I} \lambda_i \left[\sum_{j=1}^{m} \lambda_{ij} \left(\sum_{k=1}^{n} \lambda_{ijk} a_{ijk} \right) \right] \tag{1}$$

At the same time, we also use the weighted summary statistical method to process the collected information, the specific formula is as follows:

$$S = \sum_{1}^{n} Q_i S_i (i = 1, 2, \ldots, n) \tag{2}$$

Through the collection, understanding, statistics and calculation of the questionnaire information data, this paper analyzes the needs of CATA innovative design under AR technology.

3 Research and Analysis

3.1 Research Objects

The main research content of this paper is the innovative design of immersion CATA. The research object of this paper is the buyer of CATA. Through the form of questionnaire survey, this paper investigates the parents' understanding of CATA and whether their sense of CATA will become the main purchasing force.

3.2 Research Process Steps

In this paper, data mining technology is used to collect and understand children's sensitivity to the color design of teaching aids. In the innovative design of art teaching aids, high sensitivity colors are selected for use to increase the attraction of art teaching aids to children. This paper also conducted a questionnaire survey to understand the importance of teaching aids to children's art education in parents' mind, and clarify the design needs of CATA.

4   Experimental Research and Analysis on the Innovative Design of Immersion CATA Based on AR Technology

4.1   Analysis of Children's Color Sensitivity

Art teaching aids in the use of children, need to be able to attract children's attention, interact with children's senses. Children's cognition of art teaching also tends to color cognition. At this stage, children have certain sensitivity to color, and male and female children's perception of color is different. The results are shown in Table 1.

Table 1. Analysis of children's color sensitivity

	Red	Orange	Yellow	Green	Young	Blue	Purple
Boy	83.53	80.31	81.26	79.58	70.58	76.59	59.6
Girl	94.72	88.37	84.2	74.28	78.81	73.25	68.9

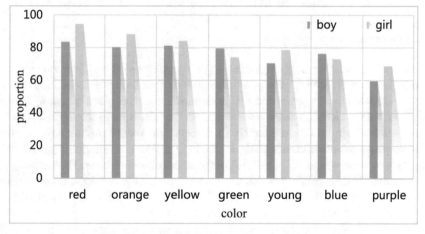

Fig. 1. Analysis of children's color sensitivity

It can be seen from Fig. 1 that children like colors with high purity, high brightness and high saturation. Girls are more sensitive to red, orange and yellow than boys, and boys are more sensitive to green and blue than girls. Therefore, when designing teaching aids for art teaching, we should pay attention to color sensitivity and carry out relative design.

4.2   Analysis of the Importance of Art Teaching Aids to Children

The art teaching aids based on AR technology make the learning content touchable and authentic. With the interest, interactivity, immersion, lifelike and creativity of AR technology, children can observe the learning content from multiple angles and actively receive learning knowledge in the learning process. In this paper, through a questionnaire survey on parents to understand the degree of parents' understanding of art teaching aids, the results are shown in Fig. 2.

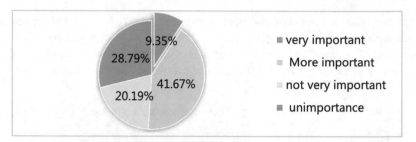

Fig. 2. Analysis of the importance of art teaching aids to children

It can be seen from Fig. 2 that 51.02% of parents think that art teaching aids are important to children, 20.19% think that art teaching aids are not very important to children, and 28.79% think that art teaching aids are not important to children.

5   Conclusions

With the rapid development of modern educational information and interactive technology, many emerging technologies have been applied to the field of art education, and new teaching ideas should be supported by corresponding teaching equipment and means. While enhancing AR technology, we should also pay attention to the intrinsic teaching content value of children's art education products. We should not excessively pursue technology and abandon content. AR technology serves for children's art education products, and is only a carrier of the teaching content of art products. The teaching content also needs to be constantly updated to keep pace with the development of the times. With the progress of technology, the immersion, interactivity and creativity provided by the new generation of AR technology provide new creative space and possibility for the creation of children's education products, and bring new vision and thinking space. The progress of AR technology and the innovation of teaching concept are the inexhaustible power to promote the continuous development of AR children's education products. New technology, new thinking and new knowledge deeply affect the exploration of AR technology in the future.

Acknowledgements. Special thanks to Jilin Province Education Science "13th Five-Year Plan" general planning topic: kindergarten creative art teaching research (GH19364) project for the support of this paper.

References

1. Gao, H.H., Fu, J.H., Xing, X.C.: Research on The design of children's toys and teaching AIDS based on sound science. In: E3S Web of Conferences, vol. 179, no. 2, p. 02025 (2020)

2. Desierto, A.J.R., Recia, A.S.A., Arroyo, J.C.T., et al.: GoonAR: a bilingual children storybook through augmented reality technology using unity with Vuforia framework. Int. J. Adv. Trends Comput. Sci. Eng. **9**(3), 3681–3686 (2020)
3. Park, H., Park, J., Kim, H., et al.: Virtual dress-fitting media art system using kinect and augmented reality. TECHART J. Arts Imaging Sci. **4**(2), 10–12 (2017)
4. Cristina, P.: When augmented reality met art: lessons learned from researcher-artist interdisciplinary work. Multimodal Technol. Inter. **2**(2), 17 (2018)
5. Plengdisakul, P., Phothisane, S., Soodsang, N.: Graphic design for children with learning disabilities based on the isaan mural painting. Acad. Jo. Interdiscip. Stud. **10**(2), 149 (2021)
6. Bajinovci, B., Bajinovci, U., Rexha, E., et al.: 5. Quality aspects of product design in artistic education. Rev. Artistic Educ. **20**(19–20), 211–217 (2020)
7. Selan, J., Potonik, R.: Art, for children's sake! At the crossroads of making, understanding and teaching visual art. Center Educ. Policy Stud. J. **10**(4), 7–11 (2020)
8. Kim, M.: A comparative study on developmental characteristics of children's digital photography and drawing. J. Res. Art Educ. **18**(1), 41–68 (2017)
9. Rice, M.F., Dunn, M.: Supporting children's writing in inclusive classrooms with arts- based strategies. Teach. Except. Child. **52**(3), 147–156 (2020)
10. Ayaka, K.: Practical research on fostering the creative imagination in art and craft teaching\n through the lower grades based on the appreciation of painting materials development—. Stud. Art Educ. **48**(1), 153–160 (2016)
11. Villarroel, J.D., Ortega, O.S.: A study regarding the spontaneous use of geometric shapes in young children's drawings. Educ. Stud. Math. **94**(1), 1–11 (2016)
12. Moon, K.A.: A study on the meaning of art activities and play through children's art therapy case. J. Act. Stud. **18**(1), 145–161 (2020)

The Innovation of Student Education Management Model Based on Internet Technology

Xiaomeng Xu[⊠]

College of Labor Relations, Shandong Management University, Jinan, Shandong, China

Abstract. With the development of society and the continuous improvement of the level of science and technology, the field of education is also undergoing earth-shaking changes. Students, as an indispensable human resource for the country's future construction, are the cornerstone of the country's development. Paying attention to student education and innovative management models are of great significance to the development of the national education system. Therefore, this article uses Internet technology to carry out innovative research on student education management mode. First, the concept of Internet education is explained, and the definition of school social work and the impact of school social work on student education mode are explained. Subsequently, the construction of the student management model was designed, and the operation effect of the model was tested. The final results show that the reliability of the student education management model under the Internet technology basically reaches about 4, the validity basically reaches about 5, and the evaluation of the use effect is consistently excellent.

Keywords: Internet technology · School social work · Student management model · Model innovation

1 Introduction

Global multi-polarization and economic globalization are two major trends in the development of today's international structure. Countries in the world have made many contributions to safeguarding world peace and common development, and they are all facing new challenges. In the 21st century, the knowledge economy is booming, the scientific and technological revolution is developing rapidly, international competition is becoming increasingly fierce, and the demand for talents is increasing, especially for innovative talents [1, 2].

Many researchers have conducted relevant research to innovate student education management models. Traditional teaching methods still occupy a prominent position in face-to-face teaching in our country. The traditional teaching model refers to a thinking and teaching model established by German educators in the early 19th century and developed by Soviet educators. Historically, it has played a positive role, and it is not useless today; however, from a development perspective, it is lagging and has become

© The Author(s), under exclusive license to Springer Nature Switzerland AG 2022
J. Macintyre et al. (Eds.): SPIoT 2021, LNDECT 98, pp. 174–182, 2022.
https://doi.org/10.1007/978-3-030-89511-2_21

a major obstacle for students at this stage to comprehensively promote high-quality innovative education. It is imperative to innovate student education management mode. The uniqueness of the intermediary teaching mode (the intermediary between teaching theory and classroom practice) has begun to arouse people's attention, and has gradually become a hot and important topic [3, 4].

In the historical process of building a well-off society in an all-round way and the great rejuvenation of the Chinese nation, innovation has become increasingly important. Education shoulders the historical mission of cultivating innovative talents and building an innovative country. Serving the development of innovative countries and cultivating innovative talents are the essential requirements of the educational age. It is for this reason that this article has innovated the student education management model based on computer technology [5, 6].

2 Overview of Student Education Management Model Based on Internet Technology

2.1 The Concept of Internet Education

Internet education is a new form of education that combines Internet technology and education with the continuous progress of science and technology. Internet education uses the Internet as a channel and platform, makes full use of Internet education resources, and uses multiple methods and expressions of information technology to provide people and users with different needs with digital education content and services anytime and anywhere. In the modern information society, the Internet has the characteristics of efficient, fast and convenient communication [7, 8].

2.2 Definition of School Social Work

School social work includes sociology and pedagogy. As for the definition of school work, it is mainly divided into two categories, one is broad and the other is narrow. In a broader sense, school social work performance goals are relatively important, including all students in the school. The narrowly defined service goals are more focused, and appropriate support and assistance are provided to grow and adapt to the school crisis. Fan Minglin emphasized, school social work is the work of social workers based on professional theories and methodology in close cooperation with teachers and school administrators, with the school as the main focus, helping students solve problems and student growth. Therefore, it is different from normal work. Educational work or psychological counseling. The main goal of school social work is students, but in order to solve student problems as soon as possible and promote better development of students, school social work should also cooperate with schools, parents and school staff to provide appropriate specific social services structure. Therefore, school social work can make full use of professional methods such as cases, groups, and communities, make full use of all available resources, link service goals with the environment, and create a family community.

2.3   Characteristics of Education Management Model

(1) Openness

The education management process presents obvious open characteristics, but it cannot be simply understood as an ineffective global opening, but a closed and conservative boundary within a certain open area. The responsibility and obligation of students is only to study, while the responsibility and obligation of teachers are to teach and educate people and their management. The most obvious feature of today's education management is its openness. The concept of openness is always implemented in education management, which directly affects the management relationship and management status. Therefore, within the framework of the modern education management system, the work attitude of education managers must remain open, and then internalize the disclosure of all management details and maintain a fair and equal status, rather than patronizing the management perspective. The subjective initiative and learning potential of students are important goals for achieving scientific education management [9, 10].

(2) Richness

The interactive activities of contemporary education management show obvious richness. They have changed the uniqueness of traditional education management activities and increased the richness of management activities. These are all important features that meet various educational needs such as learning standards, management standards, and educational development. They emphasize the interaction in the process, establish the interaction between managers and supervision mechanisms, educational theory and practice, and provide interactive action space for equal development. It mobilize the enthusiasm of participants for innovation and development [11, 12].

(3) Transformation

The transformational nature of modern education management can realize the transformation of the concept of knowledge structure, including promoting the transformation and integration of knowledge and behavior. The traditional education management model emphasizes the acquisition of knowledge and ignores the process of skill development and transformation, which has a certain gap with modern education management. In this context, the innovation and reform of contemporary education management models have become inevitable, including management methods and behaviors in the management process.

(4) Communication

In traditional education management, the educational form is too traditional and closed and lacks flexibility. The difference is that one of the communication characteristics of modern education management is flexibility. While ensuring the openness of the educational management process, optimize the regulation and allocation of resources to make the use of educational resources more reasonable.

(5) Flexibility

Modern education management also has the characteristics of flexibility. The whole process of contemporary education management can also be called the process of communication and interaction, and its organic purpose is realized in communication and interaction. Starting from the essence of the education management process, analyzing the communication and interaction process, the practical aspects of each connection fully reveal the characteristics of the flexible process. It purposefully runs through the whole process of modern education management, including the full integration of various elements of education management and the process of deep understanding of educational behavior and context. It optimize and update the education management process, fully integrate the context and process of education management, and promote the progress and development of contemporary education management.

2.4 Internet Technical Support

The PageRank algorithm focuses on old websites. It can be seen from Formula 2 that the main factor in determining the PageRank value of a website is the number of hyperlinks pointing to the website. If a website was launched not long ago, and many other websites have not yet linked to it, the PageRank value of the website calculated by Formula 1 is also very low, and it often appears in the search engine results and places it in a lower position. In this way, the new web page in the returned result is put behind instead, which is just the opposite of the user's needs. Because in many cases, users want to see the latest web pages first, such as the latest news and information. Therefore, Dr. Song believes that when calculating the PageRank value of a webpage through a hyperlink pointing to a webpage, the publication date of the webpage should be taken into consideration. To this end, he proposed a concept of date weight for web pages, namely:

$$W = C \cdot D/r \tag{1}$$

$$R(u) = c \sum 2eR(v)/N \tag{2}$$

W is the website date weight t, which is the time required for a search engine to retrieve its inventory webpage list once; H is an access cycle; T is the webpage created by the web spider (web spider is the specified number of days between the date when the file is automatically searched and loaded on the Internet and the date the webpage is placed). Finally, D is a constant whose value refers to the visit period of the web spider.

3 Construction of Student Education Management Model Based on Internet Technology

3.1 The Construction of Internet Technology on Student Education Management Mode

1) The main content of student education

   (1) Student awareness education
       Improving students' awareness is the starting point of student education. This aims to help develop an academic, correct and comprehensive understanding of students, and actively protect students' rights.

(2) Student value education

There are two aspects to consider in cultivating the value of students: First, the education of student values is a "student-oriented" education. Second, the values education of students respects the diversity of values. Since all aspects of social interests are dynamic and diverse, and tens of thousands of interests have different colors and irregularities, our requirements and standards for different objects cannot be the same. When educating students on values, we need to understand the level and relevance of education.

(3) Frustration education

There are always unsatisfactory things in the world, and setbacks are not far away from our lives. The impact of setbacks on growth is twofold. It can make those who survived failures become more courageous, turning the pressure of failure into action and ultimately achieving success. On the contrary, if you cannot bear the setbacks of life, then any setbacks may be the last straw to crush the camel. Therefore, it is necessary to carry out frustration education, help and guide people to better understand frustration, consciously avoid the negative impact of frustration, so that he can learn to adapt to frustration in time, and easily face the setbacks and challenges in life.

2) Running path

(1) Classroom education

Classroom teaching is the main position and channel of student education. Improving student education through classroom teaching activities can provide theoretical guarantees for the development of other activities, often with less effort. Integrate student education into the teaching of ideological and moral courses, basic courses and professional courses, continuously strengthen teacher professional training, improve teacher teaching quality, update teaching methods, pass on the law, enrich the appeal of education, and touch people's hearts.

(2) Group activities

With the diversified development of society, the channels for students to obtain various types of information have become more and more extensive, and the formation of students' opinions and values is affected by many factors. The Communist Youth League organization should adapt to the new situation, strengthen the development of student education, carry out various activities, and promote the realization of the unified realization of student enlightenment education through distinct themes, rich content and interesting collective learning activities.

(3) Student clubs

The existence of student clubs is based on the common hobbies or common goals of students. Therefore, students in such a group are more likely to participate in a common cause, to accept knowledge more actively, and to have a higher teamwork ability. Take the club's Dissemination of the content of student education in a form can not only increase the acceptance of education, but also facilitate the development of student mutual aid results. Students have a deeper understanding of students and life through their own subjective and active learning and practice, which will inevitably be able to shape a healthier outlook on students.

(4) Group psychological counseling

Group psychological counseling activity is a kind of heuristic training. The group psychological counseling project is divided into three units, which are closely linked and proceed step by step. The first unit "Meet and communicate with each other" aims to allow members to understand each other and build confidence and cohesion in the group; the second unit "Discover and locate the soul" aims to help students understand themselves and others, and cultivate. You have the ability to look at problems from multiple angles and plan your future life; the third unit "Heart Growth, Heart Flies", helps students according to the first two activities. The three activities are intertwined and proceed step by step, encouraging students to liberate their minds with an open mind, explore themselves, and finally gain spiritual growth, and be able to look at and accept other things with an open mind.

(5) Network mutual assistance platform

With the development of society and the advent of the Internet age, the traditional education model has obviously been unable to meet the ever-increasing demand for education. The use of online teaching methods is an extension of the current education reform, and its extraordinary benefits are gradually recognized and accepted by teachers and students. In order to carry out student education, it is necessary to increase the use of the Internet in the education process, understand the advantages of online education and the specific use of traditional methods of teaching and teaching methods, and strategically integrate the two teaching methods to complement practical teaching for students' Education opens up new ways.

3.2 Innovative Implementation of Student Education Management Model Based on Internet Technology

(1) Idea innovation

The use of advanced interdisciplinary concepts and methods to create and explore philosophical connotations for students, educational psychology acceptance mechanisms and Internet-based student education innovation models have improved the effectiveness and sustainability of education.

(2) Path Innovation

For the first time, education channels such as classrooms, classrooms, dormitories, networks, and social practices have been integrated to form a variety of effective education channels. Pay attention to cultivating students' self-education function in student education, and maximize the impact of student education.

(3) Carrier innovation

Develop a student education network platform with six functions of "self-help and mutual assistance, communication and feedback, prevention and early warning, education services, psychological evaluation and adaptation", and establish a student-based student education mutual assistance mechanism for the first time, expanding the scope of student education and student training Influence.

(4) Method innovation

Apply pedagogy and psychology experimental methods to student education and teaching reform practice, transform educational activities into experimental intervention conditions, implement student education model practice, verify model effects through comparative studies, and increase the reliability and validity of the research.

3.3 Operation of Student Education Management Model

It can be seen from Fig. 1 that the structure and operation of the student training innovation model consists of receiving subjects, receiving intermediaries and receiving objects. The receiving subject achieves educational goals through continuous processes such as attention to information, preservation of information, acceptance of information, psychological internalization, changes in cognition and influence on behavior.

The process of student education innovation model: through five educational channels including classroom teaching, collective activities, student clubs, platform networks, and campus culture. The system maintains the normal operation of the model and is a student education model with clear levels, continuous, orderly, and efficient flow.

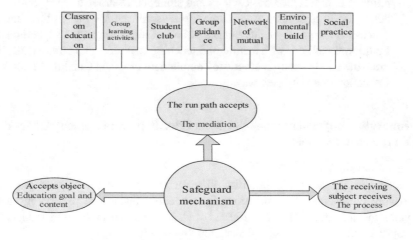

Fig. 1. Student education management mode

4   Analysis of the Effect of Student Education Management Model Based on Internet Technology

Table 1. Effect of the student education management mode

Primary coverage	Validation	Reliability	Evaluation results
Student awareness education	4	5	Excellent
Student value education	4	4	Excellent
Frustration education	3	5	Excellent

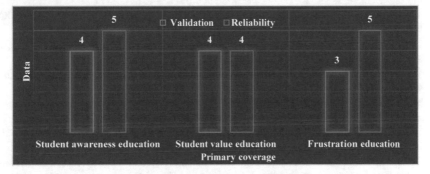

Fig. 2. Test of reliability and validity

From Table 1 and Fig. 2, it can be seen that the reliability of the student education management model under the Internet technology basically reached about 4, and the validity basically reached about 5. It can be seen that the reliability of the student education management model this time is very high. High practicability, and the evaluation of the use effect is consistently excellent, which shows that the student education management model can be promoted and practiced.

5   Conclusion

This article mainly studies the innovation of student education management model based on Internet technology. In terms of theory, it starts from the two perspectives of the construction of student education management system under the network environment and the school social work, and on this basis, combines the actual situation to find out the shortcomings and proposes future sustainable development plans and suggestions. Finally, based on the existing problems, feasible measures are given to improve our country. The current countermeasure opinions and opinions on the content and methods of student education reform in colleges and universities will provide reference for the future application of similar platform-based campus Internet technology.

References

1. Bo, G.: Research on the innovation of enterprise management in the internet era. SME Manage. Technol. **000**(026), 7–8 (2017)
2. Ting, P.: Discussion on the innovation of human resource management thinking based on the internet era. Rural Pract. Sci. Technol. Inf. **000**(007), 109–110 (2019)
3. Zhao, Y., Zhao, X., et al.: Research on the teaching mode of entrepreneurial project management general courses. Educ. Res. Front. Chin. Engl. Ed. **009**(002), 42–47 (2019)
4. Li, X., Pak, C., Bi, K.: Analysis of the development trends and innovation characteristics of Internet of Things technology – based on patentometrics and bibliometrics. Technol. Anal. Strateg. Manage. **32**(4), 1–15 (2019)
5. Yao, X., Sun, J.: Innovation and practice of educational model and method on electronic information major in polytechnic colleges. Int. J. Soc. Sci. Educ. Res. **2**(10), 16–19 (2019)
6. Pan, F., Ma, G.: Research and exploration on the development of application-oriented undergraduate education. A new generation **000**(021), 7–8,15 (2017)
7. Wu, Z.: Research on the innovation of university teaching management based on MOOCs background. Int. J. Eng. Model. **31**(1), 280–288 (2018)
8. Liu, J., Wang, C., Wu, Y.: Construction and optimization of higher education management system based on internet video online technology. Sci. Program. **2021**(1), 1–11 (2021)
9. Wang, Y., Hao, S., Zhang, X.: Research on the application of knowledge architecture in the video-based collaborative learning. C e Ca **42**(4), 1519–1523 (2017)
10. Yu, L.: Research on construction and innovation new foreign language classroom teaching mode based on the "Internet Plus." J. Intell. Fuzzy Syst. **3**, 1–6 (2021)
11. Ji, H.: Design of distributed collection model of student development information based on Internet of Things technology. Secur. Commun. Netw. **2021**(3), 1–10 (2021)
12. Chen, M., Liu, Y.: Research on college students' innovation and entrepreneurship education mode based on social participation. Revista de la Facultad de Ingenieria **32**(9), 278–284 (2017)

Influence of Chinese Language Development Based on Improved Fuzzy Mean Clustering Algorithm

Xia Zhao[✉]

Department of Mechanical and Electrical Engineering, Shandong Vocational College of Light Industry, Zibo 255300, Shandong, China

Abstract. As one of the core technologies of DM (data mining) technology, cluster analysis plays a vital role in data analysis in different fields. At the same time, this technology has attracted widespread attention, and it can be used as an independent data processing technology for people to us. FC (Fuzzy clustering) is also a kind of clustering algorithm, this algorithm is simple and easy to implement, and the DM is more efficient. In order to further improve the efficiency of clustering analysis results, people have proposed an improved FMCA (fuzzy mean clustering algorithm). This improved fuzzy mean clustering algorithm has been successfully applied in many fields, such as medical testing, food industry, data analysis, etc. The Chinese language has a long history of development, and the continuous rise of network technology in recent years has had an impact on the development of the CL (Chinese language). Therefore, in order to make the CL prosper for a long time, this article mainly studies the influence of the improvement of the FMCA on the development of the CL. First, the error rate of clustering results under different algorithms is compared and analyzed, and the results show that the clustering results of the improved FMCA are significantly better than other algorithms. The average error rate of the improved FMCA is only 21.4%, which is lower than the other three algorithms. Secondly, it analyzes the effect of the improved FMCA in promoting the development of CL. 42% of people think that improving the FMCA can standardize the development of CL. Others believe that this algorithm can speed up the spread of CL and expand the field of CL.

Keywords: Improved fuzzy mean clustering algorithm · Data mining · Cluster analysis · Chinese language

1 Introduction

With the continuous development of science and technology, the scale of data is also expanding. Traditional clustering rigidly divides the research object into a certain attribute in the classification process. This rigid division may lead to a large amount of loss of information. Improved FMCA reflects the attributes of things more truly. Compared with traditional clustering, the improved fuzzy mean clustering algorithm improves the accuracy of clustering. This clustering is currently one of the most widely

© The Author(s), under exclusive license to Springer Nature Switzerland AG 2022
J. Macintyre et al. (Eds.): SPIoT 2021, LNDECT 98, pp. 183–189, 2022.
https://doi.org/10.1007/978-3-030-89511-2_22

used and successful algorithms. In the course of the development of CL, the use of big data for cluster analysis has promoted the vigorous development of CL. Therefore, the influence of CL development based on the improved FMCA is a topic worthy of our in-depth study.

There are not a few studies on the influence of the improvement of the FMCA on the development of the CL. Geng C believes that most classic clustering algorithms accurately assign each data to a cluster, thereby giving a clear partition to the given data, but FC data belongs to different clusters. FC algorithm requires users to define the number of clusters in advance, and different clustering values correspond to different fuzzy partitions [1]. Gun-Young pointed out that clustering algorithms can be divided into two categories; hard clustering and soft (fuzzy) clustering. In hard clustering, the data is divided into different clusters, where each data element belongs to only one cluster. In soft clustering, data elements belong to multiple clusters and are associated with each element through the member level [2]. Xiao C said that data mining is the process of discovering useful data from a large amount of data. DM usually involves four types of tasks or techniques: classification, clustering, association rule mining and regression [3].

This article mainly studies the application of DM technology in regional economic analysis. The research in this paper combines the relevant theoretical basis and the results of the questionnaire to carry out the following analysis in turn: First, the error rate of clustering results under different algorithms is analyzed, and it is concluded that the clustering results of the improved FMCA are significantly better than other algorithms, and the error rate is much lower than that of other algorithms. Secondly, the effect of improving the fuzzy mean clustering algorithm on the promotion of CL development is analyzed. The results show that the improved fuzzy mean clustering algorithm is beneficial to standardizing the development of CL and expanding the field of CL.

2   Based on the Theoretical Basis of Chinese Language Development Based on the Improved Fuzzy Mean Clustering Algorithm

2.1   Data Mining

DM is the process of discovering valuable or meaningful knowledge from a large amount of data, and making decisions based on the existing knowledge [4, 5]. The main task of DM is to predict and describe. Prediction is based on the use of historical data, establishing a model to find out the law of change between the data, and then predicting future values based on the model [6]. The description task is to derive the potential patterns discovered through the data. As a modern science, DM is directly derived from probability and statistics. In other words, we often cannot get a direct and accurate expression from the real world [7, 8]. Then we can use a "relevance" idea to describe the real world.

The general process of DM includes data preprocessing, DM, and post-processing [9]. The preprocessing process is mainly a screening process to filter out those data that meet the requirements, and remove those data that have problematic formats or exist. If there is no preprocessing process, it is very easy to cause model calculation failure. Although the process of preprocessing may consume a lot of time and effort, data preprocessing is an essential step in the DM process. The next step is to build models based on data and use certain criteria to evaluate the performance of different models or combined mode. The last step is post-processing. The process of post-processing is equivalent to having discovered the pattern that you want to find. This pattern is conducive to better applying the DM results to practice [10].

2.2 Improved FMCA

Clustering is a very important analysis method in DM analysis. With the emergence and development of human society, clustering is constantly deepening [11]. Clustering is the process of distinguishing and classifying data according to its characteristic attributes. People can have a more comprehensive understanding of society and nature through clustering [12]. Cluster analysis is the process of grouping this information according to the useful information in the data. Objects in a group are related to each other, but objects in different groups are not related. The similarity within groups is proportional to the gap between groups. For objects with greater similarity, the larger the gap between groups, the better the clustering effect. The clustering effect depends on the measurement distance. The following formulas are mainly used when measuring the clustering effect:

$$D = \sqrt{\sum_{m=1}^{M} \left(x_{im} - c_{jm}\right)^2} \tag{1}$$

$$C = \sqrt{\sum_{m=1}^{M} W_m\left(x_{im} - c_{jm}\right)^2} \tag{2}$$

Improved FMCA is one of the common cluster analysis methods, the boundaries between things, some are clear, some are fuzzy [13]. Improved fuzzy mean clustering is one of the well-known unsupervised clustering techniques. However, this algorithm requires the user to pre-define the number of clusters. When the boundaries between the things involved are very blurry, the FC analysis method needs to be used. FC is derived from hard clustering. Traditional cluster analysis is a hard division with very clear boundaries. Traditional clustering technology rigidly divides the research object into a certain attribute in the classification process, but in fact most objects do not have strict attributes, This hard division may cause data loss. Compared with traditional clustering analysis, the improved fuzzy mean clustering algorithm has a larger range of membership values, and its clustering effect and data expression ability are stronger. In addition, the accuracy of FC algorithm is higher than that of hard clustering. But FC algorithm is not all advantages, it also has disadvantages. The calculation of this algorithm is relatively complicated and its ability to cluster nonlinear data is limited [14].

2.3 Chinese Language

Chinese is also called Chinese and Chinese. It is the common language among all ethnic groups in China. Chinese has a long history with thousands of years of development history. Chinese is the most spoken language in the world. At least 1.5 billion people in the world use Chinese, accounting for 20% of the world's total population. With the continuous development of the economy, many foreigners have begun to learn Chinese. Chinese has attracted widespread attention all over the world. Chinese can express massive amounts of information more easily than English and other pinyin languages. English and other pinyin languages are more time-consuming and paper-consuming than Chinese. It is easy to get started and difficult to master, and the phrases are long and inconvenient. For the massive amount of information, English is struggling to cope, but Chinese can express it easily. The conference text that records the same information in the six official languages is always the thinnest in Chinese, expressing the same information, and the shortest time in Chinese. The history of Chinese can be divided into four periods, mainly ancient Chinese, middle ancient Chinese, modern Chinese and modern Chinese. The main reason for the evolution of Chinese from ancient times to the present is communication. Cross dialect communication, mainly because of immigration and education. Migration mainly includes active migration and passive migration. Active immigration means war, being an official, doing business, studying and practicing spiritual practice, while passive immigration is mainly taking refuge or compulsory immigration after the war. Immigration will inevitably bring about changes in dialects, especially those with larger scale or higher status, which will have a greater impact on language. The influence of education on language mainly existed in ancient times. Ancient education was mainly oriented towards imperial examinations and directed at being officials. Therefore, the language used in education has always been the phonological system of Yayan or the imperial rhyme book. Where there are many scholars, the language used in education will have an impact on the dialect [15].

3   CL Development Experiment Based on Improved FMCA

3.1   Experimental Background

With the continuous expansion of the amount of information, DM technology has attracted widespread attention. Traditional cluster analysis rigidly divides the research object into a certain fixed category, and this rigid division is likely to cause a large amount of information to be lost in a certain category. The improved FMCA improves the clustering accuracy under normal circumstances, and is more suitable for the research of CL development. Therefore, this paper conducts research on the development of CL based on the improved FMCA.

3.2   Experimental Process Steps

After understanding the characteristics of the improved FMCA and the CL, this article conducts an experimental exploration between the two: Study the clustering results under four different algorithms, and list the results with specific error rates. Then 100 volunteers were convened to conduct a survey. The content of the survey was to improve the effect of the FMCA in promoting the development of CL. After the survey, the results of the questionnaire were statistically analyzed.

4   Experimental Analysis of CL Development Based on Improved FMCA

In order to improve the credibility of this article, after consulting relevant materials and combining the results of the questionnaire, we compare the error rates of several different cluster analysis results. Finally, it analyzes and discusses the effect of improving the FMCA on the promotion of CL development.

4.1   Comparative Analysis of the Error Rate of Clustering Results Under Different Algorithms

After consulting the relevant information, the improved fuzzy mean clustering algorithm and k-prototype algorithm, and the clustering error rate of the CBEST algorithm on the basic information of Chinese language lovers are sorted out as shown in the following table:

Table 1. Comparison of error rates of clustering results under different algorithms

	First time	Second	Average
CBEST	32.3%	36.5%	34.4%
Improved FMCA	20.3%	22.5%	21.4%
k-prototype	24.5%	28.3%	26.4%

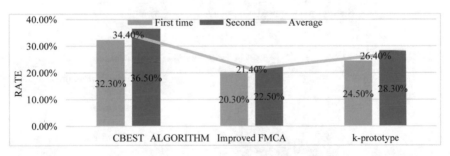

Fig. 1. Comparison of error rates of clustering results under different algorithms

Table 1 and Fig. 1 compare the clustering error rates of the improved fuzzy mean clustering algorithm, k-prototype algorithm, and CBEST algorithm on the basic information of Chinese language lovers. It can be seen from the chart that the clustering results of the improved fuzzy mean clustering algorithm are significantly better than other algorithms. The average error rate of the improved fuzzy mean clustering algorithm is only 21.4%, which is lower than the other two algorithms. And the difference between the first clustering result and the second clustering result is very small, only two percentage points apart. The two clustering results of CBEST are not ideal, the error rate of the two clusters is the highest, the average error rate is as high as 34.4%. The

K-prototype clustering result is slightly worse than the improved fuzzy mean clustering algorithm, and its average error rate is 26.4%.

4.2   Analysis of the Effect of Improved FMCA on Promoting the Development of CL

Improved FMCA can promote the development of CL, so what is its specific effect on promoting the development of CL? The results of the questionnaire survey are now organized as shown in the chart below:

From Table 2 and Fig. 2, 42% of people believe that improving the FMCA can standardize the development of CL, so that people have a correct understanding of CL, and promote the healthy development of CL. There are 27 people who think that improving the FMCA can speed up the spread of CL, which is conducive to the spread of CL. There are also 20% of people feel that improving the FMCA will help the CL to form a more complete and more specific structure and system. The remaining 11% believe that improving the FMCA can also expand the field of CL.

Table 2. The effect of promoting the development of Chinese language

	Number	Proportion
Normative development	42	42%
Perfect system	11	11%
Expand the field	20	20%
Speed up the spread	27	27%

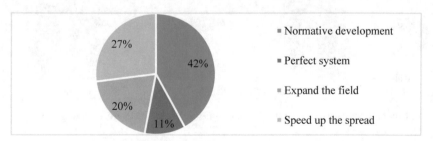

Fig. 2. The effect of promoting the development of Chinese language

5   Conclusion

As more and more data are generated in the DM process, ordinary clustering like K-prototype can no longer meet the requirements, and an improved FMCA with a lower error rate will be produced. This improved FMCA can standardize the development of

CL, speed up the spread of CL, and is beneficial to the spread of CL. Therefore, the research on the development of CL to improve the FMCA in this article is of practical significance. It will help people realize the advantages of the improved FMCA, and promote the healthy development of CL in my country by improving the FMCA.

References

1. Lei, X., Ouyang, H.: Image segmentation algorithm based on improved fuzzy clustering. Clust. Comput. **22**(6), 13911–13921 (2018). https://doi.org/10.1007/s10586-018-2128-9
2. Park, K., Hwang, G.Y.: Movement intention detection of human body based on electromyographic signal analysis using fuzzy C-means clustering algorithm. J. Korea Multimedia Soc. **19**(1), 68–79 (2016)
3. Xiao, C., Li, Y., Jing, Y., et al.: Developing the fuzzy c-means clustering algorithm based on maximum entropy for multitarget tracking in a cluttered environment. J. Appl. Remote Sens. **12**(1), 1 (2018)
4. Yu, W., Jing, C., Gao, W., et al.: Particle seed images segmentation method based on the improved fuzzy C-means clustering algorithm. Zhongbei Daxue Xuebao (Ziran Kexue Ban)/J. North Univ. China (Nat. Sci. Ed.) **39**(2), 177–182 (2018)
5. Cui, X., Yang, F., Wu, Z., et al.: Deep-sea sediment mixed pixel decomposition based on multibeam backscatter intensity segmentation. IEEE Trans. Geosci. Remote Sens. (99), 1–15 (2021)
6. Yu, C., Wang, L., Zhao, J., et al.: Remote sensing image classification based on RBF neural network based on fuzzy C-means clustering algorithm. J. Intell. Fuzzy Syst. **38**(4), 3567–3574 (2020)
7. Koryshev, N., Hodashinsky, I., Shelupanov, A.: Building a fuzzy classifier based on whale optimization algorithm to detect network intrusions. Symmetry **13**(7), 1211 (2021)
8. Hu, Q., Li, Y., Ge, Y., et al.: Clustering analysis based on improved fuzzy C - means algorithm. In: Proceedings of International Conference on Artificial Life and Robotics, vol. 23, pp. 276–281 (2018)
9. Zhou, X.G., Lu, M., Huang, X.X.: C-means clustering algorithm based on intuitionistic fuzzy sets and its application in satisfaction evaluation. J. Inf. Hiding Multimedia Sig. Process. **9**(2), 484–495 (2018)
10. Jesmon, G., Kumar, T., Job, J.: Selective brain MRI image segmentation using fuzzy C mean clustering algorithm for tumor detection. Int. J. Comput. Appl. **144**(7), 28–31 (2016)
11. Xu, H.C., Hou, R., Liu, L., et al.: The image segmentation algorithm of colorimetric sensor array based on fuzzy C-means clustering. J. Intell. Fuzzy Syst. **38**(4), 3605–3613 (2020)
12. Bai, K., Wang, J., Wang, H.: A pupil segmentation algorithm based on fuzzy clustering of distributed information. Sensors **21**(12), 4209 (2021)
13. Hidayat, S., Rismayati, R., Tajuddin, M., et al.: Segmentation of university customers loyalty based on RFM analysis using fuzzy c-means clustering. Jurnal Teknologi dan Sistem Komputer **8**(2), 133–139 (2020)
14. Kim, D.-H., Cho, H., Cho, H.-C.: Gastric lesion classification using deep learning based on fast and robust fuzzy C-means and simple linear iterative clustering superpixel algorithms. J. Electr. Eng. Technol. **14**(6), 2549–2556 (2019). https://doi.org/10.1007/s42835-019-00259-x
15. Ienco, D., Bordogna, G.: Fuzzy extensions of the DBScan clustering algorithm. Soft. Comput. **22**(5), 1719–1730 (2016). https://doi.org/10.1007/s00500-016-2435-0

Design of Flight Control System Based on Data Drive

Xu Mu[1](✉), Jingping Shi[2], and Xiang Gao[1]

[1] School of Aeronautics, Shandong Jiaotong University, Jinan, Shandong, China
[2] School of Automation, Northwestern Polytechnical University, Xi'an, Shaanxi, China

Abstract. UAVs not only have the advantages of taking off and landing at any time and flying at low speeds, but they are not subject to the normal restrictions of the driver during design. They can fly over the horizon with high intensity and perform high-risk missions. Therefore, they have a wide range of military and civilian applications. However, the UAV is a control object with complex and nonlinear mechanical structure. It is difficult to create an accurate mathematical model for this, and there is a strong connection between the various UAV channels, which has given birth to many modern control theories. These methods are difficult to apply to the UAV flight control system. The flight environment such as temperature and temperature is very sensitive, and it is easy to cause safety problems during the flight test. Therefore, it is very important to design an ideal flight control system. This paper studies the flight control system based on data drive, optimizes the flight control system based on data drive for the above problems, proposes an off-line tuning algorithm for data-driven fractional controller parameters, and performs simulation experiments on the return system. The experimental results are, the controller optimized by data-driven has better learning ability. Compared with the traditional controller, the maximum parameter is 0.2 times of them.

Keywords: Data driven · Flight control · Control system · Parameter optimization

1 Introductions

The design process of the traditional flight control system development method starts from the design of all system hardware and all system functions at this stage [1, 2]. On the basis of comprehensive inspection of system reliability requirements and corresponding software and hardware functions, the top-level structure, software unit and system hardware unit are designed [3, 4]. In the same state-of-the-art cycle, software and hardware design are carried out independently at the same time, and synthesis and testing are only carried out in the later stages of digital system simulation and integrated simulation testing [5, 6]. It can be seen from the above that the traditional system design process has many shortcomings such as information islands, increased system scale and test specifications, lack of rapid response capabilities, high maintenance costs, and time-consuming [7, 8].

© The Author(s), under exclusive license to Springer Nature Switzerland AG 2022
J. Macintyre et al. (Eds.): SPIoT 2021, LNDECT 98, pp. 190–197, 2022.
https://doi.org/10.1007/978-3-030-89511-2_23

Regarding the flight control system research, some researchers have proposed that the flight control system, as the key control unit of the unmanned helicopter, plays a key role in the normal and stable flight of the unmanned helicopter. The flight control system has two main functions: one is to control the aircraft, after completing the hovering, take-off and other flight operations, there must be good quality control; to complete the autonomous flight according to the scheduled route and other information [9]. Some researchers have proposed that the UAV is a four-input control system. In the flight control process, it is necessary to obtain an accurate system model and high-precision sensor data, and the control process is very difficult and complicated. Attitude control is the key to flight control, and it is also a hot research topic at home and abroad [10]. Various research results have shown that the secondary control simulation effect based on nonlinear control algorithms is very good, but they are based on the ideal secondary control model and have not been widely used in our practice [11]. And there have been some researchers who have designed the central control unit, which has realized the storage and expansion circuit, the system clock and the reset of the system, improved the stability of the circuit, increased the storage capacity and the data program store it. In addition, power conversion units of different voltage levels are used for power supply design to meet the power requirements of the circuit. In view of the advantages of integrating multiple input sensors into one circuit, a single i/o circuit can develop multiple input switch signals [12].

This paper conducts research on data-driven flight control systems, and has a general understanding of the composition of the flight system on the basis of related literature, and then summarizes the application of data-driven flight systems, and then on the basis of flight control the system is designed, and finally the proposed data-driven flight controller is verified by simulation experiments, and relevant results are obtained.

2 Flight Control System Research

2.1 The Composition of the Flight Control System

The flight control system must contain two most functions at the same time: one is that the flight control system must perform stable flight according to its predetermined flight altitude, speed and direction, so as to maintain the balance of the aircraft position and the accuracy of the trajectory movement, and be able to directly accept and execute instructions to ground stations and follow the instructions. The second is the management of airborne devices during the flight mission, which includes parameter collection of flight status, monitoring of airborne devices, and recording of status parameters. The function of the communication system is to be responsible for the data communication between the ground station and the flight controller to ensure the correct download and transmission of command information and point data. The main flight control chip is responsible for receiving and analyzing the data of various sensors, and sending the control signals calculated according to the sensor parameters of each steering wheel to different steering wheels. The steering wheel automatically changes the flight position according to the control signal to the corresponding deflection action.

2.2 Application of Data Driven in Flight Control System

(1) Data-driven fractional control

Extending the fractional-order controller to the flight control system can increase control flexibility and achieve better performance and dynamic response stability than traditional full-order control methods, especially for fractional-order control systems. The additional parameters of the fractional controller bring the freedom and superiority of control, but also increase the difficulty of tuning and the complexity of calculation. Taking into account factors such as non-modeling dynamics and modeling errors, traditional model-based parameter coordination methods are difficult to meet the actual control requirements. Database control theory does not depend on the exact model of the system, but its current research is applicable to integer controllers. The additional parameters of the fractional-order controller and its own non-linearity put forward higher requirements for data coordination to guide the parameters. In addition, the traditional controller configuration method also has the following shortcomings: on the one hand, when the initial configuration is unreasonable, it will cause the analysis process to fall into a local optimum, which cannot guarantee the global optimum; it increases with the operation of the system and increases the amount of calculation, which affects the real-time and convergence of the algorithm.

(2) Data-driven disturbance suppression

For the flight control system, measurement interference and unstable transmission of data will affect the quality and integrity of the data. Influencing factors include the influence of external environmental noise, the interference signal input by the sensor itself, the data conflict in data transmission, and the failure of competition with the node. When the input and output data of the system is incomplete and blocked, how to use the data actually obtained to complete the configuration suppression and improve the robustness of the controlled system is one of the key issues that need to be solved. At the same time, when the data is disturbed, how to ensure the convergence of the control algorithm and the closed-loop stability of the controlled system are also the main issues that need to be considered. In addition, changes in system operating conditions will check the adaptability of the tuning controller parameters.

3   Design of Flight Control System Based on Data-Driven

3.1   System Design Scheme

The flight control system is a multi-state, multi-parameter, nonlinear dynamic multi-task coupling system. The system design must meet the requirements of high accuracy, strong reliability, and high real-time performance. Based on the application of hardware circuits, software design and coding must consider issues such as the coordination of the sampling period and the consistency of the solution period, the inspection of overflow algorithms, and boundary conditions. The flight control system is a multi-task software system with strong real-time and high reliability. The system includes data streams from multiple subsystems. The main tasks include: collecting sensor data, calculating

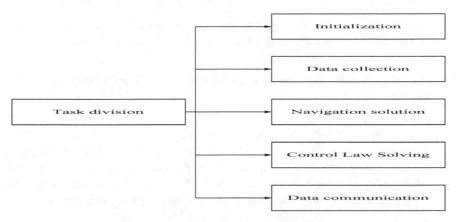

Fig. 1. Aircraft control system task division

route navigation, calculating control laws, communication point data and commands. First, ensure that the drone is stationary and correctly positioned, and can accurately fly according to the predetermined route. At the same time, be sure to accept remote commands and make changes over time. The software system of this text adopts modular design, and the main task modules are shown in Fig. 1.

3.2 System Initialization

The initialization of the flight control system is mainly the initialization of parameters, timers, communication channels, sensors and interrupt configuration. Timer preparation is mainly to configure the PWM and data calculation cycle and activate interrupts. The preparation for communication is mainly the initialization of the serial port, SPI, IIC channel and the configuration of the interrupt function. The start of the sensor is mainly the configuration of activation and sampling period IO.

3.3 Sensor Data Collection

Data collection includes accelerometer, gyroscope, magnetometer, differential pressure sensor, GPS unit and other data collection. Must ensure that the chip configuration is correct, and set the sampling period reasonably according to the design requirements. Ensuring that the data sampling period of each sensor chip is consistent is essential for behavior control and navigation calculations.

3.4 System Development Environment

(1) The leading chip of the system in this paper adopts STMicroelectronics' STM32F429 chip, and the software is realized with the built-in IAR7.2 embedded development environment. This system adopts C language development, the built-in environment supports built-in C language programming, provides a large number of ARM chip programming examples of ST, atmel and other companies. Real-time

J-link hardware debugging support, supports multiple file formats, developers can choose their favorite debugging software for simulation.

3.5 Data-Driven Off-Line Tuning Algorithm for Fractional Controller Parameters

(1) Fractional PID controller

In order to further improve the control performance of the traditional PID, the researchers introduced fractional calculus into the structure of the PID controller, thereby generalizing the form of the IOPID controller to obtain a fractional order PID (Fractional order PID, referred to as FOPID) controller, Its output expression in the time domain and the transfer function in the frequency domain are:

$$u(t) = K_p e(t) + k_i D^{-\lambda} e(t) + k D^{\mu} e(t) \tag{1}$$

$$C(s) = \frac{U(s)}{E(s)} = K_p + \frac{K_i}{S^{\lambda}} + K_d S^{\mu} \tag{2}$$

In the formula, $0 < 2$, $\mu < 2$ is the introduced fractional factor. When the FOPID controller is a high-order controller with order h and μ greater than 2, its structure is different from the traditional controller, its performance is poor, and the parameter setting is more complicated.

In order to make the fractional-order controller exert the best control effect, it is necessary to determine the best parameters of the controller through a suitable tuning algorithm to ensure the dynamic performance of the controlled object. Traditional model-based controller configuration algorithms have problems such as modeling errors and modeling dynamics, and their applications are mainly concentrated on integer controllers. Considering that the additional parameters of the fractional controller put forward higher requirements for the design of the tuning algorithm, this section studies the off-line tuning algorithm for the parameters of the fractional controller based on data.

For the system's following error control criterion, this paper uses the ideal Bode function to construct a fractional-order closed-loop system as the reference system of the controlled object, and its discrete expression is shown in the following formula:

$$M(z) = \frac{\omega_{gc}^{\alpha}}{D(z) + \omega_{gc}^{\alpha}} \tag{3}$$

In the formula, ωgc is the gain crossover frequency, and a is the order of the fractional reference system, which is the slope of the closed-loop system in the logarithmic coordinate system.

In order to establish the tuning criterion with the parameters of the fractional-order controller as the optimization variable, the system is stimulated and the corresponding input and output data set $\{u(t), y(t)\}t = 1, 2, ..., N$. Using the reference system $M(z)$ and the collected output signal $y(t)$, the virtual reference signal $r(t)$ can be calculated as:

$$r(t) = M(z)^{-1} y(t) \tag{4}$$

For the reference system M(z), this article only needs to use its discrete expression with the output signal to calculate the virtual reference signal r(t), so there is no need to assume that $M(z)-1$ physically exists, that is, there is no need to limit the reference system M (z) reversible. Furthermore, the corresponding virtual error signal e(t) is determined by the following formula:

$$e(t) = r(t) - y(t) = (M(z)^{-1} - 1)y(t) \qquad (5)$$

In order to ensure that the reference system control criterion can be solved, the ideal filter L(z) is designed to filter the related virtual error signal e(t) and the actual input signal u(t).

On the other hand, when using the filtered virtual error signal e(t) to excite the fractional controller, the output signal of the controller is:

$$u(t) = C(z, 0)e_L(t) = C(z, 0)L(z)(M(z)^{-1} - 1)y(t) \qquad (6)$$

Therefore, the parameter tuning criteria of the data-driven controller can be derived:

$$J(0) = \min_0 \frac{1}{N} \sum_{t=1}^{N} \left[u(t) - \tilde{u}(t) \right]^2 \qquad (7)$$

4 Physical Simulation Test

In order to verify the effectiveness of the flight controller based on the data designed in this paper, a simulation experiment was carried out. For the offline pre-tuning of the controller using the data-based method, the optimal fractional order is determined to be 1.3 and the initial value of the adaptive controller is obtained, which is also the parameter value of the fixed parameter controller. When t = 0.2 s, 0.4 s, 0.6 s, 0.8 s, use the self-defined data-driven algorithm based on real-time learning to update and modify the controller parameters online. The obtained controller parameters are shown in Table 1.

Table 1. Physical simulation test results

	Fixed parameter IOPI	Fixed parameter FOPI	Adaptive FOPI
Max	420.18	354.48	77.67
Standard deviation	130.91	72.24	11.21
Root mean square	175.09	85.11	20.08
ITAE	116.27	45.10	16.65

It can be seen from Fig. 2 that the maximum value of the adaptive FOPI controller is 77.67, which is 0.3 times that of the other two controllers. It can be seen from this that the adaptive FOPI controller has a strong learning ability and can be used when the operating environment of the system changes. Update the controller parameters of the correction system, so as to provide good robustness for the system to maintain the dynamic response performance of the system.

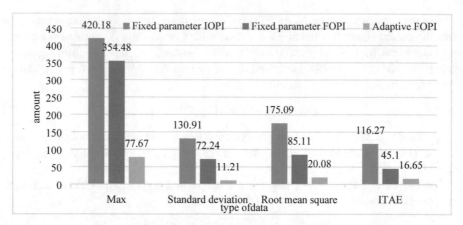

Fig. 2. Physical simulation test results

5 Conclusions

In the process of writing this article, when collecting information, it was found that the design of data-driven flight control system is not in the early stage of research; in the face of the massive information resources on the Internet, it is easy to lose sight of one another, and at the same time, due to limited knowledge, it is not possible to go deep. Or put forward a more one-sided point of view, in the future in the work and study, will continue to study related issues.

Acknowledgements. This work was supported by Doctoral Foundation of Shandong Jiaotong University (BS2018018) and Natural Science Foundation of Shaanxi Province (2019JM-163).

References

1. Zhang, K., Bi, F.H., Li, K.L., et al.: Design and implementation of a dual-IP core UAV flight control system based on Qsys - ScienceDirect. Procedia Comput. Sci. **166**, 180–186 (2020)
2. Kiyak, E.: Tuning of controller for an aircraft flight control system based on particle swarm optimization. Aircr. Eng. Aerosp. Technol. **88**(6), 799–809 (2016)
3. Xue, R., Cai, G.: Formation flight control of multi-UAV system with communication constraints. J. Aerosp. Technol. Manage. **8**(2), 203–210 (2016)
4. Zheng, N., Jin, M., Hui, H., et al.: Real-time and precise insect flight control system based on virtual reality. Electron. Lett. **53**(6), 387–389 (2017)
5. Khalesi, M.H., Salarieh, H., Foumani, M.S.: Dynamic modeling, control system design and MIL–HIL tests of an unmanned rotorcraft using novel low-cost flight control system. Iran. J. Sci. Technol. Trans. Mech. Eng. **44**(3), 707–726 (2019). https://doi.org/10.1007/s40997-019-00288-x
6. Yang, S.B., Zhang, J., Qin, X.X., et al.: Study on the flight test risk assessment of civil aircraft flight control system failure. J. Aeronaut. Astronaut. Aviat. **52**(2), 171–181 (2020)
7. He, K., Gang, L., Zhang, L., et al.: Research progress on model flight test of powered aircraft with autonomous control system. J. Exp. Fluid Mech. **30**(2), 1–7 (2016)

8. Huang, M., Wang, Z., Guo, Z.: Wind tunnel based virtual flight testing system and key technologies for the evaluation of flight control system. Guofang Keji Daxue Xuebao/J. Natl. Univ. Defense Technol. **39**(2), 1–8 (2017)

9. Goupil, P., Urbano, S., Tourneret, J.Y.: A data-driven approach to detect faults in the airbus flight control system. IFAC PapersOnLine **49**(17), 52–57 (2016)

10. Somov, Y., Butyrin, S., Somov, S., et al.: In-flight management for attitude control system with animation of satellite motion. IFAC PapersOnLine **49**(12), 833–838 (2016)

11. Yeom, W.S., Ha, S.: Method of conducting and verification for failure mode effect test of rotary automatic flight control system software. J. Korean Soc. Aeronaut. Space Sci. **48**(1), 55–62 (2020)

12. Chang, D.Y., Lee, S.D., Dong, W.J.: Analytical investigation of the influence of rotor flap dynamics on helicopter flight control system feedback gain limit. J. Korean Soc. Aeronaut. Space Sci. **48**(3), 217–224 (2020)

Solving Second-Order Nonlinear Ordinary Differential Equations Based on Improved Genetic Algorithm

Fangfang Liao and Yu Gu[✉]

Xiangnan University, Chenzhou, Hunan, China
guyu.xnu@sbcmail.com.cn

Abstract. In scientific research and production, researchers often use ordinary differential equations as mathematical modeling tools. Ordinary differential equations are often used to describe the laws of dynamic systems and have been widely used in many fields. However, there are many types of ordinary differential equations, and they are affected by many factors. In order to solve many problems encountered in the process of finding ordinary differential equation models, this paper takes the second-order nonlinear ordinary differential equations as the object to explore the improved genetic algorithm in solving the second-order nonlinear ordinary differential equations. This article first summarizes the status quo of the derivation of ordinary differential equations at home and abroad, and outlines the basic theories of genetic algorithms and ordinary differential equations. On this basis, combined with improved genetic algorithms to solve ordinary differential equations. This article systematically explained the steps of improving the genetic algorithm to solve the nonlinear equations, and used comparative analysis method, observation method and other research methods to carry out experimental research on the subject of this article. Studies have shown that when the number of individuals in the population is too large, the search space will also increase, which will directly cause the algorithm to slow down and the efficiency of the algorithm will also deteriorate.

Keywords: Genetic algorithm · Second-order nonlinearity · Ordinary differential equations · Solution analysis

1 Introduction

The use of mathematical models to predict time series is a research topic that scientific research has been paying attention to for a long time. Among them, the ordinary differential equation model is one of the more important [1, 2]. Ordinary differential equation models can show the internal connections between things, as well as possible changes and trends. By using ordinary differential equations to simulate these changes and trends, some practical problems can be solved [3, 4].

In the study of nonlinear normal differential equations, many experts and scholars have achieved good results. For example, Shin Ando et al. proposed an evolutionary

© The Author(s), under exclusive license to Springer Nature Switzerland AG 2022
J. Macintyre et al. (Eds.): SPIoT 2021, LNDECT 98, pp. 198–206, 2022.
https://doi.org/10.1007/978-3-030-89511-2_24

modeling (EM) method to establish a causal model of a differential equation system in order to model gene regulatory networks [5]. This is a method that combines genetic programming (GP) with statistical analysis. Using this method to evaluate several practical problems, good experimental results have been obtained. Xin Yi Si et al. improved the RNA genetic algorithm. Use this improved algorithm to estimate the parameters of the ordinary differential equation model [6]. And use this algorithm to deal with the problem of unknown parameters of ordinary differential equations with multiple solutions, and demonstrate the specific effects of the algorithm [7, 8].

The purpose of this paper is to propose an improved algorithm to solve the second-order nonlinear ordinary differential equations on the basis of genetic algorithm. This paper compares and analyzes the traditional solution method with the improved genetic algorithm researched in this paper, so as to make a feasibility analysis on the research theme of this paper.

2   Application Research Based on Improved Genetic Algorithm in Second-Order Nonlinear Ordinary Differential Equations

2.1   Genetic Algorithm and Ordinary Differential Equations

(1)   Genetic algorithm, the corresponding English name is GeneticAlgorithm, or GA for short. This is an adaptive exploration algorithm that combines random search and optimization.

The basic steps of genetic algorithm are as follows:

1)   Initialize the population;
2)   Design an appropriate fitness function for the research problem;
3)   Calculate the fitness value of the individuals existing in the contemporary population [9, 10];
4)   According to the calculated fitness, select individuals who are eligible to enter the next generation according to the fitness;
5)   Select individuals for crossover operations according to the pre-determined probability;
6)   Select individuals for mutation operation according to the pre-determined probability;
7)   Generate a new population;
8)   Determine whether the termination conditions are met. If the termination condition is met, proceed directly to 9) operation; if not, repeat 3)–7) operation [11, 12];
9)   Get the optimal solution.

Genetic algorithm is an algorithm that directly manipulates structural objects. This is the most prominent feature of the algorithm. It has no derivation or function continuity constraints. Genetic algorithm is an algorithm with self-adaptive characteristics. Its research has received extensive attention and attention. It is now an important algorithm in the core algorithm of computational intelligence in the 21st century.

(2) Ordinary differential equation

Ordinary differential equation is an equation that connects independent variables with unknown functions or derivatives of unknown functions. Therefore, it can be known by definition that an ordinary differential equation of order n can be expressed in the form shown in formula (1), or in the form of formula (2) or formula (3).

$$F\left(x, y, y', y'', \ldots, y^{(n)}\right) = 0 \tag{1}$$

$$y^{(n)} = F\left(x, y, y', y'', \ldots, y^{(n-1)}\right) \tag{2}$$

$$\frac{dX_i}{dt} = f_i(X_1, X_2, \ldots, X_n)(i = 1, 2, \ldots, n) \tag{3}$$

2.2 Local Search Algorithm

(1) The local search algorithm has the following characteristics:

1) The structure of the local search algorithm is universal and easy to implement. As long as the specific problem-related field is defined at the very beginning, the problem can generally be solved effectively.
2) The performance of the local search algorithm is related to the definition of the domain and the initial state of the algorithm. Different domain definitions, different selection directions, or failure to select the initial solution will all have a decisive impact on the performance of the algorithm.
3) The local optimization feature of the local search algorithm. Due to the characteristics of the local search algorithm, it is easy to fall into the local optimal solution, and it is difficult to reach the global optimal solution.

(2) Hill climbing algorithm

The hill climbing algorithm is a local search algorithm that searches in one direction. Its execution framework is as follows:

Procedure Hill Climbing choose an initial solution;
WHILE neighborhood of the current solution contains a better solution DO
choose the best solution from the neighborhood of
the current solution and move to this solution
end Hill Climbing.

2.3 Improved Algorithm for Solving Second-Order Nonlinear Ordinary Differential Equations

(1) Improved genetic algorithm

Genetic algorithm and hill-climbing algorithm have their own characteristics and shortcomings. Applying hill-climbing algorithm to genetic algorithm can improve the local path-finding ability of genetic algorithm and improve the path problem of hill-climbing algorithm.

1) Selection operator

When the algorithm is executed, the selection operation is mainly based on fitness. In the initial population, a certain percentage of individuals will be randomly selected to participate in the crossover operation through the selection operation. That is to say, the crossover operation is performed in the selected individuals. Then, among these individuals, the "parent" and "mother" individuals who perform the crossover operation are selected according to the principle of fitness selection.

The principle of fitness selection refers to the selection process based on the fitness value of an individual, but it does not completely select according to the pros and cons of the fitness, but follows the following rule: the better the fitness of an individual, the greater the chance that the individual will be selected into the next generation; on the contrary, the worse the fitness of an individual, the smaller the probability that this individual will be selected into the next generation. Because it is easy to enter the "premature" state if the selection is made completely according to the fitness value, which weakens the ability of global search.

2) Mutation operator

For each gene on the chromosome in the chromosome, there is a 50% probability that a parameter mutation will occur. The new parameter is a random number conforming to the Gaussian distribution.

When the chromosome group participating in the mutation operation has been mutated, the fitness of the chromosome group will be re-evaluated. If the fitness value after mutation is greater than the original fitness value, the chromosome group will continue to mutate, and the fitness value of the new chromosome group will be evaluated after each mutation. Until the threshold of the maximum number of mutations is exceeded, this experiment is set to 3 times. If the fitness value after 3 mutations is still greater than the original, the original individual will be retained. Through mutation operation, the diversity of individuals in the population can be maintained, and the local search of the algorithm can be enhanced.

3) Crossover operator

Because in the process of genetic algorithm, the number of individuals in each generation of population needs to remain unchanged. Therefore, after the selection, mutation, and crossover operations are completed, it will be judged whether the number of individuals in the contemporary population is equal to the number of individuals in the initial gene pool. If the number is less than the preset number, new individuals are randomly generated to keep the number of individuals in the population unchanged. The new population after one iteration will contain some new individuals generated after mutation operation, some new individuals generated after crossover operation, and some new individuals generated randomly.

The new population after one iteration is obtained through the above process, and then the new population is evaluated. If the population meets the conditions of the optimal solution or this iteration reaches the maximum number of iterations and other termination conditions, the best individual in the contemporary population is output as the result. Otherwise, continue to iterate.

4) Fitness function

The main task of this experiment is to find an ordinary differential equation model that matches the known data according to the genetic algorithm. The time course data represents the value of the variable corresponding to each moment. It can be known from the concept that this kind of data can be expressed in the form of a matrix, and this kind of data contains two elements: time and variable value. The row of the matrix represents the time, and the column represents the state of the variable at the corresponding time point. The fitness value obtained in this algorithm is obtained by calculating the square difference between the time process data generated by the ordinary differential equation model obtained from the experiment and the known data. This paper uses the following formula to measure the fitness value of a single ordinary differential equation model, as shown in formula (4).

$$fitness = \frac{\sum_{i=1}^{N} \sum_{j=0}^{T} \left(x'_{i(t_0+j\Delta t)} - x_{i(t_0+j\Delta t)} \right)^2}{N \cdot T} \tag{4}$$

(2) The steps of improving the genetic algorithm to solve the second-order nonlinear ordinary differential equation.

1) Convert the problem of solving nonlinear equations into a function optimization problem, establish a function optimization model, and determine the type of objective function.

2) Determine the strategy variables, set the upper and lower limits of the variables, determine the solution space and the required parameters of the algorithm, the length of the individual code string 1, the population size M, and the crossover probability p., Mutation probability pw, termination algebra T, generation gap G, etc.

3) Determine the encoding method, that is, determine the corresponding relationship and conversion method from the individual genotype to the individual phenotype. This article uses floating-point number encoding, and the encoding format is as described above.

4) Enter the population size M, the maximum iteration T of the population, and the error accuracy, and randomly generate the first-generation population. Calculate the fitness of individuals in the population and determine the best and worst individuals.

5) The genetic operation begins, and the optimization operation is performed on the population.

6) According to the selection probability p., Cross-generational elite selection strategy and tournament selection method, select the population and implement the optimal individual retention strategy.

9) According to the adaptive hill-climbing operator Pan, perform a local search on the offspring that have undergone genetic operations.

10) Judge the result after optimization, if the termination condition is reached, jump out of the loop and output the best individual, otherwise return to step 5).

3 Experimental Research Based on Improved Genetic Algorithm in Second-Order Nonlinear Ordinary Differential Equations

3.1 Experimental Protocol

In using the improved genetic algorithm to solve the nonlinear equations, some parameters need to be set. First, the population size must be set. After comprehensive consideration, the population size set in this paper is that the individuals are generally distributed in the local area of the coding space when the population is initialized, so to a certain extent, the search space is restricted. During the experiment, the initial population randomly generated by this program has little effect on the final result. The crossover probability and selection probability in genetic operations have a huge impact on the evolution process. In this experiment, the adaptive crossover and mutation probability is used, and the effect is more obvious.

3.2 Research Methods

(1) Comparative analysis method
This experiment compares and analyzes the improved genetic algorithm studied in this paper and the traditional algorithm in the propagation of different algebras, and conducts statistics and analysis on the results obtained. These data not only provide theoretical references for the topic selection of this article, but also provide data support for the final research results of this article.
(2) Observation method
This experiment observes the errors of the algorithm's running results in the propagation of different algebras. These data provide a reliable reference for the final research results of this article.
(3) Mathematical Statistics
Use the relevant software to carry on the statistical analysis to the research result of this article.

4 Experimental Analysis Based on Improved Genetic Algorithm in Second-Order Nonlinear Ordinary Differential Equations

4.1 Comparative Analysis of Algorithms in Different Reproduction Algebras

In this experiment, by setting different genetic algebras, and according to the experimental results, the optimal result is selected as the comparison with the traditional genetic algorithm. The final data is shown in Table 1.

Table 1. Comparison and analysis of algorithms in different reproduction algebras

		30	100	150	1000
Algorithm	X	0.861674	0.862501	0.862501	0.86250146
	F (X)	0.93276	1.0	1.0	1.0
Traditional algorithm	X1	0.65847	0.89634	0.88753	0.86153
	F (X1)	0.85321	0.92765	0.95463	0.99563

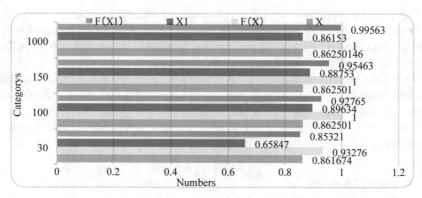

Fig. 1. Comparison and analysis of algorithms in different reproduction algebras

It can be seen from Fig. 1 that from the results of different algebraic programs, it can be seen that as the algebra increases, the results obtained are more and more accurate, and in each generation of algorithm operation, the results can be obtained, and the results are basically the same, indicating that the current number is increasing while the algorithm converges. At this time, the accurate solution of the algorithm can be basically obtained, and no iteration is required. Traditional algorithms have low accuracy when the reproduction algebra is small in the early stage, and the result of each operation is relatively divergent and cannot converge to the result.

4.2   Selection of Population Individual Number and Result Analysis

In order to see whether the number of individuals has a certain impact on the experimental effect, this paper sets up two sets of experiments according to the number of individuals, and the data obtained is shown in Table 2.

Table 2. Selection of population individual number and result analysis

		250	500	150	1000
Test 1	Best one time 1 (T1)	0.00007973	0.000424333	0.000083928	0.000000563
	Average value 1 (V1)	0.005889854	0.005198275	0.004987271	0.000396604
Test 2	Best one time 2 (T2)	0.001114025	0.000361666	0.000051593	0.000275486
	Average value 2 (V2)	0.051650176	0.034944252	0.023780158	0.003235976

It can be seen from Fig. 2 that the number of individuals in the population will have a certain impact on the experimental effect. Even when the number of individuals in the population is small, high-quality results may be obtained. The smaller the number of individuals in the population, the less likely it is to get high-quality results. However, when the number of individuals in the population is too large, the search space will also increase, which will directly cause the algorithm to slow down and the efficiency of the algorithm will also deteriorate.

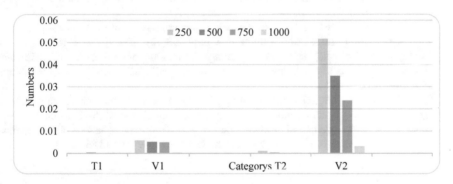

Fig. 2. Selection of population individual number and result analysis

5   Conclusion

As an excellent non-numerical algorithm, genetic algorithm has powerful global search capabilities. Its research history is earlier, and the theory formed is relatively mature, and it has been proven mathematically. It has gradually been applied to the fields of scientific research and engineering practice. It has attracted the attention of professionals in various disciplines. This paper has done some work on the application of genetic algorithm in the second-order nonlinear normal differential equation. According to the mathematical foundation of genetic algorithm, the convergence of genetic algorithm is studied, and it is theoretically proved that genetic algorithm has global convergence.

Acknowledgments. Supported by the Teaching Reform Research Project of General Institutes of Higher Education in Hunan (No: HNJG-2020-0920).

References

1. Venkatesh, S.S., Mishra, D.: Variable search space converging genetic algorithm for solving system of non-linear equations. J. Intell. Syst. **30**(1), 142–164 (2020)
2. Omar, H.A.: An integrated genetic algorithm and homotopy analysis method to solve nonlinear equation systems. Math. Probl. Eng. **2021**(1), 1–14 (2021)
3. Ahalpara, D.P.: Sniffer technique for numerical solution of Korteweg-de Vries Equation using genetic algorithm. J. Appl. Math. Phys. **03**(7), 814–820 (2017)
4. Feng, J., Lu, N.: Genetic algorithm for solving absolute value equation with Rosenbrock method. Basic Sci. J. Text. Univ. **030**(001), 69–73,86 (2017)
5. Shin, A.: Estimating the solute transport parameters of the spatial fractional advection-dispersion equation using Bees Algorithm. J. Contam. Hydrol. **203**(8), 51–61 (2017)
6. Xin Yi, S., Uddin, M., Ahmad, M.: Solving system of nonlinear equations using genetic algorithm. J. Comput. Math. Sci. **10**(4), 877–886 (2019)
7. Sawadogo, W.O., Ouedraogo, P., So, O., et al.: Identification of parameters of Richards equation using modified Hybrid Grey Wolf Optimizer-Genetic Algorithm (HmGWOGA). Eur. J. Pure Appl. Math. **12**(4), 1567–1583 (2019)
8. Hassan, O.F., Jamal, A., Abdel-Khalek, S.: Genetic algorithm and numerical methods for solving linear and nonlinear system of equations: a comparative study. J. Intell. Fuzzy Syst. **38**(3), 1–6 (2019)
9. Mangla, C., Ahmad, M., Uddin, M.: Genetic algorithm based optimization for system of nonlinear equations. Int. J. Adv. Technol. Eng. Explor. **5**(44), 2394–7454 (2018)
10. Al-Hayani, W., Alzubaidy, L., Entesar, A.: Analytical solution for the time-dependent Emden-Fowler type of equations by homotopy analysis method with genetic algorithm. Appl. Math. **08**(5), 693–711 (2017)
11. Park, D.H., Kang, D.S., Ahn, J.H., et al.: Redetermination of curve number using genetic algorithm and CN aligner equation. J. Korea Water Resour. Assoc. **49**(5), 373–380 (2016)
12. Feng, J., Lu, N.: The genetic algorithm for solving absolute value equation with the method of rosenbrock. Fangzhi Gaoxiao Jichukexue Xuebao **30**(1), 69–73 and 86 (2017)

The Application of Decision Tree and Data Analysis in Management System

Fang Mei[✉]

Chongqing College of Architecture and Technology, Shapingba District, Chongqing 400030, China

Abstract. Teaching and learning management in higher education institutions often requires handling a large amount of important data. We can apply the technologies of association analysis, classification and prediction, clustering and decision tree in data mining to help deal with the large amount of data work, such as the evaluation of students' comprehensive quality, the evaluation of teachers, the reasonable setting of courses and the selection of teaching methods, the evaluation of test papers and the quality of test questions improves the quality and efficiency of teaching and administration.

Keywords: Data mining · Higher vocational colleges · Teaching management

1 Introduction

In recent years, the rapid growth in student enrollment at institutions of higher education has raised a number of new issues in instructional management. Instructional management in higher education often deals with large amounts of data. These data relate to core faculty information, core student information, curricula, syllabi, performance evaluations, teaching evaluations, etc. By applying data mining technology in teaching management, collecting and analyzing the content of various data representations and their relationships, we can timely identify teaching management problems, improve the relevance of work, improve the quality and effectiveness of teaching management [1, 2].

2 Data Mining and Decision Tree

Data mining is the process of extracting potential and valuable knowledge (patterns and rules) from large amounts of data. It is a method of deep data analysis and an interdisciplinary discipline that includes machine learning, mathematical statistics, neural networks, databases, pattern recognition, lawsets, etc. The process of data mining includes data collection, data processing, data transformation, data mining, pattern evaluation, and knowledge representation. Data mining processes include data collection, data processing, data transformation, data mining, pattern evaluation, and knowledge representation. The above processes are not performed in one go; some steps or the entire process may have to be repeated. There are many data mining methods such as association rule mining,

J. Macintyre et al. (Eds.): SPIoT 2021, LNDECT 98, pp. 207–213, 2022.
https://doi.org/10.1007/978-3-030-89511-2_25

multi-level data aggregation and synthesis, decision tree method, neural network method, positive and negative example coverage method, rough set method, genetic algorithm, formula discovery, statistical analysis method, fuzzy theory method, and visualization techniques. The following methods are suitable for teaching and learning management in higher education institutions [3].

The decision tree is a decision analysis method that generates a probability based on known conditions, calculates the probability of net present value more than zero by constructing a decision tree, evaluates the risk of the project, and determines the feasibility of the project. This is a graphics method using the probabilistic analysis intuitively. Since such decision branches are depicted in the figure, they are very similar to tree branches and are called decision trees. In the machine learning, the decision tree is a prediction model which shows the mapping relation between object attribute and object value. The entropy = system disorder degree uses the algorithm ID3, C4.5, C5.0 to generate the tree algorithm and use entropy. This metric is based on the entropy concept in information theory.

The decision tree is a tree structure in which each internal node represents a test for an attribute, each branch represents a test output, and each leaf node represents a category [4].

The classification tree is a general classification method, and is supervisory learning, so-called supervised learning. This is the sample set. Each sample has an attribute set and category. These attributes and categories are predefined. And the classifier is acquired by learning. Machine learning which can classify a new object correctly is called "controlled learning", as shown in Fig. 1.

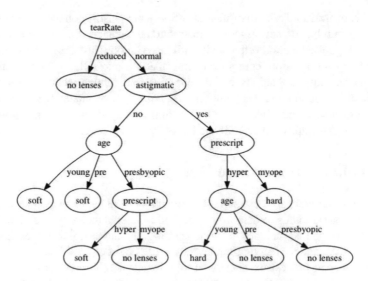

Fig. 1. Schematic diagram of decision tree and association analysis

2.1 Association Analysis

An important research method in data mining is to study relevant rules. Association rules describe the relationship between data items in database. It was originally put forward for the analysis of market shopping basket, with the purpose of discovering customers' purchase patterns. At present, association rules are widely used in other fields.

2.2 Classification and Prediction

Classification is to divide data objects into one of several given categories according to a classifier. It generally has two steps: constructing a classifier and using the classifier to classify the data. Classification is a guided learning process. The training set of samples is known and the number of categories is determined. Classification model can be used for prediction. Prediction is to use the learning model to predict the unknown class of data objects [5].

2.3 Clustering

A cluster is a process of dividing sets of data into multiple groups so that objects in the same group have higher similarity. Objects in different groups have lower similarity. Unlike classification, the cluster is a learning process that is not controlled, and in this process the sample category or category is not known beforehand.

2.4 Decision Tree Method

The decision tree retrieves the database attribute field that contains the maximum amount of information using the information theory information. Create a node in the decision tree, and build the branch of the tree according to the value of the attribute field, and repeat the process of creating the lower node and branch tree for each subunit. Classification tree is a "divide and conquer strategy", which decomposes a complex problem into several subproblems and then solves them. Moreover, divide and conquer strategy can also be recursive, that is, sub problems continue to use divide and conquer strategy until it comes down to a very basic and very simple problem. Among them, each internal node of the binary tree represents a logical judgment of a certain feature; the edge represents the result of logical judgment; the tree table node is the tab of each category; The feature sequence of leaf nodes from the root to the edge of each leaf node.

3 Data Mining in Teaching Management

3.1 Evaluation of Students' Comprehensive Quality

Comprehensive quality evaluation involves political and ideological quality, average score of academic achievement, scientific research and innovation ability, social practice ability, sports activities and so on. Teaching management departments often use artificial evaluation method, using analytic hierarchy process or fuzzy analysis method for evaluation. Manual evaluation is influenced by subjective factors, and its accuracy is not high enough. At the same time, we need to design different utility functions and give weight to each index, which is complex and difficult to promote. The decision tree classification theory can be used to solve this problem. Firstly, the decision tree is established, and the classification rules are as follows: the comprehensive quality is divided into 1, 2, 3 and 4 levels according to excellent, good, general and poor. The level of moral education is divided into 1, 2 and 3 grades according to good (top 40%), average (top 40% to 80%) and poor (bottom 20%). According to the academic level, excellent (top 20%), good (top 20%–40%), average (top 40%–60%), poor (top 60%–80%) and poor (bottom 20%) can be divided into 1, 2, 3, 4 and 5 grades. The extracurricular practical ability is divided into level 1 and level 2 according to the qualified (top 50%) and unqualified (bottom 50%). If (moral education level 1) then if (academic level 1 or 2) and (extracurricular practical ability 1) then (comprehensive quality 1).

If (academic level 1 or 2) and (extracurricular practical ability 2) then (comprehensive quality 2) if (Academic Level 3) and (extracurricular practical ability 1) then (comprehensive quality 2) else (comprehensive quality 3) if (moral education level 2) then.

If (academic level 1) and (extracurricular practical ability 1) then (comprehensive quality 1) if (academic level 1) and (extracurricular practical ability 2) then (comprehensive quality 2) if (Academic Level 2) and (extracurricular practical ability 1) then (comprehensive quality 2).

If (Academic Level 2) and (extracurricular practical ability 2) then (comprehensive quality 3) if (Academic Level 3 or 4) and (extracurricular practical ability 1) then (comprehensive quality 3) else (comprehensive quality 4) if (moral education level 3) then if (academic level 1) and (extracurricular practical ability 1) then (comprehensive quality 3) else (comprehensive quality 4) 4) According to the decision tree classification rules, the data training set is established, the existing data is input, the decision tree classification model is selected for data training, and the decision tree for predicting the comprehensive quality of students is generated.

3.2 Teacher Evaluation

In the teaching evaluation system, in order to get the law of evaluation, we must deal with many information indicators collected, some of which are key. Obtaining these indicators is helpful to the effectiveness of management and has great value in the actual teaching management. Through the investigation of teachers' class situation, we know a lot of indicators, such as teaching attitude, teaching content, teachers' use of classroom time, blackboard writing, teachers' educational background and so on. How to make an accurate evaluation with fewer indicators, and have representativeness and decision-making is the main content. Based on these, we can use rough set theory to find out the minimum set of related attributes which have the same decision-making ability as the original data in the generalization relation by calculating the importance order of attributes, so as to achieve attribute reduction and produce more concise and meaningful knowledge rules.

3.3 Reasonable Curriculum and Teaching Method Selection

Due to the connection between the courses, all the courses seem to be learning step by step, which requires prior courses for knowledge preparation. At the same time, the curriculum should be updated and adjusted according to the needs. The regularity of the curriculum setting was analyzed using the rule analysis method related to the data extraction. For example, by analyzing the relevance between course a and Course B, we can decide how to arrange the opening order of these two courses, so that students can learn according to a reasonable knowledge context. In particular, some courses do not have a very clear relationship between before and after study, but arranged to study in the same semester or different semesters, the effect may be different, at this time, you can use association rules analysis. For example, if courses a and B are set, should Course C be set. The teaching effect varies with the teaching methods. In the same course, different teachers use different teaching methods to teach, and the students' grades are also different. Through the regression analysis and correlation analysis of previous students' scores in the database, it is very important to explore the correlation between students' scores and teaching methods in order to find a suitable teaching method. This is also an advantage of data mining. In addition to the above aspects, the analysis method of association rules can also evaluate and describe students' learning, growth and teachers' teaching, and then analyze the factors that affect the teaching effect, and calculate their correlation.

3.4 Evaluation of Examination Paper and Quality

The amount of examination paper evaluation mainly focuses on the subjective questions. There is no unified standard answer for subjective questions. The marking teacher decides the score by judging the logical order of the answers and the key conclusion points. At this time, we can use the method of decision tree to evaluate. That is to construct a classified binary tree for each subjective question, each internal node of the tree represents a logical judgment of a certain feature; the edge represents the result of logical judgment; the leaf node of the tree is the label of each category; the edge from the root to each leaf node is the feature sequence of the leaf node. Generally speaking, through the statistical analysis of the distribution of students' scores, we can know the quality of the test questions. Moreover, we can calculate the kurtosis coefficient and skewness coefficient through the analysis of students' test paper scores, and then classify the test paper according to the above types, and further improve the quality of the test paper by using the method of cluster analysis.

3.5 Training and Management of Students Network Teaching Service

In terms of the cultivation and management of students in higher vocational colleges, from the original unified development to gradually pay attention to the students' personality and individualized teaching. Every student has different personality characteristics, learning interest and learning situation. To cultivate and manage different students, we need to make learning plans and management measures according to their different characteristics. The internal relationship between students can be found through students' management of database data and the use of relevant data analysis and evolutionary analysis functions; various behavior activities. First of all, the students are classified according to the association rules, and then the results are predicted. For example, according to the students' a behavior and B behavior, we can push out the C behavior that will happen, so we can push out whether to encourage or stop the C behavior. It can also track students' academic performance, make regression analysis on their assessment results in one stage, and predict the possible trend of their performance. If the forecast results show a downward trend, we need to quickly find the reasons and make up in time to avoid the decline of performance.

3.6   Network Teaching Service

Web mining is a process of extracting interesting and useful models and hidden information from WWW resources and behaviors. According to the unique flag of a visitor, find the corresponding access record set, and then divide the access record set according to a fixed time interval to find each access record set of the visitor, which is called an access transaction. Finally, sort all the access transactions in chronological order to form a mining transaction set, and each thing is equivalent to a visitor's access to the siteroute. Convert the text, pictures and other files in the web page into the available form of data mining algorithm. This paper uses statistical analysis method to count the visiting frequency, visiting time and visiting path of web pages, and to understand the visiting characteristics of visitors, so as to solve the following problems.

(1) The modification of the website has more pertinence and basis, and steadily improves user satisfaction. Modify the links between web pages according to the user access mode, and show the information users want in a faster and more effective way.
(2) View site traffic.
    Find the needs and interests of users, optimize the web pages with strong needs, and use the method of server pre storage to solve the problem of slow download.
(3) Provide personalized website.
    For different users, according to their personal interests and hobbies (user access patterns obtained by data mining algorithm), dynamic browsing suggestions are provided to users, and personalized websites are automatically provided. Personalized and intelligent online distance education is to make full use of the function of data mining technology to provide services for distance education. Its performance is: the use of student registration information, for different students, to provide different learning content and learning mode, to teach students in accordance with their aptitude, and to save students' learning records. Mining the information of learning behavior and learning records saved on the website, and combining with the information of courseware knowledge base, automatically reorganize the course content.
    Content, make it more in line with the law of teaching, and combined with the content, provide other related learning resources. Through the mining of learners' learning behavior, the user's browsing mode can be found, and the links between pages can be automatically reconstructed to meet the user's visiting habits.

(4) Find system performance bottlenecks, find security vulnerabilities.
(5) To provide important and valuable information for teachers and educational admin-
     istrators. For example, through the analysis of each student's questions, we can get
     the relevance between the questions and some other useful information, which can
     be used to guide teaching and correct the difficulty coefficient of the questions.

4 Conclusion

Data collection is used as a new technology tool that greatly affects the future of mankind
and is managed by universities and teachers. To assist teachers, find hidden mode or
mode from verifiable water quantity data to provide information support for decision
making. The quality of education is continually increased, and higher education institutions occupy an active position in intense competition. There is no future development
space.

References

1. Gong, L.: Research on the application of PDCA cycle in the teaching management of academic
    administrators in Higher Vocational Colleges. Educ. Mod. **7**(08), 164–165 (2020)
2. Li, Z.: Application of fine management in teaching management of Higher Vocational Colleges.
    Knowl. Base **12**, 191 (2019)
3. Liao, L.: Analysis of the application of big data technology in teaching management reform
    of Higher Vocational Colleges. J. Jiamusi Vocat. Coll. **03**, 286–287 (2019)
4. Zeng, H., Ge, X., Shang, X.: Research on the application of micro application based on VBA
    in teaching management informatization of Higher Vocational Colleges. Mod. Comput. **24**,
    63–67 (2019)
5. Xu, J.: Research on the application of new media in teaching management of Higher Vocational
    Education – taking a higher vocational college as an example. J. Shanxi Coll. Fin. Tax. **21**(01),
    73–77 (2019)

Research on Mathematical Modeling Optimization of Recognition Algorithm Based on Multi Feature Extraction

Peng Hao[✉]

Shanxi Mechanical and Electrical Vocational and Technical College, Shanxi 046011, China

Abstract. Data modeling optimization is widely used in the field of multi feature extraction. The recognition algorithms based on statistical features and structural features need to be implemented under the mathematical modeling optimization, which can change the input structure of the image, so as to extract the statistical features of the graphics. The image is divided by grid partition method, and then the text pixels in the grid are compared, Ensure that the text image can be recognized by fine classification pattern.

Keywords: Multi feature extraction · Function · Distinguish · Characteristic value

1 Introduction

There are two kinds of characters in the world, one is Pinyin (such as English), the other is hieroglyphic or graphic (such as Chinese characters). In the human-computer interaction, Pinyin text shows great advantages, using the keyboard can be a letter very natural and convenient input into the computer, but for graphic text, there has been no appropriate input device. Since the 1980s, a large number of microcomputers have entered our country, many of which are used for Chinese information processing, but the input device is still only the keyboard. In order to input Chinese characters into the computer, a variety of Chinese character coding schemes have emerged. The so-called coding scheme is to convert Chinese characters into characters or numbers through artificial rules to facilitate keyboard input. Using coding scheme to input Chinese characters needs memory, which brings great inconvenience to the popularization of computer in the field of life. The limitation of using western keyboard to input Chinese characters has become the bottleneck and shackle of computer popularization and application. With the development of science and technology, people realize that they need to input Chinese characters with graphic input equipment. After input, the computer will automatically recognize and transform them into internal code. As for how the computer can recognize the input Chinese characters, it is Chinese character recognition technology.

In terms of discipline, Chinese character recognition belongs to the category of pattern recognition and image processing, and also involves artificial intelligence, formal language and automata, statistical decision-making theory, fuzzy mathematics, information theory, philology, computer science, etc. Due to the large number of Chinese

J. Macintyre et al. (Eds.): SPIoT 2021, LNDECT 98, pp. 214–219, 2022.
https://doi.org/10.1007/978-3-030-89511-2_26

characters, complex structure and many similar characters, Chinese character recognition is a difficult word recognition. In recent years, with the development of pattern recognition technology and computer science and technology, Chinese character recognition has made great progress. The basic process of Chinese character recognition includes five steps: Chinese character input, preprocessing, feature extraction, classification recognition and post-processing [1].

2   Principles and Algorithms of Chinese Character Recognition

2.1   The Principle of Chinese Character Recognition

Like other pattern recognition, the basic idea of Chinese character recognition is matching discrimination. The expression forms (such as various features) representing the essence of unknown Chinese character patterns are extracted and matched with the set of standard Chinese character pattern expressions (called dictionaries) stored in the machine one by one, and the discrimination is carried out with certain criteria, and the pattern expression forms closest to the input text are found in the set of standard Chinese character pattern expressions stored in the machine, The word corresponding to the expression is the recognition result.

2.2   Chinese Character Recognition Based on Statistical Decision Method

The statistical decision-making method of Chinese character recognition extracts the features of character image to form feature vector. The feature vector does not directly represent the handwriting, but represents the characteristics of character image. Statistical character recognition method estimates the distribution of feature vectors according to the training sample set, which has excellent anti-interference ability, so it is widely used in Chinese character recognition. This method needs a function to express the sample distribution. The probability density function of multivariate normal distribution is usually regarded as an approximate function of the sample distribution, Chinese characters' rough peripheral features, Chinese characters' grid features and so on.

2.3   Extraction of Rough Peripheral Features of Chinese Characters

The outline structure of Chinese characters contains important information of Chinese characters. Rough peripheral feature (scanning from the four sides of Chinese characters to the middle, taking the stroke value of the first stroke pixel as the feature) classification is to extract the outline information of Chinese characters as the feature to classify. It is considered to be an effective rough classification method in the recognition of printed Chinese characters. The basic idea is to transform the original image, so that the internal structure is affected by the peripheral information. The specific implementation process is as follows: first, scan the text from the four borders of the text, and then set the area that can not be continuously scanned from all directions in the original text background part as the stroke part, In this way, we can get the so-called rough peripheral image. From the implementation process, we can see that the features of the original text's internal region are also affected by the contour, but because of the influence of the contour, it becomes the stroke region after transformation. Although the rough peripheral feature is still used in feature implementation, the original rough peripheral feature only considers the contour information of Chinese characters, while the improved method enhances the proportion of rough peripheral contour information in feature quantity on the basis of using the internal structure information of Chinese characters [2].

2.4   Chinese Character Recognition Based on Structural Feature Method

At present, there are two methods to extract the structural features of text. The first method is the method of obtaining the character contour based on pixels. This method does not consider the global information and structural information, so there are some problems in the results, such as the deformation of rough edges and intersections. Therefore, researchers have proposed other methods to extract the structural features, However, this makes the pixel based character recognition more complex and unstable. The second method is to use - group line segments and their relationship to describe and recognize characters, but these methods have many steps, complex calculation, difficult to achieve, and easy to be affected by boundary noise; The influence of sound and intersection.

3   Recognition Algorithm Based on Multi Feature Extraction

3.1   Analysis of Current Algorithms for Handwritten Chinese Character Recognition

In this algorithm, the strokes of Chinese characters are divided into several categories, and then according to the input sequence of all kinds of strokes, they are taken as the features of a Chinese character and stored in the character library. When the Chinese characters are input online, the stroke sequence features of the character image are obtained and compared with the features in the character library. If there is a small difference between a group of features and the image features, it can be considered as the same character. Next, we introduce the stroke classification method, and then introduce the image preprocessing and distance definition [3].

Divide the two-dimensional plane, as shown in Fig. 1.

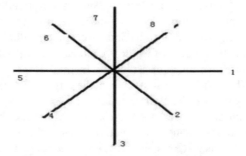

Fig. 1. Division of two dimensional plane

In the process of mathematical model design and optimization, the strokes of Chinese characters need to be divided into several categories, and then the variable analysis of each stroke is carried out, so as to form a feature of Chinese characters, and then the feature database of Chinese characters is added. Finally, through the design of mathematical function, the comparison function of stroke sequence feature and character library feature is formed, It can realize image preprocessing and structure optimization. In the process of feature extraction algorithm implementation, its direction can be subdivided by mathematical function optimization, which is usually divided into clockwise fold, counter clockwise fold and mixed direction fold. In the process of stroke classification and extraction, it will face a mathematical problem, how to design image text reading algorithm, The slope of the mathematical model will be used in the specific implementation, but it is unreasonable to use the slope in special cases. For example, there is no slope at all in directions 3 and 7. Therefore, in the design of the algorithm model, it is necessary to compare the coordinates of the first point behind the pixel with the pixel index of the point. The algorithm logic of the mathematical model is as follows:

(1) If point[a+2].y=point[a].y and point[a+2].x>point[a].x,direction 1.
(2) If point[a+2].y=point[a].y and point[a+2].x<point[a].x,direction 5.
(3) If point[a+2].x=point[a].x and point[a+2].y>point[a].y,direction 3.
(4) If point[a+2].x=point[a].x and point[a+2].y<point[a].y,direction 7.
(5) If point[a+2].x>point[a].x and point[a+2].y>point[a].y,direction 2.
(6) If point[a+2].x>point[a].x and point[a+2].y<point[a].y,direction 8.
(7) If point[a+2].x<point[a].x and point[a+2].y>point[a].y,direction 4.
(8) If point[a+2].x<point[a].x and point[a+2].y<poit[a].y,direction 6.

3.2 Mathematical Model of Handwritten Chinese Character Feature Extraction

When getting the corresponding stroke direction, we need to carry out comprehensive de-noising processing according to the direction of each pixel of the stroke, so as to eliminate the jitter problem in the process of receiving and writing, so as to obtain several pixel points, and create a mathematical model for eliminating the pixel points, In general, the distance design of mathematical model is as follows:

$$d = \frac{1}{m} \sum\nolimits_{i=0}^{m} (1 - a_i) \tag{1}$$

The optimization steps are as follows

(1) Data function control from input point can produce effective input point sequence
(2) By controlling the parameters of the input point, the de-noising process can be ensured, and the direction code sequence can be calculated through the mathematical distance function
(3) The stroke code sequence after de-noising is processed by direction code distance function
(4) According to the distance function, feature extraction and recognition of matched Chinese characters are carried out

In the process of extracting the mathematical model algorithm, it is necessary to compare the threshold value dis according to the distance. If d < DIS, the sequence code can be formed by comparing the characteristics of the font, and then the recognition optimization can be carried out according to the sequence code to ensure that the distance function can carry out the font feature recognition control and improve the overall efficiency of font recognition. For example, the stroke sequence of Chinese character "six" and "Wen" is "4134", So it can't be identified correctly [4].

4 Optimization of Eigenvalues of Mathematical Functions for Chinese Character Recognition Based on Statistical Features

In the process of statistical feature function extraction, feature variables need to be determined. Through the analysis of different feature variables, a complete feature library can be formed, which will play a guiding role in function design, In the process of feature extraction, we need to form a 4 × 4 feature value. Through the method of function transformation, we can transform a Chinese character image into a binary image, and then carry out feature processing to ensure that the feature value can meet the requirements of function image design, Generally, a Chinese character image is divided into 4 × 4 = 16 or 3 × 3 = 9 areas to control the number of points of each pixel in the statistical area, and then the total number is counted out through the function. This feature is compared with the font features in the function library, and finally the minimum character matching can be carried out, It shows that the eigenvector calculated by the eigenfunction is correct. On the contrary, the feature is not in the Chinese character feature library, so the algorithm needs to be optimized and then searched. In the process of algorithm search, two mathematical models need to be used, one is variance distance control, the other is absolute value distance control, 4 × 4 partition is adopted, and then the eigenvalues of each region are calculated. The simulation is shown in Fig. 2

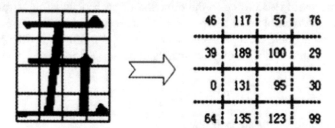

Fig. 2. Statistical feature map of Chinese character image

Figure 2 is a Chinese character image, and it is divided by 4 × 4.

Now, the feature is represented as an array, that is, test. p [16]. If the group I feature in the character library is tez [i]. p [16], then the distance between the Chinese character image to be recognized and the i-th character in the character library is:

$$d = \sum_{j=0}^{16} \left| test.p[j] - tez[i], p[i] \right| \tag{2}$$

5 Conclusion

The recognition algorithm of multi feature extraction can avoid the nonstandard problem in the process of writing Chinese characters, so as to improve the efficiency of Chinese character feature recognition. The success rate of standardized recognition of Chinese characters is as high as 90%, which can achieve the purpose of optimizing the mathematical function of multi feature extraction of Chinese characters. It has important application value for the standardized extraction of Chinese characters.

References

1. Chen, J.: Support vector machine technology and its application. Sci. Technol. Inf. (25) (2007)
2. Zhang, S., Kong, L.: Chinese character recognition and status analysis. J. Yanshan Univ. (4) (2003)
3. Du, S., Wu, T.: Support vector machine method in pattern recognition. J. Zhejiang Univ. (Eng. Edn.) (5) (2003)
4. Feng, Y., Wang, X.: Review and prospect of offline handwritten Chinese character recognition technology. Microcomput. Appl. (4) (2003)

Genetic Algorithm Image Segmentation Method Based on Membrane Computing

Yongxing Lin[1][(✉)] and Quan Wen[2]

[1] Keyi College of Zhejiang Sci-Tech University, Hangzhou 310018, People's Republic of China
[2] Zhejiang Institute of Economics and Trade, Hangzhou 310018, People's Republic of China

Abstract. A genetic algorithm image segmentation method based on membrane computing is proposed. Membrane calculation (system is a computing model abstracted from the function and structure of biological cells and tissues and organs composed of cells. The distributed, maximum parallelism, uncertainty of the system and the better adaptability and strong optimization ability of the membrane optimization algorithm have certain advantages for solving complex optimization problems. How to integrate these features and advantages of the system are applied to image processing, which is helpful for the development of image processing. It is of great significance to study the application scope of the system and new image processing algorithms. Each membrane is optimized by running evolution rules and exchange rules.

Keywords: Membrane calculation · KSW burn · Genetic algorithm · Image segmentation

1 Introduction

Image segmentation is a technology and process that divides the image into different regions with characteristics and extracts the interested objects. Image segmentation is the primary and most important step in image analysis and understanding. The quality of segmentation directly affects the subsequent recognition and understanding. Image segmentation has been widely used in many fields such as remote sensing imaging, video monitoring, industrial detection, biomedical and so on. So it is of great significance to study the segmentation methods of images.

Membrane computing, as a very young branch of natural computing, was founded in 2000 by Gheorghe paun, a Romanian scientist and academician of the European Academy of Sciences. Membrane computing is a new computational model inspired by the structure and function of living cells, as well as the tissues, organs and other higher ordered structures of organisms, such as cells in cell groups (such as bacteria). These models are often called system. The system is a new distributed and parallel computing model in essence, which has attractive characteristics for many practical applications, such as uncertainty, synchronization/asynchronism, and easy understanding (the system hierarchy); measurability and programmability; The internal separation; Nonlinear

J. Macintyre et al. (Eds.): SPIoT 2021, LNDECT 98, pp. 220–225, 2022.
https://doi.org/10.1007/978-3-030-89511-2_27

characteristics; High descriptive and high operability; Automatic and dynamic calculation characteristics; The coding characteristics and diversity of input information and calculation results. So once proposed, it attracted the attention of many researchers and became a very active research field [1].

At present, most of the research work of membrane computing is mainly focused on the system is capable of calculation and the calculation effectiveness. However, despite the system has many attractive features for image segmentation, because system research is not long and existing. The system model can not be used directly for image segmentation, so the new computing model of the system has not attracted the attention of the scholars in the field of image segmentation. The research on the image segmentation method based on membrane computing in this paper the application scope of the system, or the research of new image segmentation methods, has very important theoretical and application value.

2   Common Methods of Image Segmentation

As early as the mid-1950s, people began to study image segmentation. After half a century of development, image segmentation algorithms have emerged in endlessly [2], and many research results have been achieved. But there is still no one method to apply to all images, which is the specific analysis of specific problems. The segmentation of images is always based on specific criteria, and there is no mature theory for selecting which criteria. Until now, image segmentation is still a hot topic. The common image segmentation methods include gray threshold, edge detection, region based, mathematical morphology. This method usually uses gray histogram of image, and selects the threshold value by some criteria to divide the image into pixels set greater than and less than a specific threshold. According to the selection range and number of threshold, the following methods are used to divide the threshold value.

(1) Global single threshold method. As the name implies, this threshold segmentation method only sets a threshold for the whole image, and divides the image into two parts: target and background according to some criteria.
(2) Global multi threshold method. Because there may be multiple target objects in an image to be extracted, single threshold can not meet this requirement. Two or more thresholds can prevent the single threshold setting value from being too high to divide the target pixel into the background pixel set. Otherwise, the background pixel will be divided into the target pixel set if the threshold setting is too low.
(3) Local adaptive method. This method is generally used to deal with the segmentation of images due to uneven illumination. The whole image is segmented first, and then each block is segmented based on a local threshold.

According to the different principles or criteria, the commonly used threshold segmentation methods are: peak valley method based on gray histogram, best direct method, minimum cross Di method, maximum inter class variance method, etc.

The biggest advantage of these threshold segmentation methods based on gray level is that the algorithm is easy to implement and the calculation is small. For example, the beehive Valley method based on gray histogram can get better segmentation results quickly and intuitively for the image with double bee feature. Therefore, this threshold segmentation method can achieve better segmentation effect for the image with strong contrast between the target and background, but for noise interference, the image with overlapping region has poor effect. The limitation of this method is that the image features are too few to consider, only the gray value is considered, but the spatial features of gray are ignored. But this paper studies the image segmentation based on the membrane computing.

3   Genetic Algorithm, KSW Direct Method and System

3.1   Genetic Algorithm (GA)

Genetic algorithm is a computational model based on the evolutionary process and the mechanism of population genetics, which is based on the natural selection of biological evolution process. It takes all individuals in a group as the object to carry out random iteration and evolution to realize the optimization and efficient search of the encoded parameter space. It simulates the process of biological evolution, through hybridization, variation, selection and other ways, biological individuals keep and optimize quality, so as to promote the continuous evolution and development of the population, and finally to the optimal solution.

The basic steps of genetic algorithm are as follows:

1) Parameter coding: coding has a great influence on the performance of the algorithm, such as search ability, computational complexity and population diversity. The common codes include binary code, gray code code, real number code, symbol code, etc.

2) Generation of initial population: randomly generate a set of n individuals (i.e. initial population P (o)), which represents the possible solution domain of the problem. The calculation of genetic algorithm is to simulate the evolution process of biology to achieve the survival of fittest and finally achieve the optimization.

3) Design of fitness function: genetic algorithm only uses fitness function to update population in evolutionary process, and generally does not need external information. The fitness of individual in group P (T) is calculated according to fitness function, which is used as evolutionary basis to search. Therefore, the selection of fitness function is very important.

4) Selection: the selection operator selects the high-quality individuals from the group as the parent to generate new excellent individuals to join the next generation of groups, and the selection operation is established in the adaptive value of the individuals in the group. The main selection forms are Boltzmann selection, sorting selection, adaptive value proportion selection, etc.

5) Crossover: the selected genes of the parent individuals are exchanged and recombined to produce new individuals. Cross operation can produce new generation individuals with excellent characteristics and stronger adaptability, and improve the global search ability of the algorithm. The crossover operators are: single point crossing, double point crossing, uniform crossing, etc.

6) Variation: variation is to change one or some values on the individual coding string with a small probability of variation, and it is also the operation method to generate new individuals [3].

Group P (1) a new generation of population P (t + 1) is generated after selection, crossover and mutation operation. Each generation evolution is so iterative in a cycle that the adaptability of the optimal individual and the average fitness of the population are constantly improved. At the end of the iteration, the optimal fitness body in the population is output as the optimal solution. There are two simplest cases that indicate the end of iteration: 1) stop when the number of iterations is reached; 2) The optimal individual fitness or average fitness value in the population will not be improved for several consecutive generations, then the iteration process will end.

3.2 ksw Direct Method

Kapur [1] proposed by Kapur et al. The best method (KSW method) does not need prior knowledge, and the image with the background histogram contrast is not obvious can be segmented. For images with gray scale range of [0, n−1], the information is expressed as:

$$H = -\sum_{i=0}^{n-1} p_i ln p_i \tag{1}$$

Where, p_i is the probability of pixel points with gray value i.

$$P_i = \sum_{i=0}^{t} p_i, \; H_t = -\sum_{i=0}^{n-1} p_i ln p_i \tag{2}$$

Set the gray threshold t to divide the image into background (a) and target (b), then the probability distribution is

$$H_A(t) = -\sum_{i=0}^{t} \frac{p_i}{p_t} ln \frac{p_i}{p_t} \tag{3}$$

When $H_A(t)$ is the maximum, the threshold t is the best threshold, which can get the best segmentation effect. Therefore, it is regarded as the fitness function of the algorithm in this paper.

3.3 Point three System

The most prominent advantage of the system compared with other computing methods is the maximum parallelism. The parallel and simultaneous operation of each membrane in the system greatly improves the computing ability and efficiency of the system. There are three main types of system: cell type system, organization type system and neurotype system. Among them, cell type the system is derived from the structure and function of cell tissue, and its basic composition includes membrane structure, object multiple sets and regular sets. Cell type the system follows the principle of maximum parallelism, such as the calculation ability of the system is equivalent to that of Turing machine.

4   Experimental Structure Analysis

In the experiment, we choose the standard image Lena (512 × 512) as a test image, as shown in Fig. 1, for the proposed method based on the image segmentation experiment based on genetic algorithm and KSW multi threshold image segmentation experiment are completed respectively.

Fig. 1. Segmentation results

According to the experimental results, we can know that the optimal threshold found by genetic algorithm is very close to the other two typical algorithms, which shows that the algorithm has good feasibility. When the amount of image data is small, because the genetic algorithm needs to execute a certain number of evolution rules, compared with the other two algorithms, the running time has no advantage; With the gradual increase of image data, the parallelism of genetic algorithm will gradually reflect its efficiency [4]. The larger the amount of image data, the more obvious the advantage of running time.

The system uses two kinds of rules: the first is evolution rules, which are used to evolve the objects in the membrane, and the other is transport rules, which are used to exchange and share the objects between membranes. The system will adopt the improved operators as the evolution rules of membrane, which are selection operator, crossover operator and mutation operator. because the system is a distributed parallel computing model, and each membrane will run as an independent computing unit in maximum parallel mode under global clock control.

5   Conclusion

As a new branch of natural computing, membrane computing has become a hot research field. Many characteristics of membrane computing have attracted a large number of scholars to enter this field and achieved a lot of research results. With the development of membrane computing, more and more attention has been paid to the application of membrane computing, which has become an extremely important part of membrane computing. Based on the idea of membrane computing, this paper proposes an improved genetic algorithm (p-ga) for image segmentation the maximum parallelism of the system and the strong optimization ability of genetic algorithm. In the p-ga algorithm, each membrane completes the local optimization by executing the selection, crossover and mutation rules, and finally outputs the global optimal solution from the surface membrane through the exchange rules and global optimization. This method takes advantage of the above two models and solves the problems of low efficiency and slow convergence speed of traditional genetic algorithm. The experimental results show that the p-ga algorithm has better convergence accuracy and computational efficiency.

References

1. Zhu, Z.hen, He, Q., Zhang, B.: Research on improved genetic algorithm for large scale TT & Resource Planning. Software **34**(2), 82–83 (2013)
2. Yu, H., Lu, F.: A multi-mode and multi standard path planning method based on genetic algorithm. J. Surveying Mapping **01**, 89–96 (2014)
3. Xiaoyan, S., Shanshan, C., Dunwei, G., Yong, Z.: Weighted multi output Gaussian process proxy model based on interval fitness interactive genetic algorithm. Acta Automatica Sinica **02**, 172–184 (2014)
4. Hongwei, C., Hua, M., JieQiu, Z., Zhang Zhiyuan, X., Zhuo, W.J., Shaobo, Q.: Metamaterial optimization design based on weighted real coded genetic algorithm. Acta Physica Sinica **08**, 438–444 (2014)

Research on Brake Shoe Clearance Monitoring of Mine Hoist Based on Multi-sensor

Dan Liu[✉] and Yingying Jiang

School of Electrical and Information Engineering, Heilongjiang Institute of Technology, Heilongjiang 158100, China

Abstract. In this paper, based on the study of the influence of mine hoist braking parameters, combined with the working principle of hoist braking system, a mine hoist brake shoe clearance monitoring device is designed and implemented. The system function design, monitoring scheme, hardware design and software design are introduced. With ARM processor as the core, the monitoring device is equipped with high-precision analog-to-digital converter, multiplexer, can bus and Ethernet interface, which breaks through the traditional PLC control mode. It has more powerful data acquisition ability and superior communication mode, and is suitable for real-time monitoring in industrial field.

Keywords: Hoist brake shoe · Real time monitoring · Isolation design

1 Introduction

Mine hoist is a large and key equipment in the production process of coal and non-ferrous metal ores. The hoist is mainly used for lifting personnel, ores, coal, etc. its performance and safety reliability directly affect the production of coal and ore and the life safety of the operators. Once the accident occurs, it will inevitably lead to serious casualties and serious damage to the equipment, and the normal production of the mine is interrupted, It is known as "throat of mine" for causing significant economic losses.

The hidden brake shoe is an important factor to ensure the brake safety of the hoist. If the brake shoe clearance is too large, the brake force will be reduced, and it will not stop in time or be stationary reliably, which will cause accidents: if the brake shoe clearance is too small, the brake torque will be too large, which will lead to excessive emergency braking deceleration degree, on the one hand, the equipment will generate too large brake load, On the other hand, it will also cause harm to the person being lifted, so the hidden danger of brake shoe clearance is not to be ignored, it is an important factor related to the safety, reliability and timely action of the brake system. Therefore, it is one of the effective measures to monitor the clearance of hoist gate in real time, which is also the main purpose of the system. Through the monitoring of the brake system, the status identification and fault diagnosis of hydraulic system can be realized. Through the measurement of the working clearance of the brake, the fault identification and diagnosis of the wear degree of brake shoe, fatigue failure or fracture of disc spring can be realized, so as to determine whether the requirements of coal mine safety regulations are met. In

J. Macintyre et al. (Eds.): SPIoT 2021, LNDECT 98, pp. 226–232, 2022.
https://doi.org/10.1007/978-3-030-89511-2_28

short, the real-time monitoring system of brake shoe clearance of hoist with sound and light alarm function and high protection precision can detect the fault in time, so that the problem can be solved at the beginning, thus improving the safety and reliability of the hoist operation [1].

2   The Introduction of the Clearance Monitoring System of Hoist Brake Shoe

2.1   Working Principle of Mine Hoist Brake System

In coal mine production, the brake system of the hoist is the final means to ensure the safe operation of the hoist and to realize the normal deceleration and stop of the hoist or to perform emergency braking. The function of disc brake is: when the hoist stops normally or the working brake fails, it can quickly and meet the requirements to brake the hoist, namely, safe braking; Keep the hoist stationary when it is overhauled. In order to ensure the normal operation of the brake device, brake shoe clearance and disc spring fatigue indicator shall be set on the brake. When the brake shoe wear exceeds the specified value and when the disc shaped shell is tired or disconnected, the fault signal can be sent to input the electric protection circuit respectively, which results in that the next lifting cannot start after the lifting is finished, but the brake shoe clearance adjustment or the disc spring can be replaced before the work can be resumed.

The hidden between brake shoes refers to the distance between the brake shoe and the brake disc under the fully open state《 Article 427 of the coal mine safety regulations stipulates that the lifting equipment must be equipped with brake shoe clearance protection device, and can automatically alarm or automatically power off when the clearance exceeds the specified value< The requirements of the standard for the trial implementation of the intact mechanical and electrical equipment in coal mine include: the brake shoe and the brake wheel are in close contact, and the clearance after the sliding type is not more than 2 mm, and the upper and lower parts are equal: the angular displacement type is not more than 2.5 mm in the brake shoe center; The disc brake is not more than 2 mm. the hidden brake is an important parameter in the brake system, and it is also an important factor related to the safety, reliability and timely action of the brake system. In order to make the force of one pair of brake shoes acting on both sides of brake disc equal and synchronous, the brake clearance of each brake should be as far as possible. Moreover, the brake clearance is an active parameter, and the deflection of brake disc, wear of brake shoe, fatigue and damage degree of disc spring have an impact on it. For a long time, the brake clearance monitoring of disc brake of mine hoist has been very primitive, that is, two limit switches are used to control the brake clearance and brake skin friction. The method has a large error, and it is time-consuming and laborious to operate, and there are great safety risks. The traditional measuring tools are feeler gauge and dial gauge respectively. Stop the hoist near the cross hook position, so that the static tension of the steel wire rope on both sides of the roll is basically equal, so that the static stop of rolling after the open gate is not rotating. Close - the return oil line of some brake brakes, so that they can not open the brake and play a parking role. Open the other part of the gate, measure the clearance of each gate with feeler gauge, or fix the magnetic gauge base of dial indicator on the open brake shoe, and press the probe on the brake disc to make the indication value of 2–3 mm. Holding the brake makes the brake shoe press towards the brake disc. The indicator change value of dial gauge needle is the brake shoe clearance. Close the oil return line of the measured brake, and measure the brake shoe clearance of the other part of the brake in the same way [2].

2.2 System Control Scheme Design

According to the requirements of the standard for the trial implementation of the mechanical and electrical equipment in coal mine, the disc gate room shall not be more than 2 mm apart. Therefore, in the disc brake shoe monitoring system of mine hoist, the clearance is given to 2 mm. The clearance is detected by sensors to form closed-loop feedback, input to the controller, and the controller controls the electro-hydraulic proportional valve, adjusts the oil inlet oil pressure through the valve opening of the proportional valve, and then adjusts the brake shoe clearance through butterfly spring. The control system scheme is shown in Fig. 1:

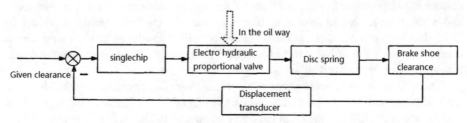

Fig. 1. Scheme diagram of control system

3   System Hardware Selection and Design

3.1   Hardware Selection

(1) Microprocessor. Microprocessor (single chip microcomputer) is the core of the whole analog gate control system, which completes signal acquisition, storage, display, communication and alarm. Its selection principle is to meet the processing requirements of the system on the premise of ensuring reliability, and to have certain flexibility and expandability. The microprocessor chip selected for this system is ATMEL's product AT89S52, which is one of the most widely used chips in 51 series single chip microcomputer. It has high performance price ratio and can be easily applied in various control fields. Moreover, many microprocessors are compatible with its pins and can be replaced in many occasions.

(2) LED display control chip. Bc7281a is an 8-bit/16 bit LED digital tube display and keyboard interface special control chip launched by bigo company. The chip can control up to 16 bit nixie tube or 128 independent LEDs through external shift registers (74HC164, 74ls595, etc.). The driving output polarity and timing of bc7281a can be controlled by software, so that it can cooperate with various external driving circuits and is suitable for any size nixie tube.

(3) Motor driver chip. Micro DC motor is widely used in small power system because of its good linear characteristics, excellent control performance and high efficiency. Considering the voltage, current level, size, appearance and other factors, L298 is used to replace the driving circuit composed of three-stage transistor.

(4) Displacement sensor. Sensor is the interface between the tested object and the detection system. It provides the system with the necessary original information for processing and decision-making, which affects the function of the system to a great extent. This design uses ST188 single speed of light reflection sampling non-contact photoelectric sensor, it is a high output infrared photoelectric diode and high sensitivity phototransistor composition, by receiving the intensity of reflected light to detect the distance from the measured object.

3.2 Control System Circuit

In this system, AT89S52, as the control core of the system, completes all the functions of detection, display and communication. The circuit diagram is shown in Fig. 2. After the system is powered on, AT89S52 first initializes the relevant peripheral devices, then circulates the detection sensor signal, displays the detection data by the nixie tube, and gives an audible and visual alarm when it detects that the gap distance is greater than the alarm value, At the same time, the command signal is sent to control the motor to rotate until the clearance distance returns to the standard range. Selecting a microcontroller with low external clock frequency can effectively reduce the noise and improve the anti-interference ability of the system [3].

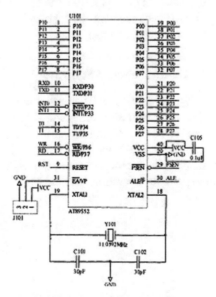

Fig. 2. Minimum system circuit diagram

4   System Software Design

4.1   Principles of Software System Design

The application software in the application system is designed according to the functional requirements of the system, and various functions of the system should be realized reliably. There are many kinds of application systems and different application software, but an excellent application system should have the following characteristics:

(1)   The software structure is clear, simple and reasonable.
(2)   Each function program is modularized and subroutined, which is easy to debug, connect, transplant and modify.
(3)   The specification of program storage area and data storage area is reasonable, which can not only save memory capacity, but also facilitate operation.
(4)   The running state realizes the symbolic management. Each function program running status, running results and running requirements are set with status flag for query. Program transfer, running and control can be controlled by status flag conditions.
(5)   After debugging, the modified program should be standardized to remove the "trace" of modification. The standardized program is convenient for communication and reference, and also lays the foundation for the future software modularization and standardization.
(6)   The overall software anti-interference design is realized. Software anti-interference is a powerful measure to improve the feasibility of computer application system.
(7)   In order to improve the reliability of operation, the self diagnosis program is set in the application software to check whether the characteristic state parameters of the system are normal before running on the system.

4.2   General Design Flow Chart of System Software

Figure 3 shows the main program flow chart. The design idea is to design the program according to the order of data transmission direction, that is, the sensor outputs the measured data, which is converted into digital signal through AD converter, and read by the microcontroller. After the relevant data processing and transformation, the data is displayed on the LED display, the alarm flag is set, the MCU receives the command signal to perform the corresponding operation, and the MCU maintains the cycle data acquisition [4].

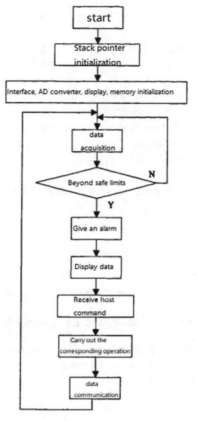

Fig. 3. Main program flow chart

4.3 Software Reliability Design

The internal arithmetic unit, controller and control register of microprocessor AT89S52 are easily disturbed. When the electromagnetic interference signal is channeled in, the CPU will execute the wrong instruction, causing the wrong operation or wrong operation result; If the information in the control register is modified by noise, it will lead to initialization error, addressing failure and even system paralysis. The experiment shows that most of the interference signals are introduced into the microprocessor by bus, and the most frequently connected with the outside world, so the most easily disturbed is the program pointer PC, which often causes fatal errors, which belongs to the object of key prevention and key correction.

The reliability design of software system plays an important role in suppressing external interference. Its function is to eliminate the voice of analog input signal (such as digital filtering technology), and make the program back on track when the program is in disorder. The main methods of reliability design of software system include: self check, software trap (program "run and fly" detection), setting program running status mark, refreshing output port, input multiple sampling, software "watchdog" and so on.

5 Conclusion

At present, hundreds of hoists are running in China's non-ferrous, black, coal, non-metallic mines. Most of these hoist brake shoe clearance monitoring systems are mechanical monitoring systems. Some mines even have no monitoring system, low automation, poor monitoring performance, weak safety protection measures, and accidents such as over rolling, Therefore, it is imperative to upgrade the monitoring system of brake shoe of mine hoist. Therefore, the promotion and application of the project will have a great impact on the progress of monitoring technology of mine brake system in China, and bring remarkable economic and social benefits.

Acknowledgements. 2020 fundamental science foundation of Heilongjiang province universities "2020-KYYWF-0518".

References

1. Wang, Q., et al.: Electric Control System of Modern Mine Hoist, p. 12. China Machine Press, Beijing (1996)
2. Editor in Chief of China Institute of Mining and Technology: Mine Hoisting Equipment. Coal Industry Press, Beijing (1980)
3. Peng, Z.: Design of Mine Hoisting Machinery. China Machine Press, Beijing (1989)
4. Hong, X., Jun, C.: Mine Transportation Improvement. China University of mining and Technology Press, Xuzhou (2005)

Design of Cooperation Platform Based on HTML5 Framework

Guiyun Chen[✉]

Changsha Medical University, Hunan 410000, China

Abstract. Through the investigation and analysis of the domestic higher education mode and enterprise talent demand, it is found that there is no effective connection scheme among universities, students and enterprises in the current market. In order to help colleges and universities transform to application-oriented and cultivate talents meeting the requirements of enterprises and institutions, it is necessary to design and develop a set of system suitable for schools, students and enterprises. The system can effectively improve the teaching quality of colleges and universities, transfer the corporate culture and ideas to teachers and students, and also help to cultivate students' good professional ethics and professional skills. By analyzing the problems existing in school enterprise cooperation, this paper expounds the design idea and function realization of school enterprise cooperation system based on HTML5 technology, which solves the problems of poor communication and lack of selection mechanism in school enterprise cooperation to a certain extent.

Keywords: School enterprise cooperation · System · HTML5

1 Introduction

The talent training mode of school enterprise cooperation is an inevitable choice for higher vocational education to strengthen specialty construction and deepen education and teaching reform. In this new mode, new changes and difficulties will inevitably arise in the management of student education. From the perspective of students' Non intelligence factors, paying attention to the psychological changes of students in school enterprise cooperation is conducive to the implementation of the new student education management strategy under the school enterprise cooperation mode, to improving the effectiveness of school enterprise cooperation, and to students' success in the enterprise environment. Because educational talents must adapt to the needs of society and market, education is a generation to cultivate successors of national undertakings and new forces. Nowadays, the teaching methods of many colleges and universities are difficult to adapt to the development of the current social situation, because most of them ignore practice and attach importance to the study of theoretical knowledge. Theory and practice of compound talents is in urgent need of our training, such talents not only have good hands-on ability, but also have a strong theoretical reserve, this is very important, can meet the needs of the unit. Although students can get some skills training through practice

J. Macintyre et al. (Eds.): SPIoT 2021, LNDECT 98, pp. 233–238, 2022.
https://doi.org/10.1007/978-3-030-89511-2_29

in school, it's just some scattered practice topics, a kind of static operating procedures, which can't be completely the same as the working environment of enterprises in reality. With the rapid development of Internet technology, the network era of education has arrived, and it is in rapid development in response to the call of the times. There are many teaching resources on the network that can be shared, providing a variety of forms of teaching activities for online network education, and realizing online communication between teachers and students, students and enterprises, teachers and enterprises through the network [1].

Through school enterprise cooperation, the school has developed, students have learned skills, and enterprises have acquired talents. In a sense, the win-win result of "resource sharing, common development, mutual benefit and complementary advantages" between the school and enterprises has been achieved. In view of this, through reading a large number of literature and investigating the mode and current situation of school enterprise cooperation of some schools and enterprises, we can understand the way of school enterprise cooperation reached by different schools and enterprises, sort out the cooperation types of different school enterprise cooperation projects, and build an online school enterprise cooperation platform to provide information release and management platform for schools and enterprises, The information of school enterprise cooperation is integrated and presented to users. With the rapid spread of network information, the school can adapt to the needs of enterprises, respond to the needs of enterprises quickly, and achieve the purpose of coordinating the resources of school enterprise cooperation, so as to promote the communication and exchange between schools and enterprises.

2 System Requirement Analysis

Although there are many attempts to cultivate excellent talents through cooperation between universities and enterprises, there are the following problems when investigating the details of the work:

(1) The stability of professional and teaching courses is relatively poor, and the demand for talents is not clear.
(2) The reorganization of teaching resources is frequent. Combined with the nature of enterprise posts, we should change the structure of teachers and the layout of venues in cooperation.
(3) The cooperation level is not deep and the level is not high, and the enterprise pursues the maximum interests.

For the project development itself, the online service platform based on school enterprise cooperation is an application platform with high business requirements. It requires developers to be familiar with the overall process, establish a correct demand analysis model, and divide the appropriate functional modules, otherwise it will bring great difficulties to the design of the whole process. On the spot investigation of both sides of the University and enterprise, the teaching and tutorial mode of the University and the management mode of the enterprise are investigated, and the detailed design scheme is formulated, and the suitable development scheme is designed.

The online service platform based on school enterprise cooperation provides a one-stop service for students and University enterprises. From the perspective of enterprises, it timely reviews the market demand, determines the enterprise employment standards, improves the students' own ability from the perspective of education, and clarifies the talent training objectives to meet the needs of enterprises [2].

After the requirements collection in the early stage of the project, according to different roles, the system requirements analysis of the project are described as follows.

Student:

(1) Be able to watch learning videos online
(2) Participate in all kinds of examinations organized by the school, check the examination results, check the recruitment information of the enterprise, and send the resume
(3) Get the interview information of the enterprise through the mobile phone or view the browsing situation of your resume
(4) Project development through intelligent data analysis.

Colleges and Universities: automatic test paper generation and examination monitoring.

Enterprise:

(1) Check the talent demand standard in the industry and upload excellent videos in the industry
(2) Release recruitment information, test and interview online
(3) Intelligent data analysis, can develop their own enterprise needs platform, through the message push notice student interview.

3 The Design Idea and Technology Selection of this System

3.1 Design Idea of the System

The design purpose of the system is to build an information transparent sharing platform, which allows schools and enterprises to freely search and publish the corresponding resources through this platform. HTML5 technology is used to realize the functions of recording users' habits and loading search results asynchronously, so as to ensure the availability of the system to the greatest extent and help both parties manage their own school enterprise cooperation projects, To promote cooperation between the two sides. The construction of the system follows the principle of openness. Both sides adhere to the principle of honesty and openness, and disclose their own resources to a certain extent. The information of the system needs to be co constructed and shared by both schools and enterprises, so as to achieve the maximum benefits of using the system, as shown in Fig. 1.

In order to meet the needs of different users, the system is mainly divided into two modules, which are search project and release project.

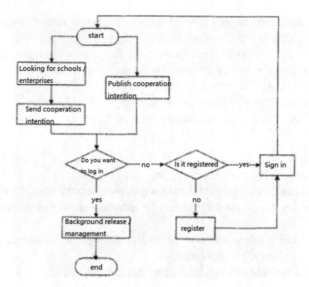

Fig. 1. Overall user flow chart of the system

The main function of the foreground system is to classify and index different projects, and set two search buttons of "find school" and "find enterprise" in the navigation. According to the differences between school projects and enterprise projects, the filtering conditions are set respectively. Through certain filtering conditions, users can search for projects that users expect to cooperate with, After login, you can use the functions of "one click to send cooperation intention" and "user-defined to send cooperation intention" to realize the function of sending intention to the project publisher [3].

The users of this system are mainly related personnel of schools and enterprises. The permissions of the two roles are roughly the same. After the authentication, they can publish school enterprise cooperation projects, manage the news trends and related contact information on their home pages, and add, delete and modify the information in the function settings, However, all the information released can only be displayed in the form of web pages through identity authentication or administrator's audit, and their respective functions are shown in Fig. 2

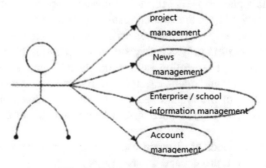

Fig. 2. System function design

3.2   System Technology Selection

The system adopts browser/server architecture, and its advantage is that it can access the system on any networked computer without installing any software. In the front-end of the system, we mainly use the latest htmls technology to realize. HTML5 mentioned in this paper is the generalized HTML5, including the sum of CSS3 and jQuery technology. In the narrow sense, HTMLs is the enhanced version of the fourth version of HTML. Compared with html4, HTML5 has great improvements, such as more descriptive tags, more user-friendly, more user-friendly, more user-friendly and more user-friendly Although W3C has not officially released the recommended version of HTML5, it claims that "HTML5 is the cornerstone of an open web platform." Based on the availability of the system in the next ten years, we decided to use HTMLs as the basic technology. The server side uses Microsoft SQL server and C #. Net for data operation and management. In order to ensure the data consistency and integrity of the whole system, we wrote some triggers for the processing of the database [4].

4   Implementation of System Function Based on HTML5

HTMLs uses local storage to realize the local storage of data in the client. Compared with cookies, HTMLs has the characteristics of large capacity, permanent storage and convenient data operation. In addition, only the current address domain can read the local data to ensure the security of the data. The local storage of HTMLs is suitable for accessing the contents only concerned by users, such as interface style, menu folding and expanding, whether to record user name and next login, etc. this kind of data does not need to be stored in the remote server, which can greatly save the communication consumption with the server. The system uses the local storage API to realize the function of recording the user login name and the usage habits of the background system. The format of data storage is "key value pair". However, for IE6 and other low version browsers, HTML5 local storage is not supported. In order to ensure the availability of the system, the system prepares the cookie back scheme, which is implemented by jQuery code as follows:

```
...
If(window .localStorage) {
localStorage. setItem(uname_str,uname);
}
else {
document.cookie(uname_ str,uname,{ expires: 10 });
}
};
$("#login_ tshi").html;
...
```

5 Conclusion

The system uses the most advanced HTMLs technology to build, and uses the latest functions of HTML5 to realize the functions of recording users' habits and searching without refreshing. While ensuring the availability and good interaction of the system, it builds a good online screening and communication platform for cooperation between schools and enterprises, which can effectively promote the cooperation between the two sides, It promotes the overall planning of school enterprise cooperation resources.

Acknowledgements. No. 811 of XiangJiaotong [2018] 436, "Exploration and Practice of the Reform of Professional Talent Training Mode of Adult Higher Education Marketing".

Construction of In-Enterprise Cooperation Mode of Non-Medical Majors from the Perspective of [2019] No. 353,19C0210.

References

1. Lee, J.: Analysis on the mechanism of school enterprise cooperation in innovation and Entrepreneurship Education. Sci. Technol. Chinese Univ. **10**, 23–24 (2016)
2. Jian, Y.: Research on enterprise applied talents training mechanism under the background of school enterprise cooperation, pp. 36–38 (2016)
3. Hogan, B.P.: HTML5 and CSS3 Case Course. People's Posts and Telecommunications Press, Beijing (2012). Translated by Li Jie
4. Zhang, Z.: Problems and Countermeasures of school enterprise cooperation. China Voc. Tech. Educ. **4**, 62–66 (2012)

Multi-objective Production Allocation Model Based on Multi-mode Technology

Xiujuan Chen[✉]

Daqing Oilfield Limited Company No. 9 Oil Production Company, Daqing 163853, China
Chenxiujuan@petrochina.com.cn

Abstract. Based on the functional simulation principle (differential simulation method and neural network method), the multi-objective production allocation optimization model of oilfield development planning is studied and established. The model successfully solves the problem of optimal allocation of the total output, the corresponding workload, cost and other development indicators to the secondary production units. Aiming at the nonlinear multi-objective programming problem, the improved Hooke Jeeves algorithm is used to solve the problem, and the model and algorithm are applied to the development planning of a domestic mid late development oilfield, which has produced good economic benefits.

Keywords: Oilfield development planning · Distribution of output · Multi objective · Optimization model

1 Introduction

Oilfield development planning refers to the corresponding measures taken to delay production decline, increase cost and maintain relatively stable production after the development of oil field enters the declining period. At present, most of the domestic oil fields (especially the Eastern oil fields) have entered the middle and later stage of development, and they are faced with technical, economic and political problems. How to maintain stable production, To meet the demand of national economic development for oil production (political factors); According to the dynamic transformation law of oil field, the development plan (technical factors) is determined; The three problems are the key problems that the oilfield development planning faces and needs to solve. So at present, all the major oilfields put the "development plan" in a very important position [1].

This paper proposes the research on the production composition of oilfield development by multi-objective programming; Based on the principle of functional simulation, the relationship among "natural production", "measure output", "new well production in old area", "new well production in New Area" and corresponding workload and other development indexes is established. The multi-objective production structure is established by using these relations. According to the concept of combination deviation degree with minimum distance from ideal point of objective function, the multi-objective production structure is established. This paper proposes an optimal harmonic algorithm for solving the nonlinear multi-objective optimization model with constraints: improved

Hooke eeves algorithm. The application of the model and algorithm to the development planning of a domestic oil field in the middle and later stage has produced good economic benefits [2].

2 Problems and Ideas

The production of oil fields is usually composed of the following four parts: natural production, measure production, new well production in old area and new well production in New Area (some oilfields may contain more production components, such as three oil recovery, etc.), and these sub production changes with corresponding cost, workload and development dynamic law, Therefore, the composition of these sub items of production should be optimized (that is, how to determine the value of production and cost and workload of each sub item, so that the whole oil field or oil production plant can not only complete the production task, but also have the lowest cost and the best benefit as possible, and also meet the development dynamic change law and the limitation of technical conditions). In essence, the cost corresponding to each sub item production should be optimized According to the actual research of a specific oil field in China, the author and the research team mainly include: the number of wells opened in the old oil field, the number of wells opened in the old oil field, the maintenance cost of natural production, the remaining recoverable reserves, the time (or the different water cut corresponding to different time), etc.; The factors influencing the production of the savings and construction include: effective wells, cost, effective period, water content, production degree, etc.; The factors influencing the production of new wells in the old area include: the number of new wells in the old area, the recoverable reserves of the new wells in the old area, the cost of the new wells in the old area, the investment in the development of the new wells in the old area, etc.; The factors of the new well production in the new area of Lhasa are mainly included; The number of new wells in the new area, the recoverable reserves of new wells, the cost of new wells in the new area, the investment in the development of new wells in the new area, etc. [3].

The three factors of establishing an optimization model are: ① determining decision variables, ② determining objectives, ③ determining constraints. The direct decision variables for the current problems are obvious, namely, natural production, measure production, new well production in New Area and new well production in old area. However, these production changes with its influencing factors most, so the real decision variables are the factors affecting these four production. Therefore, the goal can be as follows: the output is the largest, the cost is the lowest or the benefit is the best. The correlation between the four single output and its corresponding influencing factors is of course the constraint condition and the most important constraint condition. Other conditions include: the upper and lower bound of each single output, the influencing factors of each single output, such as cost, upper and lower bound constraints of workload, etc.

Generally speaking, it is more difficult to solve the extremum problem of nonlinear programming with constraints than that of linear programming without constraints; The constraint problem is transformed into an irreducible East problem; The nonlinear programming problem is transformed into linear programming problem; The multi-objective decision-making problem is transformed into a single objective decision-making problem; Hooke -Jeeves algorithm is originally an unconstrained search method.

Here we consider the appropriate improvement, taking into account the constraints. Its effectiveness is based on the fact that each priority class has multiple solutions, but the number of solutions decreases with the decrease of priority, so it is possible to search for a unique optimal solution [4].

3 Establish Optimization Model

There are many researches on the establishment of correlation between oil field production and its corresponding influencing factors, including reservoir engineering method and function simulation method. We use the correlation relationship established by the function simulation method (see (1×2) (3 differential simulation method and neural network method) to establish the multi-objective production distribution optimization model [5]. First, according to the law of historical development dynamic change, the following relationship is established by function simulation method

$$X^{(k)} = X^{(k)}\left(U_1^{(k)}, U_2^{(k)}, U_3^{(k)}, U_4^{(k)}, U_5^{(k)}, U_6^{(k)}, U_7^{(k)}, U_8^{(k)}, U_9^{(k)}, U_{10}^{(k)} \ldots \right) \quad (1)$$

k - the k-th oil production plant, $k = 1, 2, \ldots$ n has n oil production plants;

$X^{(k)}$ - the annual oil output of the kth oil production plant;

$U_1^{(k)}$ - the first factor affecting the oil production of the k-th oil production plant (produced geological reserves);

$U_2^{(k)}$ - is the second influencing factor (operation cost) of oil production of the k-th oil production plant;

$U_3^{(k)}$ - the third influencing factor of oil production of the 1kth oil production plant (total number of production wells opened);

$U_4^{(k)}$ - the fourth influencing factor (total capacity) of oil production of the k-th oil production plant;

$U_5^{(k)}$ - the fifth influencing factor (comprehensive water cut) of oil production in the k-th oil production plant;

$U_6^{(k)}$ - the sixth influencing factor of oil production of the k-th oil production plant (recoverable reserves recovery degree);

$U_7^{(k)}$ - the seventh influencing factor of oil production of the k-th oil production plant (total well times of measures);

$U_8^{(k)}$ - the eighth influencing factor of oil production (effective well number of measures) of the k-th oil production plant;

$U_9^{(k)}$ - the ninth influencing factor (total investment) of oil production of the k-th oil production plant;

$U_{10}^{(k)}$ - the tenth influencing factor of oil production of the k-th oil production plant (cumulative remaining recoverable reserves);

…

For each unit at the nianyuan level. The factors affecting the annual oil production may be the same or different, which can be analyzed and determined according to the specific situation, but this does not affect the prediction of the oil production in the planning year, because the principle is the same. The relationship between the whole oilfield production and its influencing factors can also be established similarly [6].

The output is taken as the direct decision variable (the essential decision variable is their influencing factor), the output f, the cost FZ and the investment FS are taken as the objective function. Combined with the constraints, the following multi-objective optimization model of yield composition is established

$$Max f_i = \sum_{i=1}^{n} X^{(k)} \tag{2}$$

$$X^{(k)} = X^{(k)}\left(U_1^{(k)}, U_1^{(k)}, U_1^{(k)}, U_1^{(k)}, U_1^{(k)}, U_1^{(k)} \ldots\right)$$

$$X^{(1)} + X^{(2)} + \ldots + X^{(n)} \geq A$$

$$U_2^{(1)} + U_2^{(2)} \ldots + U_2^{(n)} \leq B$$

4 Solution Algorithm

The multi-objective optimization model described above is a nonlinear optimization model with constraints. It is difficult to solve the constrained nonlinear programming extremum problem, and there is no general algorithm for nonlinear programming. In addition, the contradiction and incommensurability between multiple objectives make the solution of this kind of problem more complex. For this kind of model, an optimal harmonic algorithm is considered.

Firstly, the historical data is used to construct pairwise judgment matrix, and the weight coefficient of multiple targets can be calculated by AHP. A weighted F-norm can be used to define the minimum deviation between the objective function and the ideal point

$$d_p\left(f(u), \bar{f}\right) = \left[\sum_{j=1}^{n} w_j \left|f_j(u) - \bar{f}_j\right|^p\right]^{\frac{1}{p}} \tag{3}$$

Then the objective function of the multi-objective decision-making problem can be expressed as follows

$$Min\, d_p\left(f(u), \bar{f}\right) = Min\left[\sum_{j=1}^{n} w_j \left|f_j(u) - \bar{f}_j\right|^p\right]^{\frac{1}{p}} \tag{4}$$

The composition of each sub item production and the corresponding workload and cost composition are allocated by using this optimization result, which makes the unit cost of crude oil in 2002 be reduced by 5 yuan/T compared with the traditional method. The direct economic benefit is more than 50 million yuan after deducting financial expenses, management expenses and taxes. The optimal composition of the output, workload and cost of each sub item in the next year can be obtained by the same method. The solution algorithm of the model is an iterative algorithm, and the results are gradually approaching the optimal harmonic solution of the ideal point.

5   Conclusion

Based on the analysis of the dynamic change law of oilfield development and the correlation of development indexes, this paper establishes a multi-objective optimal production allocation model for development planning, which optimally allocates the production, cost and production of the whole oilfield to the secondary units, These optimization models are used to solve the problem of optimal allocation of the global output and the corresponding workload and cost to each secondary unit. It has been applied to Zhongyuan, Shengli and other middle and late stage oil development in China, and good economic benefits have been achieved.

According to the definition of deviation laterality of objective function value combination, the improved Hooke Jeeves algorithm is used to solve the model. Hooke Jeeves algorithm is an effective algorithm for solving nonlinear multi-objective programming problems. It combines the iterative steps of solving nonlinear programming with search techniques.

References

1. Liu, Z.: Differential simulation prediction of waterflooding oilfield. J. Southwest Pet. Univ. **15**(2), 66–70 (1993)
2. Xiao, W., Liu, Z.: Prediction of oilfield water injection performance based on neural network. Traffic Comput. **25**(2), 57–66 (1997)
3. Zhibin, L., Hui, D.: Optimization model of oil field development planning production composition and its application. Acta petrologica Sinica **25**(1), 62–65 (2004)
4. Qi, Y.F.: Method of stable production planning in middle and later stage of oilfield. Acta petrologica Sinica (1), 52–6q (1992)
5. Fan, J.: Multi objective mathematical programming method for determining the optimal number of wells in oilfield development. Acta petrologica Sinica (4), 105–108 (1996)
6. Zhang, Z.: Strategy of oilfield development system planning. J. Pet. Univ. (NAT. SCI. ED.) (2), 75–8a (1998)

Authentication and Access Control
for Data Usage in IoT

Development of E-Commerce Based on Digital Information Technology

Hexiang Huang[✉] and Zhong Chen

Fuzhou University of International Studies and Trade, Fuzhou 350202, Fujian, China

Abstract. Digital trade is a new type of trade model. Digital trade is supported by information and digital technology, deeply integrated with traditional goods and service trade, and has gradually become the mainstream growth direction of international trade. E-commerce is a combination of digital technology and commodity trade, and is an important part of digital trade. The overall development of my country's digital trade is characterized by insufficient competitiveness of high-tech cloud computing services and outstanding competitiveness of e-commerce, which is consistent with my country's current labor factor endowment and comparative advantages. This article further sorts out the problems and shortcomings of the current domestic digital trade development, and puts forward corresponding policy recommendations.

Keywords: Digital trade · E-commerce · Countermeasures

1 Preface

In the 1990s, the wave of digital communication technology application swept the world, and the digital economy developed rapidly. As an emerging trade model, digital trade became an important force in promoting economic globalization. In recent years, our country's digital trade has shown a rapid development trend, injecting new forces into the domestic economy to improve the quality and efficiency of the domestic economy, and promote the in-depth participation of small and medium-sized enterprises in internationalization. Compared with traditional trade, digital trade has new features in terms of trade methods, trade and production organization forms, and trade supervision models. Although there is still a certain gap between the scale of digital trade and traditional trade, judging from the recent rapid growth trend, digital trade has huge development potential. Digital trade is supported by information technology, enhances the circulation and sharing of economic and trade information, integrates the development of traditional goods trade and service trade, reduces trade costs, improves trade efficiency, lowers the barriers for SMEs and developing countries to participate in trade, and promotes new types of employment. However, it should also be noted that domestic digital trade is still in relatively weak competitiveness, such as low industrial added value, and weak voice in the international governance of digital trade and other issues. E-commerce is an important part of digital trade. It occupies a major share in my country's digital trade, and it is in a leading position in the world. In recent years, cross-border e-commerce has become

J. Macintyre et al. (Eds.): SPIoT 2021, LNDECT 98, pp. 247–253, 2022.
https://doi.org/10.1007/978-3-030-89511-2_31

a new growth point in the development of our country's foreign trade, with an annual growth rate of as high as 30%. It is an important starting point to gradually promote the transformation and upgrading of foreign trade. In order to maintain our country's competitive advantage in international trade, it is necessary to conduct in-depth exploration of the development path of my country's digital trade. This article aims to sort out the development trend of digital trade and e-commerce, analyze the current advantages and disadvantages of domestic digital trade, and further put forward suggestions for the development of digital trade.

2 Overview of the Development of Digital Trade and E-Commerce

2.1 Development Trend of Digital Trade and E-Commerce

In 2017, USITC made the latest definition of "digital trade". Digital trade refers to "products or services delivered through the Internet, smart phones, network-connected sensors and other related equipment", involving Internet infrastructure and networks, cloud computing services, digital Six types of digital products and services including content, e-commerce, industrial applications, and communication services. According to the "Global Service Trade Development Index Report (2018)" issued by the Institute of International Trade and Economic Cooperation of the Ministry of Commerce, digital trade is divided into three categories: digital goods trade, digital service trade and data trade. Among them, the main body of digital goods trade is e-commerce [1]. Digital trade is a broader concept than e-commerce, which includes all the production, distribution, marketing, sales or delivery of goods and services through digital means [2].

China's digital service trade grows rapidly. The "China Digital Trade and Software Export Development Report 2017" issued by the Ministry of Commerce pointed out: Currently, more than half of the world's service trade is carried out in the form of digital trade, and more than 12% of goods trade is carried out using digital technology. Among them, multinational cloud computing services are a typical representative. The international competitiveness and market share of China's cloud computing industry have developed rapidly. In 2017, the domestic online outbound travel market reached US$73 billion, doubling its size in five years. The overseas distribution market share of domestic films and games has increased significantly. In 2017, overseas movie box office revenue reached 4.253 billion yuan, and overseas game distribution reached 6.51 billion US dollars.

It can be seen from this that emerging information technologies such as cloud computing, big data, and the Internet of Things provide support conditions for the vigorous development of digital trade and promote the rapid development of digital trade. Digital trade changes the traditional trade pattern. Driven by the new generation of information and communication technologies, digital trade is showing a trend of rapid growth, and the growth rate of cross-border data trade is much higher than that of traditional trade models. The operating mode of the world economy is being reconstructed by the wave of information technology and digital technology is in the ascendant. Digital trade will have a long-term and profound impact on global economic and trade development.

2.2   Digital Trade, E-Commerce Operation Mode

Digital trade refers to trade that occurs in digital form or operations based on digital technology [3]. The characteristic of digital trade is that the international trade process uses digital information technology as a support. The application of information technology in the trade of traditional goods and services has promoted traditional trade innovation as a model to enhance economic efficiency; or online in digital form. Cross-border e-commerce is an important part of digital trade. Now take the cross-border e-commerce platform as an example to show the operation mode of digital trade. The e-commerce platform is an online trading platform as the main operating body, one end is connected to consumers, and the other end is connected to manufacturers. When consumers and production are distributed in different countries, it is difficult for consumers and producers to conduct trading activities. The emergence of e-commerce platforms provides an opportunity for trade development. The services provided by the e-commerce platform in trade activities include: 1. Information release: The e-commerce platform releases information such as commodity prices and scale, and ranks commodities through consumer feedback. 2. Payment service: Consumers pay for goods through the platform, and the platform supervises manufacturers to complete orders. 3. Other services: The platform uses information advantages to provide manufacturers with specialized market analysis services. In the case of e-commerce platform, it can be found that the innovative application of digital technology in traditional trade can promote the development of commodity transactions and extend the high value-added professional service trade. The instantaneous zero-cost transmission of information and communication technology promotes the availability of services. Digitization and commerce have promoted the optimal allocation of resources, thereby promoting the in-depth development of the global value chain [4].

The development of digital trade can reduce intermediate links, reduce international trade costs, promote trade participation of small and medium-sized enterprises, improve the efficiency of resource allocation, create new jobs, and enhance economic development [5]. Traditional trade is undergoing an important period of evolution to digital trade. Seizing the development opportunities of digital trade means grasping the initiative in the new economic era [6]. The popularity of Internet information technology is increasing, and the digital economy model based on new information technology has a solid foundation for development. Digital trade will be an important direction for the transformation and upgrading of international trade in the next phase.

3   Problems and Deficiencies

3.1   Digital Trade is at the Low End of the Value Chain

Digital trade and e-commerce have achieved rapid development. But, we should also realize that our country is at a relatively low level of competition in high-end digital service trade. This is consistent with the structure of my country's current production factors: if my country is structured, high-skilled labor factors are scarce, and low-skilled labor factors are intensive. According to the principle of H-O factor division of labor, my country's labor-intensive industries are in line with industrial advantages, and knowledge- and technology-intensive industries are at a competitive disadvantage. Judging from the performance of digital trade at this stage, developed countries such as the United States and the European Union occupy high-end digital trade positions. For example, the United States' transnational "cloud services" rank among the world's leading levels and obtain high trade added value. China wins by volume in online sales of low-end labor-intensive products. At this stage, the government attaches great importance to the development of domestic information technology and the industrial application of information technology, and through cultivating leading companies such as Huawei, Alibaba Cloud, and Tencent Cloud to drive digital trade to the high end of the industrial chain.

3.2 The Development of Domestic Telecommunications Infrastructure is Lagging Behind

Modern computer communication technology originated in the United States. Since the 1960s, the United States has promoted information technology to practical use, and in the 1990s, it has vigorously promoted the construction of the "information highway" through government forces. The mature computer technology and perfect communication infrastructure in the United States provided powerful conditions for the vigorous development of the Internet industry at the beginning of the New World. A well-developed Internet industry is the foundation for the United States to continue to lead in digital trade. China lags behind developed countries in terms of communication infrastructure. In particular, the central and western regions and rural areas generally suffer from low network density, poor access signals, and high fees. Improving communication infrastructure and establishing a complete communication network system are prerequisites for the vigorous development of Internet technology. In recent years, the government has continued to make efforts to improve communications infrastructure. Continue to introduce policies such as "speed increase and fee reduction" and "fiber access to the home" to vigorously promote the improvement and improvement of the Internet infrastructure.

3.3 Weak Voice in International Rulemaking

The development of digital trade is at an initial stage, and a mature development model and policy supervision system have not yet been formed in the world. The international regulatory rules for digital trade are still in the exploratory stage. China and developed countries in the United States and Europe have big differences in policy demands. The United States has the world's most advanced communication technology and the most competitive digital cloud output capabilities. The United States advocates a high degree of digital openness, that is, countries fully liberalize digital cross-border flow restrictions and cancel data storage localization requirements. The complete openness of the cross-border flow of data can ease the barriers to entry, circumvent regulatory barriers, and obtain broad development space for the US information service industry. The EU takes a more conservative attitude on the issue of information openness. Under the banner of "Privacy Protection", the European Union advocates the implementation of certain supervision of transnational digital providers and restricts the transnational flow of data information. The EU's policy requirements in terms of data flow, on the one hand, are concerned with domestic information network security and people's privacy; on the other hand, it also hopes that a certain degree of industrial protection will provide a more relaxed system for the cultivation and development of domestic information space.

In the field of digital trade in China, e-commerce is more internationally competitive. Through the online trading platform, domestic labor-intensive products with high quality and low price are exported to countries all over the world to obtain high additional profits on the sales side. China's policy demands for digital trade focus on countries providing convenient e-commerce customs clearance systems and open logistics systems to improve the liberalization and facilitation of customs clearance for domestic export products [7].

Based on the comparative advantages of their industries, the world's major powers put forward their own development demands in the formulation of digital trade rules. There are cooperation, conflicts and games among major countries. Relying on its super economic strength after World War II, the United States has long dominated the international economic order, and has inherent advantages in policies, rules, and regulatory systems. China has certain challenges to achieve results that are conducive to its own development in the formulation of digital trade rules.

3.4 The Domestic Intellectual Property Protection System is not Found

The degree of domestic intellectual property protection is relatively weak, which is a major obstacle to the high-quality development of digital trade. Digital trade is supported by data technology to provide cross-border professional services for goods and services. The service objects of digital service trade include knowledge-intensive industries such as finance, professional services, and R&D services. For a relatively long period of time in China, the focus of economic construction has been on the growth of scale, and there has been insufficient attention to intellectual property protection and technological innovation cultivation and the lack of necessary supporting system construction. At this stage, China's per capita disposable income has reached the international middle level. It is no longer sustainable to rely on the traditional extensive scale expansion model to develop the economy. We must attach importance to strengthening the construction of the intellectual property legal system and promoting the innovation of economic entities to maintain sustained domestic economic growth. The digital industry is a highly knowledge-intensive industry, and the development of digital trade relies on the improvement of the domestic intellectual property protection system.

4 Exploration of Development Path

4.1 Strengthen the Construction of Digital Infrastructure

On March 4, 2020, General Secretary Xi Jinping emphasized at the meeting of the Standing Committee of the Political Bureau of the Communist Party of China Central Committee to accelerate the construction of new infrastructure such as 5G networks and data centers, and accelerate new infrastructure such as 5G, big data, cloud computing, industrial Internet, and Internet of Things. The government promotes the development and application of communication infrastructure [8]. On the one hand, strengthen the construction of domestic communication networks. There is a large imbalance in the development of domestic communication networks. There is a clear gap in communication infrastructure between the more developed eastern regions and the vast central and western regions, densely populated cities and vast rural areas. By strengthening the investment and construction of communication infrastructure in the vast economically underdeveloped regions, an efficient and smooth information transmission network covering the whole country will be formed, and the advantages of the network economy will be fully utilized. The central and western regions and rural areas will be fully integrated into the tide of economic development, and the developed regions will be given a vast economic hinterland. On the other hand, strengthen the connection between the domestic network system and the international network system. The development of the digital world is inseparable from the docking and exchange of domestic and international data. Reducing the cost of cross-border data flow through international network interconnection is a necessary condition for the substantial growth of digital trade.

Fully promote the "5G" technology and take advantage of the fast data transmission speed of the "5G" technology to build an industrial IoT system. Promote the in-depth integration of traditional industries and emerging information technologies, and enhance the digitization, informatization, and intelligence level of traditional industries. For example, in the medical field, use the "5G" transmission network to develop remote diagnosis and treatment; in the agricultural field, use the advanced network communication system to develop smart agriculture, implement unmanned and intelligent management of agricultural drip irrigation and light control, and improve agricultural efficiency.

A smooth and fast communication network at home and abroad is the basic condition for fostering the competitiveness of digital trade. It is necessary to strengthen the construction of communication infrastructure to connect consumers, enterprises, and traders as a whole, realize information interconnection, and create conditions for the rapid development of digital trade.

4.2 Innovate Trade Models and Seize the Initiative to Formulate Rules

The initiative to formulate international economic and trade rules relies on strong international economic influence. The United States and other developed countries have natural advantages in economic and trade rules. How China bucks the trend and seizes the initiative in the formulation of digital trade rules is an inevitable issue for the development of digital trade.

Through the countries along the "Belt and Road" to formulate and promote digital trade rules in line with my country's development interests. The trade volume between China and the countries along the "Belt and Road" has grown rapidly. Explore diversified service methods, such as international direct mail, bonded stocking, free trade zone bonded display and sales, etc., relying on overseas warehouses, overseas marketing centers and experience store models to effectively integrate consumer product experience and after-sales service chain [9]. At the same time, my country's digital trade has certain advantages in the countries along the "Belt and Road". Through digital trade activities with countries along the route, my country's digital trade supply model can be exported and established. In the trade process, through the joint communication of the participants, appropriate digital trade supervision rules are constructed. And after improvement in practice, it will be promoted globally.

Through the domestic free trade pilot zone, innovate the digital trade supervision and regulation system. The pilot free trade zone undertakes the mission of reforming the domestic administrative supervision system and has the development advantages of pioneering trials and system innovation. Through the pilot free trade zone, innovate the supervision rules for digital cross-border flow, promote the normal and orderly flow of numbers, and strengthen supervision during and after the event. Fully integrate the actual needs of the development of the current domestic and foreign trade environment, and focus on improving the cross-border payment model and the bonded logistics operation system [10]. After the system is mature, it will be promoted nationwide.

Actively participate in negotiations and rules docking with major digital trading countries. The United States and other developed countries have accumulated rich accumulation and precipitation in the digital trade supply model and trade supervision rule system. Docking with international advanced standards can promote the convergence of domestic digital trade rules with international standards, and help realize the globalization of my country's digital trade providers. At the same time, the entry of international advanced digital trade has created a competition-promoting effect on domestic digital companies, which is conducive to domestic digital companies to enhance their competitiveness [11].

References

1. Li, G., Zhang, Q.: Thoughts on the development of digital trade in my country. Int. Econ. Coop. (01), 56–65 (2020)
2. Song, G.: Exploring a new path for the development of digital trade. Nanjing Daily **10** (A09), 23 (2019)
3. Zheng, W., Zhao, Y.: Digital trade: research on international trends and my country's development path. Int. Trade (04), 56–63 (2020)
4. Fang, Y.: Exploratory analysis of my country's digital trade development based on the OECD-WTO conceptual framework and indicator system. Cust. Econ. Trade Res. 1–13
5. Sheng, B., Gao, J.: Beyond traditional trade: the connotation, characteristics and impact of digital trade. Foreign Soc. Sci. (04), 18–32 (2020)
6. Zhang, Q., Zhou, D., Wu, S.: Research on the development situation, problems and countermeasures of my country's digital trade. Econ. Aspects (02), 106–112 (2020)
7. Zong, L., Lin, J., Wu, D.: The rise of global digital trade: the value of the times and prospects. Int. Trade (10), 58–63 (2019)
8. Zhang, X.: Symbiosis or iteration: reconsideration of cross-border e-commerce and global digital trade. Contemp. Econ. Manag. 1–15
9. Bao, H.: The status quo of the development of cross-border e-commerce in my country and relevant suggestions. Mark. Mod. (02), 49–50 (2020)
10. Niu, J., Zhang, R., Sun, F.: Discussion on the development model and strategy of cross-border e-commerce in my country. Times Financ. (24), 104–105 (2020)
11. Zhang, X., Han, Y.: Research on the energy distribution and tendency of blockchain in the e-commerce field. Int. J. Front. Eng. Technol. **1**(1), 88–97 (2019)

Big Data Clustering Algorithm Based on Computer Cloud Platform

Xiaoyun Gong[✉]

Heilongjiang International University, Harbin 150025, Heilongjiang, China

Abstract. The generation and development of big data has become a new, advanced and trending information processing technology. In the era of cloud computing, we can obtain massive and diverse user data of various types and orders of magnitude through computer network platforms. Internet information technology and mobile communication equipment are more and more widely used in people's daily life. Based on this background, how to improve the effective analysis and research of these large amounts of data is the general trend and the foundation of future development. Therefore, this paper studies the clustering algorithm of the cloud computing platform in response to this situation. This article mainly uses the data analysis method, the experimental research method and the sampling survey method to carry on the algorithm research. Experimental results show that the improved K-means recall rate and precision rate can reach 76% and 80%.

Keywords: Computer cloud platform · Big data · Clustering algorithm · Parallel computing

1 Introduction

With the increasing popularity and rapid development of the Internet, a large amount of complex and valuable information appears on all kinds of carriers in computer-readable form. The Internet has become indispensable for people to study, work and live. While obtaining information, people also face the dilemma of not being able to effectively and quickly obtain such a huge amount of information. How to finally obtain practical information from the data set has become an important direction of current research. Cloud computing provides an effective solution for processing massive amounts of data.

There are many researches on clustering algorithms for big data. Among them, scholar Xia D first introduced the types of clustering algorithms, and introduced the classic K-means algorithm and the canopy algorithm in detail. Then, combining the map reduce computing model and the spark cloud computing framework, after using the Canopy algorithm to optimize the initial value of the K-means algorithm, the parallel Canopy-K-means algorithm is introduced [1]. Some scholars believe that clustering big data in the cloud platform plays an important role in improving the efficiency of data processing. He proposed an improved fuzzy C-means clustering algorithm for large data parallel clustering methods in cloud platforms, which can be effectively applied to large-scale data processing [2]. In addition, some scholars have improved and optimized

J. Macintyre et al. (Eds.): SPIoT 2021, LNDECT 98, pp. 254–262, 2022.
https://doi.org/10.1007/978-3-030-89511-2_32

the traditional clustering mining algorithm, and implemented the parallelization of the K-means algorithm on the Hadoop cloud computing platform [3].

This article mainly studies some basic configurations of the Hadoop cloud computing platform, and at the same time studies the Spark distributed platform, and then studies the clustering algorithm in depth. From the composition of the clustering algorithm, to the improvement of the algorithm and parallel optimization. Based on theoretical learning, data experiments are carried out to detect the accuracy of the algorithm and the advantages of the improved algorithm.

2 Big Data Clustering Algorithm Based on Computer Cloud Platform

2.1 Hadoop Cloud Computing Platform

In order to solve the problem that the exponential growth of data makes the majority of Internet companies at a loss when facing massive amounts of data, Google proposed the MapReduce programming model, and developed technologies such as GFS, Bigtable, and Chubby that specifically deal with massive amounts of data, which perform when dealing with massive amounts of data [4].

Among them, HDFS is a Hadoop distributed file system. HDFS can provide high-throughput data access and is a highly fault-tolerant system designed specifically for storing large-scale data sets.

MapReduce is a programming model. The main source of this programming is the mapping and specification operations in LISP and other functional programming languages. The same MapReduce corresponds to Google's MapReduce programming model [5].

HBase is a distributed, column-based database, suitable for storing unstructured data. HBase and MapReduce need to use HDFS to store data.

In addition, Hadoop also provides a ZooKeeper subsystem to provide high-availability, distributed lock services. ZooKeeper encapsulates key error-prone services. Users can directly use interface calls to enjoy stable and efficient services. It is an open source implementation of Google Chubby [6]. As shown in the figure, ZooKeeper provides services such as unified naming, synchronization locks, leader election, queue management, state synchronization services, cluster management, and management of distributed application configuration items for the entire Hadoop platform.

After the entire Hadoop platform is successfully built, you can perform parallel programming on a specific algorithm according to the MapReduce model, and run Hadoop cloud computing applications.

Spark Distributed Platform

As a new distributed data processing platform, Spark's characteristics are mainly manifested in the following four aspects:

(1) Based on elastic distributed data set [7]. Spark uses RDD as the working set of distributed programs, forcing the use of a specific linear data flow structure in distributed programs, making full use of the advantages of distributed shared memory, and effectively improving data processing speed.

(2) Iterative calculation based on memory. The frequent data read and write of the MapReduce programming model affects the processing efficiency of the application. Spark was developed to overcome the limitations of the MapReduce cluster programming paradigm, and proposed memory-based iterative computing.

(3) Based on directed acyclic graph. The Spark platform performs task calculations in the form of directed acyclic graphs. This method is conducive to data sharing, enabling different computing tasks to access the same data set at the same time, and helping users develop complex multi-step data pipelines.

(4) Excellent fault tolerance mechanism. Spark adopts a fault-tolerant method of recording data updates. It records operations on a single execution block through RDD, and records the transformation sequence information of these RDD creation processes. Once some partition data is lost, it can be obtained by obtaining the above recorded data [8].

2.2 Clustering Algorithm

(1) Partition clustering algorithm. First, the data set needs to be divided into several piles, and then the center point of each classification is selected as the initial point, and finally the heuristic custom algorithm is used for the final data iterative analysis and calculation. Until the final algorithm termination conditions are met, the final data division is completed. The method of judging the quality of the division also has certain basis criteria, such as the error sum of squares and the weighted average square distance sum. Like K-means algorithm, K-Medoid algorithm belongs to partition clustering algorithm [9].

K-means clustering algorithm is currently the most commonly used partition-based clustering algorithm. It composes unknown object sets into multiple mutually exclusive groups or clusters, so that objects in the same group or cluster have a higher similarity, but have obvious differences with objects in other groups or clusters. As for the judgment of the similarity and dissimilarity between an object with unknown attributes and each group or cluster, it usually involves the calculation of the distance and then the division.

Assuming that the centroid can represent the group (cluster), it is denoted as v_i. The difference between the object $q \in D_i$ and the representative of the group (cluster) v_i is measured by dist(q,v_i), and dist(a,b) represents the Euclidean distance between a and b. The quality of a group (cluster) D_i can be measured by the variation within the group (cluster), which is the sum of squares of errors between all objects D_i and centroids v_i:

$$F = \sum_{i=1}^{i} \sum_{q \in D_i} \text{dist}(q, v_i)^2 \tag{1}$$

F represents the sum of squared errors of all objects (or points). q is a point in space, representing a given data object: v_i is the centroid and center of the group (cluster) (q and v_i both are multidimensional). The main goal of this formula is to find the sum of the squares of the distance between each object (or point) in each group (cluster) and the center of the group (cluster).

(2) Hierarchical clustering algorithm. The hierarchical clustering algorithm requires a tree structure to store the objects of the data set, and organizes the generation and reconstruction of the data in a tree manner [10]. The operation of hierarchical clustering is very

similar to the operation of balanced tree. The operation of balanced tree is mainly insertion and balance adjustment. Hierarchical clustering requires agglomeration operation to condense the approximate data set points together. The second is when a certain When the conditions are met, the data will be clustered and fissioned, which is equivalent to adjusting the nodes of the tree. Distance needs to be used in hierarchical clustering. There are many distance measurement methods, including maximum distance measurement, minimum distance measurement and average distance measurement methods. The algorithms of BIRCH, CURE, and Chemeleon are relatively common hierarchical clustering methods [11].

The BIRCH clustering algorithm can process more data under the same memory capacity, and it can process faster under the same data volume.

The clustering feature DE vector of a cluster is a triplet. According to a given clustering data set: M c-dimensional data points, DE is defined as follows:

$$DE = (\mathrm{m}, KW, WW) \tag{2}$$

Among them, m represents the number of data sets, KW is the linear data sum of m c-dimensional data points, and WW is the square sum of m c-dimensional data points.

The quantitative center point of the data in the cluster is:

$$y_0 = \frac{\sum_{i=1}^{m} y_i}{m} = \frac{KW}{m} \tag{3}$$

In addition, the clustering feature has an additive feature, which is one of the reasons for the fast calculation speed of the BIRCH clustering algorithm. Linear data and sum and sum of squares and sum of data are added, the formula is as (4):

$$DE = DE_1 + DE_2 = (\mathrm{m}_1 + m_2, KW_1 + KW_2, WW_1 + WW_2) \tag{4}$$

(3) Density clustering algorithm. The density clustering algorithm is based on the fact that data with a core point within the reachable range of the density belongs to a cluster, and the reachable density is a point in the neighborhood of all points in the cluster. When the density of data points in the adjacent range is greater than a given threshold, put this point in another cluster, and it must be ensured that each cluster contains at least a certain number of data points. Typical algorithms include DBSCAN algorithm and DESCAN algorithm.

(4) Grid clustering algorithm. It uses a unique grid to divide the data into grid cells according to a certain unit length. The clustering operations in the algorithm are all analyzed and executed based on these cell grids. The execution speed of the algorithm is also a major advantage of this method. The factor of the complexity of the clustering algorithm is mainly the number of cells, which is not directly related to the size of the data set. The STING algorithm is an example of a grid algorithm. The distribution of pattern data is used for grid clustering calculation and analysis. The CLIQUE proposal is a hybrid method, which mainly combines grid and density, and uses the advantages of both to gradually refine the clustering process and finally obtain the clustering result [12].

2.3 Optimization of Clustering Algorithm

(1) Data aggregation preprocessing. Faced with the compression processing of a large amount of data, this article proposes to focus a certain block of the c-dimensional data to a certain point of the data, so that a large amount of data can be simplified, and the data can even be reduced to one-fifth or less of the original.

Given a parameter point $Q(v_1, v_2, \ldots, v_c)$, for a c-dimensional data point $X(a_1, a_2, \ldots, a_c)$, the formula to gather it to a specific data point $Y(a_1, a_2, \ldots, a_c)$ is as follows:

$$A_i = v_i \left([\frac{a_i}{v_i}] + \frac{1}{2} \right), \, i = 1, 2, \ldots, c \tag{5}$$

(2) Data sorting processing. For the processing of all permutations, we adopt a recursive method. Use the K-means algorithm to cluster roughly, and then arrange all clusters according to the clusters.

2.4 Research on Parallelization of Clustering Algorithms

(1) Three basic conditions for parallel computing:

1) Parallel processor: Parallel computing must include two or more computers that implement network communication. The hardware environment and software configuration on these machines are similar.
2) The application problem must have parallelism: the application can be divided into multiple subtasks, and these subtasks have the same processing method, so that they can run in parallel without interfering with each other.
3) Parallel programming: In a distributed parallel programming environment, users only need to design parallel programs according to the parallel computing engine and business logic requirements, then run the programs, and further process to achieve the purpose of parallel solving.

(2) Parallel programming model.
Shared variables, message passing, and data parallelism are the three main ways of parallel programming.

1) Shared variables. Mainly put the data that needs to be shared in a memory that all machines can access, and then each machine must fetch data from this memory every time to ensure the unity of the data.
2) Data parallelism. The environment for parallel operation is the same, and the logic code used is also the same. The main thing is that the data used in the calculation is different.
3) Message passing programming. The communication between processes is realized by sending and receiving messages, and the collaborative operation between processes is completed by data transmission.

3   Big Data Clustering Algorithm Optimization Experiment

3.1   Experimental Environment

As an excellent distributed computing platform, Hadoop can be expanded to build a very large-scale cluster using a cheap PC, which reflects its scalability. The cluster environment used in this article is to use VMware Workstation 10 to build six virtual machines with the same configuration on a laptop as the six nodes of the cluster. This article chooses to build an HDP cluster, which is a completely open source Hadoop release version launched by Hortonworks. It uses YARN as its architecture center and retains the original Hadoop ecosystem sub-projects.

3.2   HDP Cluster Deployment

This article selects the fully open-source Ambari tool to manage and maintain HDP clusters. Compared with the initial process of manually building Hadoop clusters, HDP cluster deployment avoids the cumbersome process of manually modifying configuration files, whether in cluster configuration or cluster maintenance. The use of automated deployment to add service components not only saves a lot of work, but also avoids errors caused by manual construction and improves the accuracy of deployment.

(1) Environment configuration: modify host name, establish host mapping, modify IP, close firewall, install time synchronization.
(2) Password-free login: configure and create a key on the master node, add the key to the public key, and then overwrite the public key to other nodes.
(3) Make a local source: #mount -o loop /root/CommunityLinux7.0_86_64.iso /media/cdrom #vi/etc/yum.repos.d/CentOS-Media.repo
(4) Install JDK (all nodes): #rpm -ivh oracle-j2sdk1.7-1.7.0+update67-1.x86_64.rpm #ln -s /usr/java/jdk1.7.0_67-cloudera /usr/java /jdk #vi /etc/profile
(5) Install the database: start the service, initialize the database
(6) Install ambari-server and ambari-agent according to the local source, the installation is successful, start the service
(7) Visit the target address on the browser and install each service component in an interface.

4   Experimental Results and Analysis

4.1   Text Clustering Efficiency Test Experiment

The test data set is based on MapReduce parallel clustering processing, and then based on MapReduce's improved k-means clustering algorithm for clustering. Text vectorization and text clustering experiments were carried out on two sets of data sets on clusters constructed with 1, 3, and 6 nodes.

It can be seen from Table 1 and Table 2 that the same operation on the same data set, the more the number of cluster nodes, the less time the program runs, and the more the number of nodes increases, the faster the running time of the program decreases. In a

Table 1. Text clustering runtimes for different nodes on W0

	Vectorization processing time	Clustering time	Total running time
1	186.37	1036.48	1253.66
2	214.35	965.46	1159.68
3	135.98	465.87	632.81

multi-node cluster, when the processing data set size increases, the more the algorithm program running time consumption is reduced, the stronger the cluster computing power, indicating that the multi-node cluster is more suitable for processing large data sets. When the number of cluster nodes increases to 6, the computing power of the entire cluster is greatly enhanced.

Table 2. Text clustering runtimes for different nodes on W1

	Vectorization processing time	Clustering time	Total running time
1	252.36	1983.57	2235.79
3	286.54	1689.36	1982.33
6	98.42	990.51	1049.63

4.2 Clustering Quality Experiment

Analyze the experimental results of Sect. 4.1 clustering and count the number of texts in the final clusters. For the two sets of text vector sets W0 and W1, the precision and recall rates of the six categories are calculated according to the clustering results as shown in Fig. 1 and Fig. 2.

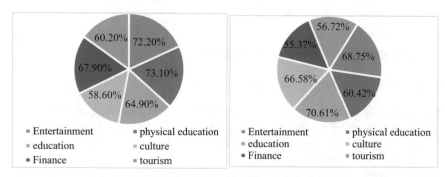

Fig. 1. The recall and precision of W0 for each clustering algorithm

From Fig. 1, we can see that the recall rate and precision rate of the data set W0 are both above 50%. And there is not much difference between the recall rate and the precision rate, and the average level remains at 66.5%.

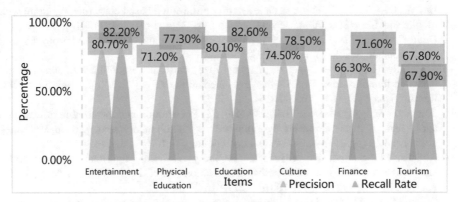

Fig. 2. The recall and precision of W1for each clustering algorithm

As shown in Fig. 2, the recall rate and precision rate of the improved k-means clustering algorithm are higher than those of the previous algorithm, and its clustering quality is better than that of the traditional k-means algorithm. When the data set size becomes larger, the recall rate and precision rate of the two algorithms decrease, which proves that the improved k-means algorithm is more stable.

5 Conclusion

Cluster analysis is a very important subject in the field of data mining. It can be used to discover the peculiar distribution of data in a data set. According to the experiments in this paper, the accuracy of the improved k-means algorithm is higher than that of the traditional algorithm. The improved k-means clustering algorithm has reduced iteration cycles, higher recall and precision. Perform text vectorization and text clustering on the same data set. The more nodes there are, the faster the program running time will be reduced, and the better the parallel performance will be. Therefore, the research of big data clustering algorithm based on computer cloud platform is still very necessary.

References

1. Xia, D., Ning, F., He, W.: Research on parallel adaptive canopy-k-means clustering algorithm for big data mining based on cloud platform. J. Grid Comput. **18**(2), 263–273 (2020). https://doi.org/10.1007/s10723-019-09504-z
2. Saghiri, A.M., Meybodi, M.R.: A distributed adaptive landmark clustering algorithm based on mOverlay and learning automata for topology mismatch problem in unstructured peer-to-peer networks. Int. J. Commun. Syst. **30**(3), e2977.1–e2977.22 (2017)
3. Begum, S., Rajesh, A.: Modified load-balanced clustering algorithm with distributed self-organization for wireless sensor networks. Int. J. Pharm. Technol. **8**(4), 23349–23356 (2016)

4. Chung, F., Simpson, O.: Computing heat kernel Pagerank and a local clustering algorithm. Eur. J. Comb. **68**(7), 96–119 (2017)
5. Bi, W., Cai, M., Liu, M., et al.: A big data clustering algorithm for mitigating the risk of customer churn. IEEE Trans. Industr. Inf. **12**(3), 1270–1281 (2016)
6. Jin, W., Cao, Y., Li, B., et al.: Particle swarm optimization based clustering algorithm with mobile sink for WSNs. Futur. Gener. Comput. Syst. **76**(nov.), 452–457 (2016)
7. Duan, Q., Yang, Y.L., Li, Y.: Rough K-modes clustering algorithm based on entropy. IAENG Int. J. Comput. Sci. **44**(1), 13–18 (2017)
8. Shang, R., Tian, P., Jiao, L., et al.: A spatial fuzzy clustering algorithm with kernel metric based on immune clone for SAR image segmentation. IEEE J. Sel. Top. Appl. Earth Obs. Remote Sens. **9**(4), 1640–1652 (2016)
9. Wang, D., Song, H., Tie, Z., Zhang, W., He, D.: Recognition and localization of occluded apples using K-means clustering algorithm and convex hull theory: a comparison. Multimed. Tools Appl. **75**(6), 3177–3198 (2015). https://doi.org/10.1007/s11042-014-2429-9
10. Anusha, M., Sathiaseelan, J.G.R.: Evolutionary clustering algorithm using criterion-knowledge-ranking for multi-objective optimization. Wirel. Pers. Commun. **94**(4), 2009–2030 (2016). https://doi.org/10.1007/s11277-016-3350-5
11. Saeed, M.: Novel linkage disequilibrium clustering algorithm identifies new lupus genes on meta-analysis of GWAS datasets. Immunogenetics **69**(5), 295–302 (2017). https://doi.org/10.1007/s00251-017-0976-8
12. Wu, W., Xiong, N., Wu, C.: Improved clustering algorithm based on energy consumption in wireless sensor networks. IET Netw. **6**(3), 47–53 (2017)

Application Analysis of Big Data Mining Based on Internet of Things Technology

Ziheng Jin[✉]

The 5th Business Division Nanjing Research Institute of Electronic Engineering, Nanjing, Jiangsu, China

Abstract. The emergence of the Internet of Things (IoT) has provided a lot of convenience for our lives. And it has become an important technical guarantee for big data mining, improving the calculation and analysis capabilities of data, and providing new directions for it. Research on the two has gradually become a hot spot in the world. This article aims to analyze the application of data mining under the IoT technology, and introduces the application of the IoT technology in big data mining from two aspects: one-dimensional data processing and image and video processing. This paper proposes the ALDCK-means algorithm, which solves the problems that the k value needs to be specified in advance, the influence of noise points and the initial clustering center cannot reach the global optimal randomly. Finally, the ALDCK-means algorithm is parallelized based on Hadoop to improve the algorithm's ability to process large-scale data, and verified the effectiveness and feasibility of the algorithm. Experimental data shows that the average clustering accuracy of K-means, LCD and ALDCK-means algorithms is 78.53%, 91.19% and 95.12%. It can be seen that the optimized ALDCK-means algorithm improves the algorithm's ability to process large-scale data and has a higher accuracy rate.

Keywords: Internet of Things · Data mining · Data processing · Data clustering

1 Introduction

Since the IoT was put forward, various fields have paid great attention to it, and great improvements have been made in the development of more than ten years. The IoT technology is used in many fields, and its scale and influence continue to expand [1, 2]. And in recent years, with the highly integrated development of big data, cloud computing, and the IoT, it has brought a lot of convenience to our lives [3, 4]. The development of data mining and IoT technology has opened up a new development space for the in-depth use of data in the database field [5, 6]. However, as far as the status quo is concerned, data mining still has certain limitations, such as the low performance of the system and the unity of mining objects. With the further deepening of research and the increasing standardization of data storage and expression, data mining based on the IoT technology has become more effective and more widely used [7, 8].

Many scholars have conducted in-depth discussions on the research of IoT and big data mining technology. For example, Rashid MM proposed that when the amount of

© The Author(s), under exclusive license to Springer Nature Switzerland AG 2022
J. Macintyre et al. (Eds.): SPIoT 2021, LNDECT 98, pp. 263–269, 2022.
https://doi.org/10.1007/978-3-030-89511-2_33

data becomes large, it is difficult for relational database systems to quickly retrieve the required data based on sensor identification or status. [9]; Liu Z believes that the method of centralized storage of perception data searches for data into a process of matching keywords [10]; Gao Y pointed out that in the IoT to identify objects, commonly used identification technologies are Sensing, recognition, and currently commonly used two-dimensional codes, etc. [11].

This article aims to analyze the application of data mining under the IoT technology, and introduces the application of the IoT technology in big data mining from two aspects: one-dimensional data processing and image and video processing. This paper proposes the ALDCK-means algorithm, which solves the problems that the k value needs to be specified in advance, the influence of noise points and the initial clustering center cannot reach the global optimal randomly. Finally, the ALDCK-means algorithm is parallelized based on Hadoop to improve the algorithm's ability to process large-scale data, and verified the effectiveness and feasibility of the algorithm.

2 Application Analysis of Big Data Mining Based on IoT Technology

2.1 Data Technology Application of the IoT

(1) One-dimensional data mining and processing
There are many forms of one-dimensional data mining and processing, such as real-time display, data fitting, alarm function, etc. [12]. The data collection is continuous, but it is not necessary to display every data when displaying, because this will display some abnormal data collected for special reasons, so the data can be fitted and processed. Display the data over a period of time as an average value, which can better show the trend of the data. Monitoring the environment is to prevent accidents and quickly deal with emergencies. Therefore, a threshold can be set for the data. When the measured data exceeds the threshold, the manager will be reminded to take the corresponding treatment, which can realize the intelligent management.

(2) Image and video data processing
The data collected by the system through cameras, medical equipment, etc. are often pictures and video data. These data can restore the condition of the monitored object. People can get the information they need by watching the pictures or videos. For example, the police can learn when the accident happened by watching the monitoring video. The circumstances and characteristics of the suspect. According to different applications, the processing methods for image and video data are also different.

(3) Application in data mining user profile
The data user portrait is a concept of "resolving into zeros". From the user's point of view, we recognize users through each label we specify. The relationship is connected. User portraits must first clarify the target and the attributes of the target. On this basis, the standards are organized through formal and non-formal methods to verify the standards. Among them, the user portrait process mainly includes the following steps:

1) Clarify the research purpose of user portraits;
2) Determine the target user based on certain characteristic attributes;
3) Randomly draw the required samples from the data set;
4) Data analysis, sorting and processing;
5) Visualization of the conclusions sought.

(4) Application of data mining in Bayesian networks

Bayesian network is a graphical model used to represent the link probability distribution of a set of variables. It provides a way to naturally represent causal information. This method is used to discover potential relationships between data. The study of Bayesian network is to find a Bayesian network model that can most truly reflect the mutual dependence of various data variables in the existing database. Use Bayesian networks to find out the potential relationships between data, the tasks that need to be completed for formal data mining. However, the main problem of Bayesian network data mining is the importance of prior knowledge. Since it is impossible for us to calculate all network structures, the special feature is that when the variables increase, the possible network structure doubles, so the network selection must be made based on existing knowledge.

2.2 K-Means Clustering Algorithm Analysis

(1) Advantages of K-means

1) K-means can handle a wide range of data, and has strong scalability. It can process text data, as well as numerical and picture data.
2) The K-means algorithm can determine whether it is a convex set, because the algorithm is more sensitive to convex sets.
3) When the K-means algorithm processes data, its scope is not limited.

(2) Disadvantages of K-means:

1) In many cases, when clustering data sets, the user initially does not know how many types of data sets should be classified, and the k value is difficult to estimate.
2) K-means algorithm is sensitive to noise and outlier data. When the data size becomes smaller, it will affect the average value of data clustering. This will lead to unstable or even incorrect results.
3) The algorithm cannot cluster clusters of arbitrary shape, and usually only finds spherical clusters with relatively uniform data.

2.3 LDCK-Means Algorithm

(1) Selection of cluster center point

In the decision diagram of the LCD algorithm, the larger local density and the larger high-density minimum distance δ can be distinguished from the normal point, which is easy to find as the cluster center point.

(2) Noise point filtering

In addition to the density extreme points having a large high-density minimum distance δ, noise points and abnormal points also have this feature. The K-means algorithm clusters the data, and LDC filters the noise by boundary density. Therefore, it is not suitable for use in the K-means method. Therefore, noise filtering is performed based on the local density ρ and the high-density minimum distance method. The specific method is to calculate the local density ρ and the minimum high density δ, and then treat the data points satisfying $\rho_i \leq \mu(\rho) - 2\sigma(\rho)$ and $\delta_i \geq \mu(\delta) + \sigma(\delta)$ as noise. Here $\mu(\rho)$ and $\sigma(\rho)$ respectively represent the mean value of ρ of all data and the standard deviation of ρ of all data.

2.4 LDC Algorithm Principle

LDC is a novel clustering algorithm that can cluster non-central point data to the nearest class. The definitions of several parameters of LDC are as follows.

(1): Local density

$$\rho_i = \sum_i x(d_{ij} - d_c)$$

(1)

$$x(i) = \begin{cases} 1, i \prec 0 \\ 0, i \geq 0 \end{cases}$$

(2)

In formulas (1) and (2), d_{ij} represents the distance between data points i to j, and d_c is the cutoff distance and is an adjustable parameter. ρ_i means the number of data points in the range of center i and the radius of d_c. The LDC algorithm is more sensitive to the value of ρ_i, so the truncation distance of d_c has a great influence on the clustering effect.

3 Application of Big Data Mining Based on the IoT Technology

3.1 Experimental Project

(1) Clustering accuracy test

For experiments on the clustering effect and accuracy after K-means optimization, this paper selects three more representative data sets, namely Aggregation, with a data volume of 750; Origin-1000 has a data volume of 1000, and Jain's data volume is 350.

(2) Clustering speedup

For computer clusters, the speedup ratio is a standard for evaluating the parallelization efficiency of algorithms and describes the execution efficiency of parallel algorithms. The calculation formula is:

$$S_p = \frac{T_1}{T_p}$$

(3)

T_1 represents the time for a single computer to execute the algorithm, and T_p represents the time for p computers to execute the algorithm in parallel.

3.2 Hadoop Experimental Environment

This experiment uses VMware Workstation Pro virtual machine to build the experimental environment. And the Hadoop cluster uses NAT for networking.

3.3 Preparation of Experimental Data Set

The experimental data to verify the feasibility of the parallel ALDCK-means algorithm are pictures of different hair colors downloaded from the Internet. These data sets consist of three colors of R, G, and B. Each picture data occupies one row, with a total of 87.1 million rows of data. Divide these data sets into 5 groups, the Dataset1 data set is the smallest, and the remaining data sets are 1.5 times the former in turn.

4 Application Data Analysis Based on Big Data Mining of Iot Technology

4.1 Clustering Accuracy of the Algorithm in Data Mining

Table 1. The accuracy of the three algorithms (unit: %)

Dataset	K-means	LCD	ALDCK-means
Aggregation	76.47	94.18	96.77
Origin-1000	85.42	90.16	93.14
Jain	73.69	89.24	95.46
Average	78.53	91.19	95.12

The clustering accuracy rates of the three algorithms are shown in Table 1: The average clustering accuracy rates of the K-means, LCD and ALDCK-means algorithms are 78.53%, 91.19% and 95.12%.

It can be seen from Fig. 1 that the accuracy of ALDCK-means clustering is the highest on different data sets, the accuracy of LDC algorithm is second, and the accuracy of K-means is the lowest. The reason for this result is that ALDCK-means optimizes the randomness of the initial clustering center selection of the K-means algorithm, the influence of noise points, and the uncertainty of the number of clusters k, meanwhile, the LDC method of processing noise points is improved. It can be seen that the clustering effect of the ALDCK-means algorithm is better.

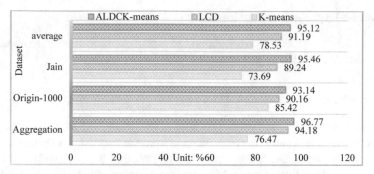

Fig. 1. The accuracy of the three algorithms

4.2 Algorithm Speedup

The speedup ratio formula (3) is used to calculate the speedup ratio of the ALDCK-means algorithm in different cluster nodes and different data sets. The final results of the calculation are shown in Table 2: when the parallel algorithm continues to increase in size of the data set, the speedup ratio of the parallel algorithm is also with the increase.

Table 2. Parallel ALDCK-means speedup ratio

Dataset	2	3	4	5	6
Dataset1	0.888	0.897	0.918	0.924	0.934
Dataset2	1.042	1.126	1.269	1.246	1.284
Dataset3	1.231	1.674	2.171	2.375	2.648
Dataset4	1.561	2.235	2.548	3.643	4.147
Dataset5	1.946	2.726	3.347	4.217	4.916

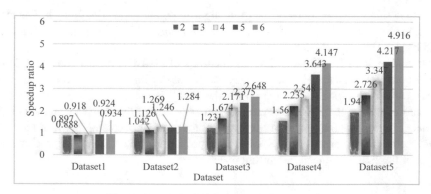

Fig. 2. Parallel ALDCK-means speedup ratio

It can be seen from Fig. 2 that the improved algorithm is executed on a data set with smaller data (Dataset 1), no matter how many nodes the cluster acceleration ratio is less than 1, that is, the time for multiple computers to execute the algorithm at the same time is less than that of a single computer to execute the algorithm. But when the scale of the computer cluster becomes larger, the acceleration ratio curve growth rate will not always be large, but will become flat.

5   Conclusion

The implementation and application of data mining technology not only brings huge benefits to enterprises, but also brings a lot of convenience to people's daily life. The development of the IoT technology has provided many technical guarantees for data mining, and the combination of the two realizes the efficient processing of large-scale data. This article aims to analyze the application of data mining under the IoT technology, and introduces the application of the IoT technology in big data mining from two aspects: one-dimensional data processing and image and video processing. This paper proposes the ALDCK-means algorithm, which solves the problems that the k value needs to be specified in advance, the influence of noise points and the initial clustering center cannot reach the global optimal randomly. Finally, the ALDCK-means algorithm is parallelized based on Hadoop to improve the algorithm's ability to process large-scale data, and verified the effectiveness and feasibility of the algorithm.

References

1. Shadroo, S., Rahmani, A.M.: Systematic survey of big data and data mining in IoT. Comput. Netw. **139**, 19–47 (2018)
2. Sattarian, M., Rezazadeh, J., Farahbakhsh, R., Bagheri, A.: Indoor navigation systems based on data mining techniques in Internet of Things: a survey. Wirel. Netw. **25**(3), 1385–1402 (2018). https://doi.org/10.1007/s11276-018-1766-4
3. Khedr, A.M., Osamy, W., Salim, A., et al.: A Novel association rule-based data mining approach for IoT based wireless sensor networks. IEEE Access **8**, 151574–151588 (2020)
4. Zhang, X., Zhu, X., Bao, W., et al.: Distributed learning on mobile devices: a new approach to data mining in the IoT. IEEE IoT J. **8**, 10264–10279 (2020)
5. Piccialli, F., Cuomo, S., Jeon, G.: Parallel approaches for data mining in the IoT realm. Int. J. Parallel Program. **46**(5), 1–5 (2018)
6. Ma, Z., Ma, J., Miao, Y., et al.: Verifiable data mining against malicious adversaries in industrial IoT. IEEE Trans. Ind. Inform. (99), 1–1 (2021)
7. Fang, Y., Chen, Q., Xiong, N.N., et al.: RGCA: a reliable GPU cluster architecture for large-scale IoT computing based on effective performance-energy optimization. Sensors **17**(8), 1799 (2017)
8. David, G., Antonio, F., Higinio, M.M., et al.: IoT: a review of surveys based on context aware intelligent services. Sensors **16**(7), 1069 (2016)
9. Rashid, M.M., Kamruzzaman, J., Hassan, M.M., et al.: A survey on behavioral pattern mining from sensor data in IoT. IEEE Access **8**, 33318–33341 (2020)
10. Liu, Z., Wang, C.: Design of traffic emergency response system based on IoT and data mining in emergencies. IEEE Access **7**(99), 113950–113962 (2019)
11. Gao, Y., Ran, L.: Collaborative filtering recommendation algorithm for heterogeneous data mining in the IoT. IEEE Access **7**, 123583–123591 (2019)
12. Zhang, S., Zhang, Y., Yin, L., et al.: Mining frequent items over the distributed hierarchical continuous weighted data streams in IoT. IEEE Access **7**, 74890–74898 (2019)

Analysis on the Application of Intelligent Data Analysis Technology of Internet of Things in E-Commerce System

Rui Lin[✉]

Department of Economics, Shanghai University, Shanghai, China

Abstract. The traditional Internet of Things model has been unable to meet the expanding needs of the current e-commerce platform, and it is urgent to upgrade to the direction of intelligence. With the continuous development of technologies such as artificial intelligence and big data analysis, IoT intelligent data analysis technologies are increasingly used in e-commerce systems to optimize and upgrade e-commerce systems. The intelligent data analysis technology of the Internet of Things enables the entire logistics system to collect and process information in real time like a human brain, filter out invalid information, and achieve the optimal layout of the entire industry, ultimately enabling all participants in the logistics system to perform high-quality, high-efficiency, and low-cost Cooperation. This article first reviews the application history of IoT intelligent data analysis technology in e-commerce systems. This development trend has gone through a period of initial contact, a period of gradual penetration, and a period of widespread application, supporting the upgrade of traditional logistics to smart logistics in the e-commerce system, and fulfilling the fundamental requirements of the e-commerce model to reduce costs and increase efficiency. Secondly, from the perspective of the supply chain, comparing the traditional e-commerce system with the e-commerce system after applying the intelligent data analysis technology of the Internet of Things, it demonstrates the positive effect of the intelligent data analysis technology of the Internet of Things. In each link of the e-commerce supply chain process, the core modules of the intelligent data analysis technology of the Internet of Things are introduced in detail. Finally, from the perspective of the specific connection between e-commerce supply chain logistics and big data systems, this article analyzes how the intelligent data analysis technology of the Internet of Things matches the various needs generated in the process of e-commerce supply chain, and realizes the overall optimal allocation of resources.

Keywords: Internet of Things · Intelligent data analysis · E-commerce · Supply chain

1 Introduction

China's e-commerce is in the golden age of development, and the development speed of the Internet of Things model that connects customers, merchants and logistics companies

© The Author(s), under exclusive license to Springer Nature Switzerland AG 2022
J. Macintyre et al. (Eds.): SPIoT 2021, LNDECT 98, pp. 270–275, 2022.
https://doi.org/10.1007/978-3-030-89511-2_34

is difficult to match the development speed of e-commerce, so this imbalance of matching poses a huge challenge to the traditional Internet of Things model [1]. At present, China's Internet of Things model cannot effectively cope with the logistics hockey effect brought by e-commerce festivals such as "Double Eleven". Under the e-commerce system, the huge transaction volume easily leads to the low operating efficiency of the Internet of Things system, high maintenance costs, and low customer service satisfaction, which seriously hinders the further application of the Internet of Things in the e-commerce system [2]. However, with the gradual maturity of the Internet of Things and cloud computing information technology, more and more e-commerce platforms invest in intelligent data analysis technology and use the Internet of Things field. This change has realized the information sharing and coordinated operation of all links in the logistics industry, improved efficiency, promoted the transformation of traditional logistics to the intelligent Internet of Things model, and better served the e-commerce system [3].

The intelligent data analysis technology of the Internet of Things refers to the use of advanced technologies such as big data processing, cloud computing, and artificial intelligence in the e-commerce logistics system. Based on the connotation of IoT intelligent data analysis, this paper sorts out the development trend of IoT intelligent data analysis technology based on big data cloud computing in e-commerce system, analyzes how to apply intelligent data analysis technology in e-commerce system, and introduces electronics The Internet of Things in the business system provides companies with functions such as data collection, processing, analysis, and prediction. These functions can help reduce business operating costs, realize effective allocation of resources, and provide better services for e-commerce systems [4].

2   The Development Trend of Intelligent Data Analysis Technology of Internet of Things in E-Commerce System

In the first stage (initial contact period), the intelligent data analysis technology of the Internet of Things comes from its own mechanical accumulation, the logistics data is low in informationization, the information among enterprises in the e-commerce platform cannot be shared, and many links of logistics do not have intelligent characteristics. In the second stage (gradual penetration period), the intelligent data analysis technology of the Internet of Things began to use barcode, RFID, GPS/GIS, EDI and other technologies for data interaction and processing. However, the ability to process and analyze a large number of e-commerce platform data is limited, and semi-structured and unstructured data cannot be identified. Unable to get real-time, comprehensive and effective logistics data of e-commerce platform, resulting in the application fault of big data cloud computing technology [5]. In the third stage (widely used period), the intelligent data analysis technology of Internet of Things has effectively realized the digitalization of e-commerce business, such as Alibaba's rookie network. In warehousing, the system has deeply excavated the order data of "200,000 × 200,000/day" in the e-commerce platform, and built a data cube based on time series. Extract the stable and credible commodity correlation degree, and realize the optimization of warehouse commodity layout [6].

3   Analyze the Advantages of Intelligent Data Analysis Technology of Internet of Things Under E-Commerce Mode

The supply chain system of e-commerce includes the whole process of goods from original factory production to sales. Each part of the traditional e-commerce supply chain system is independent of each other, and each completes its own work, and the relationship is not close. The application of intelligent data analysis technology of the Internet of Things links enterprises in all stages of the e-commerce supply chain. Specifically, the logistics of the entire e-commerce supply chain needs the help of IoT technology. In the raw material procurement stage, companies use technology to store raw material information and pass the information to manufacturing companies through some channels; manufacturing companies evaluate the quality, quantity and source of raw materials based on warehousing information and logistics information, and inspect qualified raw materials. It is retained and processed into finished products, and the relevant information is stored in the RFID system, and then transmitted to the retail enterprise; after the retail enterprise obtains the finished product, it must further check the quality of the finished product and verify the product information of qualified quality. Enter, and finally sell qualified products to consumers through the e-commerce platform. Consumers can check the raw materials, finished products, retailers, logistics and other information in the entire process on the e-commerce platform to ensure that they purchase suitable and high-quality products [7].

Specifically, the core modules of the IoT intelligent data system under the e-commerce model mainly include three sections, namely, the communication layer, the isolation layer, cloud computing, and big data. Figure 1 describes the composition of the core module in detail. First, the communication layer aggregates a large amount of transaction data on the entire e-commerce platform, which usually does not have a regular pattern, and then the communication layer transmits these data information to the system. The usual ways of transmitting information are microwave communication and satellite communication, because these two methods have fast transmission speed and huge traffic. Secondly, the isolation layer in the core module is very important, and the isolation layer ensures data security. There are two commonly used isolation layers. One is to isolate the network. Its working principle is to set up a firewall between the external network of the communication layer. In this way, harmful attacks can be resisted; One is data isolation, which uses data encryption technologies such as node encryption, endpoint encryption, and link encryption, as well as data authentication technology. These two points can ensure that the data in the e-commerce platform is not illegally obtained. Third, the daily trading volume of e-commerce platforms is huge, with massive amounts of data. Storing these data requires certain infrastructure and procedures, and the emergence of cloud savings has just solved this problem. A large number of distributed computers have implemented large-capacity data storage tasks, and then use cloud computing analysis technology to identify the importance of the data, and then sort them from high to low. Complicated chaotic information is sorted out, useless information is eliminated, and effective information is retained. Finally, the key to big data processing lies in the programmer's use of artificial intelligence and programming languages. Programmers usually conduct secondary data mining on data sources and extract key information through multiple screenings. Then, the collected key information is processed through deep learning and natural language processing. Through the above methods, the intelligent processing of the Internet of Things data on the e-commerce platform is realized, and the optimal production decision is obtained. The R&D personnel transfer the communication layer of the optimal resource allocation method to each platform under the intelligent logistics mode to help each platform realize the improvement of sales efficiency [8].

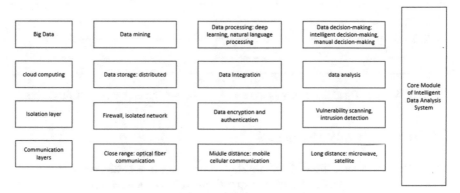

Fig. 1. Core framework of intelligent data analysis

4 How to Match the Demand of E-Commerce Supply Chain Process with Intelligent Internet of Things

From the specific link between E-commerce supply chain logistics and big data system, in the process of continuous transfer of e-commerce supply chain logistics, logistics related demand will be generated, as show in Fig. 2. First of all, in the process of purchasing raw materials, purchasers can input relevant raw material information into big data system, The big data system stores a large number of goods storage information and related business information. Through intelligent analysis and decision-making of Internet of things data, the optimal supplier and optimal storage information of relevant raw materials can be automatically fed back [9]; Secondly, in the process of transportation, suppliers, manufacturers and retailers have demand for warehousing, loading and unloading, means of transport, freight drivers, etc. Input relevant requirements into the big data system, and intelligently analyze the data of the Internet of Things, so as to make intelligent decisions on the supply of related human and material resources, and feedback the optimal transportation tools, driver candidates, warehousing and loading and unloading choices [10]; Thirdly, in the process of sorting and distribution, it links with the big data platform according to the RFID technology information embedded in the goods. After comprehensive analysis of factors such as transportation route, commodity size and precious degree, intelligent sorting is carried out, and then the optimal distribution route is intelligently planned according to commodity distribution information and stock resource information, so that the whole distribution process is efficient and timely, and the optimal allocation of resources is realized as a whole [11].

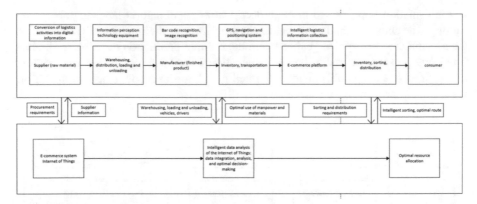

Fig. 2. Smart IoT and the matching of demand in the e-commerce supply chain process

5   Conclusions and Prospects

In the information age, the field of big data processing is a battleground for military strategists and the core of corporate competitiveness. E-commerce platforms also need intelligent data processing technology to enhance their core competitiveness. Intelligent data analysis technology can help e-commerce companies realize information sharing, and on this basis, integrate the supply chain and realize the coordinated development of upstream and downstream companies. At the same time, because of the good matching of the needs and supply of upstream and downstream companies, it improves resource availability. Usage efficiency. The above methods are conducive to the advance planning of logistics. Therefore, the use of intelligent data analysis technology not only brings a comprehensive update and upgrade to the logistics model under the e-commerce system, but also brings a product that combines new technology with the times. Of course, smart logistics under the e-commerce system also requires continuous innovation of high-end technologies such as cloud computing, big data, and artificial intelligence to meet the ever-changing e-commerce logistics needs and customer needs, and to promote the further development of smart logistics. At present, the application of intelligent data analysis technology of the Internet of Things in e-commerce systems is still restricted by some factors, such as social environment, high-end technology and other factors. On this basis, this article puts forward several suggestions for its future development direction:

5.1   Internet of Things Application Upgrade

In the future, with the further development of the Internet of Things technology, the identification of objects and people will be more accurate, which can be widely used in warehousing, transportation and distribution [12]. First of all, the Internet of Things can capture all the States of people, equipment, orders, inventory, etc. in warehousing, and realize warehousing optimization and manpower sorting optimization through big data analysis; Secondly, Through the dynamic identification of people and things in logistics and transportation, all the information in the process of logistics and transportation can be collected to better allocate responsibilities; Thirdly, match the goods with the delivery personnel, and realize the tracking of the whole logistics industry chain of goods.

5.2 Application of Blockchain Technology

The essence of blockchain is a traceable ledger for distributed data storage, which has strong advantages in data storage, transmission and confidentiality. It can be combined with intelligent logistics platform to optimize information transmission and ensure data security. Based on the anonymity of blockchain, each enterprise in the intelligent logistics platform is regarded as a single node. In the application of blockchain, information exchange between enterprises protects the private information of enterprises [13]. Based on the characteristics of data traceability, the data information transmitted by each node is traceable, which is conducive to preventing data loss and establishing responsibility allocation mechanism. Based on asymmetric encryption technology, the security of data transmission process can be well guaranteed.

References

1. Li, J., Xie, Q.: Talk about how to alleviate the hockey stick effect of "Double Eleven" logistics. Pract. Foreign Econ. Relat. Trade (4), 89–91 (2014)
2. Zhang, J., Chen, Y.: Summary of research on logistics "last mile." China Circulation Econ. **29**(4), 23–32 (2015)
3. Huang, H., Wu, Z.: The awareness of logistics information security in the era of big data needs to be strengthened urgently. Logist. Technol. (22), 18–23 (2013)
4. Dai, D.: Talk about Internet of Things and intelligent logistics again. China Logist. Purch. (23), 36–38 (2010)
5. He, L.: China's smart logistics development trend. China's Circulation Econ. (6), 3–7 (2017)
6. Wang, X.: Big Data and Smart Logistics, pp. 16–17. Beijing Jiaotong University Press, Tsinghua University Press, Beijing (2016)
7. Manyika, J., Chui, M., Brown, B., et al.: Big Data: The Next Frontier for Innovation, Competition, and Productivity. McKinsey Global Institute, Chicago (2011)
8. Zhang, W., SEAO: Research on big data feature mining technology in cloud computing environment. Mod. Electron. Technol. **41**(20), 161–164 (2018)
9. Sun, B., Wang, D.: "The Belt and Road Initiative" construction of big data system in logistics center. China Circulation Econ. **31**(8), 32–40 (2017)
10. Tao, J., Pan, L., Chu, Y.P.: Research on reconstruction of urban and rural logistics network in big data era. China Circulation Econ. **30**(11), 22–32 (2016)
11. Fu, P.: Intelligent logistics model construction based on big data. Logist. Technol. **37**(1), 135–139 (2018)
12. Zhang, X., Yuan, Z.: Research on China's "smart cloud logistics" platform system and collaborative operation mode in the age of online shopping. China Sci. Technol. Forum (7), 99–104 (2013)
13. Susan, W., Wang, H., Qin, X., Zhou, X.: Architecture of big data: challenges, present situation and prospects. Chin. J. Comput. **34**(10), 1741–1752 (2011)

Application of Image Processing Technology in Target Recognition by UAV

Donghao Cui[1], Lifu Chen[2], and Chaomin Ou[1(✉)]

[1] School of Systems Engineering, National University of Defense Technology, Changsha 410073, Hunan, China
[2] PLA Army Academy of Artillery and Air Defense, Nanjing 211100, Jiangsu, China

Abstract. With the continuous development of electronic and information technology, application based scenarios based on image processing and machine vision have been universally seen in all the fields. With the help of image and visual recognition technology, humans have solved many problems. As a key element of national security and crisis management, the capability of military unmanned aerial vehicles (UAVs) in target identification have a critical impact on the performance of their functions. Starting from the requirements on target recognition of military unmanned aerial vehicles (UAVs), this paper conducts discussions and researches on the application of image processing technology in the field, conceiving the application of image processing technology in UAVs field, hoping to provide some foundation for the research and application in related fields.

Keywords: Image processing · Target identification · Unmanned aerial vehicle (UAV)

1 Introduction

With the development of electronic information and computing science, human life has undoubtedly entered an era of information and intelligence. As a typical information processing and computing technology, image processing has become an important basic link of machine vision. In the civil field, many other goal-based algorithms have been applied in the fields of automatic driving, industrial protection, robotics, smart transportation, biomedicine and so on [1]. In the field of public security and military, especially in the field of battlefield reconnaissance, surface-to-air strike and crisis management, the use of UAV carrying target identification system can accurately identify the target and closely track the target to ensure the successful completion of missions in detection, guard, defense and attack. Based on the above factors, it is particularly necessary to study the UAV target recognition and image processing technology. This article is designed with this field as the starting point, and the application is taken in to consideration.

2 Image Processing Technology and Commonly Used Target Recognition Algorithms

Image processing technology refers to the use of a variety of numerical calculation methods for processing of the digitized image, so that it can match many algorithms

J. Macintyre et al. (Eds.): SPIoT 2021, LNDECT 98, pp. 276–283, 2022.
https://doi.org/10.1007/978-3-030-89511-2_35

to achieve special purposes. The target recognition algorithm is a special application based on a variety of image processing technologies combined with specific recognition algorithms [1].

2.1 Mainstream Image Processing Methods

2.1.1 Image Graying Processing

Image grayscale, in essence, is to convert the color picture to black and white picture, the color of which is converted to gray value for performance. Grayscale is essentially pixel brightness, and its value varies between [0, 255]. On the one hand, this method can significantly reduce the size of the image, and on the other hand, it can retain the features of the image to some extent [2]. Assuming that the image is an RGB image, and each pixel can be represented as Rx(I,j), Gx(I,j), and Bx(I,j), then the gray value can be represented as gx(i,j). Commonly seen grayscale algorithms include:

(1) The mean value method

In essence, the mean value of the three channels of the pixel point is taken as the gray value of the pixel point. The algorithm is made as:

$$g_x(i, j) = \left[R_x(i, j) + G_x(i, j) + B_x(i, j)\right]/3$$

(2) Maximum method

The highest value in the three-channel pixel is taken as the gray value. It can be shown as:

$$g_x(i, j) = MAX[R_x(i, j) + G_x(i, j) + B_x(i, j)]$$

(3) Weighted average method

The weighted average based rule means to carry out the weighted calculation on the three channels to finally get the mean value, which is used as the gray value of the pixel. It can be shown as:

$$g_x(i, j) = \left[W_R R_x(i, j) + W_G G_x(i, j) + W_B B_x(i, j)\right]/3$$

Where, W_R, W_G and W_B are the weights of channels of R, G and B respectively.

2.1.2 Image Denoising

In terms of image filtering, denoising is essentially for smooth filtering, the commonly seen filtering covers mean filtering and median filtering. Average filtering is a linear filtering algorithm. The core idea is to take the mean value of the neighborhood of the noise reduction area as the filtering basis to remove individual sharp pixels in the image and achieve smoothing processing. This method is simple in calculation and efficient in algorithm, but its effect may lead to the disappearance of key information and loss of clarity due to the large target area [3].

Median filtering is achieved based on the median value of the image pixel in the target template region. This filtering algorithm is mainly used to weaken excessive or too dark pixel points.

2.1.3 Binarization Processing of Image

binarization processing, in effect, converts each pixel of the image to 0 or 1. The commonly usded processing method is to adopt gray processing for the image at first, and then set a threshold value for the grayscale value. Based on this threshold, the binary processing is carried out. Therefore, the core of the method is to select the appropriate threshold. The processed image will hide a lot of details, which only highlight some subject targets [4]. The visual effect of binarization can be seen in Fig. 1.

Fig. 1. Image binarization

2.1.4 Image Enhancement

The essence of image enhancement is to improve the contrast of the image to distinguish the main body of the image and the background in a significant manner. Common image enhancement algorithms include histogram equalization and wavelet transform. The former is the most commonly used image enhancement algorithm with simple implementation process and low computational complexity, which can effectively realize image enhancement processing. Based on histogram equalization, there are many optimization algorithms, such as double histogram or regional histogram enhancement, which can meet different demands of image enhancement [4].

2.2 Target Recognition Technology and Algorithm

2.2.1 Common Target Identification Technologies and Processes

Target recognition algorithm can be divided into visible light, infrared and laser according to the acquisition method and source of target image recognition [5].

Image target recognition based on visible light is essentially achieved on the basis of photoelectric conversion technology to record the visible light image for gray processing and processing in noise reduction. From the perspective of processing methods and characteristics, there is little difference between visible light target recognition technologies in civil and military scenes, including automatic material and information recognition, face recognition, object recognition and so on.

Target recognition based on infrared spectral imaging can perform capture based on the infrared wave emitted by different objects, which can detect and identify targets in the case that visible light is not easy to capture. On some airborne reconnaissance equipment, infrared spectral imaging can be effectively adopted to find some camouflage targets, providing auxiliary support for battlefield reconnaissance and accurate strike.

Laser-based target recognition is achieved for construction of the target shape and even the internal structure based on the reflection of the laser, as a result, its accuracy and penetration is stronger. This method is one of the most accurate detection methods at present. In the military field, target recognition based on laser is capable of finishing the followings such as the detection of surface or underground fortifications, the capture and lock of hidden targets, and so on.

The essence of target recognition is to find the target in the image and keep continuous tracking the target. Therefore, target identification and tracking are usually carried out based on the following process (see Fig. 2):

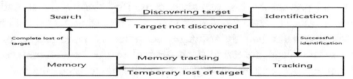

Fig. 2. Target recognition process

2.2.2 Target Recognition and Tracking Algorithm

Based on the above process, it can be found that recognition and positioning are the core of target recognition and tracking algorithm. As the target positioning is conducted, the geometric coordinates of the target in the graph are calculated by using the closed outer contour of the target. In this process, common algorithms include Canny edge detection and Hough transform [6].

The former one is mainly based on the color for the transition distinction, and then the descending speed of gray value is adopted to judge whether the boundary point is reached, which can be marked. Later on, the adjacent boundary points are linked to form the edge image of the target. In order to realize the above process, Gaussian filtering is firstly used for image denoising. Then, based on the grayscale value of the image, the partial derivative of the image in all directions is processed to obtain the gradient amplitude of each pixel. Then edge information is obtained based on non-maximum suppression. Finally, the false edge caused by noise or other factors is eliminated by setting a threshold value to realize target recognition.

The latter one is essentially an image information detection algorithm. The image is transformed to the parameter space, in which the parameter conditions conforming to the boundary point are defined, and the accumulator is used to accumulate the parameters one by one until the peak value is approached. At this time, the detection target is obtained, which is, in essence, based on the physical form of the object to carry out parameterized evaluation to infer the position of the target.

The two algorithms above mentioned enjoy their own advantages over each other, the former one is capable of quickly extracting the edge, while the latter one can provide higher anti-jamming performance for target detection.

3   Application Research of Target Recognition of Military UAV

3.1   Overview of Application of Military UAV in Target Recognition

In practical application, military UAV is mainly used to identify armed targets on the ground, including armored vehicles, personnel, armed facilities and so on. The battlefield information is changeable, the environment is complex, and the enemy frequently adopts a variety of camouflage, jamming and other means to prevent the detection and attack of UAV. Therefore, it is particularly necessary to study the target recognition technology based on vision for UAV.The target identification of military UAV should be optimized from the following aspects. The first is image acquisition. By improving the resolution of the visual acquisition camera, it can effectively capture the details of the investigation environment and prevent the loss of targets caused by clarity. The second is image transmission. Advanced network and data transmission technology should be used to ensure the integrity and reliability of image data transmission to people for decision-making. The third is image processing, the battlefield environment is optimized based on the current image processing technology. The fourth is target recognition, optimization in target tracking is conducted in the battlefield environment and scene switching, and the algorithm for fast search after lost can be improved to avoid the temporary target lost for a long time due to the shelter.

3.2   Framework of Target Recognition System

The framework of target recognition system is mainly divided into four modules, including image acquisition, image processing and recognition, image transmission and command control. This is shown in Fig. 3.

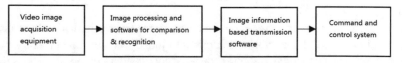

Fig. 3. Framework of target recognition system

The image acquisition module selects visible light and infrared spectrum for visual acquisition. In terms of image processing and recognition, target recognition algorithm and reference database are provided [7]. On the one hand, target identification and control are carried out, and on the other hand, real-time tracking is carried out according to the control results. The information transmission module provides high level encryption and transmission, which is equipped with a server for data processing and computing services. The command system focuses on the operational control of UAV, including strike command, tracking command and target change command.

3.3 Target Identification Process of UAV

According to the previous analysis of the logic and process of target identification, the UAV target in the battlefield environment can be designed for other processes. See Fig. 4 for details.

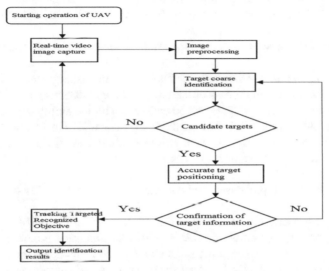

Fig. 4. UAV target recognition process

The first is visual image acquisition. In the process of high-speed movement, a high-resolution camera is used to extensively record the battlefield environment and undersurface images, and the collected information is transmitted frame by frame to the image recognition module for processing [8]. The image recognition module will prepro-cess the image first, including gray processing, noise reduction, binarization processing and other ways to provide material for the subsequent target recognition module. The target recognition stage is divided into two steps. Firstly, rough recognition is carried out. Canny and Hough transform algorithm is used to capture the target edge, and the interference edge information is supplemented, and finally the closed edge information is extracted. After that, the existence of candidate targets is confirmed by comparing and identifying databases [9]. If selected, the target position will be tracked and positioned to lock the target and output the result.

3.4 Application Value and Prospect of UAV Target Identification

UAV target identification technology is one of the typical development directions in the current military UAV field. Efforts in the direction of target identification can effectively improve the individual and comprehensive capabilities of UAV in the battlefield, which is of great help to guarantee China's homeland security and enhance the ability of crisis management [10]. In particular, its value can at least be summarized as follows:

(1)   Improving the operational capability of unmanned aerial vehicles

UAVs with powerful target recognition capabilities can more efficiently identify and track targets in complex battlefield environments for targeted and precised strike. The integration of this technology can achieved the transformation of the UAV from the traditional single control source to the multi-dimensional automatic intelligent control level, so that its single combat and reconnaissance capabilities can be further improved.

(2)   Strengthening the strike ability of the armored forces in the battlefield of the UAV

The target recognition technology itself is defected in misjudgment of risks, but combined with the comparison process of the back-end target database, the UAV equipped with the target recognition module can be endowed with the ability to identify the current common military facilities. Among them, the edge and visual features of armored vehicles are the most typical. Therefore, this technology can effectively strengthen the UAV's ability to attack the armored forces.

(3)   Enhancing the comprehensive and coordinated combat capability of the PLA

As well as carrying out combat missions, drones are often used for aerial reconnaissance. Improving the ability of UAVs to identify and track ground targets can be beneficial for other units in providing accurate movements of enemy forces and targets. On the one hand, it can effectively reduce the intensity of ground reconnaissance troops, and on the other hand, it can make up for the deficiency of ground reconnaissance and build a three-dimensional battlefield environment detection system to provide support for the comprehensive and cooperative operations of troops.

4   Conclusion

As the basis of the target recognition algorithm, the selection of various image processing technologies plays a decisive role in the application effect of the target recognition algorithm. The application of target recognition algorithm and related technologies in the field of UAV can significantly improve the comprehensive cooperative combat capability of the UAV, which is of significant significance for safeguarding national territorial security and improving the ability of crisis management BY integrating algorithm design and optimization based on the working scenes and environmental characteristics of military UAV.

Acknowledgements. Special appreciation goes to the National Natural Science Foundation of China (NSFC) project 2018, Dynamic Evolution and Regulation Strategies Based on Multi-level Network Terrorist Cooperation, Project No.: 71774168, for the support of this paper.

References

1. Min, F.: Research on target recognition technology of SLR system based on image processing. Institute of Seismology, China Earthquake Administration (2019)

2. Zhu, S.: Research on object recognition method of UAV remote sensing image based on deep learning. Beijing University of Civil Engineering and Architecture (2018)
3. Li, P.: Research on Target Recognition and Tracking Technology for Moving Vehicle. North University of China (2016)
4. Lu, J.: Application of image processing technology in license plate recognition. Digital Space **12**, 114 (2018)
5. Yu, X.: Application of digital image processing technology in agricultural weed identification. Qiqihar University, Heilongjiang Province, 02 September 2018
6. Szpak, Z.L., Tapamo, J.R.: Maritime surveillance: tracking ships inside a dynamic background using a fast level-set. Expert Syst. Appl. **38**, 6669–6680 (2017)
7. Sakhamuri, S., Sri, K.V.: An overview on prediction of plant leaves disease using image processing techniques. IOP Conf. Ser. Mater. Sci. Eng. **981**(2), 022024 (2020)
8. Liu, Q., Peng, H., Chen, J., Yang, S.: Face detection based on open Cl design and image processing technology. Microprocess. Microsyst. **81**, 103745 (2021). (prepublish)
9. Xi, J.: Research on Target Recognition and Grabbing Control of Intelligent Wheeled Vehicle with Manipulator. Tianjin University of Science and Technology (2015)
10. Kunlin, Y.: Research on UAV target recognition technology based on computer vision. J. Changsha Aeronaut. Vocat. Tech. Coll. **18**(04), 47–50 (2018)

The Application of Big Data Technology in University Library Management Service

Jianan Zhong(✉)

Hezhou University, Hezhou, Guangxi, China

Abstract. At present, the amount of information stored in university libraries is increasing, and readers' needs for information and document types are becoming more and more detailed, personalized and complex. In order to better manage university libraries and provide readers with better services, it is necessary to vigorously develop personalized information services. This is also a service currently being carried out by various university libraries in our country. Based on this, the purpose of this article is to study the application of big data technology in university library management services. This article first summarizes the basic theory of big data, and then extends its core technologies such as data mining. Combining with the current situation of university library management in our country, analyze the existing problems and shortcomings. On this basis, data mining and other technologies are used for application research. This article systematically expounds the composition and design of the library management service system based on data mining and other technologies, and uses comparative analysis and observation methods to study the subject of this article. Experimental research shows that compared with the traditional university library management service system, the university library management service system based on data mining technology studied in this paper is 5.9% higher in the new book recommendation module, 11.4% higher in the resource link module, and course reference book recommendation. The module is 4.8% higher, and the recommended classic book module is 5.8% higher. Fully reflects the feasibility of the subject of this article.

Keywords: Data mining technology · University library · Management service · Applied research

1 Introduction

With the continuous development of big data technology, all walks of life are experiencing a wave of digitization. As an information resource center, the library also follows the trend of informatization and moves forward. In the process of informatization, the library has accumulated many digital information resources [1, 2]. Among so many information resources, how can readers find the information they need conveniently and quickly is a big problem facing libraries at present [3, 4].

In recent years, economically developed countries such as the United States, Britain, France, and Japan have attached great importance to the construction and development of digital libraries, and have invested heavily in research on this project [5, 6]. In terms

© The Author(s), under exclusive license to Springer Nature Switzerland AG 2022
J. Macintyre et al. (Eds.): SPIoT 2021, LNDECT 98, pp. 284–291, 2022.
https://doi.org/10.1007/978-3-030-89511-2_36

of personalized customized services for digital libraries, these four countries are also at the forefront [7, 8]. For example, the My Library@LANL system developed by the Los Angeles International Research Laboratory and the My Library system of the University of Toronto Library were developed earlier, with mature structures, and promoted in schools and communities, and achieved very good results [9, 10].

The purpose of this article is to improve the library's abilities, and to use data mining and other technologies for its application research. The system structure, data processing, mining implementation algorithm and application of mining results are elaborated in detail. And combined with the actual work, it has a greater guiding significance for the actual work.

2 The Application of University Library Management Service Based on Data Mining Technology

2.1 Data Mining Technology

Through data mining, interesting knowledge and laws hidden in massive data can be discovered from the database. These laws or knowledge can be used in business areas such as guiding decision-making, process control, sales promotion, medical diagnosis, etc. The data mining system can also It is convenient to browse and store these laws and knowledge, to facilitate our research and study the regularity from them [11, 12].

2.2 Analysis of Existing Problems in Management Services of University Libraries

(1) Did not actively explore the needs of readers

Digital services are based on the user's personal characteristics and needs as the premise to screen dynamic information to provide digital services. At present, the existing digital services at home and abroad mainly stay on the initiative of users, provide demand information according to users' requirements, and then determine the goals of digital services. This method is difficult to intelligently and actively obtain user information and dynamically track changes in user needs. There is still a long way to go for quantitative analysis and accurate description of user characteristics.

(2) Not fully considering user needs

Existing libraries do not provide services based on personal interest information. Due to the lack of pertinence, there are still a lot of inappropriate content in the information provided to users. In addition to the lack of maturity in information mining technology, the lack of consideration from the user's point of view is also one of the important reasons why the existing library digital service system does not meet the digital needs of users.

2.3   Design of University Library Management System Based on Data Mining

(1) Analysis of ease of operation

The main groups of university libraries are divided into two categories, one is teachers; the other is students. The knowledge and skills of the two groups are different, which determines their different feelings during operation. The users of the copy are ordinary readers. For them, they like simple and quick operation. Therefore, the various requirements of the reader should be considered when designing, that is, the reader's operation is convenient.

(2) System function analysis

1) Collect user information behavior record module

The recording module mainly includes two parts: book information collection module and reader information collection module. The book information collection module mainly provides comprehensive book information for later data processing, including the title of the book, call number, Chinese library classification number, author, publication, etc., and can be provided to readers during the borrowing and recommendation stage detailed book information for reference. The reader information collection module mainly collects the reader's personal registration information and other attribute information that can fully describe the reader's personality.

2) User model analysis

The main function of this module is to generate a user model, that is, according to the analysis of the above-mentioned collected information, the readers are divided according to its implicit rules, and the readers' borrowing behavior patterns are found according to the division situation.

3) Recommended modules

This article uses recommendation based on association rules, because there is a large amount of borrowing data in the library, and this type of data has only one characteristic, that is, which readers have borrowed which books, and there are only borrowing and borrowing between books and readers. No, there is no other form of content.

2.4   Data Association Analysis Based on Books

The relevance analysis of books is mainly to perform relevance analysis on the books that readers have borrowed. The frequency of a given data set can be determined by the degree of support, and the degree of confidence is used to determine how frequently y appears in a transaction that contains x. The form of the two measures of support (S) and confidence (C) is defined in the following formulas (1) and (2).

$$S(x \rightarrow y) = \frac{\sigma(x \cup y)}{n} \tag{1}$$

$$C(x \rightarrow y) = \frac{\sigma(x \cup y)}{n} \tag{2}$$

2.5   Implementation Analysis of Digital Service Data Mining

(1) Selection of data sources for library data mining

In the university digital library system, not only many book resources are stored for readers to use, but also a large amount of book circulation data is generated every day, that is, reader borrowing records. These data are used to record the reader's borrowing situation, in general. The download is only used to make some regular simple business statistics, and the unknown, valuable, and useful knowledge hidden behind these data is not known, and the discovery of this knowledge can help the library to a large extent. Make decisions to increase the reader's attendance rate, and at the same time, develop targeted services after understanding the reader's borrowing behavior.

(2) Processing of borrowing data

Data is the foundation of data mining and analysis. If there is no data, there is no source of mining and analysis. Because the data we obtain does not necessarily meet the conditions and format of the mining tool processing, this requires us to clean, transform, and organize the data so that the subsequent work can proceed normally. In order to be able to fully understand the reader's borrowing situation, and to proceed smoothly in the future work, three data sets need to be extracted from the library management system: readers' basic information, borrowing history information, and collection information. Among them, the most common data collection of collection information includes: title, collection number, call number, collection location, storage time, etc. These information libraries have complete record formats, we only need to extract them, but in order to be able to connect with the information in other tables and prepare for future data mining, some attributes need to be generalized before they can be used.

(3) Data cleaning

There are high requirements on the correctness and reliability of the data in each part of the data table, because the data is a database backup file, and its storage form is suffixed to the. Backup data type, so this work uses the SQL Server 2008 database to change Import the backup file and process it with the ETL tool integrated in the database.

(4) Data conversion

In the process of data cleaning, the accuracy and reliability of the data required for data mining have been achieved, but the form of storage fields in the three information tables to be processed by data mining cannot meet our needs. While setting goals, we It is necessary to know the category of each book borrowed by readers, which is the book category that readers prefer. Therefore, each book should be classified according to the Chinese Library Book Classification Method. For the table name, field name, time, etc., further processing is also needed to facilitate the subsequent work.

(5) Data integration

1) According to the needs of the data mining model, select data from the business data and store it in the data set to be mined.

2) According to the relationship between tables and various aspects of information, fill in empty values, delete useless data information and attributes, and form the direct information table required for mining.

3) Count the attributes that need to be counted in the data table. The number of times the book has been borrowed and the number of readers has borrowed.

3 Experimental Research on the Application of Big Data Technology in University Library Management Services

3.1 Experimental Protocol

In order to make this experiment more scientific and effective, this experiment went deep into a university library in a certain place. And use questionnaire survey to investigate and analyze its management services. This experiment compares and analyzes the traditional university library service system and the university library management service system based on data mining. To judge the feasibility of the subject of this article. On this basis, through face-to-face interviews with 9 library managers, we discussed and analyzed the performance of the university library service system based on data mining studied in this article. And use the mathematical statistics method for statistics and analysis of the obtained results.

3.2 Research Methods

(1) Questionnaire survey method

In this experiment, a targeted questionnaire was set up by asking relevant experts. In addition, a semi-closed method is used to conduct a questionnaire survey on library borrowers, the purpose of which is to promote the correct filling of the surveyed personnel.

(2) Field research method

In this research, we go deep into a university library in a certain place, investigate and analyze its management service system and collect data. These data provide a reliable reference for the final research results of this article.

(3) Interview method

This article organizes and analyzes the recorded data by conducting face-to-face interviews with library managers and recording data. These data not only provide theoretical support for the topic selection of this article, but also provide data support for the final research results of this article.

(4) Mathematical Statistics

Use related software to make statistics and analysis on the research results of this article.

4 Application Experiment Analysis of Big Data Technology in University Library Management Service

4.1 Comparative Analysis of Library Management Service Systems

In order to make this experiment more scientific and effective, this study conducted a questionnaire survey and analysis of borrowers. The data obtained is shown in Table 1.

Table 1. Comparative analysis of library management service system

	New book announcement	Resource link	Course reference book recommendation	Recommended classic books
Data mining	68.7%	72.1%	69.1%	70.4%
Traditional	62.8%	60.7%	64.3%	64.6%

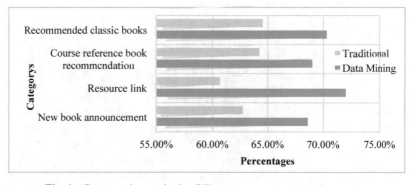

Fig. 1. Comparative analysis of library management service system

As can be seen from Fig. 1, compared with the traditional university library management service system, the university library management service system based on data mining technology studied in this paper is 5.9% higher in the new book recommendation module and 11.4% higher in the resource link module. Course reference the book recommendation module is 4.8% higher, and the classic book module is 5.8% higher, fully reflects the feasibility of the subject of this article.

4.2 Performance Analysis of University Library Management Service Based on Data Mining Technology

In order to further conduct experimental analysis on the research results of this article, this research conducted face-to-face interviews with relevant library staff, and the collated data are shown in Table 2.

Table 2. Analysis on the performance of university library management service based on data technology

	High efficiency	Convenience	Information security	Others
1	7.21	7.06	6.45	5.69
2	7.68	7.00	6.25	5.47
3	7.49	6.98	6.39	5.18
4	7.76	6.48	6.18	5.66
5	8.23	6.74	6.48	5.82
6	7.99	7.12	5.59	5.69
7	7.43	7.35	6.21	5.98
8	8.03	6.58	6.81	6.48
9	6.48	6.99	7.06	6.27

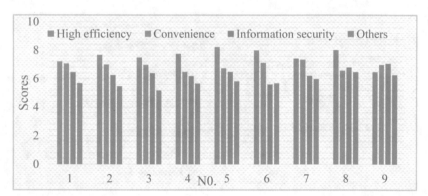

Fig. 2. Analysis on the performance of university library management service based on data technology

It can be seen from Fig. 2 that the evaluation of the university library management service system based on data mining studied in this paper is above 5, which shows that this research is feasible. Among them, the evaluation of the efficiency of this research system is the highest, which fully reflects the excellent performance of the university library management service system based on data mining studied in this paper.

5 Conclusion

With the continuous development of information technology, all walks of life are experiencing a wave of informatization. As an information resource center, the library serves as an information resource center to meet the individual needs of readers, innovate library service models, and conform to the trend of the times. The function should change from passive to active. We need to use new scientific and technological achievements to serve the library. Coincidentally, another technology, the emergence and development of data mining technology, has opened a new chapter for library service functions, making it possible to transform library service functions from passive to active, and has created more for libraries to carry out digital services.

References

1. Madhusudhan, M., Singh, V.: Integrated library management systems: comparative analysis of Koha, Libsys, NewGenLib, and Virtua. Electron. Libr. **34**(2), 223–249 (2016)
2. Ming, Z., Xizhu, L.: Theoretical construction of internalization of university library management system—exploratory research based on grounded theory% theoretical construction of internalization of university library management system—exploratory research based on grounded theory. Libr. Inf. Serv. (5), 32–41 (2019)
3. Wangchun, Y.: Research on the role of big data technology in the construction of university library management system. Sci. Technol. Inf. Dev. Econ. **004**(005), 16–19 (2019)
4. Liu, C., Sheng, M.: Design of library management system. Open Access Libr. J. **05**(12), 1–8 (2018)
5. Dwiyantoro, D.: Evaluasi software aplikasi Senayan Library Management System (SLIMS) di lingkungan Universitas Lancang Kuning Pekanbaru menggunakan pendekatan ISO 9126 usability. Al-Kuttab Jurnal Kajian Perpustakaan informasi dan kearsipan **2**(1), 21–31 (2020)
6. Bwalya, T., Akakandelwa, A.: Challenges of using Koha as a library management system among libraries in higher education institutions in Zambia. Desidoc J. Libr. Inf. Technol. **41**(2), 82–87 (2021)
7. Muhamad, S.S., Darwesh, A.M.: Smart university library management system based on Internet of Things. UHD J. Sci. Technol. **4**(2), 63 (2020)
8. Thiruvengad Am, S., Gokulakrishnan, S.: Library management system using RFID. Int. J. Comput. Appl. **175**(10), 31–33 (2020)
9. Malipatil, N., Roopashree, V., Sanjana Gowda, R.H., Shobha, M.R., Sateesh Kumar, H.C.: RFID based library management system. Int. J. Eng. Sci. Technol. **3**(7), 112–115 (2020)
10. Shanmugam, A., Ramalakshmi, A., Ganeshan, S., et al.: Library management system. Xi'an Jianzhu Keji Daxue Xuebao/J. Xi'an Univ. Archit. Technol. **12**(11), 743–753 (2020)
11. Naveed-e-Sehar, Tariq, H., Burney, S.: Proficient automated library management system (palms): a new model for public libraries of Pakistan. Int. J. Adv. Res. **7**(4), 505–522 (2019)
12. Rehman, H.A., Soomro, A.S., Surahio, F.A., et al.: Implementation of library management system using radio frequency identification technology in Sindh libraries. Int. J. Comput. Sci. Inf. Secur. **14**(7), 419–422 (2019)

Project Management Mechanism Research of Distribution Network Based on Big Data

Wei Song[1], Lei Xue[1(✉)], Yao Wang[1], Jiwu Liu[1], Yuming Zheng[1], Jiahong Wei[1], Huiwen Zheng[1], Zhenbo Xu[1], and Siyuan Cheng[2]

[1] Economics and Technology Institute of State Grid, Shanxi Electric Power Company, Taiyuan, Shanxi, China
[2] State Grid Yuncheng Power Supply Company, Taiyuan, Shanxi, China

Abstract. There are many kinds and large quantities of distribution network projects, and the traditional manual control method is inefficient. Power grid construction presents the trend of intellectualization, digitalization and informatization. At the same time, the scale of modern power grid construction based on sensor detection technology, communication technology, machine learning technology and remote control technology is also gradually expanding. Therefore, the application of power big data in power grid construction is increasingly widespread, and the effect is remarkable. This paper studies the project management mechanism of distribution network based on big data mining method. The general research ideas are as follows: firstly, describe the digital transformation path of power grid enterprises; Secondly, the characteristics and difficulties of control of distribution network engineering projects are analyzed. Finally, a fine management approach of distribution network engineering project under the application of big data is proposed. The research results can highlight the role of big data technology in realizing related and dynamic management and control of distribution network engineering projects, and promote the construction and management level of distribution network.

Keywords: Distribution network construction · Big data · Project management · Fault early warning

1 Introduction

Emerging digital technologies such as "Big Cloud, Thing and Intelligence Shift" have brought about all-round changes in thinking, culture and mode in the economic field, providing technical support for the digital transformation of enterprises. As a new factor of production, data effectively breaks the system barriers and builds a data bridge for each relatively closed and independent system, enabling enterprises to quickly respond to changes in the demand end and reconstruct the entire value chain process. At present, the power grid operated by State Grid Corporation of China and China Southern Power Grid Corporation has become the largest Internet of Things in the world. In the daily operation and maintenance process, massive monitoring data of power grid equipment operation, power grid operation data and power grid management data will be generated [1]. Since much starker choices-and graver consequences-in, electric power development by focusing on the" access to electricity "to safeguard" with good electricity, power distribution network as a bridge that crosses from the transmission network with the user, is a guarantee service for national economic and social development and improve people's livelihood the important public infrastructure, is the guarantee of our country economic growth and structural transformation, the important guarantee to promote the development of urbanization, Its engineering project has the characteristics of point more area wide, size, and the network connection is complex, different kinds of equipment, operational data is complex, the data information is difficult to fully explore the, caused the current situation of "data rich, lack of information, the urgent need to digital and modern methods to promote the management change, management the whole process of real-time perception, visual control, lean and efficient, we will promote overall improvement in the quality, efficiency and effectiveness of development.

Traditional distribution network engineering project management, mainly concentrated in the project construction cost, safety and operation scheduling management, literature [2, 3] analyzed the changes in the distribution network project construction technology, and the distribution network project cycle, construction environment, personnel management system, such as Angle analyses the reasons of the safety management system cannot fall to the ground. Literature [4] based on the requirement of technological upgrading projects not power distribution network operation, using benchmarking management, target management, standardized management, pareto chart management tools such as optimization of operation management mode, operation process, safety control, to solve the distribution network engineering do not power operation strength is weak, basic unit enthusiasm is low, the process control problems such as low efficiency and security; Literature [5–7] aims at achieving precise investment in distribution network, and proposes to improve the scientific rationality and effective control of construction fund planning arrangement by realizing the modularization of distribution network project cost.

Although the above research has further improved the management accuracy of distribution network in terms of management system and technical means, it lacks of data coordination analysis and fails to make full use of data information. Data analysis of the current distribution network system and information mining is mainly using the classical model in data mining, large artificial tear open data and data mining, mining results accurate rely too much on original data, fails to reflect the thought of big data, ignored the data, the relationship between the distribution network construction management level and application level of the large data long-term improvement. Therefore, this paper firstly analyzes the status quo of whole-process control of distribution network project. Secondly, the digital transformation path is proposed based on the digital transformation goal of power grid enterprises. Finally, based on the characteristics of the era of big data, the innovation management mechanism of distribution network engineering projects is proposed to promote the application of big data technology in the management of distribution network engineering and accelerate the process of digital management.

2 Digital Transformation Path of Power Grid Enterprises

The digital transformation of power grid enterprises is mainly carried out from three aspects: enterprise itself, industrial structure and national mission. Enterprises themselves need to be guided by digital information flow, manage and control their production and operation activities, develop into "ubiquitous and efficient" digital enterprises, use data flow to resolve the uncertainty of complex systems, optimize the efficiency of resource allocation, and build new competitive advantages [8].From the perspective of the energy industry system, power grid enterprises need to transform and upgrade the traditional energy industry chain to the energy industry groups that meet the needs of social development, and become the Internet platform economy at the core of the national digital energy system (see Fig. 1). From the perspective of Digital China, power grid enterprises need to carry out the national mission in the new social form of Digital China and become one of the most important basic Internet platforms in the information society, namely the "National New Energy Internet Cloud Platform" [9, 10].

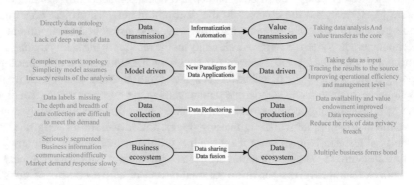

Fig. 1. New paradigm of power grid enterprise digital management

3 Innovation of Distribution Network Project Management Mechanism

3.1 Analysis of Difficulties in Power Distribution Network Project Management and Control

(1) The uncertain factors faced by the construction and control of distribution network are complicated and their relationship is chaotic

Distribution network of quantities large and complicated, which involves many professional, contractors, construction cycle is long, the construction unit of qualification and different quality of construction. At the same time, because the distribution network project is close to the residential load density point, there is a multilevel logical relationship between the management elements of the distribution network project.It makes it more difficult to sort out uncertain influencing factors and identify key control factors in the process of distribution network engineering management.

(2) The control emphases of the whole process of the distribution network project are different, so flexible management mode is needed

At present most of the distribution network construction projects lack of scientific, standardized, professional, comprehensive management method, management relative lag, and lack of information management, this makes the project organization, planning, implementation, coordination and control is difficult to effectively carry out, leading to the actual execution of the project is difficult to implement. In addition, the management mode is relatively single, the coordination between departments is not coordinated, unreasonable staffing, progress management and cost management is not scientific phenomenon is common, it is difficult to achieve unification in a short time, seriously affect the implementation of the project, but also greatly reduce the work efficiency and quality of engineering project management.

(3) Lack of data decision technology

Due to distribution network big data project is multifarious, lead to low efficiency, data processing analysis as a whole analysis is more difficult, and distribution network construction project implementation process involves a large number of wasting resources and not wasting resources, resource consumption in the change of time and space dimensions are hard to match the big data project lead to distribution network project construction management decision analysis response capacity is insufficient, lack of scientific decision support.

3.2 The Status Quo of Big Data Management of Distribution Network Engineering Projects

The evaluation and decision of early distribution network project usually adopts the method of expert rating, and the system planner evaluates the rationality and feasibility of distribution network planning scheme according to the relevant guidelines and expert experience of power network. In recent years, with the promotion of big data strategy, the power grid has widely carried out the construction of "three rates of integration" system projects ("three rates" refers to the completion rate of construction progress, the completion rate of investment progress, and the completion rate of account entry progress), and the statistical analysis system for the whole process of power grid project investment has been improved day by day. "Unity of three" system using technology of data mining huge amounts of data resource value, basic can achieve grid infrastructure project milestones plan execution, physical quantities form, closely tracking the financial cost accounting process, improve the investment statistics to submit the scientific nature, data accuracy and management effectiveness, enhance the level of project the lean controls the whole process. In the process of distribution network project construction, the control of quality, cost and schedule is not separated, but coordinated and synchronized. However, at present, big data technology is more used in the benefit analysis of distribution network operation stage, while in the construction stage, a large number of data are scattered in different work, process and system, and there is a lack of scientific and standard data force analysis and evaluation, which cannot support the decision-making of distribution network construction.

3.3 Application of Big Data in Distribution Network Engineering Projects

The application of big data in power distribution network engineering projects has effectively broken the internal ecology of enterprises with independent governance and numerous barriers, which is conducive to the establishment of effective communication and mutual cooperation management mode and the improvement of management efficiency. The full sharing of data and information can ensure the smooth connection of management at all stages of the project. Relying on scientific analysis methods can effectively solve cross-professional, cross-stage, cross-department affairs. Based on the management of key data resources, the original professional barriers can be broken, and the network collaborative management mode of vertical professional management and horizontal resource management can be formed.

(1) Breaking the information barriers of each link of project management through data integration

Through the comprehensive integration of data level, application system, user interface and business process, the enterprise's business process, public data, application software, hardware and various standards are combined to achieve seamless integration between different business systems and enable them to conduct business processing and information sharing as a whole. Big data analysis technology is used to analyze and process the data of administrative office and operation management of enterprises, forming analysis reports, supporting traditional enterprise operation and management with data, and providing objective and rational basis for decision-making.

(2) Alleviating the friction in each stage of the project by using digital management means

The development of digital technology reduces the time cost and economic cost of information communication. In the information age of data sharing and resource exchange, traditional project managers can use big data technology to find the business links with low efficiency and risks and constantly optimize the management process. Through learning and using relevant technologies to improve business and innovate business, using big data to solve problems in the way of thinking to improve project professional skills, improve project management efficiency.

(3) Replacing the management process based on experience judgment and manual operation by using digital technology

Using digital technology to analyze the system data of the production and operation links of the distribution network project, the real-time operation status can be obtained more intuitively, and the conventional fault situation and typical operation scenes can be judged according to the accumulated large amount of system data. Artificial intelligence, deep learning and other algorithms are utilized to realize the self-learning and self-decision-making of equipment, significantly improving the production and operation efficiency of enterprises.

4 Conclusion

Large and complicated distribution network project construction period is longer, the actual situation was complex, the entropy value is high, the management has a point, face more wide, the current management system is given priority to with log data, the lack of the analysis and feedback of information, or just stay in and the historical data contrast, compared with the standard level, not to exceed the time limit, does not conform to the standards of the subsequent impact of record of early warning, It cannot reflect the correlation between all links and nodes in the construction of distribution network engineering projects, and lacks intuitiveness, which cannot meet the current lean management requirements of distribution network engineering projects. The introduction of big data analysis can realize the full mining of data value, strengthen the refined management of data, subdivide data categories, improve data sources, and effectively improve the efficiency of distribution network project management. At the same time, using the data flow of information, through the data system, implement the data linkage, the relationship is beneficial to strengthen the management system, the deviation of the data in a timely manner with the expected warning, the scattered data index, depicting the relationship between each target, looking for the logic of error transmission line, provide a reference for subsequent management focus.

References

1. Chen, J., Shi, W., Liu, Y., et al.: Big data analysis supports the research and pilot application of distribution network planning technology. IOP Conf. Ser. Earth Environ. Sci. **632**(4), 042047 (2021)
2. Yang, Z., Yu, J., Li, H., et al.: The relationship between management and control indexes of distribution network construction based on Apriori algorithm. E3S Web Conf. **237**(10), 1–14 (2021)
3. Song, Y., Wang, G., Li, Z.: Research on terminal management technology based on distribution management system. J. Phys. Conf. Ser. **1550**(3), 032116 (2020)
4. Mueller, H., Makarov, A.Y., Nikitina, E., et al.: Efficiency of advanced smart grid technologies applied in Ufa's distribution network in Russia. In: International Etg Congress, Die Energiewende-Blueprints for the New Energy Age. VDE (2016)
5. Angioni, A., Lipari, G., Pau, M., et al.: A low cost PMU to monitor distribution grids. In: 2017 IEEE International Workshop on Applied Measurements for Power Systems (AMPS). IEEE (2017)
6. Liu, Y., Ma, Q., Xu, C., et al.: Investment optimization model for distribution networks projects with risk and efficiency constraints. In: 2020 IEEE 4th Conference on Energy Internet and Energy System Integration (EI2). IEEE (2020)
7. Liu, Z., Lin, F., Lin, Y., et al.: Visualization system design for investment and planning of active distribution networks. In: 2019 IEEE Innovative Smart Grid Technologies - Asia (ISGT Asia). IEEE (2019)
8. Yue, H., Li, W., Yue, Y., et al.: Research on the current status and transformation strategy of digitalization at the State Grid Jibei Electric Power Company. IOP Conf. Ser. Earth Environ. Sci. **585**(1), 012107 (2020)
9. Zhang, L., Zhang, Z., Li, J.: Digital management of power transmission and transformation works based on building information model. In: IOP Conference (2018)
10. Nardelli, P., Baptista, M.S., Pomalaza-Raez, C., et al.: Network models for the structures of modern electric power grids (2014)

Big Data Information Technology Application in Logistics Cost Management and Control

Huachang Miao(⊠)

Applied Technology College of Soochow University, Suzhou, Jiangsu, China
mhc.suda@longmail.cn

Abstract. China's logistics system has achieved rapid development, but there is still a certain gap compared with Western countries. However, from the beginning of nothing to the present integration of Chinese logistics companies, information technology has played an key role in promoting the establishment of the logistics system. With the information technology rapid development, big data technology is imperceptibly affecting people's lives. Logistics is closely related to people's lives. It not only promotes the development of various industries, but also brings many conveniences to people's lives. As the basis of logistics cost management, big data information technology has promoted the development of the logistics industry, but it has also increased the pressure of logistics cost management invisibly. This article starts with the conceptual analysis of logistics cost management in the era of big data, and conducts an in-depth analysis of the role of big data information technology in logistics cost management.

Keywords: Big data · Information technology · Logistics · Cost control

1 Introduction

In the 1990's, the concept of logistics cost management in China has developed rapidly for decades. Especially with the information technology development, the big data application technology in logistics cost management optimizes China's logistics cost management model [1]. The content of big data technology mainly includes network technology, barcode technology and computer technology, etc. These big data technologies play an important role in the logistics cost management of enterprises, but there are still many problems in the big data application in logistics cost management. Therefore, the current problems in logistics cost management are analyzed, and measures to optimize big data logistics cost management are discussed [1].

2 The Connotation of Big Data

Big data has attracted the attention of many foreign business scholars since its emergence. Compared with traditional data analysis software and tools, big data has unparalleled advantages in obtaining information, storing information, and analyzing and managing

J. Macintyre et al. (Eds.): SPIoT 2021, LNDECT 98, pp. 298–304, 2022.
https://doi.org/10.1007/978-3-030-89511-2_38

data information [1]. The significance of using big data in logistics cost management lies in the ability to process data professionally, rather than using big data as a means to obtain economic benefits. If big data is used as a means of logistics cost management, it needs to add value. As the connection between cloud computing and big data is getting closer, big data also creates more value for mankind [2]. The foundation of big data development is big data theory, and big data processing methods mainly include distributed processing and storage technology. The modern society development is closely related to big data. Big data can dig more useful information for human beings and bring more convenience to people's lives. In the development of enterprises, big data has become the main factor in the competition of modern enterprises. Companies use big data to provide high-quality products and precise marketing. Some enterprises will also use big data for transformation. Traditional enterprises can use big data to balance the value, cost and decision-making of the enterprise during the transformation process [2]. For example, they can develop the problems encountered in the development of the enterprise in time and solve them in time. It will save a large amount of manpower, material and financial resources during the operation of the enterprise to a large extent, so that the information analyzed by big data can be used to deliver the optimal logistics route for express information, and to formulate scientific and reasonable preferential information according to customer habits, so as to avoid the emergence of fraud.

3 The Impact of Big Data Information Technology on the Logistics Industry

Therefore, the current big data information technology is actually a good opportunity for the industry development for the logistics. The role of big data information technology for the logistics industry is mainly reflected in the following aspects:

3.1 Improve the Intelligentization of the Logistics Industry

The logistics industry is a highly procedural industry, and big data technology is a rigorous data analysis technology. Combining the two, through the tracking and logistics data analysis, big data can be used to make intelligent decisions and recommendations for logistics companies. In logistics decision-making, big data can also analyze the environment, match logistics supply and demand, and optimize the allocation of logistics resources [3]. In the logistics industry, various transportation, warehousing, handling and other procedures can be greatly improved compared to before, and the intelligence of the logistics industry can be improved.

3.2 Can Promote the Industry Development

In the logistics industry, goods transportation, warehousing, and loading and unloading are all controlled and completed by people, and they are relatively backward in efficiency. And big data technology can complete the iterative arrangement of the whole process of the logistics industry as a whole [3]. Moreover, in the analysis of the competitive environment of the industry, for maximize the benefits of the brand, predict their behavior and trends, so as to find the deficiencies of the brand in the industry and the areas that need improvement, as shown in the Fig. 1. In this way, big data can quickly complete the promotion of the industry and accelerate the industry development [4].

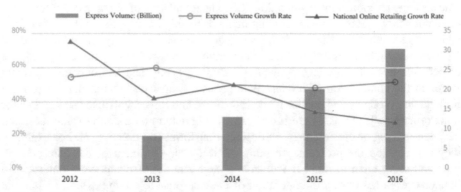

Fig. 1. Big data information technology promotes the development of the logistics industry

3.3 Improve User Experience

At the end of the logistics industry, products need to be delivered to users. With the big data support, compared to manual delivery, big data technology can provide users with a better logistics experience and increase user stickiness [4].

4 Advantages of Big Data Information Technology in Logistics Management

4.1 Master the Information About the Operation of the Enterprise

The traditional data analysis model cannot adapt to the development of modern logistics enterprises. Therefore, logistics cost management need big data supported, to increase the optimization and integration of the information of each node in the logistics operation process, and then analyze and process the collected data through the data center [5]. Transform this information into valuable information for cost management, so as to grasp the logistics operation mode that arise in the operation.

4.2 Provide a Basis for Logistics Companies to Make Correct Decisions

Traditional logistics companies generally make decisions based on market research and personal experience, which no longer to meet the needs of a modern information society. The market needs to truly reflect the data analysis in the logistics enterprises operation, so as to help logistics enterprises make scientific and effective decisions through this information [5].

4.3 Avoid Losing Customers

Through the information mining and analysis by big data, the results of these analyses can be reasonably and effectively used in logistics cost management, so as to optimize and consolidate in logistics relationship, the trust level of the company can fundamentally prevent the loss of customers [6].

5 The Big Data Application of Information Technology in Logistics Cost Management

The application process of data in the logistics industry, first of all, can effectively collect and process massive data and huge amounts of information from logistics links such as transportation, warehousing, packaging and processing, and establish mathematical models to mine high-quality information and perform mathematical analysis; Secondly, apply data to practice to optimize the allocation of resources, improve transportation and distribution efficiency and timeliness, reduce logistics costs, quickly respond to market demand, realize data feedback, and help logistics companies avoid market risks and make reasonable and satisfactory decisions. To maximize profits [6]. Figure 2 shows the general process.

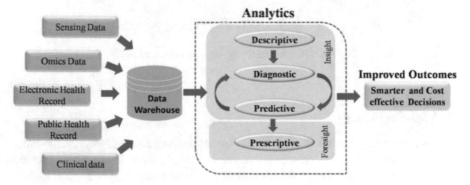

Fig. 2. The data mining process in the logistics industry

5.1 Intelligent Commercial Property Forecasting (Analysis of Goods-Routing Planning)

The logistics market is highly dynamic and random. In the big data environment, real-time supply and demand relationships in the market, the flow and flow of different product categories and other information can be mined from it, so as to realize reasonable adjustment of resource allocation, optimize logistics routing and transportation, and quickly respond to the supply and demand market [7]. Up to now, many logistics companies only focus on the fast and efficient delivery of goods from the starting point to the end point, rather than the goods themselves. In fact, logistics companies can use their own data mining to sell market information to related service target companies, expand business segments, and expand their sources of income [7]. The source of information and protecting consumer privacy, logistics operators carefully collect and mine this valuable information, and report it to the enterprises that need this information, as shown in Fig. 3. In addition, logistics companies can also use big data analysis technology to effectively plan distribution routes, predict traffic conditions and accidents, and balance the load of the logistics chain [7].

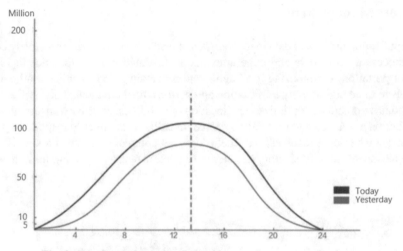

Fig. 3. Big data information technology predicts commodity items

5.2 Logistics Pricing

The modes of transportation of goods include road transportation, railway transportation, air transportation and ocean transportation. Take the freight market of road transportation as an example to discuss the big data application analysis technology to support logistics pricing. As the freight market has the characteristics of low degree of informationization, fragmentation, and non-standardization, while the logistics price fluctuates greatly, the price war between logistics companies is also intensifying, which seriously affects the stability of the logistics market [8]. Logistics prices are based on cost calculations, and various factors must be considered, such as the type of goods, the cost of transportation capacity, geographical factors, traffic network conditions, and so on. At present, freight pricing in the logistics market is only determined based on simple linear analysis of the delivery mileage, quality, volume, toll, vehicle warranty, and fuel price of freight items, and there is a lack of solutions [8]. The logistics pricing method based on big data is as follows: The first step is to collect cost information data that affect logistics pricing, and establish a logistics freight database. The cost information data includes the type of goods, the cost of transportation capacity, geographical factors, traffic network conditions, etc.; through data preprocessing Feature engineering processing of massive data, including data cleaning, data integration, data reconstruction, data conversion, data association and fusion. The second step is to establish a data model and use the calculation module and the big data calculation engine to perform multi-criteria decision calculations to obtain the optimal logistics price. In the third step, the high-level managers of the logistics company refer to the guide prices in different regions supported by big data and make reasonable and satisfactory decisions [8].

5.3 Smart Terminal Distribution (Hub Network Layout + Smart Express Cabinet)

The "last mile" has been widely concerned as a terminal distribution problem in the logistics industry. A research conducted by iResearch shows that the "pain points" between city users and couriers are due to the mismatch of their respective time. The emergence of terminal hubs and smart express cabinets provides consumers with more efficient services [9]. However, the location of outlets and smart express cabinets is difficult to make decisions. On the one hand, big data can collect a lot of information and data related to decision-making such as land rent, transportation network, radiation area, market demand, and competitors' locations. It can help companies make optimal decisions. On the other hand, the compartment layout of most smart express cabinets on the market is uniform, but in fact, packages delivered to different regions have different sizes and different characteristics. For example, the packages in central CBD office buildings and colleges and universities are mostly letters and documents, the packages in industrial areas are more parts and machines, and the packages in residential areas are more life products [9]. Logistics companies can rely on big data technical support to set up different express cabinet compartment layouts in areas with different needs, so as to save costs, improve customer experience, and upgrade efficiency.

5.4 User Satisfaction Analysis

In logistics customer management, big data can analyze the internal connections and laws of massive customer data and commodity data, and provide a basis for correct decision-making for enterprise development [10]. By analyzing user information and user behavior, using multi-disciplinary knowledge and practical experience, the data is transformed into actual problems, based on the actual situation of the enterprise, to improve user satisfaction, try to retain old customers while mining as much as possible Potential customers, mobilize customer enthusiasm, improve user experience, expand market share, and win word of mouth. There are various applications of big data analysis in practice. While attaching importance to it, we must continue to practice, through data mining, drive efficiency, help companies reduce costs and increase efficiency, and provide strong support for logistics companies [10].

6 Conclusion

With the modern information technology development, the big data scope in the management of logistics enterprises has become more and more extensive. According to the relevant survey reality, there are still many problems in China's modern logistics cost management. At present, the big data application information technology in logistics cost management is still in the preliminary development stage. Therefore, modern logistics enterprises should pay attention to the big data application information technology in logistics cost management, so as to improve the efficiency of modern logistics distribution, to promote the sustainable development of enterprises while providing customers with better services.

References

1. Li, R.B.: Research on HB tobacco company's logistics cost management system optimization based on big data. Jilin Univ. (2017)
2. Liu, W.L., Yin, J.: Research and implementation of a currency logistics cost management system for inquiry and tracking currency logistics under the background of big data era. Wuhan Finance **8**(04), 58–60 (2016)
3. Ye, B., Huang, W.F., Yu, Z.: Research on the Big data application in Logistics Enterprises. Logistics Technology **33**(15), 22–24 (2014)
4. Yang, S.: Research on intelligent logistics model in the big data era. Enterp. Technol. Develop. **11**(05), 112–113 (2018)
5. Guo, B., Meng, W.L.: Thoughts on informatization education and teaching in the big data environment. J. Hexi Univ. **10**(02), 109–113 (2017)
6. Luo, L., Xu, D.W.: On the construction of technology platforms in the era of big data. China's New Technol. New Prod. **18**, 23–24 (2019)
7. Li, G.W.: Big data application in agricultural product logistics cost management system. Logistics Technol. **11**(10), 97–100 (2013)
8. Zhu, Y.F.: A preliminary study on the optimization of the professional structure of logistics cost management in higher vocational education from the perspective of "Big Data." J. Ningbo Vocat. Tech. College **10**(04), 10–13 (2016)
9. Wang, B.Y.: Research on the construction and construction countermeasures of logistics information platform in the big data era. Inf. Sci. **12**(03), 120–123 (2016)
10. Ma, L.Y.: Analysis of the opportunities and challenges faced by logistics management companies in the era of big data. Chin. Market **13**(22), 256–259 (2017)

Exploration on the Application of GIS Technology in the Field of Ecological Protection

Xiang Zhang[1,2(✉)]

[1] College of Ecological Environment and Resources, Qinghai Nationalities University, Xining 810007, Qinghai, China
[2] Qinghai Tibet Alpine Wetland Restoration Engineering Technology Research Center, Xining 810007, Qinghai, China

Abstract. The ecological environment system is a complex system related to the sustainable development of society, economy and nature, including biological resources, land, water resources and climate resources that affect the survival and development of human beings. The security status of the ecological environment system affects the sustainable development of the entire ecosystem, and ensuring ecological security has become the key to coordinated regional development. The red ecological protection line is the boundary line of the ecological protection area. Delineating the ecological protection red line can not only achieve the purpose of protecting the ecological environment and protecting natural resources, but also promote the sustainable development of the city. Therefore, this article aims to study the application of GIS technology in the field of ecological protection. On the basis of analyzing the functions of GIS and the advantages of GIS in ecological protection, make full use of the powerful spatial data processing and analysis functions of GIS to quickly and efficiently obtain county areas. Eco-environmental sensitivity relies on data that is not easy to obtain by conventional methods, builds an ecological environmental sensitivity assessment model, and completes the ecological environmental sensitivity assessment. The evaluation results show that through statistics of important areas, insensitive area accounts for 22.92%, general sensitive area accounts for 37.73%, sensitive area accounts for 28.21%, and extremely sensitive area accounts for 13.15%.

Keywords: GIS technology · Ecological protection · Red line division · Sensitivity assessment

1 Introduction

In the past ten years, the speed of social development has gradually accelerated, the number of people has increased rapidly, the scope of urban construction has been gradually expanded, the ecological environment has been destroyed year by year, the resources and environment have been deteriorating, the ecosystem service functions have degraded, and the environmental problems in my country have become increasingly prominent. The problem of ecological environment is becoming more and more serious [1, 2].

J. Macintyre et al. (Eds.): SPIoT 2021, LNDECT 98, pp. 305–313, 2022.
https://doi.org/10.1007/978-3-030-89511-2_39

In order to ensure the sustainable development of the ecological environment and social economy, my country first proposed measures to delineate the ecological protection red line in 2011, and later proposed a series of policies to support the implementation of the ecological protection red line system, with the purpose of improving the ecological environment. The system service function and ecological environment quality alleviate the deterioration of the ecological environment caused by the rapid economic and social development and the continuous advancement of urbanization, the continuous occupation of ecological land, and the increasingly prominent problem of the contradiction between the city and the ecological environment [3, 4]. In order to achieve the maximum utilization of resources and energy, and at the same time coordinate the relationship between resources, environment and population, so that the three can form a new pattern of mutual benefit, promote the balanced development of social, economic and ecological benefits, and achieve. While the economy is speeding up, it is necessary to ensure that the ecological benefits can be steadily improved. The delineation of the ecological protection red line is not only conducive to the country's ecological security work, but also has a certain role in promoting the sustainable development of the country's ecology [5, 6].

Based on the analysis of the functions of GIS and the advantages of GIS in ecological protection, this paper makes full use of the powerful spatial data processing and analysis functions of GIS to quickly and efficiently obtain the sensitive data of the county ecological environment and rely on conventional methods to build ecological environment. The environmental sensitivity assessment model has completed the ecological environmental sensitivity assessment.

2   Exploration of the Application of GIS Technology in the Field of Ecological Protection

2.1   Functions of GIS

(1) Data collection.
   Map data, measurement data, statistical data and text data are input into the map database, and the data is often sorted and analyzed by the vector method, so that GIS can identify and analyze the data [7, 8].
(2) Data processing.
   Data processing is the process of transforming, integrating and maintaining spatial data. Generally speaking, the geometric images and spatial features of the collected spatial data often contain errors and imperfections, which must be corrected by subsequent processing.
(3) Spatial analysis of data.
   ArcGIS spatial analysis function is an important part of ArcGIS and is essential in ecological applications. According to the type, it can be divided into vector data and raster data [9, 10].

2.2   Advantages of GIS in Ecological Protection

Remote sensing technology, as one of the current mainstream methods for studying regional environment, has unique advantages in the delineation of ecological protection red lines in counties; however, with data and images, the development of specific ecological protection red lines will analyze data processing and analysis. It is also essential. Geographic Information System is a tool that can efficiently and accurately process and analyze the data and images obtained by RS technology, and can provide a scientific and objective reference for the process of delineating the ecological protection red line.

GIS is a comprehensive technology of multiple disciplines such as geography, research, cartography and computer science. The core is computer science, and the basic technology is data analysis, mapping and spatial analysis [11, 12]. The computer's high-speed operation and powerful logic judgment function can provide multiple design comparisons in a short time, which increases the rationalization of designs and design schemes. In addition, the computer can automatically create various design drawings, tables and reports, and use the database to facilitate deletion and update, so that the red ecological protection line can also be dynamically monitored and dynamically designed.

3   Experiment

3.1   Data Processing

This paper uses the combination of spatial information technology to quantitatively evaluate the ecosystem service functions of a certain city. After collecting a large amount of data related to the study area, the data is systematically analyzed based on GIS technology, and the ecological environment of the study area is of good quality and ecosystem functions. Under the sound premise, the factor of ecological environment sensitive area is selected as the basic index of the county ecological protection red line. The data processing involved mainly includes the preprocessing of remote sensing data, image enhancement and remote sensing interpretation.

3.2   Environmental Sensitivity Assessment of Soil Erosion

(1)   Construction of environmental sensitivity model for soil erosion.
      According to the city's basic data and remote sensing data, the city has a good ecological environment index, abundant water resources, large changes in topography and landforms, and high vegetation coverage. This article combines the above characteristics of the study area and summarizes the main factors affecting the city's soil erosion. Elevation, vegetation coverage, hydrological factors, slope and aspect. The above single factor is superimposed and calculated by using GIS technology:

$$A = \sum\nolimits_{i=1}^{n} P_i \cdot C_i \tag{1}$$

Sensitivity to soil erosion mainly refers to the degree of impact of changes in external pressure on ecological factors under the premise that there will be no loss of the ecological environment or the ecological quality is stable. This assessment combines the characteristics of the regional natural ecological environment to determine the geographic unit ecological sensitivity level. This evaluation selects elevation factors, slope factors, hydrological factors, vegetation factors, and ecologically sensitive area factors as the single factors for the evaluation, and analyzes the spatial distribution of the ecological sensitivity of the single factors, and then comprehensively evaluates the ecological sensitivity of the ecological unit through GIS technology and calculates comprehensive index of environmental sensitivity of soil erosion in each geographic unit.

(2) Model evaluation factor extraction.

1) Elevation factor.
    The similarities between mountain ecosystems and hilly ecosystems are: as the altitude increases, its temperature will gradually decrease, and the material circulation speed will gradually slow down. Therefore, the higher the altitude, the higher the ecological sensitivity.

2) Slope factor.
    In areas with large slopes, water conservation is low and soil erosion is more likely to occur, so the ecological sensitivity is high. If the vegetation cover is destroyed by human interference, it is very easy to cause exposed mountains and natural disasters such as mudslides.

3) Vegetation factors.
    The forest ecosystem is the most important ecosystem on the land and has important ecological functions such as maintaining ecological balance. Vegetation coverage reflects the status of the forest ecosystem. Areas with high forest coverage have important ecological value and belong to areas with high ecological sensitivity.

(3) Classification of evaluation results.
    According to the dynamic conditions of soil erosion, the types of soil erosion can be divided into water erosion and wind erosion. The sensitivity of soil erosion, mainly wind erosion, will be evaluated based on the sensitivity of soil desertification. Due to the abundant water resources in the study area, this paper mainly conducts a systematic evaluation of soil erosion sensitivity based on hydrodynamics. According to the basic principles of the general soil erosion equation, relevant indicators such as slope length, slope slope, and sediment erosion are selected. Use geographic information system technology to multiply single factor rating data, the formula is as follows:

$$SS_i = {}^4R_i \times K_i \times LS_i \times C_i \tag{2}$$

3.3 Assessment of Geological Hazard Sensitivity Model

(1) Construction of the framework of geological hazard sensitivity assessment model. Since this article has considered the sensitivity of soil erosion from the perspective of the importance of ecological functions in the ecological environment sensitivity assessment, according to the definition of the ecological protection red line, the construction of the geological disaster sensitivity assessment mainly focuses on ensuring the safety of the ecosystem itself and the impact of the ecosystem. The

safety of human services is considered, so the identification of ecological assessment factors is also analyzed from the above two aspects.

(2) Sensitivity assessment of geological disasters.

Geological hazard sensitivity refers to the probability of geological hazards occurring in the regional ecosystem under the combined action of external forces such as precipitation, volcanic earthquakes and other factors. Based on the theoretical basis of historical documents and expert opinions, this research evaluates the city's current geological hazards. On this basis, the evaluation factors are selected and weighted with reference to relevant technical guidelines such as geological hazard planning, and GIS technology is used for geological analysis. The disaster sensitivity evaluation determines the evaluation factors and weights of the geological disaster sensitivity in the study area. The evaluation formula is:

$$RR_j = \sum_{i=1}^{n} C_{ij} \times P_j \tag{3}$$

4 Discussion

4.1 Evaluation Results of Soil Erosion Sensitivity Based on GIS Overlay Analysis Model

According to the calculation based on the GIS overlay analysis model, the evaluation results of the city's soil erosion and geological disaster sensitivity assessment model are obtained, as shown in Fig. 1 and Fig. 2.

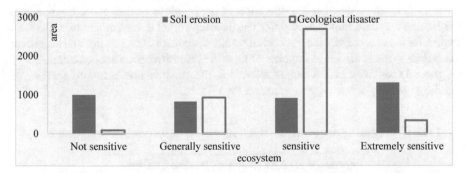

Fig. 1. Results map of sensitivity assessment of soil erosion and geological disasters

According to the constructed soil erosion and geological disaster sensitivity model based on the GIS overlay analysis model, it can be seen that the sensitivity of soil erosion is mainly extremely sensitive, accounting for 32.38% of the city's total area, followed by insensitive areas, accounting for the 24.7% of the city's total area. The city's geological hazard sensitivity is mainly sensitive. The statistical results show that the city's geological hazard sensitivity is mainly sensitive, with an area of 2692.6 km2, accounting for 66.23%; the second is general sensitivity, which accounts for 22.98% of the city's total area.

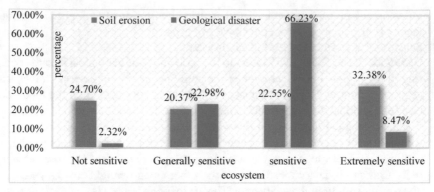

Fig. 2. Percentage map of soil erosion and geological disaster sensitivity evaluation

4.2 Comprehensive Assessment of Ecological Environment Sensitivity

Select the city's soil erosion sensitivity and geological disaster sensitivity to comprehensively evaluate the city's ecological environment sensitivity. By assigning weights to the sensitivity of soil erosion and geological disasters, the weighted raster images can be obtained, and then the weighted raster images are superimposed using GIS technology to obtain a comprehensive map of ecological environmental sensitivity, using natural discontinuities. The method divides the obtained comprehensive analysis results into four levels: insensitive, general sensitive, sensitive and extremely sensitive.

Through consultations with 3 experts with research basis on the sensitivity of ecological environment and scoring of the above two types of ecological environment sensitivity, after comprehensive consideration of the opinions of 3 experts, the weight of soil erosion was determined to be 0.67. The disaster weight is determined to be 0.33. See Table 1 for the comprehensive evaluation results of ecological environmental sensitivity. According to the statistics of important areas, the insensitive area accounts for 22.92%, the general sensitive area accounts for 37.73%, the sensitive area accounts for 28.21%, and the extremely sensitive area accounts for 13.15%.

Table 1. Comprehensive evaluation result table of ecological environmental sensitivity

Importance rating	Area (Km²)	Percentage of total area (%)
Not sensitive	850.75	20.92%
Generally sensitive	1533.23	37.73%
Sensitive	1147.01	28.21%
Extremely sensitive	533.98	13.15%

4.3 Ecological Protection Management Countermeasures

(1) Implement management measures for the ecological protection red line area.

The red line area of ecological protection mainly includes areas with extremely important value and areas with high ecological sensitivity. This area is sensitive to external disturbances and human activities. The stability of the ecosystem is not high. It is difficult to recover after destruction, and ecological problems are prone to occur. Therefore, strict management and protection of the red line area of ecological protection must be carried out. Strengthen ecological space management and strictly prohibit all forms of construction and development activities. Important water sources, water areas and meadows shall be used for key protection of the nature reserves within the red line of ecological protection. Strict use of extensive human resources has caused serious damage to the ecological environment, which is serious for economic growth. At the cost of destroying natural resources, the economic development and environmental protection should be coordinated, and on the basis of protecting the natural environment from damage, develop the economy, prohibit energy harvesting, and maintain the balance of the ecological system in the region.

(2) Establish a supervision system based on GIS platform.

Establish the relationship between the geographical location of the ecological red line area and the data that can reflect its current situation, and establish a red line management system based on GIS. Monitor the ecological environment through the data layer, create a systematic ecological protection system, and provide accurate data support for the work of environmental management departments.

(3) Change the mode of economic development.

Strengthen the adjustment of energy structure, eliminate equipment and processes with high energy consumption, heavy pollution, and backward technology in accordance with the law, and accelerate the pace of closure of enterprises that do not comply with industrial policies; strictly restrict the development of polluting industries and relocate or close them in phases and batches, so that the relocated companies. The reconstruction of the city is combined with the planning assumptions of suburban industrial districts to form a new industrial area with high standards and reasonable configuration.

Promoting the development of circular economy is the general trend of national development. It encourages the development of green and low-carbon industries, maximizes the utilization of ecological resources, and shifts the development direction of enterprises to energy conservation and environmental protection, so that the development of enterprises mainly depends on technology. Investment in innovation and technology, rather than destroying the environment as the price of enterprise development.

(4) Ecological publicity and education.

Through multiple channels and various positions, we will vigorously carry out multi-form and multi-level publicity and education of the city's ecological environment with the goal of popularizing ecological environment knowledge and enhancing environmental protection awareness. In the city's primary and secondary schools, vigorously promote various forms of ecological education with ecological environment and ecological city construction as the content, instill the awareness of ecological development, and cultivate a new generation of ecological awareness. Vigorously improve the ecological awareness and ecological concepts of leading cadres at all levels, and enhance the sense of urgency and responsibility for building an ecological city.

5 Conclusions

The construction of ecological culture is the basis for the coexistence and coordinated development of mankind and the natural environment. It is also an important manifestation of the spiritual and material development and material relationship of people in the relationship between the social environment and the natural environment. This is the driving force for a country to promote the construction of an ecological market. The coordinated and sustainable development of population, resources, and environment of a country and a city is an important prerequisite for promoting ecological and cultural construction. Delineating the red line of ecological and environmental protection and promoting the construction of the ecological environment will better help our country build a new type of city that interacts and develops in harmony with technology, enterprises and knowledge, smart culture, and smart economy.

References

1. Fletcher-Lartey, S.M., Graziella, C.: Application of GIS technology in public health: successes and challenges. Parasitology **143**(4), 401–415 (2016)
2. Abdalla, R.M., Alharbi, S.M.: The role of GIS technology in assessing critical infrastructure resiliency during emergencies in the City of Jeddah, Saudi Arabia. Int. J. Emergency Manage. **13**(2), 183 (2017)
3. Liu, Y., Fang, X., Cheng, C., et al.: Research and application of city ventilation assessments based on satellite data and GIS technology: a case study of the Yanqi Lake Eco-city in Huairou District. Beijing. Meteorol. Appl. **23**(2), 320–327 (2016)
4. Rodolfo, P., Giuseppe, P., Enrico, M., et al.: The application of two approaches using GIS technology implementation in forest road network planning in an italian mountain setting. Forests **9**(5), 277 (2018)
5. Kong, H.: Prediction and spatial correlation analysis of standard grain theoretical capacity for land improvement based on GIS technology. IOP Conf. Series Earth Environ. Sci. **514**(2), 022064–10 (2020)
6. Al-Rojaie, Y.: Mapping perceptions of linguistic variation in Qassim, Saudi Arabia, using GIS technology. J. Linguist. Geogr. **8**(1), 9–30 (2020)
7. Shaikh, H., Deng, X., Li, Z., et al.: Projections of future land use in bangladesh under the background of baseline, ecological protection and economic development. Sustainability, **9**(4), 505 (2017)
8. Gao, H., Zhang, H.: Study on coordination and quantification of ecological protection and high quality development in the Yellow River Basin. IOP Conf Series Earth Environ. Sci. **647**(1), 012168–5 (2021)
9. Wang, X., He, et al.: Thoughts and practice of ecological protection and restoration of mountains, rivers, forests, farmlands, lakes and grasslands. J. Landscap. Res. **12**(03), 8–11 (2020)

10. Liu, X.: Thoughts on talent gathering under the background of ecological protection and high quality development in the yellow river basin. Open J. Soc. Sci. **09**(1), 378–387 (2021)

11. Wang, N., Ai, F.: Coordination mechanism of marine ecological protection and its impact on Island tourism resources development. Arab. J. Geosci. **14**(6), 1–16 (2021). https://doi.org/10.1007/s12517-021-06714-2

12. Peiheng, Y.U., Feng, M., Liu, B., et al.: Demarcation and administration of watershed ecological protection red line considering the ecological security pattern-A case of the Qilu Lake watershed, Yunnan Province. J. Lake Sci. **32**(1), 89–99 (2020)

Remote Mechanical Monitoring Electronic Technology Based on KNN Optimization Algorithm

Fan Li[✉], Guoxiu Qin, Wenping Zhou, and Weizhe Li

College of Energy and Power, Shenyang Institute of Engineering, Liaoning, China

Abstract. With the further combination of Internet and artificial intelligence technology, remote monitoring system based on artificial intelligence and Internet technology emerges as the times require. The operation and control of modern mechanical equipment is gradually developing in the direction of automation, refinement, and integration. Therefore, to enhance the level of automation technology for remote mechanical monitoring (RMM) has been widely concerned by enterprises. The research on electronic technology of RMM has a deeper significance. RMM electronic technology is optimized on the basis of traditional mechanical monitoring, which makes it more convenient and fast, and realizes automatic monitoring fault detection at the same time. This paper mainly studies the RMM electronic technology based on KNN optimization algorithm, expounds the structure and working principle of the condition monitoring system, and analyzes the function of RTU and LabVIEW software platform. For the problem of time delay that may appear in the electronic technology of RMM, this paper proposes the time delay processing of RMM, uses buffer strategy to keep the random time delay consistent, and diagnoses the RMM system based on KNN optimization algorithm. In this paper, an intelligent fault monitoring model is established based on KNN optimization algorithm. According to the actual situation, five mechanical sample data are selected for experiments. Through the experimental training model, the mechanical pressure, oil temperature, vibration peak and engine speed of the sample machinery are observed. The experimental results show that in the fault monitoring training samples, the mechanical pressure of sample 1 is significantly abnormal compared with other samples, and the value is 0.98. The oil temperature of sample 3 is as high as 0.93, and the engine speed of sample 4 is significantly lower than other samples, which is only 0.43, sample 5 has problems in both oil temperature and engine speed. The RMM system combined with KNN optimization algorithm can observe a number of indicators in fault monitoring, and it is easy to deal with the fault phenomenon of one cause and multiple effects and one cause and multiple effects.

Keywords: KNN optimization algorithm · Remote monitoring · Mechanical monitoring · Monitoring system

J. Macintyre et al. (Eds.): SPIoT 2021, LNDECT 98, pp. 314–322, 2022.
https://doi.org/10.1007/978-3-030-89511-2_40

1   Introduction

At present, with the rapid development of Internet technology, the popularity of mechanical automation is gradually increasing. While saving manpower and material resources, mechanical automation needs to be monitored to ensure its normal operation [1, 2]. With the continuous development of economy, the continuous development of enterprise scale, the continuous innovation of technology, and the increasing number of mechanical automation equipment, it is imperative to carry out remote monitoring to ensure its use parameters and the use safety of large machinery [3, 4]. Large scale mechanical equipment is a key part of mechanical automation. In case of failure and shutdown, it will cause stagnation of production work, even more serious mechanical accidents, which will bring huge losses and even social reputation impact to enterprises. Therefore, it is necessary to monitor large-scale remote machinery to ensure that the accident risk is strangled in the cradle [5, 6]. Remote monitoring system refers to the system in which the monitoring personnel are far away from the controlled object, can also know the current working status of the controlled object in time, and can change the current equipment operation status by sending instructions when necessary [7, 8]. The remote monitoring system enables the staff to monitor the operation of the current working equipment in real time only in the monitoring center, make accurate judgment on the sudden abnormal conditions at the first time, and avoid the occurrence of safety accidents. It integrates a variety of advanced technology levels in the new century, and has a long-term development prospect [9, 10].

In the research on the application of remote monitoring electronic technology, many scholars at home and abroad have studied it, and have achieved some research results. Hyde P and others pointed out that the research on remote monitoring in China started later than that in foreign countries. Although some practical systems have been developed, from the perspective of software system development, these systems have weak ability to cope with demand changes, and many domain knowledge and design experience cannot be effectively accumulated and inherited [11]. Orlosky J and others pointed out that at present, remote monitoring system is widely used in transportation, construction, electric power and other industries. In modern industrial production management, a large number of process parameters need real-time detection and automatic control, which has a wide range of applications [12]. These researches on remote monitoring electronic technology provide the corresponding theoretical support for the research of RMM electronic technology in this paper.

This paper mainly studies the RMM electronic technology based on KNN optimization algorithm. By studying the RMM system, the structure and working principle of the condition monitoring system are expounded. At the same time, the function of RTU and LabVIEW software platform is analyzed. For the problem of time delay that may appear in the electronic technology of RMM, this paper proposes the time delay processing of RMM, uses buffer strategy to keep the random time delay consistent, and diagnoses the RMM system based on KNN optimization algorithm. This paper establishes an intelligent fault diagnosis model based on KNN optimization algorithm, studies the monitoring performance index of the remote monitoring system, and tests the relationship between the vibration peak and the machine condition according to the vibration peak index.

2 RMM Electronic Technology Based on KNN Optimization Algorithm

2.1 RMM System Technology

The remote monitoring and analysis system of large-scale mechanical equipment combines the traditional condition monitoring and analysis technology with computer network technology, embedded system technology and virtual instrument technology. It is mainly composed of RTU and upper computer LabVIEW software platform. RTU itself as a server, its composition structure is remote terminal unit, analog signal sensor, signal conditioner, high-speed A/D converter, switch input circuit, switch output relay, network interface circuit and other peripheral circuits, which can realize data acquisition and remote transmission.

Remote Terminal Unit RTU can collect, process and store the analog electrical signal converted by vibration acceleration sensor and signal conditioner. While collecting the vibration signal of mechanical equipment, RTU unit can also collect the switching signal of equipment and control various switches of mechanical equipment through the switching value acquisition and control circuit. Through the serial port, it communicates with other devices on site to obtain the required data, and communicates with the remote host computer through the network to transmit the collected data to the host computer. Through Remote Terminal Unit RTU, complete parameters of large mechanical equipment running state can be obtained.

LabVIEW platform of remote host computer is used as the client. According to the requirements of data acquisition, instrument control, signal analysis and data processing, the design provides rich and perfect function icons, and controls the system through interactive graphical panel. The LabVIEW platform of remote host computer mainly receives the data collected by RTU of remote terminal unit to the host computer through TCP/IP transmission module, and displays the waveform of the signal. It uses the file management module to store and read the data, and uses the amplitude range analysis module to display the characteristic parameters of the signal and the out of limit alarm. LabVIEW platform uses the time domain analysis module to analyze the data signal in time domain, and do autocorrelation analysis to determine whether the signal contains periodic signal.

The working principle of the system is that the processor of RTU converts the analog electrical signal converted by sensors and signal conditioner into ad, collects the data, and then packages the collected data according to the required format. The software module of RTU uses embedded operating system to realize multi task real-time scheduling, including TCP communication task, electrical signal acquisition task, analog signal A/D conversion task, system parameter download task, data processing interrupt program, data package upload task and serial communication task. The upper computer platform communicates with the RTU through the network communication module to obtain all kinds of mechanical equipment data online and in real time. At the same time, it can also set the remote sampling parameters through the network. LabVIEW platform displays and analyzes the collected signal. If the signal exceeds the specified range, alarm information will be generated. At the same time, the signal is further analyzed by means of time domain and frequency domain analysis to extract the characteristic signal.

2.2 Function of RMM System

In the RMM system, the server uses GPRS network to collect and process the real-time operation data of construction machinery equipment, and creates the communication thread between the server and the client through socket mode to communicate. The user completes the operation of the system through the client according to his own requirements for the system. Therefore, the establishment of communication protocol, the server accurately process the data of the lower computer and timely respond to user requests, and the client clearly display the results of user requests become the focus of system software design.

(1) Server function analysis

Because the mechanical equipment has high requirements for data security, and the customer also expects to always monitor the equipment when monitoring the equipment, so in the design process of the system, the server adopts TCP/IP protocol for the communication between the equipment and the client. The server mainly realizes the following functions:

1) Realize the communication with the construction machinery equipment terminal in TCP mode: the server establishes the communication connection with the lower computer, and receives the data of the current equipment status regularly sent by the lower computer to the server. After receiving the data, on the one hand, the data is converted into internal unified standard for storage, on the other hand, the processed data is sent to the client monitoring the device.
2) Realize the communication with the client in TCP mode: the server establishes the communication service. After receiving the data request from the client, it first judges the validity of the client's data. If it is legal, it will respond to the corresponding request of the client.
3) Complete the connection and access to the database: the server receives the data sent by the lower computer, and stores the corresponding data to the SQL Server database through ADO.

(2) Client function analysis

The object of the client program is all users of the system, so the friendly interface design is the primary consideration of the client. Secondly, the functions of the system can be reflected in the relevant operations of the client. According to the relevant analysis of the demand analysis of the system, the main functions of the client are as follows:

(1) System user management. The system user management module mainly includes user login, system user adding and modifying functions. For the design principle of security, the user of this program must have the legal user name and password of the system to carry out relevant operations, so the legitimacy of the user's

identity must be checked before entering the system; for practical design principles, administrators can manage users accordingly, including adding users, deleting
users, modifying permissions and other functions.

(2) Equipment management of lower computer. In this monitoring system, the data
types of monitored objects are different, so the monitoring interfaces displayed by
different equipment types are inconsistent in the monitoring process. Therefore,
the monitoring pages of each equipment type should be customized in the system
design process.

(3) System equipment monitoring management. The function of system monitoring
management module mainly includes real-time data monitoring, historical data
query and alarm information statistics.

2.3 Delay Processing of RMM

Because of the randomness and uncertainty of the network, we can only find a way
to reduce the impact of the delay on the real-time performance of the RMM system
according to the rules and characteristics of the network. In the RMM, software is used
to measure and calculate the network traffic in real time. According to the traffic size
and network stability, the network bandwidth is effectively used to improve the control
accuracy. Accordingly, it is required to continuously develop the test software with good
performance in order to measure the network traffic more accurately.

In this paper, a buffer strategy is designed to keep the random time delay consistent.
Because the time delay is constant, the remote controlled machine can receive control
information at the same time interval. The size of the buffer can fully ensure the continuity
of command transmission. The command sent by the controller through the network is
first received by the buffer of the server. When the buffer is full, the control command
will be sent to the server according to the required transmission cycle, and the command
sent by the client will be continuously stored in the buffer, In this way, the random time
delay will be converted into a fixed time delay. But there is a case that must be considered,
that is, if all the commands in the buffer are sent to the server and the client's commands
have not yet arrived, the control system will be in the interrupt state. Therefore, it is
necessary to design a buffer management module, which calculates the network delay
in real time to determine the size of the buffer. It is the adjustment function of the buffer
management module that makes the control system not affected by the time delay. In
addition, we can also mix TCP and UDP to reduce network delay.

2.4 KNN Optimization Algorithm Monitoring and Diagnosis

Artificial intelligence neural network has a good learning ability. It can learn from the
condition monitoring data of the field system, and process and supplement the unknown
knowledge. However, neural network can not make full use of a large number of language
of domain experts to express knowledge. Therefore, this paper uses KNN neighbor
algorithm to combine neural network technology with KNN algorithm to broaden the
range and ability of neural network to deal with data information, so that it can not only
deal with accurate information, but also deal with fuzzy information or other imprecise

information, so that it is easy to deal with those fault phenomena with multiple causes and multiple causes.

KNN neighbor algorithm in determining the classification decision only depends on the nearest one or several samples to determine the category of the samples to be divided, mainly depends on the surrounding limited adjacent samples, rather than the method of identifying the class domain to determine the category, so KNN method is more suitable for remote mechanical fault monitoring than other methods. The KNN algorithm model is expressed as follows: let the data have n samples and each sample has m attributes, then the data can be expressed as:

$$Z_{n*m} = \begin{bmatrix} x_{11} \ x_{21} \ ,..., \ x_{1m} \\ x_{21} \ x_{22} \ ,..., \ x_{2m} \\ ... \ ... \ ,..., \ ... \\ x_{n1} \ x_{n2} \ ,..., \ x_{nm} \end{bmatrix} = [X_1, X_2, \ldots, X_m] \tag{1}$$

$$X_j = (x_{1j}, x_{2j}, \ldots, x_{nj})^{\tau} (j = 1, 2, \ldots, m) \tag{2}$$

Among them, $[X_1, X_2, \ldots, X_t](1 < t < n)$ is benign application, $[X_{t+1}, X_{t+2}, \ldots, X_m]$ is malicious application. When the similarity between the samples of monitoring values in the test data set and the training data set is not correct, the test sample is a malicious application, that is, it has a fault problem and needs to be repaired, otherwise it is a benign application.

3 Research and Analysis

3.1 Research Objects

This paper mainly studies the RMM system based on KNN optimization algorithm. This paper establishes an intelligent fault diagnosis model based on KNN optimization algorithm, studies the monitoring performance index of the remote monitoring system, and tests the relationship between the vibration peak and the machine condition according to the vibration peak index.

3.2 Research Process Steps

In order to study the application of KNN optimization algorithm in the RMM system, this paper compiles a program for sample training, establishes an intelligent fault monitoring model based on KNN optimization algorithm, and selects five mechanical sample data according to the actual situation for experiments. Through the experimental training model, the mechanical pressure, oil temperature, oil temperature and other parameters of the sample machine are observed the vibration peak value and engine speed and other performance indicators. This paper also takes 8 different mechanical engines as samples to test the relationship between vibration peak and machine condition.

4   Experimental Research and Analysis of RMM Electronic Technology Based on KNN Optimization Algorithm

4.1   Remote Monitoring System of KNN Optimization Algorithm

The application of KNN optimization algorithm in RMM system makes the judgment of mechanical fault more accurate and can deal with fuzzy information or other imprecise information. According to the fault monitoring process, this paper takes five remote mechanical samples as an example to establish an intelligent fault diagnosis model based on KNN optimization algorithm. Through the experimental training model, the mechanical pressure, oil temperature, vibration peak value, engine speed and other performance indicators of the sample machinery are observed. The results are shown in Table 1.

Table 1. Training analysis of RMM system

	Mechanical pressure	Engine-oil-temperature	Vibration peak	Engine speed
1	0.98	0.63	0.36	0.54
2	0.84	0.73	0.43	0.52
3	0.83	0.93	0.45	0.68
4	0.86	0.75	0.48	0.43
5	0.84	0.38	0.49	0.71

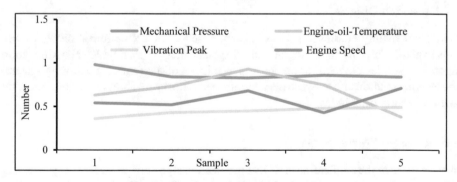

Fig. 1. Training analysis of RMM system

It can be seen from Fig. 1 that in the fault monitoring training samples, the performance indexes of remote mechanical samples are detected. It can be observed that the mechanical pressure of sample 1 is significantly abnormal compared with other samples, and its value is 0.98. The oil temperature index of sample 3 is as high as 0.93, and the engine speed of sample 4 is significantly lower than other samples, which is only 0.43, sample 5 has problems in both oil temperature and engine speed. The RMM system combined with KNN optimization algorithm can observe a number of indicators in fault monitoring, and it is easy to deal with the fault phenomenon of one cause and multiple effects and one cause and multiple effects.

4.2 Remote Monitoring of Vibration Peak

Among the fault monitoring indicators of the samples, the comparison of the indicators is from the standard source database. Through the comparison of the normal data in the database, the abnormal signal of the machinery is fed back. The vibration peak value of the mechanical sample is analyzed, the strength of the mechanical sample is assessed, and the mechanical vibration peak value is judged. In order to show the relationship between the peak vibration and the machine condition, eight engine samples from different machines are selected to monitor the peak vibration. The data results are shown in Table 2.

Table 2. Study on vibration peak of engine sample

	Vibration peak	Standard value	State judgment
1	0.12	0.25	Benign application
2	0.25	0.25	Benign application
3	0.51	0.76	Benign application
4	0.59	0.76	Benign application
5	1.39	1.27	Malicious application
6	2.78	2.52	Malicious application
7	3.97	2.52	Malicious application
8	5.42	5.21	Malicious application

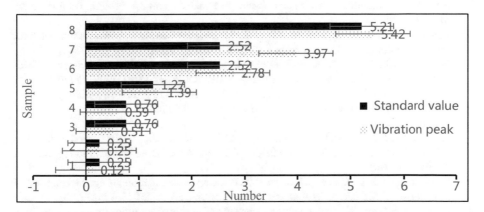

Fig. 2. Study on vibration peak of engine sample

It can be seen from Fig. 2 that the smaller the vibration peak value is, the better the condition of the machine will be. If the vibration peak value is within the data standard value, the mechanical data is a benign application, while if the vibration peak value is beyond the data standard value, the mechanical data is a malicious application, and the monitoring index is the vibration peak value.

5 Conclusions

With the further development of mechanical automation in China, the remote monitoring technology of mechanical equipment will be further improved. In this paper, the RMM electronic technology based on KNN optimization algorithm is studied. Taking the RMM system as the research object, the structure and composition of its state monitoring system are studied this paper deals with the problem of time delay which may appear in the electronic technology of RMM, and diagnoses the RMM system based on KNN optimization algorithm. In short, the remote monitoring technology of mechanical equipment must rely on the coordination and integration of multi-disciplinary algorithms, and continuous improvement can adapt to the development of mechanical automation and play its role.

References

1. Battista, L.: A new system for continuous and remote monitoring of patients receiving home mechanical ventilation. Rev. Sci. Instrum. **87**(9), 57–1153 (2016)
2. Jung, J., Sang, B.L., Lim, C., et al.: Electrical monitoring of mechanical looseness for induction motors with sleeve bearings. IEEE Trans. Energy Convers. **31**(4), 1 (2016)
3. Minh, N.T., Zorin. V.: Methods of remote monitoring of operability of mechanical systems. IOP Conf. Ser. Mater. Sci. Eng. **918**(1), 012080 (4 p.) (2020)
4. Gn, Y.M., Moses. R., Vyas, A.: S58 the use of remote monitoring to assess ventilator adherence and outcomes within a regional home mechanical ventilation service. Thorax **71**(Suppl. 3), A34.2–A35 (2017)
5. Blanik, N., Heimann, K., Pereira, C., et al.: Remote vital parameter monitoring in neonatology – robust, unobtrusive heart rate detection in a realistic clinical scenario. Biomed. Tech. **61**(6), 631 (2016)
6. Marcelli, E., Capucci, A., Minardi, G., et al.: Multi-sense CardioPatch: a wearable patch for remote monitoring of electro-mechanical cardiac activity. ASAIO J. **63**(1), 1 (2017)
7. Rose, M., Schettler, S., Klemm, F., et al.: Mechanical properties of remote-laser cut CFRP and thermographic laser-process monitoring. Mater. Sci. Appl. **11**(8), 560–575 (2020)
8. Aldosari, M., Al-Rawabdeh, A., Bullock, D., et al.: A mobile LiDAR for monitoring mechanically stabilized earth walls with textured precast concrete panels. Remote Sens. **12**(2), 22 (2020)
9. Radogna, A.V., Siciliano, P.A., Sabina, S., et al.: A low-cost breath analyzer module in domiciliary non-invasive mechanical ventilation for remote COPD patient monitoring. Sensors **20**(3):653 (2020)
10. Ng, Q.A., Loo, N.L., Chiew, Y.S., et al.: Mechanical ventilation monitoring: development of a network data acquisition system. IFAC-PapersOnLine **53**(2), 15916–15921 (2020)
11. Hyde, P., Defossez, F., Ulianov, C.: Development and testing of an automatic remote condition monitoring system for train wheels. Intel. Transp. Syst. (IET) **10**(1), 32–40 (2016)
12. Orlosky, J., Theofilis, K., Kiyokawa, K., et al.: Effects of throughput delay on perception of robot teleoperation and head control precision in remote monitoring tasks. Presence Teleop. Virt. **27**(2), 226–241 (2020)

The Design of Household Intelligent Health Care Products Based on Internet Technology

Xia Zhou[✉]

Department of Mechanical and Electrical Engineering, School of Hope,
Southwest Jiaotong University, Chengdu, Sichuan, China
zxswjtu@yeah.net

Abstract. Health has always been the focus that people pay attention to. Under the background of "Internet +" various kinds of household intelligent health care products have emerged in endlessly in the market, and have become the design achievements with great economic and social value in the health care industry. In the context of the Internet, combined with the development of "Internet +" network and Intelligent high-tech, this paper discusses the innovative design practice of household intelligent health care products, summarizes the design criteria of such products, and solves the remaining technical barriers among countries, product homogeneity and uneven quality issues to promote innovative home intelligent health care products can better serve people's healthy lives.

Keywords: Internet technology · Intelligentization · Household health care products · Innovative design

1 Introduction

Because of the difficult situation caused by the COVID-19 epidemic, all kinds of intelligent health care products based on health theme are more welcome and valued by people. With the support of "Internet +" thinking and technology, various kinds of smart home health devices have become the necessary products for people's health care at home due to the integration of advanced Internet and intelligent technology, bringing a diverse user experience to people [1].

2 Product Features

As early as 2006, there was a surge in the development of smart electronic hardware in the United States. With the development of Internet technology, IT presents the trend of low power consumption, miniaturization and low cost. The related products on the market are dazzling, dazzling. Taking the European and American markets as an example, it has two characteristics: First, it has developed a wide range of smart and healthy household products, ranging from common household blood pressure meters and body fat meters to hats and clothing, and second, it has an early start in its intelligent industry,

© The Author(s), under exclusive license to Springer Nature Switzerland AG 2022
J. Macintyre et al. (Eds.): SPIoT 2021, LNDECT 98, pp. 323–331, 2022.
https://doi.org/10.1007/978-3-030-89511-2_41

the development pattern is becoming mature, and the large number of fans and high consumption level of the products can provide effective feedback for product research and development. The quality of the intelligent health devices produced by some enterprises can even reach the level of medical use [2, 3]. Compared with the European and American countries, our country starts late but develops rapidly. After 2010, Internet companies such as Xiaomi have stepped into the field of health-related smart products, with a number of smart health hardware companies investing between 2013 and 2014, pushing the industry to upgrade. Coupled with the growing emphasis on health management among consumers, this trend has stimulated both innovative product development and a booming market [4].

Although domestic intelligent health products are developing rapidly under the action of various resultant forces, there are still many problems, summarized as follows: First, the lack of humanization in design [5]. The attention to the physical and mental health of users is not comprehensive enough. Some products simply superimpose their functions, or load their functions directly onto mobile APPS, without taking into account the detailed needs of customers. In addition, some enterprises abuse the concept of intelligentization, leading users to misdirect their consumption, leads to the product Salable Heat Reduces; third, the product supply way is unitary. Most smart health devices are happy to sell online in order to make a name for themselves, but due to the lack of user experience, some older people can not be very good access to and trial products on the sidelines. The lack of offline promotion greatly limits market development [6].

The innovative development and market development of household intelligent health products can realize sustainable development better only by facing these problems squarely and actively studying countermeasures.

3   Definition and Countermeasures

3.1   Definition of Smart Home Health Care Products

Home intelligent health care products are highly intelligent products which can be used by individuals or family members to maintain and keep fit under the two-way drive of "internet +" thinking and technology.

3.2   Innovative Design of Double Drill Design Pattern

In 2005, the British Design Association (BDA) put forward the idea of "dual-dispersion -- focusing design pattern", which holds that designers should start from questioning the problems of products, and go deep into the root causes behind the problems, and focus on one of the core issues, take a scaled-up solution, and then, from there, find a set of optimal solutions [7]. This design concept is known as the "twin-drill design pattern". I think this design concept still has an important reference value for the development and design of household intelligent health care products (see Fig. 1).

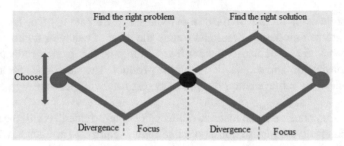

Fig. 1. Dual drill design pattern

From the thinking level, "Internet plus" has changed people's habitual thinking mode and life style, and also refreshed people's old cognition of product consumption, such as: people stay indoors, you can buy goods on smartphone apps such as Taobao, Pinduoduo and Jd.com, and you can have the chance to get a free gift to experience the product, and at the same time, make people enthusiastic about the online experience and cultural consumption, more emphasis is placed on the emotional communication brought about by human-computer interaction, so that the design of home health care products should not rest on the simple technology stack, but instead present a colorful cultural, entertaining, easy-to-use and emotional interaction of the products, and popular with the public [8].

At the technical level, smart product design, which integrates various Internet technologies such as mobile interconnection, Smart Sensor, big data and cloud computing, offers more innovative possibilities. Designers and users have developed their preference for products from "form-following function" to human-computer interaction experience, the co-existence and integration of shape, function, color, technology, material, interaction and emotion in product design will be the focus of future research [9]. The application of big data analysis technology can collect, analyze and sort out the massive data produced by the users'unconscious behavior, which not only reduces the tedious data collection work intensity in the product research stage, also through the big data technology for in-depth analysis, combined with the "dual-drill design model", can obtain profound and economic value of user demand information, insight into which important product feedback, improve the accuracy and humanization of product design to achieve more practical product innovation and service.

Therefore, high and new Internet technology is like a magic finishing touch. Designers can understand the real needs of users through big data analysis, extract valuable data, and make use of unique design language, infuse the product with entertaining, interactive, emotional and other experience elements, and transform into data visualization, personalized prescriptions, functional intelligence, and process simulation of intelligent selling points. Let the intelligent product design more innovative and more perfect!

3.3   The Essentials of Innovative Design of Three-Level Theory

Donald, American scholar. In his Book Design Psychology, Norman a put forward the three-level theory of product, that is, the design level of instinct, behavior and reflection.

User's different demand level, to the design request is different [10]. The level of instinct focuses on product modeling, color and sound; the level of behavior focuses on product operation behavior and human-computer interaction; and the level of reflection focuses on user identification and satisfaction with the product. The same goes for smart home health products. To achieve better design, I suggest following these design guidelines:

(1) A two-prong design at the instinctive level. "people-oriented" is still the first principle of innovative design of household intelligent health care products. All intelligent products should be based on the premise of satisfying people's psychological and physiological needs. Great products are based on good user experience with minimal design. Therefore, the designer should first carry out market research according to the needs, behaviors and usage of the target consumers, and then construct different modules according to the needs of the target consumers to carry out targeted detailed design, the end user experience to meet the user's level of instinct requirements, which is the most intuitive, most attractive to users. For example, the design of the portable John Field Health Care Bracelet, the appearance of the lower end of the arc broke the characteristics of the traditional circular shape, with five colors can basically meet the needs of the user diversity. The bracelet especially hides the opening setting, which makes the form integrity of the product design more prominent. Function button is slightly prominent in the surface of the bracelet, reducing user error probability, simple and easy to use design (see Fig. 2).

Fig. 2. John field

(2) Intelligent operation at behavior level. Under the background of "Internet" era, the usability of products based on user behavior is one of the standards to test whether product design is excellent. Intelligent products are Internet technology products based on behavior design. Home intelligent health care products as a modern hot selling digital products because of the integration of high-tech Internet technology easy to use, easy to use quality products. For example, at present, in order to facilitate middle-aged and elderly people to seek medical treatment, many communities use network technologies such as big data sharing to combine medical treatment with the Internet and achieve Internet medical treatment. This industrial model makes use of the advantages of the Internet to make up for the shortcomings of traditional

medical treatment. Users can monitor their own health status in real time by using all kinds of intelligent health care devices, such as intelligent detectors for blood pressure, heart rate, sleep quality, etc., without leaving home. Residents can transmit the data to the medical center through the medical APP on their smart phones, and doctors can obtain the user's physical health data. In this way, users can easily and quickly complete the monitoring or diagnosis and treatment at home, greatly shortening the diagnosis and treatment time. The intelligent system of community medical network can also develop personal electronic health records for residents in the community, which can relieve the problem of hospital congestion to a certain extent, making it very convenient for users to seek medical treatment at home (see Fig. 3 and Fig. 4).

The pleasure of using the product is the core level that users pay attention to, and it is the real reason that consumers buy or use the product. High-level behavior design level will bring higher product satisfaction. Intelligent products generally use the human-computer interaction technology, the greatest extent to enhance the level of product intelligence. Through information refinement, Ai Association and analysis technology, the product can acquire a large number of users' habitual interactions in the process of use, and build the "ideas" of intelligent products through big data technology, finally through the cloud computing technology to complete the interactive behavior of the machine. In the future, the ultimate version of intelligent health products for home use will realize closed-loop automatic control, and truly become studious and diligent expert-type intelligent health care products. Such as: multi-sensory system and sensor information integration of home health care robot, for digital home health care can provide medical monitoring, remote diagnosis and consultation services, can be functional, all four aspects of behavioral design, such as ease of use and physical sensation, have evolved to be as good and perfect as a dedicated family doctor.

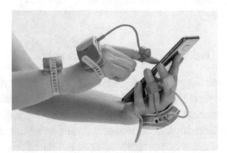

Fig. 3. Wearable smart medical device

(3) A reflective big data platform. The user's instinct and behavior level is about the product shape design and the use time real feeling, is relatively short, but the user's reflection level duration is long, through uses the product, the user will think the person and the product relations, this level determines whether the user is willing

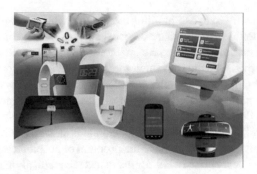

Fig. 4. Home intelligent physical examination system

to buy the product again, that is, brand loyalty. If the product is good and easy to use, people's satisfaction and recognition of the product is high, it is possible to buy again, and vice versa. However, the reality is that in the face of tens of millions of users in the network product feedback data, various interest groups have set up their own exclusive healthy open platform, each in its own way, leading to a serious fragmentation of data collection, this situation is extremely detrimental to improving the user experience of the product. Obviously, multi-device multi-dimensional collaborative work needs a unified data collection and analysis platform, so as to facilitate cross-device cross-system collaboration, can achieve multi-user product experience. Therefore, the value focus of network product feedback data is embodied in the construction of the unified big data platform of "everything interconnected".

4   Design Trends for the Future

At present, home intelligent health care products have the characteristics of automatic intelligent interactive products. Generally speaking, the design trends of such products in the future are as follows:

4.1   Human Computer Interaction Data Visualization

Data visualization refers to the process of interaction between products and users through the electronic screen can directly display the user's heart rate, heart rate, blood pressure and other medical information (Fig. 5). Relying on the Internet, Smart Sensing, cloud computing and other modern high tech products to achieve human computer interaction data visualization, so that users can directly see personal health data, thereby improving the user experience, at the same time, it can also encourage users to monitor their physical condition in real time and keep exercising indoors. For example: Millet Home Smart Blood pressure meter is a home for the elderly can be real-time detection of personal blood pressure of a classic product. This product powerful data processing ability can realize "blood pressure high and low automatic detection", among them, the 4.3 inch Chinese color screen and live audio player let middle aged and elderly users know about their health without having to consult a manual. The product is easy to use, can connect directly to Wifi, wechat interaction. Users only need to wear the cuff, the product can complete automatic pressurization, automatic measurement, automatic analysis, automatic broadcast and other functions, the whole process only need to press the key once, simple and practical operation.

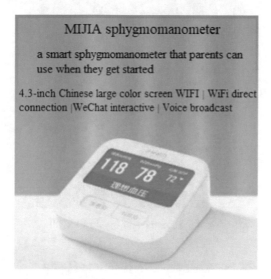

Fig. 5. Millet Home Smart Blood pressure meter

4.2   Sports and Fitness Entertainment

Sports and fitness entertainment refers to the intelligent health care equipment to meet the needs of users at home sports, but also to provide users with interesting and novel way of fitness. This kind of intelligent products use internet technology, can make the original boring and difficult to adhere to the fitness exercise process become rich and interesting, the experience of the exercise process is full of all kinds of surprises, much favored by users. For example: Yi Jian Intelligent treadmill 8009 (Fig. 6) with ingenuity and intelligence skills, after 7 technological breakthroughs, 54 optimization upgrades, finally has 18 grade automatic adjustment of the lifting double motor running table, automatic refueling system, patented six-shock Absorption Technology, 55 cm luxury running belt and 15.6-inch color screen. Through the Internet, with its own wide-screen display online entertainment programs, so that users at home sports at the same time, enjoy audio-visual entertainment, happy body and mind.

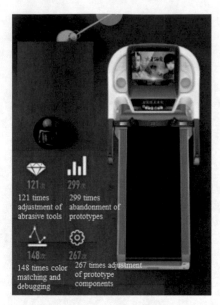

Fig. 6. Yi Jian smart treadmill 8009

4.3   4Digitalization of Ways to Maintain Health at Home

Home health care is convenient means that users at home can use the Internet technology under the support of intelligent health care equipment, easy to achieve a variety of user beauty, health care needs. Such as: left-point AI 2Max SMART moxibustion box, simple product shape, can connect the Xiaomi APP, intelligent products, easy to use, accurate temperature control, 5-mode free adjustment, to meet the needs of different users. At the same time, the product breaks through the tradition, no fire, no smoke, safe and effective, so that the fast-paced modern people at home can experience the power of traditional moxibustion medicine, feel the progress of digital technology (see Fig. 7).

Fig. 7. Left: Ai 2Max SMART moxibustion box

5   Conclusion

Under the dual driving conditions of "Internet" thinking and technology, the upgrading of intelligent products has been stimulated. People's yearning for a healthy life and the demand for home fitness expand the marketing market of household intelligent health care products. This paper puts forward the point of view that the innovative research and development process of household intelligent health care products will be more perfect under the double-drill design mode. Through the case analysis, it summarizes the design essentials of realizing innovative design of intelligent products under the three-level product theory, and points out the design trend of household intelligent products based on the "Internet". The research results have a high reference significance for the innovative design of intelligent household health products in the future. At the same time, the subsequent relevant practice and exploration and the intelligent products on the market will further verify the existing research conclusions and continue to explore new ideas and new methods for the design of such products.

Acknowledgments. Provincial Project Fund: Research on training mode of diversified industrial design talents in private universities (MBXH19YB025).

References

1. Huang, W., Yi, Z., Xu, H., Wu, J.: Research on innovative design of home fitness products based on internet +. Packag. Eng. **11**(6)–(10) (2017)
2. He, S.: Research on the design of healthy home products in the intelligent age. Design. 2018. 120 Zhao, Y., Liu, Y.: Analysis of home health product design in the intelligent age. Packag. Eng. **11**, 6–10 (2017)
3. Zhao, Y., Liu, Y.: Analysis of home health product design in the intelligent age. Packag. Eng. **11**, 6–10 (2017)
4. Zhang, S.: Industrial design in the age of artificial intelligence. Dual Use Technol. Prod. 196–197 (2017)
5. Yang, M., Hu, Y.: Internet + and the innovation of traditional household appliances. Packag. Eng. **37**(12), 139–142 (2016)
6. Zhang, R.: Product innovation design in the big data era. J. Heihe Univ. **7**(08), 49–50 (2016)
7. Qin, J.: Big interaction design in the age of big data. Packag. Eng. **36**(8) (2015)
8. Gong, G., Zhu, Q.: Research on technology innovation and development prospect of sports fitness products. Tech Startup Monthly, **29**(1), 106109
9. Wang, Y., Qiu, X.: Experience-oriented interactive packaging design. World Packag. **108**, 109 (2010)
10. Xue, X.: Research on interactive experience packaging design practice under Internet thinking. Packag. Eng. **37**(16), 146–148 (2016)

Improvement Strategy of Distributed Delay Task Based on Timing Wheel Algorithms

Ping Wu[✉]

College of Information and Control Engineering, Weifang University, Weifang, Shandong, China

Abstract. Delay task are widely used in project development and can be implemented in various ways. For distributed and independent delayed task processes, there are often no appropriate modules. Based on the existing Timing Wheel algorithm, this paper improves it by adding the concept of round to the link list node of the timed task, and designs a strategy to implement the delay task with precision of 1 s, which can realize the distributed multi-process deployment and load balancing conveniently, and make the process occupy as little memory as possible.

Keywords: Timing Wheel · Distributed delay task · Round

1 Introduction

Delay task, just as its name implies is point to the task that is delayed for a period of time. Delay task are different from scheduled tasks. Timed task often have a fixed cycle and a definite trigger time. Generally, delay task do not have a fixed start time. It is often triggered by one event, and triggers another event for a period of time after that event. Delay task are widely used in project development. For example, if the order is not paid within 15 min after the order is generated, the system automatically closes the order; a user reminder message sent before an event starts; if the order is not answered within the prescribed time, it will be set to overtime; a periodic consistency check, and so on.

There are many ways to implement delay task. A simple delay task function can be easily implemented by ticker in Go language. But if we want to achieve distributed and independent delay task, there is no corresponding module. This paper describes a delay task implementation strategy based on Timing Wheel algorithm with accuracy of 1 s, which can realize distributed multi-process deployment and it can easily achieve load balancing.

2 Timing Wheel Algorithm

Typically, delay task are implemented by Timing Wheel, which can be regarded as an efficient delay queue or timer. As shown in Fig. 1, the time wheel consists of eight slots, each slot is associated with a linked list, and the timer triggers the rotation of a slot and performs the corresponding tasks.

© The Author(s), under exclusive license to Springer Nature Switzerland AG 2022
J. Macintyre et al. (Eds.): SPIoT 2021, LNDECT 98, pp. 332–338, 2022.
https://doi.org/10.1007/978-3-030-89511-2_42

There are a lot of delay operations in Kafka. The delay operation of Kafka is a relatively independent component. Its main function is to manage the delay operation. The bottom layer is implemented through the time wheel provided by Kafka. Instead of using JDK's own Timer or Delay Queue to implement delay, Kafka customizes a system Timer based on the time wheel [1]. The average time complexity of insertion and deletion of JDK's Timer and Delay Queue operations is O(nlog(n)), which can not meet Kafka's high performance requirements. The time complexity of insertion and deletion operations can be reduced to O(1) based on time wheel.

Timing Wheel in Kafka is a circular queue that stores timing tasks. The bottom layer is implemented by an array. Each element in the array can store a Timer Task List. Timer Task List is a circular two-way linked-list. Each item in the linked-list represents a Timer Task Entry, which encapsulates the real Timer Task [2, 3].

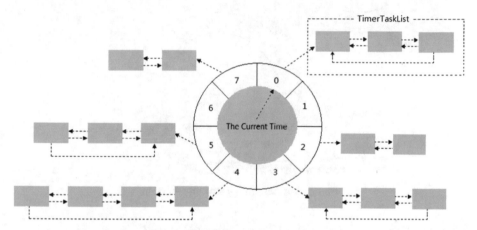

Fig. 1. Timing Wheel structure diagram

Timing Wheel provides the concept of hierarchy, because the time span is relatively large and the number is large, the single-layer timing wheel will result in a large round of tasks, and the linked-list of individual lattices is long. Therefore, Kafka adopts the concept of multi-tier timing wheel. When the expiration time of the target delayed task exceeds the time range indicated by the current timing wheel, it is necessary to create a higher-level timing wheel and try to add the task to the higher-level timing wheel. As shown in Fig. 2, in general, the time span of the first floor is the smallest, and the time span of the second floor is relatively large. If the task time exceeds the cycle of the lower wheel, it will be put into the upper wheel in turn. However, the multi-layer wheel needs to move the corresponding task with time, which is more complicated in management [7].

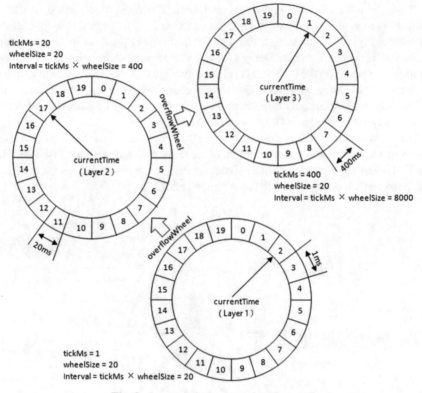

Fig. 2. Layered timing structure diagram

3 Improvement Strategy of Distributed Delay Task

The delay task module delaytask designed in this paper adopts the idea of adding round to the linked-list node of the timed task. That is to say, the current task will be executed after running round cycles. Its definition is as follows:

```
type TaskNode struct
{
    Runner
    round uint64        // Execution after running rounds
    // Pool interface
    worker IWorker      // Perform the worker pool of task
}
type runnerInfo struct
{
    task *TaskNode      // TaskNode of runner
    slot int64          // Runner's slot in wheel
}

type Wheel struct
{
    ticks time.Duration
    // Number of slots in time wheel
    count int64
    // Index
    index int64
    // Slots
    slots []*Node       // Time wheel slots, each containing a TaskNode
    // linked-list and taskNode count
    // Save the index of task and task
    runnerMap map[int64]*runnerInfo
}
```

The following key points need to be guaranteed in order to realize the delay task: (1) Guarantee the time accuracy of tasks and avoid excessive delays. (2) Memory occupancy. When there are more tasks and longer delays, the process occupies as little memory as possible. (3) Distributed deployment. (4) Recovering current situation. When the execution module is down, recover current situation as soon as possible [8].

3.1 Accuracy of Tasks

In order to ensure the execution of timed tasks, timed tasks are usually switched to a separate goroutine pool to execute; at the same time, in order to ensure the sequential process of adding tasks and triggers of timers. Therefore, the time wheel is put into a separate goroutine, and the added tasks and timer events are switched to the timewheel goroutine to execute, which not only ensures the sequentiality, but also avoids the use of locks [2, 4–6].

3.2 Memory Occupancy

If there are few tasks, put them directly in memory; but if there are millions of tasks, especially those with long latency, put them all in memory, it will lead to low native memory utilization. Therefore, the delay task module designed in this paper has built-in a threshold (integral multiple of the time wheel round time) according to the timewheel. When the execution time is longer than this threshold, the task is serialized and put into redis, and the waiting task is managed by sorted set; the time wheel goes to redis every round time to extract the tasks to be executed. Thus the process occupies as little memory as possible.

3.3 Distributed Deployment

Through redis pub/sub, the client serializes the task and sends it to redis, where delaytask subscribes to the task. If you want to achieve load balancing, you can do it through subscribed channels, such as delaytaskchannel-0, delaytaskchannel-1, delaytaskchannel-2; the client sends the task to the specified channel through a certain load balancing algorithm [9].

3.4 Recovering Current Situation

The task entered into timewheel is serialized into redis ongoing task list, and the corresponding task is deleted from ongoing when task is completed; if the system crashes, the task is read from ongoing task when the system is restarted, and if the task is not timed out, it is put directly into timewheel to execute.

4 Analysis of Test Results

The tester uses Apple MacBook. It has 4* Intel (R) Core (TM) i7-4870HQ CPU @ 2.50 GHz and 16 G memory.

After the tester performs HTTP get Baidu tasks 50,000 times (part of it is timeout failure by setting 1 s timeout), the memory consumption is about 8M, and the peak value is about 12M. The test results of concurrent execution of more than 500 tasks per second show that the error between the planned execution time ToRunAt and actual execution time RunAt of tasks is less than 1 s.

The number of workers in the goroutine pool can be changed according to the type of task. For IO tasks, the number of workers can be increased appropriately, the default value is CPU number. CPU performance test flame diagram can be obtained by using Go pprof performance analysis tool, as shown in Fig. 3.

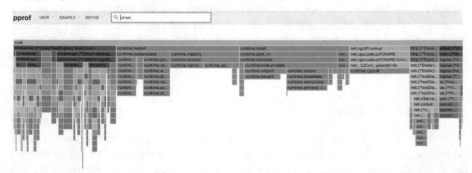

Fig. 3. CPU performance test flame diagram

Through the flame diagram, we can easily find where the CPU spends most of its time, and we also know how to optimize it.

We can use the top command of pprof to get the CPU consumption statistics as shown in Fig. 4. Through analysis, it is found that besides the occupancy of system scheduling, logrus and redis operations take up a lot of time, so if log are removed, CPU occupancy should be reduced and running speed should be improved.

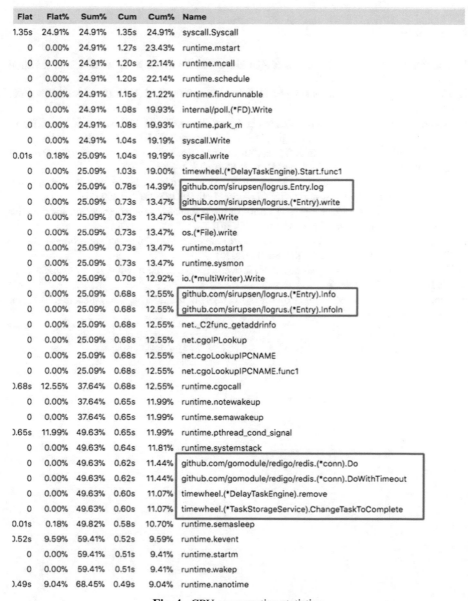

Flat	Flat%	Sum%	Cum	Cum%	Name
1.35s	24.91%	24.91%	1.35s	24.91%	syscall.Syscall
0	0.00%	24.91%	1.27s	23.43%	runtime.mstart
0	0.00%	24.91%	1.20s	22.14%	runtime.mcall
0	0.00%	24.91%	1.20s	22.14%	runtime.schedule
0	0.00%	24.91%	1.15s	21.22%	runtime.findrunnable
0	0.00%	24.91%	1.08s	19.93%	internal/poll.(*FD).Write
0	0.00%	24.91%	1.08s	19.93%	runtime.park_m
0	0.00%	24.91%	1.04s	19.19%	syscall.Write
0.01s	0.18%	25.09%	1.04s	19.19%	syscall.write
0	0.00%	25.09%	1.03s	19.00%	timewheel.(*DelayTaskEngine).Start.func1
0	0.00%	25.09%	0.78s	14.39%	github.com/sirupsen/logrus.Entry.log
0	0.00%	25.09%	0.73s	13.47%	github.com/sirupsen/logrus.(*Entry).write
0	0.00%	25.09%	0.73s	13.47%	os.(*File).Write
0	0.00%	25.09%	0.73s	13.47%	os.(*File).write
0	0.00%	25.09%	0.73s	13.47%	runtime.mstart1
0	0.00%	25.09%	0.73s	13.47%	runtime.sysmon
0	0.00%	25.09%	0.70s	12.92%	io.(*multiWriter).Write
0	0.00%	25.09%	0.68s	12.55%	github.com/sirupsen/logrus.(*Entry).Info
0	0.00%	25.09%	0.68s	12.55%	github.com/sirupsen/logrus.(*Entry).Infoln
0	0.00%	25.09%	0.68s	12.55%	net._C2func_getaddrinfo
0	0.00%	25.09%	0.68s	12.55%	net.cgoIPLookup
0	0.00%	25.09%	0.68s	12.55%	net.cgoLookupIPCNAME
0	0.00%	25.09%	0.68s	12.55%	net.cgoLookupIPCNAME.func1
0.68s	12.55%	37.64%	0.68s	12.55%	runtime.cgocall
0	0.00%	37.64%	0.65s	11.99%	runtime.notewakeup
0	0.00%	37.64%	0.65s	11.99%	runtime.semawakeup
0.65s	11.99%	49.63%	0.65s	11.99%	runtime.pthread_cond_signal
0	0.00%	49.63%	0.64s	11.81%	runtime.systemstack
0	0.00%	49.63%	0.62s	11.44%	github.com/gomodule/redigo/redis.(*conn).Do
0	0.00%	49.63%	0.62s	11.44%	github.com/gomodule/redigo/redis.(*conn).DoWithTimeout
0	0.00%	49.63%	0.60s	11.07%	timewheel.(*DelayTaskEngine).remove
0	0.00%	49.63%	0.60s	11.07%	timewheel.(*TaskStorageService).ChangeTaskToComplete
0.01s	0.18%	49.82%	0.58s	10.70%	runtime.semasleep
0.52s	9.59%	59.41%	0.52s	9.59%	runtime.kevent
0	0.00%	59.41%	0.51s	9.41%	runtime.startm
0	0.00%	59.41%	0.51s	9.41%	runtime.wakep
0.49s	9.04%	68.45%	0.49s	9.04%	runtime.nanotime

Fig. 4. CPU consumption statistics

5 Conclusions

This paper designs an improved strategy for distributed delay task based on Timing Wheel algorithm. By adding the concept of round to the linked list nodes of the timed tasks, and improving the algorithm of several key points of the delay task, the distributed multi-process deployment can be realized, the load balancing can be easily realized, and the process occupies as little memory as possible.

References

1. Gao, G., Wang, N.: A software timer algorithms based on time wheel mechanism. J. Ezhou Univ. **25**(5), 104–106 (2018)
2. Xu, L., Zheng, J.: Improvement of the task-delay management in μC/OS-II. Inf. Secur. Technol. **7**(5), 67–69 (2016)
3. Wang, Z., Zhao, Y., Wang, G.: Research and implementation of load balancing algorithm in heterogeneous Flink cluster. J. Nanjing Univ. Nat. Sci. **57**(1), 110–120 (2021)
4. Wu, P.: Ad-hoc detector a high-efficient ad-hoc synchronization detector for multi-threaded applications. ICIC Express Lett. part B Appl **6**(10), 2661–2666 (2015)
5. Zhu, H., Li, T., Yan, J., Zhang, Z.: Research on a distributed task scheduling algorithm based on dynamic load balancing. Chin. High Technol. Lett. **24**(12), 1261–1269 (2014)
6. Liu, Y., Kao, Y., Gu, S., Karimi, H.: Soft variable structure controller design for singular systems. J. Franklin Inst. **352**(4), 1613–1626 (2015)
7. Yan, X., Zhang, G., Qiu, X.: Improved partition overload optimization algorithm in Kafka. Comput. Technol. Develop. **30**(12), 88–91 (2020)
8. Wu, P.: Efficient self-tuning strategy for hardware transactional memory. ICIC Express Lett. **10**(10), 2391–2396 (2016)
9. Fan, C., Su, R.: Task scheduling optimization algorithm based on load balancing. Comput. Eng. Design **38**(6), 1532–1535 (2017)

Application of VR Virtual Simulation Technology in Teaching and Learning

Wenye Yu[1](✉) and Zhenyu Chen[2]

[1] NanChang Institute of Science and Technology, Nanchang 30108, Jiangxi, China
[2] Jiangxi Tourism and Commerce Vocational College, Nanchang 330108, Jiangxi, China

Abstract. With the rapid development of modern technology, the promotion and popularization of computer multimedia technology and the updating of teaching concepts and methods, a variety of new teaching methods are emerging, and virtual simulation technology as a new teaching media began to appear in the field of education and teaching. The purpose of this paper is to investigate the application of VR virtual simulation technology in teaching and learning. Based on this, this paper designs a high school geography flipped classroom model based on VR virtual simulation technology. This paper mainly uses the questionnaire method to conduct VR virtual teaching experiments in a middle school in this city, and then conducts questionnaires and analyzes the data on students. The experimental results showed that students' interest in the new teaching model increased dramatically. The average score of senior students rose from 74.21 in the previous test to 82.62, and the average score of sophomores also rose by 5.58, while the average score of seniors in geography also rose by 7.7, and their performance also improved significantly.

Keywords: VR · Virtual simulation technology · Teaching learning · Flipped classroom

1 Introduction

In recent years, with economic development and social progress, more and more people have realized that the success or failure of education determines not only a person's destiny, but also the country's future [1, 2]. One of the important ways to update teaching methods is to use modern information technology for teaching [3, 4]. At present, with the continuous advancement of virtual reality technology, virtual reality has involved all aspects of our lives, and it has changed people's views on themselves, the world, space and time. As for the education field, virtual reality has also emerged [5, 6].

Virtual reality technology provides an ideal effective teaching method for educators, and is currently widely used in military teaching, physical training and medical teaching abroad [7]. Virtual reality technology originated in the United States, and many related technical research institutions are also in the United States. Therefore, it can be said that the level of virtual reality technology in the United States can represent the overall level of virtual reality technology in the world [8]. The U.S. education field attaches great importance to the use of virtual technology. In 1994, the University of Washington and the Westinghouse Science Foundation organized a mobile teaching plan. A mobile teaching vehicle equipped with virtual equipment was used to educate students in various schools. Teaching services of virtual reality technology [9]. Zhejiang University has applied virtual planning and virtual design in architecture [10].

By designing a new teaching model, this paper first summarizes the challenges of carrying out practical teaching in the VR era, then proposes countermeasures for practical teaching, and finally designs a flipped classroom model for high school geography using VR virtual simulation technology, and conducts a questionnaire survey on students and analyzes the collected data.

2 Application Research of VR Virtual Simulation Technology in Teaching and Learning

2.1 Challenges of Carrying Out Practical Teaching in the VR Era

The technical problems of virtual reality helmets. From the results of related research, the resolution is usually required to be 4K and above in order to produce high-quality images. However, at this stage, many of these products have not reached the standard, and it is difficult to completely solve the dizziness problem in the use of the equipment [11, 12].

First, the threshold of teaching development technology remains high. It is necessary to combine a large number of technology and teaching resources to carry out virtual time teaching work, especially to solve core technical problems such as interactive technology and policy technology, then it means to strengthen the research and development investment in order to accurately understand and master the basic content of teaching and basic methods.

Second, instructional design is encountering a violent impact under the new situation. The application of VR technology to teaching activities has brought great convenience to teachers' teaching. In the new situation, how to plan the teaching content, how to fully reflect the interactivity in the learning process, and how to evaluate the effectiveness of learning are all problems that need to be solved in this type of teaching.

2.2 Countermeasures for Practical Teaching

Based on the VR industry, it should develop from the following points:

First, continue to upgrade equipment, improve and optimize current VR equipment, accelerate the development of new products, and enter the stage of product upgrading as soon as possible;

Second, strengthen technology research and development, continuously improve key technologies in this field, such as visual sensing, tactile feedback and other technologies, to break through the shackles of two different worlds, thereby enhancing user experience.

3 Design of Classroom Flipping Teaching Mode Based on VR Virtual Simulation Technology

3.1 Design Purpose

In order to verify the role played by VR virtual simulation technology in teaching and learning, this paper uses VR virtual simulation technology to design a high school geography flipped classroom model, and takes a high school student in this city as the experimental object, uses questionnaires to collect data, and finally analyzes the data, explore whether this kind of classroom flipped teaching model can stimulate students' interest, enhance the initiative of learning, and thus improve students' performance.

3.2 Design Process

The content of the flipped classroom teaching design based on VR virtual simulation technology mainly includes three parts: pre-class knowledge acquisition, classroom knowledge internalization and post-class reflection and summary.

(1) Pre-class pre-study
   Pre-class knowledge acquisition mainly includes analysis of teaching objectives and learner characteristics, then comprehensive teaching resources and design of teaching contents, while students learn independently before class and give feedback on independent learning results. The main process is shown in Fig. 1.

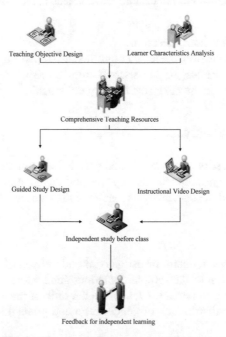

Fig. 1. Flowchart of pre-course preliminaries

(2) Internalization of classroom knowledge

Firstly, students report and display the results of their independent learning before class, and then the teacher determines the important and difficult points to be discussed in this class based on the students' report and display as well as the feedback of the Internet distance learning results before class. All learning that takes place on the Internet during the lesson can be permanently archived for the teacher to access and students to review later. The main flow of the in-class session is shown in Fig. 2.

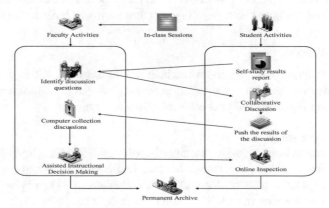

Fig. 2. Flow of the interactive session in the lesson

(3) Summary of after-school reflection

In the post-lesson session, the teacher quizzes the students again to get a full picture of their learning. Reflect on the shortcomings and summarize the experience to provide reference for the next time to carry out teaching.

4 Experimental Analysis

The first questionnaire survey was conducted before the implementation of the teaching mode, and the second questionnaire survey was conducted after the implementation of the teaching mode, and the data analysis results of the two surveys are as follows.

4.1 Sample Analysis

The first part of the questionnaire investigated the identity, according to which we can get the basic information of the sample, including gender and grade level. In terms of gender, the ratio of male to female is 1:1.2, and the ratio of male and female selected is more appropriate. The distribution of gender ratio and grade distribution are shown in Fig. 3.

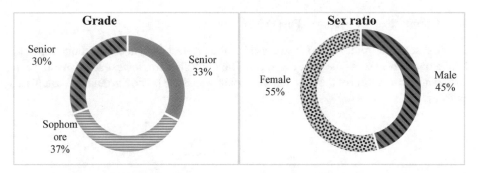

Fig. 3. Sex ratio distribution and grade distribution

4.2 Interest Analysis

First of all, students' interest in geography class in high school was analyzed, and the interest level of students in the three grades before and after the change of teaching mode is shown in Fig. 4.

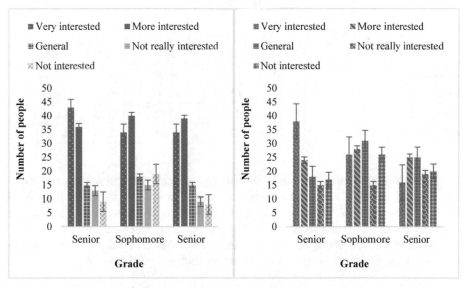

Fig. 4. Students' interest in geography class before and after changing the teaching mode

It is obvious from Fig. 4 that after using the flipped teaching mode based on VR virtual simulation technology, the students' interest in geography course has increased significantly, and the number of students who are very interested has increased from 80 to 111, up by 9.1%; from 22.4% to 33.5%. The number of students who are interested in the course has increased from 80 to 111, an increase of 9.1%. In general, the interest of students in all grades in geography courses has increased significantly after using the new teaching mode. This shows that the use of VR virtual simulation technology can improve students' interest.

4.3 Comparison of pre- and Post-performance

The first semester midterm exam was used as the pre-test (X1) and the second semester final exam was used as the post-test (X2). The highest score, lowest score, mean score and standard deviation of the students were mainly analyzed. The formula was calculated as follows.

Sample variance:

$$s^2 = \frac{\sum_{i=1}^{n} (x_i - \bar{x})^2}{n - 1} \tag{1}$$

Sample standard deviation:

$$s = \sqrt{s^2} = \sqrt{\frac{\sum_{i=1}^{n} (x_i - \bar{x})^2}{n - 1}} \tag{2}$$

Table 1. Test score

	Senior		Sophomore		Senior	
	X1	X2	X1	X2	X1	X2
Minimum	47.00	54.00	46.00	49.00	43.00	51.00
Maximum	94.00	95.00	95.00	96.00	93.00	96.00
Mean	74.21	82.62	73.54	79.12	73.53	81.23
Std. deviation	13.43	8.51	12.35	11.64	12.54	9.34

As can be seen from Table 1, after the change in teaching methods, students' scores improved in all three grades, starting with the mean scores, which increased from 74.21 in the previous test to 82.62 for seniors, 5.58 for juniors, and 7.7 for seniors. The standard deviation of students in all grades also decreased, indicating that the degree of dispersion is also getting smaller. Taking all the performance data together, the use of VR technology in the teaching process was helpful in improving students' performance.

5 Conclusions

The development and application of virtual simulation technology in education has been very popular and popular in countries with developed education, and it has also been widely used in some fields in China, which has attracted great attention from the education community. With the rapid development of computer software and hardware technology, network technology, and the continuous improvement of the level of education development, virtual simulation technology will be more widely used in education. The simulation effect will be better and the fidelity will be higher, creating good conditions for promoting the improvement of education level.

References

1. Little, J.: Using virtual simulation to increase deep learning in radiography students. Radiol. Technol. **92**(4), 324–330 (2021)
2. Chang, W., Zhang, Q., Xia, H., Yang, Y.: Construction of virtual simulation teaching platform for elevator control. J. Phys. Conf. Ser. **1848**(1), 012120 (2021)
3. Fang, M., You, F., Yao, R.: Application of virtual reality technology (VR) in practice teaching of sports rehabilitation major. J. Phys. Conf. Ser. **1852**(4), 042007 (2021)
4. Li, Z., Cao, Y., Luo, J.: Application of virtual simulation technology in chemistry teaching. E3S Web Conf. **267**, 02067 (2021). https://doi.org/10.1051/e3sconf/202126702067
5. Wang, G., Zhou, J., et al.: Civil engineering teaching research based on the virtual simulation technology. IOP Conf. Ser. Earth Environ. Sci. **267**(4), 42103–42103 (2019)
6. Xiao, X., Liu, X., Xiao, Z.: Construction and application of computer virtual simulation teaching platform for medical testing. J. Phys. Conf. Ser. **1915**(4), 042074 (2021)
7. Chen, J., Liang, M., Liu, H.: The Application and Discussion of virtual simulation platform in the teaching of ship handling and collision avoidance course. IOP Conf. Ser. Mater. Sci. Eng. **719**, 012075 (2020)
8. Shen, C., Qi, A.: An adaptive learning mode of "public psychology" based on creative thinking with virtual simulation technology. Int. J. Emerg. Technol. Learn. (iJET) **15**(23), 131 (2020)
9. Gu, Y., Huang, W.: Analysis of the application of virtual simulation software in food testing technology course with atomic absorption spectrum1–taking the example of determining calcium in drinking water by atomic absorption spectrometry. IOP Conf. Ser. Earth Environ. Sci. **680**(1), 012085 (2021)
10. Peng, Y., He, X., Huang, Y.: A virtual simulation experiment system for requirement analysis. J. Phys. Conf. Ser. **1757**(1), 012194 (2021)
11. Zhao, J., Ying, F.: Research on the construction of virtual simulation experiment teaching center based on computer-aided civil engineering in colleges and universities. J. Phys. Conf. Ser. **1744**(3), 032115 (2021)
12. Hong, H.: Research on the application of virtual simulation practice teaching in engineering cost specialty. E3S Web Conf. **253**, 02017 (2021)

Implementation of Security Technology of Cloud Computing Platform Based on Hadoop

Haijun Huang[✉]

School of Intelligent Systems Science and Engineering, Yunnan Technology and Business University, Kunming, Yunnan, China

Abstract. In recent years, cloud computing has been widely researched and applied, and the key factor that affects the popularity or acceptance of cloud computing is the security of cloud computing. As an open source basic cloud computing framework, Hadoop has become more and more widely used in the corporate world. Research on the security technology of cloud computing platform based on Hadoop is of great significance for better research on cloud computing security and promotion of Hadoop. The purpose of this paper is to study the security technology of cloud computing platform based on Hadoop. This article first summarizes the basic theories of cloud computing security and the core framework of Hadoop, and then researches and designs the cloud computing file management system based on Hadoop, and conducts research and analysis on its security solutions. This paper systematically expounds the architecture design, function design and detailed design of the cloud computing file management system, and uses comparative analysis and observation methods to study the subject of this paper. Experimental research shows that the choice of k (nearest neighbor range) has little effect on the detection rate. It is inferred that the detection rate is mainly affected by the length of the subsequence. However, as the value of k increases, the detection rate decreases slightly, but the false detection rate is proportional to the value of k, that is, the detection effect is relatively reduced.

Keywords: Hadoop · Cloud computing · Security technology · Research and implementation

1 Introduction

As an efficient open source platform, Hadoop is not only widely used in the field of cloud computing, but also more and more popular in the fields of massive data processing, data mining, and scientific computing. However, the security mechanism of Hadoop is very weak, which makes users have concerns when deciding whether to use Hadoop. At the beginning of its design, the issue of platform security was not considered.

Cisco announced that it will provide a new cloud computing framework CloudVerse in 2012, and plans to extend its SecureX security architecture to CloudVerse. SecureX includes the use of tagging technology in the Cisco firewall to identify information about user network applications, such as applications, devices, locations, and time of day, so

J. Macintyre et al. (Eds.): SPIoT 2021, LNDECT 98, pp. 346–352, 2022.
https://doi.org/10.1007/978-3-030-89511-2_44

that security decisions can be made in a familiar way [1, 2]. Rising launched a "cloud security" strategy to create an Internet environment of mutual trust. In response to the complex Internet environment, Rising believes that the Internetization of viruses and web browsing have become the main channels for the spread of various viruses, and more than 90% of the virus transmission history web pages are in line [3, 4].

The purpose of this paper is to improve the security of cloud computing, and to study the security technology of cloud computing platform based on Hadoop. By comparing the security technology of traditional cloud computing platform with the security technology of cloud computing platform based on Hadoop researched in this paper, This article analyzes the research theme of this article.

2 Security Technology Application of Cloud Computing Platform Based on Hadoop

2.1 Cloud Computing

Cloud computing is not an emerging technology, but a mode of IT resource integration under the existing network environment. Through the Internet or local area network, users can obtain resources such as software as a service, platform as a service, and hardware service in an on-demand and easily scalable manner [5, 6].

2.2 Design of File Management System Based on HDFS Cloud Computing

(1) HDFS

HDFS is a sub-project under the Hadoop project, and is a popular open source distributed file system in a variety of cloud computing platforms [7, 8].

(2) System architecture design

On the whole, the system is divided into three parts. The client is mainly responsible for interacting with the user, which is an important part of the user experience of the entire system. Various service requests of the user are directly transmitted from the client to the control center Masterservice [9, 10]; the client client uses the control center to respond to the message MetaData to fragment the file data and implement the pipeline pipeline to copy the data to the corresponding storage node according to the set block size according to the HDFS block data redundancy strategy; the user downloads the block data scattered in different data nodes on the system, through the client, according to the metadata mapping information provided by the control server, the block data is merged into a complete file [11, 12].

(3) System function design

This system, the cloud computing file management system, mainly consists of four functional modules: user management functions, directory management functions, file management functions, and shared management functions similar to public cloud features.

2.3 Safety Plan Design

(1) User audit management

User audit management is set for system administrators and is not visible to ordinary users. Its operation interface is also based on the login interface of the cloud computing file management system. The system administrator is a built-in user such as super_cloud_admin. The administrator logs in to the system, conducts a preliminary review of user information, enables, disables, and deletes users to achieve the effect of controlling users as a whole.

(2) DataNode identity authentication

When a user requests access to file data, malicious and illegal users can obtain metadata information about the file from the complex network environment, communicate directly with the DataNode without interacting with the NameNode, and illegally steal the relevant data, which brings irreparable damage to the user.

In order to prevent this security hazard, each time the client interacts with the DataNode, the DataNode authenticates the client's identity to ensure that the client's access is permitted by the main server NameNode to ensure that the NameNode controls the file system. This article uses an IBE-like system Identity authentication scheme.

IBE system parameter initialization

Step1. Obtain a finite field Fp characterized by a prime number p, and select field elements a, b on the finite field Fp, so that the field elements a, b satisfy the equation of the elliptic curve E(Fp):

$$y^2 = x^3 + a \cdot x + b (mod\ p). \tag{1}$$

Which satisfies the conditions:

$$4 \cdot a^3 + 27 \cdot b^2 \neq 0 (mod\ p). \tag{2}$$

Step2. Select a base point G on the elliptic curve E(Fp), and calculate the order q of the base point G, and then calculate the two q-order groups Gl and G2 of point G respectively. G1 and G2 meet the following bilinear mapping conditions:

$$G_1 \times G_1 \rightarrow G_2 \tag{3}$$

The client completes the initialization of the system parameter T according to the above-mentioned system parameter initialization rule, and the client maps the user identification, such as the email address EIDu, to a point EIDp on the elliptic curve E(Fp). Get the private key du = s.EIDp corresponding to the public key EIDu of the user.

3 Experimental Research on Security Technology of Cloud Computing Platform Based on Hadoop

3.1 Experimental Program

In order to make this experiment more scientific and effective, the abnormal data based on the sudden change of Hadoop cluster resource consumption was tested. Due to the detection efficiency of the k-nearest neighbor local anomaly algorithm based on the sub-sequence pattern, it is mainly composed of two parameters It is determined that the length of the subsequence seqLen and the neighboring range k, so in the laboratory, the parameters of these two parameters were trained separately.

3.2 Experimental Method

(1) Comparative analysis method

In this experiment, a comparative experiment was conducted by setting different sequence lengths and nearby ranges to analyze the feasibility of the research on the security technology of the Hadoop-based cloud computing platform in this paper.

(2) Mathematical Statistics

Use relevant software to make statistics and analysis on the research results of this article.

4 Experimental Analysis of Security Technology of Cloud Computing Platform Based on Hadoop

4.1 Change the Test Result of Sequence Length

The following is the parameter training for the sequence length seqLen and the neighbor range k respectively. The following table shows that k = 10; winLen = 20; f = 4, and the experimental results obtained by setting different seqLen are shown in Table 1.

Table 1. Change the test result of sequence length

	Detection rate	Error rate	Not detected
30	0.868	0	0.132
20	0.737	0.135	0.128
10	0.592	0.276	0.132

It can be seen from Fig. 1 that within a certain range, the longer the sequence length, the higher the detection rate. Because for the full abnormal sub-sequence (that is, the data points in the sub-sequence are abnormal data points), the more points there are, the deeper the degree of abnormality and the more obvious the anomaly, so the distance from other normal sub-sequence patterns It is also farther away, so that the fewer data points in its neighboring range, the higher the anomaly coefficient.

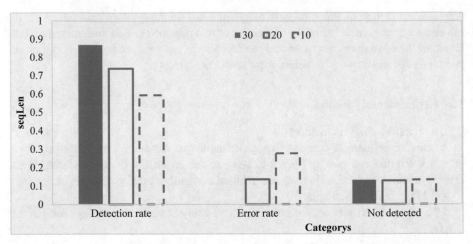

Fig. 1. Change the test result of sequence length

4.2   Change the Test Result of k Value

When taking different values of k, according to the order of appearance of the abnormal subsequences, the abnormal coefficients of the first three abnormal sequences are obtained. The experimental results obtained by setting different values of k are shown in Table 2.

Table 2. Change the test result of k value

	Detection rate	Error rate	Not detected
10	0.868	0	0.132
13	0.882	0.092	0.026
15	0.829	0.145	0.026

It can be seen from Fig. 2 that the choice of k value has little effect on the detection rate, so it is inferred that the detection rate is mainly affected by the length of the subsequence. However, as the value of k increases, the detection rate decreases slightly, but the false detection rate is proportional to the value of k, that is, the detection effect is relatively reduced.

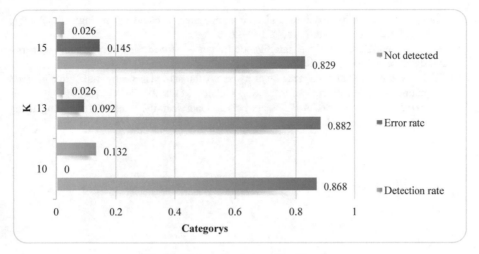

Fig. 2. Change the test result of k value

5 Conclusions

With the advent of the Internet era and the rise of the era of big data, the information processing technology derived from it has become critical. Cloud computing just makes up for the shortcomings of high concurrency processing data in reality, and it meets the needs of resource integration. And whether this new model can be quickly popularized and applied, an important factor is whether cloud computing security can meet the demand. Therefore, it is particularly necessary to study a comprehensive security protection system in the cloud computing environment.

References

1. Hu, G., Yang, Y., Li, L.: Education technology cloud platform framework establishment and security. J. Comput. Commun. **04**(7), 7–14 (2016)
2. Hema, A., Visalatchy, S., Mallika, C., et al.: A survey paper on cloud computing security and outsourcing data mining in cloud platform. Int. J. Adv. Sci. Res. Dev. (IJASRD) **4**(11), 75 (2017)
3. Kritikos, K., Kirkham, T., Kryza, B., et al.: Towards a security-enhanced PaaS platform for multi-cloud applications. Futur. Gener. Comput. Syst. **67**(2), 206–226 (2017)
4. Wei, W., Zhang, Y., Liu, Y., et al.: FRP: a fast resource placement algorithm in distributed cloud computing platform. Concurrency Comput. Pract. Experience **28**(5), 1399–1416 (2016)
5. Adejo, O.W., Ewuzie, I., Usoro, A., et al.: E-Learning to m-Learning: framework for data protection and security in cloud infrastructure. Int. J. Inf. Technol. Comput. Sci. **10**(4), 1–9 (2018)
6. Malomo, O., Rawat, D.B., Garuba, M.: A survey on recent advances in cloud computing security. J. Next Gener. Inf. Technol. **9**(1), 32–48 (2018)
7. Youssef, A.: A Delphi-Based security risk assessment model for cloud computing in enterprises. J. Theor. Appl. Inf. Technol. **98**(1), 151–162 (2020)
8. Su, Y., Shen, J., Wang, W.: Reconfigurable design and implementation of nonlinear Boolean function for cloud computing security platform. Int. J. Inf. Comput. Secur. **11**(2), 145 (2019)

9. Anjana, S.A.: Security concerns and countermeasures in cloud computing: a qualitative analysis. Int. J. Inf. Technol. **2018**(4), 1–8 (2018)
10. Gong, S., Lee, C.: Cyber threat intelligence framework for incident response in an energy cloud platform. Electronics **10**(3), 239 (2021)
11. Ali, M., Malik, S., Khalid, Z., et al.: Security issues, threats and respective mitigation in cloud computing -A systematic review. Int. J. Sci. Technol. Res. **9**(8), 474–484 (2020)
12. Thirukrishna, J.T., Arvind, A.B.: R Associate. Data security in cloud computing using transparency service model. Int. J. Technol. Manage. **9**(6), 3131–3236 (2019)

Exploration of Ecological Restoration Test Simulation Based on Simulation Virtual Technology

Shuming Ju[1,2], Pin liu[1], Kuimei Qian[1], Chengfang Yang[1], Li Guo[1], Mingxia Gao[1], Jiankun Zhang[1], and Kuiqing Chen[1(✉)]

[1] Xuzhou Institute of Technology, Xuzhou 221018, Jiangsu, China
[2] Jiangsu Laboratory of Pollution Control and Resource Reuse, Xuzhou 221018, Jiangsu, China

Abstract. In order to solve the problem of the long time, large site, non-repeatable, high cost of experiment on landscape ecological restoration of damaged site, by using advanced information technology, this experiment is transformed into a virtual simulation experiment without the restrictions of space, time and funds. On the basis of elaborating the background, necessity and purpose of project development, the content and construction method of virtual simulation experiment are introduced. The application of this project improves students' practical ability of ecological restoration engineering, and promotes the cultivation of application-oriented talents in environmental and ecological majors, and actively serve the needs of ecological restoration and environmental improvement.

Keywords: Coal mining subsidence area · Landscape ecological restoration · Virtual simulation · Experimental project

1 Construction Background of Virtual Simulation Experiment

The coal mining subsidence area refers to the area caused by the coal mining activity of the displacement or deformation within a certain scope. In the process of coal mining, the ecological environment is deteriorated [1, 2]. At present, 69 cities in China are resource-exhausted cities. The area of coal mining subsidence is about 2 million hectares in the whole country, while it is still increasing at a rate of 27,000 hectares every year [3]. The ecological restoration of coal mining subsidence area has important strategic significance.

Ecological restoration project is an important course for environmental ecology related majors, which has strong practicality and applicability. Its experimental teaching is an indispensable part of the teaching system, which plays an important role for cultivating students' practical, engineering practice ability and the spirit of innovation [4–6]. The ecological restoration of damaged sites is an important part of the teaching and research of ecological restoration engineering. However, because of the reasons such as, the huge cost, long duration and non-reproducibility of ecological restoration projects, it is extremely hard to carry out physical experiments. At present, the development of

J. Macintyre et al. (Eds.): SPIoT 2021, LNDECT 98, pp. 353–360, 2022.
https://doi.org/10.1007/978-3-030-89511-2_45

information technology has provided technical conditions for solving practical teaching problems, while the construction of national virtual experimental projects has provided an opportunity for the construction of virtual simulation experimental projects of ecological restoration projects [7–9]. Combining the regional urban economic transformation with the professional characteristics, it takes the national demonstration case of ecological restoration of damaged sites in the coal mining subsidence area in Pan'an Lake as an example, it has established a virtual simulation experiment platform for the landscape ecological restoration project of the coal mining subsidence area, combined with the virtual and real, which not only expanded the depth and breadth of the experiment, but also increased the systematicness, openness and immersion of the experiment, which is beneficial for the students to cultivate practical and innovative abilities.

2   Teaching Content Design of Virtual Simulation

The teaching content of the virtual experiment teaching platform is designed according to the technical route of ecological restoration engineering, which mainly consists of on-site survey, engineering treatment, landscape ecological restoration and restoration effect evaluation (Fig. 1).

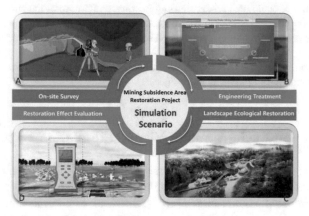

Fig. 1. The teaching content of virtual simulation experiment of landscape ecological restoration project in coal mining subsidence area. A: On-site Survey; B: Engineering Treatment; C: Landscape Ecological Restoration; D: Restoration Effect Evaluation.

2.1   On-site Survey

Through the artificial intervention of ecological restoration technology, it has performed the restoration of the ecosystem of coal mining subsidence area. The main points involved are: (1) the main content of on-site survey in coal mining subsidence area (2) the ecological damage characteristics of coal mining subsidence area.

2.2 Engineering Treatment

In the process of mining subsidence area restoration. First of all, the damaged surface needs to be renovated, to make it similar to the natural landform and coordinate with the surrounding landscape, which is beneficial for the landscape reconstruction and the conservation of water and soil, and also meets the requirements of land reuse. The main points involved are: (1) the engineering treatment principles of coal mining subsidence areas (2) the engineering treatment and construction techniques of coal mining subsidence areas.

2.3 Landscape Ecological Restoration

The landscape construction has become an important ecological restoration method in the coal mining subsidence area. According to the various characteristics of different site scenes after the engineering treatment, it requires students to carry out the ecological restoration design of the site landscape. The main points are: (1) the landscape design of the restoration site (2) the screening and configuration of functional plant.

2.4 Restoration Effect Evaluation

The environmental detection and effect evaluation after restoration in the recovery process of coal mining subsidence area is an essential part of the ecological restoration project. It shall carry out site environment testing, analyze the repair effect, and discuss the current existing problems. The main points are: (1) Soil ecological quality evaluation principles and methods for the evaluation of key indicators; (2) Principles and methods for the evaluation of key indicators for environmental quality assessment of water ecosystems.

In the process of mining subsidence area restoration, the key technologies include the site exploration, engineering treatment, landscape ecological restoration and remediation effect evaluation, which form a closed-loop relationship the whole restoration project.

3 Framework Construction of Virtual Simulation Teaching Process

According to the acceptance process of knowledge from perception → cognition → practice → application of human beings, the virtual simulation experiment platform sets six modules of perceptual learning, cognitive learning, practical operation, comprehensive application, online testing and comprehensive evaluation (Fig. 2 and Fig. 3).

3.1 Perceptual Learning

Through the video display of ecosystem restoration engineering process, students can experience the ecological restoration of coal mining subsidence area with a vivid, concrete and direct experience of sensory organs, master the process of landscape ecological restoration in coal mining subsidence area, enhance the understanding effect. On the basis of obtaining a lot of perceptual knowledge, the next step is to carry out in-depth cognitive learning.

3.2   Cognitive Learning

This part mainly includes five parts of "Project background", "Site exploration", "Engineering treatment", "Landscape ecological restoration" and "Repair effect evaluation". And it introduces the contents of each part through the text, pictures, etc. In the "Project background" part, it introduces the current situation of coal mining subsidence area, the purpose and significance of ecological restoration, and the demonstration project of Pan'an Lake Wetland Park. In the "Site exploration" part, it introduces the contents of the site exploration in coal mining subsidence area, as well as topography, vegetation type and hydrogeological characteristics. In the "Engineering management" part, it introduces the common principles and technologies of different site engineering treatment in coal mining subsidence areas. In the "Landscape ecological restoration design" part, it introduces the principles and methods of ecological slope protection design and plant species screening and configuration. In the "Repair effect evaluation" part, it introduces the determination of key indicators for water and soil ecological quality evaluation, principles and methods of phytoplankton collection and quantitative analysis, evaluation criteria and methods. Through cognitive learning, students can master the theory and technology of landscape ecological restoration in coal mining subsidence area, and lay a solid foundation for completing the task of practical operation module.

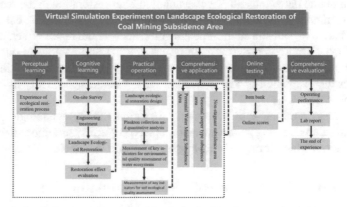

Fig. 2. Virtual simulation experiment teaching link of of landscape ecological restoration project in coal mining subsidence area

3.3   Practical Operation

The practical operation module includes two parts of "Landscape ecological restoration design" and "Repair effect evaluation". Choose to enter the "Landscape ecological restoration design experiment", load the simulation environment, complete the experiment according to the prompts in the taskbar, select the scene, select the appropriate revetment type according to different site characteristics, combine the engineering standards, design the appropriate revetment type, and combine the plant ecological habit

Fig. 3. Frame module interface of virtual simulation platform of landscape ecological restoration project in coal mining subsidence area. A: Perceptual learning; B: Cognitive learning; C: Practical operation; D: Comprehensive application; E: Online testing; F: Comprehensive evaluation

and function, students shall carry out the design and configuration of waterfront landscape. The project requires students to fully understand and master the principles, purposes, methods and steps of the landscape ecological restoration experiment. The design concept varies according to different students, and the design effect diagram shall be transmitted to the experimental report module. Students can design and evaluate the landscape according to the design content.

The "Repair effect evaluation" includes three experiments of "Measurement of key indicators for environmental quality assessment of water ecosystems", "Plankton collection and quantitative analysis" and "Measurement of key indicators for soil ecological quality assessment". Select to enter the experimental project, load the simulation environment, each experiment includes two operation modes of demonstration mode and operation mode, students can perform demonstration experiments and then enter the operation experiment in order to complete the experimental tasks. The project requires students to fully understand and master the experimental principles, objectives, methods and steps of the experiment. The experimental results shall be imported into the experimental report, and combine with the data of the fixed-point timing measured before and after the ecological restoration as shown in the experimental report, the students can analyze the repair effect.

3.4 Comprehensive Application

In order to improve students' ability of solving complex practical problems, entering the "comprehensive application" module. Select the "Comprehensive application" button, enter the virtual environment. Three site types were set up in the experiment: perennial water-collecting coal mining subsidence area, seasonal water-collecting coal mining subsidence area and non-water type coal mining subsidence area. Students can randomly select and enter the experimental scene to carry out landscape design, carry out suitable plant screening and configuration for the selected coal mining subsidence area, and achieve the goal of coupling environmental restoration and landscape design.

3.5 Online Testing

It has been set up various questions in the online testing, which covers the theoretical knowledge and technical points in cognitive learning and practical operation. Students can detect the theoretical and technical master situation by answering questions online.

3.6 Comprehensive Evaluation

Explore multi-evaluation model, the computer automatically scores the experimental operation, comprehensive application and online testing. Based on the student's experimental report, the teacher can understand the master situation of the students' theory and practice of ecological restoration in the coal mining subsidence area. Students can also access the experimental records online through the platform.

4   Construction Method of Virtual Simulation Experiment

The experimental platform has the ability of providing 1,000 students to access online at the same time without any waiting for login. The platform adopts B/S architecture, and the computer system is Windows 7 and even latest. The project is a web version, and users do not need to download any special plugins when using it. System development mainly includes two parts of multimedia resource production and software programming. It adopts 3D simulation development technology and development tools such as Unity 3D, 3D Studio Max, Maya, Adobe Flash, Animate CC, and Visual Studio and other development tools, and constructs the interactive virtual simulation experiment platform. The website platform adopts JAVA development language, Eclipse development tools and MySQL database to build an integrated management mode, for which the client can access the virtual simulation experiment platform efficiently and quickly through the browser.

5   Characteristic Innovation of Virtual Simulation Experiment

5.1   Case Relied on of the Project Development - Advanced, Typical and Exemplary Case

It has used advanced restoration concepts and techniques in the repair process of the Pan'an Lake coal mining subsidence area, and the repair effect was fully affirmed, which has become a national model of "regional ecological restoration engineering in coal mining subsidence area" [10]. The landscape ecological restoration project of the coal mining subsidence area in Pan'an Lake has become a typical case of the ecological restoration engineering teaching of the damaged site.

5.2 Necessity of Virtual Simulation Experiment Project Construction

Which is mainly based on the following two points: first, the ecological restoration project is an important teaching content of the environmental ecology majors. It takes a long time and the space span is large during the practice process, and it is impossible for students to practice repairing while it is hard for students to imagine the change process with their knowledge reserve; second, the landscape ecological restoration project is artificially intervened, which is generally costly. In traditional teaching, students can only understand the repair project through the materials, and cannot systematically carry out the actual practice.

6 Conclusion

The ecological restoration project and the wetland landscape construction project of Pan'an Lake in the coal mining subsidence area, are in accordance with the aim of cultivating the application engineering talents who master advanced restoration concepts and technologies, with modern information technology, such as advanced virtual reality technology and computer network communication technology to develop diversified virtual simulation teaching resources and achieve resources sharing. The experimental project plays an important role in promoting the cultivation of ecological restoration talents, scientific research and development, and achievements transformation.

Acknowledgements. This work was supported by the Jiangsu Virtual Simulation First-class Curriculum Construction Project.

Conflicts of Interest:.   The authors declare that they have no conflict of interest.

Software copyright registration number: Virtual simulation software for ecological restoration and landscape construction in coal mining subsidence area (code: 2019SR086574).

References

1. Zhou, X., Zhou, Y., Zhai, Z.H.: Practice and thinking on comprehensive control of coal mining subsidence area. China Land. **7**, 42–45 (2018)
2. Pang, J., Song, X.H., Zhou, M.J.: Comprehensive management model for coal mining subsidence area: a case study of Heze city. J. Shandong Univ. Finance Econ. **4**, 109–119 (2018)
3. National Development and Reform Commission. Guidelines on the compilation of transformation plans for resource-exhausted cities. China (2013)
4. Hu, S.H., Xu, Z.M., Ma, D.D., Cui, K.P.: Exploration on teaching construction of ecological rehabilitation engineering course based on professional certification. Educ. Modernization **5**(10), 103–106 (2018)
5. Yang, F.G.: Teaching reform and practice of ecological restoration engineering course. Sci. Technol. Innov. Herald **14**, 202–203 (2019)
6. Du, Q.P., Chen, L.C., Shi, Y., Yin, G.C., Li, D.M.: Practical research of training innovative and application-oriented talents of eco-environmental engineering under the background of emerging engineering education. Res. Explor. Lab. **8**(197–200), 223 (2019)

7. Li, P.: Promote the application of virtual reality technology to improve the education quality in universities. Res. Explor. Lab. **37**(1), 1–4 (2018)
8. Liu, Y.F., et al.: Exploration on opening and sharing strategy for virtual simulation teaching resources. Exp. Technol. Manag. **33**(12), 137–141, 145 (2016)
9. The CPC Central Committee and the State Council. Outline of the national program for medium - and long-term education reform and development. China (2010)
10. The Xinhua News Agency. Xi jinping is visiting xuzhou, jiangsu province. China Development Observation, vol. 24, no. 2 (2017)

Fault Identification of Power Grid Based on BP Neural Network

Mingjiu Pan[1], Zhou Lan[1], Kai Yang[1], Zhifang Yu[1], Huaiyue Luo[2], and Di Zheng[2(✉)]

[1] Economic and Technical Research Institute of Zhejiang Electric Power Corporation, Hangzhou, Zhejiang, China
[2] China Ji Liang University, Hangzhou, Zhejiang, China

Abstract. This paper presents a data-driven fault identification method for distribution network. It is mainly based on the distribution network fault when each electrical quantity will change, and different fault types will have different electrical quantity changes, so as to select the fault characteristic quantity. Then, the distribution network fault simulation model is built to simulate different types of faults when changing the phase to phase resistance, grounding resistance, load and fault location. The fault feature quantity is collected and the feature sample library is formed, which is used as the data basis to complete the distribution network fault identification.

Keywords: Data-driven · Fault identification · BP neural network

1 Introduction

The existing power grid fault identification often depends on relay protection device, and its principle is mostly based on electrical mechanism. There are 10 types of power grid fault, which can be roughly divided into phase to phase short circuit fault and ground fault, of which 4 are phase to phase fault and 6 arc ground fault. In this paper, a power grid fault model is built on MATLAB to simulate different kinds of faults.

This paper presents a method of power grid fault identification based on BP neural network, introduces the algorithm and parameters of BP neural network, and introduces how to apply BP neural network algorithm to realize power grid fault identification. Finally, the simulation results are presented and compared [1, 2].

2 Power Grid Fault Simulation and Characteristics

In this paper, Matlab software is used to establish the simulation of power grid faults, and different fault types are simulated by changing various parameters of power grid simulation, and the relevant three-phase current voltage and zero-sequence voltage are collected. Then, based on the collected data, BP neural network is used to complete the training, thus realizing the fault identification of power grid.

© The Author(s), under exclusive license to Springer Nature Switzerland AG 2022
J. Macintyre et al. (Eds.): SPIoT 2021, LNDECT 98, pp. 361–367, 2022.
https://doi.org/10.1007/978-3-030-89511-2_46

2.1 Fault Type

Grid fault and large ground fault and short-circuit fault is given priority to, in this design will be considered A phase ground fault, B phase ground fault, C phase ground fault of three single-phase earth fault and A, B phase ground fault, A and C phase ground fault, B, C phase ground fault three double phase ground fault and A, B and short circuit, A, C phase short circuit and B, C phase short circuit three kinds of two-phase short circuit fault and A, B, C three-phase short circuit fault these 10 kinds of power grid fault types.

2.2 Fault Simulation

Figure 1 is the power grid fault model required by the scheme built in Simulink of MATLAB software. This model is composed of three-phase power supply, three-phase voltage and current measurement module, three-phase fault module, three-phase PI line and three-phase series RLC load.

The design of the three-phase fault module to achieve 10 types of fault simulation, by changing the module each resistance parameters to simulate different situations.

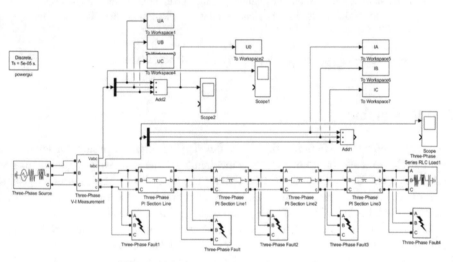

Fig. 1. Power grid fault simulation model

2.3 Fault Feature Selection

Three - phase current, three - phase voltage and zero - sequence voltage are selected as characteristic variables to identify 10 fault modes.

In this paper, the interphase fault resistance is 40 Ω, the grounding fault resistance is 20 Ω, the total length of the line is 400 km, and the load is rated at 10 kW. The total duration of system simulation is 0.25 s, and the fault time is 0.1–0.15 s. In order to facilitate observation and sampling, 0.25 s is divided into 5000 points, taking single-phase grounding fault as an example, as shown in the abscissa of Figs. 2 and 3.

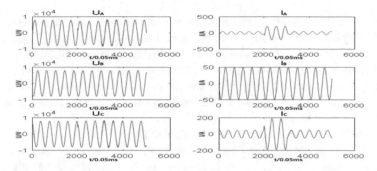

Fig. 2. A Three-phase current and three-phase voltage for ground fault phase

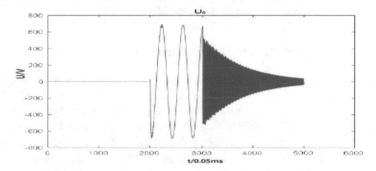

Fig. 3. A Zero sequence voltage for phase ground fault

During the fault the voltage amplitude of A phase is about 7.813 kV, that of B phase is 7.966 kV and that of C phase is 7.961 kV. The zero-sequence voltage changes dramatically during the fault, and the amplitude can reach about 696 V, which decreases slowly after the fault [3–8].

3   Fault Identification of Power Grid Based on BP Neural Network

3.1   BP Neural Network

Artificial neural network can be trained to get a network, the network in accordance with certain rules, when you give a certain input, the network will be trained according to the network to get an output close to the expected value. And BP neural network, also known as back propagation algorithm, is the most widely used neural network model. BP neural network is a multi-layer feedforward network, which can transfer information forward. Data of the input layer will pass through the hidden layer and output the actual output, which will produce a certain error with the expected output of the design. Then, the error will be transmitted in reverse to adjust the weight and threshold of the hidden layer, so as to continuously improve the network and make the result close to the expected value.

BP neural network has certain advantages and disadvantages. BP neural network has good learning ability, and the reverse transmission of error can constantly adjust the weight and threshold value, so that the network model can learn the characteristics of data and complete pattern recognition. BP neural network can obtain strong nonlinear mapping ability through a large number of sample data training, and this mapping does not need explicit mathematical expression. BP neural network has a certain generalization ability, and can prepare mapping output for untrained input data. In this design, all the test samples are different from the training samples, so as to reflect the generalization ability of BP neural network.

In this paper, the required data will be collected through the above simulation experiments, and part of the collected training and test data will be presented. BP neural network will train the data, and transfer the error to the input end, constantly changing the parameters of the hidden layer and reducing the error between the output value and the expected value. The last two types of output are 0 and 1, which can clearly explain the fault type according to the actual output [9, 10].

3.2 BP Neural Network Training Set and Test Set

For the network training set, this design changes the parameters of each component on the grid fault simulation model shown in Fig. 1. The earth fault resistor is selected as 5 Ω, 10, 20 and 40 V. Interphase fault resistance 10 Ω, 40 and 100; The fault locations were selected as 0%, 25%, 50%, 75% and the end of the line. The load was taken at 10 kW, and 15 groups of samples were taken for the fault-free mode to strengthen the difference between the fault-free mode and the fault-free mode. For the change of interphase fault resistance, there are 4 (interphase fault resistance) *5 (fault location) *10 (different fault types) = 200 sets of training samples. There is an additional 4*5*6 (type of ground fault) *2 (different ground fault resistances) = 240 sets of training samples for ground fault. In order to strengthen the generalization ability of the training set, 160 fault samples were taken by changing the transition resistance. With 15 trouble-free samples, there are 615 groups of samples, which constitute the neural network training set samples.

Table 1 shows the partial data presentation after the normalization of the training set. It can be seen that the comparison of similar data after normalization is also sharper, and they are all between 0 and 1.

Epochs: the maximum training times of the network, namely the number of training steps specified for BP neural network. An Epoch represents that all training samples have completed a training in the neural network, and the training will stop until the network training reaches the specified Epoch or the expected error range. In this design, the step size is set as 1000. After several measurements, the design can achieve the minimum mean square error (MSE) in about 700 times. In order to ensure the reliability of the results, the step size is set as 1000.

Goal: the accuracy achieved by network training. In this design, the Goal is set at 0.002.

LR: The learning rate of the network, the default value is 0.01. The higher the learning rate, the higher the learning efficiency, but too large a learning rate is not conducive to the convergence of the network. In this design, LR is set as 0.005, because enough samples are collected in this design, and the learning rate can be lowered to achieve a higher accuracy.

Table 1. Normalized results of data of group 11–20 of BP network training set

Feature	The sample number									
	11	12	13	14	15	16	17	18	19	20
U_0/V	0.24	0.36	0.58	0.00	0.00	0.00	0.18	0.22	0.13	0.00
U_A/V	0.80	0.97	0.97	0.79	0.00	0.85	0.76	0.05	0.85	0.27
U_B/V	0.91	0.87	0.97	0.00	0.91	0.84	0.04	0.95	0.81	0.33
U_C/V	0.93	0.97	0.88	0.93	0.85	0.00	0.95	0.82	0.04	0.32
I_A/A	0.16	0.01	0.01	0.84	0.85	0.01	0.87	0.83	0.01	0.99
I_B/A	0.02	0.20	0.02	0.86	0.02	0.87	0.83	0.02	0.90	0.99
I_C/A	0.00	0.00	0.16	0.00	0.87	0.86	0.00	0.89	0.83	0.99

Netff function is used to create BP neural network. This design mainly considers the above parameters, and the default values are selected for the maximum number of faults and minimum gradient requirements.

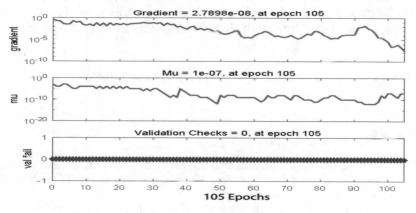

Fig. 4. BP neural network training results

Figure 4 shows the training process of the neural network, in which the gradient is generally on the decline and the feedback error is getting smaller and smaller, indicating that the prediction accuracy of the network is continuously improving with the progress of EPOCH.

4 Conclusions

(1) Based on BP neural network, this paper realizes power grid fault identification, and takes three-phase power supply, three-phase voltage and current measurement module, three-phase fault module, three-phase PI line and three-phase series RLC load

as main units to build power grid fault simulation. The fault simulation under different conditions is realized by changing the fault resistance and ground resistance as well as the fault location and load, and the characteristic quantity is collected. With the increase of the characteristic quantity, the accuracy of the network is gradually improved, and 10 fault types can be identified without the influence of load, fault ground resistance, interphase fault resistance and fault location.

(2) This paper also compares the fault identification of power grid based on support vector machine, and finds that: in the case of a small number of samples, the support vector machine method is more suitable. When the number of samples reaches tens of thousands of groups, it takes a lot of time for support vector machine to search for the optimal solution, which is why the use of BP neural network can quickly get better identification results.

(3) The installation of shunt reactor is mainly to eliminate the error caused by overvoltage of power frequency. Due to the existence of transition resistance, the measured voltage and current will change, resulting in changes in the measured impedance. The simulation results show that when the shunt reactor is installed outside the measuring point, that is, on the bus, the accuracy is higher than that of the shunt reactor installed inside the measuring point. As the transition resistance is pure resistance, it will affect the value of the measured voltage and current, thus affecting the calculation of the fault distance. It can be seen from the simulation results that the simulation accuracy is higher when the transition resistance is 0Ω.

Acknowledgements. This work was supported by SGZJJY00JJJS2100058.

References

1. Zhang, C., Song, N., Li, Y.: Incipient fault identification of distribution networks based on feature matching of power disturbance data. Electr. Eng. (2021). https://doi.org/10.1007/s00202-021-01232-6
2. Li, Y., Zhang, Y., Liu, W., et al.: A fault pattern and convolutional neural network based single-phase earth fault identification method for distribution network. In: 2019 IEEE Innovative Smart Grid Technologies - Asia (ISGT Asia). IEEE (2019)
3. Jamali, S., Bahmanyar, A., Ranjbar, S.: Hybrid classifier for fault location in active distribution networks. Protect. Control Modern Power Syst. 5(1), 1–9 (2020)
4. Nadour, M., Essadki, A., Nasser, T.: Improving low-voltage ride-through capability of a multimegawatt DFIG based wind turbine under grid faults. Protect. Control Modern Power Syst. 5(1), 1–13 (2020)
5. Mishra, S.K., Tripathy, L.N.: A critical fault detection analysis & fault time in a UPFC transmission line. Protect. Control Modern Power Syst. 4(1), 24–33 (2019)
6. Xiao, Y., Ouyang, J., Xiong, X., Wang, Y., Luo, Y.: Fault protection method of single-phase break for distribution network considering the influence of neutral grounding modes. Protect. Control Modern Power Syst. 5(1), 1–13 (2020)
7. Zhang, F., Mu, L.: New protection scheme for internal fault of multi-microgrid. Protect. Control Modern Power Syst. 4(1), 1–12 (2019)

8. Jain, T., Ghosh, D., Mohanta, D.K.: Augmentation of situational awareness by fault passage indicators in distribution network incorporating network reconfiguration. Protect. Control Modern Power Syst. **4**(1), 1–14 (2019)
9. Madhiarasan, M.: Accurate prediction of different forecast horizons wind speed using a recursive radial basis function neural network. Protect. Control Modern Power Syst. **5**(1), 1–9 (2020)
10. Hamdi, H., Regaya, C.B., Zaafouri, A.: A sliding-neural network control of induction-motor-pump supplied by photovoltaic generator. Protect. Control Modern Power Syst. **5**(1), 1–17 (2019)

Intelligent Traffic App Operation Mode and Prediction of People Based on Big Data

Li Tao[✉]

Urban Vocational College of Sichuan, Chengdu 610000, Sichuan, China

Abstract. The flow prediction plays an important role in the operation and management of urban rail transit system. It is an important prerequisite for the rational deployment of transportation resources, the optimization of train operation plan, the guidance of daily transportation organization and the evaluation of economic benefits. This paper mainly studies the operation mode and flow prediction of smart transportation APP based on big data. This paper mainly takes the city's public transport as an example, through the installation of public transport passenger flow collection equipment, collect each station on and off the passenger flow data, and take it as the sample data, analyze the time distribution characteristics of the station passenger flow, summed up the passenger flow change rules. Then the time series model & modeling and forecasting of passenger flow series are adopted, and the random influencing factors of statutory holidays and emergencies are introduced to remove the interference of external factors and improve the robustness of the passenger flow prediction model of the station.

Keywords: Big data · Intelligent transportation · Traffic operation · Crowd prediction

1 Introduction

Public transportation is a transportation system that provides services to travelers through a group travel system. The system is usually managed as planned, runs on a given route and charges passengers a fee based on the status of each trip. Urban public transportation system, as one of the core systems of urban transportation system, is an important and main component of urban transportation network [1]. Due to the huge number of passengers, the bus system will produce a large number of bus card swiping data every day. How to organize traffic operation according to the big data of card swiping has become one of the focus issues of public transportation system. With the progress and development of science and technology, the data accumulated by all walks of life expands and becomes larger. The analysis and mining of massive data through models and algorithms gradually transforms into new economic growth points. The analysis and calculation of a large number of data expand people's perspective on the problem and provide new ideas and data support for solving the problem. As the traditional pillar industry of the country, the transportation industry has also accumulated a large number of passenger, vehicle and road-related data. A large number of traffic industry data

J. Macintyre et al. (Eds.): SPIoT 2021, LNDECT 98, pp. 368–375, 2022.
https://doi.org/10.1007/978-3-030-89511-2_47

reflect the state of national and city traffic, people's travel mode and characteristics and possible traffic problems. Through the analysis and mining of a large number of traffic data, we can more accurately predict the traffic volume, fully understand the behavioral characteristics and preferences of traffic pedestrians, and better solve the corresponding traffic problems. It is a complex system engineering to fully understand passengers' demands for public transport based on their travel data and optimize the public transport operation organization plan [2, 3].

With the continuous development of urbanization, the traffic situation in many cities has become more complicated. Research on urban transportation mainly involves many scientific fields such as urban planning, transportation, environment, computer science, information management, behavioral science and systems engineering [4]. From the 1950s to the present, scholars around the world have made a lot of theoretical basis and relevant application work on transportation related theories, demand forecasting, traffic management and planning, environmental science, safety assurance, simulation and intelligent system [5]. However, due to the lack of in-depth understanding of the dynamic balance mechanism of urban traffic and the lack of in-depth analysis of the causes of urban traffic congestion, most of the studies on the alleviation of urban traffic congestion only stay at the "symptom" level, and there are still passivity and limitations in the effectiveness of solving urban traffic problems [6].

With the development and application of new technologies, an important direction for the intelligent transportation operation system to serve the general public is to provide more timely, accurate, perfect and intelligent transportation services for the public. With the provision of these public services will further accelerate the integration of the transportation industry and other industries.

2 Bus Operation and People Flow Prediction Based on Big Data

2.1 Bus Operation Organization Process and Bus Data

Public transportation operation organization is a complex system engineering including people, vehicles, lines and stations. The bus operation organization usually optimizes the bus operation plan based on the travel data of passengers and aims to solve the travel demands of passengers by comprehensively considering various influencing factors, and then reasonably arranges the personnel and vehicles according to the operation plan [7]. Among them, the bus operation plan design and schedule preparation is the basis of personnel and vehicle organization, is also the core link of the bus operation organization, the study of this paper mainly for the bus operation organization in the passenger travel demand analysis to the bus operation plan design schedule preparation stage. The general process of bus operation organization and the main research stages and methods of this paper are as follows:

(1) Traffic information collection
    In this stage, the server interface calls and automatic data collection are used to obtain relevant bus data from the bus card swiping data, operation data and map APP.

(2) Traffic demand analysis

In the second stage, through big data technology, passenger travel characteristics analysis, single line passenger flow prediction, travel time calculation, multi-line passenger flow prediction, etc.

(3) Theory of bank opening scheme

In this stage, through the analysis results of the second stage, the optimization theory is used to analyze the needs of regular passengers, transfer passengers and the company's operation needs. The bus operation plan considering the situation of district buses and the bus operation plan considering the transfer of coordinated buses.

(4) Sales personnel scheduling

The fourth stage is to set up the scheduling plan according to the bus cost plan.

The bus card swiping data is the core data of the bus operation organization. The data is collected by the card swiping device and contains the information of each passenger on and off the bus. The data volume is huge but the dimensions are few, so other data are needed to effectively expand the dimensions of the data [8]. In this paper, bus operation data and point of interest data are selected as auxiliary data, and bus card swiping data is fused with bus operation data and point of interest data to improve data availability.

2.2 Big Data Acquisition and Preprocessing Technology

Big data can be collected through sensor data, RFID radio frequency data, mobile Internet data and social network interaction data, including various types of structured, semi-structured and unstructured mass data. Data collection techniques are also different in different ways of data generation [9, 10].

The file log data is generated by the company's business platform. Mining the potential value of the log data of the company's business platform can provide reliable data support for the company's decision-making and the performance evaluation of the company's backstage server platform [11]. The most commonly used logging data collection technologies are Apache Flume and Scribe. Apache Flume is designed to efficiently collect, aggregate, and move large amounts of log data, with strong fault tolerance. Scribe improves the fault tolerance of log collection systems by providing persistent message queues. In addition, databases can also be used to collect business data of enterprises and industries. The commonly used databases include MySQL, Oracle, Redis and Mongod, etc. [12].

Network data can be obtained from the website through the public API and web crawler technology provided by some websites. Web crawler systems (such as Apache Nutch, Crawler4J, Scrapy and other frameworks) can extract large amounts of unstructured and semi-structured data from Web pages, clean it up and convert it into structured data.

Big data preprocessing technology can transform chaotic data into relatively simple and easy to process structured data and improve data quality. ETL, a data warehouse technology, can preprocess data through extraction, Transform, Load and other processes. First of all, the required data can be extracted from the original format of the complex data, some unimportant fields can be discarded, and the data can be cleaned to filter and eliminate the incorrect data. The data can be converted into the corresponding format for specific application scenarios and loaded into the data warehouse through the predefined data warehouse model. Typical ETL tools include Informatica, DataStage, Microsoft DTS, and so on.

2.3 Traffic Flow Prediction Modeling

(1) Preprocessing of pedestrian flow data

According to the forecast of the daily passenger flow of the station in this paper, the daily passenger flow of the station is basically in line with the time series model with the cycle of week. This time series model is affected by the influence of statutory holidays, working days and weekends, and temporary emergencies. Therefore, it is necessary to adjust the time series model. Then the forecast model of SARIMA(P,D,Q) (P,D,Q) S is used to forecast the daily passenger flow of the station.

The modeling process in this paper is based on the central limit theorem. If the sequence samples do not conform to the characteristics of contract distribution, the prediction result is not reliable. Firstly, the daily passenger flow data containing legal holidays in the original data were adjusted to remove the influence of the later holidays. The mathematical expression is as follows:

$$Y = X * \frac{W_d(i)}{W_f(j)} \tag{1}$$

Where, X represents the original passenger flow Wd(I) represents the weight factor of week I, and Wf(j) represents the weight factor of legal holidays j.

The STL algorithm is used to decompose the time series, and the trend component, periodic component and residual term of the time series are observed. If the time series fluctuates around a fixed value without an obvious upward or downward trend, the time series is likely to be stable.

Whether the time series is stable or not can be determined by observing the trend of the trend component manually. However, this method is not logical and has some misconducts. When there is no significant change in land use rate and land use classification around the site, the change trend of future passenger flow is positively correlated with the change trend of previous years.

With the development of econometrics, the American statistician USES unit root to determine whether a autocorrelation coefficient is 1 the stability of the method of testing sequence, through constant scrutiny, finally the method is defined as ADF test, the first sequence of null hypothesis is not stable, test results by the test statistic and some of the confidence interval of the critical value, if the test statistic is less than the critical value, then the invalid hypothesis is not valid and the sequence is stationary. Meanwhile, the closer the stationary probability value of time series is to zero, the more stationary the sequence is. ADF unit root detection was performed on the time series to determine whether the series was stable.

(2) Selection of time series model

After the daily passenger flow time series of the station is stable, the order of the SARIMA model can be determined according to the trailing characteristics of the auto-correlation function and partial correlation function, and the data fitting is completed. The fitting operation is carried out on the basis of the pretreatment of the original data series, so after the fitting, the reverse operation of the pretreatment is also needed.

According to the truncation characteristics of autocorrelation function and partial autocorrelation function graph, three models $(p,q) = \{(1,0),(0,1),(1,1)\}$ are preliminarily determined through observation. According to the experience, the values of p and q are not greater than 2. Usually, AIC and BIC are used to select the best model from a number of alternative models. The principle is that the best trade-off is between the ability of the model to fit the data and the complexity of the model.

AIC is commonly used to measure the degree of fit of a model, and model complexity is added as a penalty term to suppress overfitting. Its definition is as follows:

$$AIC = 2k - 2\ln(L) \qquad (2)$$

Wherein, K represents the number of model parameters and L represents the likelihood function.

The difference between the two models is mainly reflected in the likelihood function term. If the likelihood function term difference is not obvious, then the model with fewer parameters is the best, that is, the model with less complexity.

BIC Bayesian information criterion is similar to AIC, but it has a large penalty term for model complexity. It not only takes into account the number of model parameters, but also introduces the number of samples N, which is defined as follows:

$$BIC = k\ln(n) - 2\ln(L) \qquad (3)$$

Assuming that the values of P and Q are no more than 2, all possible combinations are traverses and the AIC and BIC values of each model are calculated. The order of the final model is determined according to the principle of model optimality with minimum values. The AIC and BIC values of Arimax $(1,1,1) \times (0,1,1,7)$ model are the smallest. Therefore, the SARIMA(1,1,1)(0,1,1)7 prediction model is used to fit the original passenger flow data for the prediction of the station's daily passenger flow time series.

3 Human Flow Prediction Simulation Experiment

3.1 Data Sources

This paper takes the bus data of a certain road in July in the city as an example to verify the results. The whole month of June is divided into a training set, and the data of July is divided into a test set. Based on the SARIMA model, the traffic flow of the station is predicted.

3.2 Evaluation Indicators

In this paper, the mean absolute error (MAE), mean relative error (MRE), mean square error (MSE), mean square percentage error (MSPE) and equality coefficient (EC) are selected to analyze the passenger flow forecast results.

The mean absolute error (MAE) represents the average of the absolute deviation between all the predicted values and the true values, which can better reflect the actual situation of the error.

The mean relative error (MRE) represents the percentage between the absolute error generated by the prediction and the true value, which can reflect the degree of confidence of the measurement.

The mean square percentage error (MSPE) represents the deviation between the predicted value and the actual value to some extent.

Equalization coefficient (EC) represents the degree of fit between the predicted value and the true value, and the prediction effect is in direct proportion to this value. It is generally believed that when EC value is greater than 0.9, the prediction result is relatively satisfactory.

4   Simulation Experiment Results and Analysis

4.1   Prediction Value Error

Table 1. Actual and predicted values

	Actual value	Predicted values	Absolute error
Day 1	234	239	5
Day 2	243	257	14
Day 3	338	330	8
Day 4	604	616	12
Day 5	611	609	2

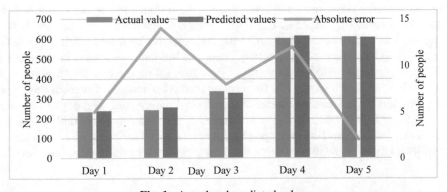

Fig. 1. Actual and predicted values

As shown in Table 1 and Fig. 1, this paper randomly selects the predicted values of five days in July to compare with the actual data. In the five days, the maximum absolute error was on the second day, which reached 14. The minimum absolute error was on the fifth day, which was only 2.

4.2 Predictive Error Indicators

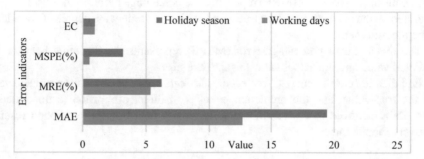

Fig. 2. Prediction error index of working day and holiday population

As shown in Fig. 2, the average absolute error MAE remains at a low level, the average relative error MRE and the mean square percentage error MSPE are within 6%, indicating that the prediction accuracy of the flow prediction system proposed in this paper can reach more than 90% in working days and holidays, with high robustness. And the results of EC are all greater than 0.9, indicating that the prediction results are relatively ideal.

5 Conclusions

The application of big data technology provides opportunities for the development and innovation of intelligent transportation systems, and also poses new challenges to the current passenger flow forecasting methods. For 4 units, this paper analyze site traffic forecast, consider site 4 main influence factors, the introduction of the statutory holidays impact factor and days rest day impact factor, adjust the original passenger flow sequence, and choose according to AIC and BIC adaptive optimal model, improve the robustness and accuracy of the models. There are still some problems in the passenger flow statistical method and passenger flow forecasting method which need to be improved in the future work. The time series prediction model adopted in this paper only considers the influencing factors of the official holidays and weekday rest days on the passenger flow of the station. For the impact of emergencies on the daily passenger flow, it only puts forward the corresponding concepts. In the next step, more data should be collected to study the impact degree of emergencies on the daily passenger flow.

References

1. Baccarelli, E., Cordeschi, N., Mei, A., et al.: Energy-efficient dynamic traffic offloading and reconfiguration of networked data centers for big data stream mobile computing: review, challenges, and a case study. Comput. Chem. Eng. **91**(2), 182–194 (2016)
2. Wibisono, A., Jatmiko, W., Wisesa, H.A., et al.: Traffic big data prediction and visualization using Fast Incremental Model Trees-Drift Detection (FIMT-DD). Knowl. Based Syst. **93**, 33–46 (2016)
3. Xu, F., Lin, Y., Huang, J., et al.: Big data driven mobile traffic understanding and forecasting: a time series approach. IEEE Trans. Serv. Comput. **9**(5), 796–805 (2016)
4. Tang, M., Niemeier, D.A.: Using big data techniques to better understand high-resolution cumulative exposure assessment of traffic-related air pollution. ACS ESAndT Eng. **1**(3), 436–445 (2021)
5. Song, M., Li, R., Wu, B.: A novel prediction model of traffic accidents based on big data. Int. J. Model. Simulat. Sci. Comput. **10**(04), 600–606 (2019)
6. Xie, K., Ozbay, K., Kurkcu, A., et al.: Analysis of traffic crashes involving pedestrians using big data: investigation of contributing factors and identification of hotspots. Risk Anal. **37**(2), 1459–1476 (2017)
7. Yap, M., Munizaga, M.: Workshop 8 report: big data in the digital age and how it can benefit public transport users. Res. Transport. Econ. **69**, 615–620 (2018)
8. Hu, S., Liu, W.: Research on the application of big data in intelligent transportation system. Rev. Facult. Ingenier. **32**(5), 517–524 (2017)
9. Zhang, X., Zhao, Y., Xie, J., Li, C., Hu, Z.: Geological big data acquisition based on speech recognition. Multim. Tools Appl. **79**(33–34), 24413–24428 (2020)
10. Braun, M.T., Kuljanin, G., Deshon, R.P.: Special considerations for the acquisition and wrangling of big data. Organ. Res. Methods **21**(3), 633–659 (2018)
11. Castillo, C.: Big Crisis Data (Social Media in Disasters and Time-Critical Situations) ‖ Volume: Data Acquisition, Storage, and Retrieval (2016). https://doi.org/10.1017/CBO978131 6476840(2):18-34
12. Tonnang, H., Balemi, T., Masuki, K.F., et al.: Rapid acquisition, management, and analysis of spatial maize (Zea mays L.) phenological data—towards 'big data' for agronomy transformation in Africa. Agronomy **10**(9), 1363 (2020)

Attack Methods and Prevention Strategies of Computer Network Security

Yinghuan Wei[✉]

Guangxi Software Management Center (Guangxi Software Testing Center),
Nanning, Guangxi, China

Abstract. Since entering the 21st century, computer Internet technology has made breakthrough progress. However, in the context of network resource sharing, computer network security issues have become more and more important. The endless network intrusions and attacks such as web viruses and hacker attacks have caused huge losses to the social economy and corporate costs. Therefore, the prevention of computer network security has become an urgent problem to be solved. The purpose of this article is to study the attack methods and prevention strategies of computer network security. This article first summarizes the common computer network security attacks from the aspects of information leakage, denial of service attacks, and hidden attacks, expounds related defense strategies, and designs a computer network security attack defense model based on stochastic differential games. This paper verifies through experiments that the strategy designed in this paper improves the detection accuracy and detection speed of network security intrusion. Experimental data shows that for nodes S1, S2, and S3, the traditional detection method takes 0.39 us, 0.52 us, and 0.53 us respectively; the node verification calculation time under the stochastic differential game model is 0.24 us, 0.46 us, and 0.41 us. It shows that the stochastic differential game model is beneficial to improve the detection speed of attacking nodes in network security.

Keywords: Network security · Network attack · Attack prevention · Offensive and defensive game

1 Introduction

Nowadays, the Internet has become the primary medium connecting the world, constantly changing people's study, work, travel methods and other aspects [1, 2]. Just as the network is a double-edged sword, while it brings development to society, there are also many potential threats and insecurity factors [3, 4]. The direct economic losses caused by network intrusions to individuals, businesses, and even the entire country are immeasurable. With the complexity of the network environment and the diversification of attack methods, traditional defense measures can only statically identify and block known types of attacks. Due to the limitations of traditional defense methods, it is easy to misclassify new data [5, 6]. Therefore, judging from the current development, salt-baked, a preventive strategy for computer network security, still has a high research value and development space in the industry [7, 8].

© The Author(s), under exclusive license to Springer Nature Switzerland AG 2022
J. Macintyre et al. (Eds.): SPIoT 2021, LNDECT 98, pp. 376–384, 2022.
https://doi.org/10.1007/978-3-030-89511-2_48

Regarding the research on network security, many scholars have conducted in-depth discussions. For example, Peng M analyzed the protection costs and benefits in the system on the basis of dynamic risk assessment, and obtained protection strategy selection recommendations through genetic algorithms [9]; Stojanov analyzes the threat of internal attacks, uses a probabilistic attack graph model to describe the uncertainty caused by internal attacks, uses a greedy algorithm to select information security protection schemes, and conducts effectiveness verification experiments in a real environment [10]; Yang L uses the Bayesian attack graph model to analyze the system information security status and uses the ant colony algorithm to obtain the optimal security protection strategy [11].

The purpose of this article is to study the attack methods and prevention strategies of computer network security. This article first summarizes the common computer network security attacks from the aspects of information leakage, denial of service attacks, and hidden attacks, expounds related defense strategies, and designs a computer network security attack defense model based on stochastic differential games. This paper verifies through experiments that the strategy designed in this paper improves the detection accuracy and detection speed of network security intrusion.

2   Computer Network Security Attack Methods and Prevention Strategies

2.1   Common Computer Network Security Attacks

(1) Information leakage

Information leakage is extremely easy to occur in network communication channels. Attackers use software and hardware vulnerabilities, and packet capture software to illegally obtain information through bypass or backdoors, and then obtain system operating data and control information. These are unauthorized operations. Information leakage is a prerequisite for the realization of most network attacks. Monitoring, scanning, counting, destruction, infection and even advanced persistent threats are effective ways for attackers to obtain information.

(2) Denial of service attack

Denial-of-Service (DoS, Denial-of-Service) is a type of malicious attack that makes the server (controller) unable to provide services. It is one of the commonly used attack and destruction methods by hackers [12]. In CPS, DoS attacks are mainly aimed at the physical system. Its purpose is to prevent the information interaction between the controller and the execution device, and use malicious programs to consume communication bandwidth to prevent the control instructions and feedback information between the controller and the execution device. The transmission, cut off the information exchange between the execution device and the controller. The main reason is that malicious attacks lead to disconnection or a large number of invalid service requests occupy routing and server resources.

The DoS phenomenon will cause the control loop and the feedback loop to be disconnected and disconnected, and the controller cannot get the feedback information in time, which in turn makes the system out of control, and eventually its performance deteriorates or even crashes. During a DoS attack, there is neither sending nor receiving information in the channel.

1) Hidden attack

Concealment attacks are a type of malicious attacks that are concealed and difficult to detect. The attack mode and behavior characteristics are the main research content of the communication and computer profession in the field of network security (especially communication security). Hidden attacker is on the controller, simulating the condition of the local closed-loop control system without network, which is equivalent to a device that plays a "cover" role.

(3) Replay attack

A replay attack is a type of malicious behavior that interferes with or disrupts the normal operation of the system by means of complete interruption/dissemination of non-current data. Replay attacks generally do not work alone, and require the cooperation of some other attack programs or behaviors to accomplish the purpose of the attack. Initially, this attack used network monitoring or other methods to steal authentication credentials, usually cookies or some authentication session sessions, after certain processing, and then resend it to the authentication server, such as attacking the Stuxnet of Iranian nuclear power plants. This type of attack is more harmful. Although encryption can effectively prevent the plaintext from being monitored, it cannot prevent replay attacks. This is mainly because the attacker can carry out the attack only by obtaining the uncracked ciphertext, and does not need to clarify the original content of the ciphertext. In the field of information security, the conventional solutions are "challenge response", timestamp, serial number, etc.

(4) ICMP Flood attack

ICMP Flood uses large traffic packets with ICMP packets to send service requests to the target server, consuming the server's system resources, making the server unable to respond to normal service requests, and then becoming paralyzed. However, many firewalls on the network now have the ability to filter ICMP packets. Therefore, it can maximize the occurrence of ICMP Flood attacks, so this type of attack is currently relatively rare on the Internet.

1) SQL injection attack

SQL injection attacks are currently one of the lowest cost of entry, the most harmful, and most destructive vulnerabilities based on application layer attacks. The method of SQL injection is usually, at the WEB application layer, fill in data into the form and submit SQL attack data by submitting the form, or assemble SQL attack statements through URL links with parameters to achieve the purpose of attacking the server. SQL injection is an operation performed by the attacker on the SQL statement according to his own purpose, and executed on the related SQL statement through the server.

2.2 Preventive Strategies for Computer Network Security

(1) Chinese wall model

This model is mainly used to solve the problem of conflict of interest in business, and the purpose is to prevent the occurrence of conflict of interest. The principle of the Chinese Wall Model to achieve access control technology is: whether the visitor has the access right, the visitor who already has the access right can access the data normally. When the visitor makes the visit, the data that can be accessed is the same as the one known. The information does not have a conflict of interest. Moreover, the meaning of the Chinese Wall is: in the initial state, the visitor can access any data information, once the visitor has accessed a certain data information, he will not be allowed to access any information that is known to have conflicting interests in the previous state.

(2) PPDR model

The primary role of the PPDR model (Policy& Protection& Detection& Response, security strategy, protection, detection, response) is to provide the attack defense timetable of the network security information system. The view it agrees with is that the network security protection strategy is based on time. Once the protection time of the protection measures expires, there is a chance of being damaged by attackers.

(3) WPDRRC model

On the basis of the PDRR model, the WPDRRC model is added with early warning and counterattack functions. Among the three elements of the WPDRRC model, personnel are the core, strategy is the bridge, and technology is the guarantee. Only in this way can the six links of the model be implemented, making it a truly effective security strategy model. The advantage of the WPDRRC information security model is that it is more suitable for China's national conditions when compared with other information security models. When constructing a system to deal with network security issues, in order to achieve security benefits, it is necessary to implement early warning, protection, detection, response, recovery, and counterattack. The strategy of infiltrating and implementing the personnel in the six links to improve the skills of the personnel is a more scientific and effective security strategy method, which can fully protect the security of the network environment.

2.3 Network Attack and Defense Stochastic Differential Game Model Design

The evolutionary game model of network offense and defense transforms the original two players of the game into a player space where both offense and defense have multiple decision makers, which can be expressed as a 4-tuple, ADEGM = (N,S,P,U).

(1) $N = \{N_A, N_D\}$ is the participant space of the evolutionary game, N_A and N_D respectively represent the attacker and the defender in the offensive and defensive game, and there are multiple decision makers on both sides.

(2) $S = \{S_A, S_D\}$ is the game strategy space, S_A is the attacker's optional strategy set, and S_D is the defender's optional strategy set. Both parties have multiple strategies to choose from.

(3) p = {p, q} is the game belief set, p represents the probability set of the attacker's choice of different strategies, and q represents the probability set of the defender's strategy choice.

(4) $U = \{U_A, U_D\}$ is the set of game utility functions, and the profit functions of players U_A and U_D in the network attack and defense game are represented by N_A and N_D respectively.

In the offensive and defensive game process, use p_i to indicate the probability of the attacker's decision maker choosing attack strategy AS_i,q_j to indicate the probability of the defending party's decision maker choosing defense strategy DS_j, $a_{ij} = U_A(AS_i, DS_j)$ to indicate the attacker's revenue, and $b_{ij} = U_d(AS_i, DS_j)$ to indicate the defender's revenue.

2.4 Network Security Attack Detection Model Based on Support Vector Machine

Network security attack detection is essentially a classification problem, which is to accurately distinguish between normal data and abnormal data in a data set. The support vector machine is a two-classifier. In order to accurately classify Normal, Dos, Probe, U2R, R2L in the KDD99 data set, SVM can be used to build a hierarchical detection structure, that is, construct multiple SVM classifiers, and treat other attack types as the same when detecting a certain type of attack, and then remove the detected attack type from the data set, and then detect the remaining types in the same way.

(1) Principal component analysis

The combination of principal component analysis (Principal Component Analysis, PCA) and linear discriminant analysis (Principal Component Analysis)

Analysis and Linear Discriminant Analysis (PCA-LDA), Fuzzy Principal Components Analysis (FPCA), and Kernel Principal Component Analysis (KPCA) are used in network intrusion detection. The calculation process of principal component analysis is as follows (the angle of matrix transformation):

1) Let the standardized and normalized data be X, (i = 1,2,...,m;j = 1,2,...n)

Among them, m represents the number of data in the data set, and n represents the number of features of each data, that is, the number of data dimensions.

2) Solve the correlation coefficient matrix:

$$Z = \frac{x^T x}{m - 1} \tag{1}$$

3) Solve the eigenvalues of Z:$\lambda_1 \geq \lambda_2 \geq \dots, \geq \lambda_n$

4) Generate the principal component data set S to replace the original data set:

$$S_i = \beta_{1i}X_1 + \beta_{2i}X_2 + \dots + \beta_{ni}X_n (i = 1, 2, \dots n)(n' \prec n) \tag{2}$$

Replace the original data set with the newly generated data set S, which greatly reduces the dimensionality of the data and improves the detection speed.

3 Experimental Research on Computer Network Security Attack Methods and Defense Strategies

3.1 Experimental Environment

This article divides the local area network into three parts: the trust zone, the isolation zone and the Internet as the experimental environment.

3.2 Experimental Equipment

The realization platform of the simulation program is MATLAB2016a, the test platform is windows system, the CPU is quad-core 2.3 Ghz, the simulation coefficient is $L = 10$, $q = 28$, and $m = 50$.

In the simulation experiment, the attacker invaded through social engineering to obtain the administrator authority of the network management server, switch, interlocking communication board, upper computer, maintenance machine and other equipment or cause the denial of service. The consequences of the functional safety domain mainly include partial function failures, information leakage, and equipment failures.

3.3 Evaluation Index

This paper uses three indicators of detection accuracy and detection time to verify the comprehensive performance of the detection model.

Detection risk (Accuracy): Refers to the risk value of network attack nodes detected in an experimental environment, and whether the risk nodes can be accurately captured.

Testing time (Testing_ time): Refers to the time spent by the model verification node.

4 Computer Network Security Attack Methods and Defense Strategy Data Analysis

4.1 Risk Situation

Table 1 is the comparison result of the risk situation in the experimental environment. For nodes S1 and S2, the risk value captured by the experiment increased from the original 70.2% and 64.3% to 100% and 95%.

As shown in Fig. 1, for nodes S2, S3, and S6, the prior probability increases from 64.3%, 52.6%, and 63.7% to 95.4%, 72.6%, and 89.7%, respectively. The overall risk of the network system is on the rise. Under the premise of detecting the attack event, the risk value of the target node is increased by more than 25%. Compared with the traditional risk analysis technology, the network security risk situation can be analyzed more accurately.

Table 1. Comparison of risk conditions in the experimental environment (unit: %)

Attribute node	Node's risk value when there is no network attack	Capture the node's risk value during network attacks
S1	70.2	100
S2	64.3	95.4
S3	52.6	72.6
S4	70.1	100
S5	70.0	100
S6	63.7	89.7
S7	43.6	65.9
S8	50.8	85.9
S9	44.9	70.4
S10	60.7	85.8

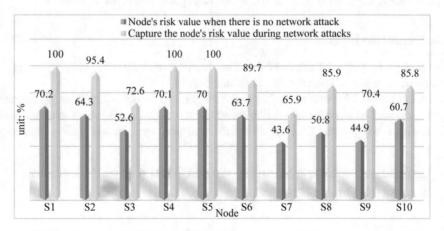

Fig. 1. Comparison of risk conditions in the experimental environment (unit:%)

4.2 Node Verification and Calculation Time-Consuming

Check the received message at the intermediate node, perform simple matrix inner product operation, and obtain the node verification calculation time-consuming data as shown in Table 2: for nodes S1, S2, and S3, the traditional detection method takes 0.39, 0.52 us and 0.53 us; the node verification calculation time under the stochastic differential game model is 0.24 us, 0.46 us, and 0.41 us.

It can be found from Fig. 2 that the stochastic differential game model's detection time for network attack nodes is controlled within 0.50 us, and the detection time is shorter than traditional detection methods, and the efficiency is higher.

Table 2. Node verification calculation time-consuming

Time(us)	Traditional detection methods	Stochastic differential game model
S1	0.39	0.24
S2	0.52	0.46
S3	0.53	0.41
S4	0.62	0.50
S5	0.36	0.32
S6	0.47	0.40
S7	0.45	0.39
S8	0.49	0.41
S9	0.34	0.30
S10	0.39	0.21

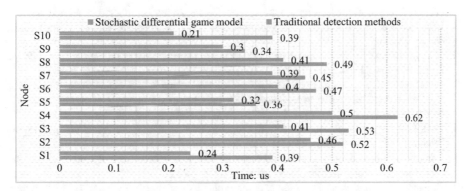

Fig. 2. Node verification calculation time-consuming

5   Conclusion

In recent years, with the increasing popularity and application depth of computer networks, Internet applications have expanded to many new fields such as medical treatment, education, and industrial manufacturing, which have added new impetus to the
development of social productivity and also brought potential safety hazards. A series of
cyber attacks such as cyber violence and cyber fraud have caused immeasurable losses in
social value. The methods of cyber attacks have also developed from traditional server-
side intrusion to mobile application-side intrusion, showing a diversified development.
The purpose of this article is to study the attack methods and prevention strategies of
computer network security. This article first summarizes the common computer network
security attacks from the aspects of information leakage, denial of service attacks, and
hidden attacks, expounds related defense strategies, and designs a computer network

security attack defense model based on stochastic differential games. This paper verifies through experiments that the strategy designed in this paper improves the detection accuracy and detection speed of network security intrusion.

References

1. Sudar, K.M., Deepalakshmi, P.: A two level security mechanism to detect a DDoS flooding attack in software-defined networks using entropy-based and C4.5 technique. J. High Speed Netw. **26**(2), 1–22 (2020)
2. Ahlawat, P., Dave, M.: A cost-effective attack matrix based key management scheme with dominance key set for wireless sensor network security. Int. J. Commun. Syst. **31**(12), e3713.1-e3713.24 (2018)
3. Li, J., Wang, D., Wang, Y.: Security DV-hop localisation algorithm against wormhole attack in wireless sensor network. IET Wirel. Sens. Syst. **8**(2), 68–75 (2017)
4. Kwon, S., Yoo, H., Shon, T.: IEEE 1815.1-based power system security with bidirectional RNN-based network anomalous attack detection for cyber-physical system. IEEE Access **99**, 1 (2020)
5. Xu, X., Weicui, Z.T.: Security dynamic performance evaluation method of ad hoc network based on anti-attack model. Acta Technica CSAV (Ceskoslovensk Akademie Ved) **62**(1), 45–56 (2017)
6. Sun, L.: Discussion on computer network security prevention based on big data era. J. Phys. Conf. Ser. **1848**(1), 012103 (2021)
7. Kizza, M.J.: [Computer communications and networks] Guide to computer network security. Syst. Intrusion Detect. Prevent. (Chapter 13), 275–301 (2017). https://doi.org/10.1007/978-3-319-55606-2
8. Gao, J.: A support vector machine model for computer network security technology. Boletin Tecnico/Tech. Bull. **55**(12), 564–568 (2017)
9. Peng, M., Jiang, J., Dong, Y., et al.: Research on computer network security monitoring based on extreme learning machine. Paper Asia **2**(2), 176–179 (2019)
10. Stojanov, Z., Dobrilovic, D., et al.: Solving problems in a physical laboratory for computer networks and data security: a conceptual framework with students experiences. Int. J. Eng. Educ. **32**(6), 2517–2530 (2016)
11. Yang, L.: An optimized prefix span algorithm for computer network security and intrusion detection system simulation. Boletin Tecnico/Tech. Bull. **55**(11), 277–284 (2017)
12. Bourekkache, S., Kazar, O., Aloui, A.: Computer and network security: ontological and multi-agent system for intrusion detection. J. Digit. Inf. Manag. **17**(3), 133 (2019)

Control System of Quad-Rotor Aircraft Based on EDA

Xu Mu[1](✉), Jingping Shi[2], and Li Ma[1]

[1] School of Aeronautics, Shandong Jiaotong University, Jinan, Shandong, China
[2] School of Automation, Northwestern Polytechnical University, Xi'an, Shaanxi, China

Abstract. In recent years, estimation of distribution algorithms (EDA) has attracted widespread attention in the international academic community, making this algorithm as a popular tool widely used in the actual optimization of flight controllers. The flight control system is an important part of the aircraft design, and its control performance determines the performance of the aircraft to a large extent. The purpose of this paper is to study the quadrotor control system based on the EDA. This article introduces the working principle of the quad-rotor aircraft control system, improves the EDA structure, designs the quad-rotor aircraft control system, and finally conducts experiments on the aircraft control system. The experimental data shows that the maximum error between the actual measured altitude of the quadrotor and the expected hovering altitude value is 0.16 m, and the aircraft basically moves along the expected trajectory preset in the experiment. This shows that the system altitude has good control performance.

Keywords: Estimation of distribution algorithms · Flight control law · Aircraft control system · Flight trajectory

1 Introduction

The EDA is an effective tool and method for solving practical engineering problems [1, 2]. Entering the 21st century, with the continuous update and progress of microelectronics technology, control theory methods, and high-tech equipment, quadrotors have become an important part of the aircraft field [3, 4]. Among them, the research on the quadrotor control system has become the focus of researchers. Therefore, the research on the quadrotor control system based on the distribution estimation algorithm has very important practical significance [5, 6].

Regarding the research of aircraft control system, many scholars at home and abroad have conducted multi-faceted discussions. For example, Y Yang, uses VC++ to establish flight dynamics and aircraft motion models, and realizes the visual simulation of the flight process of the aircraft through professional visualization simulation platform Multigen Vega and other software [7]; Chen J uses backstepping algorithm and automatic anti-jamming algorithm to achieve stable flight of the aircraft [8].

This paper mainly researches the control system of the quad-rotor aircraft based on the EDA. This article first introduces the working principle of the quadrotor control

J. Macintyre et al. (Eds.): SPIoT 2021, LNDECT 98, pp. 385–392, 2022.
https://doi.org/10.1007/978-3-030-89511-2_49

system, then introduces the improved EDA structure, analyzes the operation process and related modules of the quadrotor control system, and finally conducts experiments on the aircraft control system.

2 Quadrotor Control System Based on EDA

2.1 Working Principle of the Flight Control System

The main power source of a quadrotor is only the lift generated by the four propellers, and the four rotors can achieve six degrees of freedom of movement, so it is an under-driven, strongly coupled system [9, 10]. Moreover, quadrotors are generally small in size, so they are very sensitive to external forces, so an excellent control system directly determines whether the aircraft can complete the function of stable flight. The main task of the aircraft control system is: to maintain the stability of the three-axis take-off angle, to be precise, to ensure that the take-off angle of the aircraft is maintained at the required angle; and to swing to a fixed position according to the program design.

2.2 Quadrotor Control System Based on EDA

(1) Operation process of quadrotor control system based on EDA

After the system is started, the flight control related modules first receive the initialization data from the main control module. For example, the instruction solution module needs to receive the bound flight track information. After that, the navigation parameters are received from the aircraft modeling module, and after the rationality judgment is made, they participate in the altitude calculation, the command calculation, the control law calculation, and the engine control calculation. After the flight control calculation is completed in each cycle, the flight control commands and parameters need to be sent to the main control module and the auxiliary analysis module [11, 12].

(2) Wireless communication module

The wireless module of this system is realized by the NRF24L01 produced by NORDIC. At the remote control terminal, the control signal is sent out based on the high-frequency circuit and the encoding circuit, and sent to the aircraft through the PPM encoding. After the aircraft receives it, the CPU decodes and analyzes it. Then it is sent to the motor drive to complete the control and complete the communication between the remote control and the aircraft.

In the process of aircraft control, the specific working principle is that the microchip of the chip outputs PWM signals to control the opening and closing of different field effect transistors. The CPU calculates the static state according to the sensor data, and then generates different PWM waves, which will generate different voltages after adjustment, so that the aircraft's motors produce different speeds, thereby completing the control and output of the motors.

(3) Navigation information collection module

The range method is used to judge the rationality of the sky acceleration, longitude, and latitude in the navigation data; the range + increment method is used to judge the rationality of the north speed, east speed, and altitude.

(4) Height calculation module

The input of this module is two heights: theoretically calculated height without error and simulated height with actual error. This module needs to select the height mode used by the flight controller according to the initial settings.

(5) Instruction solving module

According to the complexity of the command system task, the software structure divides the entire command control system into three layers:

The first layer is the task decomposition layer: it mainly receives and decomposes this flight task before the aircraft takes off to activate different task management layers. Simulink receives the task set data packet from the main control module when it is running, and then sends it to the instruction module for processing. The received information includes the location of the launching point, the launching direction, the ambient temperature, the trajectory planning, and the location of the target area. The task decomposition layer is responsible for sending the setting information to different modules for program initialization.

The second layer is the management and control layer, which mainly performs various state switching and scheduling during the flight, and handles various abnormal situations in the process to generate different instructions. For example, set the boost phase control and cruise phase control switch before and after the booster is separated according to power working conditions, aircraft flight attitude, speed and other conditions; set the track according to the relationship between the current position of the aircraft, the flight trend and the set track point scheduling and switching, etc.

The third layer is the command issuance layer: it calculates the module corresponding to the command issuance timing and issuance conditions according to the commands of the management control layer, and generates target parameters as the input for switching the flight control mode.

(6) Control law solution module

The core idea of the control strategy of the quadrotor is to control the flight attitude of the aircraft, that is, to control the deflection angle of the aircraft on the three axes of space (X-axis, Y-axis, and Z-axis). According to the idea of PID control algorithm, the angle can be used to design a single closed-loop PID control module. The calculation expression of the dual closed-loop PID control algorithm of the attitude angle-angular velocity is:

$$
\begin{cases}
\text{Angle}PID\_u\ (t) = K_p(e_1(t) + \frac{1}{T_i}\int_0^t e_1(t)dt + Td\frac{de_1(t)}{dt} \\
\text{AngleRate}PID\_u\ (t) = K_p'(e_2(t) + \frac{1}{T_i}\int_0^t e_2(t)dt + Td\frac{de_2(t)}{dt}
\end{cases}
\tag{1}
$$

Where AnglePID_$u(t)$ is the output value of the outer loop, AngleRatePID_u is the output value of the inner loop, $\frac{de_2(t)}{dt}$ is the error value between the desired attitude angle and the actual attitude angle, and $e_1(t)$ is the error value between the outer loop output value and the actual angular velocity.

The default values of the corresponding parameters of the roll angle, pitch angle and yaw angle have been set in the control module program. In the process of continuous experimentation, the parameters are adjusted through the aircraft control effect to obtain the best value. When calculating the output value of each attitude angle and the remote control signal, the addition and subtraction relationship of each attitude angle will be determined according to the parameters set by Motor A, Motor B, Motor C and Motor D relative to the position of the nose. Since the obtained throttle output value may exceed the threshold value, it needs to be compensated and normalized. The processed throttle output value is passed to the signal output module, and the throttle output value is analyzed by the module to obtain the periodic waveform and the signal duty ratio are transmitted to the drive motor to adjust its speed.

2.3 EDA

EDA has good global search capabilities, but EDA still lacks local optimization capabilities. For simpler optimization problems, such as low-dimensional one-way optimization problems, EDA has its own unique functions. However, when it comes to complex multi-dimensional optimization problems, it is easy to fall into a local minimum. Adding chaotic mutation sequence to traditional EDA allows the algorithm to quickly escape from the local extreme area and find the global optimal solution, which is of great help in solving such problems.

(1) Perform statistical analysis on the distance information between different individuals in the population. The sum of the distance between the sth individual and other individuals is:

$$d(s) = \sum_{l=1}^{k \times NP} \sqrt{(X_{S,G} - X_{l,G})^2} \tag{2}$$

(2) Normalize the obtained d(s) to get:

$$d'(s) = \frac{d(s)}{\underset{l=1,2,\ldots k \times NP}{\mathrm{maxd}(l)}} \tag{3}$$

(3) Obtain the individual concentration:

$$\rho(s) = \frac{1}{d'(s)} \tag{4}$$

(4) Adjust the fitness value $fit(s) = fit(s)/p(s)$ of the individual according to the individual concentration, sort the individuals in descending order of the fitness value, and select NP as the next-generation population.

3 Experimental Research on Quadrotor Control System Based on EDA

3.1 Software System Construction of Quadrotor Flight Platform

Embedded system: μC/OS-III supports an unlimited number of task priorities, which has a greater effect on the later expansion of the program. Comprehensive considerations, μC/OS-III is a suitable choice.

After the system is powered on, according to the programming sequence of the program, the corresponding modules, I/O interfaces, interrupts, ADCs and variables will be initialized first. When data is transferred, an interrupt will be triggered, and the system interrupt program will be read sensor data, perform attitude calculation and data fusion on the data, use the corresponding filtering algorithm to achieve accurate attitude calculation, and then implement the final output PWM wave according to the corresponding control algorithm, which is passed through the ESC control the speed of the motor so that it returns to the interrupt receiving state after completing a cycle.

3.2   Experimental Project

(1) No wind field interference

  The Backstepping controller is optimized by the nonlinear control design (Nonlinear Control Design for short as NCD) toolbox in MATLAB, and different expected hover height values are set, and the hover control effect of the controller is simulated and verified.

(2) Whether there is wind field interference

  Add interference torque at t = 30 s to test the position change and error value of the aircraft.

4   Data Analysis of Quadrotor Control System Based on EDA

4.1   No Wind Field Interference Position Changes and Errors

In the absence of wind disturbance, the set flight trajectory of the quadrotor is shown in Table 1: the maximum error between the actual measured altitude and the expected hovering altitude is 0.16 m.

Table 1. The position change and error data of no wind field interference

Time (s)	Measured value (m)	Setting value (m)	Error
0	−0.16	0	0.16
10	7.5	7.5	0
20	10	10	0
30	5	5	0
40	−3	−3	0
50	−5	−5	0
60	7.48	7.5	0.02
70	10	10	0

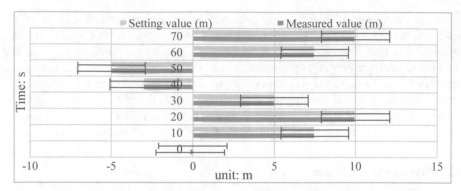

Fig. 1. The position change and error data of no wind field interference

It can be found from Fig. 1 that the aircraft basically moves along the desired trajectory preset in the experiment. From the perspective of position changes and errors, the aircraft's position curve has basically not changed, which shows that the algorithm based on the distribution estimation algorithm has relatively stable aircraft control performance under the condition of no wind field interference.

4.2 Location Changes and Errors with Wind Field Interference

After that, the wind field interference item is added during the flight of the quadrotor, and the wind speed interference is increased at T = 30.0 s with its initial state being zero time. Use the control algorithm to construct the attitude control module, and carry out the simulation experiment of the stability control of the aircraft against wind field interference, and record the experimental results of the control algorithm on the aircraft within 70 s, as shown in Table 2:

Table 2. Location change and error data with wind field disturbance

Time (s)	Measured value (m)	Setting value (m)	Error
0	0	0	0
10	7.5	7.5	0
20	10	10	0
30	6.2	5	1.2
40	−1.8	−3	1.2
50	−3.8	−5	1.2
60	8.7	7.5	1.2
70	11.2	10	1.2

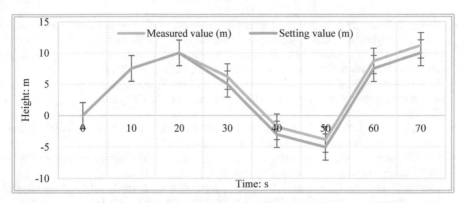

Fig. 2. Location change and error data with wind field disturbance

Looking at Fig. 2, it can be found that in the case of wind field disturbance factors, the actual flight path of the aircraft deviates significantly from the preset expected trajectory, and in the subsequent flight process, although the flight trajectory form is similar to the expected trajectory, the actual position is different. Not at the corresponding position of the desired trajectory, but higher than the desired position.

5 Conclusion

With the rapid development of MEMS devices and embedded processors, the emergence of quadrotor aircraft has become possible. There is no doubt that aircraft will become an indispensable assistant to mankind in the near future. And as the demand increases, the aircraft control system becomes more and more complex. This paper mainly conducts related research on the quadrotor control system based on the EDA, introduces the working principle of the quadrotor control system and the structure of the quadrotor, and improves the EDA, and designs the quadrotor control system. Through testing, the system has good control performance.

Acknowledgements. This work was supported by Doctoral Foundation of Shandong Jiaotong University (BS2018018) and Natural Science Foundation of Shaanxi Province (2019JM-163).

References

1. Fei, G., Bozhko, S., Asher, G., et al.: An improved voltage compensation approach in a droop-controlled DC power system for the more electric aircraft. IEEE Trans. Power Electron. **31**(10), 7369–7383 (2016)
2. Chen, Y., Wickramasinghe, V., Zimcik, D.: Active control of a hybrid actuation system for aircraft vertical fin buffet load alleviation. Aeronaut. J. **110**(1107), 315–326 (2016)
3. Botez, R.M., Kammegne, M., Grigorie, L.T.: Design, numerical simulation and experimental testing of a controlled electrical actuation system in a real aircraft morphing wing model. Aeronaut. J. **119**(1219), 1047–1072 (2016)

4. Wang, X., Rui, L., Shi, C., et al.: Linear extended state observer-based motion synchronization control for hybrid actuation system of more electric aircraft. Sensors **17**(11), 2444 (2017)
5. Khalid, S.: Optimized aircraft electric control system based on adaptive tabu search algorithm and fuzzy logic control. J. Acoust. Soc. Am. **122**(2), 1102–1110 (2016)
6. Keller, D., Rudnik, R.: Numerical investigations of aerodynamic properties of a propeller blown circulation control system on a high wing aircraft. CEAS Aeronaut. J. **7**(3), 441–454 (2016)
7. Yang, Y., Chen, S., Sheng, C., et al.: Study on coupling performance of turbo-cooler in aircraft environmental control system. Energy **224**(20), 120029 (2021)
8. Chen, J., Chen, S., Ma, C., et al.: Fault detection of aircraft control system based on negative selection algorithm. Int. J. Aerosp. Eng. **2020**(2), 1–10 (2020)
9. Miazga, T., Iwański, G., Nikoniuk, M.: Energy conversion system and control of fuel-cell and battery-based hybrid drive for light aircraft. Energies **14**(4), 1073 (2021)
10. Fang, Q., Chen, W., Zhao, A., et al.: Control system designing for correcting wing–fuselage assembly deformation of a large aircraft. Assem. Autom. **37**(1), 22–33 (2017)
11. Ijaz, S., Yan, L., Hamayun, M.T., et al.: Active fault tolerant control scheme for aircraft with dissimilar redundant actuation system subject to hydraulic failure. J. Franklin Inst. **356**(3), 1302–1332 (2019)
12. Yang, Y., Gao, Z.: A new method for control allocation of aircraft flight control system. IEEE Trans. Autom. Control **65**(4), 1413–1428 (2020)

Software Development of Intelligent Pension Monitoring System Based on Ant Colony Optimization Algorithm

Yizhi Wu[✉]

Guangzhou College of Technology and Business, Guangzhou, Guangdong, China

Abstract. Since the 21st century, the problem of China's population aging has intensified, and accelerating the development of aging industries to ease social pressure has become an important issue to be solved in China. The development of the intelligent elderly care industry has shared the demand pressure faced by nursing homes to a certain extent. Therefore, in a Chinese society where the economy is developing rapidly, people's consumption level is rising, and the problem of population aging is increasingly prominent, the intelligent elderly care monitoring system has received increasing attention. The purpose of this paper is to study the software development of intelligent elderly care monitoring system based on ant colony optimization algorithm. This article analyzes the functional requirements of the intelligent elderly care monitoring system based on the ant colony optimization algorithm, introduces the key functional modules of the software, and explains the overall system architecture. This article debugs and tests the functions of the system "blood oxygen saturation test" and "old man fall detection" to complete the system reliability test. In the experiment, the results measured by the blood oxygen saturation detection module of the system are compared with the oxygen saturation measured by the traditional blood oxygen detector. The maximum relative error between the two is 1.13%, and the minimum relative error is 0. It is visible that this blood oxygen saturation test module of the system has high test accuracy and has reached the design goal of the system.

Keywords: Smart elderly care · Ant colony algorithm · Smart elderly care monitoring · Environmental monitoring

1 Introduction

With the development of my country's economy and technological progress, on the one hand, people's living standards have continued to improve, on the other hand, the problem of population aging has become more and more serious [1, 2]. In addition, young people are under great work pressure and have no time to take care of the elderly. Our country's smart elderly care equipment has gradually developed and become a new development trend. And various elderly care equipment have been emerging on the market [3, 4]. Using artificial intelligence to monitor the lives of the elderly can avoid hidden dangers when the elderly live alone and provide them with a safe and

© The Author(s), under exclusive license to Springer Nature Switzerland AG 2022
J. Macintyre et al. (Eds.): SPIoT 2021, LNDECT 98, pp. 393–400, 2022.
https://doi.org/10.1007/978-3-030-89511-2_50

comfortable living environment [5, 6]. With the rapid development and application of machine learning, the combination of ant colony optimization algorithm and intelligent pension monitoring system can achieve certain practical effects [7, 8].

Regarding the research of smart elderly care, many scholars at home and abroad have conducted multi-faceted and in-depth discussions on it. For example, Yang SR conducted research on the remote control system of the Internet of Things for smart home elderly care based on DTMF [9]. Udupa P on smart elderly care the wireless sensor of furniture is designed and researched to improve the convenience and intelligence of elderly furniture [10]. Chiang KF applies smart mobile devices in smart elderly medical care to help patients with chronic diseases in home care [11]. It can be seen that the research on intelligent elderly care has always attracted much attention. Therefore, this paper proposes a new idea of combining the ant colony optimization algorithm with the software of the intelligent elderly care monitoring system.

This paper first analyzes the functional requirements of the intelligent elderly care monitoring system based on ant colony optimization algorithm, and then designs the software module in detail according to the requirements analysis, and realizes the use methods of the computer application end and mobile phone client of the system. Finally, the overall function of the system is debugged and tested to complete the test of system reliability. Through the analysis of the test results, it is concluded that the system software is feasible.

2 Software Development of Intelligent Elderly Care Monitoring System Based on Ant Colony Optimization Algorithm

2.1 Demand Analysis

(1) Safety protection-fall detection

The behavior of the elderly who unconsciously loses their balance and thus falls is difficult to control. Especially for the elderly, a fall can cause a series of injuries that are difficult to recover, and in severe cases, it can lead to paralysis or death, which greatly endangers the health and life safety of the elderly. Therefore, it is necessary to add a detection function to the fall behavior of the elderly in the intelligent elderly care monitoring system to ensure that the alarm and processing are performed as soon as possible after a fall.

(2) Medical health monitoring-heart rate and blood pressure test

Currently, there are many smart bracelet products for the elderly on the market. The auxiliary equipment connected with the system has a bracelet for measuring heart rate and blood pressure. Through the measurement, it can be determined whether the heart rate and blood pressure health of the elderly is normal or there is a hypertension disease, and the elderly provide corresponding medical-grade care and early warning measures.

On the basis of health detection, complete health files can be formed, including constitution files, blood pressure files, blood glucose files, blood oxygen files, ECG files, blood uric acid files, blood lipid files, all-in-one machine files and hospital cases. After determining the target file, you can view the details of the file.

(3) Health management

In the health management part, the system can display data information related to health detection indicators. The statistical information includes blood sugar, blood pressure, ECG, blood lipids, body temperature, etc., and the system can feed back information such as the number of detection indicators and the degree of abnormality to the elderly. The feedback results can be displayed in the form of bar graphs, tables, circular graphs, etc.; the number of tests includes all-in-one device testing, home testing, etc. The specific indicators include BMI, blood sugar, blood pressure, blood oxygen, ECG, body temperature, blood uric acid, four items of blood lipids, medical records, all-in-one report, etc. The statistical results of the test situation include the number of tests, the percentage of normal, etc.

2.2 Software Development and Design of Intelligent Elderly Care Monitoring System Based on Ant Colony Optimization Algorithm

(1) System network architecture

Users can log in to the system through the computer application and mobile client. The computer application user sends a request to the system through a web browser. After processing by the main server, data server and other components, the processing result is returned. It is filtered by firewalls, switches and other components after processing, the user accesses the Internet to obtain it. After the user logs in to the mobile phone, they need to access the wireless network to make a request and obtain the final result.

(2) Software function module design

1) Fall detection module

The sensor is used to collect human activity data, and the processor will process and analyze the collected data, and then determine whether the elderly has fallen down according to a certain algorithm. Once the system determines that the human body has fallen, it immediately sends out the corresponding alarm signal. Here, the ant colony optimization algorithm is used to optimize the alarm decision. Among them, at a certain decision time t, the probability transition calculation formula is as follows:

$$P_{ij}(t) = c_{ij} \cdot \frac{[\tau_{ij}(t)]^m \cdot [\eta_{ij}]^n}{\sum_{j \in Vf} [\tau_{ij}(t)]^m \cdot [\eta_{ij}]^n}, \forall j \in V_f, i \in V \tag{1}$$

Where m and n are the weight coefficients of pheromone and heuristic parameters respectively; $\tau_{ij}(t)$ is the pheromone amount of road section (i, j) at time t; η_{ij} is the heuristic parameter value of road section (i, j); c_{ij} is a binary variable.

The Q matrix records the value of each state-action combination, and the Q matrix is updated every time the algorithm takes an action, that is, the Q learning algorithm is a one-step Q learning algorithm (one-step Q learning), which updates the value of the state-action combination and the calculation is shown in formula (2):

$$Q'(s, a_s) \leftarrow Q(s, a_s) + \theta \cdot (r_{sa_s} + \gamma maxQ(s', a_{s'}) - Q(s, a_s)) \tag{2}$$

In the formula, $\theta \in [0, 1]$ is the learning rate, $Q(\cdot)$ is the value of the current Q matrix, and $Q'(\cdot)$ is the updated matrix value. At the beginning of the iteration, the Q matrix is initialized to a zero matrix, and gradually converges as the algorithm continues to proceed [10].

The sensor of the fall detection module can also upload the raw data collected by the acceleration and gyroscope to the computer application terminal through the wireless network, and then the data can be preprocessed, feature extraction, and training model.

2) Blood oxygen measurement module

When measuring blood oxygen saturation, the main task is to control the driving circuit through the main controller to ensure that the photoelectric sensor converts the intensity of transmitted light into electrical signals, then transfer these data to the processor for corresponding sampling, and finally calculate the relevant data of blood oxygen saturation according to the formula. The detailed process is as follows: after the power supply for the system is completed, the main control system first initializes, and then transmits the probe detection pulse on time to detect the shape of the blood oxygen probe. Then clamp the detection equipment on the finger clamp. After the system detects the finger, it will regularly send the driving timing control signal to automatically measure the blood oxygen saturation of the elderly. That is, the timing control signal is transmitted on time, the light emission of the diode is controlled, sampled, and then the information of blood oxygen saturation is packaged through filter control, and then transmitted to the main controller through serial port for display in the software interface.

3) View the real-time status of the elderly

The main interface of the computer application or mobile phone will display the current status of the elderly. The current state of the elderly is obtained through real-time analysis of the received event sequence. The main interface display content includes: user login status (online, not logged in), MQTT connection status (connected, not connected), edge device online status (online, offline), date and time, indoor population (no one, single, multiple person, unknown), sleep state (falling asleep, waking up, unrecognized), currently activated abnormalities and duration (abnormal analysis only in the first 24 h).

4) Personal health management module

The first realization of the personal health management module is to store the basic information of the elderly, including the contact information of the family and the health files of the elderly. When the elderly use the monitoring system for the first time, they must fill in personal information through the registration page, and then send it to the server. After the server confirms, the registration is successful, and the elderly can enter the login system through the newly registered account. Taking into account the particularity of the system for the elderly, in order to facilitate the elderly to log in to the system and prevent the elderly from forgetting their password or account number, this software sets a specific digital label for each elderly person. The elderly only need to remember their own center. The number can be logged into the system, eliminating the need for complicated user names and passwords.

The health management module mainly includes health records and historical data query. The health record module can store basic information and corresponding physiological parameters of the elderly, so that the elderly can understand their physiological conditions more conveniently.

The historical data query module allows the elderly to have a clearer understanding of recent changes in physiological parameters. For historical data query, the system will store the data of each test in the database, just check the part of the information you want to know, and the system will call the recent test data and date of the information from the database, and display on the interface.

3 Experimental Design

3.1 Elderly Fall Experiment

The use of the fall detection function is the majority of elderly people, but because the elderly are relatively fragile, they are not suitable for violent experimental testing activities such as jumping and falling. Therefore, the 50 testers selected for this experiment are adults with a height between 154 cm and 165 cm, a weight between 45 kg and 80 kg, and the age between 30 and 40 years old.

During the experiment, the tester wears the detection equipment at the waist and simulates the activity characteristics of the elderly as much as possible. The testers performed 40 falls, 40 times of bending, sitting and standing, jumping, walking, running, and up and down stairs respectively.

3.2 Spo2 Module Test

In order to test the accuracy of the blood oxygen saturation test of the system, ten adults were selected for testing, and the results measured by the system were compared with the oxygen saturation results measured by a traditional blood oxygen detector.

4 Analysis of Test Results

4.1 Blood Oxygen Measurement Data

In order to test the accuracy of the blood oxygen saturation detection results of this system, the results measured by the blood oxygen saturation detection module of this system are compared with the oxygen saturation measured by the traditional blood oxygen detector. The test results are shown in Table 1. It shows that the maximum relative error between the two is 1.13%, and the minimum relative error is 0.

Observing the blood oxygen saturation test results of the 10 testers in Fig. 1 can be concluded that the blood oxygen saturation test module of this system has a high test accuracy and has reached the design goal of the system.

Table 1. Blood oxygen saturation test results (%)

Experiment	System blood oxygen saturation monitoring data	Blood oxygen saturation monitoring data of oximete	Relative error
1	97.6	97.6	0
2	96.3	96.0	0.31
3	97.2	97.1	0.10
4	97.4	97.4	0
5	97.3	97.2	0.1
6	98.1	97.0	1.13
7	97.4	97.6	0.20
8	96.2	96.3	0.10
9	95.7	95.4	0.31
10	97.3	97.0	0.31

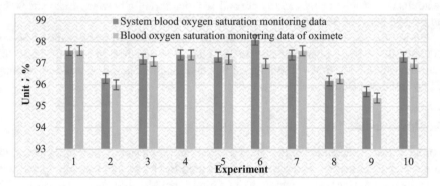

Fig. 1. Blood oxygen saturation test results

4.2 Analysis of the Test Results of the Fall Experiment

The results of the elderly fall experiment are shown in Table 2: the correct recognition rate of backward and left fall detection is 100%, and the correct recognition rate of forward and right fall detection is 97.5%.

It can be seen from Fig. 2 that the system has high accuracy in detecting the fall behavior of the elderly, and can distinguish falls from other daily behavior events more accurately. Using video to monitor the entire test process, it is found that false alarms occur when sitting or standing up. The reason is that the tester's motion acceleration when lying on the bed is large, and the inclination angle of the body after lying on the bed also changes significantly. These motion changes feature it is similar to the fall of the human body, which is easy to make the system misjudgment. And the false alarm rate is low, so the experiment proves that the system has good reliability and meets the design requirements.

Table 2. Test results of the fall test

Event	Number of experiments	Number of alarms	Number of false negatives	Recognition rate (%)	False report rate (%)	False alarm rate (%)
Fall forward	40	39	1	97.5	2.5	
Fall backward	40	40	0	100	0	0
Fall to the left	40	40	0	100	0	0
Fall to the right	40	39	–	97.5	0	0
Sit down and stand up	40	2	–	95	0	5
running	40	1	–	97.5		2.5
walk	40	0	–	100		
Down stairs	40	0	–	100		
jump	40	2	–	95	0	5

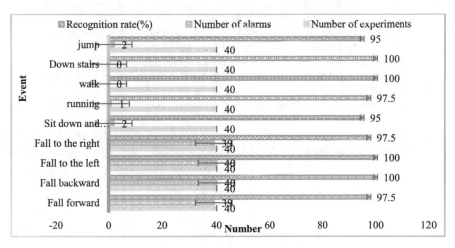

Fig. 2. Test results of the fall test

5   Conclusion

With the development of population aging, the market of pension products is expanding rapidly. With the rapid development of intelligent technology, its application to the elderly monitoring system is conducive to improve the service and supervision level of the elderly and ensure the health of the elderly. Therefore, it is necessary to develop a new intelligent elderly care monitoring system software based on ant colony optimization algorithm. Through research, this paper completes the following work: analyzes the functional requirements of the intelligent elderly care monitoring system based on ant colony optimization algorithm, introduces the key functional modules of the software, and explains the general architecture of the system. Through the system test, the reliability of the system software is verified.

References

1. Wang, S., Liu, T., Wei, W.: Design of intelligent pension platform based on Internet of Things. J. Phys.: Conf. Ser. **1744**(4), 042240 (2021). (5pp)
2. Cui, Y., Zhang, L., Hou, Y., et al.: Design of intelligent home pension service platform based on machine learning and wireless sensor network. J. Intell. Fuzzy Syst. **40**(2), 2529–2540 (2021)
3. Xia, M.: Analysis on the development status and prospects of the intelligent pension model in Tianjin. Human. Soc. Sci. **7**(1), 34 (2019)
4. Yassine, A., Singh, S., Alamri, A.: Mining human activity patterns from smart home big data for health care applications. IEEE Access **5**(99), 13131–13141 (2017)
5. Vanus, J., Machacek, Z., Koziorek, J., et al.: Advanced energy management system in Smart Home Care. Int. J. Appl. Electromagnet. Mech. **52**(1–2), 1–8 (2016)
6. Huang, F.L.: 57A predictable smart home integrated with cloud computing and long-term care. J. Investigat. Med. **64**(Suppl 8), A20 (2016)
7. Marjan, A., Jennifer, R., Uwe, K., et al.: An ontology-based context-aware system for smart homes: E-care@home. Sensors **17**(7), 1586 (2017)
8. Kang, H., Kim, B., Kwon, G.H.: The smart Servicescape framework in smart home healthcare service experience. Design Manag. J. **14**(1), 50–59 (2019)
9. Yang, S.-R., Yuan, S.-C., Lin, Y.-C., Yang, I.-F.: DTMFTalk: a DTMF-based realization of IoT remote control for smart-home elderly care. Mobile Netw. Appl. (2020). https://doi.org/10.1007/s11036-020-01641-0
10. Dupa, U., Yellampalli, S.S.: Smart home for elder care using wireless sensor. Circuit World **44**(2), 69–77 (2018)
11. Chiang, K.F., Wang, H.H.: Nurses' experiences of using a smart mobile device application to assist home care for patients with chronic disease: a qualitative study. J. Clin. Nurs. **25**(13–14), 2008–2017 (2016)

Design of Computer Network Security Monitoring System Based on Programming Language

Wei Zhang[✉]

School of Information Engineering, Lanzhou Vocational and Technical College of Resources and Environment, Lanzhou 730021, Gansu, China

Abstract. With the rapid development and increasing popularity of the Internet, information security has attracted more and more attention. Although people install firewalls and intrusion detection systems on servers to prevent network intrusions, intruders may continue to steal or forge confidential information through illegal means. Therefore, timely detection and control of illegal activities has become an urgent need for network security. This paper studies the computer network security monitoring system based on the programming language. Through some existing shortcomings of the system, the overall framework of the system is proposed, and then each module proposed is designed in detail, and some existing problems of the system are attempted to be solved. And through the experimental detection system, the test results show that the company's network security problems have been well resolved after the system is used, especially in the area of iP conflict, which has been reduced from 10 to 2, but from the evaluation of the use of the system, it can be concluded that the system still has some shortcomings.

Keywords: Programming · Computer network · Network security · Monitoring system

1 Introductions

Internet technology, especially the rise of mobile Internet, is fundamentally changing the traditional information science and technology industry [1, 2]. With the development and popularization of Internet information technology and mobile Internet, information exchange has become faster and more convenient, and the requirements for information confidentiality and data security have been improved [3, 4]. In recent years, researchers have done a lot of research work in information encryption technology, such as public keys, symmetric encryption algorithms, network access and control (such as firewalls), and electronic computer system security management and network security management, and have achieved a lot of research results. Moreover, network surveillance technology is also an urgent problem for many Internet users [5, 6]. The scientific research results of network security and management-related technologies often fail to fully consider the requirements of such technologies. As a result, the current network security

J. Macintyre et al. (Eds.): SPIoT 2021, LNDECT 98, pp. 401–408, 2022.
https://doi.org/10.1007/978-3-030-89511-2_51

and management-related technologies may not meet specific network monitoring users [7, 8].

Regarding the research of computer network security, some researchers have proposed that traditional manual data storage can no longer meet the needs of the current network age, so the technology of encrypting data transmission and storage to the network has emerged, but there is a certain degree in the process of data encryption transmission. Security issues, so using data encryption technology to analyze the factors that affect network security, it is proposed that the application of data encryption technology to computer network security monitoring can improve data security [9]. Some researchers have proposed that user authentication is one of the most important links in a network security system. Therefore, a user authentication based on image sequences is proposed, and the authentication based on image sequences is more secure than text passwords through experiments. The security of authentication is high [10]. Researchers have also designed a network security device that includes a security detector to detect the security of data transmission between the host and the sub-network, as well as to monitor some malicious attacks on the host, and to generate a predictive data model [11]. There are also researchers who combine artificial intelligence with computer network security to propose a Trojan horse monitoring model based on artificial intelligence, and use experiments to verify that the artificial intelligence-based monitoring model can accurately and quickly detect Trojan horse programs, and there are fewer false positives and false negatives [12].

This article studies the computer network security monitoring system based on the programming language. On the basis of relevant literature, it summarizes some of the shortcomings of the computer network security monitoring system, and then analyzes the application of the programming language development in the system. Xiyong's shortcomings put forward the overall framework of the system, and carried out a detailed design for each function, and finally tested the system designed in this paper, and obtained relevant conclusions through the test results.

2 Research on Computer Network Security Monitoring System

2.1 Defects of Computer Network Security Monitoring System

(1) Data collection issues in high-traffic environments. The current processing capacity of the system is generally between 1 billion, but the current network traffic on the Internet is usually more than 1 billion. How to solve the problem of data collection in a high-traffic environment is a major challenge faced by the network content security monitoring system.

(2) Distributed development issues and collaborative processing. Because backbone nodes use routing and motion balance algorithms, specific messages or conversations cannot completely pass through specific collection points, so incomplete information will be received at specific collection points, but cannot be submitted for processing, and needs to be completely edited again, and the system needs to be distributed type deployment, the information in different locations needs to be integrated and coordinated for data mining.

(3) Data analysis and knowledge discovery. The current system has simple content control and filtering algorithms, but practical applications require relevant and intelligent data analysis. In the analysis, a correct judgment should be made through a comprehensive analysis of various relevant information collected on the network. Just getting a certain level of information sporadically will bring great uncertainty to information analysts.

(4) The servers inside the network now only install traditional firewalls and intrusion detection systems, and can only detect viruses and Trojan horses in the usual sense. If a low-privilege user reads, modifies, or deletes viruses and Trojan files on the host, the traditional network monitoring system is powerless.

2.2 Application of Programming Language Development in System Design

Java is very suitable for the Internet or corporate network environment, so it has become one of the most popular and important programming languages on the Internet. Compared with C++, Java has deleted many unused features, including those features that have more advantages than disadvantages, such as simplicity, object-oriented, distributed, structurally neutral, portability, high performance, interpretability, reliability, security, multithreading, dynamics, etc., at the same time it allows clients of any processor to run and stream on the Internet.

2.3 Evaluation of Computer Network Security and Health

After analyzing the fluctuation of the data under different pressure conditions, combining industry experience, formulating a reasonable calculation method, and selecting the final index to reflect the performance of the server. The specific index calculation method is as follows:

(1) Calculation method of disk usage

$$dis = \frac{\sum_{i=1}^{n} d_i}{n} \times 100\% \tag{1}$$

d_i represents the disk in the disk group, the average utilization of i.

(2) CPU usage calculation method

$$cpu = (1 - \frac{\%ide}{\%ide + \%_{5y} + \%us + \%wai}) \times 100\% \tag{2}$$

% Sy represents the percentage of time that the program running in kernel mode uses the CPU, % Us represents the percentage of time that the program running in user mode uses the CPU, % wai represents the percentage of CPU idle time, and % ide represents the percentage of CPU idle time.

3 Computer Network Security Monitoring System Design Based on Programming Language Development

3.1 The Overall Framework of the System

Through the analysis of the defects of the computer network security monitoring system, the overall framework of the system is proposed, including: data acquisition module, data analysis module, network equipment security monitoring module, and visualization module.

3.2 Data Acquisition Module

Data collection consists of infrastructure and collectors in the business network information platform. Infrastructure is the hardware foundation of the system, which mainly includes routers, switches, servers, terminals, firewalls, IPS, RSAS, and communication links. Collectors mainly include active detectors, detectors and customer detectors. Different collection methods are designed to collect data from different types of equipment.

3.3 Data Analysis Module

Data analysis is the pre-layer of the control system, which mainly includes equipment in the business network information platform, resource statistics, operation status analysis, and business capability evaluation. Use Pandas, Numpy and other common data analysis technologies to analyze the historical status of the equipment and analyze the operating status of the equipment in real time according to the company's security policy. Business capability evaluation includes comprehensive evaluation of alliance network operations and Web server service capability evaluation. The results of the comprehensive evaluation will serve as the basis for the company's evaluation. Evaluating the serviceability of the Web server will provide a basis for company employees to optimize the network topology. The above analysis results will generate an analysis report for users to download.

3.4 Network Equipment Security Monitoring Module

The main task of network equipment security monitoring is to monitor the network equipment and realize human-computer interaction, that is, the user interface, which can correctly reflect the received messages and errors, and correctly display all the activities of the equipment in the network tree. Check the machine list to display its status in color, or move it to the bottom of the bar where the received message is displayed. At the same time, in order to prevent the system load from increasing a lot of messages, the network equipment security monitoring has an automatic deletion message bar. At the same time, ensure that all basic functions are correct and activate the corresponding correct functions. The second is the operation of the data center monitoring station. Add and delete tests in the data center, add, delete, and change administrator attributes and permissions, all tests must display relevant results. It can also realize the remote control of the monitoring station. In the test, the dynamic transmission of the remote control desktop should be ensured to be quite smooth. Regardless of the resolution and color, it can accurately reflect the current desktop of the remote control in time. During the remote mouse control process of the remote desktop operation, it may be exactly the same as the operation on the machine. Various operations can be done on the remote desktop through the mouse to achieve the effect of remotely controlling the user's desktop.

3.5 Visualization Module

The visualization function is to present the real-time operating status of the network equipment with a graph. If the operating status is normal, it will be displayed in green, and if the operating status is abnormal, it will be displayed in red.

3.6 System Implementation

Although the structured b/s application system in the network has the advantages of simple management, client development, operation and maintenance costs, etc., a structured b/s application system in the network may bring a lot of pressure on the network stream data processing in the network. At the same time, especially when a structured b/s application system is used on a network, compared with a structured c/s application system, it will bring dozens of times the number of concurrent tcp sessions to the firewall system of the entire network. Therefore, this system chooses the cs structure.

4 Computer Network Security Monitoring System Monitoring Based on Programming Language Development

4.1 System Performance Test

Randomly select 3 companies of the same size in this city and let them use them for 3 months. The comparison of the security problems that often occur in the office network before and after use is shown in Table 1:

Table 1. System performance test results

	Before the implementation of the plan	After the plan is implemented
IP address conflict	10	2
Multiple computers poisoned	1	0
The virus caused the computer system to crash	2	1
Server poisoning causes access interruption	2	2

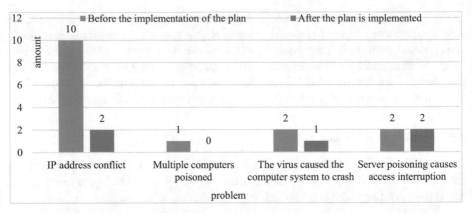

Fig. 1. System performance test results

It can be seen from Fig. 1 that after using this system, the company's computer network security situation has improved significantly. The most obvious problem is the IP conflict, which has dropped from 10 to 2.

4.2 System Use Evaluation

After the system is used, a questionnaire survey will be conducted on the company's operation and maintenance staff to collect the evaluation of the system's use. This paper recruited 45 people, so 45 questionnaires were issued and 45 were collected. Effective questionnaires were obtained through sorting out 43 copies, and the evaluation results of the system are shown in Table 2:

Table 2. System usage evaluation results

	The function of the system	Human-computer interaction of the system	System response time
It is good	35%	36%	15%
Better	36%	37%	26%
Generally	10%	10%	33%
Have no idea	19%	17%	22%

It can be seen from Fig. 2 that the overall evaluation of the system is good. The functional evaluation of the system accounts for more than 71%, and the human-computer interaction evaluation of the system accounts for about 73%, but the system's response time is not very good.

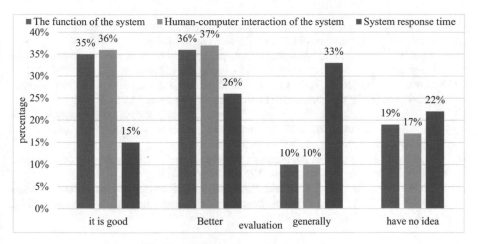

Fig. 2. System usage evaluation results

5 Conclusions

In this paper, a computer network security monitoring system developed based on a programming language is researched. The system is designed according to the existing problems of the system at this stage, and the problems are optimized in the design process. Then the system is verified by experiments to optimize the problem. The experimental results show that the system designed in this paper can monitor the company's network security very well. After use, the emergence of network security problems has been significantly reduced. However, it can be concluded from the evaluation of the use of the system that the system designed in this paper still has certain defects, which are mainly manifested in the response time of the system.

Acknowledgements. Gansu University Innovation Fund Project: Research on a computer communication network wiring device, Project No.: 221b-424.

References

1. A.D.A, B.A.M., C.A.O.: Network security and contagion. J. Econ. Theory **166**(3), 536–585 (2016)
2. Liyanage, M., Abro, A.B., Ylianttila, M., et al.: Opportunities and challenges of software-defined mobile networks in network security. IEEE Secur. Priv. **14**(4), 34–44 (2016)
3. Li, Y., Hua, N., Song, Y., et al.: Fast lightpath hopping enabled by time synchronization for optical network security. IEEE Commun. Lett. **20**(1), 101–104 (2016)
4. Ramos, A., Lazar, M., Filho, R.H., et al.: Model-based quantitative network security metrics: a survey. IEEE Commun. Surv. Tutor. **19**(4), 2704–2734 (2017)
5. Xiang, W., Shi, W., Yang, X., et al.: Efficient network security policy enforcement with policy space analysis. IEEE/ACM Trans. Netw. **24**(5), 2926–2938 (2016)
6. Saxena, N., Grijalva, S., Chukwuka, V., et al.: Network security and privacy challenges in smart vehicle-to-grid. IEEE Wirel. Commun. **24**(4), 88–98 (2017)

7. Moreira, R., Moreno, R., Strbac, G.: Value of corrective network security for distributed energy storage applications. IET Gener. Transm. Distrib. **10**(7), 1758–1767 (2016)
8. Luong, N.C., Hoang, D.T., Wang, P., et al.: Applications of economic and pricing models for wireless network security: a survey. IEEE Commun. Surv. Tutor. **19**(4), 2735–2767 (2017)
9. Sharma, P.K., Singh, S., Park, J.H.: OpCloudSec: open cloud software defined wireless network security for the Internet of Things. Comput. Commun. **122**, 1–8 (2018)
10. Jin, Y.H., Shen, Y.J., et al.: The model of network security situation assessment based on random forest. In: 2016 7th IEEE International Conference on Software Engineering and Service Science (ICSESS), pp. 977–980. IEEE (2016)
11. He, J., Yang, J., Ren, K., et al.: Network security threat detection under big data by using machine learning. Int. J. Netw. Secur. **21**(5), 768–773 (2019)
12. Nezhad, S.M.T., Nazari, M., Gharavol, E.A.: A novel DoS and DDoS attacks detection algorithm using ARIMA time series model and chaotic system in computer networks. IEEE Commun. Lett. **20**(4), 700–703 (2016)

Environmental Monitoring and Evaluation of 10 KV Distribution Station Room Based on Internet of Things

Changsong Ni[✉], Yanjun Sun, Fengkai Niu, and Jingwen Cai

State Grid Dalian Power Supply Company, Dalian 116001, Liaoning, China

Abstract. With the increase in user load and the increase in the rate of cable distribution in the distribution network, the distribution station room has gradually become one of the main power distribution facilities in the urban distribution network. Due to the large number of equipment, wide distribution, changeable geographical environment, and susceptibility to user capacity expansion and urban construction, the 10 KV power distribution station room makes it difficult to implement safety and defense measures, environmental monitoring measures and inspections for the power distribution station room. Therefore, it is of great significance to establish a 10 KV distribution station room environmental monitoring system based on the IoT. The purpose of this article is to study the environmental monitoring and evaluation of the 10 KV distribution station room based on the IoT. This research uses IoT sensor technology to intelligently monitor the environment of the 10 KV distribution station room. Through real-time monitoring, analysis and calculation of related data, the system provides management and service functions such as real-time status monitoring, alarm linkage, and statistical reports. The monitoring system designed in this study is compatible and expandable, and can adapt to the ever-increasing smart demand of smart grids. The system function test data shows that the minimum relative error of the data collected by the sensor is 0 and the maximum is 1.68%. It can be seen that the humidity measurement of the sensor is more accurate and the error is small, and it can realize the monitoring of the environmental data of the 10 KV power distribution station room.

Keywords: Online monitoring · Distribution station environmental monitoring system · System communication · Internet of Things

1 Introduction

With the increase of electricity load, the environment of the distribution station has become more and more complicated, which has caused the instability of equipment and voltage to increase [1, 2]. In the past, the main method of monitoring the environment of the distribution station room was manual inspection, but this method not only cannot reflect the environment of the distribution station room in real time, but also consumes a lot of human resources, and the efficiency is extremely low [3, 4]. With the development of the IoT technology, its application in the 10 KV power distribution station room

© The Author(s), under exclusive license to Springer Nature Switzerland AG 2022
J. Macintyre et al. (Eds.): SPIoT 2021, LNDECT 98, pp. 409–416, 2022.
https://doi.org/10.1007/978-3-030-89511-2_52

environmental monitoring and evaluation system is very helpful to realize the real-time grasp of the 10 KV power distribution station room environmental status.

Regarding the research on the IoT and smart grids, many scholars at home and abroad have conducted in-depth discussions on them from multiple angles. For example, Mortaji H studied the role of the IoT in load shedding and smart direct load control in smart grid response management [5]; Al-Rubaye S studied the industrial IoT driven by the SDN platform for smart grids [6]; Li Y introduced and designed smart options for smart grids in detail, and proposed the concept of broadband IoT [7]. It can be seen that since the development of the IoT and smart grid technology, research on its application has always been the focus of the academic circle. Therefore, the combination of the IoT and the environmental monitoring of the distribution station room under the smart grid has important research significance for the development of the smart grid.

The purpose of this article is to study the environmental monitoring and evaluation of the 10 KV distribution station room based on the IoT. This article first analyzes the functional requirements of the system, including management functions, communication functions, real-time status monitoring and visual display functions, report statistics and analysis functions, and alarm linkage functions. Then the system is designed and related modules are introduced. Finally, collect and test the temperature and humidity data of the system from the sensor. Through the analysis of the experimental data, the feasibility of the application of the system in the environmental data monitoring of the 10 KV distribution station room is verified.

2   Environmental Monitoring and Evaluation of 10 KV Distribution Station Room Based on the IoT

2.1   Demand Analysis of the 10 KV Distribution Station Room Environmental Monitoring and Evaluation System Based on the IoT

(1) Management function
   According to actual needs, different workers have different levels of authority, and the scope of their operations on the system is also different. Managers can add or delete users, set security attributes, and give different personnel different operation permissions, effectively preventing non-management and developers from misoperation of the monitoring system.

(2) Communication function
   The communication function ensures that the environmental data collected in the distribution station rooms at different locations are transmitted to the main interface of the monitoring system through the network, and the main interface of the monitoring system completes the conversion of the transmitted data from analog to engineering quantities.

(3) Real-time status monitoring and visual display function
   This system can realize the functions of real-time power generation viewing, abnormal collector status, and abnormal inverter operation status monitoring. This is to allow managers to intuitively and quickly view the operation of the entire system on real-time monitoring of the distribution station room environment.

In order to facilitate the staff to understand the situation of the distribution station room in real time, the system dynamically displays the data. The way of dynamic data display is: dynamic display of data in the form of output value output, and the use of trend curves to intuitively reflect changes in monitoring data.

(4) Report statistics and analysis functions

The environmental status data in the power distribution station room can use the report function in the system to call the report function to realize the data analysis of any time period. Through data analysis, it is possible to determine the areas that are more prone to problems in the distribution station room, and provide scientific guidance to the staff.

The system stores and queries the collected environmental data in the distribution station room, and combines the external database with the system's SQL database function to design the database [8, 9].

In the report analysis part, through communication with front-line operation and maintenance personnel, the key points of daily work are clarified, and daily reports are designed. Commonly used reports include power generation statistics, power generation time statistics, new installed capacity statistics, power station distribution statistics, user distribution statistics, communication interruption statistics, etc. This method eliminates the manual collection and statistical information process of operation and maintenance personnel, and directly opens the system report page for viewing and analysis, which improves efficiency.

(5) Alarm linkage function

When an environmental indicator in the power distribution station room exceeds the threshold, the system needs to prompt the alarm in real time, and notify the staff who are not in the monitoring center by means of text messages and emails.

2.2 Design of the 10 KV Distribution Station Room Environment Monitoring and Evaluation System Based on the IoT

(1) Sensor acquisition module

1) Temperature acquisition circuit

This system adopts digital temperature sensor DS18B20, DS18B20 supports multi-point networking function, multiple DS18B20 can be connected in parallel on the only three-wire, to achieve multi-point temperature measurement, its conversion rate is relatively high [10, 11]. The detailed technical parameters of DS18B20 are as follows:

Power supply voltage: 3.3 VDC (carrying range 3–5.5 V);

Interface definition: power input terminal, power ground and signal input/output terminal;

Temperature measurement range: $-55\ °C$–$+125\ °C$.

2) Humidity acquisition circuit

This system selects the humidity sensitive resistor HR202, the detailed technical parameters of the humidity sensitive resistor HR202 are as follows:

Rated voltage: 1.5VAC (Max, sine wave);

Operating humidity: below 95%RH;

Response time: moisture absorption $\leq$ 20 S, dehumidification $\leq$ 40 S;

Stability: $\leq 1\%RH/year$;

Humidity detection accuracy: $\leq \pm 5\%RH$.

3) Infrared monitoring module

There are a large number of electrical equipment in the 10 KV distribution station room, and there are a large number of joints and electric shocks in the conductive loop. When a line fault occurs, it will cause a local temperature increase, resulting in aging of the insulation layer and overheating of the wire, and electrical fires are prone to occur. The infrared monitoring module monitors the infrared radiation energy emitted by these fire hazards. When the infrared radiation exceeds the standard, the system will send an alarm signal to remind the staff.

(2) Data monitoring platform

The modified platform includes front-end monitoring interface, back-end database and Windows service. The front-end monitoring interface is compiled by Visual Studio software, and it is the window for the relevant personnel to interact with the host computer. This interface can realize direct dialogue with the monitoring platform, and retrieve and query real-time data and historical data from the back-end database.

The back-end database uses SQL Server, which can store all collected environmental data [12]. Windows service program is C/S mode software written by Visual Studio. After the host computer is started, the service program starts to work independently. Its main function is to use the Socket connection to read the data packets sent by the network, analyze and verify the data content, classify and group, and finally save the data in the database.

(3) ZigBee wireless network data transmission

The nodes in the ZigBee wireless sensor network include: terminal nodes, router nodes and coordinator nodes. The terminal nodes are installed around the solar panels to collect and upload environmental information around the distribution station room, and at the same time receive commands issued by the platform. The routing node clusters the information collected by all the terminal nodes according to the relevant algorithm. Each routing node will connect the terminal nodes within its coverage area and communicate with the coordinator to complete the wireless sensor network of the mesh topology to ensure the collected information can be effectively transmitted to the coordinator node in real time.

(4) Remote monitoring and abnormal diagnosis management module

After the real-time environmental data of the distribution station room is collected and transmitted, the monitoring and diagnosis management module analyzes and diagnoses the collected data, and handles potential abnormal problems in a timely manner, so as to realize the function of equipment status maintenance in the distribution station room.

In the functional design of the detection terminal, advanced modern dynamic interactive technology is used to process dynamic information. On the basis of wavelet analysis and BP neural network analysis, a complete data analysis and diagnosis module is constructed to analyze the collected equipment operating state parameter information in the power environment of the 10 KV power distribution station room. Work managers can also log in and access the entire system server through the client, execute the functional applications of the system, monitor the power environment of the 10 KV power distribution station room in real time through specific instructions, and perform status diagnosis management.

3 Experimental Design

3.1 Temperature Acquisition Test

In order to ensure the accuracy of the experiment and to avoid the inconsistency of the experimental results due to accidental errors, the data volume of 5 days is selected as the sample to test the success rate of temperature monitoring.

The significance of this inspection is to check whether the device is capable of on-line monitoring the temperature of the distribution station. After turning on the device for 5 days, the overall statistical inspection of the success rate of device temperature collection was performed in the database.

The temperature sensor is placed in a changing temperature field, and each data detection endpoint collects the ambient temperature information in a fixed period, and displays it on the OLED synchronously. Data is recorded every 5 min, therefore, the amount of data should be: $24 \times (60 \div 5) = 288$, and the measurement data of OLED and professional tools are recorded at the same time.

3.2 Humidity Collection Test

In the laboratory, the controller circuit board is set up, the humidity sensor is placed in different experimental environments, the sensor performance is tested, and the professional measurement tools are used to synchronize the test, and then the test results are analyzed to optimize the system function. The professional measurement tool is an industrial-grade humidity detector with a humidity measurement range of 0.1–99.9%RH and a response time of less than 10 s.

3.3 Test Error Analysis

In the error analysis, the following two statistical indicators are used:

(1) Mean absolute error $\overline{\delta}MAE$

$$\overline{\delta}MAE = \frac{1}{N} \sum_{i=1}^{N} \left| P_i - P_i' \right| \tag{1}$$

(2) Average percentage error $\overline{\delta}MAPE$

$$\overline{\delta}MAPE = \frac{1}{N} \sum_{i=1}^{N} \frac{\left| P_i - P_i' \right|}{P_i} \tag{2}$$

In formulas (1) and (2), P_i and P_i' are the measured values of professional tools and prototypes respectively; N is the number of data measurements.

4 Experimental Data Analysis

4.1 Analysis of Temperature Acquisition Data

The sensor temperature check results are shown in Table 1: the five-day temperature collection success rate is 98.61%, 99.65%, 98.96%, 98.26%, and 99.31% respectively.

Table 1. Temperature collection data table

Experiment day	Amount of data due	Actual amount of data	Success rate (%)
First day	288	284	98.61
The next day	288	287	99.65
The third day	288	285	98.96
The fourth day	288	283	98.26
Fifth day	288	286	99.31
Total	1440	1431	99.38

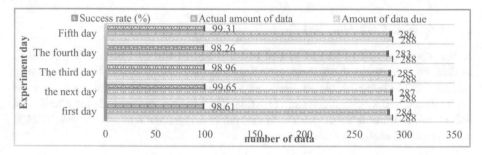

Fig. 1. Five-day temperature collection data

It can be seen from Fig. 1 that the success rate of temperature acquisition by the sensor is as high as 99.38%, which is sufficient for the task of online temperature monitoring and meets the requirements of the relevant technical standards of the state grid corporation. The collected temperature data was checked by time record and all met the requirement of 5 min of collection frequency. There were no problems such as overheating or slow speed during use, which proved that the device can meet the frequency of once every 5 min.

4.2 Analysis of Humidity Test Data

In order to reduce the error, the average value is calculated according to the arithmetic average method within a sampling period, which can reduce external interference and improve the measurement accuracy. Keep the ambient temperature at 25 °C and place the humidity sensor in a changing humidity environment. Each data detection endpoint collects the humidity information of its surrounding environment in a fixed period, and the data records are displayed on the OLED simultaneously. This experiment is set to sample 8 times, and the recorded data is shown in Table 2: The relative error of the data collected by the sensor humidity is at least 0 and the maximum is 1.68%.

Table 2. Sensor humidity test data record table

Serial number	Collect humidity (RH %)	Actual humidity (RH %)	Error (%)
1	42.6	42.4	0.47
2	44.1	43.8	0.68
3	43.7	44.2	1.13
4	42.4	41.7	1.68
5	38.1	37.6	1.33
6	38.6	38.2	1.05
7	37.5	37.5	0
8	39.3	39.3	0

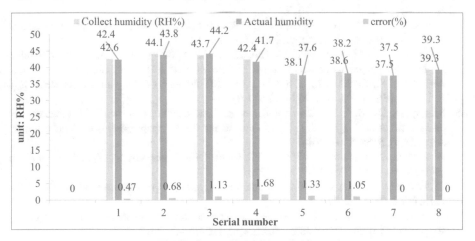

Fig. 2. Sensor humidity test data

It can be seen from Fig. 2 that the sensor humidity measurement is more accurate and the error is small, which proves that the system is more reliable and can monitor the environmental data of the 10 KV power distribution station room.

5   Conclusion

The management and maintenance of power distribution station equipment play a vital role in ensuring reliable power supply for users. Therefore, it is very meaningful to design a new environmental monitoring system for the distribution station room by adopting sensor technology and IoT technology to adapt to the development requirements of the smart grid, to ensure the safe operation of the distribution station. The research in this paper is a new application of the IoT technology in the smart grid, which realizes the intelligent perception, identification, monitoring, centralized management and traceability of the environmental parameters of the distribution station room. The system is compatible and expandable, and can adapt to the smart grid. The ever-increasing demand for intelligence is of great significance to the production, operation and maintenance of smart grids. This monitoring system is compatible and expandable, and can adapt to the ever-increasing smart demand of smart grids.

Acknowledgements. This paper is funded by the Science and Technology Project of State Grid Liaoning Electric Power Company Ltd (2018YF-38), Research on dynamic environment monitoring and evaluation technology of 10 KV distribution station building based on Internet of things.

References

1. Collier, S.E.: The emerging Enernet: convergence of the smart grid with the IoT. IEEE Indust. Appl. Magaz. **23**(2), 12–16 (2017)
2. Viswanath, S.K., Yuen, C., Tushar, W., et al.: System design of internet-of-things for residential smart grid. IEEE Wirel. Commun. **23**(5), 90–98 (2016)
3. Chen, H., Hui, H., Zhou, S., Fang, D., Hui, Y.: Real-time pricing strategy based on the stability of smart grid for green internet of things. Mobile Inf. Syst. **2017**, 1–11 (2017)
4. Al-Rubaye, S., Kadhum, E., et al.: Industrial IoT driven by SDN platform for smart grid resiliency. IEEE IoT J. **6**(1), 267–277 (2019)
5. Mortaji, H., Ow, S.H., Moghavvemi, M., et al.: Load shedding and smart-direct load control using internet of things in smart grid demand response management. IEEE Trans. Indust. Appl. **53**(6), 5155–5163 (2017)
6. Al-Rubaye, S., Kadhum, E., Ni, Q., et al.: Industrial IoT driven by SDN platform for smart grid resiliency. IEEE IoT J. **99**, 1 (2017)
7. Li, Y., Xiang, C., Yang, C., et al.: Smart choice for the smart grid: narrowband IoT (NB-IoT). IEEE IoT J. **99**, 1 (2017)
8. Guan, Z., Li, J., Wu, L., et al.: Achieving efficient and secure data acquisition for cloud-supported IoT in smart grid. IoT J. IEEE **4**(6), 1934–1944 (2017)
9. Javier, M., Francisco, R., Pedro, M., et al.: A new approach to detection of systematic errors in secondary substation monitoring equipment based on short term load forecasting. Sensors **16**(1), 85 (2016)
10. Hajikhani, M., Labeau, F., Agba, B.L.: Power allocation for a self-sustainable power substation monitoring system using wireless transfer of energy. IEEE Access **7**, 141456–141465 (2019)
11. Zhang, Y., Yang, Q., Xie, S., et al.: Mechanism and application of arrester block voltage division to lightning transient voltage monitoring in substation transformers. IEEE Trans. Electromagn. Compat. **61**(3), 689–696 (2019)
12. Ballal, M.S., Jaiswal, G.C., Tutkane, D.R., et al.: Online condition monitoring system for substation and service transformers. IET Electr. Power Appl. **11**(7), 1187–1195 (2017)

Natural Language Processing in Language and Literature Based on Target Detection Algorithm

Qianyu Ma[⊠]

Northwest Minzu University, Lanzhou 730030, China

Abstract. Chinese language and literature is broad and profound, which contains a long history and profound cultural connotation of the Chinese nation, is the treasure of the national culture of the Chinese nation. Understanding and learning Chinese language and literature can enrich our knowledge system, expand our knowledge, and improve our aesthetic level and cultural connotation. This is very helpful for the development of humanistic education and aesthetic education. Based on the target detection algorithm, this paper summarizes the aesthetic factors of Chinese language and literature, so as to further explore the learning path of knowledge points and aesthetic factors in Chinese language and literature.

Keywords: Chinese language and literature · Aesthetic factors · Learning path

1 Introduction

Chinese language and literature is an excellent cultural tradition in China. It is a treasure of traditional Chinese culture. For example: the beauty of language, image and emotion in Chinese literary works, the beauty of thought and philosophy in literary works, etc. by studying these aesthetic factors, we can establish a perfect aesthetic standard and lay a solid foundation for the improvement of aesthetic level. In the actual teaching process. Most colleges and universities are too casual, such a teaching method may not be conducive to the improvement of students' professional quality and aesthetic level. As a kind of aesthetic education, aesthetic education is relatively unique. Aesthetic education is very important for our students. The improvement of aesthetic level can not only improve people's perception and imagination, but also enrich people's spiritual world. If we can learn Chinese language and literature, be influenced by it, and cultivate our aesthetic consciousness, it will have a far-reaching impact on the improvement of our living standards and their comprehensive quality in the future [1]. Based on the target detection algorithm, this paper puts forward some suggestions on the penetration and integration of aesthetic factors in Chinese language and literature, hoping to have a certain reference value for some experts or scholars in the actual operation of aesthetic factors and learning path of Chinese language and literature.

J. Macintyre et al. (Eds.): SPIoT 2021, LNDECT 98, pp. 417–423, 2022.
https://doi.org/10.1007/978-3-030-89511-2_53

2 Target Detection Technology

The main applications of hyperspectral for low probability exposed targets of specific interest are anomaly detection and target recognition. Foreign field detection: to locate and recognize unknown but interested low probability targets from a blind image. Target recognition: according to the prior knowledge of the image, it can be the target endmember spectrum extracted from the image or the optical data obtained from the spectrum library for accurate target recognition. Both belong to target detection, which is to find a few pixels with certain spectral characteristics in the image and answer the question whether the target has "yes" or "no" for the pixels.

2.1 Target Detection Algorithm

According to the understanding of target information, target detection algorithms can be divided into three categories: supervised sub-pixel detection, unsupervised sub-pixel detection and automatic sub-pixel detection, as shown in Fig. 1.

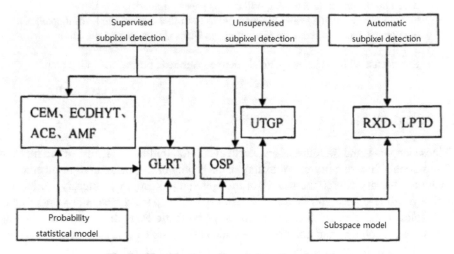

Fig. 1. Classification of target detection algorithms

Supervised subpixel detection needs to know the information of the target to be detected, including the spectral information or abundance information of the target to be detected. For example, the algorithms that only need to know the spectral information of the target: ACE (adaptive consistent spectral angle estimator), AMF (adaptive coherent matched filter), CEM (constrained minimum energy operator), CMF (constrained minimum energy operator) Ecdhyt (detector based on hyperbolic threshold with elliptical contour background); At the same time, we need to know the spectral information of target and background: OSP (orthogonal subspace projection detection) and GLRT (likelihood ratio detection without significant characteristics). Unsupervised sub-pixel detection does not need to know the target information, the algorithm directly extracts the required target information from the image, such information is called posterior target information, such as: utgp (unsupervised target generation processing). Automatic sub-pixel detection does not need any target information, and detects abnormal information directly from the image, such as RXD (anomaly detection algorithm) and LPD (low probability detection algorithm) [2].

According to the mathematical model, the target detection algorithm can be divided into: detection algorithm based on probability and statistics model and detection algorithm based on subspace model. There is a certain correlation between the algorithms based on the same model. Ace and AMF are algorithms based on likelihood ratio rule, ACE is the spectral angle of target vector and sample pixel in whitening space, AMF is the measurement of their distance in whitening space, CEM is based on Fisher criterion, which can highlight the target and suppress the interference of background information; The previous algorithms are based on the assumption that the scene satisfies the Gaussian distribution. ECD is based on the "ellipse contour model" algorithm, which uses non Gaussian model to describe the scene distribution more reasonably, highlighting the goal: RXD is the calculation of the inner product of the sample vector itself in the whitening space, and the greater the inner urgency, the more likely it is to be abnormal; OSP is an algorithm based on subspace model, which projects the sample vector and the target vector into the background subspace to make the inner product to maximize the signal-to-noise ratio. Both GLRT models have their own ideas.

Because of the similarity between algorithms, this paper mainly analyzes the detection effect of CEM, ACE, AMF, ecdhyt, RXD, OSP and GLRT on small geological targets.

2.2 Evaluation of Algorithm Performance

The performance of the algorithm is evaluated by calculating the signal to clutter ratio and drawing receiver operating characteristic (roc'o) curve.

Receiver operating characteristic (ROC) curve is composed of many corresponding points of PD and PFA. PD = number of pixels correctly judged as target/total number of target pixels, PFA = number of pixels wrongly judged as target/total number of background pixels. By changing the threshold value, the corresponding false alarm probability and detection probability are obtained. The higher the threshold, the lower the detection rate and false alarm rate. As shown in Fig. 2: the decision-maker needs to select a threshold value through the ROC curve, so that the detection result can obtain the maximum detection rate and the minimum false alarm rate. For the same scene, the more the ROC curve obtained by various detection algorithms warps to the upper left corner, the better the performance of this detection algorithm.

The result of detection can be regarded as a decision statistic with floating-point accuracy. SINR = variance of interested signal/(variance of non interested signal + noise) reflects the separation degree of interested target and background. The greater the separation degree is, the better the detection effect of the algorithm is. Because of the low probability of the target, the intensity can be ignored relative to the whole image, so the signal to clutter ratio is approximately equal to the variance of the signal of interest/the variance of the whole image.

$$SINR = E\left\{(y_d - u)(y_d - u)^l\right\}/D\{y\}$$

$D\{\}$ denotes variance, y_d is the decision statistic of the target, u is the mean of the decision statistic of the target. Signal to clutter ratio (SCR) is an important index to evaluate the performance of detection algorithm. In the case of the same image quality, the detection algorithm with higher SCR decision statistics often gets better detection effect.

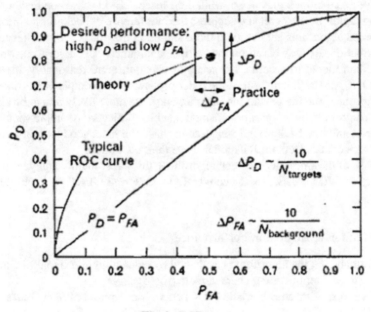

Fig. 2. ROC curve

3　The Current Situation of Language and Literature Education in China

In China's long history, culture has a long history. As a traditional discipline with a long history, Chinese language and literature is an indispensable basic discipline in Colleges and universities. The major of Chinese language and literature has the advantages of strong historical foundation, wide range of employment and relatively large social demand. So. Chinese language and literature has always been a relatively traditional major. However, with the advent of the information age, traditional disciplines have suffered a certain degree of impact.

　　The traditional teaching method of language and literature may be difficult to create an information age. The obvious problems are as follows. First, it is the problem of curriculum setting in universities. Colleges and universities do not pay enough attention to the subject of Chinese language and literature, and set up relatively few related courses. Therefore, due to the problem of curriculum schedule, it is difficult for many students to form a relatively complete knowledge system and reduce their interest in Chinese language and literature. Second, the mode of Chinese language and literature education is relatively single, and the method of education is relatively backward. To a certain extent, it reduces the students' interest in learning Chinese language and literature. We must try our best to change this cramming teaching method. Take students as the main body to meet their personalized needs. Third, the backward teaching concept. Nowadays, the evaluation of the educational effect of Chinese language and literature mainly focuses on the comparison of students' evaluation. Too much emphasis on achievement will naturally make students lack of practical experience. Students and teachers are only after results. And ignore the learning process of Chinese language and literature. Fourth. It's the students' subjective factors. In this materialistic society, students do not realize the importance of Chinese language. These emotional and cognitive defects are difficult to generate great interest in Chinese language and literature, which is the most direct reason for students' low interest. To a certain extent, the emotional problems of students make it more difficult to achieve the goal of Chinese language and Literature Teaching [3].

4 This Paper Analyzes the Aesthetic Factors Embodied in Different Literary Forms of Chinese Language and Literature

Chinese language and literature is extensive and profound. Its content covers all aspects of ancient Chinese social life, contains rich humanistic spirit, and embodies the aesthetic standards and pursuit of ancient Chinese working people. This aesthetic pursuit is reflected in different literary forms of Chinese language and literature, such as poetry, novels, essays, plays, songs, etc., The following is a detailed analysis of the three literary forms of Chinese language and literature which can best reflect the aesthetic factors.

4.1 Aesthetic Factors Embodied in Chinese Characters

Chinese characters are the unique invention of the Chinese nation. Since the birth of ancient times, Chinese characters have been adopted and applied by people. In modern society, Chinese characters are still the most effective way for us to communicate and exchange in our daily life. Chinese characters not only have great practical effect, but also are an art and a culture. In the long period of the formation and development of Chinese characters, the working people in ancient China constantly improved them according to their own aesthetic tendency, making them more and more beautiful. Chinese characters not only became a tool for people to communicate and spread culture, but also became an important carrier of aesthetic education and a concrete embodiment of people's aesthetic pursuit. Calligraphy extended from the writing of Chinese characters has also become an important part of Chinese traditional art. The birth of calligraphy also confirms the aesthetic factors contained in Chinese characters and their words. Each square character is a unique style of language and literature; Every word contains rich aesthetic factors. Chinese characters and the words composed of Chinese characters often have their rich connotations in shape, sound and meaning. The shape, sound and meaning of Chinese characters together build up the world of Chinese characters, which embodies the aesthetic factors of Chinese characters everywhere.

4.2 Aesthetic Factors Embodied in Poetry

Poetry is one of the most important literary forms of Chinese language and literature, and it is also one of the forms of Chinese language and literature with rapid development, large scale and far-reaching influence. People's love for poetry has continued from ancient times to the present. One of the key factors is that poetry contains rich aesthetic factors. The threshold of poetry creation is high: because of its own formal characteristics, poetry has higher requirements for the level of poetry creators, not only requiring the creators to have skilled skills, rich life experience, unconstrained imagination, moving thoughts and emotions, but also requiring the creators to strictly abide by the rules of poetry in the process of pretending, It not only has rich connotation and emotion, but also has concentrated language form, cadence and catchy rhythm, which is convenient for people's recitation and secondary artistic creation. Therefore, poetry as a literary form must have a high degree of musical beauty and rhythmic beauty; Ancient people's poems were often created to express their inner feelings, so they also had emotional beauty; Poetry is the essence of highly condensed language, and poetry must contain language beauty. Poetry often contains the poet's life perception, so poetry has the beauty of thought and Philosophy [4].

5 The Specific Implementation Strategy of Integrating Aesthetic Factors into the Education of Chinese Language and Literature

In the education of Chinese language and literature, the education of aesthetic factors has always been inseparable. According to the characteristics of Chinese language and literature and aesthetic education, combined with the current situation of the development of Chinese language and literature and the existing problems, this paper puts forward the following effective strategies:

First, deeply explore the role of aesthetic factors in Chinese language and literature. In the teaching of Chinese language and literature, the aesthetic factor is the most important, which is the basis for the successful teaching of Chinese language and literature. Chinese language and literature contains a lot of traditional culture essence, including a large number of classic works. Chinese language and culture itself is also a kind of "beauty". We should understand the feelings and thoughts in Chinese language and literature, and realize the beauty of Chinese language and literature. So as to improve the students' own aesthetic level.

Second, starting from the students themselves, improve students' learning enthusiasm. In the teaching of Chinese language and literature, it is very important to improve students' learning autonomy. Interest is the best teacher. Aesthetic appreciation of Chinese language and literature. Everyone's appreciation of literature is different. A thousand people have a thousand Hamlets. Therefore, we need to pay attention to the uniqueness of students in the teaching of Chinese literature.

6 Conclusion

The extensive and profound Chinese language and literature not only contains rich historical and cultural connotations, but also embodies many aesthetic factors. The study of Chinese language and literature can not only enrich our knowledge, expand our scope of knowledge and improve our Chinese performance, but also gradually form aesthetic consciousness and improve our aesthetic level in the process of learning Chinese language and literature. Therefore, the study of Chinese language and literature is not meaningless, but worthy of our serious study. We should timely adjust the learning concept of Chinese language and literature, formulate learning strategies and learning methods in line with our own actual situation, so as to better learn Chinese language and literature, so as to improve our Chinese performance, form aesthetic consciousness and improve our aesthetic level.

References

1. Wang, M.: On aesthetic factors and learning strategies in Chinese language and literature. Motherland **22**, 193–193 (2017)
2. Gao, L.: Information Enhancement and Feature Extraction in Hyperspectral Remote Sensing Target Detection. Institute of Remote Sensing Applications, Chinese Academy of Sciences, Beijing (2007)
3. Yang, X.: Analysis of aesthetic mood of language in Chinese language and literature. Young Writer **36**, 70–71 (2017). https://doi.org/10.3969/j.issn.1002-2139,2017.36.053
4. Xiong, J.: On the effective development of aesthetic education in Chinese language and literature teaching. Knowl. Econ. **9**, 166–166 (2016)

Application of Cluster Analysis Algorithm in the Construction of Education Platform

Bin Chen[✉]

Fuzhou Melbourne Polytechnic, Fuzhou 350001, China

Abstract. With the rapid development of the Internet, people use the Internet to carry out social and learning needs to continue to improve. Like other industries, the education field is more and more aware of the importance of education platform, and constantly applies its new functions to education and teaching. Clustering algorithm provides an effective evaluation method for the evaluation of students' personality and learning status in distance education. It can provide targeted guidance and learning strategies according to the clustering analysis results. This paper studies the construction of English teaching education platform based on clustering analysis algorithm, aiming at exploring new methods of English teaching by using new network means, and trying to solve some problems existing in junior middle school English teaching, and puts forward some new ideas for improving English teaching.

Keywords: Clustering analysis algorithm · English teaching · Education platform

1 Introduction

With the rapid development of information technology and the popularity of mobile intelligent terminals, the era of micro technology has come. MOOCS and micro classes are rising, changing the traditional teaching methods. In the micro era, knowledge is generated through the interaction and collaborative construction between people, people and situations, and the learning community based on the micro environment is also formed in this interactive process. The ubiquitous network and mobile intelligent terminals promote the development and application of Micro Technology in teaching, so that students can "learn from time to time and everywhere". With the great changes in the media and ways of communication, College English teaching should follow the trend of the times and carry out the research and practice of educational informatization.

Teaching and education platform is one of the important teaching methods based on modern network technology. Teachers and students maintain a state of separation in the form of educational resources or as learning promoters. With the compression of College English class hours, students are required to have strong self-learning ability, do a good job in preview and review, how to position students' learning status through the network education system, and provide more targeted guidance to achieve good learning results, which is an urgent problem to be solved in the teaching process. In this paper,

© The Author(s), under exclusive license to Springer Nature Switzerland AG 2022
J. Macintyre et al. (Eds.): SPIoT 2021, LNDECT 98, pp. 424–430, 2022.
https://doi.org/10.1007/978-3-030-89511-2_54

data mining technology is applied to network education, using the improved k-meana clustering algorithm combined with educational statistics to analyze and evaluate the students' academic performance information, so that the teaching staff can make specific adjustments according to the students' learning characteristics in time, and improve the efficiency of teaching and learning [1].

2   Cluster Analysis

Data mining is the whole process of discovering knowledge from large-scale data. Data mining focuses on designing efficient algorithms to find the knowledge needed from massive and miscellaneous data. It is a process of extracting hidden, unknown and potentially useful information or knowledge from large database. The flow chart is shown in Fig. 1.

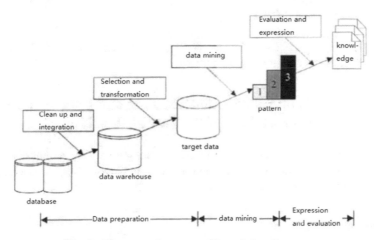

Fig. 1. The general process of knowledge discovery

As an important research branch of data mining, cluster analysis is a process of dividing data object set into several clusters (clusters) composed of similar data objects through certain methods and rules. The data objects in the same cluster are similar enough, and the data objects in different clusters are not similar enough. Cluster analysis is one of the research contents of statistics, which has a deep mathematical theoretical foundation. However, most of the statistical cluster analysis methods are based on the probability assumption of the distribution characteristics of data objects, and less consider the characteristics and differences of data objects in practical application. With the rapid development of data mining technology, clustering analysis has also received unprecedented attention.

The definition of cluster analysis is as follows:

Clustering is the most basic activity of human cognitive world. It is a common data mining function and an unsupervised learning process. The fundamental difference between clustering and classification is that classification needs to know the characteristics of the objects and define the categories in advance. The number of categories remains unchanged, while clustering has no predetermined categories and the number of categories is uncertain. In short, it is the process of dividing the data object set into several clusters or classes composed of similar objects. The cluster generated by clustering is a set of data objects. The data objects in the same cluster have higher similarity and smaller distance, while the data objects in different clusters have smaller similarity and larger difference [2].

The formal description of clustering model is as follows: given the data set $X = \{x_1, x_2, ..., x_n\}$. For $\forall_i \in \{1, 2, ..., m\}$, $x_i = \{x_{i1}, x_{i2} ... x_{im}\}$ is an object of X. for $\forall_j \in \{1, 2, ..., m\}$, x_{ij} is the attribute of x_i. According to the inherent characteristics of data object, X can be divided into k sets (clustering) $C_1, C_1, ..., C_K$, which satisfy the following three conditions:

(1) $C_i \neq \varphi, i = 1, 2, ..., k$
(2) $C_1 \cup C_2 \cup ... \cup C_k = X$
(3) $C_i \cap C_j = \varphi, i \neq j, i,j = 1, 2, ..., k$

Let $C = \{C_1, C_2, ..., C_k\}$, $K = \{X, C\}$ be a clustering space, and C_i is called the class i (cluster) of the clustering space.

3 The Feasibility of Applying the Teaching and Education Platform Based on Clustering Analysis to English Teaching

Based on Constructivism and social interaction learning theory, the author discusses the feasibility of the application of educational platform in College English teaching based on the characteristics of English teaching.

3.1 Constructivism Theory

Constructivism theory holds that effective teaching is not to teach students "knowledge in one way, but to provide students with a variety of learning options, broaden their learning channels, and help them establish a knowledge system that conforms to their own cognitive characteristics by creating certain teaching situations, and then promote the construction of significance of the knowledge they have learned.". As a kind of micro-technical medium, education platform includes text, pictures, images, sounds and other elements, and has the characteristics of one-to-many, one-to-one, many to many information push, real-time interaction, etc. it has a positive role in the construction of learning community for teachers and learners in the teaching process, and can provide learners with a channel for deep learning with the goal of common knowledge construction, Thus, it helps learners to establish knowledge system in learning community and construct meaning of the knowledge they learn [3].

3.2 Theory of Social Interaction Learning

The theory of social interaction learning emphasizes the real-time interaction between the teachers and learners in the teaching process, and the effective improvement of the learners' interpersonal interaction ability. It has a certain effect on creating a teaching environment of easy and pleasant, real-time interactive follow-up learning and establishing harmonious relationship between teachers and students. At the same time, it also helps teachers to construct the micro environment of teaching based on the education platform, and promotes the students to deepen their understanding of the knowledge learned and to learn deeply in the interaction and discussion between teachers and students. It can be seen that the teaching and education platform is of great significance to the theory of social interaction in promoting the interpersonal interaction, harmonious development between teachers and students, and promoting students' deep learning. There's some overlap on it.

3.3 The Characteristics of College English Teaching by Using Platform

The English teaching based on the teaching platform breaks the limitation of traditional teaching methods in time and space, and provides convenience for the smooth implementation of teaching activities. At the same time, it also makes the students make full use of their free time to learn independently, which helps them to choose the appropriate learning content according to their learning progress, learning habits and learning ability, and join in the creation, thinking and evaluation of knowledge, and obtain support from other learners, and establish their self identity in the process of forming consensus knowledge.

This also reflects that the network community is the key to improve learning efficiency. Therefore, the teaching platform based on clustering analysis algorithm enhances the interaction of learning community and is feasible for the teaching of College English courses [4].

4 The Application of Cluster Analysis in Teaching

As an important tool of the information age to assist college English teaching, the platform plays a certain role in establishing a new digital learning mode based on the education platform, which is helpful to the formation of College Students' English learning group, the real-time sharing of English learning resources, and the formation and development of micro environment in College English classroom teaching. The author discusses the application of the education platform in College English teaching.

4.1 Building Educational Learning Platform

Because of the autonomy of the students' learning groups, the diversification of learning methods and the fragmentation of learning time, College English teachers should not inherit the traditional teaching mode to instill "English knowledge into students" one-way, but should actively use micro technology to help students build a micro environment for English learning and build an English Education learning platform. The platform is widely used and even relied on by college students, which provides a favorable opportunity for College English teachers to construct a micro environment of curriculum teaching. English teachers can build learning platform, gather English learners and form learning groups, and then make use of wechat platform to learn deeply in exploration and cooperation.

In the process of building the learning platform of clustering analysis algorithm, College English teachers can be the administrators of wechat learning group, who are responsible for uploading curriculum teaching plans, teaching resources, arranging learning tasks, checking students' learning situation and conducting teaching evaluation. At the same time, students can be guided to establish English film group, English song group, spoken English group, English writing group, etc. according to their own learning interests, and can also join the corresponding group as members to communicate with students in real time.

4.2 Design of Teaching Plan of College English Course

Teaching plan plays an overall role in guiding teaching activities. Based on the cluster analysis teaching platform, the design of College English curriculum curriculum includes three stages: before class, in class and after class. In the specific content, the teaching plan should include not only the learning objectives, the key and difficult points, the subject of the course discussion, but also the language points to master, but also the specific thinking quality, cultural character, learning ability and so on. Meanwhile, students should be asked to make use of wechat platform to preview and review themselves in time according to the specific requirements of the curriculum teaching plan, master the key and difficult knowledge, and strive to improve their core English literacy.

4.3 Track Students' Progress

In order to understand the application effect of the platform in College English Teaching in time, College English teachers should follow the progress of students' learning. It can be divided into three tracking stages: preparation before class, classroom learning and consolidation after class. Prepare before class, ask teachers to control the students' Preview situation immediately before starting a unit of classroom teaching. The specific steps are as follows: send the learning tasks to the students, and ask them to preview the specific tasks; Encourage students to collect information related to the course in the form of group cooperation; Classroom learning, let students around their understanding of teaching content, through group discussion activities, in the activities to share learning experience, in the discussion of thinking gradually build a knowledge framework and knowledge system; After class consolidation, a large number of contents related to the course will be uploaded to the group, providing students with a variety of learning options, realizing the digital assisted teaching in the Internet era, and expanding the classroom teaching space.

The clustering based data mining technology can find out the internal factors, such as students' learning attitude, learning habits, weak chapters, and after-school practice, according to the students' usual and final exam results. The summary analysis is applied to guide students' learning and daily teaching, which not only saves a lot of practice time but also gets good learning effect.

Figure 2 describes the basic structure of an education system based on clustering analysis algorithm. Among them, the intelligent tutoring function module based on clustering is added in the performance analysis module, which is used to analyze the test results of students. The results include suggestions for selecting courses, recommended bibliography, parameters of group papers, characteristics of categories and performance. The system gives the learning suggestions and guidance according to the above results, and modifies the parameters of the paper to give the test questions which conform to the students' learning level and characteristics; Students can also study according to the analysis results, thus improving the intelligence of the whole system.

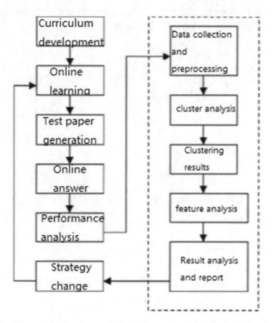

Fig. 2. Cluster analysis module in the structure of education system

5 Conclusion

The English teaching and education platform has a large number of English learning resources, which brings convenience to the improvement of the effectiveness of College English classroom teaching. With the popularity of mobile intelligent terminal in Colleges and universities, the research and practice of educational platform in College English teaching are becoming more and more mature. However, there are many problems to be solved in promoting effective learning and cultivating their comprehensive English application ability in the information environment. In English teaching, how to use the platform to enhance students' learning enthusiasm and participation, and how to do a good job in teaching management is particularly prominent, which requires the majority of English teachers to continue to explore and study.

Acknowledgements. Discussion on the Teaching Mode of Interdisciplinary Cooperation between Chinese and Foreign Teachers in International Schooling. No: JAS180380.

References

1. Dong, P.: Application of cluster analysis in teaching evaluation. Journal of Hunan Institute of Technology: Natural Science Edition 35–38 (2001)
2. Tang, D.: Cluster analysis and its application. University of Electronic Science and Technology, Chengdu (2010)
3. Wang, P.: Construction and application of wechat mobile learning platform. Modern Educational Technology (5), 88–95 (2014)
4. Fan, W., Ma, Y., Qiu, B.: Research on flipped classroom supported by wechat in mobile learning environment. Open Education Research (3), 90–97 (2015)

Based on Reconfigurable Modular Intelligent Snake Robot and Three-Dimensional Multi Motion Simulation

Zhenmin Yang, Xiaodong Wang[✉], Hongbo Li, and Yuying Feng

Kunming Metallurgy College, Yunnan 650300, China

Abstract. With the rapid development of human life and social economy, nowadays, the application of robot is more and more widely, and it has penetrated into almost all fields, so robot has become an irreplaceable necessary equipment in all fields of society. Modular snake like robot adopts the modular reconfigurable design concept, which can be combined into different configurations according to the special environment to complete the corresponding tasks. It is very suitable for the operation in unknown environment. As an important branch of robot field, modular snake robot has many advantages, such as diverse assembly, good robustness, low price and so on. This paper focuses on the research and design of the mechanical structure of the snake like robot, the control principle and control algorithm analysis of the snake like robot, the motion performance of the snake like robot, and the three-dimensional multi motion simulation.

Keywords: Modularization · Snake like robot · 3D multi motion simulation

1 Introduction

With the continuous expansion of robot applications and the complexity of operation tasks, fixed-point operation in structured environment can no longer meet the needs of human beings. Human beings hope that robots can complete more complex and dangerous tasks. The unknown working environment and constantly changing tasks require the robot to have a variety of motion modes and good terrain adaptability. The traditional wheels, tracks and legs can not meet the unknown and changeable working environment. It is this demand that makes the research of bionics get great development. Bionics is a new and challenging research topic in the field of robotics. Bionic robots are very suitable for high-risk and complex tasks. Snake like bionic robots with multi joints and multi redundant degrees of autonomy have become an indispensable part of the bionic robot family, It has aroused wide attention of scholars at home and abroad. The snake like robot has realized the limb less movement like a snake. The snake like robot has the advantages of small size, low center of gravity, flexible movement mode, good adaptability to the environment, and good stability. It is very suitable for working in complex and high-risk environment and narrow space, such as emergency rescue and disaster relief, interstellar exploration, nuclear radiation, dust, toxic and earthquake collapse environment On the battlefield, reconnaissance and attack, elimination of explosives, pipeline

© The Author(s), under exclusive license to Springer Nature Switzerland AG 2022
J. Macintyre et al. (Eds.): SPIoT 2021, LNDECT 98, pp. 431–437, 2022.
https://doi.org/10.1007/978-3-030-89511-2_55

maintenance, disease diagnosis and treatment, archaeological exploration and other non structural environment of autonomous operation. Due to the characteristics of modular structure, high reliability and maintainability, and the joints at the module connection are not exposed, which can be well encapsulated with protective layer, so it is very suitable for deep-sea operation. Because of its own advantages of multiple redundant degrees of freedom, it can not only move in two-dimensional space, but also realize the flexible movement in three-dimensional space. It can replace human beings to work in dangerous environment. For example, it can replace human beings to complete dangerous tasks, clamp objects, drag objects and so on. Because of the typical characteristics of snake like robot and its wide application prospect, the research and design of snake like robot has sprung up in recent years [1].

2   Characteristics and Structure Design of Modular Snake Machine

2.1   Characteristics of Modular Snake Robot

Modular snake like robot has a wide range of theoretical and practical application value, which has been proved in practical application in recent years. As an important branch of robot research, modular snake like robot has the advantages of diverse assembly, good robustness and low price. Modular snake like robot uses its own modular structure, which can be combined into different configurations according to different environment conditions to complete the corresponding tasks, so it is very suitable for the operation in unknown environment; It also has the typical feature of self-healing. When one of the modules fails, it can throw away the damaged module and replace it with other functional modules, so that it can continue to complete the scheduled task; Due to the unity of module structure, mass production of the same module will greatly save the cost of research and design of snake like robot. Modular design makes snake like robot have these typical advantages, which is also the significance of domestic and foreign scholars to study it [2].

2.2   Mechanical Structure Design

The modular snake like robot can be easily formed by using gz-i unit module. Because the left and right bottom plates of the module can rotate around the rotation axis of the steering gear, and other modules can be connected to the left and right bottom plates, the adjacent modules are linearly connected, which can form a modular snake like robot. The rotation direction of each unit module is centered on the pitch axis, The modular robot, which is composed of several unit modules rotating around the pitch axis, is called the pitch modular snake robot. The snake like robot with this structure can make continuous motion changes in various forms in the vertical direction, and can realize two-dimensional motion. Therefore, the snake like robot, like a caterpillar, can make forward and backward sine E-string crawling motion on the ground. This kind of snake like robot has a certain speed of crawling back and forth. It can not only crawl on flat ground, but also be used in narrow straight line or narrow pipe environment. This function is incomparable with other traditional robots. As shown in Fig. 1, this structure is based on the pitch snake like robot. By further changing the connection mode of adjacent modules, the snake like robot can have more kinds of motion forms and realize three-dimensional motion. For the snake like robot of this kind of mechanism, the adjacent modules are linearly connected with each other in the way of rotation angle of 90°. A unit module rotates around the pitch axis, and its adjacent unit modules rotate along the yaw axis. Therefore, the snake like robot of this kind of structure is called pitch yaw modular snake like robot [3].

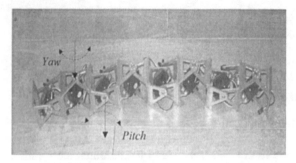

Fig. 1. Pitch yaw modular snake like robot

3 Control System and Control Algorithm

3.1 Control System

The snake like robot uses mscc20 actuator control chip as the main control system. The rotation of each actuator is driven by PWM signal. The number of actuators controlled by mscc20 actuator control chip can reach up to 20. The control chip mscc20 of the steering gear receives the motion command sent by the PC through the RS-232 serial port, and generates the corresponding PWM signal to drive the steering gear to rotate. The motion command received by the steering control chip is generated by the corresponding control algorithm of the main control central computer. The external power supply is used to drive the servo control chip and the whole snake robot. The structure of the whole control system is shown in Fig. 2.

In the actual control operation, the steering gear control system operating box is mainly used. It is mainly composed of mscc20 actuator control chip, switch power supply for chip power supply, and switch power supply for driving power supply of steering gear, including input power line, RS-232 serial port line and data line for output PWM signal.

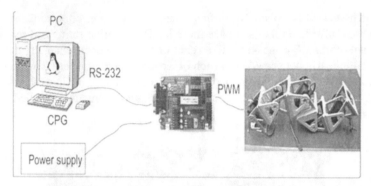

Fig. 2. Control system structure diagram

3.2 Control Algorithm

The modular snake robot can move normally, and a good control algorithm is needed to control. First, this good control algorithm can also control the movement of single module unit. The control of the rotation of a single module means that it can control its free movement within the maximum rotation angle allowed by the physical mechanical structure. The maximum rotation range of GZI unit module is 180°. Because the mechanical structure of the module is symmetrical, it can be considered that its self-output rotation range is $-90 \sim 90°$. The control algorithm also has a timing effect on the rotation control of the module, because in the process of controlling the movement of modular snake robot, the adjacent modules rotate in a certain order. That is to say, in the process of continuous motion of modular snake robot, every module that constitutes the snake structure is doing regular periodic rotation.

Considering the above consideration of the unit module motion control, the control algorithm of center pattern generator (CPG) is proposed. This algorithm is a model simulating the movement mode of snakes in the biological world. The medium Maple mode generator is the neural network which can generate rhythm signals in the spine. The control algorithm in this paper uses the mathematical sine function generator to simulate the central mode generator. Each sine function generator in mathematics can generate rhythmic motion signals to control the motion of a single module. Then a group of sine function generators can control the movement of a modular snake robot. Each sine function generator can be expressed in formula (1):

$$\varphi(t, i) = A_t \sin\left(\frac{2T}{T} + \phi_i\right) + O_i, i = \{1, 2, ...n\} \tag{1}$$

4   Analysis of 3D Multi Motion Simulation

4.1   Gait Analysis

In this paper, the pitch snake like robot composed of multiple modules is simplified. It is assumed that the modular snake like robot is linear connected with N modules (n >

2), the length of each module is set to 2 h, and the number of modules needed to form a motion waveform is assumed to be n = 4. Next, take the motion waveform composed of four modules as an example to analyze the motion process of the snake like robot in a cycle, and so on, the transmission principle of the motion waveform composed of more than four modules is the same [4].

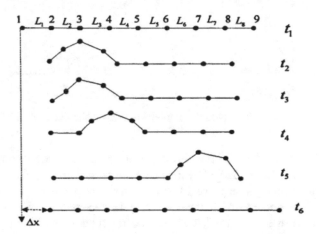

Fig. 3. Gait of pitch snake like robot in one cycle T

As shown in Fig. 3, the whole process of waveform transmission in the whole movement process of snake robot is described. The initial state of the snake like robot is assumed to be a horizontal line, that is, the rotation angle of the embedded actuator in each module is at zero degree, and the maximum rotation range of the actuator between adjacent modules is −90 ~ +90°. In Fig. 3, the waveform transmission process of the pitch snake like robot in one cycle is described. The motion waveform is transmitted from the tail to the head of the snake in turn along the snake body.

4.2 Mathematical Model Analysis

A simplified mathematical model of the multi module pitch snake like robot in the process of motion waveform transmission is constructed; The motion curve part is composed of N modules (n > 2). The length of each module is set to 2 h, and the parameter a is the angle between each module and x-axis, β It is the angle between two adjacent modules. Assuming that the steering axis of the module joint rotates clockwise to be positive and counterclockwise to be negative, the snake like robot's motion depends on the sine wave formed to advance forward in turn. The simplified mathematical model of multi module pitch snake like robot in the process of motion waveform transmission is shown in Fig. 4.

4.3 Experimental Results

In order to verify the feasibility of the above-mentioned pitch snake like robot gait, we use a snake like robot model composed of eight modules to carry out a simple experimental

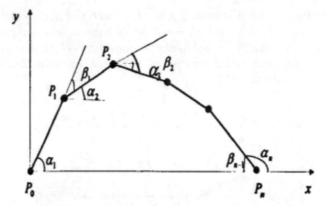

Fig. 4. Wave propulsion model of pitch snake like robot

verification. The snake like robot is controlled by a manual two-way switch. Because the movement displacement is affected by the phase difference, after repeated experiments, when the steering gear rotation phase difference is zero and positive or negative 180°, the snake like robot does not move and the movement displacement is zero; When the phase difference is between 80° and 160°, the snake like robot has high efficiency; If the phase difference is less than 50°, the snake like robot's motion is unstable and its efficiency is very low, so it is the least desirable. Therefore, through experimental analysis, when the phase difference is 120°, the snake like robot has stable motion and fast motion speed. The experimental results are shown in Fig. 5.

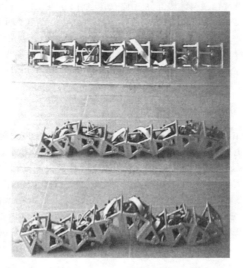

Fig. 5. Motion process of pitch snake like robot

5   Conclusion

Through the theoretical and experimental research of the modular snake like robot technology, it has guiding significance for its future application in the theoretical research of experimental teaching, social life services, manufacturing, battlefield and other high-risk environment, medical diagnosis and treatment, deep-sea exploration, planetary exploration and other fields. In particular, it can provide some help for the research of modular snake like robot technology.

References

1. Wang, L., Zhou, H.: Research status and development direction of bionic robot. J. Shanghai Normal Univ. **36**(6), 58–62 (2007)
2. Chen, L., Wang, C.Y., Li, B.: Research status and progress of snake like robot. Robot **24**(6), 684–687 (2002)
3. Ma, J., Zhang, H., Liu, J.: Design and research of a new modular reconfigurable robot. Autom. Manufact. Ind. **26**(9), 35–43 (2004)
4. Li, B.: Research on snake like robot and its application in fire rescue. Robot Technol. Appl. **21**(3), 22–26 (2003)

Cloud Storage for Management System

Yan Zhao[✉] and Wenjie Li

Shandong Medical College, Jinan, Shandong, China

Abstract. The development of education informatization has been highly valued by our government. It conforms to the development of education in the intelligent environment and is of great significance to promote the development of education informatization in our country. Based on the Internet platform and information technology and equipment, education management information system can fully reflect the development level of education informatization, and the establishment of a perfect education management information system is of great significance to promote the development of education informatization and improve the national quality. This paper studies the development process and current situation of education management information system in the process of education informatization at home and abroad, and makes a comparative analysis. Based on the relevant research of education informatization and education management information system at home and abroad, this paper analyzes the problems in the development of education management information system in China from two aspects of information system and project management, Put forward to establish a public network learning platform and exchange forum.

Keywords: Education informatization · Education management information system · Mobile education

1 Introduction

In today's society, in the information age with developed network, the application and popularization of management information system in all walks of life has brought us many unexpected convenience and quickness, so that people who use it correctly can obtain the information and services they need through the information system, so that managers can improve the efficiency in management, and even provide the basis for the decision-making of the country and managers, It has created tangible and intangible value for many enterprise groups and government units. Therefore, the establishment of a perfect education management information system is not only convenient for the students, teachers and administrators of the school to learn, educate and manage, and serve the public, but also convenient for the national education department to unify and efficiently manage the personnel of various schools and departments, and provide a reliable basis for the education reform by using the analysis of education information data [1].

In recent years, with the continuous development and popularization of the Internet, the development of education informatization at home and abroad continues to deepen

© The Author(s), under exclusive license to Springer Nature Switzerland AG 2022
J. Macintyre et al. (Eds.): SPIoT 2021, LNDECT 98, pp. 438–442, 2022.
https://doi.org/10.1007/978-3-030-89511-2_56

and expand, the degree of education informatization is also higher and higher, various countries and organizations pay more and more attention to education informatization, and gradually establish the relevant education network and system platform, so that people's learning and education management is more and more efficient. With the advent of 5g era, people pay more and more attention to mobile education. There are more and more education management and learning software on the mobile platform, which makes the mobile terminal an important part for people to obtain the required knowledge and information in school, life and work. At the same time, online learning platform and exchange forum have also become a paradise for people to obtain information and exchange. With the full coverage of the Internet in China, computers, tablet computers and smart phones are widely used in people's daily life, study and work, which makes people's online education and autonomous learning unprecedented hot and makes people's learning more convenient and efficient. In the information age, network education and learning make the development of education in our country more balanced in the region, age, gender, occupation and other classes and categories, so that people can learn what they need anytime and anywhere, promote people's lifelong learning, facilitate people's career planning, and greatly improve the population quality and education level of our country.

2 Research on the Development of Education Management Information System Under Education Informatization 2.0

The content of education informatization 2.0 action plan includes its general requirements, objectives and tasks, implementation actions and safeguard measures. In the process of the implementation and development of educational informatization 2.0, the overall requirements and objectives need to be implemented and deepened through the implementation of eight major actions and five major safeguard measures. However, based on the current development status of education informatization and education management information system, the eight action rules of education informatization 2.0 implementation action are difficult to implement in the implementation process, and many problems will be encountered in the process of education management information system construction and project management [2].

2.1 Analysis on Information System

In the process of promoting the implementation of the 2 action plan of education informatization, it is particularly important to actively promote "Internet plus education", and to build an educational system of networking, digitalization, intellectualization, individuation and lifelong education, and to build a learning society that everyone can learn, learn everywhere and learn from time to time. It is of great significance to realize more open, more suitable, more humanistic, more equal and more sustainable education. The construction of the Internet plus education platform can lay a solid foundation for actively promoting the "Internet plus education", and will promote the popularization and education of digital resources, standardize the digital campus, cover the learning space of the network, promote the development of intelligent education and intelligent education.

It is worth our deep thought and Research on how to build a platform for "Internet plus education" and how to achieve its goal better.

The calculation method of distribution density is to first take v consecutive line segments from i to j, and then calculate the distribution density of effective data distributed on them, as shown in formula (1).

$$\rho_{ij} = \frac{\sum_{p=i}^{j} N_P}{\sum_{p=i}^{p=i} 1_{mp}} (i = j - v, 0 < j \le m - v) \tag{1}$$

If the maximum linear density is extracted from the corresponding interval, then the normal value R is estimated. See formula (2):

$$R_0 = \frac{1}{m} (\sum_{i=i_0}^{i_0+v} \sum_i rs) \tag{2}$$

In the formula, m refers to the value of the effective sampling points in the corresponding interval of the maximum linear density. It refers to not only the value of one point, but the value of all the effective sampling points.

2.2 Project Management Analysis

The construction and improvement of the system platform, database, hardware facilities and other supporting facilities in the education management information system cannot do without a large number of funds and personnel investment, and the huge amount of funds required plays a decisive role, which is a strong guarantee for the construction of the education information team, and has a huge role in accelerating the construction and improvement of the education management information system and accelerating the modernization of education. Although the capital investment in the construction of educational management information system mainly depends on the investment of national modernization construction in education, compared with the development requirements of educational informatization and educational modernization, the investment in the construction of educational management information system in national educational funds is bound to be a drop in the bucket, The operation and maintenance of the system platform, the expansion and security of the education database, the development and application of the mobile platform all need a lot of money to invest and support, and the guarantee of the education informatization funds is widespread.

3 The Countermeasures of Developing and Perfecting Education Management Information System

Combining online and offline, on the basis of education management information system, using the office space, facilities, books, digital information resources that can be shared by various education departments and schools, we can organize the establishment of education resources integration and sharing, construction institutions and education information collection and processing, publishing and sharing institution system, so as to realize the real education resources sharing and education information sharing. Educational management information system can provide people with management services, educational administration and educational information inquiry services [3]. At the same time, Internet plus education platform system should be provided for people to provide educational resource sharing knowledge learning service and public network exchange platform, so as to provide learning and communication places for scholars. The foundation of the Internet plus education platform is the sharing and interconnection of the national educational information resources on the Internet platform, including the construction of the national public network library system, the specialized teaching video website platform, the knowledge information resources database, the public network exchange forum, the education public information inquiry service platform and other integrated education service system. It mainly provides educational information and resource sharing and exchange platform for people from all walks of life in China, and solves the problems of school teachers and students, sociologists, enrollment and other education related information query, as well as the lack of learning resources, educational resource sharing, knowledge learning and exchange in the process of learning and exchange.

3.1 Promote the Development and Application of Mobile Education Management Information System

In order to realize the leap forward development of school education and lead school education into the fast lane of rapid development, it is necessary to use the information highway to establish the management information system and basic database covering all levels of education administrative departments and all kinds of schools, so as to provide technical and data support for strengthening education supervision, supporting education macro decision-making, and comprehensively improving education public service ability, It is of great significance to realize the modernization of education management. Making full use of the advanced educational resources at home and abroad, sharing the teaching achievements of advanced schools, and giving full play to the advantages of the main position of promoting the informatization construction of schools are the key to quickly improve the level of running schools [4].

3.2 Increase Capital Investment and Make Efficient Use of Capital

By increasing investment and improving investment channels to increase investment in education management information system, at the same time, planning, supervision and management of the use of funds can make the use of funds more efficient. The construction and operation of education management information system, the equipment and update of hardware facilities, the construction and improvement of security system need a lot of money, especially the salary of construction and operation management personnel, information technology personnel, and infrastructure construction. It is suggested that the national government should increase the investment in the construction of education management information system, ensure the annual investment in the support and operation and maintenance of education management information system, strengthen the cooperation with social enterprises and organizations in education informatization, encourage and support social charity funding and investment in various industries, and make effective use of resources and funds from all walks of life, Optimize the organization and management structure of education management information system construction and operation, and supervise the use of management funds.

4 Conclusion

The realization of education modernization cannot do without informatization, which should serve education modernization. The state and education departments should plan and organize the construction of education informatization in a unified way. All fields of the education system should make extensive and in-depth use of modern information technology to improve education management, promote mutual exchanges between schools and all sectors of society, facilitate people to learn and obtain the required information, and promote the more balanced and comprehensive development of China's education. At the same time, educational informatization is a gradual and continuous development process. The state should increase capital and resource investment, and make use of the strength of all sectors of society to continuously improve the level of educational informatization in China.

References

1. Jiang, D.: General design of national education management information system. Chin. Inf. Circles (12), 76–83 (2013)
2. Luo, F.: Construction practice and Prospect of China's national education management information system. China's Educ. Informatization (1), 7–8 (2016)
3. Li, W.: Design and optimization of wireless campus network. Electron. Technol. Softw. Eng. (06), 13 (2017)
4. Xiao, N.: Design and implementation of mobile educational administration management system based on Android. Southeast University (2015)

Fitting Approximation Algorithm for Applied Mathematical Model

Xiaoxiao Ma(✉)

School of Traffic and Transportation, Chongqing Vocational College of Transportation,
Chongqing 402247, China

Abstract. Objective in the field of computer aided design (CAD), the property of progressive iterative approximation (PIA) of curve or surface is widely used in interpolation and fitting. If PIA method is used to fit all the data points directly, it will lack the flexibility of Y1 determination when fitting large-scale data points. In order to further improve the flexibility of incremental iterative approximation method in fitting large-scale point sets, an applied mathematical model based on fitting approximation algorithm is proposed.

Keywords: Fitting algorithm · Function approximation · Mathematical model

1 Introduction

Approximation and fitting of scattered data points is a basic and important research topic in computer aided geometric design (CACD). The progressive iterative approximation (PIA) method has been widely concerned in recent years because of its stable convergence, self adaptability and obvious geometric significance. The asymptotic iterative approximation method was proposed by Qi Dongxu and de boor (2). Lin Hongwei et al. Studied it in detail and extended it to the case of standard totally positive basis (NTP). For the configuration matrix is not standard totally positive, Chen Jic et al. Proved for the first time that the curve generated by general mixed basis function has the generalized PIA property when it satisfies certain conditions. On this basis, Zhang Li and others pointed out that said ball bases over triangular domains have generalized asymptotic iterative properties. Delgado and Pena, the Spanish scholars, compared the convergence rates of the curves (surfaces) generated by them using PIA method, and found that B-spline curves (surfaces) have the fastest convergence rate. In order to accelerate the convergence speed of PIA, Lu Lizheng proposed a weighted progressive iterative approximation (wpia) method, which can speed up the convergence speed by weighting the adjustment vector. Liu Xiaoyan et al. [9] proposed a Jacobi PIA algorithm based on non-uniform cubic B-spline interpolation. In 2010, Lin Hongwei proposed a local progressive iterative approximation (LPIA) method, which only adjusts a subset of the initial control vertices, so it is more flexible than the global PIA method. When there are many given data points, Lin Hongwei puts forward a generalized PIA method (EPIA), that is, by grouping the given set of points to calculate the adjustment vector in parallel, so as to improve the iterative efficiency and make it have great advantages

J. Macintyre et al. (Eds.): SPIoT 2021, LNDECT 98, pp. 443–448, 2022.
https://doi.org/10.1007/978-3-030-89511-2_57

in fitting large-scale data points. In 2014, Deng Chongyang and others further proposed the least squares based incremental iterative approximation method (lspia), which makes the fitting limit curve/surface be the least squares fitting curve/surface of a given set of points. After that, the method is extended to the case with different weights. In 2015, Lin Hongwei reviewed PIA from both theory and application to show its wide application. This paper presents an applied mathematical model based on fitting approximation algorithm to improve the flexibility of PIA method in fitting large-scale point set [1].

2   Algorithm Model

2.1   Fitting Model

The objective function of numerical optimization is to minimize the sum of squares of residuals Z (Y-Y). In the operation of genetic algorithm, the smaller the objective function value of the chromosome, the better the fitness of the chromosome. Because the parameter setting of genetic algorithm has a great influence on the actual operation results, this study uses trial matching method to determine the parameters according to the final results. The specific parameter selection is: population size n = 30, crossover probability p = 0.85, mutation probability p = 0.05, the maximum evolution algebra is 100, using real number coding. From the iterative results, it can be seen that the fitness of the initial generations of chromosomes is low, and the objective function value is large. With the increase of evolution algebra, the fitness of chromosomes increases rapidly, and the objective function value decreases. After 15 generations of evolution, the objective function value has been basically stable at about 234. Finally, when the model parameters are 110.060 and 36.115 respectively, the sum of residual squares reaches the minimum value of 234.33.

According to the final running results of genetic algorithm, the optimal fitting equation is as follows.

$$\hat{Y} = 110.060 + 36.115 lgX \tag{1}$$

2.2   Fuzzy Value Function Fitting

The least square method and algebraic interpolation method are commonly used in traditional function fitting. However, when the fluctuation of observation data is very complex and chaotic, any precise function fitting is meaningless. For example, in the following two graphs, one is the scatter diagram observed for variables (x, y), and the other is the complex curve graph [2].

From the scatter distribution in Fig. 1, in the section with smaller x, the scatter is relatively concentrated near a certain curve, and the law is also very obvious. When x is large, the scatter distribution is very scattered and the correlation between variables is very small. Generally, an explicit function y = f (x) is used to approximately describe the relationship between variables X and y, but it can not depict the data uncertainty presented by scattered points. If the fuzzy value function y = f (x) is used to describe the relationship between the variables reflected by the scatter point, the change trend

Fig. 1. Scatter plot of data on plane

Fig. 2. Complex curve graph

implied by the scatter point can be given, and the uncertainty degree of the change trend at any x point can also be given. Similarly, it is more natural to use a fuzzy value function to describe the motion law of variables implied in Fig. 2 than to use an accurate number. We call the method of constructing fuzzy value function by using observation data the problem of fitting fuzzy value function.

For the convenience of description, the linear regression problem of one variable is taken as an example. Suppose that there is a certain correlation between the random variable and the variable x, and the observed data is (x, y) (x2) y.. (x, y). Further assume that for every value of X, the random variable obeys the normal distribution, that is, y ~ n (a + bx, O), where a, B and σ 2 are unknown parameters independent of X. The regression equation is y = i + bx, and the parameters can be estimated by traditional methods

$$\hat{b} = \frac{\sum_{i=}^{n}(x_i - \overline{x})(y_i - \overline{y})}{\sum_{i=}^{n}(x_i - \overline{x})} \tag{2}$$

3 Approximation Algorithm

3.1 Background and Significance of Approximation Algorithm

From the 18th century to the early 19th century, some problems of optimal approximation of specific functions have been involved in the research of mathematicians such as L. Euler, P. - S. Laplace, J. - B. - J. Fourier, J. - V. pencailles. These problems are from the following aspects: Cartography, geodesics, geodesics, and the best approximation of specific functions The paper proposes the concept of best approximation and studies the properties of the best approximation element when the approximation function class is n-order polynomial, The characteristic theorem which can judge polynomial as the best approximation element is established. He and his students have studied the problem of polynomial with minimum deviation from zero, and obtained many important results.

In 1885, German mathematician K. (T.W.) weierstras proved a theorem when he studied the problem of approximation of continuous functions by polynomials, In principle, this theorem affirms that any continuous function can be approximated uniformly in the definition interval of a function with any pre specified precision. Although it is not pointed out how to select polynomials to approach the best, Chebyshev and weierstras are the founders of the modern development of approximation theory [3].

There are a lot of nonlinear problems in the field of natural science and science and technology. In recent years, people have obtained a series of profound results in the study of numerical and functional approximation and computer aided geometric design. With the emergence of high-performance and large-capacity computers, the problems that have been difficult to realize in the past have become possible, So there is great potential for the theoretical research and application of function approximation.

3.2 Approximation Algorithm and Function Space

In numerical calculation, the function values are often calculated, such as the basic elementary function and other special functions calculated on the computer. These involve approximation of the given function by simple functions such as polynomial, rational fraction or piecewise polynomial, etc., In order to describe more accurately mathematically, some basic concepts and preparatory knowledge are introduced.

In mathematics, it is often called to introduce certain certain definite relations into various sets as giving a set some spatial structure, and such a set is called space. For example, in linear algebra, all real n are composed of linear space in real number field, which is composed by vector addition and multiplication of vector and number, is written as R", and called n-dimensional vector space. Similarly, the paper introduces the concept of" linear space "in the real number field by adding vector and multiplication of vector and number, For all real coefficient polynomials with no more than N, the addition of the common polynomials and multiplication of numbers and polynomials also constitute a linear space on the number field R, expressed by H, called polynomial space. Also, for example, all the continuous functions defined on interval [a, b], and the linear space on the number field R is formed by adding and multiplying the number and function by function, and it is recorded as c[a, b] as function space.

Definition: set s is a linear space on the number field P, $x1, x2,\ldots xn \in s$, if there is a number not all zero….a. $\in$ P makes

$$a_1x_1 + a_2x_2 + \ldots a_nx_n = 0 \tag{3}$$

4   Applied Mathematical Model

4.1   Significance of Model

In the era of big data, a large amount of data emerges every day. These data structures are complex and high-dimensional, and they have many redundant information and noise. This not only brings great challenges to processing work, but also increases the corresponding cost for storage. Therefore, the first mock exam is proposed. It is to use the low dimension part of high-dimensional data to find a suitable low rank tensor instead of the original one. This low rank tensor has the main characteristic information of the original complex tensor. However, there is still a major challenge in the application of the model, that is, how to describe the rank of tensors accurately. The kernel norm is used to calculate the rank of tensor. However, because the kernel norm can treat singular value equally, that is, the punishment for singular value with different size is the same, which leads to the application effect in practical problems is not particularly good, and the optimal solution is not very accurate [4].

Therefore, in order to avoid the above-mentioned kernel norm, this paper proposes an applied mathematical model based on fitting approximation algorithm. Based on the matrix rank defined by singular value decomposition of tensor, the model describes the rank of tensor by constructing a convex problem of nonconvex penalty function. The non-zero singular value is estimated more accurately by using the parameterized non convex penalty function than kernel norm. In this chapter, it is proved theoretically that under certain assumptions, the global optimal solution of the objective function can be obtained. The MATLAB is used to test the method proposed in this paper and several common low rank tensor approximation methods.

4.2 Model Establishment

The singular value of known matrix has decreasing property. For any third order tensor x'e rngms, its singular value decomposition can be written as x = u * s * V, because inverse discrete Fourier transform gives

$$S(i, i, 1) = \frac{1}{n_3} \sum_{j=1}^{n_3} \overline{S}(i, i, j) \tag{4}$$

Therefore, the term on the diagonal of the first positive slice $S(i, i, 1)$ of tensor s also has the same decreasing property,

Let d be the low rank tensor polluted by noise, X be the potential target tensor, e be the tensor containing Gaussian noise, then y = X + ε, As shown in Fig. 3. So the low rank tensor approximation model can be written as follows:

$$argmin\left\{\Psi(x) = \frac{1}{2}\|y - x\|_F^2 + \lambda \sum_{t=1}^{k} \varphi(S(i, i, 1)), a\right\} \tag{5}$$

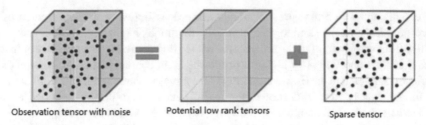

Observation tensor with noise   Potential low rank tensors   Sparse tensor

Fig. 3. Low-rank and sparse tensor decomposition from noisy tensor observations

5 Conclusion

This method is suitable for fitting large-scale data point sets. If this method is used in fitting a small number of data point sets, not only the flexibility of the algorithm will be greatly limited, but also the overall calculation amount of the algorithm will be increased compared with the pia method due to the influence of the splicing algorithm. As the future work, we will use this method to study some complex curves and surfaces.

References

1. Leader, J.J.: Numerical analysis and scientific computing. Tsinghua University Press, Beijing (2008)
2. Yuan, X.: Numerical analysis. Dalian University of Technology Press, Dalian (2010)
3. Li, Q.: Numerical analysis. Tsinghua University Press, Beijing (2008)
4. Lu, Z.: Analysis and comparison of approximation ability of several function approximation methods. J. Wuzhou Univ. (2008)

Multi-sensor Technology for Construction Safety Early Warning System of Bowl Buckle Full Hall Support of Highway Bridge

Weiping Li[✉]

Shandong Transport Vocational College, Weifang 261206, China

Abstract. Bowl buckle full framing is widely used in highway and bridge construction because of its strong versatility, high bearing capacity, less auxiliary equipment and low cost. However, the safety accidents caused by its instability often occur, which seriously endanger the life safety of on-site workers and bring huge economic loss and adverse social impact to all parties of the project. This paper studies from the technical route and management route respectively, and develops the construction safety early warning system of bowl buckle type full framing for highway bridge girder. It carries out real-time monitoring on the construction scaffolding, analyzes the monitoring data of the early warning system, and prevents the occurrence of scaffolding accidents, which has great practical and economic significance.

Keywords: Bowl buckle type full hall support · Safety warning · Early warning system

1 Introduction

With the development of national infrastructure construction, the construction of high way and bridge infrastructure develops rapidly. Full framing system has been widely used in highway and civil engineering due to its various vertical and horizontal spacing, low foundation bearing capacity and low cost. However, with the rapid development of social economy, the national highway network has been continuously expanded and optimized, the construction of highway bridges and other infrastructure has developed rapidly, and modern large-span bridges and tall buildings have emerged, In order to meet the construction requirements, the height of the formwork support is constantly improving, and the technical requirements are also becoming higher. The collapse accidents of the formwork support system occur frequently. On the one hand, it causes serious economic losses and increases the cost of the project. On the other hand, the collapse accidents seriously endanger the safety of the construction personnel.

At present, a series of researches on full framing have been carried out at home and abroad, which are mainly divided into three types: Research on the stress of framing system; Research on the support force and shape control, the relationship between the deformation of the structure after the force; Support construction safety monitoring research, but the monitoring content is mainly support internal force, foundation settlement, support displacement and other single or several single combination of research. There is no research project to monitor internal force, deformation, settlement, environmental temperature and humidity in all aspects at the same time, and there is no research content to directly link the mechanism of safety accidents with monitoring data, and stage early warning to directly guide on-site safety management [1].

Due to the lack of real-time warning and timely feedback of support safety information, support collapse and other safety accidents are caused. This paper systematically analyzes the use characteristics of the support in the construction process, explores the mechanism of safety accidents, and uses a variety of sensors, equipment and software to build a complete and reliable full support safety early warning system for real-time monitoring of the support safety status. The research of full support safety monitoring technology and early warning system has an important impact on solving support safety monitoring problems and preventing support safety accidents.

2 Bowl Buckle Full House Support

Bowl buckle type steel pipe support frame is a new type of support developed on the basis of portal support and fastener support. Its reasonable joint structure, easy construction and high bearing capacity can meet the construction requirements of buildings and long-span bridges. The most significant difference between bowl buckle type and fastener type steel pipe support bracket is that bowl buckle type support bracket is a fixed structure. In other words, the spacing between columns, step distance and row to row spacing are no longer arbitrarily determined. The specific size of support structure is determined by its structural elements, and can only be referenced by some standard sizes, Adjust it appropriately. At present, in the construction of overpass and Viaduct in some cities, bowl buckle steel pipe support has been popularized. Bowl buckle steel pipe support has been used instead of fastener type steel pipe support, and remarkable results have been achieved [2]. The structure of bowl buckle joint is shown in Fig. 1.

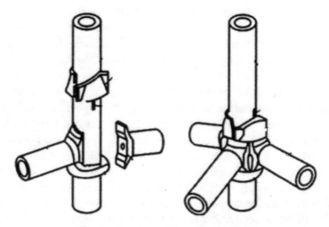

Fig. 1. Structure of bowl buckle joint

Socket type support is a form of single pipe support. Its structure is basically similar to fastener type steel pipe support. It is mainly composed of vertical pole, horizontal pole, inclined pole, adjustable base, etc., but the connection between main pole, horizontal pole and inclined pole is not made by fastener, but by welding socket on main pole, welding plug on horizontal pole and inclined pole, and inserting plug into socket, various sizes of support can be assembled.

3   Analysis of Influencing Factors of Full Support Accident

3.1   Reasons for the Construction Materials of Fastener Type Steel Pipe Support

At present, steel pipe is the most important material in China's construction industry, and fastener type steel pipe full framing has become the most frequently used way of steel pipe support in China's construction engineering field. The main components of full framing are support column, horizontal and vertical horizontal bar, cross bracing, horizontal support construction and wall connecting bar. In addition, there are three kinds of components: right angle, butt joint and rotation. In the manufacturing process of support bar, it needs to meet the relevant standard value, but in practice, due to a variety of reasons such as technology, personnel, equipment and so on, the bar often has some defects. Such as the initial eccentricity, initial bending and other geometric defects in the manufacturing process of the rod, the initial stress in the production process of the rod, and the quality problems of the rod material [3]. Some products do not meet the requirements, steel pipes and connectors have great problems and defects, which are "unqualified" products when they are shipped in the factory.

3.2   The Design Model of Full Framing is Unreasonable

No matter what kind of use the full framing is used for, it should be optimized in structure design and checked accurately in calculation. The strength, stability, stiffness and other checking calculation of full framing should be considered in its optimization design: in addition, the anti sliding of connectors, the anti bearing capacity of foundation, the strength of transverse and longitudinal horizontal bars and the strength of connectors with wall should be checked. The calculation of the strength and rigidity of the tie rod, the pulling rope, the base and the limit erection height of the cantilevered full framing, etc.

The design model of full support is seriously inconsistent with the calculation condition, the various parameters of the support are not calculated empirically, the strength and stiffness are not ensured to meet the needs of the actual situation, and the construction process operation is not standardized, which can cause some unnecessary disasters.

3.3   Nonstandard Construction of Full Framing

In the process of full support construction, the indicators that affect its stability mainly include: the setting of the sweeping pole, the step distance of the vertical pole, the distance between the vertical poles, the length of the vertical pole extending out of the top horizontal pole, the setting of the horizontal reinforcement layer, the setting of the vertical cross bracing, and the foundation stiffness. Whether these building indexes meet the design specifications directly determines the bearing capacity level of full framing. Among them:

(1) The step distance of the bottom broom bar of the fastener type support should be limited to 0.2 m. Setting the broom bar according to this specification can significantly improve the bearing capacity of the full support;

(2) In practical engineering application, there are many reasons for uneven load distribution and local instability of full support, among which the number and location of scissors support and horizontal reinforcing layer are not enough and inappropriate. In addition, diagonal bracing will also affect the strength and stiffness of support system, Therefore, in the construction of the support system, attention should be paid to the setting of diagonal bracing to avoid excessive lateral deformation of the support, resulting in the instability of the system;

(3) In order to effectively improve the bearing capacity of the full support, the distance between the vertical pole and the top horizontal pole should be reduced.

4   Research on Construction Safety Early Warning System of Bowl Buckle Full Framing

4.1   System Function Requirement Analysis

The functional requirements of bowl buckle support construction safety early warning system are as follows:

(1) Real time data acquisition, display, storage and access of bowl buckle bracket are realized
The data collection and real-time display of bowl buckle type full hall support are used to guide the construction of site construction workers, save the data for easy viewing, and then realize remote data transmission and access through wireless communication technology, so as to achieve the purpose of remote monitoring.

(2) It can forecast the data collected by the system and has the function of early warning
By collecting the stress state parameters of bowl buckle full support, comparing with the theoretical calculation value and buckling load value, and combining with the prediction algorithm to predict the follow-up data trend, judge whether the monitoring rod is in a safe state, form an early warning function for the collapse of bowl buckle support, and inform the relevant responsible personnel

4.2   System Structure

After signal conditioning, the electrical signals collected by various sensors are converted into digital signals by the AD conversion module, which are transmitted to the embedded computer security early warning system to judge and save the collected data, and then sent to the communication module according to the communication protocol. The data acquisition system mainly includes all kinds of sensors, signal conditioning module, a / D conversion module, embedded computer, GPRS DTU module, power supply and so on. The structure of data acquisition system is shown in Fig. 2.

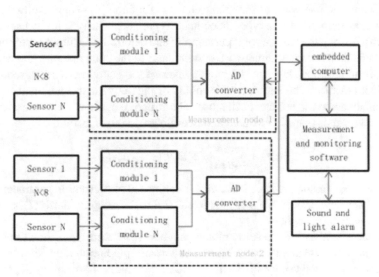

Fig. 2. Structure diagram of data acquisition system

4.3   System Functions

The design of safety early warning system takes into account the complexity of the construction site and the universality and safety of the system. Most of the hardware uses common components to enhance the universality of the system. The software adopts modular design, which is easy to transplant and upgrade. The main functions of the system are as follows:

(1)   Support stress, displacement, inclination and other sensor data acquisition and monitoring. The system supports stress sensor, displacement sensor, temperature sensor, inclination sensor and so on, and can select monitoring according to the demand, which improves the application scope of the system.

(2)   It has good data access function. The monitoring data can be transmitted to the monitoring host for real-time display, and can also access the database and edit the relevant parameters according to the SMS instructions edited by the user to obtain the real-time status information of the stent system, and the system can add multiple managers to facilitate multi-party monitoring and management.

(3)   Alarm conditions can be set at each monitoring point. Due to the different stress and early warning values of each member in the support system, the system can set independent early warning values for each member to better ensure the safety of the support.

4.4   Research on Real-Time Monitoring Pre-processing Based on Wavelet Neural Network

Wavelet neural network is the combination of wavelet analysis and neural network. Its basic idea is to use wavelet function to replace function [4]. There are two types of wavelet neural network: loose type and compact type. The loose type firstly uses wavelet transform to extract the features of input samples and determine the input vector of neural network, and then uses neural network to train data; In the compact type, the excitation function from the input layer to the hidden layer of the neural network is replaced by the wavelet function. The learning and training of the network still adopts the gradient descent method and the pre feedback mode. The structure of the wavelet neural network (the unified form of one-dimensional and two-dimensional) is as follows

$$f(x) = \sum_{i=1}^{N} \omega_i \phi[D_i R_i(x - t_i)] + \bar{f} \tag{1}$$

Where, the input vector $x \in R^n$; t is the vector composed of translation parameters; D_i is a diagonal needle composed of scale parameters; R is the rotation matrix; $\bar{f}$ is the mean value of f; ω_i is the weight.

This paper studies the real-time monitoring and early warning of bowl buckle full house support, transplants the wavelet neural network prediction algorithm to predict, compares the predicted value with the measured value to judge the accuracy of prediction, and compares the predicted value with the alarm value to judge whether the alarm condition is reached. When the length of measured data meets the requirements of prediction calculation, the short-term prediction is started. In order to fully reflect the latest state of full support, improve the accuracy of prediction, and ensure the efficiency of prediction, the system uses the cyclic vector to change the array structure at any time. After the completion of the first prediction, the measured data collected in the next time is added and the first data of the current time series is removed, and then the second prediction is started.

4.5   Monitoring Information Alarm

Monitoring is divided into three alarms and one verification, which also includes monitoring safeguard measures under abnormal conditions.

(1) Foundation settlement overrun alarm:
    The settlement is within the warning value, the settlement is normal, the green flashing light works, and no sound is emitted. If the settlement exceeds the first level warning value, the warning alarm will be given. The light is yellow and the warning alarm will sound. If the settlement exceeds the second level warning value, the danger alarm will be given. The light is red and the danger alarm will sound.
(2) Alarm when the vertical force at the bottom of the vertical pole exceeds the limit (the force on single rod of Jieneng mould base coil buckle is $n = 17$ tons)
    When the force at the bottom of the pole is below the warning value, the bracket works normally, and the green flash lamp works without sound. When the force on the bottom of the vertical pole exceeds the first level warning value or the force

on the bottom of the vertical pole is lower than the initial value of the support installation, a warning alarm will be given. If the force on the support is large or the uneven force on the support is serious, the alarm light will be yellow and the alarm sound will be given. When the force on the bottom of the vertical pole exceeds the level II warning value or the force on the bottom of the vertical pole is far lower than the initial value of the support, the danger alarm will be given. The alarm light is red and the alarm sound will be given.

5 Conclusion

In this paper, the reasons for frequent accidents of full support are analyzed, and the key technologies for preventing serious accidents of full support are studied. The forms and causes of collapse accidents of full support are analyzed systematically. The optimization and monitoring methods of full support early warning system are studied, and the early warning system of full support is developed to realize phased and hierarchical early warning, The early warning information of the system is displayed by numbers, curves, pictures, sound and light, which can be divided into three levels: normal, warning and danger. The scene lights are green, yellow and red respectively.

References

1. Chen, X.: Real time monitoring and early warning method for external scaffold safety based on RFID technology. Harbin Institute of Technology (2012)
2. Zhu, X.: Research on real time monitoring method and application of external scaffold safety based on mobile it. Harbin Institute of Technology (2014)
3. Zeng, S., Wang, Y., Chen, P.: Scaffold safety management and control technology of Jinping High and steep slope. Constr. Technol. **42**(05), 44–47 (2013)
4. Research on safety state early warning of shibaolu high formwork support structure. Constr. Technol. (45–11), 12–14 (2016)

Service Innovation of Intelligent Library Based on 5G + AI

Yuqin Huang[✉]

Zhangzhou Health Vocational College, Fujian, China

Abstract. according to the requirements of library information management system, this paper uses Java language and diango framework to build library information management system, introduces the system module, optimizes the design, tests the system functionality and stability, and provides beneficial reference for library management. Innovation is a common behavior in human activities, which exists in every field of human life. Therefore, there are innovation problems in every link and function of management activities. With the coming of the era of knowledge economy, it will inevitably shake the management ideas, management systems and management methods of all kinds of organizations. Modern library management is no exception. In a certain sense, this is the essence of modern library management. In the face of the rapid development of science and technology, the huge increase of knowledge and information, and the drastic changes of the market in the new century, who can feel agile, seize the opportunity, make a quick decision, make a quick response, and strive to take the lead everywhere, will win in the competition.

Keywords: Computer technology · Library information system · Java language

1 Introduction

With the rapid development of social economy, it promotes the development and upgrading of electronic information technology and Internet technology. Information processing system relies on the advantages of high security and high efficiency. It is widely used in most fields. In the information processing system, computer data processing and integration are the advantages of technology application. Information management system can not only improve the management efficiency, simplify the workload, but also maintain the security and privacy of user information. According to the survey of the library, the library uses computer technology to manage books, providing a single docking mode for readers. This kind of phenomenon has a great impact on the enthusiasm and convenience of readers, and also increases the waste of human and financial resources. Perfect information management system, to a certain extent, can replace the traditional management, so in this paper, we focus on optimizing the design of library information management system [1].

After entering the information age, people gradually realize the importance of information data, and with the rapid development of science and technology and network

© The Author(s), under exclusive license to Springer Nature Switzerland AG 2022
J. Macintyre et al. (Eds.): SPIoT 2021, LNDECT 98, pp. 456–463, 2022.
https://doi.org/10.1007/978-3-030-89511-2_59

technology, people get more and more new ways, access speed is also faster and faster, no longer like before what information and information need to run to the library, to read a thick book, now through the Internet technology needsAny information and materials can be found quickly, so the development trend of library service innovation should be towards the direction of new service concept, high-tech management means and service technology equipment, content keeping pace with the times, innovation of access methods, large-scale improvement of library storage capacity, and establishment of knowledge and information sharing platform, so as to break the shackles of traditional library service mode,We should develop an open service mode, improve the ability of knowledge collection and sharing, give full play to the extensive and profound role of the library, and provide better services for knowledge seekers [2].

Carry out efficient and convenient services. With the development of network information technology, the whole society is gradually stepping into the information age. Network information technology should also become an important innovation point of library work and service. The network management tool of library has its own characteristics. For example, to meet the needs of readers, we should create special information database, integrate network information resources, develop information columns on the basis of the initial library website development, improve the efficiency of literature retrieval, strengthen video publicity and establish electronic books and periodicals.

Hardware facilities are the foundation of library construction. If the foundation is not firm, other work can not be carried out. The hardware construction of the library includes collection resources construction, human environment construction and so on. The library should ensure the systematicness, comprehensiveness, continuity and timeliness of the collection, increase the quantity of the collection, improve the quality of the collection, optimize the structure of the collection, and realize sustainable development; the relaxed and elegant humanistic environment will greatly enhance students' interest in information inquiry and autonomous learning [3, 4].

Information resources are the core of library construction. We should focus on the long-term interests, increase the investment in library information resources construction, introduce advanced equipment and professional talents, and strengthen the management of information library. At the same time, in order to realize the comprehensiveness, systematicness and integrity of university library information, we should integrate social forces and realize the co construction and sharing of information resources. Information resource sharing can make the information resources reasonably allocated, build a relatively complete system together, avoid repeated construction, meet the different needs of readers in all aspects to the maximum, improve the utilization rate and service rate of the overall literature information, save manpower, material and financial resources, and expand the function and space of Library services.

The basic work of the library is the fundamental work of serving readers. Opening and closing the library on time, warm service and meticulous work are our most basic common sense and discipline. In this era of knowledge explosion, readers put forward diversified and personalized demands for information. Only by understanding the needs of readers can we meet the needs of readers to the greatest extent. The library should regard the needs of readers as the standard and purpose of service innovation. With the development of society, readers' demand for information is constantly changing.

The library should study the needs of contemporary readers, understand the direction they are interested in, and establish a book collection system and service mode suitable for modern needs. The characteristics of openness in this era determine that the closed characteristics of Libraries in the past can not conform to the trend of the times. Libraries should change the self satisfaction view of "large and complete" and "small and complete" and establish the concept of resource sharing. Strengthen the communication and cooperation with other libraries, so that readers can get more information and services, and provide as much information and services as possible with the least funds.

The innovation of service concept is to continuously optimize the overall function of the management system with new ideas, new measures and new methods, so as to maintain a state of the best effect. From the current situation of our country, the key of management innovation is the change of ideas, that is to focus on the development of people's ability, the mobilization of enthusiasm and the stimulation of creativity. In order to survive and develop better, the library must sublate and choose the traditional management ideas and methods, and establish a new management operation mechanism through reform and innovation, so as to meet the requirements of social development.

2 Overview of Computer Technology

The technical methods and means used in the computer field are relatively advanced, and the comprehensive characteristics of computer technology are remarkable. It can combine mathematics, modern communication, mechanical engineering, applied physics and electronic engineering, and promote the development of computer technology accordingly. Computer technology is a complete system technology, involving system structure, maintenance, application, management and other technologies.

2.1 System Architecture Technology

TThrough the system structure technology, it can improve the efficiency of computer system problem solving and ensure the highest cost performance. The rapid development of electronic devices, the development of microprogramming and solid-state engineering technology, has driven the development of operating system, virtual memory technology, programming language technology, which will greatly affect the system structure technology, and closely combined with computer software, firmware, hardware, involving computer science theory, microelectronic engineering and other technologies.

2.2 System Management Technology

Through the operating system, the automation of computer system management can be realized. The operating system can efficiently use computer hardware and software resources, strengthen the machine throughput capacity, operation convenience, disintegration timeliness, etc., and improve the system reliability, so as to reduce the cost of solving problems.

2.3 System Maintenance Technology

The computer system has the technology of automatic maintenance and diagnosis. The function inspection program and automatic diagnosis program are the main maintenance and diagnosis automation software. Function check program can realize the micro function of computer system. Through strict data graphics and actions, it can realize inspection and test, compare the correctness of the results, and ensure the normal operation of components.

2.4 System Application Technology

Computer system application standard, program design automation, software engineering technology, and technology application are closely related. Program design automation can use computer to design program automatically, so as to promote computer technology. The traditional computer system, relying on manual and machine instructions to write programs, takes a lot of time and energy, is prone to errors, and is difficult to read, debug and modify.

3 Management Information System and Its Development

3.1 Management Information Systems

Management information system is human-oriented, through computer hardware, software, network communication equipment, office equipment, at the same time do a good job in information collection, transmission and processing, comprehensively improve the efficiency and efficiency of the enterprise, at the same time can support the enterprise high-level decision-making, middle-level control, highly integrated human computer system.

3.2 Understanding of System Development

The new management concept and computer information system need to take the management system as the basis, based on the material demand theory, which is quite different from the traditional inventory concept and method. The time division is introduced to the basis to comprehensively reflect the bill of materials, and to deal with the inventory management and production control problems, which can be realized by using the order point method. It takes a long time for material supply, so it is not possible to replenish the inventory after the material inventory is completely consumed. It is necessary to accurately judge the time and replenish in advance. The above problems can be solved by order point method.

4 Demand Analysis of Library Information System

Library information management subsystem is an important information management mode. Therefore, it has specific information management system and functions, information security, human-computer interaction experience, data scalability, retrieval methods, paper resources and electronic resources synchronous management. The information management system takes the library as the management object, so we must comprehensively analyze the particularity of the library. Usually, the service objects of library can be divided into staff, administrators and readers. The management personnel mainly maintain the system information, detect the system status, summarize the special borrowing situation according to the actual needs, and generate the information report file. Staff is responsible for the management of books, according to the book borrowing situation, generate reports. Readers mainly input and modify personal information, query book information, query personal borrowing, etc. The requirement of library information system involves data flow chart and data dictionary. Data flow analysis can separate the internal flow of the organization, remove the organizational structure, processing tools, information carrier, material and so on. Through the data flow process, the business data processing mode is investigated.

5   Design of Library Information System

The traditional library mainly adopts the way of providing resources and literature. In the new era, digital library can provide knowledge to customers. The design of library information system can effectively combine pictures, images, multimedia and database through knowledge unit to provide high-quality service for customers. The retrieval of metadata and content can promote the new development of technology. In the new era, library information system can provide open knowledge structure. In the development of information processing technology, library information system can effectively solve the problem of readers' knowledge and information, and realize the one-time satisfaction of information data. Therefore, the following details of library information system design related issues.

5.1   Overall Design

The function of the library management system will be affected by the overall design of the system. System application services, relying on diango framework. The framework has the functions of template engine and object relation mapping. The above functions help to improve the efficiency of the system and speed up the development of the system. In addition, the system framework can provide web page and API interface to the outside. In the system database table, we need to complete the object writing through the object relational mapping mode. Through the management system, it can deal with the problem of manual construction of traditional database, and significantly improve the efficiency and accuracy. Diango framework involves "model", "template" and "router", where "model" corresponds to database and "template" corresponds to system file.

5.2   Database

5.2.1   Database Conceptual Design

According to the system function analysis, when the permissions are different, the system users can be divided into system administrator, book operator and borrowing operator.

The work content of the borrowing operator is as follows: issuing and reporting the lost book card, paying the fine information, urging the return of books, viewing the reader information, book information, saving information, so it is the role of control authority. The book operator is responsible for maintaining the book information, and also belongs to the role of control authority. In this kind of system users, the system administrator has the highest authority and can apply system functions at will. Before different types of users log in to the system, they should check the effectiveness. If the input is invalid, you must reenter it. After that, according to the requirements of the library and the borrowing and returning of books, corresponding treatment should be done. According to different situations, maintain the book system, issue a fine, etc. Through the effectiveness check, it can show the readers, books, statistical tables and so on.

5.2.2 Database Logic Design

In the design of logical structure, the conceptual structure needs to be transformed into logical data model and optimized. According to the previous analysis, the design drawings are obtained, and the characteristics of structured query language are modified to obtain the main relational database tables. The system data flow is shown in Fig. 1.

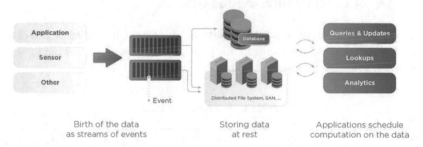

Fig. 1. System data flow

5.2.3 Physical Design of Database

The physical design belongs to the second half of the database. The given logical structure is put into the specific environment. The logical data model selects the specific working environment and provides the data storage structure and access method. It belongs to the physical design of the database. Physical structure is more dependent on hardware system and database management system. Designers should master the storage structure, internal characteristics and access methods of database management system. The physical design of database involves determining the physical structure of database and evaluating the time efficiency and space efficiency. For determining the physical structure of the database, the following contents are involved: determining the data storage structure, storage path, storage location, system configuration, etc. During the physical design of database, we should balance the space efficiency, time efficiency, maintenance cost and user requirements, and choose the best scheme as the physical structure. In the physical design of database, objects should be stored and retrieved centrally.

5.3 Input and Output Design

The input and output design is the visualization of the organization system, which can build a high-quality working environment for users, ensure that users can actively participate in the work, conform to the user's operating habits, ensure that users can accept the target system, and provide users with easy to understand information.

(1) Input design. The input interface belongs to the management information system and the user interaction link. The design task is to define the input form according to the business requirements, so as to ensure that the information system can obtain the management tool information. Therefore, the input design helps to improve the input efficiency and reduce input errors.

(2) Output design. The output design task can ensure that the system can meet the needs of users. Through the output design, it can correctly reflect the information needed for management and ensure that the information meets the needs of users, which has a great impact on the use effect of the system and the success rate of system design.

6 Testing of Library Information System

In order to test the stability and functionality of the system, it is necessary to comprehensively test the system functions. In this paper, the functions of some modules of the system are tested, and the results are shown in Table 1.

Table 1. System test results

Modular	Operation	Expected results	Actual results
System login	Password modification	Success	Success
User management	User name and password recovery	Success	Success
	User information modification	Success	Success
Library management	Name of publishing house editor	Success	Success
	Return books	Success	Success
	Add new books	Success	Success
	Borrowing new books	Success	Success

7 Conclusion

To sum up, through the investigation of library information management, this paper puts forward the requirements of library information management system. Based on the overall design of information system, this paper puts forward the content of database design, introduces the problems related to the system design in detail, and analyzes the typical functional modules, puts forward the overall design of the system, database design, input and output design, and expounds the realization of library information system, which involves the system login module, user management module, book information management module and borrowing information management moduleFinally, the library information system is tested to verify the functionality and stability of the library information system. Through the research and analysis of this paper, it provides reference value for library management.

Acknowledgements. Funded by the Sub-committee of The Guidance Committee of Library and Information Work in Higher Vocational Colleges of the Ministry of Education in 2020 (TQ-2020-29, TQ-2020-46); 2020 Academy-level research project of Zhangzhou Health Vocational College (Project No: ZWYS202009).

References

1. Gong, J.: Application of human resource management in information system project management - taking the construction of military literature and art digital library as an example. Wirel. Internet Technol. **16**(13), 155–156 (2019)
2. Luo, K.: Practice of mobile library system based on Android platform - taking Shaoyang University Library as an example. Electron. Technol. Softw. Eng. (08), 79–80 (2016)
3. Yang, M.: Dynamic simulation of mobile library community service system based on information ecosystem perspective. Inf. Sci. **38**(01), 153–161 (2020)
4. Ming, J., Cao, H., Tu, R., Yang, Y.: Research on the elements and structure of social media information ecosystem - taking Library microblog as an example. Mod. Inf. **39**(11), 128–136 (2019)

Evaluation of Low Carbon Economy Development Level Based on Fuzzy Synthesis Algorithm

Weihua He[✉]

Beijing Jiaotong University, Beijing 100044, China

Abstract. With the rapid development of the world economy, the development of low-carbon economy has become more and more important, using the evaluation of low-carbon economy development index data, according to the evaluation results of the first-level ambiguous comprehensive evaluation model, provinces and cities, evaluate the development of low-carbon economy, provide a basis for accelerating the development of low-carbon economy, and promote the harmonious and sustainable development of the national economy and environment. and sustainable development of national economy and environment.

Keywords: Fuzzy comprehensive evaluation model · Low carbon economy · Membership function · Condition assessment

1 Introduction

With the continuous development of economy and society, people's awareness of ecological environment protection is becoming weaker and weaker, the problem of economic and environmental coordination and sustainable development is becoming more and more serious, and the contradiction between them is becoming more and more serious. As a famous scholar, it is loved by scholars all over the world.Developed countries abroad have also taken the lead in many "low-carbon actions", such as reducing the discharge of waste gas from factories, limiting the use of vehicles, encouraging "single cycle travel" and other behaviors.

The theory have not been completely systematic, but due to various "low-carbon behaviors" put forward by various countries, a reasonable and comprehensive evaluation system of needed for the sustainable development of the world economy and society.On the other hand, the actual production and development of each province and province are different, the economic environment, geographical environment and other states are different, and the development relationship between provinces and cities is different, which will affect their evaluation, so the comprehensive evaluation of low-carbon economy has certain fuzziness. This paper uses fuzzy comprehensive evaluation method and model, and uses some index data obtained by investigation to measure. We introduced the current situation of Chinese provinces and cities developing low-carbon economy, and submitted reasonable opinions and suggestions. It provides scientific basis for each province and province to develop the regional economy of cities more continuously [1].

© The Author(s), under exclusive license to Springer Nature Switzerland AG 2022
J. Macintyre et al. (Eds.): SPIoT 2021, LNDECT 98, pp. 464–469, 2022.
https://doi.org/10.1007/978-3-030-89511-2_60

2 The Level of Evaluation of Low Carbon Economy Development and Analysis of Fuzzy Factors

Stage 1: the economic essence of low-carbon economy includes two meanings. First, to develop a low-carbon economy according to the principles and institutions of the market economy, and to develop a low-carbon economy, people's living standards and welfare levels cannot be lowered. We will oppose extravagant consumption and energy waste and improve people's living standards.

Stage 2: the technical nature of low-carbon economy, that is, through technological progress, energy efficiency is improved, and the emission intensity of greenhouse gases such as CO_2 is reduced. That is, on the premise of minimizing energy consumption, we can obtain the equivalent economic output; Or to obtain more effective economic output than before with the same amount of energy consumption.

Stage 3: The goal of a low-carbon economy is to achieve relative stability of greenhouse gas concentrations in the atmosphere, increase global temperatures, affect human survival and development, and achieve harmonious development of people and nature.

Reviewing the previous literature on low-carbon economy, it is found that there are a series of fuzziness in the evaluation of the development level of low-carbon economy. One is the fuzziness related to the evaluation criteria of low-carbon economy.

The judgment of qualitative index to the target subject is determined by the subjective judgment of the judge. In essence, there is interval ambiguity in itself. The second is the fuzziness related to the evaluation process of low-carbon economy. Because the final result of the evaluation is to analyze the evaluation results of experts comprehensively, and the evaluation index is interval fuzzy quantity, the evaluation process can only use fuzzy method to get more objective results. Based on the above reasons, the author thinks that the fuzzy comprehensive evaluation method combining AHP and fuzzy set theory is more reasonable to evaluate the development level of low carbon economy, which can not only deal with the fuzziness of evaluation index but also distribute the weight of evaluation index scientifically [2].

3 The Establishment of Decision Model of Fuzzy Comprehensive Evaluation

Comprehensive principle of fuzzy relations. The idea of fuzzy mathematics is the degree of subordination. The key to building a mathematical model with fuzzy mathematics is to establish the degree of subordination that meets the actual requirements, and the basic idea is the fuzzy linear change point. Using the exchange rate principle and the maximum subordination principle, considering the relevant factors affecting the evaluation object, quantify the indicators, allocate different weight values according to different benchmarks and the influence degree of the evaluation object, and finally reasonably evaluate each evaluation object. Carry out high and comprehensive comprehensive evaluation. When there are few evaluation indexes, the first-level fuzzy comprehensive evaluation is adopted. When the problem is complex and there are many indexes, the multi-level fuzzy comprehensive evaluation model can be used to improve the accuracy of evaluation. Many achievements need to be evaluated, so the multi-purpose fuzzy comprehensive evaluation decision model is used for evaluation, and then the construction process of the first-class ambiguous comprehensive evaluation model and the multi-purpose ambiguous comprehensive evaluation decision model is introduced [3].

3.1 A Fuzzy Comprehensive Evaluation Model of First Level

(1) Set of determining factors and comment

Comprehensively should be made from many aspects, such as harmless treatment of domestic waste, per capita support, number of private cars, forest coverage, etc. All these factors constitutes the evaluation standard system. That is, the parameter set, with N, it is recorded as follows:

$$U = \{u_1, u_2, ...u_N\} \tag{1}$$

For each evaluation index, there are different evaluation values, so different grades will be formed. For example, the forest coverage rate of various provinces and cities is 29.63% in Beijing, 13.95% in Shanxi and 8.16% in Shanghai. There are high and low, so the evaluation will be good, good, medium and poor. The evaluation values or comments of all these evaluation indexes constitute a set of comments. If there are m, it is recorded as:

$$V = \{v_1, v_2, ...v_M\} \tag{2}$$

(2) Determine the weight of each factor

As the factors influencing low-carbon economic development differ, so do their roles. The overall assessment result depends, to a certain extent, on the role of different factors on low-carbon economic development [4]. This evaluation indicator has a significant impact on the overall evaluation result.

$$A = \{a_1, a_2, ...a_n\} \tag{3}$$

(3) Constructing fuzzy comprehensive evaluation matrix R

For each evaluation index u, the membership of each evaluation value is a fuzzy subset on the evaluation set v. The evaluation of evaluation index is recorded as

$$R_i = \{r_{i1}, r_{i2}, ...r_{in}\} \tag{4}$$

The fuzzy comprehensive evaluation matrix of each evaluation index is as follows

$$R = \begin{pmatrix} r_{11} & r_{12} & ... & r_{1m} \\ r_{21} & r_{22} & ... & r_{2m} \\ ... & ... & ... & ... \\ r_{n1} & r_{n2} & ... & r_{nm} \end{pmatrix} \tag{5}$$

3.2 Decision Model of Multi-objective Fuzzy Comprehensive Evaluation

When two or more objects need to be evaluated, the method of selecting the optimal object which meets the requirements is called the multi-objective fuzzy comprehensive evaluation decision-making method. The specific process of evaluation: 1. the first level fuzzy comprehensive evaluation model is applied to each object to evaluate; 2. quantify the evaluation set and calculate the priority of each object. The priority of each object is to set the quantitative set of fuzzy comprehensive evaluation comments as Q

$$N_k = B_k Q^I = \left(B_k^1, B_k^2, \ldots, B_k^m \right) \cdot (Q_1, Q_2, \ldots, Q_m)^T \tag{6}$$

4 Comprehensive Evaluation and Countermeasures of Low Carbon Economy

4.1 Comprehensive Evaluation of Low Carbon Economy Development Level

In the comprehensive evaluation, there are three types of generalized fuzzy operators: the main factor determining type, the main factor prominent type and the weighted average type. From the fuzzy evaluation matrix, we can see that the evaluation indexes of low-carbon economy have a certain degree of influence on the rating domain, and the weighted average type is closer to people's experience judgment

$$M = (\vee \cdot)b_x = \wedge \left\{ 1, \sum n = I(\alpha_i, r_{ik}) \right\} \tag{7}$$

The comprehensive evaluation result matrix is obtained

$$B = W \times R = \begin{vmatrix} 0.28 \\ 0.17 \\ 0.22 \\ 0.08 \\ 0.09 \\ 0.12 \\ 0.04 \end{vmatrix} \times \begin{vmatrix} 1.53\% & 23.7\% & 55.8\% & 5.2\% \\ 25.5\% & 34.6\% & 21.8\% & 18.1\% \\ 17.1\% & 22.5\% & 43.2\% & 16.9\% \\ 19.8\% & 27.1\% & 38.3\% & 14.8\% \\ 15.18\% & 50.1\% & 26.13\% & 7.19\% \\ 35.1\% & 46.2\% & 16.8\% & 1.89\% \\ 5\% & 39.4\% & 5.28\% & 50.3\% \end{vmatrix}$$

$$= (19.7432\%, 31.265\%, 36.554\%, 12.3209\%)$$

Then we can get the comprehensive evaluation table of the development level of low-carbon economy, as shown in Table 1.

Table 1. Comprehensive evaluation value of low carbon economy development level

Evaluating indicator	Very high	High	Higher	Low
Comprehensive evaluation value(%)	19.7432%	31.265%	36.554%	12.3209%

4.2 Countermeasures and Suggestions

To develop low-carbon economy, take economic development as the basic premise, low-carbon market as the short-term goal, building a low-carbon society as the ultimate goal, phased implementation as the development path, take the phased transformation path of high carbon, medium carbon and low carbon, and clarify:

(1) Under the requirements of the industrialization of a resource-saving and environ-ment friendly society and "standing in an important position of the modernization development strategy, strengthening the ability to respond to climate change and making new contributions to world climate protection", firmly establish the concept of low carbon, and the 17th generation of leaders of the party will firmly take the road of sustainable development, Formulate the important concept of the 12th Five Year Plan for low-carbon economic development, guide governments at all levels to change the one-sided pursuit of GDP growth, and take low-carbon economic devel-opment as an important goal and strategy for economic and social development during the 12th Five Year Plan. As value.

(2) Vigorously develop low-carbon industry, vigorously develop emerging energy industry, vigorously develop wind power generation, stabilize bioenergy, accel-erate solar energy industry, actively promote the construction of nuclear power generation, cultivate and expand low-carbon demonstration enterprises, parks and economic zones, and accelerate the development of low-carbon industry.

(3) Innovate low-carbon technology and track R & D; D. The development direc-tion of low-carbon technology in developed countries refers to the successful expe-rience of low-carbon technology application, and studies the development status of low-carbon technology as a regional key field; To develop the international com-munity, especially the low-carbon economy, we must strengthen cooperation and exchanges with good countries, actively introduce, digest and introduce mature technologies from advanced countries, and improve our ability to innovate.

(4) Establish a low-carbon market production technology and low-carbon production and living equipment consumption market, innovate financial instruments, cultivate a low-carbon financial market trading system, develop green carbon funds, build a carbon venture capital system, and develop a variety of low-carbon financial financing channels. open.

(5) Building a low-carbon society requires a comprehensive low-carbon production, life and consumption of the whole society, involving everyone's clothing, food, housing and transportation. Social citizens practice their lifestyle and consumption habits with a low-carbon production mode, so as to form a social atmosphere proud of low-carbon.

5 Conclusion

At present, China has entered the era of rapid industrialization. However, under the dual constraints of resources and environment, the extensive mode of economic growth will seriously affect China's industrialization process. To maintain a balance between the requirements of industrialization and environmental energy, developing low-carbon economy is an inevitable strategic choice.

References

1. Shuili, C., Jinggong, L., Xianggong, W.: Fuzzy Set Theory and Its Application, pp. 187–223. Science Press, Beijing (2005)
2. Si, S., Sun, Z.: Mathematical Modeling Algorithm and Application, pp. 375–377. National Defense Industry Press, Beijing (2016)
3. Xihai, Z.: Application of Fuzzy Comprehensive Evaluation in Network Security Evaluation. Nanjing University of Technology, Nanjing (2006)
4. Wang, W.: Technological and economic paradigm and path of low carbon economy development. Yunnan Soc. Sci. **4**, 114–117 (2009)

Cloud Computing in Computer Digital Media Art in Animation Design

Ju Shen[✉]

School of Art, Wuhan Business University, Wuhan 430056, Hubei, China

Abstract. In recent years, with the rapid development of digital media art, the animation industry has entered an unprecedented new field with the support of digital media art. This paper analyzes the development status of animation design industry, and discusses the basic connotation and characteristics of digital media art. The application of digital media in animation design is analyzed.

Keywords: Digital media art · Animation design

1 Introduction

Since the birth of computer technology, it has been widely used in several fields. The development of computer technology has produced many emerging industries, including animation industry. In the process of the development of animation industry, digital media art has been used flexibly.which helps to realize the interaction from static to dynamic and from one-way to two-way, and helps to improve the visual experience of the audience, and ushers in a good development wave of animation industry.

2 The Concept and Characteristics of Digital Media Art

Digital media art is developing rapidly in China. With the support of digital media art, the world animation industry has entered an unprecedented new field.With the emergence of new media means, the professional talents in the animation industry design and production industry should not only have rich professional theoretical knowledge, but also have excellent practical operation ability.

2.1 Concept and Development of Digital Media Art

Xu Peng talks about the new media art theory, which changes the artistic form in creative, carrying, spreading, appreciating and criticizing, and then realizing the artistic feeling, experience and thinking of aesthetics. Computer science and technology, the development of network media and the formation of new art field. Information technology is the necessary premise for the formation of digital media art and an important part of the content and form of digital media art.

J. Macintyre et al. (Eds.): SPIoT 2021, LNDECT 98, pp. 470–475, 2022.
https://doi.org/10.1007/978-3-030-89511-2_61

This paper holds that in the environment of rapid development of science and technology, the development of all technologies must be accompanied by changes in art forms and new art forms. Such changes are closely combined not only in material aspects, but also in spiritual aspects, science and technology, art, culture and aesthetics, and there are several new ideas. New technologies and new technologies are closely combined, and new perspectives promote new art forms into people's lives and bring unprecedented convenience to people. In these technologies and thoughts, a unique new art form of digital media art has been born. It is a product of the combination of science, culture, art, media and other fields. Its emergence broadens the vision of the masses and closely connects our world. Digital media art is a natural science. It embodies the concept of "science, art and humanities", a multi-disciplinary comprehensive discipline integrating social science and humanities. The digital technology in this term reflects the basis of its science and technology. The media emphasized the foothold of the media industry. It defines all application fields such as the creation of artistic works and the artistic design of digital media products [1].

However, all new art forms are new ideas. Driven by new technologies and new perspectives, the development of digital media art presents rapid and powerful vitality and broad development prospects. Digital media art also has such performance. With the development of computer technology, the emergence of digital media art, and the development of computer and digital technology after the Second World War, Digital media art has also begun to take shape.

That is mainly divided into three stages of development. The 1960s is the stage of digital media art which has just begun to develop. Everything is from scratch. The digital media in this stage can only be called digital art. At this time, most artists are constantly exploring, researching and creating in the laboratory to obtain new art resources; The 1970s–1980s is the most significant decade for the development of computer images and 3D computer animation. In this decade, the digital media art has undergone qualitative changes. Computer painting software has begun to appear, making more and more artists begin to use computers, so computer animation and Movie Stunts appeared in the entertainment market at that time; The 1990s is the stage of Internet appearance and multimedia technology maturity. It has diversified forms, rich content and strong expressive power. To some extent, it has surpassed our traditional art. Digital media art has entered all walks of life in this period. Digital media art has entered and affected our entire human society step by step [2].

2.2 Essential Characteristics of Digital Media Art

(1) Technology and digital media art are the products of science and art. Technology is the pillar of digital media art. The form and thought of digital media art works change with the development of science and technology. The development of science and technology is new to the form and thought of digital media art works. When new technologies appear, with the changes of art forms, in a word, digital media art is the perfect combination of rationality and sensibility, and the product of the development of the times.

(2) Interactivity. An important measure of digital media art is the user interaction experience. It is a field of multi-disciplinary and cross-disciplinary technical expression. The interactivity of the work brings uncertainty to the display results of the work, which enhances the curiosity, freshness and exploration spirit of the users.

(3) Artistry. Digital media art has its unique artistic characteristics. Besides the basic attributes of traditional art, its uniqueness mainly lies in "technical art", such as immersion Art, virtual reality art and network art are art forms closely related to technology. With the continuous development of digital technology, digital media art has undergone more changes [3].

3   The Influence of Digital Media Art on Animation Design

Due to its application in the animation design of computer digital art, the effect of film exceeds people's imagination. With the continuous development of film and television, people can use digital art to display visual images and action effects perfectly, which makes the effect of film and television constantly exceed people's imagination.

3.1   Diversified Performance Style

Whether it is new media or mixed media, one of the most obvious features is the diversity of collage. The characters, images, paintings, patterns, prints and images, etc., through digital transformation into digital images, break the limitations of traditional animation media, make the visual effect of the picture more powerful, and strengthen the expressive force and dynamic of the picture.

3.2   Animation Design Enters a New Era of Development

Because the content of digital media is more and more abundant. The production of flash memory tools is the result of the integration of digital media art and film. Also has many advantages, such as good sharing, opening up, and so on. Through the combination of information technology and mass media technology, the mass media will develop towards the direction of "small masses" and "public separation", and its communication mode is more purposeful. Through the advantages of digital art, animation technology has achieved a high market recognition.

3.3   The Content of the Work has More Space for Expression

Due to the limitation of time, material and human resources, the development space of traditional art is limited. Its application in art creation greatly enriches the content of art creation. By using information technology, not only can we successfully display the modern and ancient pictures, but also show the content of film and television from the micro and macro perspectives.

3.4   The Efficiency of Creation has been Greatly Improved

Using digital media art to design animation can improve the creation efficiency of animation. Compared with the creation of traditional animation, this new animation creation technology has many advantages. With the development of information technology, the creation efficiency of animation is higher and higher. The application of digital media art in animation design not only optimizes the animation production process, but also make the promotion and publicity process of animation works more smooth, and bring animation technology to the public view [4].

4   The Application of Digital Media Art in Animation Design

4.1   Application of Digital Media Art in 2D Animation

The birth art has brought new impetus to the design and production of animation. It is widely used in the design and 3D animation, and the picture effect and pattern have been significantly improved. In the design of two-dimensional animation, digital media art has a strong advantage. Animation design, when digital media art is adopted, only the image needs to be processed automatically, and the ideal picture will be automatically generated in the system, and the convenience of color painting is greatly improved, which makes the whole operation simple. In addition, digital media art is characterized by key frames. There are only small (sub-pixel) displacement and rotation between images. Only the motion model between them needs to be estimated. In fact, the original observation data obtained is used in reconstruction. In fact, there are large deformation, scaling, rotation and translation between the image data obtained at different times, Through reconstruction of frequency domain or spatial domain, high (super) resolution data with uniform sampling is generated. The technical flow is shown in Fig. 1.

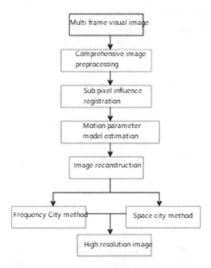

Fig. 1.   High resolution image fusion technology process

4.2   The Application of Digital Media Art in Film and Television Animation

Digital media art is the creation of films and TV dramas, which has been widely used in production and popularization, which not only makes people feel the entertainment, but also promotes the production and dissemination of films and TV. Mainly, the production and dissemination of films and TV are reflected in the development and release, film production, Reduce the film production cost, provide convenient production technology, greatly improve the film production efficiency, and cannot shoot in reality through digital media art. It can display scenes and improve the artistry and visibility of film and television works. With the development of film and television industry, digital media art is more and more widely used in film and television production. Now we can use communication technology to distribute these films at the same time, improve the distribution efficiency of films and provide a stronger guarantee for the interests of films. In addition, in the process of film distribution, through digital film chain, the film can maintain the integrity of image quality in the process of continuous reproduction and maintain the integrity of image quality in the process of film reproduction for a long time. Can be saved. Compared with previous methods, this method not only has new images with high image quality, but also realizes interactivity and provides users with the greatest convenience.

4.3   Adhere to the Development of Animation with Chinese Characteristics

At present, the development level of China's film industry is not high, and there are few excellent animation works. In contrast, there are many excellent foreign animation film and television works, such as European and American countries in animation production resources are relatively rich, and also produced a lot of excellent works. In the process of developing our own animation industry, we can reasonably learn from foreign animation technology, and on this basis, combine animation works. Therefore, in the process of developing animation technology, we should not only ensure that the animation industry is loved by the public, but also combine with the cultural characteristics of our country.

4.4   Pay Attention to the Cultivation of Animation Digital Technology Talents

With the better meet people's needs, the animation industry continues to abandon backward ideas and technologies, make full use of new digital art and improve the diversity of animation works. Therefore, the application of digital media art can bring vitality to the development of animation industry and promote the healthy development of China's animation industry.

5   Conclusion

With the development of digital media art, the production of artistic works has reached a new level, and the methods and means of artistic creation have also changed. Under the background of modern information technology, digital media art plays a more and more important role in the design and production of two-dimensional and three-dimensional animation, digital animation technology, digital media art provides more convenience for animation designers, but in the actual animation design and production process, the requirements for the essence of animation art can not be ignored. I am emerging for this. We must make full use of the artistic design and animation of digital media, organically combine them, promote and develop them on the basis of strictly abiding by the basic requirements of animation.

References

1. Sida, L.: Introduction to Digital Media Art, vol. 07, p. 6. Tsinghua University Press, Beijing (2012)
2. Liruoyan: The characteristics and ontology of digital media art. Art Exploration **1**, 39–41 (2012)
3. Wanliang, W., Shangshang, Z., Yuping, Y., Hanshanshan, Z., Xu, Z.: Interdisciplinary talent training mode of digital media integrating software and art. Teach. Chinese Univ. **03** (2011)
4. Xu, M., Changhe, T.: Seeking the way of cultivating special talents and seeking the way of innovation of education mode - from the construction of digital media art major. Comput. Educ. **03** (2010)

Experiments, Test-Beds and Prototyping Systems for IoT Security

Optimal Tourist Route Optimization Model Based on Cloud Data

Shuqing Xiao[✉]

School of Modern Service Management, Shandong Youth University,
Jinan 250103, Shandong, China

Abstract. In recent years, with the continuous evolution of Internet of Things, cloud computing, mobile Internet, wireless communication network and other technologies, "smart tourism" makes use of Internet and other technologies and portable terminal devices to customize tourism products for tourists and improve users' travel experience. This paper mainly studies the optimal tourism route optimization model based on cloud data. In this paper, a tourism route optimization algorithm based on genetic algorithm is proposed at first. Then, according to the characteristics of cloud computing and cloud data technology, the optimization algorithm model is changed to a parallel genetic algorithm adapted to the characteristics of cloud computing. The effectiveness of the proposed algorithm is verified by comparing the tourism benefit values of the other two algorithms.

Keywords: Cloud data · Cloud computing · Travel routes · Model optimization

1 Introduction

With the rapid development of social economy, tourism not only plays a more and more important role in the national economy and society, but also grows into a force that cannot be ignored in the social, cultural and foreign communication construction. With the development of economy, people's living standard is also improving continuously. People have gradually changed from simple material needs to the pursuit of life experience. In addition, in recent years, the national investment in infrastructure construction continues to increase, and the scope of people's travel activities continues to expand, which also creates a good environment for the development of tourism. While giving full support to the tourism industry, the state insists on expanding the intensity of opening up to the outside world and enhancing the intensity of international economic cooperation, which creates new opportunities for the development of tourism [1]. However, traditional travel planning applications have various shortcomings in actually solving the problem of tourist travel planning, researchers need to use new methods to solve the problem of travel planning. Considering the need to select suitable tourist attractions and transportation arrangements for tourists during travel planning, researchers first need to obtain scenic spot information and transportation information of tourist cities. At present, cloud data provides researchers with real and reliable data and can effectively provide relevant information [2, 3].

© The Author(s), under exclusive license to Springer Nature Switzerland AG 2022
J. Macintyre et al. (Eds.): SPIoT 2021, LNDECT 98, pp. 479–486, 2022.
https://doi.org/10.1007/978-3-030-89511-2_62

The rapid development of the Internet has accelerated the construction of tourism informatization, and tourism information services have developed from the original traditional artificial paper to the later electronic and digital [4]. Informatization has brought revolutionary changes to the development and value of tourism again and again. From the perspective of travel service platforms, after travel service providers such as hotels, airlines, and travel agencies transfer their original offline travel service business to the Internet, a single travel service platform has been formed, subsequently, one-stop service platforms such as food, lodging, travel, travel, shopping, entertainment, etc. have flourished everywhere at home and abroad, all of which depend on the support of network technology [5]. The progress of science and technology, make the information transfer and service experience is becoming better and better, therefore, tourism service providers also have generous returns, at the same time, tourism also has made a great contribution to the country's economic development, the rapid development of tourism industry and tourism information is inseparable, the evolution of the tourism industry closely related to the key high and new technology, it has become an important tool and way to strengthen tourism competition [6].

At present, what needs to be solved in tourism planning is the comprehensive problem of the coverage of interest points and route planning. However, the research in this aspect is insufficient in terms of the quality and efficiency of solutions. Therefore, how to effectively plan users' travel routes based on their personalized preferences and multiple resource constraints is the main research direction of this paper.

2   Tourism Route Optimization Model Based on Cloud Data

2.1   Route Optimization Based on Genetic Algorithm

In the greedy heuristic algorithm based on the shortest path and the greedy heuristic algorithm based on the scoring time ratio, the greedy strategy is adopted as the point of interest search strategy. However, the greedy algorithm always makes the best choice based on the current state and does not take into account the overall optimal case of the travel path. Therefore, this paper proposes a genetic heuristic algorithm based on maximum benefit value. Compared with the greedy strategy algorithm, the genetic heuristic algorithm can cover a larger number of candidate paths, reduce the risk of being trapped in local optimization, and is more conducive to the selection of the optimal solution from the global perspective [7].

In the genetic heuristic algorithm based on the maximum benefit value, there are mainly the following four steps:

(1) Initialize the population.
    The first step of the algorithm is to initialize the genetic population and generate individuals randomly in the initialized genetic population. When the tourists specify the set of interest point types, they first randomly select one interest point from each interest point type, add the starting point and the destination, and then combine it into a new individual. The same work is then repeated to construct an initialized genetic population of sufficient size.

(2) Constructing evaluation function.

After the genetic population is established, the evaluation function is constructed in the second step of the algorithm to calculate the individual fitness of the population. According to the relevant factors that affect the value of travel benefit, the following functions are adopted as the evaluation function.

$$f_i = \frac{s(tp_i)^\lambda}{t(tp_i)} \tag{1}$$

Where, TPI is a certain tourist route, S represents the quality score of the points of interest of the route, and T represents the travel time of the points of interest of the route. The parameter λ is used to adjust the proportion of the influence of line quality and line time on the travel time. The higher λ is, the more the line quality can affect the size of fi.

(3) Selection, crossover and variation.

It is an important step of the algorithm to generate individual population with higher fitness through crossover and mutation. According to the evaluation function in Step 2, individuals with high fitness were selected for crossover and mutation. In the process of individual crossover, two individuals with high fitness are randomly selected, the location of crossover is randomly selected, and then the first half of an individual is combined with the second half of another individual to generate a new individual. Individual cross section, select the part stage between the type of the third interest point and the type of the fourth interest point, and split and combine the two individuals. In the process of mutation, an individual with high fitness is selected randomly in the population at first, and then the position requiring variation is selected in the individual. Since the types of interest points in the same position are the same, new interest points are selected from the same type of interest points in the mutation process to complete individual mutation.

In each generation of inheritance, crossover and mutation behaviors occur with their own probabilities. Through continuous inter-individual crossover and individual variation, a sufficient number of population size can be generated in each generation of population, and then individuals with low fitness can be removed from the population and new individuals can be selected for inheritance of the next generation [8, 9]. Considering the actual genetic process, the probability of individual crossover is generally greater than the probability of individual variation. Therefore, in this algorithm:

$$P(cross) + P(mutate) = 1, P(cross) > P(mutate) \tag{2}$$

P (cross) represents individual crossover, and P (mutate) represents the probability of individual variation.

(4) Genetic termination.

According to the limitation of genetic effect and time, the running algebra of genetic algorithm needs to be limited. After satisfying the running algebra, the algorithm selects the optimal solution in the genetic population as the travel planning path.

2.2 Route Planning in Cloud Data Environment

(1) Data processing in cloud computing environment.

In the process of using cloud computing technology, the number of users increases rapidly, so the difficulty of data processing in the cloud will increase accordingly. Therefore, this paper has to find a technology to ensure user satisfaction [10].

Map Reduce is the typical technology proposed by Google to solve this problem, which guarantees the processing of large data sets. First of all, it is conducive to his parallel programming mode. Developers only need to write the Map function and Reduce function according to their own development requirements, and then they can complete the operation and processing of large data sets. At the same time, it also provides a convenient and quick interface for developers [11].

In this paper, Baidu's Map Reduce (BMR) is used for data analysis. BMR is a fully hosted Hadoop/Spark cluster that builds and deploys the environment according to the user's needs. Users only need to focus on the data processing reports generated by the system, rather than on the process.

BMR has many advantages: one is the ability of automatic operation and maintenance, the creation of the cluster is no longer based on the personal ability of the developer, but automatically created, the service time is humanized 24x7; second, the security of the data is high, BMR adopts the form of private network, and is no longer the common use of public management, so as to escort the user's data security information; third, the management interface is clear and clear, which facilitates the user's information view and task management; fourth, the flexible pricing method takes the form of charging on demand to ensure the cost expenditure of users [12].

The implementation of business logic layer adopts Spark framework, which has excellent computing speed and compatibility. For the order information with a large amount of data, it can be calculated and processed quickly. And the service layer is to provide external API computing interface, convenient for users to use and view.

(2) Design of genetic algorithm in cloud computing environment.

The solving process of the improved model in this chapter is as follows: the population is divided according to the number of processors selected, which is decomposed into MXN, the number of small groups is m, and the number of individuals in each small group is n. Corresponding to this m a small groups, each group separately to iterate, iteration to set the number of times for the exchange of small groups of excellent individual operation, ensure the diversity of population and good sex, make the quality of the solution in the whole solution space as much as possible, to a certain extent to jump out of local optimal algorithm, to find the global optimal. And the m subgroups correspond to the number of processors we select, so the more processors we select, the smaller the group will be, and the less difficult the task will be. In each processor, the process of genetic iteration is carried out in parallel with each other. In other words, the selection operation, crossover operation and mutation operation are carried out independently, and then the selection of the next generation population is carried out through the calculation of the fitness function value.

The master-slave model, as the name suggests, divides the processor into two parts: the master and the slave. The former is mainly to control the large undecomposed population, and is a large main cycle, responsible for the overall monitoring and calculation. The latter mainly carries out genetic operation on each individual in a small population, and feeds back the final result to the former through the calculation of fitness value and the selection of individuals. This model is the most direct way to parallelize the genetic algorithm. Its advantage is that it can effectively reduce the communication workload, but its disadvantage is also obvious. The imbalance between the two processors is inevitable and needs to be improved.

Fine-grained models require more processors, because they need to occupy a processor for each individual in a small group. These processors will calculate fitness values and select individuals, which have high requirements on communication ability, and are generally used in practical cases with a large scale.

Coarse-grained model is the improvement of master-slave model and fine-grained model, which divides the population, assigns tasks to the processor, and the processor will carry out independent genetic operation and select excellent individuals according to fitness value. In order to make up for the problem of high demand for communication volume, in the process of small group iteration, the migration operation of excellent individuals is set artificially, so that the program running time and the solution result are significantly improved, and the situation of falling into local optimum is also greatly reduced.

3 Simulation Experiment Settings

3.1 Data Sources

This paper collects and sorts out the specific information of 300 points of interest in a location service website and classifies the 300 points of interest into 10 types of points of interest. In terms of the type attribution of interest points, this paper classifies each interest point to a certain type of interest point, that is, the same interest point belongs to a certain type of interest point type set. When tourists are planning their travel routes, they only need to give specific starting points, destinations and select an ordered set of interest points from a given set of interest point types.

According to the collection of interest points in Foursquare, the interest points in the collection are substituted into the Amap open platform to collect the time information between the interest points. In particular, the algorithm does not need the path between specific points of interest in calculation, and only needs to judge the path according to the time information. After the final result is planned according to the time information, the specific travel route is provided according to the result.

3.2 Experimental Environment

The experimental environment of this paper is Win10 64-bit operating system, based on x64-bit processor. The processor is Inter(R) Core(TM) i7–10400 CPU@ 2.90GHz, and the memory is 16.00GB. All the simulation experiments in this paper use Eclipse software to run three travel route planning algorithms. Among them, Algorithm 1 and Algorithm 2 respectively represent the greedy algorithm based on the shortest path and the greedy algorithm based on the scoring time ratio.

4   Simulation Experiment Results

4.1   Relationship Between Travel Benefit Value and Travel Interest Point

As shown in Table 1 and Fig. 1, when the number of interest points keeps increasing, the travel benefit value of the travel paths searched by the three algorithms keeps increasing. This is because when the number of interest points increases, the search scope of the algorithm is larger, which is more conducive to searching for a better path. Among them, the travel path planning based on algorithm 1 has the lowest benefit value, while the greedy algorithm based on algorithm 2 is better than the previous one. Based on this paper, the genetic algorithm planning result under cloud data is the best. As can be seen from the figure, when the number of points of interest is greater than 200, the travel benefit value of the route planned by the three algorithms tends to be stable.

Table 1. Travel benefit value/travel interest point relation chart

	100	150	200	250	300
Algorithm 1	10.3	10.9	13.2	13.8	14.2
Algorithm 2	10.7	12.8	18.6	18.9	19.3
Our algorithm	11.1	13.4	20.7	20.7	21.1

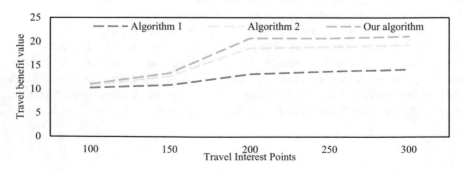

Fig. 1. Travel benefit value/travel interest point relation chart

4.2   Relation between Travel Benefit Value and Designated Point of Interest

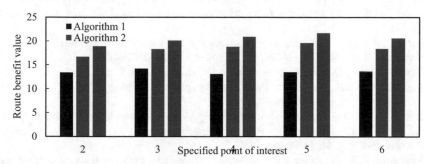

Fig. 2. Graph of travel benefit value/number of types of designated points of interest

Figure 2 shows the calculation results of the travel benefit value of the planned route when the types of interest points specified by the three algorithms change. It can be seen that the benefit value of the travel route planned by the first algorithm is the lowest, which is generally lower than 15. The second algorithm has a significant improvement, and the third genetic algorithm based on cloud data has the highest travel benefit value for planned routes. When the types of designated points of interest of tourists change, the benefit value of the planned route of the third algorithm is higher than that of the first two algorithms.

5   Conclusions

Traveling route planning is a kind of service that selects suitable travel routes for tourists according to their individual needs. When tourists give different restrictive conditions, travel route planning needs to search the most suitable travel route for tourists under the condition of meeting the restrictive conditions. This paper runs the improved genetic algorithm based on cloud computing technology, and verifies the effectiveness of the algorithm under cloud computing technology and the help of cloud computing technology to the running speed of the algorithm. After comparing the efficiency of the three algorithms with different number of interest points in the spatial database and different number of tourists' designated interest points, it is proved that the proposed algorithm can plan the best travel route for tourists.

References

1. Peng, L., Chen, Y.: Tourism safety monitoring information service system based on internet of things and block-chain. J. Intell. Fuzzy Syst., 1–7 (2021). https://doi.org/10.3233/JIFS-219101
2. Guo, Q., Jian-Xia, L.I., Yu-Ming, W.U.: Evaluation of intelligent tourism information service capabilities: A case study of Hanzhong City. Ecol. Econ. **01**, 4–15 (2017)

3. Schegg, R., Stangl, B. (eds.): Information and Communication Technologies in Tourism 2017. Springer, Cham (2017). https://doi.org/10.1007/978-3-319-51168-9
4. Marchenkov, S.A., Vdovenko, A.S., Korzun, D.G.: Enhancing the opportunities of collaborative work in an intelligent room using e-tourism services. Tr Spiiran **1**(50), 165–189 (2017)
5. Kim, T.: A Study on the Security Technology of the Location based Tourism Information Service. J. Korea Soc. Digital Ind. Inf. Manag. **12**(2), 25–29 (2016)
6. Wei, C., Wang, Q., Liu, C.: Research on Construction of a Cloud Platform for Tourism Information Intelligent Service Based on Blockchain Technology. Wirel. Commun. Mob. Comput. **2020**(2), 1–9 (2020)
7. Chung, B.D., Kim, B.S.: A hybrid genetic algorithm with two-stage dispatching heuristic for a machine scheduling problem with step-deteriorating jobs and rate-modifying activities. Comput. Ind. Eng. **98**, 113–124 (2016)
8. Sant'Anna, L.R., Filomena, T.P., Guedes, P.C., Borenstein, D.: Index tracking with controlled number of assets using a hybrid heuristic combining genetic algorithm and non-linear programming. Ann. Oper. Res. **258**(2), 849–867 (2016). https://doi.org/10.1007/s10479-016-2111-x
9. Kaweegitbundit, P., Eguchi, T.: Flexible job shop scheduling using genetic algorithm and heuristic rules. J. Adv. Mech. Design Syst. Manuf. **10**(1), JAMDSM0010-JAMDSM0010 (2016)
10. Li, F., Qu, Z., Li, R.: Medical cloud computing data processing to optimize the effect of drugs. J. Healthc. Eng. **2021**(9), 1–15 (2021)
11. He, Y., Zhang, X.: Key technologies of massive concurrent data processing in smart city based on cloud computing. Int. J. Comput. Appl. **4**, 1–8 (2020)
12. Luo, M., Zhang, L.-J. (eds.): CLOUD 2018. LNCS, vol. 10967. Springer, Cham (2018). https://doi.org/10.1007/978-3-319-94295-7

Practical Application Exploration of Animation Design Based on Short Video Technology

Yu He[✉]

Huizhou Economics and Polytechnic College, Huizhou, Guangdong, China

Abstract. With the rapid development of the mobile Internet, human society has entered a new era of visual communication. As an important visual communication method, short video has gradually become popular with the development of the mobile Internet and is loved and sought after by the general public. Based on the background of the Internet era, this article analyzes the development status of short video technology in the practice of animation design, unique advantages, the relevant situation of short video animation consumers, and the age distribution and occupation distribution of short video animation fans. Therefore, the innovative strategies and application measures of animation design are proposed.

Keywords: Short video technology · Animation design · Short video · Application exploration

1 Introduction

With the rapid development of the mobile Internet, human society has entered a new information age. In this fast-paced information society, the way the public receives information is gradually replaced by color images and videos. The public began to use graphics and images to identify the world, clarify opinions and convey emotions [1]. In the context of the new era of visual communication, short video technology, as an important visual communication method in the era of visual communication, has gradually become popular with the development of the mobile Internet, and has been loved and sought after by the general public. Short video technology often brings joy and relaxation to busy people with a new attitude, and gradually breaks the original state of the animation industry; at the same time, short video technology overcomes the limitations of traditional animation, and brings new development opportunities to the animation industry [2]. The animation industry needs to broaden the audience's positioning, improve core technology research and development, deepen the local cultural imprint, create high-quality animation IP layout pan-entertainment, and develop micro-anime models to enhance comprehensive competitiveness, form a competitive advantage, and create a unique Chinese animation brand. Enhance the soft power of Chinese animation [3].

J. Macintyre et al. (Eds.): SPIoT 2021, LNDECT 98, pp. 487–495, 2022.
https://doi.org/10.1007/978-3-030-89511-2_63

Chinese scholar Lei Ao analyzed the short video function in animation software, understood the current situation of short video function, and explored the possibility of increasing user enthusiasm and stimulating the function to a more personalized and healthy system. Through in-depth comparison and research and analysis, it is concluded that it is necessary to strengthen the social diversification of short video functions, reduce the complexity of user operations during creation, and combine product personality with user behavior characteristics to create animations that conform to product personality [4]. Qianhong Meng pointed out that in the context of fragmented short video technology, social groups and time are being segmented, and the animation industry is facing new opportunities. Universities are important production channels for industry talents and need to actively adapt to the market environment. For this reason, college animation education with the characteristics of animation size should innovate teaching methods, expand teaching content, expand the scope of courses, improve the content of practical activities, and further establish a new type of talent training mechanism [5].

Although text and images can convey information, they have no dynamic content records. Video can not only convey a large amount of information in a timely and fast manner, but also present the development and changes of real-time content in front of the audience. The short film not only contains images, text and speech, but also combines these expressions in a dynamic way [6]. This concise and clear form of content expression can meet the needs of public expression in all aspects and bring people's performance and sharing. Based on the background of the Internet era, this article analyzes the development status of short video technology in the practice of animation design, unique advantages, the relevant situation of short video animation consumers, and the age distribution and occupation distribution of short video animation fans. Therefore, the innovative strategies and application measures of animation design are proposed.

2   Short Video Technology and Animation Design

2.1   Basic Characteristics of Short Video

Short video is a new type of video that can be completed within a few seconds and relies heavily on mobile smart terminals to complete instant recording and editing, and can be directly shared on social media platforms. When creating short videos, the length of most short videos is in seconds, and most of them are within 1 min [7]. Later, as short videos were accepted and used by more and more viewers, the duration of the videos also increased. Currently, short films in China are basically less than 20 min in length. No need to rely on professional equipment, and not limited by time or place. Relying on smart terminal equipment, you can download, edit and upload immediately. Once a short video is released, it is easy to be liked, followed and shared by the audience due to its innovative content and interesting pictures, and its social functions have been further enhanced [8]. For short videos, the short duration, low traffic consumption and instant viewing capabilities make users very satisfied. Short recording time and low data consumption are the number one reason for the rapid spread of short videos. In addition, video playback does not need to be based on a dedicated video site, and can be viewed at any time on any social platform or information platform, which is conducive to the looping of small videos and creates a good environment for their rapid dissemination. This is the rapid dissemination of short videos. The second largest reason.

2.2   The Unique Advantages of Short Video Technology in Animation Design

Short image and text editing can be viewed as an animated version with images and videos. The requirements for making animation videos are also very low for audiences who like animation. They only need to have basic cognitive skills to understand the production methods. It is different from pure animation production. The audience is required to have professional production capabilities for the entire animation production, and the production technology is not easy to accept. Therefore, short animated videos are very popular among the public, and are more suitable for the fast-paced lifestyle of modern people, and can emotionally resonate with the audience [9]. In recent years, the rapid development of short video animation is inseparable from the update and upgrade of short video applications. The various short video software in the app store can not only be downloaded with one click, but also can you can also customize filters according to personal preferences, add text and music when downloading to create personal short videos, and encourage users to continuously extract and generate new content as needed. In terms of spreading influence, since the production of short video animation content is mainly for individual users, the number of creations is very large, and those short videos with more than one million views often have high-quality original content and interesting pictures. This is also the key to whether short video animation can be recognized by the public. In this era of panoramic entertainment, no audience is willing to be a pure audience. What they need most is the feeling of display and being paid attention [10, 11]. The audiovisual effects of short video animation are eye-catching, expressive and attractive. Compared with other types of videos, short video animation uses a large number of auxiliary elements to convey the atmosphere, enhance emotional background music, etc., and combine it with warm and healing The story immediately brings the audience into this deep emotional communication.

2.3   Animation Design Combined with Artistic Expression of Short Video

The artistic level and influence of short video animation determine the popularity of short video animation. Short video animation should show as much as possible various phenomena in social life, resonate emotionally with a certain group, and bring warmth and care to them. As a new form of video, short video uses the self-media platform as the main communication channel. It satisfies the audience's reading style and psychological needs in the new media environment, and has obvious social networking attributes. Each issue of short video animation has a specific theme. Love, emotion and friendship in life can be output and shared in the video, which is the charm of short video. As a short video animation, it is necessary to be excellent in the light and shadow of the picture, but also to be smooth in narrative expression. No matter how good the animation design is and how coveted, if the work does not have impressive ideological connotations, it should not be regarded as excellent animation. Whether it is a short video animation or an anime TV program, it is through the video content to communicate with the audience. A good creator will definitely make the video unique, because letting the audience watch is not the fundamental purpose. The most important thing is to communicate with the audience spiritually; at the same time, the sense of identity is satisfied. The biggest difference between online short video animation viewers and traditional animation program viewers is whether they have the initiative to choose. Viewers watching short video animations can choose what they want to watch, and only short video animations with real emotions can win everyone's favor.

3   Research and Analysis of Short Video Animation Fans

This article randomly selects 200 short video animation enthusiasts and investigates their individual attributes such as age and occupation. The relevant analysis results are now presented as follows:

3.1   Age Distribution of Short Video Animation Enthusiasts

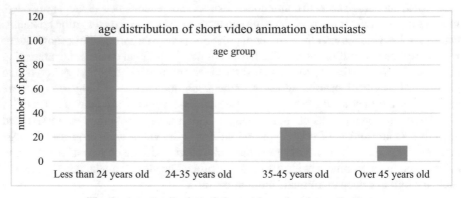

Fig. 1. Age distribution of short video animation enthusiasts

According to Fig. 1, among the 200 short video animation enthusiasts, 103 are younger than 24 years old, accounting for 51.5%; 56 are between 24–35 years old, accounting for 28%, and 28 The age is between 35–45 years old, accounting for 14%; 13 people are older than 45 years old, accounting for 6.5%. Through the analysis of the data, it can be known that most of the short video animation enthusiasts are people under the age of 24. There are also many short video animation enthusiasts in the 24–35 year old youth group, but middle-aged people over 35 years old are not. I don't really like watching short video animations. Therefore, as the age increases, the number of short video animation fans will gradually decrease, and short video animation consumption objects should be under 35 years old.

3.2 Professional Distribution of Short Video Anime Lovers

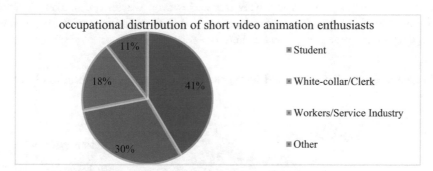

Fig. 2. Occupational distribution of short video animation enthusiasts

According to Fig. 2, among the 200 short video animation enthusiasts, 83 occupations are students, accounting for 41.5%; 61 occupations are white-collar workers/employees, accounting for 30.5%; 35 occupations are workers/. The service industry accounted for 17.5%; 21 occupations were other occupations, accounting for 10.5%. By analyzing the data, it can be known that the majority of short video animation fans are students and white-collar workers, while workers and short video animation fans engaged in the service industry are relatively few. The reason may be that workers and groups engaged in the service industry are busy with their work and leisure time. less. Therefore, when using short video APP algorithm to promote animation, students and white-collar workers should be regarded as accurate users.

3.3 Gender Distribution of Short Video Animation Lovers

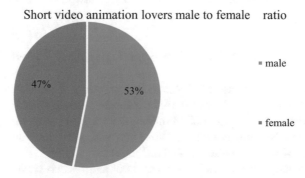

Fig. 3. Gender distribution of short video animation fans

According to Fig. 3, among the 200 short video animation enthusiasts, 106 are men, accounting for 53%; 94 are women, accounting for 47%. Through the analysis of the data, it can be known that the gender gap among short video animation fans is not large. Male short video animation fans are slightly higher than women by 6%. This data shows that the male otaku group is relatively large. However, the female market cannot be ignored, so how to better take into account the common preferences of the two genders when designing anime images and content is a very key point.

3.4   Distribution of Educational Background of Short Video Animation Fans

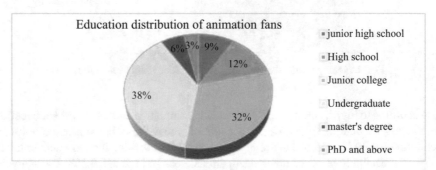

Fig. 4. Educational background distribution of short video animation enthusiasts

According to Fig. 4, it can be seen that the educational background of 200 short video animation enthusiasts is distributed. 18 are junior high school students, accounting for 9%; 24 are high school students, accounting for 12%; 64 are junior high school students, accounting for 32%; 76 are undergraduates, accounting for 38%; 12 are graduate students, Accounted for 6%; 6 people are PhD or above, accounting for 3%; through data analysis, the proportion of undergraduates among the fans of animation short video is the highest, followed by junior college students. Decrease gradually. Therefore, short videos should be designed based on the interests of undergraduates and junior college students when designing anime.

4   Animation Design Innovation Based on Short Video Technology

Driven by innovations in industry fields such as new technical means, creative methods, platform carriers, etc., in recent years, the field of animation design in China has promoted diversified exploration in the field of animation design with pioneering and innovative development ideas. Not only has more practice in traditional animation design methods, but also researched new animation design techniques.

Through the early investigation and analysis of short video crowd, the short video market survey has a clear target. Therefore, the short animation short video application crowd is mainly students and white-collar workers under the age of 35. Affected by the pressure of study and employment, it will virtually cause emotional vacancies for students and white-collar workers who education background is undergraduate and junior college. Because of the relatively high level of education, there are certain requirements for the art and connotation of animation design, and how to design a form that both sexes love in the direction of gender is an important design parameter. Based on the above information, it can be inferred that in the animation design, the cute and soft animation images are more in line with the needs of the above-mentioned consumers.

4.1 Emotional Appeal of Animation Design

If you simply pursue efficiency in animation design, it will appear very impersonal. Therefore, it is necessary to create a relaxed and happy mood and environment in terms of emotional needs, which not only helps to reduce stress, but also makes it easier to awaken the audience's empathy, form a kind of "positive energy" feedback, and enhance mutual emotional connection and sense of belonging. Therefore, insisting on emotional design as an animation design concept can not only effectively enhance the fun, but also enhance the interaction with consumers.

4.2 Animation Modeling Design Innovation

In the well-known IPs in short videos, such as the images of pig little fart, Ari, Yichan monk, etc., in the animation design, dots, lines and planes are used to form objects. When you want to express cute things, the contrast between curved lines and tough lines will make more use of soft lines, so that the finished product designed will be more lovely, and the shape will use more spheres instead of cubes, and chamfers. The relationship with the right angle is also worth noting. After being chamfered, many shapes will have a lovely effect, which makes adults more protective. Designing the attributes of cuteness, the head-body ratio plays a very important role in the overall shape. From the psychological feelings of the audience and creators, it mainly reflects the gradual and emotional aesthetics of the configuration. The key lies in the head-body ratio. Items with smaller body proportions will look more cute. At the same time, it is necessary to consider the relationship between the height of the main character and the supporting role. Different heights, different body shapes and ratios are the manifestation of personality differences, so that different groups of people have a stronger sense of substitution.

4.3 Animation Color Design

The contemporary American visual art psychologist Bloomer believes: "Colors evoke various emotions, express feelings, and even affect normal psychological feelings." The color design of animation characters is an important factor that impresses consumers. The beauty, purity, and purity of animation characters are Highly saturated colors can create fashion, personality, cuteness and full of vitality. The emotion of color is reflected in people's preference for high-purity, high-brightness colors, and it can also be reflected in the different interpretations of color emotions by people of different genders and different cultural backgrounds. For example, in China, people prefer warm colors because they are warm and festive. The saturation and lightness are increased on the basis of warm colors. The colors are mainly red, which represents passion, and the new color system supplemented by blue, can make The audience will have more pleasure after watching.

4.4   Animation Scene Design

Animation scenes refer to elements that create situations other than animation characters, such as scenes, decorations, light and shadow, etc.. Animation scenes can be used to create an atmosphere that people can quickly substitute, such as the Buddhist epiphany of a Buddhist monk, which can nourish people in such a situation. The type of plot determines the style of animation works to a certain extent, and the style can be reflected in the background and character shape. The style of the character and the background should be unified to make it easier for the audience to accept. The props that the character needs when necessary should have dual characteristics, one is the external simple functional props, and the other is the externalization of the character's personality. In order to narrow the distance with the user and form a relaxed and interesting atmosphere, appropriate addition of props Enliven the atmosphere of the picture, enrich the activities between the audience and the cartoon characters.

4.5   Animation Text Design

How to make cartoon cute fonts? First, let's analyze the representative Chinese character " 永", and divide it into horizontal, vertical, abbreviated, napped, tick, lift, and dot strokes of Chinese characters. You will find that the basic strokes of Chinese characters are covered by this "永". From another point of view, strokes in font design are actually a combination of basic stroke structure and modeling. Similarly, when designing cute fonts, the solution is to build the basic stroke structure, and the method adopted is "first square and then round". First use a square to build a stroke structure and then gradually round it. Secondly, in the process of rounding, adjust the center relationship, strokes and local details of the entire font, and finally get the desired cute and soft effect.

5   Conclusions

Short video is a short and powerful medium, which can effectively fit the characteristics of animation design. It can not only help animation companies to vividly display animation products in a short period of time, but also spread animation brands at the fastest speed and optimal effect., So as to gain more recognition from potential customers.

Acknowledgments. This work has been supported by Teaching reform and practice of productive training in schools based on modern apprenticeship——Taking the animation design and production major of Huizhou Economic Vocational and Technical College as an example (project number: 2020hzzjkt24).

References

1. Tao, G.: Analysis on the application of short video shooting skillless transition technology. Digit. Des. **009**(002), 238 (2020)
2. Lili, X.: The expression of graphic art style in animation video production. Res. Commun. Power **3**(072(24)), 42–43 (2019)

3. Sansan, S., Xinyue, Y.: Research on countermeasures to improve the competitiveness of my country's animation industry. J. Northeast Normal Univ. (Philos. Soc. Sci. Edn.) **1**, 183–188 (2019)
4. Lei, A.: comparative research on short video functions of animation app products. New Media Res. **6**(125(09)), 45–47 (2020)
5. Qianhong, M.: Exploration of college animation teaching in the fragmented context. Educ. Modernization **5**(28), 126–127 (2018)
6. Yan, L.: The creative way of "Stationery Story" animation short video. Art Grand View **63**(27), 83–84 (2020)
7. Yang, F.: MG animation short video and the spread of meteorological science knowledge. Sci. Technol. Commun. **11**(249(24)), 192–193 (2019)
8. Mengmeng, H.: Research on the application of digital special effects in film and television animation production. Digit. User **25**(014), 281 (2019)
9. Dejie, L.: Exploring the role of video animation art in teaching. Knowl. Libr. **457**(09), 124–124 (2019)
10. Le, Z.: On the application of digital special effects in animation production. Ind. Technol. Forum **018**(006), 59–60 (2019)
11. Yikun, S.: Research on the dubbing and commentary of the public account of short animation videos. News Res. Guide **10**(04), 85–86 (2019)

The Multi-module Professional Service Training Platform Based on the Internet and Information Technology

Hua Xin, Jing Xu, and Zhongnan Wang[✉]

College of Humanities and Science of Northeast, Normal University, Changchun, China

Abstract. With the emergence of a new generation of information technology, new solutions have been brought to different scenarios in the elderly care industry, and the service content and methods have also become rich and advanced with the advancement of technology. The purpose of this article is to study the construction of a multi-module professional service training platform with home care as the core in the information age. Explains the current domestic and foreign research progress in home care for the elderly; for the home care services for urban retired elderly, analyzes the credibility measurement of the current home care services; explores the application of cloud computing technology to home care services, In-depth analysis of the advantages of introducing cloud computing in intelligent home care services. Taking the elderly in this city as the survey object, through interviews and questionnaires, the operation effect of the multi-module professional service training platform with home care as the core is analyzed, to find out the reasons that affect the operation effect of the service, and to learn from the beneficial international experience On the basis of this, it discusses the countermeasures to improve the service level and path of home care for the aged.

Keywords: Home care · Training platform · Multi-module specialization · Information age

1 Introduction

The aging of the population continues to escalate, and the demand of the elderly population continues to increase, which puts forward higher requirements for social elderly care services [1, 2]. Among the three models of home, institution, and community care for the elderly, the home care model more highlights its own vitality and advantages [3, 4]. Under the background of a healthy China and an aging population, this project focuses on home care, aims to improve the quality of life of the elderly at home, and starts from meeting the elderly's personalized and diversified care needs and services. Internet+ thinking and big data technology support, build an education, teaching and senior care training simulation platform that matches the needs of home care services, build a service and management team, establish service standards and specifications, cultivate professional senior care service personnel, and deliver services for community senior care the team provides decision support for the government [5, 6].

J. Macintyre et al. (Eds.): SPIoT 2021, LNDECT 98, pp. 496–503, 2022.
https://doi.org/10.1007/978-3-030-89511-2_64

With the aging of our country's population, the elderly population will continue to increase, and the concept of home care for the elderly is still the mainstream. It is very important and urgent to build and improve the home care service system [7, 8]. Wu MP combines the practice of home care services to establish an embedded home care service framework analysis, and builds an urban home care service model, service objects and service platforms from three aspects: the interaction between the main body supply and the service platform, and the interaction between the main service and the service platform. The embeddedness, the interaction between the supplier and the service object [9]. Stome L N promotes performance evaluation through the analysis of the third-party evaluation results, thereby promoting the government's purchase of home care services. Use Stata 13 to analyze and process data. The evaluation of the purchase of home care services by the B city government shows that the satisfaction of home care services paid by the government is high, but the degree of awareness is low, indicating that there is a cognitive bottleneck in the service process and provision process [10].

"Smart elderly care" has begun to use Internet of Things technology in many fields in China, and has been implemented as a new type of telecommunications communication business. This article focuses on home care for the core information age, and studies the construction of multi-module professional service training. The elderly care system platform, and the concept of credibility is introduced to evaluate the home care platform services. Based on the questionnaire survey, it deeply analyzes the service modules, maintenance modules and education and training modules in the construction of a multi-module professional service training platform with home care as the core, and measures the reliability of home care.

2 Construction of a Multi-module Professional Service Training Platform with Home Care as the Core

2.1 Multi-module Professional Service Training Platform for Home Care

The system design should fully consider the system progress and classification, adopt B/S structure, DONET framework technology and SOA framework design concept, this structure has started data mining technology, actively analyze the needs of the elderly, and combine advanced communication technology to make the whole system Become a unified platform.

Modularization is beneficial to improve the degree of knowledge coding, can transform individual local knowledge into unique knowledge, and improve the degree of structure of personal knowledge. Professional service industry can improve knowledge network structure and promote knowledge sharing through knowledge coding. The specific contents of the three modules included in the entire platform are education and training modules, service modules and maintenance modules.

Education and training module: build an education and teaching training platform supported by big data technology. The platform provides data support for education and training of medical and health care, rehabilitation training, physical examination and nursing, spiritual comfort (psychological assistance), special populations, nursing care services, and old-age health technology.

Service module: Build data services and data management for day care, economic support, life care, housekeeping services, and legal aid supported by big data technology.

Yiyang module: Constructing data services and data management for the education and practice of leisure and entertainment, art fitness, calligraphy, painting and other art forms supported by big data technology.

2.2 Credibility Measurement of Home Care Services

Trustworthiness is the degree of user trust in the service, that is, a subjective feeling formed by the user in the process of using the service. If the service behavior is always consistent with expectations, it is credible. The concept of credibility is introduced to evaluate the home care platform services, which provides a guarantee for the elderly users to choose the home care service platform.

The home care service provided by the home care service platform is S, and its published QoS attribute set is a(a1, a2, ...an), where n is the number of attributes, and the preference set of the home care service user Req is w(w1, w2, ...wn), where w1 is the normalized value of preference. The home care service provider initially publishes it on the platform. The credibility of the home care service without usage records is the initial credibility:

$$T_{init}(Req, S) = \sum_{i=1}^{n} w_i a_i \tag{1}$$

According to the historical records of direct contact between home care service users and home care services, credibility is the direct credibility derived from multiple historical interactions between home care service users and home care services.

If the home care service user interacts successfully with the home care service, the credibility is as follows:

$$T_x = 1 - \left(\sum_{i=1}^{n} \frac{w_i (p_i - a_i)^2}{n} \right)^2 \tag{2}$$

Conversely, if the home care service user fails to interact with the home care service, the credibility of the punishment:

$$T_y = -p/T_x \tag{3}$$

3 Experiment on the Construction of a Multi-module Professional Service Training Platform with Home Care as the Core

This paper uses the questionnaire survey method to investigate the elderly in this city, and obtains the whole content of the three modules of the specific service items of the home care service, namely, the education and training module, the service module, and the maintenance module. And using SPSS software for statistical analysis, using a combination of quantitative and qualitative methods to determine the level of importance of the home care service project system, which has important guiding significance for the orderly development of home care service business. The survey adopts a snowball sampling method, using offline and online methods to directly survey the elderly or indirectly through the elderly's family members and managers who provide home care services to investigate the evaluation information of the elderly on the importance of home care services. A total of 180 questionnaires were distributed. Of which 150 valid questionnaires were collected.

4 Experimental Research on the Construction of a Multi-module Professional Service Training Platform with Home Care as the Core

4.1 Service Module

There are two "very important" items in the survey results of the service party and the provided service party on day care, economic support, life care, housekeeping services, and legal aid in the service module: day care and economic support. Accounted for 43% and 33%. Generally speaking, the elderly pay more attention to physical health, and the elderly at home respond positively to physical exercise. Due to the limited economy, they have demand for venues and sports facilities, and they are more recognized for the collective activities of the morning team.

The service module can centrally manage the information subscription of each elderly person, and can make accurate service information subscription recommendations for each elderly person according to the elderly's attention to various types of information, so that the service information seen by the elderly is personalized and Suitable for your own situation.

4.2 Maintenance Module

Among the survey results on the education and practice of leisure and entertainment, art fitness, calligraphy, painting and other art forms of the service provider and the service provider in the nursing module. The proportion of the first four answers "very important" is relatively high, as shown in Fig. 1. Leisure and entertainment (28%), art fitness (22%), calligraphy, painting and other art forms education (38%) and tourism organized by service platforms Activities (10%), as shown in Table 1. It can be seen that the elderly have a relatively consistent understanding of the needs and importance of cultural leisure, indicating that the elderly have relatively high spiritual and cultural needs. Home care services should increase the intensity of community cultural and leisure services, and use information technology to strengthen the spiritual civilization of community care services.

The maintenance module can manage and maintain the classification of active care services, such as adding, editing and deleting service categories. In different regions, services can be classified into different categories to respond to the special needs of the local elderly. Active care service is a kind of warm service for the elderly living alone, empty nest, and lonely. Because this type of old people do not live with their children, and their children seldom communicate with the old people, this type of old

Table 1. The demands of the elderly on the service items of pension activities

Variable	Frequency (number)	Percentage (%)
Entertainment	42	28
Art fitness	34	22
Calligraphy and painting	58	38
Tourism activities	16	10

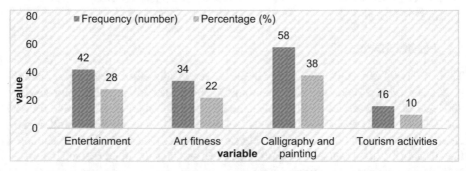

Fig. 1. The demands of the elderly on the nursing module of pension activities

people are withdrawn and unwilling to contact the outside interpersonal circle socially, and the outside world cannot know the status of the old people. This function is used to implement regular, active and personalized care services for each elderly person. Active care services for the elderly mainly include: birthday reminders, event notifications, news broadcasts, weather forecasts, etc.

4.3   Education and Training Modules

From the education and training module, the survey results of service providers and service providers on medical and health, rehabilitation training, physical examination and nursing, mental comfort (psychological assistance), special populations, nursing services, and old-age health technology are shown in Fig. 2. The proportions of choosing physical examination nursing and old-age health care technology accounted for 41% and 36%, respectively, as shown in Table 2, which also shows that medical needs are the most needed by the elderly who care at home. Nowadays, with the improvement of living standards and medical technology, more and more elderly people are suffering from chronic diseases, and they urgently need attention in nursing and rehabilitation. Professional medical services with better service quality are what the elderly people strive for. It can be seen that the education and training module should strengthen the training in the provision of professional medical services, and the home care service platform should vigorously promote medical services tailored to the characteristics and differences of the elderly.

Table 2. The demands of the elderly on community medical services for the aged

Variable	Frequency (number)	Percentage (%)
Health care technology for the aged	55	36
Nursing service	31	20
Physical examination nursing	62	41
Special groups	2	1

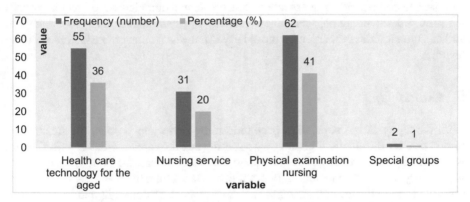

Fig. 2. The demands of the elderly on community medical services for the aged

Education and training modules should pay attention to physical examination and nursing, establish complete, dynamic, and life-long health files for the elderly, record the elderly's health history, and dynamically record the elderly's physical examination data, expert diagnosis, disease treatment records, test results, medication and identity health the development trend of the virus is timely imported into the health file for relevant users to read and view at any time. At the same time, health file information can also be provided to doctors for the elderly when they are undergoing disease diagnosis and treatment, so that doctors can have a more comprehensive understanding of the physical condition of the elderly, and more timely and accurate targeted treatment.

The education and training module must establish a medical and health database. The medical and health data sources come from various wearable devices and home health testing equipment used by the elderly. The main functions include physical data monitoring, health data monitoring, and health assessment early warning. Through wearable devices such as health smart watches, the pulse, exercise volume, sleep and other physical signs of the elderly can be collected in real time, and continuous physical sign monitoring information can be formed in the time dimension. Through intelligent home health testing equipment, such as smart blood pressure monitors, smart blood glucose meters, smart blood oximeters, smart electrocardiographs, smart body fat meters, etc., the relevant health data of the elderly are collected regularly and uploaded to the cloud service platform in real time. According to the different physical conditions of the elderly, different health testing equipment can be configured in the family to achieve targeted measurement. The corresponding blood pressure, blood sugar, blood oxygen saturation, body fat, electrocardiogram and other data will form continuous health monitoring information in the time dimension. Elderly physical sign data and historical health monitoring data collected through front-end systems (including wearable devices and intelligent home health testing equipment, etc.), combined with the elderly's personal disease history, family disease history, personal living habits, etc. Carry out comprehensive analysis, evaluation, and early warning. Based on the results of the health assessment, personalized health management plans are proposed for abnormal situations and different risk factors, and corresponding intervention measures and a complete set of health services are provided, including physical fitness evaluation, psychological analysis evaluation, and nutritional status evaluation for the elderly. Analysis of unfavorable factors affecting health.

5 Conclusions

With the aging of the population becoming more and more serious, the function of family care for the elderly is gradually weakening and the care facilities are insufficient. Restricted by the level of social and economic development, the issue of pensions in my country has become increasingly prominent. Based on a large amount of literature, this article investigates and analyzes the development of domestic and international home care service models, as well as the status quo and needs of home care service models. Aiming at home care services for urban retirees, this paper analyzes the current application status of elderly care information under the background of cloud computing and the Internet of Things, discusses the combination of cloud computing technology and home care services, and deeply analyzes the advantages of cloud computing in smart home care services. Through the design and testing of the questionnaire, basic statistics and cross-analysis were carried out on the needs of the group for the elderly. Based on the results of the analysis, this paper proposes a design plan for a multi-module professional service platform for the aged at home that is suitable for my country's national conditions and meets the objective needs of urban retired elderly groups.

References

1. Wang, S., Tang, Q.: Construction of guiding system for growth and development of college students under the student-oriented concept. Asian Agric. Res. **10**(05), 93–95 (2018)
2. Solov'Ev, V.P., Pereskokova, T.A.: Management of professional training. Russian J. Ind. Econ. **12**(3), 356–376 (2019)
3. Susomrith, P., Coetzer, A., Ampofo, E.: Training and development in small professional services firms. Eur. J. Train. Dev. **43**(5/6), 517–535 (2019)
4. Wilson, C.M., Arena, S.K., Adcock, K., et al.: A Home-Based Older Person Upstreaming Prevention Physical Therapy (HOP-UP-PT) program utilizing community partnership referrals. Home Healthc. Now **37**(2), 88–96 (2019)
5. Iezzoni, L.I., Gallopyn, N., Scales, K.: Historical mismatch between home-based care policies and laws governing home care workers. Health Aff. **38**(6), 973–980 (2019)

6. Lamine, E., Bastide, R., Bouet, M., et al.: Plas'O'Soins: an interactive ICT platform to support care planning and coordination within home-based care. Innov. Res. Biomed. **40**(1), 25–37 (2019)
7. Hronová, T., Souralová, A.: Managers, consumers, visitors: roles of caring relatives in emerging home-based eldercare in the Czech Republic. Socialni Studia **16**(2), 7–24 (2019)
8. Adinan, J., Adamou, B., Amour, C., et al.: Feasibility of home-based HIV counselling and testing and linking to HIV services among women delivering at home in Geita, Tanzania: a household longitudinal survey. BMC Public Health **19**(1), 1–8 (2019)
9. Wu, M.P., Huang, C.M., Sun, W.J., et al.: The promotion of resources integration in long-term care service: the experience of Taipei City hospital (in Chinese). Hu Li Za Zhi **65**(1), 24–32 (2018)
10. Stome, L.N., Norrud, A., Fjordholm, M., et al.: OP97 digital home services: overcoming critical barriers in early decision making. Int. J. Technol. Assess. Health Care **34**(S1), S36 (2018)

Interaction Rituals Performed by Fans Using New Media in China Against the Background of the "Internet+"

Juan Cheng[✉]

College of Literature and Journalism, Sichuan University, Chengdu, Sichuan, China
chengjuan88@xueshumail.cn

Abstract. In the context of the increasing openness of the Internet and the sharing of information, various fan communities on the Internet are nourished by idol culture and have come into being. Many fans gather on social media because of their shared love for a person, and jointly build online fan communities based on "fan sites". Interaction rituals are very common within Internet fan communities, such as daily ranking for idols, meet-ups, and gatherings to listen and discuss electronic albums that their idols newly released. Fans can obtain emotional energy from these rituals. Long-lasting high emotional energy can encourage fans to participate in interaction rituals related to their idols for a long time; frequent low emotional energy will cause fans to unfollow the idol. This paper attempts to systematically examine how the interaction ritual chain gathers the power of fan communities as the development of the fan community matures, and how Internet fans obtain emotional energy through online interaction rituals. In this way, this paper will evince a clear understanding of the psychology and behavior of fans using new media.

Keywords: Internet+ · New media · Fans

1 The Theory of Interactive Ceremony in the Context of "Internet+"

The theory originates from the Interaction Ritual Chain Theory by Durkheim, which became a framework after being improved by American sociologist Collins. The core mechanism of Interaction Ritual Chain Theory is that a combination of high degrees of mutual attention with high degrees of emotional connection, forms a sense of membership associated with cognitive symbols that brings emotional energy to the participants, rendering them confident, enthusiastic and willing to participate in morally permitted activities. In the context of the Internet [1], there are four main components of an interaction ritual: The first is the gathering of two or more people in the same location. The second is restricted access to outsiders. The third is that people's attention be focused on the same person or activity. The last is sharing common emotions or emotional experiences among the members. When the four components are effectively integrated and

© The Author(s), under exclusive license to Springer Nature Switzerland AG 2022
J. Macintyre et al. (Eds.): SPIoT 2021, LNDECT 98, pp. 504–510, 2022.
https://doi.org/10.1007/978-3-030-89511-2_65

accumulated to a high degree of mutual attention and emotional sharing, interaction rituals against the Internet background will produce four main results: Group unity, a sense of membership; individual emotional energy (a feeling of confidence, elation, strength, enthusiasm, and initiative when taking action); symbols representing the group, such as visualized icons, words, and gestures. These symbols make members feel that they are related to the group. They are the "sacred object" according to Durkheim; and safeguard a sense of justice in the group. Moreover, the symbols fuel respect for group symbols, a sense of morality that prevents the group from being violated, and the sense of sin or impropriety when betraying the group's unity or symbols. In the context of the Internet [1], the core of interaction rituals is a process in which participants develop a common focus and feel each other's micro rhythms and emotions [1] (see Fig. 1).

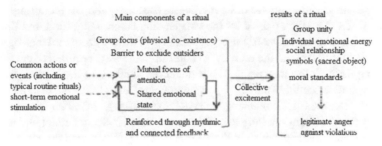

Fig. 1. The model of Collins' interaction ritual theory [1]

2 Fans' Performance in Interaction Rituals Using New Media Against the Background of the "Internet+"

"The purpose of interaction rituals is to awaken emotions [2]." In fan activities, interaction rituals are common. Fans form emotional resonance through interaction rituals to achieve collective excitement and obtain emotional energy [1]. Fans can follow a star for a long time because fans can obtain continuous and high emotional energy from idol meeting activities on the Internet. Fans' emotional energy is obtained from interaction rituals.

Collins divided the interaction rituals into formal rituals and natural rituals. A formal ritual is a formal ceremony, carried out through a generally recognized ceremony procedure. It has a set of programmed actions, such as reciting the ceremony rules, singing, performing traditional postures, wearing traditional clothes, and so on. A natural ritual is to establish mutual attention and emotional bonds without formal formalization procedures. In the context of the Internet, the core elements, processes, and results of formal and natural rituals are the same, and the key elements of ritual operation are the focus of mutual attention [2].

Both formal rituals and natural rituals exist in fan activities. Offline support activities held by fans are mostly formal rituals. Fans' often wear the same clothes during offline support activities; or share the same support items and the same slogans. The official rituals also take place in fans' online activities. For instance, Cai Xukun Global Fan Club will organize fans to complete the daily tasks that must be done, including completing charity tasks, votes for him in Oriental Billboard, QQ Music, UNI music, Paopao circle of IQIYI, and earn member loyalty points in Super Topics [3]. Such rituals are spontaneously carried out by Cai Xukun's fans on the new media platform of the Internet. Such rituals have fixed procedures and rules. It requires active learning of relevant tutorials to complete the tasks specified in these procedures. Therefore, these rituals impose restrictions on outsiders. Cai Xukun's fans express their support for him by completing tasks day by day. In another example, the online fan meeting of the TV series The Romance of Tiger and Rose is also a formal ritual. On June 5, 2020, Tencent Video held a live internet fan meeting for the TV series The Romance of Tiger and Rose. The number of people attending this activity exceeded 400,000. In the context of the Internet, fans will focus their attention on the crew members, especially the CP (coupling) [4], Han Shuo and Chen Qianqian, played by the two leading actors Ding Yuxi and Zhao Lusi. This online fan meeting set restrictions on outsiders. Those who are not fans of the show could not get notified about the activity. In addition, fans were required to get tickets for the meeting through the Internet. In addition, some symbols in the show were used in the activity, such as calling Han Shuo and Chen Qianqian as "deal" (Chengjiao) couple. In the show, Han Shuo's nickname is "banana" (Xiangjiao), and Chen Qianqian's nickname is "orange" (Chengzi). People who haven't seen this show cannot understand the meaning of these language signs. In this online meeting, fans shared the same emotions: Happiness came the first. Then they felt "sweet" from the interaction between the actor and actress. During the entire Internet meeting, fans strongly called for season 2 of the show. This online meeting was conducted in accordance with a fixed process, which was a formal procedural ritual.

However, in the new media fan circle in the context of the Internet, there is more mutual attention and emotional connection without a fixed program, which is called natural rituals. For example, Zheng Shuang's fans boycotted her starring in the movie "The Journey of Flower" on Weibo. It is actually an interactive ritual. The main character in "The Journey of Flower" starring by Zhao Liying is a classic image. When Zheng Shuang, a better-looking celebrity whose acting skills were not on the same level as Zhao Liying, was reported to play the character of Zhao in "The Journey of Flower", Zheng Shuang's fans persuaded Zheng Shuang to cherish her "feather" (fame) on Weibo. This topic received more than 100 million readings and more than 80,000 discussions. The advice and discussion of the fans formed an interactive ritual without a fixed procedure. Many people participated in it, with all their attention on the news of her starring. Fans had a common sentiment and emotion admonishing Zheng Shuang to cherish her fame and reject "poisonous cake" (bad resources). In order to persuade Zheng Shuang, fans listed five problems and four foreseeable consequences [5]. This ceremony placed restrictions on outsiders. For example, fans use phrases such as "feather", "flower road", and "poisonous cake" that outsiders generally can't understand. Fans are required to be familiar with the film, Zheng Shuang's characteristics as an actress, and the rules of the domestic film circle so that they could list the five major issues and four consequences in order to discourage Zheng Shuang from taking on the filming.

Another example is the release of Jay Chou's new song Mojito on the Internet. It is another interactive ritual among fans. On June 8, 2020, Jay Chou shared a prelude to Mojito in Kuaishou, creating momentum for the single's release. Then he announced the news that the single will be launched at 0:00 on June 12. Before the release of this song, more than 3 million people made reservations on QQ music. On June 12, the song premiered on QQ Music. When it officially released, the platform was paralyzed for a while due to too many purchasers. There were fans discussing about this song on Internet platforms such as Weibo, Moments of WeChat and QQ music. They gathered on the Internet and listened to the song together [6]. Many great fans gathered on Internet platforms such as QQ Music, focusing on buying, appreciating, and commenting on the song Mojito, sharing common emotions such as their love for Jay Chou and the youthful memories of listening to Jay's songs. Since people was asked to pay for the song, this ritual was restricted to outsiders. The collective excitement of the fans led to the paralysis of QQ music. The results of this ritual are:

(1) Group unity. The fans are proud to be Jay Chou's fans who has been one of the most popular singers for so many years. The netizen Honghuang girl said: "The new album is here. The street is too crowded, too many people waiting for the album. The taste of saury, Jay Chou's album, cat and I want to know. Where is the rainbow tell me, can you return my album to me. I will be in a daze, then smile, and then urge you to release the album. Give him two more minutes to make the album into MV. Azure is waiting for misty rain, I'm waiting for the new album. The most beautiful day is not the rainy day, but the day you release your new album. Our beginning was a very long album. It has been delayed for four years, and I am still thinking about it. How can I sleep without seeing your album. The song is very nice, but the album still needs to be reminded. What an idol who always stands fans up." (She reedited Jay's lyrics).

(2) Individual emotional energy. Jay's fans spontaneously bought the single on Internet music platforms, forwarded and shared the song on Weibo, Moments and other online platforms. There are some other fans covered the song. #Mojito covered by fans#, #Mojito in Spanish#, #6-year-old girl sings Mojito by guitar" became popular topics on Weibo. The mojito MV posted on Bilibili is full of Danmaku. These activities are actively driven by the fans' passion for Jay Chou. It is precisely because being a fan of Jay Chou can bring them long-term high emotional energy, that the fans are willing to continuously invest their emotions in Jay Chou and actively participate in rituals related to Jay Chou.

(3) Symbols represent the group. Jay Chou and his music have become a symbol. His music is the youth memory of people who were born in the 80s and 90s. Every release of his new song will set off a wave of people's recollection of youth. Some fans even said that only Jay Chou can make them pay for music. A fan on the Internet once said: "Jay Chou is at the forefront in the music industry, facing the vast universe alone. His creation is amazing, every new song is a surprise, and every album is a classic. When there was no one to learn from and no one to compete with, he understood the true meaning of making music. He gives people warmth, touch and beauty again and again. I like Jay for a long, long time. As his big fan,

I will always adhere to this belief and let his unparalleled music accompany me through my youth and continue to accompany me to a farther life."

(4) Sense of morality. Fans will get angry when someone criticizes this song, saying that Jay Chou lost his talent, the song is terrible, or even that the song is plagiarized. Reasonable anger is a collective emotional outburst towards people who offended and desecrated the "sacred object" [1]. Jay Chou's fans argued with anti-fans on Internet social platforms, which is a legitimate expression of anger. For example, @vic love musical box best said: "Those who say it is bad, maybe they don't know what Latin music is. That is why they can't appreciate this kind of music. Speaking of plagiarism, let's figure out how to define plagiarism in music theory first! Well, I have loved football since I was a kid, and it also gave me a chance to get in touch with the world, to feel and understand other cultures! Mojito is super good. I also want to wear fancy beachwear and walk on the beach [7]."

3   Features of Interaction Rituals of Fans Using New Media Against the Background of Internet+: Presence Without Physicality

Collins believes that a ritual is a process of physical experience. The ritual process begins with people gathering in the same place. Personal presence is necessary for rituals. Being physically present makes it easier for people to perceive the signals and physical performance of others. People can have the same rhythm, absorb the gestures and emotions of others, and focus on a core of mutual attention. Being physically on a mutual locale also makes participants in the ritual more likely to share common emotions. On the contrary, interaction rituals carried out through remote media will make people lack a sense of unity, lack of respect for common symbols, and reduce enthusiasm and personal motivation. Collins believes that although programs through remote media can provide certain shared concerns and emotions, and make people aware of a sense of membership, only adequate physical contact will have a strong effect. Only when the electronic media is designed to simulate people's physiology, makig the nervous system directly and remotely produce emotional bonds, will the rituals through electronic media be the same as being on the spot [1].

However, new media in the context of the Internet provides technical support for people to carry out interaction rituals. The new media emerging with the Internet has realized real-time interactive communication, combined with the characteristic of anonymity present in some Internet media enables people to speak daringly, enabling participants to express their true feelings. For fans, both online and offline rituals are important. In China, Weibo has become the main venue for fan activities. Daily rituals on Sina Weibo are common. In addition, WeChat group chats and video Danmaku are all ways to carry out interaction rituals. Chen Yu regards the virtual community of fans as a special ritual place. He believes that this ritual venue breaks through the limitations of space and time: new media technologies such as real-time images and virtual reality can reproduce scenes that are thousands of miles apart or even imagined. Virtual reality technology enables non-real-time images to be presented in real time, making linear time a virtual non-linear node. This ritual venue is also an interactive venue where fans can enter at any time to witness and share events, express and shape personal opinions. Therefore, the experience of "presence" and "participation" of fans in virtual communities in the context of the Internet shows the characteristics of breaking through the limitations of time and space. Although fans are scattered in the vast "cloud space", with the help of mobile phones and applications, they can feel each other's existence. What's more, through real-time interaction, they can feel the joy of the interactive rituals. Especially the appearance of memes has realized the restoration of physically being on the spot. Before memes were used, netizens could not see each other's facial expressions and actions because they were not face to face. The expression of emotions could only be done through non-intuitive text descriptions. Memes can imitate people's expressions and gestures to the maximum, so that people can chat immersively. Memes are a kind of physical expression. They use facial expressions or body movements to intuitively represent emotions. Memes have replaced the physical body anchored in reality with an "online body", so that fancs can be on the spot [8]. The emergence of memes allows fans to capture expressions and gestures and achieve emotional resonance [9, 10]. In addition, since praying to symbols can also awaken emotional energy, through symbols, people can recall the collective excitement of the past when they were physically present together. Therefore, the awakening of emotional energy is no longer entirely dependent on the co-presence in physical space [2]. However, whether online rituals can produce such strong emotional resonance as offline rituals requires the use of measurement tools.

4 Conclusion

The continuous development of Internet technology has enabled the establishment of online communities based on virtual communities and has developed into a new type of social relationship. The emergence of new media technologies on the Internet has made the forms of idols more diversified. The scene where everyone is an idol and everyone is a fan may also exist. Fan culture will have an increasingly significant impact in the online society. Based on the Theory of Interaction Ritual Chains, this paper conducts in-depth research on the fan culture communication mechanism in the era of new media of the Internet, constructs an interaction ritual chains model for fans using new media, and points out that the interaction rituals of Internet fans have the characteristics of being present without physicality.

References

1. Collins, R.: Interaction Ritual Chains, pp. 79, 85–89, 93–106, 184, 186. The Commercial Press, Beijing (2009)
2. Turner, J.H., Stets, J.E.: The Sociology of Emotions, pp. 63, 65–66. Shanghai People's Publishing House, Shanghai (2007)
3. Cai Xukun Global Fan Club: Daily tasks (2018). https://weibo.com/2839635900/J70rVzefN? ref=collection
4. Shao, Y. (ed.): Book for Breaking the Wall, pp. 194–198. SDX Joint Publishing Company, Beijing (2018)
5. Tencent: Zheng Shuang will star in the movie version of "The Journey of Flower", but fans strongly oppose it, enumerating five major problems and four major consequences (2015). https://xw.qq.com/cmsid/20200609A0503Q00

6. News: Jay Chou brings a new song Mojito, is the digital music economy having an explosion? https://3g.163.com/war/article_cambrian/FF7DA50L05118DKC.html
7. vic love musical box best: #mojito plagiarism ##mojito unpleasant to hear# (2017). https://weibo.com/2159854590/J6Ms8nBz2?refer_flag=1001030103_
8. Liu, H.: Meme culture: body expression and identity construction under power transformation. Soc. Sci. Yunnan **01**, 32 (2017)
9. Chen, Y.: Sharing rituals and exchange gift game-a study on the virtual fan community based on Baidu Tieba. Contemp. Commun. **06**, 04 (2013)
10. Borcan, I.: The role of dynamic capabilities, business model and organizational culture in the digital transformation of a traditional organization. Manage. Market. J. **19**, 55–57 (2021)

Recommendation System Based on Semantic Web

Wenyue Chen[(⊠)]

Tianjin University of Commerce, Tianjin 300134, China

Abstract. The rapid development of the Internet has greatly facilitated people's work and life, but at the same time, the information overload has also made people feel at a loss in the face of mixed data. The search engine and recommendation system created for this can help users filter and filter what they need. Information, but as people's requirements for personalized services are getting higher and higher, traditional recommendation methods cannot meet more precise needs. Therefore, this article comprehensively explains the new recommendation system that introduces the concept of the Semantic Web, and introduces its origin, implementation, and finally, the successful application of semantic recommendation system in four different fields is cited.

Keywords: Recommendation system · Semantic web · Ontology

1 Introduction

With the rapid development of today's technology, the amount of data is increasing every day. Faced with massive amounts of data, people are increasingly feeling helpless. People have no clear and specific expectations and goals for their needs, or their goals are vague, and users only know them. The general demand direction, so there are often choices made by herd mentality, which leads to more and more projects with a large number of people selected, while projects with a small number of people are always selected. It is common for people to judge the quality of a certain good based on the number of other people consuming that good [1].

Initially, this phenomenon was referred to as the "long tail". Used to describe the business and economic models of websites such as Amazon, Netflix, and Real.com. The long tail effect refers to the phenomenon that the sales volume is small but the total amount of many products or services that were not originally evaluated is huge, and the cumulative total revenue exceeds that of mainstream products especially in the Internet field. [2] Therefore, the long tail theory can explain the existence of the search recommendation system—The recommendation system filters out the user's personalized information and provides a way to understand their taste and recommend appropriate things to them by considering the user's preferences and the level of various things [3]. Meanwhile, the recommendation system can also provide exposure opportunities for all items appropriately, so as to tap the potential profits of long-tail projects. Although the recommendation system has a history of twenty-six years since its inception, and predecessors have done a lot of related research in this field, there is still a lot of room for improvement in the recommended methods and effects until now. For example, cold-start optimization, [4] application of recommendation methods to new scenarios, protection of data involving user privacy, etc., in particular, new concepts can be introduced to make the recommendation system better provide personalized services.

According to the different data sources of the recommendation system, Thomas divides the personalized recommendation system into three categories: demographic filtering, content-based filtering, and collaborative filtering [5]. Demographic filtering is an earlier recommendation method. It is classified according to the user's demographic data. For recommendation systems with increasing personalization requirements, this method is too rough. Therefore, this article will not dwell on demographic filtering, but mainly analyze the other Two recommended methods:

1) Content-based filtering describes the attributes of the item itself through tags, and recommends items with higher similarity to the user based on the user's previous preferences and behavior habits. Since the content-based filtering method uses keywords and tags to describe item metadata, the analysis of item similarity only relies on the characteristics of the item itself, and does not consider people's attitudes toward the item.

2) Collaborative filtering, which is the recommendation method with the most applications currently. In Marko Balabanović's opinions, collaborative recommendation recommends items other similar users have liked, rather than recommending items because they are similar to items a user has liked in the past. And it computes the similarity of the users, instead of calculating the correspondences of the items [6]. This filtering method is mainly based on historical data. Its recommendation result depends on whether the user's historical preference data is sufficient and accurate, so users with special requirements cannot get very suitable suggestions. In addition, this feature also leads to the inability of the system to flexibly modify the results according to the user's usage after capturing and modeling user preferences.

In summary, in the domain of recommendation systems, the attribute tags used to describe items usually come from third-party calculations. The extracted content is only based on text information and does not correctly define user interests with certain association relationships. Moreover, the use of text functions involves a large number of ambiguities due to natural language, which cannot understand the real needs of the user. Therefore, the modern recommendation system that connects users to users, users to items, and items to items can enhance the semantic information of data to further boost the accuracy of suggestion, which has important research significance.

2   Semantic Recommender System

2.1   Semantic Web

As a new computer concept, the Semantic Web is favored by major Internet companies. Its data storage and calling methods that are more inclined to artificial intelligence can achieve more precise and personalized functions. The explicit representation of Web meta-information and domain theory (i.e. ontology) will make it possible to provide websites with a new level of quality. It will weave a huge human knowledge network and supplement it with machine machinability [7]. Two important technologies for the development of the Semantic Web are already in place: Extensible Markup Language (XML) and Resource Description Framework (RDF). XML allows everyone to create their own tags-hidden tags, and allows users to add arbitrary structure to their documents, but does not explain the meaning of the structure. The meaning is represented by RDF, which encodes it in the form of triples, and each triple is very similar to the subject, verb, and object of a basic sentence. Subjects, objects, and verbs are all identified by Universal Resource Identifiers (URIs), and their usage is similar to web links. URIs allows anyone to define a new concept or a new verb by defining a URI somewhere on the Web [8].

Currently, the Semantic Web is mostly used in Internet search engines. There is no greater advantage for pushing knowledge questions and answers with the recommendation system. However, the personalized data technology provided by the Semantic Web will substantially change the intelligence and accuracy of the recommendation system. As Wanaskar believes, the description displayed by traditional text mining methods based on lexical and syntactic analysis can be understood by users, rather than computers or recommendation systems. That is the reason that it is essential to create new text mining techniques based on semantic analysis [9]. And the recommender system with this technology is called a semantic-based recommender system.

2.2 Recommended Generation Methods

According to the representation of the knowledge graph in the recommendation process, the recommendation generation methods of the semantic recommendation system can be divided into two categories: based on ontology and based on linked open data (LOD).

1) Recommendation generation based on ontology. In the field of knowledge representation, ontology is one of the main manifestations of knowledge graphs. It is a sct of entities, relationships, and examples that clearly describe the concept of the domain. It emphasizes the hierarchical relationship between concepts, with logic Based on semantics, computers can simply and automatically reason about data based on the ontology. CHANG Liang considers that the main idea of ontology-based recommendation generation is: use the fine-grained description of the concept hierarchy in the ontology, and use the fine-grained classification description of the concept to more accurately represent the entity features, so as to dig out the deep-level information contained in the facts [10]. The concept of ontology can be integrated into existing content-based recommendation and collaborative filtering-based recommendation methods to generate recommendations based on the user's current context information. For example, Marco de Gemmis summarized the shortcomings of content-based recommendation systems, proposed to integrate ontology knowledge into more complex domains, and use unstructured or semi-structured encyclopedia knowledge resources (such as BabelNet) to semantically represent content [11]. In terms of ontology generation, we can obtain the corresponding semantic web model through a certain extraction algorithm based on existing text documents

and content. Benjamin Adrian introduced a recommendation system to tag or anno-
tate documents with concepts of a called ConTag [12]. Semantic Web ontology. It
provides services of text and term analysis functions by Web 2.0, such as phrase
extraction, dictionaries, thesaurus, classification and term association, and then uses
the concept of semantic web ontology to process this information. This method integrates the information provided by Web2.0 services and Web3.0 (Semantic Web)
ontology, enabling us to recommend semantic document tags. As shown in Fig. 1.

Fig. 1. Web3.0 (Semantic Web) ontology

2) Recommendation generation based on LOD. For link data, Christian Bizer believes
that its core idea is to use HTTP URIs to represent data about these entities in the
form of Resource Description Framework (RDF), which can not only complete the
semantic analysis of user text content, but also construct arbitrary real-world entities. Generate recommendation results that are more in line with actual needs [13].
In this method, the recommendation system uses a large amount of interrelated data
in the LOD to measure the similarity between resources in a more refined manner,
mine user preferences, and finally generate recommendation results based on contextual information. At present, the relatively complete development link databases
include DBpedia, LinkedMDB, Freebase, etc. The DBpedia data set alone provides
more than 1.95 million information, including at least 80,000 people, 70,000 venues,
35,000 music albums, 12,000 movies, etc. Data and links [14]. Chang Liang considers that the advantages of LOD-based recommendation generation technology are:
strong data association expression, strong logic and certain reasoning ability, can
automatically discover some implicit semantic information, and can improve the
accuracy of recommendation [10]. However, the recommendation process also has
problems such as excessive reliance on external knowledge bases and large amount
of calculation.

3 Applications

At present, recommendation systems based on the Semantic Web are used in practical operations in many fields and have achieved considerable results. This article lists several typical application cases.

1) News recommendation. Traditional news recommendation is based on the TF-IDF algorithm and Hermes personalized semantic recommendation framework. Wouter IJntema proposes Athena on the basis of the existing Hermes, which can mine information from the user's personal data and the news items browsed by the user [15]. The method of TF-IDF and cosine similarity is used to calculate content similarity, and then three semantic-based methods are used-concept equivalent to find the concepts in the news items browsed by users, and binary cosine and Jaccard are used to calculate articles and individuals Semantic correlation between data, and then proposed a new method called ranking recommender, which uses related concepts in user data to create extended user profiles. Finally, the semantic recommendation system ends with method evaluation. This system's extremely high accuracy, precision and recall rate are superior to other methods that are considered to be based on ontology.

2) E-government. Modern government business decision makers mostly search for business partners and establish new partnerships through the Internet. Qusai Shambour has developed a semantic recommendation system that provides personalized G2B electronic services, mainly for business partners to recommend electronic services to Australian SMEs [16]. The hybrid semantic recommendation method adopted combines the project-based collaborative filtering similarity and the project-based semantic similarity technology, and is implemented in the prototype BizSeeker called the "intelligent business partner locator recommendation system", which can provide information to individual business users (such as Exporters) recommend relevant business partners. The significance is that it can help e-government agencies to provide more personalized electronic services to corporate users.

3) E-learning. With the increase of online learning content, it has become quite difficult and complex for online learners to find suitable high-quality courses. Compared with other fields, online learning has not yet established a suitable semantic recommendation system. Saman Shishehchi proposed a semantic recommendation system for e-learning, which helps learners find courses and learning materials that suit their interests and needs based on their knowledge level and personal information [17]. The system construction includes two subsystems: a semantic system and a rule-based system. The semantic system uses ontology to represent learning content, and the rule-based system uses OWL rules to use inference engine rules based on recommendations. The most important feature he put forward is the profile of learners, including learning history, knowledge and personal characteristics, some of which will change in the learning process, and learner information will be updated in real time in the system.

4) Drug recommendation. At present, there are many useful drugs and knowledge resources related to Adverse Drug Events (ADE) available from the Internet, which can be provided to patients who are not familiar with drugs and ADE terms to

find drugs through simple keyword-based search, and their potential Interactions and side effects. Nidhi Kushwaha uses semantic web technology and data mining for disease diagnosis [18]. According to the records of the patient's personal label, medical history, medications taken, etc., a list of drugs that can cure a specific disease without any interaction between drugs can be obtained through semantic query from the existing Linked Open Drug Data (LODD), and finally The doctor makes the final decision based on the generated suggested drug tuples, and the patient can also obtain relevant explanations and precautions for the drug.

4 Tag

This article provides a basic overview of the origin and development status of recommendation systems based on the Semantic Web. In the introduction section, it starts from the recommendation system and introduces the existence value of the recommendation system in combination with the long tail effect, as well as two types of current mainstream recommendations with different data sources. System, based on the shortcomings of the traditional recommendation system to elicit the concept of the semantic web. In the part of semantic recommendation system, this article firstly cites literature to give a basic explanation of the concept and implementation method of the semantic web, then introduces two semantic web-based recommendation generation methods, and presents the current application status of the two methods. In the application part, this article selects current successful cases in the four fields of news, politics, education, and medical treatment, which reflect the current development status of semantic recommendation systems in many aspects.

Of course, the rapid development of the Semantic Web will inevitably bring some instability factors, and there are still many problems that need to be overcome. For example, CHANG Liang believes that typical problems include multi-source information fusion, the mining of user characteristics in the knowledge map, and the analysis of potential needs. Multi-dimensional recommendation, combination with time and space reasoning, user privacy protection, etc. In addition, the recommendation system can also combine popular concepts such as deep learning and system inquiry to achieve a more efficient and accurate recommendation system that is more in line with the trend of the times.

References

1. Spicer, M.P.: Following the Herd: An Economic Analysis of the Effects of Herd Mentality on the US Housing Bubble (2011)
2. Anderson, C., Andersson, M.P.: Long tail (2004)
3. Das, D., Sahoo, L., Datta, S.: A survey on recommendation system. International J. Comput. Appl. **160**(7), 6–10 (2017)
4. Schein, A.I., Popescul, A., Ungar, L.H., Pennock, D.M.: Methods and metrics for cold-start recommendations. In: Proceedings of the 25th Annual International ACM SIGIR Conference on Research and Development in Information Retrieval, pp. 253–260, August 2002
5. Hess, T.: Recommender Engines Seminar Paper, 1 February 2009

6. Balabanović, M., Shoham, Y.: Fab: content-based, collaborative recommendation. Commun. ACM **40**(3), 66–72 (1997)
7. Davies, J., Fensel, D., Van Harmelen, F.: Towards the Semantic Web. Wiley, Chichester (2003)
8. Berners-Lee, T., Hendler, J., Lassila, O.: The semantic web. Sci. Am. **284**(5), 34–43 (2001)
9. Wanaskar, U., Vij, S., Mukhopadhyay, D.: A hybrid web recommendation system based on the improved association rule mining algorithm. arXiv preprint arXiv:1311.7204 (2013)
10. Chang, L., Zhang, W.T., Gu, T.L., Sun, W.P., Bin, C.Z.: Review of recommendation systems based on knowledge graph. CAAI Trans. Intell. Syst. **14**(2), 207–216 (2019)
11. de Gemmis, M., Lops, P., Musto, C., Narducci, F., Semeraro, G.: Semantics-aware content-based recommender systems. In: Ricci, F., Rokach, L., Shapira, B. (eds.) Recommender Systems Handbook, pp. 119–159. Springer, Boston (2015). https://doi.org/10.1007/978-1-4899-7637-6_4
12. Adrian, B., Sauermann, L., Roth-Berghofer, T.: Contag: a semantic tag recommendation system. Proc. I-Semant. **7**, 297–304 (2007)
13. Bizer, C.: The emerging web of linked data. IEEE Intell. Syst. **24**(5), 87–92 (2009)
14. Auer, S., Bizer, C., Kobilarov, G., Lehmann, J., Cyganiak, R., Ives, Z.: DBpedia: a nucleus for a web of open data. In: Aberer, K., et al. (eds.) ASWC/ISWC -2007. LNCS, vol. 4825, pp. 722–735. Springer, Heidelberg (2007). https://doi.org/10.1007/978-3-540-76298-0_52
15. IJntema, W., Goossen, F., Frasincar, F., Hogenboom, F.: Ontology-based news recommendation. In: Proceedings of the 2010 EDBT/ICDT Workshops, pp. 1–6, March 2010
16. Lu, J., Shambour, Q., Xu, Y., Lin, Q., Zhang, G.: BizSeeker: a hybrid semantic recommendation system for personalized government-to-business e-services. Internet Res. Electron. Netw. Appl. Policy **20**(3), 342–365 (2010)
17. Shishehchi, S., Banihashem, S.Y., Zin, N.A.M.: A proposed semantic recommendation system for e-learning: a rule and ontology based e-learning recommendation system. In: 2010 International Symposium on Information Technology, vol. 1, pp. 1–5. IEEE, June 2010
18. Kushwaha, N., Goyal, R., Goel, P., Singla, S., Vyas, O.P.: LOD cloud mining for prognosis model (Case study: Native app for drug recommender system). In: Advances in Internet of Things (2014)

Short-Term Power Load Forecasting Method Based on Cloud Computing and Intelligent Algorithms

Linan Hu, Chao Yang$^{(\boxtimes)}$, Xianwei Meng, and Peng Pang

Heilongjiang University of Technology, Jixi 158100, Heilongjiang, China

Abstract. In recent years, with the widespread application of computer technology, smart power grid has developed rapidly. In order to ensure the safe and economic operation of the system, the staff of the electric power sector put forward higher requirements for the stability, accuracy and efficiency of the short-term load forecasting results. This paper analyzes the practical application background of short-term load forecasting and summarizes the current research status in this field at home and abroad. Due to the complex operation of the combined prediction model, the difficulty of operation is greatly increased. In order to solve the problem of insufficient single computing resources, cloud computing is introduced in this paper to parallelize the combined prediction model, which improves the big data processing ability of the prediction model and enhances the practical application effect of this new model. By introducing cloud computing, the parallel computing performance of the prediction model is improved, the computing speed is accelerated, and the working efficiency of the staff is improved.

Keywords: Cloud computing · Intelligent algorithms · Power load forecasting · Combination models

1 Introduction

Power load the electricity users of electricity equipment at some point to the sum of the electrical power system access, has been the power sector coordination scheduling priority, if we can accurately predict the future for a period of time will allow for coordination of power load play a significant supporting role, greatly reduce the energy waste caused by excessive power generation, therefore, it is necessary to forecast the power load [1]. Power load forecasting refers to exploring the change rule of historical data and its influence on future load change, and seeking the internal relationship between power load and various related factors according to the historical data of power load, economy, society and meteorology. In the power system, the overall planning, specific design and real-time operation of security, economy, social and other aspects will be involved in the problem of power load prediction. The main purpose of the power sector is to provide all users with low price, high quality electric energy. Therefore, it is necessary to make as accurate a prediction as possible for the characteristics and changes of power loads,

J. Macintyre et al. (Eds.): SPIoT 2021, LNDECT 98, pp. 518–525, 2022.
https://doi.org/10.1007/978-3-030-89511-2_67

which is also an important reason why people continue to explore the load prediction theory of power system [2, 3].

Foreign scholars began to foresee the importance of power load forecasting very early, so foreign research on power load forecasting started earlier. At the beginning of the 21st century, foreign power load prediction methods develop more rapidly, and the combination technology of neural network algorithm and power load prediction is also in continuous innovation. Japanese scientists use ANN artificial neural network method to establish a similar daily power load prediction method. In the prediction, the future power load data can be predicted by learning the power load data under a large number of similar conditions. The final results prove that this method can predict the trend of power load several hours in advance [4, 5]. In order to cope with the challenge of prediction accuracy brought by complex environment, the prediction method combining swarm intelligence algorithm with single neural network prediction method has gradually become a hot topic [6].

In this paper, the historical data samples were carefully analyzed, and the factors affecting the load size were classified in detail. By adopting targeted prediction models for different types of influencing factors, a combined prediction model with better prediction performance was finally obtained.

2   Combined Prediction Model Based on Cloud Computing and Intelligent Algorithm

2.1   Power Load Forecasting

Power load forecasting technology is to use the historical data of the load and the influence factors of the historical load, combined with a certain regression method or intelligent algorithm such as data mining method, establish a load forecasting model, and use the model to forecast the power load in a certain period of time in the future. The precision of power load prediction is the important foundation of power system planning, operation and dispatching. Load forecasting can be divided into medium - and long-term forecasting, short-term forecasting and super-short-term forecasting according to the length of forecasting cycle. The forecast period of medium and long term load forecast is generally in a few months or even a few years. This kind of forecast combines the growth rate of national economy and the support of regional government and other factors, and plays a reference role in the future power construction planning. Short-term load forecasting is the most widely used, and the forecast period is generally the next 24 h. It is used by power plants to make power generation plans, start and stop of units, etc., and power supply companies to make scheduling arrangements and set peak and valley electricity prices, etc. Ultra-short-term load forecasting is generally a minute-level forecasting period, which is used for online safety checking and optimization adjustment of inner-day and real-time generation plan and dispatching plan of the power grid [7].

In order to improve the core competitiveness, the power supply service providers should provide the corresponding power supply service strategies according to the different types of customer loads. With the increasing popularity of electric vehicles and high-power electrical equipment for residential users, the impact of residential load on the power system is getting higher and higher. Power supply service providers should formulate corresponding demand response strategies based on the actual load prediction [1]. At the same time, in recent years, in addition to the influence of the temperature, humidity, air pollution effects on people's life and work and power load is higher and higher, especially for industrial load, affected by the fog is very high also, according to the situation of industrial load forecast, formulate the corresponding response strategies, reduce the economic loss, reduce pollution emission is one of the important research direction in future industrial enterprises.

2.2 SVM Predictive Model Construction

When using the particle swarm optimization algorithm to optimize the least square support vector machine, it is found that the particle swarm will fall into the local optimum, so the particle swarm optimization is carried out to improve the accuracy of the algorithm. When the particle swarm reaches prematurity, the solution space is redistributed to guide the particle swarm out of the local optimal accelerated convergence. At the same time, the nonlinear variation of particle swarm weight coefficient is introduced to replace the fixed value or the general linear inertia factor replacement to optimize the particle swarm optimization algorithm.

The updating mode of nonlinear inertia factor is as follows:

$$\omega_g = \omega_{min}(\omega_{max}/\omega_{min})^{1/(1+10\ Count\ LoopC)} \tag{1}$$

In the formula, ωmax and ωmin are the maximum and minimum values of inertia factors respectively; Count and LoopC are the current iteration times and the upper limit of iteration respectively.

In order to solve the problem that many data are linearly inseparable, the kernel function is used to reduce the difficulty of data processing. The Gaussian kernel function, which is simple in form, radially symmetric and can guarantee the existence of arbitrary derivatives, is selected for load forecasting.

The Gaussian kernel function can be expressed as:

$$K(x, x_i) = \exp(-\|x - x_i\|^2/2\sigma^2) \tag{2}$$

Where, x is the input vector, xi is the center of the kernel function, σ is the parameter of the kernel function.

The only SVM parameters selected by particle swarm optimization are penalty parameter γ and kernel parameter σ. The flow of LS-SVM short-term power load prediction model based on particle swarm optimization is shown below

(1) The parameter regularization parameter and kernel parameter that need to be optimized are mapped to particle swarm.
(2) Initialize the parameters (such as acceleration constant, population size, iteration evolution number, etc.).
(3) Select the update mode of inertia factor.

(4) Calculate the fitness value of each particle:

$$f(x) = \sum_{i=1}^{n} (y_i - y_i)^2 \tag{3}$$

Obtain the optimal position of the ith particle and the optimal position of the population.

(5) Modify the particle position and velocity according to the formula to obtain a new population X(t).

(6) The adaptive value of the new position of each particle is obtained and compared with the historical optimal position and the historical optimal position of the population. The new optimal value is generated by judging whether to replace according to whether the new value is optimal.

(7) Judge whether the result is up to the standard. If the set threshold is met, the optimal solution will be returned; if not, it will return to (4) to continue.

2.3 Combined Predictive Model Based on Cloud Computing

(1) Principle of cloud computing

Cloud-computing refers to the service that provides users with dynamic and easily extensible virtualized storage of computing resources through the Internet, and is a new mode of use and delivery of Internet services [8]. Users can connect their computers to the data center of the cloud platform through personal computers and provide corresponding computing services according to their own needs. Cloud computing technology can provide users with a super-fast computing speed of 10 trillion cycles per second. Technically, cloud computing technology is inseparable from the emergence of big data. When the processing of big data is limited by the lack of single computing resources, the emergence of cloud computing becomes an inevitable result of the development of big data [9].

(2) Cloud computing construction

All the major cloud computing service providers have developed in combination with their own business characteristics and have their own advantages in various fields. In view of the characteristics of power system short-term load forecasting, this paper chooses Windows Azure cloud computing platform provided by Microsoft as the big data processing platform of the combined forecasting model. First of all, Windows Azure Visual Studio and Visual Environment tools have good applicability for the staff using C++ and SQLServer. In addition, the high quality services of Windows Azure platform have been widely praised by users. It has the advantages of always online, flexible and open, unlimited storage, two-way interaction and complete functions [10]. Platform as a Service is a platform where users can deploy their own personalized requirements without considering platform management and control. This paper utilizes the operating system and code runtime provided by Windows Azure platform. The short-term load forecasting combined forecasting model application is deployed to the Windows Azure cloud computing platform for calculation in strict compliance with the computing specifications of the platform [11]. In this platform, you can detect and modify your own work at any time, giving full play to the powerful parallel computing ability of the cloud computing platform.

After the parallelization, the combined prediction model application greatly improves the computational efficiency of the prediction model and can adapt to the big data processing architecture of the cloud computing platform. The combined prediction model parallel application is deployed to Visual Studio application development platform, and migrated to Azure project under Windows 7 operating environment for project initialization. It mainly includes project attribute Instance, virtual computer Size VM Size and virtual machine geographical location. After the project is deployed to the Azure platform, the preparation of the computation is basically complete.

3   Simulation Experiment of Combined Forecast Model Based on Cloud Computing

3.1   Experimental Environment

The experimental cluster system in this paper is set up in the computer network information security laboratory, and the cluster contains eight computers. Since the laboratory computers are not equipped with a special Linux system, virtual machines are installed for each computer in the experiment, and the algorithm execution process is completed in the virtual machine environment.

3.2   Sample Selection

The traditional short-term load forecasting is the theoretical basis of the extended short-term load forecasting. Therefore, the traditional load forecasting model can be retained and introduced into the construction process of the forecasting model based on the actual demand. For this reason, we introduce the following two sample sets in this chapter:

The first group, real-time monitoring sample set. The data is the load size after each time node of the day. The purpose of using this data is to provide a rigorous and accurate data basis for the subsequent revision of the forecast results.

The second group is the sample set of the load forecast results of the day. Data is the load size before each time node of the day and the load size at the time node. The data construction of the second group provides data reference for the subsequent model building.

4 Simulation Experiment Conclusion

4.1 Computing Time of Different Platforms

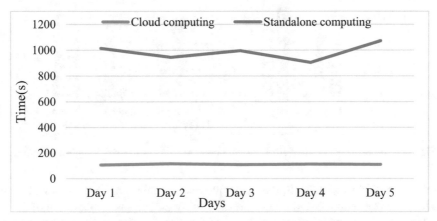

Fig. 1. Single computation time of stand-alone computing platform and cloud computing platform

As shown in Fig. 1, when the combined prediction model is run on a single computer in the laboratory, the operation time is relatively long, which is basically maintained between 900 s and 1200 s. However, the introduction of cloud computing into the calculation process of the combined prediction model greatly shortens the calculation time and improves the calculation efficiency, and the prediction time is basically maintained between 100 s and 120 s. The results show that cloud computing not only solves the problem of big data processing for smart grid short-term load forecasting and improves the calculation speed, but also considers more influencing factors to improve the accuracy of the calculation results without sacrificing the calculation accuracy, and well solves the contradiction between speed and accuracy.

4.2 Error Analysis of Different Platforms

Table 1. Comparison of relative errors of prediction results of different computing platforms

	Day 1	Day 2	Day 3	Day 4	Day 5	Mean value
Cloud computing	1.20%	3.74%	1.14%	3.02%	2.78%	2.37%
Standalone computing	2.84%	5.13%	0.61%	2.04%	3.15%	3.15%

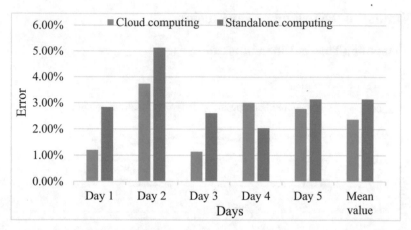

Fig. 2. Comparison of relative errors of prediction results of different computing platforms

As shown in Table 1 and Fig. 2, the model proposed in this paper has a relatively small prediction error, with a five-day mean error of only 2.37% and a maximum of 3.74%. The two-day mean error of the model was 3.15% and the maximum was 5.13%. Through the comparison experiment of daily peak load prediction simulation, it is proved that the intelligent combination short-term coincidence prediction model based on cloud computing proposed in this paper is feasible and effective.

5   Conclusions

A large number of studies show that power load is chaotic, and it is difficult to accurately predict by conventional forecasting methods. This paper first explores the detailed load sample, and combined with the characteristics of short-term load forecasting, using the analysis of the influencing factors, select the appropriate influencing factors as the training sample. On this basis, the comprehensive prediction model based on cloud computing is further constructed, and the determination of uncertainty is used. Finally, the influencing factors are further divided into two basic categories, namely, uncertainty factors and deterministic factors. For the uncertain factors, Cloud-SVM is selected to improve the accuracy of the prediction results, and particle swarm optimization SVM method is used to obtain the final load prediction results after the completion of the weighted processing.

Acknowledgments. Research on ultra short term load data analysis and forecasting method of smart grid" research results, the subject comes from the special fund project of basic scientific research business expenses of Heilongjiang Provincial undergraduate universities in 2020, subject number: 2020-KYYWF-0515.The name of the project is "Research on data analysis and forecasting method of ultra-short-term load in smart grid".This paper is supported by Natural Science Foundation Project of Heilongjiang province (Grant No.LH2019F050). The name of the project is "Research on the control system of intelligent discrimination of electricity stealing and omission". Research achievement of "Research on propagation velocity model and defect detection of internal stress wave in trees", which is from the natural science foundation of Heilongjiang Province, subject number: LH2020C091.

References

1. Zhu, Z., Bo, Y., Wu, P., et al.: Short-term power load forecasting model based on QPSO-RBFNN. Nanjing Li Gong Daxue Xuebao/J. Nanjing Univ. Sci. Technol. **40**(1), 97–101 (2016)
2. Quan, H., Srinivasan, D., Khosravi, A.: Short-term load and wind power forecasting using neural network-based prediction intervals. IEEE Trans. Neural Netw. Learn. Syst. **25**(2), 303–315 (2017)
3. Zhao, H., Guo, S.: An optimized grey model for annual power load forecasting. Energy **107**(jul.15), 272–286 (2016)
4. Hu, R., Wen, S., Zeng, Z., et al.: A short-term power load forecasting model based on the generalized regression neural network with decreasing step fruit fly optimization algorithm. Neurocomputing **221**(C), 24–31 (2017)
5. Li, C., Li, S., Liu, Y.: A least squares support vector machine model optimized by moth-flame optimization algorithm for annual power load forecasting. Appl. Intell. **45**(4), 1–13 (2016)
6. Jurasz, J., Mikulik, J.: Day ahead electric power load forecasting by WT-ANN. Przeglad Elektrotechniczny **1**(4), 154–156 (2016)
7. Khuntia, S.R., et al.: Forecasting the load of electrical power systems in mid- and long-term horizons: a review. IET Gener. Transmission Distrib. **10**(16), 3971–3977 (2016)
8. Xia, Z., Wang, X., Zhang, L., et al.: A privacy-preserving and copy-deterrence content-based image retrieval scheme in cloud computing. IEEE Trans. Inf. Forensics Secur. **11**(11), 2594–2608 (2017)
9. You, C., Huang, K., Chae, H.: Energy efficient mobile cloud computing powered by wireless energy transfer. IEEE J. Sel. Areas Commun. **34**(5), 1757–1771 (2016)
10. Abbas, H., Maennel, O., Assar, S.: Security and privacy issues in cloud computing. Ann. Telecommun. **72**(5–6), 233–235 (2017). https://doi.org/10.1007/s12243-017-0578-3
11. Hameed, A., et al.: A survey and taxonomy on energy efficient resource allocation techniques for cloud computing systems. Computing **98**(7), 751–774 (2014). https://doi.org/10.1007/s00 607-014-0407-8

Digital Hotel Unified Management Platform Under Artificial Intelligence

Yan Li[✉]

Guangdong Vocational College of Innovation and Technology, Dongguan 523960, Guangdong, China

Abstract. Along with the development of information technology and the change of emerging technologies, AI is driving a new wave in the field of platform design with new trendy technical concepts and industry applications, and the theme of profitability of hotels in the process of operation is transformed from single mode to multiple modes. With the rapid development of China's economic system reform, people's income is increasing, the concept of consumption is also improving, and the requirements for service quality are more stringent. Many regions have tourism as the local pillar industry. Hotels as people rest and entertainment places and has been rapidly developed. How to provide more accurate and timely service has become the key to hotel competition. Therefore, hotel information construction has become a standard for hotel customer service providers and become the basis of information system. This paper designs and develops a unified management platform for digital hotels under AI, in which the platform contains the common tasks in the daily operation and management of hotels, and the platform has novel and convenient features. Experiments show that the unified platform has significantly reduced training and customer service costs compared with the traditional platform, saving software development costs, while the security and stability of the system can be guaranteed. The operation modules are relatively independent and correspond to the actual business needs, and the business logic is clear, which makes the whole solution configuration flexible and convenient for users to choose.

Keywords: Artificial intelligence · Digital hotel · Unified management · Platform design

1 Introduction

The hotel industry is a promising and competitive industry. Due to the rapid development of China's tourism industry especially in the last decade, the scale of the industry has been expanding, the number of employees has been increasing, the business system has been innovating, the business environment has been improving, and the hotel industry has been completely opened to the outside world [1, 2]. At this time, the hospitality industry is facing unprecedented opportunities and challenges [3, 4]. Therefore, to ensure the improvement of the quality and efficiency of the work of hotels as well as to promote the development of tourism in China, it is necessary to adopt advanced computer information

© The Author(s), under exclusive license to Springer Nature Switzerland AG 2022
J. Macintyre et al. (Eds.): SPIoT 2021, LNDECT 98, pp. 526–533, 2022.
https://doi.org/10.1007/978-3-030-89511-2_68

technology and AI to change the current business model of the hotel industry, and it has become a necessity to realize the automation of the business management model of the hotel industry [5, 6].

Hotel management information system was developed in foreign countries in the 1970s. In the 1980s, with the expanding application of computer technology and hotel management information systems in the hotel industry, hotel management information systems have been continuously developed and improved [7]. EECC hotel system in the United States, HIS hotel system and Fidelio hotel system in Germany, etc., the overall model of these hotel management systems has been basically finalized, the technology is more mature, the functions are also more complete and basically realized the hotel front desk business management of room reservation, room arrangement, cashier checkout, restaurant management, night audit management and other functions, but at that time, personal computers were not popular, the deployment of the whole system was based on the centralized minicomputer/terminal management model [8, 9].

This chapter introduces the design of the digital hotel unified management platform under artificial intelligence (AI), which mainly includes the architecture design of the unified management platform, the design of the system workflow and the functional design of the specific business system and the comprehensive query system in the unified management platform, and proposes the evaluation algorithm of the hotel operation status, and finally the layout of the overall interface of the system is designed in a unified manner.

2 Research and Design of Digital Hotel Unified Management Platform Under AI

2.1 Overview of Digital Hotel Unified Management Platform Under AI

The world enters the era of informationization in the twenty-first century, and the rapid development of high technology brings opportunities and challenges to hotel management informationization, which will promote the development of hotel management in the direction of networking, intelligence, virtualization and digitalization, and it is an inevitable trend to build a digital hotel unified management platform under AI [10]. However, looking at the design and use of the current digital hotel platform, there are the following main problems.

(1) digital hotel platform both design and application should be a whole, but in the implementation of hardware and software separate, not only the development is separate, and engineering implementation is also independent of each, the result is to cause the application of the overall low compatibility, affecting the use of efficiency [11].

(2) In the same hotel management software, there are different development periods and the fruits of labor of different developers, resulting in the developers of information management software, streaming software, e-commerce software to do their own thing, the lack of unified design specifications and concepts, the result is that the digital hotel platform in the application, the transfer and collection of data is cumbersome, affecting the accuracy and timeliness of data.

(3) In the process of hotel management process changes are difficult, in the previous hotel management system in a variety of management processes are generally "hard-coded" to achieve, in the future application, to change the workflow, you must modify the software code, which brings the problem of software maintenance difficulties [12].

In summary, the establishment of a unified digital hotel management platform with AI, the unified integration of the loose and independent software resources of hotels into the network management platform, and the construction of a complete digital hotel unified management platform is the problem that needs to be solved, and this problem is also the core of the research in this paper.

2.2   System Functional Requirements

Through the demand analysis of the operation business in hotel management, its main functional requirements are: to complete the front desk management of hotel reservation, reception, cashier, checking, room, sales, report, audit, etc.; to complete the peripheral cashier system management of catering, bathing, entertainment, etc.; to complete telephone billing, accounts receivable management, inventory management, system management, data query, personnel salary management, general manager comprehensive query, financial accounts management and other background management; complete the connection with membership card, attendance machine, second generation card, SMS cat, call box, face recognition, handheld machine, touch screen, fingerprint instrument, Android tablet ordering and other peripheral systems; realize the purpose of hotel industry management informationization to save manpower, optimize configuration, improve efficiency, control costs, data analysis, decision-making, meet the needs of hotel operators and solve the current traditional hotel industry business difficult problems.

2.3   Design of Digital Hotel Unified Management Platform Under AI

(1) Architecture design

   The digital hotel unified management platform under AI is an integrated platform, which contains the main work in the daily activities of each hotel operation site, and it can be simply said that almost all the work related to hotel management can be completed on the digital hotel unified management platform.

   Each system in the digital hotel unified management platform under AI can run independently without affecting each other, but when data sharing is needed, data sharing can be effectively realized between each system to avoid duplication of data entry. In the implementation of the digital hotel unified management platform, you can also install different systems according to the needs of the user, these systems are module independent, when the user needs, you can install a system, if the user in order to save operating costs, you can also temporarily not install a system.

(2) Assessment of the hotel's operating conditions

The assessment of the hotel's operating condition is mainly based on the existing equipment and facilities in the hotel's business management activities, the degree of impact on the guests, and the satisfaction of the guests with this activity or project. It allows the general manager to have a general understanding of the overall operation of the hotel on a macro level, and also provides an early warning of the crisis that may occur in the hotel's operation in the future.

Hotel evaluation index vector.

$$V_1 = W_1 \times T_1 \tag{1}$$

Where W1 represents the weights of each evaluation index, T1 represents the specific value matrix of each evaluation index, and V1 represents the specific evaluation results.

(3) User interface design

Digital hotel unified management platform, in the development of the software interface design is required to be unified, beautiful, and easy to use, convenient for user information input. Whether the user interface is friendly or not reflects the programming level of the designer to a large extent, and also directly affects the efficiency of user use and the authenticity of the data generated. In this paper, for the design of each software in the unified management platform, the main points to consider in the design of the user interface are as follows.

(1) interface unity. Different systems to maintain a unified operating style, so that users can use a system, other systems can also achieve the degree of "will operate".

(2) Simple interface. All the systems in this platform adopt the design style of Windows forms, and put different information management into different Windows forms, so that users can understand at a glance in the process of use.

(3) Avoid duplicate data entry. The more information entered, the greater the chance of error, the guest information needed in the hotel management information system will be more, in the unified management platform, a variety of data as much as possible to do data sharing, if the data exists in the unified management platform, should be obtained directly from the system, without requiring the user to re-enter.

(4) clear prompt information, when the user enters the information wrong, should give a clear prompt information, but not give the error information given in the process of program modulation, as far as possible the error prompt informations are able to let the user see.

(5) The design of dynamic service permission management model based on role assignment.

In simple terms, the role-giving operation is to extend the runtime code provided by AspectJ, so that the type that does not have a specific function can dynamically support this function and achieve the purpose of code-level enhancement. This enhancement is dynamic and can be determined and changed at runtime.

The dynamic service permission management module based on role endowment has the following key points, namely.

1) The source type is dynamically converted (or extended) to the target type by dynamically adding attributes and functions to the operation to complete the dynamic adaptation of the user to the role user over.

   This process is different from the common object-oriented IS-A and Has-A relationships, but a new form of object transformation, which is described as "A AS B". Therefore, the authors define this operation as an AS operation.

$$User\ user \tag{2}$$

$$user.as(CheckInRole) \tag{3}$$

2) Permission verification: before using AS operators, it is necessary to verify that the instance has this role permission, and if not, the transformation cannot be performed.
3) Service invocation: After dynamic role assignment using AS operator, expect to invoke the function intuitively using the following way.

$$user.as(CheckInRole).changeRoom(aRoom) \tag{4}$$

3 Platform Deployment and Testing

3.1 Deployment Environment

In the development of the platform, its core modules are realized by Java and related technologies, and its subsystems are realized by Delphi, VC++ and other languages according to the development situation at that time. Its data integration uses Web Service technology. The message channel, using JMS, is implemented. Its platform deployment environment is as follows.

Tomcat: core platform components.
Microsoft SQL: core platform database, general manager integrated query database.
Apache Axis: open source Web Service engine.
Apache Active MQ: open source JMS implementation.

3.2 Platform Operation

Hotel front desk management is one of the most important system modules in the hotel business, it needs frequent interaction with users, in the reception business of the front desk can be divided into casual guests, groups and companies, for different types of guests using different reception processes.

(1) Casual guest reception
   Select the room you want to register in the available rooms, enter the casual guest check-in interface, and then you need to fill in some information about the guest. After confirmation, the user's casual check-in function will be completed and a record will be inserted in the database check-in information table, at which time the system will also set the room to occupancy status.

(2) Group reception

Similar to casual reception, group reception can complete the task of opening rooms for groups of guests. There is no limit to the number of rooms, and different types of rooms can be added at the same time.

(3) Guest Checkout

This function will count all the consumption amount and payable amount incurred by each guest while staying in the hotel, and complete the related checkout and cashier operation.

(4) Room Reservation

The function realized by this module can complete the functions of adding reservation, modifying reservation, deleting reservation, querying reservation status, etc.

(5) Business Inquiry

The function of this module can be realized to query the billing that has been checked out according to the checkout time, guest name, bill number and room number.

4 Test Analysis

4.1 Platform Operation Efficiency

Since the platform uses SOAP requests based on HTTP protocol, this architecture can effectively meet the needs of distributed and Internet applications, but it is foreseeable that its execution efficiency will be slightly slow compared with the traditional stand-alone mode system software. In this paper, the average response time of several major functions of the hotel front desk management module is summarized in Table 1.

As can be seen in Table 1 and Fig. 1, the average response time has increased substantially due to the use of HTTP network and Web Service requests, and the average response time is basically the same as the time using Web software. However, due to the significant decrease in the cost of hardware resources at this stage, the increase in response time is still within the acceptable range for users. Moreover, due to the unified data exchange method used, the business logic is also clearer and the resource utilization is more reasonable, so this reasonable delay is acceptable.

Table 1. Comparison of the response time of several functions of the guest room management module

	Traditional software	Unified platform
Room reservation (load)	35	127
Room reservation	123	237
Guest Statistics	431	1974

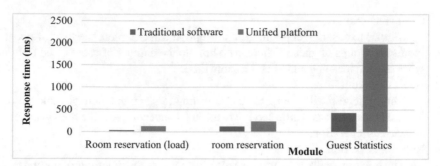

Fig. 1. Response time comparison of several functions of the guest room management module

4.2 Training Implementation and Customer Service Evaluation

Since the new platform is erected on the existing software, the user's operation habits and ways can be retained through the platform, so that the average training cost for platform users is significantly reduced, as shown in Fig. 2.

The system platform is analyzed by hundreds of users' training and customer service tracking data in terms of training course time, consultation call time, number of non-repeat questions and bug feedback. From Fig. 2, we can see that the unified platform has significantly reduced the training and customer service costs compared with the traditional platform.

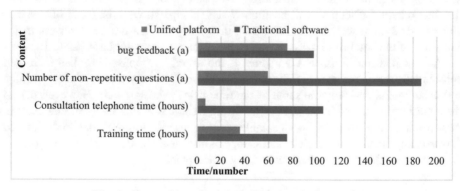

Fig. 2. Comparison of training and customer service costs

5 Conclusions

With the rapid development of China's tourism service industry, the number of hotels is increasing, the service level of hotels is gradually improving, and the scope of hotel business services is gradually expanding, which requires the level of informationization of hotel management to keep pace with the development of hotel service industry. In the daily management of hotels, there are many business management systems, and there are many business processes in these systems, and these business processes are relatively similar, if each one is developed separately, the workload is relatively large, and it is difficult to maintain in the operation process; at the same time, the processing of a business process may span multiple management information systems, which requires the sharing of business processes between systems data between systems. The workflow management technology is used to establish a unified workflow management platform, and all workflows are extracted from specific business application systems for separate development, and workflows are allowed to be added dynamically by users in the process of application, and users are free to modify and delete workflows, so that all workflows are "customized" by users. This not only saves development cost, reduces the workload of program design, but also increases the scalability and maintainability of the system business process.

References

1. Zheng, L., Wang, H., Mei, L., et al.: AI in digital cariology: a new tool for the diagnosis of deep caries and pulpitis using convolutional neural networks. Ann. Transl. Med. **9**(9), 763 (2021)
2. Sabri, M.S.: Role of AI in digital marketing. SSRN Electron. J. **10**(5), 1–6 (2021)
3. Lutz, C.: Digital inequalities in the age of AI and big data. Human Behav. Emerg. Technol. **1**(2), 141–148 (2019)
4. Lai, Z., Yu, L.: Research on digital marketing communication talent cultivation in the era of AI. J. Phys. Conf. Ser. **1757**(1), 012040 (2021)
5. Gorodetsky, V.I., Laryukhin, V.B., Skobelev, P.O.: Conceptual model of a digital platform for cyber-physical management of a modern enterprises. Part 2. Digital services. Mekhatronika Avtomatizatsiya Upravlenie **20**(7), 387–397 (2019)
6. Schneider, I.: democratic governance of digital platforms and AI: exploring governance models of China, the US, the EU and Mexico. JeDEM – eJ. eDemo. Open Gov. **12**(1), 1–24 (2020)
7. Huo, D., Chen, Y., Hung, K., et al.: Quality management of E-business: a key node analysis of ecological network in digital economy by using AI. Romanian J. Econ. Forecast. **22**(2), 166–179 (2019)
8. Bunti, L., Dami, M., Duevi, I.: AI in business models as a tool for managing digital risks in international markets. SHS Web Conf. **92**(2), 03005 (2021)
9. Ma, X., Cheng, J., Qi, Q., et al.: AI enhanced interaction in digital twin shop-floor. Procedia CIRP **100**(1), 858–863 (2021)
10. Leontev, M., Magera, T.: Unified digital platform for the transport industry and human factor. IOP Conf. Ser. Mater. Sci. Eng. **918**(1), 012070 (2020)
11. Sirait, A., Murdianingrum, S.L.: Management decision in joining hotel network on digital marketing-based for financial and non-financial impacts (study: 4 and 5-star hotels in yogyakarta special region). Open J. Bus. Manag. **08**(2), 754–762 (2020)
12. Duan, L., Ma, S., Li, Y., et al.: Unified management method for multi-dimensional basic power resources. Procedia Comput. Sci. **183**(15), 827–832 (2021)

Analysis on the Application of Pulse Acquisition Technology in Intelligent Crutches System

Guokun Xie[✉] and Ningning Wang

Electrical Engineering Institute, Xi'an Traffic Engineering Institute, Xi'an, Shaanxi, China

Abstract. As the aging of our country's population continues to increase, the number of elderly people in our country continues to increase, and the demand for aging products will also continue to increase. At the same time, with the continuous innovation of technologies such as sensors and artificial intelligence, the pulse acquisition technology has been developed in the field of integration, standardization, and intelligence in the integration, standardization, and intelligence of the intelligent crutches system designed for the elderly and the integration of pulse acquisition technology in the research process. This article focuses on the application of pulse acquisition technology in the intelligent crutches system. This paper designs the overall structure of the pulse signal acquisition system, and completes the design and implementation of the hardware platform of the pulse signal acquisition system. The hardware part of the system includes signal acquisition circuit and FPGA platform. This article introduces the design and implementation process of the two functions of the intelligent crutches system using pulse acquisition technology, namely, the fall detection algorithm and the heart rate detection algorithm, and the feasibility of the system is verified through tests. In order to test the reliability and accuracy of the device and algorithm, a fall detection experiment was carried out. The average detection accuracy of the five experiments was 97.332%. It can be seen that the fall detection device of the smart crutches can correctly distinguish most fall actions, and can achieve the expected design effect.

Keywords: Pulse collection · Smart crutches · Heart rate detection · Smart products

1 Introduction

With the improvement of people's living standards, people pay more and more attention to health. Pulse is an important indicator of human life activities, and pulse signals are rich in physiological and pathological information [1, 2]. The accurate measurement and analysis of pulse signals can effectively monitor the health of the human body and realize early detection and early prevention of some cardiovascular diseases [3, 4]. Especially for the elderly, the collection of pulse is even more important [5]. Contemporary technology is changing with each passing day, and the update speed is staggering. The use of pulse acquisition technology to design a smart crutches for the elderly has important research significance [6, 7].

© The Author(s), under exclusive license to Springer Nature Switzerland AG 2022
J. Macintyre et al. (Eds.): SPIoT 2021, LNDECT 98, pp. 534–541, 2022.
https://doi.org/10.1007/978-3-030-89511-2_69

Regarding the research of pulse acquisition and intelligent systems, many scholars have launched in-depth discussions on it. For example, DeAbreu comprehensively
applies B-ultrasound technology and MRI to obtain pulse information, including the
relative position, size, length, and movement of the pulse [8]; Tofield A researched and
developed a pulse acquisition device that integrates 3 pressures and 9 photoelectric sensors. At the same time, a pressure mechanism is set to continuously adjust the pulse
pressure to collect photoelectric pulse signals and pressure pulse signals under different
pressures [9]; Zhang SB has developed and manufactured three pulse collection devices,
which detect and analyze different pulses by using different sensing devices [10].

This article focuses on the application of pulse acquisition technology in the intelligent crutches system. This paper designs the overall structure of the pulse signal acquisition system with FPGA as the core chip, and completes the hardware platform design
and implementation of the pulse signal acquisition system. The hardware part of the
system includes signal acquisition circuit and FPGA platform. This article introduces
the design and implementation process of the two functions of the intelligent crutches
system using pulse acquisition technology, namely, the fall detection algorithm and the
heart rate detection algorithm, and the feasibility of the system is verified through tests.

2 Application of Pulse Acquisition Technology in Intelligent Crutches System

2.1 Pulse Acquisition Module Design

(1) Filter circuit

The frequency of the power frequency noise in this system is around 50 Hz,
which is greater than the highest frequency of the pulse signal. A low-pass filter
can be used to filter the power frequency noise mixed in the pulse signal [11, 12].
Therefore, when designing the conditioning circuit, a low-pass filter with a cut-off
frequency of 30 Hz can be set.

The gain calculation formula of the filter:

$$K = 1 + \frac{R_{10}}{R_{11}} \tag{1}$$

The filter cutoff frequency calculation formula:

$$f_0 = \frac{1}{2\pi \sqrt{R_{10} * R_{11} * C_5 * C_6}} \tag{2}$$

Take $R_{10} = 0$, $R_{11} = \infty$, and directly connect the inverting input end of the
operational amplifier U7D with the output end, $C_6 = 2.2u$, $C_5 = 1u$, after calculation:
$K = 1$, $f_0 = 30.1216$ Hz. Therefore, its gain is 1, and its cut-off frequency is 30 Hz.

(2) Amplification and voltage boost circuit

Because the signal sent by the pulse sensor is weak and has a small amplitude, and various interferences from the environment will be mixed into it,
and high impedance exists between the skin and the sensor, the purpose of the
pre-amplification circuit in this part is clear and needs to meet the following Claim:

1) Very high input impedance means that the input terminal must have a particularly high input impedance, which can avoid a large part of the unstable signal caused by the poor circuit connection, so as to enhance the anti-interference effect.

2) It must have a particularly high common-mode rejection ratio, that is, it can well suppress the common-mode signal generated by the circuit.

3) The selected chip must have fairly small noise, good linearity, and negligible drift.

4) The selected chip must have a frequency range suitable for the signal characteristics of the entire acquisition system.

Based on the above requirements and analysis, and in various comparisons carefully selected, this study chooses AD620 as the core chip of the preamplifier.

It is well known that the gain of the amplifier circuit is proportional to its noise, so the amplification factor of AD620 is not easy to be too large. This article uses a potentiometer R7 with a size of 0–5K. Different gains can be adjusted when doing experiments. At the end of this article choose a magnification of 20 times.

In addition, in order to better eliminate the influence of various signals of the human body, a compensation circuit that resists various interference signals is added to this circuit. The chip used in the compensation circuit is OP07, and the specific method is to design a negative feedback between the signal source and the feedback terminal of the amplifying circuit.

(3) Serial port circuit

A serial port function module is designed in this system, and the user can transmit the collected pulse signal to the PC as needed. The standard of RS-232 interface level is adopted in the design. When FPGA uses serial communication, this system chooses MAX3232 as the level conversion chip.

2.2 System Software Design

(1) AD conversion

This system selects ADC0809. The function of ALE is to latch the address. It works when it is 1, and the function of START is to indicate the start of the conversion start signal. It works when it is 1, and the chip will be autonomous at the rising edge of the signal. Reset, start conversion at the falling edge, and keep the low level state until the end of the conversion. EOC is the judgment point for stopping the AD process. The conversion time is about 100 microseconds. After the AD process stops, EOC becomes 0. The function of OE is whether to enable the signal when data is output. After the current signal AD is completed, it is 1. Then the data in the tri-state buffer will be output from the bus.

(2) Serial communication

The second important function of the main core chip 89S52 is to transmit the signals processed and transmitted by AD to the host computer through the serial communication module in real time and accurately. According to the introduction in this article, the MCS-86 optically isolated USB to TTL serial port module is selected. It is regarded as a serial port in the application, so 89S52 can send data to the host computer through its own TXD and RXD through this serial port module. Write the program of the main core chip according to the functions realized above.

(3) PC software design

Choose MATLAB in the host computer to process and display the final pulse diagram. The data sent by the single-chip microcomputer through the serial port module is used in this article to receive and save the data using the serial port assistant, and then send it to MATLAB for processing and display. The signal data collected by the serial port assistant is binary data that has been converted from analog to digital. Therefore, this data must be processed in advance before being sent to MATLAB for data processing. The purpose is to display the data as the essential image of the original pulse signal. It is conceivable that the process of analyzing and processing the data can be completed next.

2.3 System Function

(1) Fall detection

1) The system first detects the current acceleration of the crutches, and compares the detected value with X_1. When the acceleration value of the crutches is detected to be less than X_1 for a period of time, it is judged that the crutches are weightless.

2) Continuing, when the system detects that the crutch has experienced weightlessness, it will continue to detect whether the angular velocity of the crutch is greater than Y_1. In normal life, only the acceleration vector and reduction cannot determine the accurate posture of the human body. Normal human activities such as bending and squatting cannot have a great impact on the crutches angular velocity sensor. Therefore, if and only when the system determines that the current crutches motion angular velocity value is too large and exceeds Y_1, it will further initiate the fall determination process.

3) When the system determines that the crutches' motion state meets the above two conditions at the same time, the system will continue to detect whether the crutches has a collision with the ground after falling to the ground, that is, whether the crutches' motion acceleration exceeds the threshold x_2. After a fall, the crutches will collide with the ground. The collision process is generally severe. The elderly are injured mostly during this period. Therefore, the crutches will undergo a large acceleration change during this process, and at the same time this is also a very obvious feature of the entire process of falling.

4) When the system determines that an impact action has occurred, it will further detect whether the crutches have also taken up a standing action after the fall. If the detected angular velocity does not exceed the threshold y_2, it indicates that the crutch has maintained a relatively static state for a short time after falling to the ground. When the detected angular velocity exceeds the threshold y_2, it indicates that the ward can still lift the crutches by itself in a short time, and the angular velocity exceeds the threshold, indicating that it is not a valid fall, and may be caused by the crutches accidentally falling to the ground.

5) If the crutch does not move up, the system will continue to detect whether the crutch is in a static state for a long time, and set a threshold of X_3. When the acceleration of the crutches cannot exceed the threshold X_3 for a long period of time, it indicates that the ward moved before the occurrence, you may have lost consciousness, or you may be conscious but unable to save yourself; if the crutches remain stationary for a long time and the angular velocity does not change, it is determined that this is an effective fall.

(2) Abnormal heart rate monitoring

Since each body's constitution and physical condition are different, and the target group of smart crutches is the elderly, the elderly themselves are often accompanied by some chronic diseases, which may have a certain impact on their heart rate range. Therefore, this study adopts a personalized approach. The guardian can set the normal heart rate range according to the ward's own situation, and if it exceeds the range, it is regarded as abnormal heart rate. However, there is a certain difference in the range of heart rate between people during exercise and when they are stationary. The method for judging abnormal heart rate designed in this article is as follows:

1) Set the resting heart rate threshold S_{min}, S_{max}, which respectively represent the minimum and maximum heart rate of the ward at rest, and settings M_{min} and M_{max} respectively represent the minimum and maximum heart rate of the ward under exercise.

3   Experimental Research on the Application of Pulse Acquisition Technology in the Intelligent Crutches System

3.1   Body Posture Monitoring Function Test

In this experiment, 5 tests were performed for normal walking, falling forward and backward, and falling left and right. A total of 15 tests were carried out for normal walking, falling forward and right, and the rest were carried out 10 times for a total of 65 tests to determine whether the system a fall warning will be issued.

3.2   System Heart Rate Monitoring Function Test

Due to the large gap between the age of the tester and the target group of actual crutches users, they can only imitate the actions of the elderly in their posture, setting $M_{min} = 96$, $M_{max} = 144$, $S_{min} = 60$, $S_{max} = 100$. In order to test the reliability and accuracy of the heart rate test, the following three experiments were done:

(1) Walk normally;
(2) Falling down;
(3) Test within 6 min after exercise;
(4) More than 6 min after exercise.

4 Application Data Analysis of Pulse Acquisition Technology in Intelligent Crutches System

4.1 Fall Inspection Test Results

In order to verify the reliability and accuracy of the device and algorithm, a fall detection experiment was performed. The experimental results are shown in Table 1. The average detection accuracy of the five experiments is 97.332%.

Observing Fig. 1 can be concluded: the fall detection device can correctly distinguish most fall actions, and can achieve the expected design effect.

Table 1. Test results of fall inspection

Behavior	Number of experiments	Determine the number of falls	Judgment normal times	Correct rate
Walk normally	15	0	15	100%
Fall forward	15	14	1	93.33%
Fall backward	10	10	0	100%
Fall on the left	10	10	0	100%
Fall on the right side	15	14	1	93.33%
Average	–	–	–	97.332%

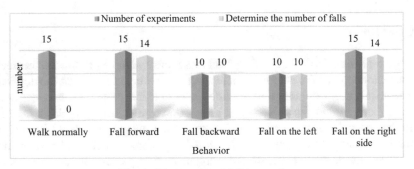

Fig. 1. Test results of fall inspection

4.2 System Heart Rate Monitoring Function Test

Use the heart rate test bracelet to compare with this system. Each experiment is tested 4 times, and the results are averaged. The experimental results are shown in Table 2: The crutches test heart rate is 92, 155, 128 during normal walking, fall, and 6 min after exercise; and the heart rate of the bracelet test is 88 bpm, 154 bpm, 129 bpm.

Table 2. Heart rate monitoring function test results

Behavior	Crutches to test heart rate	Bracelet test heart rate
Walk normally	90	88
Fall down	155	154
Within 6 min after exercise	128	129
More than 6 min after exercise	70	69

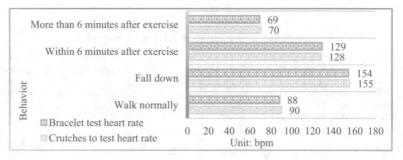

Fig. 2. Heart rate monitoring function test results

It can be seen from Fig. 2 that the heart rate detection device of the system can basically accurately measure the real-time heart rate in various situations faced by crutches users, and can achieve the expected design effect.

5 Conclusion

As the problem of aging in our country becomes more and more prominent, the demand for smart products for the elderly is increasing. This article focuses on the application of pulse acquisition technology in the intelligent crutches system. This paper designs the overall structure of the pulse signal acquisition system, and completes the design and implementation of the hardware platform of the pulse signal acquisition system. The hardware part of the system includes signal acquisition circuit and FPGA platform. This article introduces the design and implementation process of the two functions of the intelligent crutches system using pulse acquisition technology, namely, the fall detection algorithm and the heart rate detection algorithm, and the feasibility of the system is verified through tests.

References

1. Rui, J., Guan, R., Zhang, J., et al.: Design of Information acquisition system for high voltage pulse power supply. J. Phys. Conf. Ser. **1894**(1), 012094 (2021)
2. Zhou, B., Wang, X., Li, Q., et al.: Human pulse signal acquisition system based on PVDF piezoelectric film. J. Phys. Conf. Ser. **1924**(1), 012019 (2021)

3. Rui, W., Lu, S., et al.: Design and implementation of a Chinese pulse condition acquisition system. Tsinghua Sci. Technol. **04**(v.21), 89–98 (2016)
4. Xue, S., Hao, Z., An, Y., et al.: The research of three regions acquisition and analysis system of pulse based on flexible sensor. E3S Web Conf. **271**(10), 03056 (2021)
5. Wang, J., Lu, Y., Xi-Bo, Q., et al.: Design of body temperature and pulse data acquisition system based on simpliciti wireless network protocol. Int. J. Fut. Gener. Commun. Netw. **10**(1), 97–110 (2017)
6. Chen, C., Li, Z., Zhang, Y., et al.: A 3D wrist pulse signal acquisition system for width information of pulse wave. Sensors (Basel, Switzerland) **20**(1), 11 (2020)
7. Chen, Y., Napoli, D., Agrawal, S.K., et al.: Smart crutches: towards instrumented crutches for rehabilitation and exoskeletons-assisted walking, pp. 193–198 (2018)
8. DeAbreu, C.: Material database for resistive and pulse magnets. McNair Schol. Res. J. **5**(1), 5 (2018)
9. Tofield, A.: CardioPulse: a new european society of cardiology research database: atlas of cardiology. Eur. Heart J. **37**(10), 801 (2016)
10. Zhang, S.B., Hobbs, G., Russell, C.J., et al.: Parkes transient events. I. Database of single pulses, initial results, and missing fast radio bursts. Astrophys. J. Suppl. Ser. **249**(1), 14 (2020)
11. Cahan, A., Lerner, U., Fishbain, B., et al.: Abstract P2053: blood pressure variability is higher when pulse rate is lower, in a large outpatient database. Hypertension **74**(Suppl_1) (2019)
12. Lee, J., Lee, S., Hyun, J.M., et al.: Robotic scanning technology for laser pulse-echo inspection. Electron. Lett. **56**(19), 1027–1029 (2020)

The Impact of Audiovisual Network Media on Remodeling and Policies for College Students' Values

Wentao Yin[✉]

School of Marxism, Xi'an Fanyi University, No. 123 Taiyigong, Chang'an District, Xi'an, Shaanxi, China

Abstract. Audiovisual network media is an important carrier of university popular culture and has increasingly become an important source of information and distributing centre of ideas for college students. To deal with the contradiction between some value distortions and establishment of college students' correct values, we need to examine the role of audiovisual network media in reshaping college students' values, use audiovisual network media as the basic communication law for popular culture, and explore a method of improving college students' value accordingly. Finally, we can provide a reference for reshaping college students' values.

Keywords: Audiovisual Network media · Value · Remodeling

Under the strong promotion of new technologies and new applications, the audiovisual network media, which is represented by live broadcast, online video, network audiovisual, short video, and more, has developed rapidly and is gradually becoming an important source for information acquisition and distribution of ideas for college students. In addition, it has become a main form of college student consumption. The audiovisual network media caters to the individual needs of college students and satisfies their emotional appeals. The popular cultural elements carried by the audiovisual network media directly affect the thinking habits, living habits, communication methods, and values of college students. Vulgarity, pornography, fraud, and other chaos have become alienated forms of audiovisual network media. Platforms such as Kuaishou and Huoshan are alleged to be the anchors of underage pregnant women, underage mothers, and teenage second-born mothers. A large number of negative elements, such as "diaosi culture", "Otaku culture", "corrupt culture", and "declining culture" in online drama deviate from the traditional culture, distort the aesthetic value of college students, and weaken the cultural identity of college students. This breeds the values of college students' consumption supremacy, affecting their correct value judgment.

© The Author(s), under exclusive license to Springer Nature Switzerland AG 2022
J. Macintyre et al. (Eds.): SPIoT 2021, LNDECT 98, pp. 542–548, 2022.
https://doi.org/10.1007/978-3-030-89511-2_70

1 Audiovisual Network Media is an Important Carrier of University Popular Culture

With the advancement of China's Internet technology and the rise of new media, the audiovisual network media, which is an important part of the Internet media, has developed rapidly. It incorporates many characteristics, such as Internet, literature, diversity, multimedia, and so on. With the representatives, such as live broadcast, online video, online audiovisual, and short videos, audiovisual network media has greatly enriched the transmission channels of audiovisual programs. In the Internet + era, with the rise of memberships and mobile payments, the consumption of audiovisual network media has become more diverse and convenient. The rise of audiovisual network media not only leads to the transformation and upgrade of the traditional audiovisual communication pattern, but also quietly realizes the inter-generational transformation of netizens' lifestyles and the collective identity of cultural psychology [1].

Audiovisual network media is an important carrier of university popular culture. As an important part of popular culture, college students' online culture has also become diverse, such as Aidou culture, constellation culture, otaku culture, buzzword culture, runaway comic culture, subtitle group culture, barrage culture, animation culture, and network divine songs in various styles. These popular culture forms in college students' campuses can be also presented through audiovisual network media.

College students are not only active followers but also communicators of audiovisual network media. As an important carrier of popular culture, audiovisual network media features diversified content, entertaining form, and younger audience. With the increasing college students using Internet, the Internet public opinion becomes vital. Review the role of audiovisual network media in reshaping college students' values, and explore the reshaping of college students' values from the perspectives of operation mechanisms, value communication, and life influence of audiovisual network media. This is of great practical significance.

2 Audiovisual Network Media Remodels College Students' Values

As an important carrier and expression of popular culture, audiovisual network media itself has the universal characteristics of popular culture, such as popularity of the audience, secularity of the content, and entertainment of forms [2]. It is the existence of these characteristics that the audiovisual network media plays an increasingly important role in the life of college students, satisfying the desire of extensive participation and the need to relieve stress, thereby affecting the behavior patterns, value judgments, and consumption concepts of college students. Therefore, it plays an increasingly important role in shaping the values of college students.

2.1 Positive Meanings

Audiovisual network media is the product of the Internet + era. In the process of communication, audiovisual network media contains the spirit of the times and the value pursuit of college students, and therefore has positive significance for reshaping college students' values. It promotes the Chinese culture diversity and meets the growing cultural needs of the public. Its core connotation should be a unique expression in the context of multiple social values [3].

The audiovisual network media updates and enriches the values of college students. Through its vast amount of audiovisual information, the audiovisual network media spreads diverse values and caters to the current value demands of college students. The audiovisual network media steps into the forefront of campus popular culture because of its novel form, rich content, and diverse audience. The content of audiovisual network media covers the knowledge elements of many disciplines and conveys the most advanced ideas of the society. It is in line with the psychological characteristics of the students who are active in new knowledge, new ideas, and new concepts. In the blending and collision of different ideologies and cultures, the values of college students have been re-evaluated and considered. This helps increase college students' knowledge, broaden their horizons, and enhance comprehensive quality.

The audiovisual network media is conducive to stimulating and cultivating the innovative consciousness of college students. The popular culture represented by the audiovisual network media has cutting-edge characteristics, which can closely follow the pace of the times and reflect the current trend of fashion. Through the audiovisual network media, college students can accept new ideas and be active thinking, which helps them to get rid of the restraint of conformity for knowledge authority and mindset. This is conducive to the thinking integration and cultivation of reverse and divergent thinking. As a result, help college students cultivate creation and innovation of college students.

The audiovisual network media is helpful for formulating college students' health psychology. The entertainment forms of audiovisual network media are suitable for the cultural taste of college students, which can help college students get rid of pressure and troubles in real life, relieve their pressure in learning and life, and stimulate psychological needs. The popular characteristics of audiovisual network media can meet the general aesthetic appeal of college students, adapt to their modern life philosophy, and help cultivate their health psychology.

Network audiovisual media helps college students to accumulate and form their own cultural character. The popular culture form experienced by college students through the audiovisual network media makes them more quickly, comprehensively contact with the social ecology and therefor forms a more comprehensive understanding of the outside world. In the process of experience, college students cultivate their own cultural character by integrating their understanding and cognition of social life and promoting the formation of their own social consciousness and earlier socialization. Under the extensive participation of audiovisual network media, college students can further enrich their individuality and contribute to the cultivation of democratic participation consciousness.

2.2 Negative Meanings

Shoddy at will, vulgar and malicious transmission, negative intensification, and irrational emotions are common occurrences, which mislead audiences, confuse right and wrong, and lead to lower social recognition of some audiovisual network programs [4]. For example, in April 2018, TouTiao's Neihan Duanzi app and official account were permanently closed by the NRTA. The reason was that it spread vulgar content. This highlights the shortcomings of audiovisual network programs in China on content recognition.

The audiovisual network media pursues the click-through rate and attracts audience by gimmicks, which makes "cheap laughter, amusing to death and bottomless hype flooding". Vulgar, violent, anti-social and other content and shots are common. The audiovisual network media itself also becomes the "communicator" of vulgar content. The negative effects of audiovisual network media are mainly on college students' value judgments, moral values, and codes of conduct. These effects include imperfect valuation, discrimination at will, vulgar malicious communication, negative speech, and irrational emotions. The tendency of mediocrity and entertainment dwarfs university spirit and ideological realm of college students. For example, package hype stars' children, which in turn cause violations of the rights and interests of minors, contributes to the society's exaggerated atmosphere and game mentality. The "ghost" culture in audiovisual network media is not simply the category of emotional expression. "Demotivational culture" continues to display and ferment in cyberspace and social media by virtue of texts, pictures, videos and other carriers, and has gained group identity and emotional resonance among some young people [5].

Audiovisual network media can easily lead to utilitarianism in values. Part of the popular culture of audiovisual network media encourages individual success while traditional values with the core of ideals and beliefs become increasingly weak. This changes the passive cultural reception status of college students in the past, and increasingly tends to individuals, utilitarian and realistic in terms of values [6].

Audiovisual network media weakens the cultural identity of college students. Its content of cultural identity is particularly rich, but its core is to recognize the ideals, beliefs and values contained in the culture [7]. Colleges and universities are the key areas for the infiltration of western culture. As an important carrier of cultural communication, the audiovisual network media has been penetrated by western ideology and spreads western decadent ideas, lifestyles and values. In the wave of globalization, the western world, led by the United States, strongly advocates cultural integration, promotes their cultural patterns through audiovisual network media, and encapsulates capitalist values as universal values to promote their cultural hegemonism. This greatly impacts the values of college students, weakening the cultural identity of some college students.

Western values prevail in college students, such as the admiration of western festivals among college students. Cultural factors that infiltrate western values, such as Japanese anime, Korean flu, and American drama are sought after among college students. The proliferation of negative values such as nihilism, hedonism, money worship, and even extreme individualism, are flooding. They all reflect the weakening of the mainstream values of college students. It is necessary to face up to the impact of western values on the ideological concepts of college students, adopt targeted measures, integrate socialist core values into the education of college students' cultural identity, and create a good cultural atmosphere with good virtues.

Audiovisual network media have induced the perceptual values of college students. The popular culture carried by audiovisual network media is often appealing through sensory shocks. "Attracting the eye" is a common feature of audiovisual network media nowadays, and rarely asks the minds of college students. Consumer culture is in power. It's simple and intuitive form caters to the low-level needs of college students, and never cares about education. Fast-food entertainment, vulgar, kitsch, and even ugly beauty spread, rarely touching the impetuousness and anxiety of college students. Facing the intuitive experience of audiovisual network media, college students ignore the exercise and cultivation of rational thinking and indulge in obtaining sensory sensations from various perceptual images. This induces college students' perceptual values and therefore is not conducive to the exercise and cultivation of their rational thinking.

Audiovisual network media breeds the value of college students' consumption supremacy. Under the influence of such mass consumption concept, college students have immediate pleasures. The pursuit of personalized consumption has become the mainstream of college student consumption. Under the influence of such consumption concept, college students indulge in vulgar entertainment and boring pastimes. They hope to temporarily avoid the responsibility of daily life and their own studies. The traditional concept of arduous struggle has been greatly challenged [8].

Audiovisual network media leads to the utilitarianization of college students' value choices. The popularizations of audiovisual network media content and realistic themes have influenced college students to become more rational in choosing value, which is easy to form the trend of profitable value choice. In terms of learning, they are eager to test based on the principle of usefulness and ignore the improvement of comprehensive quality. In interpersonal communication, they pay attention to interests and ignore emotions by taking benefit as principle. In the life, they pay attention to enjoyment and ignore dedication. In choosing a career, they take self-interest as the principle, pay attention to self-development, and ignore the needs of the country and social responsibility [9].

3 Audiovisual Network Media Strengthens the Recognition of College Students' Values and Helps Explore the Value Remodeling Method

We can truly realize cultural identity. The core values of socialism are the essence of socialist culture only by strengthening the value recognition of college students. The core value of socialism is the essence of socialist culture. Incorporate the core values of socialism into the cultural life of college students, spread the traditional culture and socialist cultural forms that are more acceptable to students in audiovisual network media, and enhance the cultural identity of college students and cultural self-confidence. If young students learn to understand life and society with rational thinking at the beginning of their education, it will be more conducive to their mental maturity and becoming social talents with a happy life [10].

Taking the communication, penetration, consumption, and emulation of popular culture as the starting point, examines the role of audiovisual network media in the remodeling of contemporary college student values. The youth popular culture is still a subculture in the final analysis, and still needs to be built and developed under the broad framework of socialist culture. Therefore, the primary direction of the youth popular culture is also to strengthen cultural confidence [11]. Aiming at the contradiction between the value distortion in audiovisual network media and the establishment of the correct values of college students, find out the law of its spread as a popular culture, explore specific ways to enhance the values of college students, and provide a reference for remodeling college students' values.

Adhere to the core values of socialism to lead the reshaping of university students' values. Persist in the attitude of "tolerance, openness, and truth-seeking" towards audiovisual network media, and create a harmonious campus culture that meets the needs of students' development. In the Internet + era, enhance the diversity of education forms of socialist core values and the effectiveness of education using the characteristics of informatization of audiovisual network media. Strengthen the construction of campus culture, explore the hidden education mode, correct people's value orientation and ideological deviation, and improve people's spiritual realm.

Play a role in leading the Communist Youth League. Think about the role of audiovisual network media from the perspective of meeting the needs of students' spiritual growth. Focus on social hotspots that students pay attention to, create audiovisual network products, and improve college students' social observation capabilities. In the struggle of public opinion on the network, we must dare to face the problem, actively promote the main theme, and spread positive energy. Attach importance to the construction of media and enhance the public place where students gather, such as live platforms, to increase their influence. Pay attention to the way and skills of thought-leading content dissemination and enhance the professionalism of content dissemination.

Enrich the carrier of ideological and political education on the network and build the educational role of great ideological and political education. On the basis of the existing ideological and political education carrier, integrate traditional campus media with audiovisual media to create a comprehensive ideological and political education media matrix. Using internet audiovisual media as the thinking carrier open all ideological and political education activities in colleges and universities to solve practical problems in life and learning. Cultivate campus audiovisual network media opinion leaders to improve the persuasion of network ideological and political education and public opinion guidance through peer education. Innovate a group of audiovisual network media programs that are independently operated by students, close to the lives of college students, and cultivate their correct values. Absorb the liveliness and interest of popular pop culture and innovate the discourse methods of ideological and political education in universities. Form healthy sentiments and correct Aesthetic consciousness to lay the foundation for firm ideals and beliefs.

4 Epilogue

Audiovisual network media is an important carrier of popular culture. It not only contains values but also carries values, and affects the generation of college students' values. On the other hand, the shaping of college students' values affects the development direction of popular culture. The two penetrate and interact with each other.

Education on college students' core values is a necessary way for them to grow up healthily. During the prevailing period of popular culture, audiovisual network media actively accept and try to learn resources that are useful for the education on college students' values, and realize the benign interaction between the audiovisual network media and education on core values of college students. On the basis of correctly handling the audiovisual network media and reshaping of college students' values, the students' rational thinking and discerning ability are cultivated accordingly.

Acknowledgements. 2021 Major Research Project of Xi'an Fanyi University. Project name: Research on the Theory Construction and Practice Mode of Labor Education (Project No. 2021Z05, Host: Wentao Yin).

References

1. Lin, H.: Insight into and understand the four dimensions of audiovisual network media communication. China Radio TV Acad. J. **10**, 57–59 (2019)
2. Huang, C.: Analysis of the influence of popular culture on college students' values. High. Educ. Exploration **4**, 143 (2011)
3. Chen, W.: Analysis of Contemporary Internet Popular Culture, p. 32. China Social Sciences Press, Beijing (2015)
4. Xiaoxian, S.: Analysis on new development trend and long-term mechanism of audiovisual network programs under great integration. China Publ. J. **22**, 38 (2017)
5. Shen, J.: Evolution and conflict of contemporary youth popular culture. J. Chinese Youth Soc. Sci. **06**, 80–87 (2018)
6. Jiang, H.: Influence of mass culture on the values of young college students and countermeasures. Morality Civ. **4**, 84–85 (2007)
7. Luo, D.: Education of college students' socialist core values from the perspective of cultural identity. Stud. Ideol. Educ. **2**, 106–109 (2014)
8. Xu, W.: Research on the Impact of Mass Culture on College Students' Values and Countermeasures. Shandong University, vol. 34 (2013)
9. Chang, Z., Lei, Y.: Research on the influence of popular culture on the ideology of college students and countermeasures. J. Hubei Normal Univ. (Philos. Soc. Sci.) **2**, 109 (2016)
10. Li, R., Sun, J.: Cultivation of young students' rational spirit from the perspective of civic culture. Ideol. Polit. Educ. Res. **4**, 39–42 (2017)
11. Su, H., Jia, R.: Social representation and reflection of online "Demotivational Culture" from the perspective of post-subculture. Modern Commun. **5**, 12–19 (2019)

Innovative Design and Theory of Urban Environmental Protection Art Network Under Information Technology

Yingqi Zhang[✉]

Beijing Institute of Technology, Beijing 100081, China

Abstract. With the development of social economy and the advancement of urbanization in China, urban environmental art has made great progress. This paper mainly studies the relationship between modern urban environmental art and environmental protection. On the basis of discussing the relationship between urban environmental art design and environmental protection, this paper introduces that urban environmental art is the product of the development of society and the times. It highlights the important factors of urban environmental protection and plays an important role in improving people's quality of life. This paper discusses the necessity of strengthening the network innovation of art and urban environmental protection art in engineering colleges, analyzes the current situation of the network innovation of art and urban environmental protection art in engineering colleges, and puts forward some suggestions for the network innovation of art and urban environmental protection art in the new century. The results show that: the standard deviation of a is 0.55 higher than the mean, the standard deviation of B is 0.94 higher than the mean, the standard deviation of C is 2.44 lower than the mean, and the standard deviation of D is 2.56 higher than the mean.

Keywords: Information · Urban environmental protection · Art · Online innovation

1 Introduction

With the continuous progress of computer technology, information technology has become an indispensable part of our life. There are different degrees of competition between cities in terms of economic growth and environmental protection. It is of great theoretical and practical significance to study how urban competition affects the improvement of ecological efficiency for the realization of regional sustainable development.

With the continuous development of science and technology, many experts have studied urban environmental protection. For example, some domestic teams have studied the construction and optimal design of urban green infrastructure network, introduced the latest research results of environmental protection department in urban rainwater and confluence overflow pollution control, and elaborated the main contents of urban runoff

© The Author(s), under exclusive license to Springer Nature Switzerland AG 2022
J. Macintyre et al. (Eds.): SPIoT 2021, LNDECT 98, pp. 549–557, 2022.
https://doi.org/10.1007/978-3-030-89511-2_71

pollution control scheme. It includes problem definition, user assistance tools, management alternatives and technology transfer, including some highlights of the future development direction of the project and the products of several research projects. Through case analysis, the paper understands the application of green infrastructure construction in practice, establishes two sets of evaluation systems of macro whole and micro part, and determines the index weight by using analytic hierarchy process; analyzes the network construction of green infrastructure by using morphological spatial pattern analysis (MSPA) combined with landscape index method and cost path method; and advances the network construction of green infrastructure The control and treatment cost of rainwater and combined pipe are compared. This paper identifies and explains the attitude of urban residents towards tourism development, and groups the attitude of urban residents towards tourism development. This paper outlines the process of developing social indicators for regional resource management plans. Insights and lessons are drawn from the experience of developing social water quality solution indicators for non-point source pollution. The main lessons learned from this process include: the value of building the capacity of stakeholders to use social data; the importance of developing a conceptual framework to guide the selection of indicators; and the importance of selecting a few core indicators. The design is based on site-specific horological data and uses data from nearby river sections with known priority ecological habitat functions [1]. Some experts studied the diversified ways and promotion of urban environmental protection publicity system, elaborated the diversified performance of environmental protection knowledge publicity system and the source, purpose and significance of promotion research, discussed the development status of environmental protection knowledge publicity at home and abroad, elaborated the definition, background of environmental protection and various measures taken by the government in environmental protection. This paper points out the necessity of environmental protection knowledge publicity and education, and analyzes the development, changes and effects of environmental protection publicity means in environmental protection cities. This paper mainly discusses the formation method, design method, purpose and function of the environmental protection publicity system, compares the advantages and disadvantages of the current environmental protection publicity system and the environmental protection publicity system from three aspects of interaction, vision and initiative, and puts forward the advantages of the environmental protection publicity system. Through the construction of multi supplier participation incentive platform, the establishment of sustainable capital investment mechanism and resource saving mechanism, the construction of incentive supply mechanism. To build an incentive oriented supply supervision mechanism, it is necessary to establish an incentive mechanism in supervision and improve the incentive supervision system. Through the construction of incentive mechanism for decision-making, supply and supervision of environmental protection public products in urban communities, we can promote the supply of environmental protection public products in urban communities, and constantly meet the diversified needs of community residents, so as to achieve the balance between supply and demand of environmental protection public products in urban communities [2]. Some experts have studied the design and application of "green products" under the information network, analyzed the previous research results in green products, and found that the research on green products and green design is very rich, but based on the background of network development, the research on the changes of green products under the network environment and how to design and apply green products is less. This paper defines the network green products, and discusses the nature and characteristics of the network green products. Through the analysis of the application of network green products in various fields, it is found that the application of network green products in efficiency products can bring positive impact on network users and network environment [3]. This paper analyzes the design method of network green products in energy efficiency products, and finds that there are some problems in the design of energy efficiency products, such as the interface style is not unified, and it is difficult for users to adhere to the use. Through the research on the modular design and safety design of the urban wind energy device, the integrated design of the wind energy device is formed: Modular part of the structure of the wind energy device to improve the utilization rate of the wind turbine; modular the monitoring system and alarm system to adapt to the development of the wind power device and the use environment of the wind energy device. In order to improve the urban ecological efficiency and realize the regional green development, we should adopt the differentiated restraint mechanism according to different urban types, adhere to the classified guidance and characteristic development. Although the research on urban environmental protection is fruitful, there are still some deficiencies in the network innovation of urban environmental protection dramatization.

In order to study the online innovation of urban environmental protection, this paper studies the public goods theory of urban environmental protection infrastructure, and finds the measurement method of high pressure and low suction. The results show that information technology is conducive to online innovation of urban environmental protection.

2 Method

2.1 Public Goods Theory of Urban Environmental Protection Infrastructure

(1) Public goods theory of urban environmental protection infrastructure

Public goods refer to non-competitive and non-exclusive goods with consumption function, which can be divided into private goods and public goods. Exclusive and competitive goods are called personal goods, such as food [4]. The reason why they become private goods is that once they are consumed, others cannot consume them at the same time [5]. Therefore, we can see its two characteristics: one is exclusiveness, which limits the use of goods or services by those who pay the price, while others cannot enjoy them at the same time; the other is competitiveness, which means that if the payer has consumed a certain kind of goods or services, others have no right to consume [6]. The corresponding public goods refer to the products and services provided by the government that private people are unable or unwilling to produce or provide [7]. Contrary to the characteristics of private goods, non-competitive and non-exclusive consumption are the most important characteristics of public goods [8]. No matter the number of people who consume certain public goods increases or decreases, the utility of the users of the goods will not be reduced, and the cost will not increase or decrease [9]. In other words, a person's consumption of public goods cannot detract from others' consumption of the goods, and the marginal cost of the goods provided by the increase in the number of consumers is equal to zero, which is the so-called noncompetitive [10].

(2) Green products

Green products refer to products that have little or no negative impact on the environment during the whole life cycle of products [11]. Environment has become the most important factor to define green products, but environment not only refers to macro natural material elements such as water, air, soil, animals and plants, microorganisms, but also includes non-material elements such as ideology, concept, policy guidance, etc. green products under information network environment can be defined as green products under network environment. Network environment can be understood as an environment that connects computers all over the world, communicates with each other using established protocols, and shares software and hardware. From the macro point of view, the network environment can be understood as a collection of virtual space. Green products are generally communicable, but there are differences between traditional green products and network green products in communication channels. Traditional green products mainly rely on offline channels established by ordinary products, and carry out large-scale product promotion and concept dissemination through distributors and other channels. Green products have high ecological value, and the products themselves are also very green, which have made great contributions to the optimization of the world environment. Through the green empowerment of the product itself, it also brings economic value to the product itself, and the R & D investment in the product also provides the value of accumulating intellectual property for the long-term development of the enterprise.

(3) Urban public facilities

The design of modern public facilities is not only a simple product design, its design requires designers to fully consider the shape of the product and the color of the environment, so that public facilities can be perfectly integrated with the surrounding environment of the city. The classification of public facilities is mainly divided into two categories. The first type is the design of single facilities, which is also the most basic, core and important part of public facilities; the second type is the system planning and design, which is the systematic design and planning of single public facilities to form the overall public facilities design coordinated with the social environment. With the rapid development of society, the design of public facilities is constantly improving and developing, more and more types of facilities will be divided into more detailed.

2.2 Measurement of High Pressure and Low Suction

The measure of spatial spillover is further refined, and the spatial spillover values of two adjacent regions in opposite directions of local high pressure and low suction are calculated. Sohit (Solit) is used to express high pressure (low suction), that is, the economic growth (ratio) of neighboring cities. If GDP and ER are higher (lower) than I, we can calculate the ratio between I and J. The specific calculation method ($v =$ ratio GDP or $V = $ ER) is shown in formula (1–2):

$$SOH_{it} = \sum_{j \in J_i} w_{ij(t-1)}\left(V_{j(t-1)} - V_{i(t-1)}\right), if V_{j(t-1)} \geq V_{i(t-1)}, j \in \tau \qquad (1)$$

$$SOH_{it} = \sum_{j \in K_i} w_{ij(t-1)}\left(-V_{j(t-1)} + V_{i(t-1)}\right), if V_{j(t-1)} < V_{i(t-1)}, j \in \tau \qquad (2)$$

Considering the geographical distance and population size between the two cities, the specific calculation formula is shown in formula (3):

$$w_{ij(t-1)} = \frac{Pop_{j(t-1)}/D_{ij}}{\sum_{j=\Delta i} Pop_{j(t-1)}/D_{ij}} \qquad (3)$$

Assuming that the number of DMUs is n, DMUs can be divided into H (H1) group according to the heterogeneity. The number of DMUs in group H is shown in Eq. (4):

$$\sum_{h=1}^{H} N_N = N \qquad (4)$$

3 Experience

3.1 Experimental Object Extraction

In order to save energy, reasonable shading design can avoid high indoor temperature in hot summer. In the hot summer and cold winter area, for ordinary residential buildings, if the sunshade energy-saving design is not adopted, in the summer thermal environment, in order to achieve the purpose of cooling, the house will consume a lot of energy. There is a long history of building sunshade in hot summer and cold winter areas in China, and there are many kinds of design forms, such as large eaves building, Hangnail arcade, stilts in the Yangtze River Basin, etc. If we use the popular sunshade decoration in the last century for building sunshade design, and keep the consistency of sunshade components, we cannot achieve good sunshade effect, affecting the normal lighting and ventilation of residential buildings. From the angle of sun height and azimuth, the East and West windows should choose vertical sunshade and shutter sunshade, the south window should choose horizontal sunshade, and the North window should choose unilateral temporary vertical sunshade. Choosing proper sunshade design for windows with different orientations can save the project cost, make the residents do not need to add additional sunshade facilities in the later period, and avoid damaging the building facade effect due to the residents' use of sunshade.

3.2 Experimental Analysis

The energy saving design of building shading in hot summer and cold winter area is divided into design steps. First of all, the building needs to determine the shading conditions. Building shading must start from the biological climate comfort of human body, analyze the climate and environmental conditions of buildings, and then determine the design strategy. Due to the difference of climate in hot summer and cold winter areas, when the solar radiation reaches a certain condition, the building must take shading measures. Secondly, determine the season and time of shading, and finally calculate the angle and size of shading. After determining the shading mode, angle and size, the designer should design from the perspective of vision, ventilation and lighting, and then adjust the shading components to obtain the best shading form. In order to facilitate the exhaust of hot air in hot summer and cold winter areas, the sun shading surface is usually made into the form of louvers and openings. Although the opening on the board can induce ventilation and heat dissipation, the current construction technology has some difficulties and the effect is not good. The installation position of sunshade plays an important role in indoor ventilation and lighting of residential buildings. Choosing a good installation method can guide the wind.

4 Discussion

4.1 Variable Selection and Data Description

Four factors affecting eco efficiency are selected: population density, foreign capital utilization, technological progress and industrial structure. The utilization of foreign capital refers to the ratio of the total industrial output value of Foreign-invested Enterprises above Designated Size to the GDP; the technological progress refers to the proportion of the utilization of foreign capital, and the number of employees engaged in scientific research, technical services and geological exploration at the end of the year; the industrial structure is represented by the proportion of the secondary industry to the GDP of each city. As shown in Table 1.

Table 1. Variable meaning and descriptive statistics

Variable name	Mean value	Standard deviation	Number of samples
A	3.72	4.27	4.83
B	4.73	5.67	6.78
C	6.72	4.28	3.82
D	3.27	5.83	4.73

As can be seen from the above, the mean value of a is 3.72, the standard deviation is 4.27, and the number of samples is 4.83; the mean value of B is 4.73, the standard deviation is 5.67, and the number of samples is 6.78; the mean value of C is 6.72, the standard deviation is 4.28, and the number of samples is 3.82; the mean value of D is 3.27, the standard deviation is 5.83, and the number of samples is 4.73; the concrete presentation results are shown in Fig. 1.

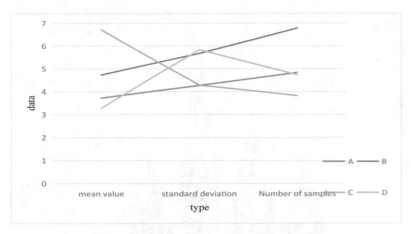

Fig. 1. Variable meaning and descriptive statistics

It can be seen from the above that the standard deviation of a 0.55 higher than the mean; the standard deviation of B is 0.94 higher than the mean; the standard deviation of C is 2.44 lower than the mean; the standard deviation of D is 2.56 higher than the mean.

4.2 Analysis on the Composition of GSCM

Through the analysis of the composition of green supply chain management, it is concluded that the implementation of green supply chain management includes four measurement dimensions: enterprise internal environmental management, green cooperation with the upstream and downstream of the supply chain, enterprise internal environmental management and enterprise internal environmental management, green design and green recycling. Therefore, the hypothesis A1 is supported, as shown in Table 2.

Table 2. GSCM composition hypothesis test results

First order factor	Path coefficient
Enterprise internal environment management	2.48
Green cooperation with upstream and downstream of supply chain	3.89
Green design	1.74
Green recycling	2.58

It can be seen from the above that the path coefficient of enterprise internal environmental management is 2.48; the path coefficient of green cooperation with upstream and downstream of supply chain is 3.89; the path coefficient of green design is 1.74; and the path coefficient of green recycling is 2.58. The results are shown in Fig. 2.

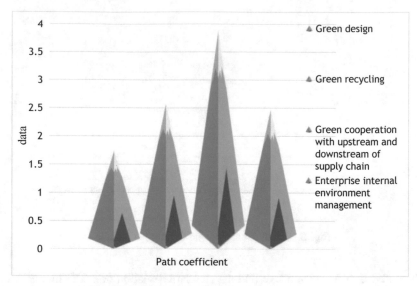

Fig. 2. GSCM composition hypothesis test results

It can be seen from the above that the path coefficient of green cooperation with the upstream and downstream of the supply chain is the maximum of 3.89, and the path coefficient of green design is the minimum of 1.74.

5   Conclusion

In recent years, more and more complex urban functions challenge urban environmental management. Based on the concept of sustainable development, the project of reusing old industrial buildings combined with creative industries has played a positive role in urban environmental protection and promoted the sustainable operation of urban environmental management. A new mode of urban environmental management in China, mufti network management mode of sustainable development, is proposed. The task is divided into multi-access, mufti block parallel working mode, and transformed into a mufti member, multi-access coordination grid model. The focus of power should be gradually transferred to the grassroots. At the same time, the political game among government, non-governmental organizations, developers and market mechanism is introduced in urban planning to cope with the changing requirements and improve the efficiency of urban environmental management.

References

1. Link, A. N., Morris, C. A., Van Hasselt, M.: The impact of public R&D investments on patenting activity: technology transfer at the U.S. Environmental Protection Agency. Econ. Innov. New Technol. **28**(5–6), 536–546 (2019)
2. On the innovation and development path of provincial TV stations in China. Art Res. Lett. **10**(1), 1–6 (2021)

3. Konieczny, K., Wszelaka-Rylik, M.E., Macherzyński, B. Membrane processes innovation in environmental protection: review. Arch. Environ. Prot. **45**(4), 20–29 (2019)
4. Man, W.: Impact of environmental pollution on the construction of urban brand in Hangzhou and its protection strategies. J. Environ. Prot. Ecol. **20**(4), 1863–1869 (2019)
5. Mena Mejía, I., González, Á.B., Vargas López, S., et al.: Evaluación de la Condición Ecológica del río Zahuapan. Revista Internacional De Contaminacion Ambiental, **33**(1), 7–19 (2017)
6. Bendixen, M., Overeem, I., Rosing, M.T., et al.: Promises and perils of sand exploitation in Greenland. Nat. Sustain **2**(2), 98–104 (2019)
7. Jiang, T.: Urban public art and interaction design strategy based on digital technology. Clust. Comput. **22**(2), 3471–3478 (2018). https://doi.org/10.1007/s10586-018-2194-z
8. Krupnova, T.G., Rakova, O.V., Mashkova, I.V., et al.: Health risk assessment of metal (LOID)S exposure via indoor dust from urban area in Chelyabinsk, Russia. Int. J. Geomate **16**(55), 1–7 (2019)
9. Martin, C. J., Evans, J., Karvonen, A.: Smart and sustainable? Five tensions in the visions and practices of the smart-sustainable city in Europe and North America. Technol. Forecast. Soc. Change **133**(AUG.), 269–278 (2018)
10. Bajinovci, B.: Design and art guidelines for complex structures. J. Sci. Humanit. Arts - Josha **5**(10), 1–11 (2018)
11. Research on the model and mechanism innovation of NGO's participation in the Urban Lake Environmental Governance. Adv. Environ. Prot. **10**(6), 887–896 (2020)

Application of Computer Virtual Simulation Technology in Landscape Design

Jing Zhang$^{(\boxtimes)}$ and Xiaoxiao Ma

Wuhan Donghu University, Wuhan, Hubei, China

Abstract. The use of virtual simulation technology to help students visualize the landscape space in front of students is of great help to students' spatial cognitive learning. This paper briefly introduces the landscape design and virtual simulation technology, and analyzes the benefits of applying virtual simulation technology in landscape space cognitive teaching. Finally, the construction of virtual simulation platform and its specific application in landscape space cognitive teaching are discussed.

Keywords: Virtual simulation technology · Landscape design teaching · Landscape teaching · Virtual simulation platform

1 Introduction

First of all, multimedia technology into teaching, through sound, image and other forms of expression, so that students can grasp the knowledge more firmly. Multimedia courseware is rich in pictures, pictures, sound and images. It adds infinite charm to classroom teaching by using vivid pictures, pleasant music and easy to understand explanation. It is conducive to arouse students' interest in learning and fully mobilize their enthusiasm for learning; It is helpful to optimize the teaching process, improve teaching efficiency, break the key and difficult points of teaching materials, and show some general and micro phenomena; It is conducive to students' perception, imagination and understanding, and absorb all students to participate in classroom learning activities.

Secondly, the computer demonstration reduces the teachers' language transfer and writing on the blackboard. Especially in the case of a relative shortage of teachers, the original teachers in teaching to do a lot of demonstration, explanation, by the computer agent, the teacher as long as timely to the computer instructions. This not only reduces the labor intensity of teachers in teaching, but also increases the teaching density. Through the computer demonstration, the students can change the abstract into the image and master the knowledge quickly.

Third, in the process of transforming examination oriented education into quality education, the infiltration of computer technology undoubtedly provides a new and lively way of teaching and learning, and changes the boring "cramming" teaching method in the past. It has changed the traditional educational ideas, methods and means, and greatly improved the teaching efficiency and level. In mathematics class, the scientific

J. Macintyre et al. (Eds.): SPIoT 2021, LNDECT 98, pp. 558–565, 2022.
https://doi.org/10.1007/978-3-030-89511-2_72

application of computer-aided teaching is more suitable for the development of students' personality, which is conducive to eliminating students' learning obstacles, making the mathematics teaching reform more purposeful and targeted, and improving the quality of mathematics teaching in a large area.

Virtual simulation technology has developed rapidly in recent years. As a technology that can simulate the real system in all directions, it plays a very important role in the development of various fields in our country. Among them, the education industry has begun to explore the use of virtual simulation technology to improve its teaching results, and the teaching tasks that have been tested have achieved very good results. Especially in the teaching of landscape space cognition related to three-dimensional images, the application of virtual simulation technology is remarkable.

2 The Content of Space Cognition in Landscape Design

2.1 A Place to Present the Landscape

First of all, the spatial cognition of landscape is the cognition of the place that constitutes the landscape. In the process of landscape design, the designer must divide a whole area, or through the combination of different areas to help the whole landscape design space to get the most efficient use. All the spaces that make up the final landscape are connected by different elements, and finally get the effect of harmony and unity. For example, in landscape design, designers will use different plants, sculptures, fountains and other landscape structures to separate the large landscape space into several small spaces, which is to connect the visible space and invisible space of people [1]. In addition, designers can also recombine the small space to present a new landscape effect. In the actual landscape design, designers often through the scientific use of space and layout, help urban residents have a local characteristics, can gather, entertainment, leisure space. As shown in Fig. 1.

Fig. 1. Spatial cognition of landscape

2.2 Centralization of Space

Space is a three-dimensional environment. If we want to have a stronger perception of it, we need a center as a reference point, so that other activities can be further implemented around this center. The centralization of landscape space is to further ensure the security of the space of landscape design for designers and users. It emphasizes the region of landscape and the direction of space. In addition, the centralization of landscape space is also to be able to get along more harmoniously with the people who are carrying out activities in this landscape area, and help the whole landscape design to be more methodical in artistic processing. The centralization of landscape space makes the landscape show a richer sense of hierarchy in the overall presentation, highlighting the dominant position of the central space. After feeling the overall charm of the landscape, urban residents can pay more attention to the deep connotation of the spatial structure of the landscape itself. As shown in Fig. 2.

Fig. 2. Centralization of landscape space

2.3 Virtual Space in Landscape Composition

Landscape space in the process of design can present an open type, can also be enclosed. Whether it is open or enclosed, it is composed of actual space. The virtual space is the use of plants, light and shadow or architectural structure to form a virtual effect, so that people have the illusion of space. For example, the trees in the landscape design will produce scattered projections under the night light, forming an independent shadow space, which gives people a sense of space extension. In the landscape design with fountain, after the fountain is opened, a water curtain performance space is formed in the void, which helps the environment of the whole landscape space have more aesthetic feeling.

3   Virtual Simulation Technology

Virtual simulation technology is a kind of simulation technology with epoch-making significance. It simulates the real world into the virtual space by using computer technology, and brings people the most real feelings. To some extent, virtual simulation technology is a representation of the "real world", but the "real world" may be the past, the present and the future. Users of virtual simulation technology can also interact with the virtual world through their own senses [2]. Virtual simulation technology involves many advanced science and technology fields, such as computer technology, computer vision, visual psychology, physiology, simulation technology and so on. Virtual simulation technology in the current stage has achieved certain research results, and began to be applied in different industries.

4 Virtual Simulation Technology for Landscape Design Teaching

4.1 Deepen Students' Understanding of Space Scene

The application of virtual simulation technology in the teaching of landscape space cognition will make students understand the space scene more deeply and comprehensively. In the teaching of landscape space related knowledge, because space itself is the existence of three-dimensional concept, if only with the help of books and blackboards, students can only see the two-dimensional plane composition, and then according to some known information, they can associate and conceive the three-dimensional space in their mind, so they can't control the details of space accurately. But virtual simulation technology can help students list all the elements of landscape design one by one, and arrange them regularly according to the teaching content, so that students can more intuitively feel the spatial structure of the whole landscape, and even the gradual effect of color, light and shadow in the landscape space can be clearly presented in front of students, Let the students not only have the feeling of being in the scene, but also examine the landscape space from the perspective of bird's-eye view.

4.2 Help Students and Teachers Have Better Interaction

In the teaching of landscape space cognition, it is not easy for teachers to convey their ideas and thoughts clearly to students. Because the teacher's thinking nerve is not connected with the students, so many landscape space theories which cannot be intuitively displayed cannot be accurately accepted by the students in the process of teaching. But with the help of virtual simulation technology, teachers and students seem to set up a communication bridge [3]. In the process of explanation, teachers can let students participate in the process of explanation through the interactive function of virtual simulation platform. Teachers construct the theoretical knowledge of landscape design, and students use the theoretical knowledge to carry out the design practice. Once there is a difference in thinking between the two, teachers can also find it in time and achieve effective communication with students through direct modification. Moreover, the virtual simulation platform also provides more practical functions for teachers and students. For example, teachers can release students' training tasks through the platform, and students can also conduct skill training and ability evaluation through the platform, which greatly facilitates the communication of landscape space cognition between teachers and students.

4.3 Stimulate Students' Creative Thinking

Virtual simulation technology for students of landscape design teaching also has a great benefit, is to help stimulate students' creative thinking. The purpose of students learning landscape design is to help them better control their spatial structure in the future landscape design, so that the final landscape effect can be appreciated. And design is a field that needs students' creative thinking most. Using virtual simulation technology, students can realize the inspiration of landscape design scheme in their mind at any time, and further adjust the landscape as a whole by examining its effect, so as to enhance its feasibility and aesthetics. This kind of pleasure of turning inspiration into reality can stimulate students' enthusiasm for landscape design learning, constantly exercise students' thinking sensitivity in creation, and make students have a broader development.

5   Construction of Virtual Simulation Training Base

5.1   Allocation of Hardware Equipment

To apply virtual simulation technology to landscape design teaching, it is necessary to build virtual simulation training base to help students create conditions and environment for learning relevant knowledge by using virtual simulation technology. The virtual simulation training base consists of hardware facilities, software and database. Generally speaking, the construction of virtual simulation training base must be able to bring students immersive experience, in order to give full play to its help students improve the ability of landscape design. So in the vision, touch, hearing, taste, smell need to apply the corresponding equipment to achieve one by one [4]. The hardware required for the whole virtual simulation training can be roughly divided into four parts. The first part: the equipment used to build the simulation system, including computer equipment, 3D scanner and so on. The second part: display the whole simulation system; make it present in front of the students. For example, VR helmet, VR glasses, virtual reality display system, etc. The third part: to help students with voice interaction equipment, such as audio, voice recognition equipment. The fourth part is for students to operate a variety of devices to facilitate interaction with the system, such as joystick, tactile feedback device.

5.2   Support of Corresponding Software

The construction of virtual simulation training base also needs the support of corresponding software. For the teaching of landscape space cognition, the following functions need to be realized in the process of classroom teaching and after class practice with the help of virtual simulation technology: teachers should be able to manage the learning of students by using virtual simulation platform, and also achieve certain exchange with students on the course content. Therefore, according to this function, the virtual simulation training base needs to be equipped with education management, online education corresponding software, which helps students and teachers log on to the virtual simulation platform through personal accounts, and realize the division of authority. Landscape design teaching involves most of the design drawings and 3D models of various landscapes. Therefore, it also needs software to identify drawings and software for landscape simulation construction according to drawings and related parameters. Students' inspiration needs to be verified by virtual simulation platform. So we need some drawing software to help students turn their ideas about landscape space into reality.

5.3 Construction of Data Material Database

A good virtual simulation platform must be supported by some basic materials and data to help students use more smoothly. Therefore, after completing the hardware construction and software configuration of the virtual simulation platform, the teachers related to landscape spatial cognition need to cooperate with the professionals who carry out virtual simulation technology research to establish a basic data base for the whole platform. For example, we should provide some small tools to use in the data material database, various trees, pavilions, plant racks, fountains, sculptures, etc. commonly used in the construction of landscape space, and some basic graphics. With the help of technical personnel, all kinds of landscape materials drawn on the plane are transformed into three-dimensional virtual existence, so that students can use the platform more conveniently.

6 The Content of Virtual Simulation Technology in Landscape Design Teaching

6.1 Theoretical Teaching of Design Courses

There are many aspects when applying virtual simulation technology in landscape design teaching. For example, in the landscape space of the enclosure change, infiltration relationship, scale ratio, spatial sequence, if the teacher tells the students directly, the students not only have difficulties in understanding, but also are easy to be absent-minded because the course is too boring. So we can use the virtual simulation platform to make these abstract concepts vivid. Through the simulation of three-dimensional landscape space, we can make students have more intuitive feelings for different concepts related to it. Through the dynamic teaching process, we can make students more interested in the classroom, so as to improve the learning efficiency of students' theoretical knowledge of landscape space.

6.2 Practice of Engineering Courses

In the study of landscape design, there is a large part of the content that students need to understand and verify through practice. But the traditional teaching of students to practice often need to spend more resources, so the opportunity to really practice is very few. But with the virtual simulation technology, students can simulate the practice process on the virtual simulation platform, saving the cost of students' practice [5]. Moreover, in the actual engineering practice, a landscape from its design to the formation of the final space needs not short time, and the virtual platform can directly save this time, students can carry out more practice simulation in the limited learning time, greatly enhancing the efficiency of students' experience accumulation [6].

6.3 Perception of Landscape Space

In the teaching of landscape design, it is difficult for students to express the landscape in three-dimensional space through a group of numbers [7]. After the application of virtual simulation technology, students have more time to observe the landscape space structure in the virtual system, and can observe the landscape space 360 degrees from different angles. With the help of specific virtual degree, students can see the construction process of the whole landscape space from scratch, and have a more accurate understanding of its parameters and construction steps, And through the observation of many cases to have a deeper understanding of the sequence of landscape space, and then more quickly and firmly grasp the relevant knowledge, but also help teachers save a lot of teaching time [8].

6.4   Verification of Landscape Design Scheme

The best way for students to determine the height of their landscape design is to design a set of landscape scheme and realize it, and then observe whether the landscape effect is consistent with their expectations. This is difficult to achieve in the past landscape cognitive teaching. But virtual simulation technology is just the expert in this field [9]. Students scan their design drawings into the virtual simulation platform, set the construction technology, construction materials and other parameters, and then use the simulation software to turn the whole scheme into reality in the virtual space. If there is a gap between the presentation effect and the original design intention, we can also optimize the design scheme by modifying the parameters, so as to help ourselves find the deficiencies in spatial cognition and improve them [10].

6.5   Further Sharing of Network Teaching Resources

The application of virtual simulation technology in landscape design teaching can also make the network teaching resources realize further sharing [11]. Take the various elements of landscape composition for example, whether teachers or students in the process of using the virtual simulation platform, if they have the inspiration to make new landscape elements, they can store them in the material library of the virtual simulation platform, so that other people can use it in their own landscape space [12]. Moreover, all kinds of courseware used by teachers in daily teaching can also be stored in the virtual simulation platform. If students encounter problems that cannot be solved in the process of after class review and daily training, but they cannot communicate with teachers immediately, they can independently find solutions to problems through viewing these courseware.

7   Concluding Remarks

When virtual simulation technology is applied in landscape design teaching, it can help students understand the theory more intuitively, have more opportunities and time for practice, and have easier cognition of landscape space composition. It can also help students test their own spatial cognition level through virtual verification of landscape design scheme, the teaching resources about landscape design can also be further shared on the virtual simulation platform.

Acknowledgements. This work was supported by teaching research project of Wuhan East Lake University in 2016.

References

1. Zhang, H., Bao, R.: Application of virtual simulation technology in landscape design course teaching. Grand View **208**(04), 87–88 (2020)
2. Wang, S., Wu, D.: Application of virtual reality technology in the teaching of landscape architecture design. Chin. For. Educ. **037**(003), 51–55 (2019)
3. Chen, Y.: Application of virtual simulation technology in landscape architecture experiment teaching. Art Sci. Technol. **32**(009), 17–18, 21 (2019)
4. Yujia, Z.: Construction and application of landscape spatial scale sense teaching module based on immersive virtual environment technology. Chin. Gard. **034**(004), 68–73 (2018)
5. Yu, Q., Guo, Y., Zhang, Y.: Application of virtual simulation technology in space science experiment teaching. Lab. Res. Explor. **39**(02), 137–141 (2020)
6. Brzozowski, S., Kandrack, M., Oermann, M.H., Dangerfield, C., Virginia C, S.M.: Virtual simulation to reinforce nursing staff resuscitation responses. J. Nurses Prof. Dev. (2021)
7. Dolan, H., Amidon, B.J., Gephart, S.M.: Evidentiary and theoretical foundations for virtual simulation in nursing education. J. Prof. Nurs. **3** (2021). (prepublish)
8. Mak, V., Fitzgerald, J., Holle, L., Vordenberg, S.E., Kebodeaux, C.: Meeting pharmacy educational outcomes through effective use of the virtual simulation MyDispense. Currents Pharm. Teach. Learn. **13**(7), 739–742 (2021)
9. Qiu, C., et al.: Brain structural plasticity in visual and sensorimotor areas of airline pilots: a voxel-based morphometric study. Behav. Brain Res. **6**(411), 113377 (2021)
10. Chang, H.Y., Wu, H.F., Chang, Y.C., Tseng, Y.S., Wang, Y.C.: The effects of a virtual simulation-based, mobile technology application on nursing students' learning achievement and cognitive load: randomized controlled trial. Int. J. Nurs. Stud. **120**, 103948 (2021)
11. Zhang, Z.: Research on key technology of virtual simulation of minimum processing unit. Int. Core J. Eng. **7**(6), 237–243 (2021)
12. Fauziyah, S., Pramono, N.A., Anggraini, R.T., Hidayat, A., Supriana, E., Ali, M.: The increase of students' critical thinking abilities on optical instrument topic through pbl-stem with virtual simulation media. J. Phy. Conf. Ser. **1918**(5), 052067 (2021)

Improved Method of Parallel DC-DC Boost Converter Based on Main Current Control

Lirong Wang[✉]

Beijing Information Technology College, Beijing, China

Abstract. With the development of modern power electronics technology, traditional AC converters have some problems in power conversion, such as low efficiency and large losses. Therefore, this paper proposes an inner loop control method based on the main current in the DC-DC parallel system. The method is based on the SVC controller to improve the system performance by improving and perfecting it; at the same time, the multivariable nonlinear equation root is used to determine whether the value meets the set range and the optimal solution is obtained as the final goal to solve the grid voltage fluctuation, Network loss is too large.

Keywords: Main current control · Parallel DC · DC Boost converter · Current improvement

1 Introduction

DC-DC converters are widely used in various portable consumer electronic products and wearable smart electronic devices due to their advantages of high conversion efficiency, fast response speed, and small size. With the continuous improvement of user needs, these electronic products are also developing in the direction of low power consumption and miniaturization. Therefore, how to improve the efficiency of DC-DC converters, reduce system power consumption, and extend the service life of electronic products has become a hot research topic [1, 2].

In recent years, with the rapid development of digital signal processors (DSP) and programmable logic devices (FPGA), digital controllers have also been rapidly developed, and at the same time have promoted the corresponding academic research on digital power in related fields. With the increasing recognition of digital power in the industry, digital power has also developed to a higher level. Texas Instruments has also been committed to the research of digital power products in recent years, providing a wide range of digital power solution products and supporting multiple communication protocols. Its digital power products include PTD08A006W, PTD08A010W, PTD08A015W, UCD3020, UCD3028 and UCD3040 series products [3, 4].

Power converters are classified according to the conversion method, and can be divided into linear power supplies and switching power supplies. Among them, the linear power supply has the advantages of good stability, fast transient response, high reliability, high output voltage accuracy, relatively simple circuit structure, and small output voltage

© The Author(s), under exclusive license to Springer Nature Switzerland AG 2022
J. Macintyre et al. (Eds.): SPIoT 2021, LNDECT 98, pp. 566–574, 2022.
https://doi.org/10.1007/978-3-030-89511-2_73

ripple. However, it also has the advantages of large volume, low operating frequency, and conversion. Therefore, designing a parallel DC-DC Boost converter controlled by the main current is the key to improving the above problems [5, 6].

2   Working Principle of Parallel DC-DC Boost Converter Based on Main Current Control

2.1   Modulation Mode

In switching power supplies, common modulation methods include pulse width modulation (Pulse Width Modulation, PWM) and pulse frequency modulation (Pulse Frequency Modulation, PFM).

Pulse width modulation means that the frequency of the output pulse remains unchanged, and only the duty cycle is changed in each cycle, thereby changing the time ratio of the high and low levels of each cycle to achieve regulation; pulse frequency modulation refers to the output pulse. The empty ratio is constant, that is, the time ratio of the high and low levels in each cycle is fixed, and the frequency of the pulse is changed to achieve the adjustment effect.

In power converter applications, compared with pulse frequency, pulse width modulation has many advantages, such as easy implementation, constant operating frequency, low noise, demodulation can be achieved through a simple low-pass filter, etc. [7, 8].

(1)   Pulse-width modulation:

The control strategy for pulse width modulation is shown in expression (1):

$$s(t) = \begin{Bmatrix} 2, & V_f(t) < V_{saw}(t) \\ 1, & V_f(t) \geq V_{saw}(t) \end{Bmatrix} \tag{1}$$

Among them, vf(t) is the control signal output by the control module, and vsaw(t) is the periodic signal. According to the above control strategy, the pulse width modulation module generates the pulse drive signal s(t) by comparing the magnitude of the control signal vf(t) and the signal vsaw(t).

(2)   Pulse frequency modulation

The DC-DC converter adopts the pulse frequency modulation mode, by adjusting the pulse frequency, and then controlling the on and off of the switch tube S1 to maintain the stability of the output voltage. Pulse frequency modulation mode has the advantages of low static power consumption and high conversion efficiency in light load mode, but because it works in variable frequency mode, it is prone to electromagnetic interference, so additional circuits are needed to prevent ringing.

2.2   Control Mode

According to the different control principle of feedback control loop of DC-DC converter, the control mode can be divided into voltage mode control and current mode control, and current mode control includes peak current mode control, average current mode control and hysteretic current mode control. This section will introduce voltage mode control and peak current mode control in detail. Average current mode control and hysteretic current mode control will no longer be discussed in detail [9, 10].

(1) Voltage mode control

The voltage mode control developed earlier, and it has emerged as early as the 1960s. The voltage mode control uses a feedback resistor to sample the output voltage in the power stage circuit, and amplifies the error signal through the error amplifier, and adjusts the duty cycle according to the error signal. The voltage mode control technology has a single feedback loop, simple structure, simple debugging, and good anti-noise margin. However, when the input voltage of the external input DC source changes suddenly or the load impedance changes suddenly, the transient response of the DC-DC converter is slow, and because the voltage control mode does not need to sample the inductor current, additional additional overcurrent protection circuit [11, 12].

(2) Peak current mode control

Peak current mode control is usually called current mode control, and it has been developed rapidly since the early 1980s. Peak current mode control is a dual-loop control, including two control loops, an external voltage loop and an internal current loop. The internal current loop is used to sample the inductor current and then control and adjust the inductor current waveform. The external voltage loop is to sample the output voltage. The duty cycle is controlled through a series of actions. Suppose the charging phase time in peak current mode is Won, the discharging phase time is Woff, and a complete clock cycle is WS = Won + Woff, then the duty cycle D = Won/WS can be defined. In the charging phase, the power supply voltage VIN is all added to the inductor to charge it. The increase in inductor current during the time Ton is:

$$\Delta I = \frac{V}{L} \cdot D \cdot W_{\mathrm{s}} \tag{2}$$

Peak current mode control has the following advantages: first, peak current mode control needs to sample inductance current, and overcurrent protection can be realized only by limiting the control signal, so there is no need for additional overcurrent protection circuit. Second, the peak current mode control directly controls the inductance current, which increases the function of feedforward compensation and improves the response speed when the input voltage changes or the load impedance changes. Third, compared with the voltage control mode, the bandwidth is increased, so the transient response is also improved. Fourth, the peak current mode control adjusts the inductance current according to the error of the error amplifier output, and then determines the duty cycle change, so there is a certain constraint relationship between the inductance current and the load current, which improves the stability. Although the peak current control mode has so many advantages, it also has its shortcomings. For example, if the duty cycle is more than 50%, if the disturbance occurs, the system can not automatically eliminate the disturbance and achieve self-stability, which leads to the increasing disturbance and affects the stability of the system, so it is necessary to increase the slope compensation circuit. In this paper, the voltage mode is used to control the output voltage, only the output voltage is sampled, the digital control loop is adjusted, a new duty cycle is generated, the power level circuit is controlled, and the output voltage is adjusted.

2.3 Implementation

At present, the implementation of DC-DC converter is divided into two ways: analog implementation and digital implementation, which have their own advantages and disadvantages. The power level circuits of analog control technology and digital control technology are the same, but the difference is that the realization of feedback control loop is different. Analog implementation is rising earlier and the technology is more mature. Digital implementation is a new technology in recent years, which has the advantages that analog implementation does not have. According to the voltage control mode, this section mainly discusses the difference between analog implementation and digital implementation.

(1) Simulation implementation

The realization of analog control has developed earlier, and has been rapidly developed as early as the 1970s, and the technology has been very mature. The implementation of analog control usually uses feedback resistors to sample the output voltage. The error amplifier module obtains the error amount according to the sampled voltage signal and the reference voltage. The comparator then generates the PWM signal according to the error amount and the sawtooth wave, and the gate-level drive signal controls the PWM After the signal is processed, it is sent to the gate of the switch tube in the power stage circuit to control the on and off of the switch tube S1, and then adjust the output voltage to stabilize the output voltage. The feedback loop in the analog implementation mode is designed to build a circuit with components. The circuit design and layout design require a very large amount of work, and the circuit modification is relatively complicated, which affects the whole body, and the circuit needs to be modified. The schematic and layout of the circuit must be manually modified, so the development cycle of the analog implementation method is relatively long. Because the circuit in the analog implementation mode is built by components, and the parameters of the components are easily affected by temperature and humidity, and the aging of the components will also change the parameters, which will lead to parameter changes, which will affect the performance of the circuit. Therefore, the performance of the DC-DC converter implemented by analog is more dependent on the parameters of the components. It is difficult to implement complex algorithms in analog implementation. Although the implementation of analog control has these shortcomings, the implementation of analog control has been developed for a long time and the technology has been very mature. Therefore, in the case of relatively small environmental changes and relatively single performance requirements, the DC-DC converter implemented by analog control have more advantages.

(2) Digital implementation method

The application of digital control technology in DC-DC converters has risen rapidly in recent years, and many companies have introduced digital power products.

The digital controller of the DC-DC converter realized by digital control is usually realized by a digital signal processor (DSP), a programmable logic device (FPGA) or a microcontroller. The feedback loop of the DC-DC converter implemented by traditional digital control usually includes the following modules: sampling module, analog-to-digital conversion (ADC) module, digital proportional integral derivative (digital PID) module and digital pulse width modulation (DPWM) Module. Digital control is usually realized by sampling the output voltage through a sampling module. The sampling signal is converted into a digital signal by an analog-to-digital conversion (ADC) module. This digital signal is compensated by a digital PID module to obtain a control signal. Digital pulse width modulation (The DPWM module then generates a gate drive signal according to the control signal, and then sends it to the gate of the switch tube in the power stage circuit to control the on and off of the switch tube to realize the adjustment of the output voltage. The real-time sampling of the DC-DC converter realized by digital control is different from the real-time sampling of the analog realization mode. The output voltage is sampled only once in each switching cycle, thus reducing the power consumption of the system. The digital control implementation transfers the hardware circuit design issues in the analog implementation to the Verilog programming, and can use the automatic placement and routing tools for layout design, so its development cycle is short. The implementation of digital control is less sensitive to changes in the parameters of the components, so the system performance is relatively less affected by the changes in the parameters. The implementation of digital control shifts the focus of algorithm implementation to programming, which greatly reduces the difficulty of implementing complex algorithms. Although the digital control implementation has the advantages described above, it also has its shortcomings. For example, the analog-to-digital converter in digital control technology has a large delay, which affects the response speed of the DC-DC converter; digital control implementation due to the limitation of character length, the quantization effect will affect the accuracy; the discretization process in the implementation of digital control has caused some information loss; and the digital signal processor (DSP), programmable logic device (FPGA)) and microcontrollers are expensive, which increases the cost. Even if the DC-DC converter realized by digital control has the above shortcomings, for the case of higher functional requirements, the digital power supply is more advantageous than the analog power supply. And with the analysis and research of digital power in related fields, digital power technology will become more mature and gradually overcome its shortcomings. Therefore, the integrated realization of the DC-DC converter realized by digital control is its inevitable development trend.

2.4   Design of Parallel DC-DC Boost Converter Based on Main Current Control

2.5   Principle of Parallel Main Current Control

Parallel main current control, one power module is set as the main module, and the current loop of the other module (slave module) is based on the output current of the main module as a reference. The main module is a voltage control inverter, which ensures that the output frequency of the system synchronizes with its own frequency reference through the phase-locked loop PLL, and makes the output voltage stable through the feedback link. The slave module is a current control inverter, which is usually designed as a current following property. The control module outputs the current to track the reference current communicated by the main module to achieve load sharing. The current control inverter can set current feedforward to speed up the current response.

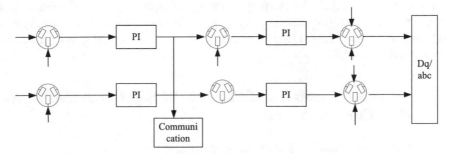

Fig. 1. Parallel-type main current control block diagram

Figure 1 is the schematic diagram of the parallel system with parallel main current control technology. The converter adopts voltage and current double closed loop control to ensure the output voltage waveform quality of the parallel system. The output signal of the voltage outer loop of the converter is given as a reference from the current inner loop of the converter. As long as the parameters of the current inner loop of each module are consistent, the phase consistency and amplitude of the output current of each parallel module can be realized.

2.6 Design Method of Parallel DC DC-Boost Converter Circuit for Main Current Control

In this article, this article will design a DC-DC converter circuit, you can manually control the circuit with a button. When the direct current in the circuit turns on and off (the component uses a push button switch to do this), the inductor deposits current, and each time current flows through the capacitor. Therefore, every time the operation is performed, the voltage will reach a peak value. The size of the output amplification depends on many factors, which includes the inductance value. In this circuit, this component uses an inductance of 100 mH. Therefore, when the inductance value becomes a lower value, a smaller output voltage can be generated. If the inductance value is changed to a higher value, a higher voltage can be generated. Therefore, you can use the inductance value to understand how it affects the output voltage. When using a higher inductance, the resistance of the inductance is greater and prevents the current from flowing, so the charging time may be longer, but it will eventually produce a higher output voltage. The rated voltage of electrolytic capacitors is approximately 50 V or higher. The voltage across the capacitor exceeds its rated voltage and may explode. Since the converter circuit generates extremely high voltages, it must far exceed the usual 25 V of capacitors in order to achieve the highest possible load capacity.

This circuit uses diodes as resistors. When the button is released, there is an interruption at the button connection, so all current is discharged through the diode and flows to the capacitor. The working principle of this circuit is that when the button is pressed, current flows through the inductor and is grounded through the button. After releasing the button, the node on the button becomes an open circuit, so no current flows through the node. The inductance consumes all the current accumulated on it and transfers it to the capacitor through the diode, thereby charging the capacitor, at which point voltage starts to build up across the capacitor. As the component continues to press and release the button, the load increases further, thereby further increasing the voltage. Until the tension reaches its peak.

3 Experimental Simulation of Parallel DC-Boost Converter Based on Main Current Control

3.1 Experimental Platform and Parameter Setting

This paper builds an experimental platform with a rated power of 1 kW, which is mainly divided into the main experimental platform, control board and expansion board. The expansion board is used to fine-tune the phase shift angle to make the power reach the target value. In the main experimental platform, the primary side capacitor bridge arm has a relatively large current, which requires multiple thin film capacitors to shunt in parallel; the secondary side capacitor bridge arm capacitor uses three thin film capacitors in parallel to reduce voltage ripple. The DC power supply used in the experiment is a programmable DC power supply from Magna-power electronics. The load is replaced by a sliding rheostat. The input and output power is measured with a FLUKE digital multimeter. Table 1 is the data sheet of this experiment.

Table 1. Experimental data sheet

Input voltage/V	Transmission power/W	Fifty-four	Fifty-five	Fifty-six	Fifty-seven	Fifty-eight	Fifty-nine	Sixty
Ten		0.7450	0.7575	0.7725	0.7850	0.8000	0.8150	0.8245
Twenty		0.7450	0.7575	0.7725	0.7850	0.8000	0.8150	0.8275
Thirty		0.7450	0.7575	0.7725	0.7850	0.8000	0.8125	0.8275
Forty		0.7450	0.7575	0.7725	0.7850	0.7975	0.8100	0.8275
Fifty		0.7425	0.7575	0.7700	0.7825	0.7975	0.8100	0.8250
Sixty		0.7425	0.7550	0.7700	0.7825	0.7950	0.8075	0.8225
Seventy		0.7425	0.7550	0.7675	0.7800	0.7925	0.8050	0.8200
Eighty		0.7400	0.7525	0.7675	0.7800	0.7900	0.8025	0.8175
Ninety		0.7450	0.7525	0.7650	0.7775	0.7900	0.8000	0.8125
One hundred		0.7450	0.7500	0.7625	0.7750	0.7875	0.7975	0.8100

3.2 Efficiency Analysis of the Parallel DC-Boost Converter

Figure 2 shows the experimentally measured conversion efficiency curve of the DC-DC Boost converter with output power under different input Vin and load R. When Vin = 26 V and the output power is 60 W, the maximum efficiency of the proposed prototype is about 96%. The efficiency of about 91% at 180 W full load can still reach the design index of the DC-DC Boost converter. From the change trend of the curve, it can be seen that the output power increases correspondingly after the input voltage increases. When the input voltage increases, the input current and the duty cycle decrease, and the on-off loss of the power switch tube and the diode decreases, so that the electromagnetic loss of the coupled inductor during the charging and discharging process is correspondingly reduced, thereby improving the efficiency.

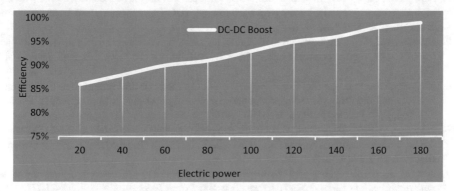

Fig. 2. DC-DC-BOOST converter efficiency curve

4 Conclusion

This article studies and summarizes the current application and research results of DC/DC converters in new energy power generation systems, and discusses key technologies to improve converter boost ratio and conversion efficiency in parallel. Based on this premise, two kinds of interleaved parallel high-gain Boost converters that can be applied to the requirements of new energy power generation systems are studied, and their functional realization and performance characteristics are discussed in depth.

References

1. Shebani, M.M., Iqbal, T., Quaicoe, J.E.: Modified droop method based on master current control for parallel-connected DC-DC boost converters. J. Electr. Comput. Eng. **2018**(5), 1–14 (2018)
2. Du, H., Jiang, C., Wen, G., et al.: Current sharing control for parallel DC-DC buck converters based on finite-time control technique. IEEE Trans. Ind. Inf. **15**(4), 2186–2198 (2019)

3. Zhang, X., Min, R., Lyu, D., et al.: Current tracking delay effect minimization for digital peak current mode control of DC-DC boost converter. IEEE Trans. Power Electron. **34**(12), 12384–12395 (2019)
4. Thounthong, P., Mungporn, P., Pierfederici, S., et al.: Robust hamiltonian-energy control based on lyapunov function for four-phase parallel fuel cell boost converter for DC microgrid applications. IEEE Trans. Sustain. Energ. **12**, 1500–1511 (2021)
5. Rana, N., Banerjee, S.: Development of an improved input-parallel output-series buck-boost converter and its closed-loop control. IEEE Trans. Industr. Electron. **67**, 6428–6438 (2019)
6. Karimi, M., Pichan, M., Abrishamifar, A., et al.: An improved integrated control modeling of a high-power density interleaved non-inverting buck-boost DC-DC converter. Mil. Oper. Res. **15**(6), 688–699 (2018)
7. Harryson, R., Germàn, G., Carlos, T., et al.: LMI control design of a non-inverting buck-boost converter: a current regulation approach. Tecciencia, **12**(22), 79–85 (2017)
8. Kawakami, T., Harada, T., Yamamoto, M., et al.: Proposal for the new control method having high responsiveness and robustness for the boost-type DC-DC converter. IEEJ Trans. Electron. Inf. Syst. **138**(4), 395–404 (2018)
9. Chincholkar, S.H., Jiang, W., Chan, C.Y.: An improved PWM-based sliding-mode controller for a DC-DC cascade boost converter. IEEE Trans. Circ. Syst. II Expr. Briefs **65**, 1639–1643 (2017)
10. Mohanty, P.R., Panda, A.K.: Fixed-frequency sliding-mode control scheme based on current control manifold for improved dynamic performance of boost PFC converter. IEEE J. Emerg. Sel. Top. Power Electron. **5**(1), 576–586 (2017)
11. Hassan, M.A., Su, C.L., Chen, F.Z., et al.: Adaptive passivity-based control of DC-DC boost power converter supplying constant power and constant voltage loads. IEEE Trans. Ind. Electron. (99), 1–1 (2021)
12. Ahmad, S., Ali, A.: Active disturbance rejection control of DC-DC boost converter: a review with modifications for improved performance. IET Power Electron. **12**(8), 2095–2107 (2019)

Identification of Traditional Chinese Medicine Based on Internet of Things Technology

Chen Sun[(✉)] and Jing Wang

Shaanxi University of Chinese Medicine, Xianyang, Shaanxi, China

Abstract. The development of the Internet of Things (IoT) has become one of the focuses of attention in the IT field and society at home and abroad. The Internet of Things is an emerging next-generation network that integrates radio frequency identification technology, wireless data communication technology and computer technology on the basis of the Internet. It has broad application prospects and high research value. Our country has used Chinese medicinal materials to treat diseases for thousands of years. In long-term medical practice, Chinese medicine has formed its unique theoretical system, which has good therapeutic effects and small side effects. As China's national treasure, Chinese medicinal materials are an indispensable material foundation for the people to prevent and treat diseases, and they are also a very important part of my country's medical industry. There are many types of Chinese medicinal materials, but the number of fake and inferior products on the market is not uncommon, and it is difficult for non-professionals to correctly identify them, which seriously hinders the sound development of the Chinese medicine industry. Therefore, it is extremely urgent to develop and establish a systematic identification technology for Chinese medicinal materials. However, traditional identification methods rely too much on manual intervention, which is subjective and costly. Secondly, the existing computer-aided identification methods are mainly based on machine learning algorithms for research, and their identification accuracy is low and time-consuming. In response to the above problems, this article mainly uses the Internet of Things technology to identify Chinese medicines. This article is based on the Internet of Things technology image classification and recognition algorithm for Chinese medicinal materials, and compares with traditional machine learning algorithms, optimizes the training of the neural network and constructs independent Chinese medicinal materials images the library ultimately improves the accuracy and objectivity of the classification and identification of Chinese medicinal materials.

Keywords: Internet of Things · Chinese medicine recognition · Deep learning · Image fusion

1 Introduction

In the process of fighting against diseases for thousands of years, the working people of our country have gradually accumulated a wealth of medical knowledge through practice and continuous understanding [1, 2]. Since writing was not developed in the

J. Macintyre et al. (Eds.): SPIoT 2021, LNDECT 98, pp. 575–582, 2022.
https://doi.org/10.1007/978-3-030-89511-2_74

ancient times, this knowledge could only be dictated by teachers. Later, when writing was available, it was gradually recorded and medical books appeared. These books have played a role in summarizing previous experience and facilitating the circulation and promotion [3, 4]. Chinese medicine has a history of thousands of years, it is the extremely rich experience of our people in the long-term struggle against diseases, and it has made a huge contribution to the prosperity of the Chinese nation [5, 6].

The development of traditional Chinese medicine in my country has a long history. It originated in the ancient times of "Shen Nong tasted herbs" and originated in the Yellow River Basin. Since the initial inheritance was passed down by word of mouth, many current Chinese medicinal materials are not detailed and accurate, and many of them are of the same category. The external morphology of the medicinal materials is often the same, so there is an applied discipline to identify the types and pros and cons of Chinese medicines-Chinese medicine identification [7, 8]. Traditional Chinese Medicine Identification is an applied discipline that uses modern scientific and technological knowledge to analyze and record the source, category, efficacy, characteristics and quality of Chinese medicinal materials, so as to formulate its quality standards and discover new varieties of medicinal materials [9]. For the identification of Chinese medicinal materials, in addition to traditional identification methods such as the empirical identification method of "looking, smelling, tasting, touching, water, and fire", and the microscopic identification method of observing the cell characteristics of medicinal materials, the currently commonly used methods mainly include molecular identification method, Chemical identification method, morphological identification method, biological effect identification method and other methods [10].

In this paper, in view of the complex model and time-consuming problem in the image identification method of Chinese medicinal materials, a computer-aided classification and identification algorithm of Chinese medicinal materials combined with deep convolutional neural network is proposed. The advantage of this algorithm is that it is based on the self-learning of traditional convolutional neural network, and combined with the unique advantages of traditional image processing technology and deep separable network, the accuracy and classification rate in the classification and identification of traditional Chinese medicine images have been improved to a certain extent.

2 Identification of Chinese Medicine Based on the Internet of Things Technology

2.1 Construction of Chinese Medicinal Materials Data Set Based on the Internet of Things Structure

In this paper, an ordinary optical microscope combined with computer equipment is used to collect images of Chinese medicinal materials such as fruit seeds. Due to the environment and the proficiency of the operation during the collection process, it will interfere with the identification results of the final Chinese medicinal materials. Therefore, in order to improve the accuracy of identification and make the follow-up work more convenient, this paper needs to preprocess the image of Chinese medicinal materials. First, to deal with the image duplication and naming confusion problems in the image acquisition process of the tablet computer, the image of the Chinese medicinal materials was sorted; then, the noise generated during the shooting and preservation of the Chinese medicinal materials was denoised through the Gaussian filtering algorithm; finally, due to the deep learning network requires a large amount of data, and the data set in this article is relatively small, so the image data is augmented [11, 12].

(1) Image collation of Chinese medicinal materials

Using image processing and machine learning to identify and classify Chinese medicinal materials requires relatively high data sets, which makes the algorithm more efficient. Since the current images are all self-collected, there are certain problems in the naming method, image efficiency, etc., so the collected original Chinese medicinal material images are processed as follows:

1) Organize the sample images one by one: due to the small depth of field of the ordinary optical microscope, in order to obtain all the features of the sample in the subtle state, it is necessary to take multiple shots to obtain a multi-focus image. For the convenience of future work, separate each of the same categories the images of the samples are organized in a unified manner, and each sample has a folder.

2) Image size: At the beginning of image acquisition, the resolution cannot be set specifically, resulting in different resolutions of 1600*1200 and 1400*1080. The image resolution is too large to have a certain impact on the image fusion in the later stage, but it has an impact on traditional Chinese medicine. The impact of material classification and identification is relatively small. Considering the fusion effect, fusion rate, and classification training time and effect, the resolution is unified to 800*600.

3) Delete duplicate images: In the process of image acquisition of Chinese medicinal materials, the adjustment of the depth of field is relatively small, which leads to image redundancy and has a greater impact on the efficiency of subsequent work. Therefore, the redundant photos of Chinese herbal medicine images are deleted, and there are 13,547 images remaining in the processed image data set.

(2) Establishment of data set of Chinese medicinal materials

By processing the collected images, a unified data set of Chinese medicinal materials is obtained, which lays the foundation for the subsequent identification of Chinese medicinal materials.

The data set of Chinese medicinal materials constructed in this paper includes 11 kinds of medicinal materials, namely, mustard seeds and yellow mustard seeds, perilla seeds, white sage seeds, wild small sesame seeds, scorpion vines, south scorpion vines, north scorpion scorpion seeds, and small flowers Sugar mustard, Psyllium, Psyllium, Ping plant and Chushizi, a total of 13,547 sheets.

2.2   Convolutional Neural Network

With the development of science and technology, deep learning technology has achieved good results in image and speech recognition, text processing, and smart recommendation. As a network model with a high degree of attention and a wide range of applications in deep learning, convolutional neural network (CNN) is developed on the basis of multi-layer perceptrons, and the output of the previous layer is the input of this layer, there is no information exchange and feedback between the same layer, which belongs to the feed forward neural network.

The calculation formula of the feature map x_i generated by convolution is as follows:

$$x_i = \frac{x_{i+1} + 2p - k}{s} + 1 \tag{1}$$

K is the convolution kernel, s is the moving stride, and p is the number of extra layers beyond the edge when the convolution kernel moves.

The back propagation process is similar to the feedback system in the circuit, which is a process of optimizing and fine-tuning the network according to the error. The first step of backpropagation is to obtain the error value, which is to find the error between the actual output value a(n) of the output layer n and the initial target value y, where f(z(n)) is the derivative function of the excitation function, the formula for:

$$\delta(n) = -(y - \alpha(n))f(z(n)) \tag{2}$$

2.3   Design of Chinese Herbal Medicine Identification Algorithm

(1) Image fusion

Since the traditional Chinese medicinal material image acquisition system used in this article uses an ordinary optical microscope, its depth of field is relatively small. When the distance between the medicinal material and the lens in the scene is large, it is impossible to obtain a clear image with full focus, and the multi-focus fusion technology can be used in In the synthesis process, the information content is expanded or enhanced, so it is important to obtain clear images of all targets, which is of great significance for image recognition and classification. In this paper, a guided filter-based image fusion algorithm (GFF) is used. First, the image is decomposed in two scales through the mean filter, then the image is reconstructed, and finally the base layer and the detail layer are fused by the guided filtering and weighted average method.

(2) Image feature extraction

Image features are complex and diverse. Feature extraction is the core operation in the identification of Chinese medicinal materials. The quality of feature extraction has a greater impact on the final identification effect. Among them, the following points should be paid attention to obtain the characteristic information of the micro-character images of Chinese herbal medicines.

1) Discrimination requirements, in terms of feature description, this image can be distinguished from other images;

2) Robustness is required to maintain good characteristic information when faced with external environmental influences such as light changes, visual angles and scale changes;

3) Local requirements, to tolerate the lack of information to a certain extent, and keep the expression of the micro-character image information of Chinese medicinal materials unaffected.

(3) Feature fusion

There are many types of Chinese medicinal materials, and there are endless emergence of fake products, and the medicinal materials are small in shape and difficult to distinguish with the naked eye, but their texture, color and shape are different. Therefore, the color feature, morphological feature, and texture feature of the extracted micro-character images of Chinese medicinal materials are serially fused. At the same time, in order to prevent the influence of large-value features on small-value features, different types of features are normalized to [0, 1] interval.

3 Chinese Medicine Identification Experiment

3.1 Experimental Environment

In order to verify the performance of the algorithm, this article conducts experiments in the MATLABR2016a environment, and the specific configuration is as follows:

The operating system is: WIN10, 64-bit.

Processor: Intel(R) Core (TM)i7-6700CPU@3.40 GHz.

Memory: 8.00 GB.

Graphics card: GT730.

3.2 Evaluation Index

This article mainly classifies and distinguishes Chinese herbal medicines and their confounding products, and uses several commonly used evaluation indicators, Accuracy, Recall, Precision, Comprehensive Evaluation Index (F1_Score), and Confusion Matrix to analyze the model.

4 Analysis of the Results of Traditional Chinese Medicine Identification Experiments

4.1 Authenticity Identification Experiment

The white mustard seed and its traditional product, yellow mustard seed, perilla seed and its counterfeit wild small sesame seed, were classified into two categories. Among them, there are 250 micro-character pictures of each medicinal material, each of which takes 165 as the training set, and the remaining 85 as the test set. The size is uniformly normalized to 128*128*3; various types of Accuracy, Recall. Precision and F1_ Score are calculated respectively.

Table 1. Two-classification experiment results

Data	Model	Accuracy %	Precision %	Recall %	F1_Score
Mustard	Morphological characteristics	62.17	62.47	62.84	0.62
	Color characteristics	78.14	80.52	78.29	0.78
	Texture feature	90.54	91.71	90.51	0.91
	Morphology + color	83.62	84.67	83.49	0.83
	Morphology + texture	91.54	91.32	91.47	0.91
	Color + texture	92.56	93.07	92.79	0.92
	All features	95.48	95.67	95.47	0.95

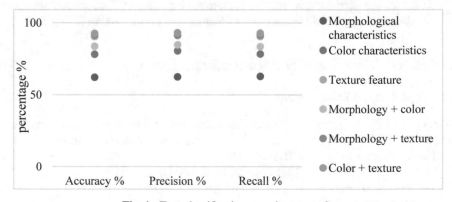

Fig. 1. Two-classification experiment results

According to Table 1 and Fig. 1, the recognition method proposed in this paper has a good effect on the accuracy and precision of the two-classification of authenticity and falsehood of traditional Chinese medicinal materials. Under normal circumstances, precision and recall rate affect each other and are inversely proportional. In disease drug monitoring research, the recall rate is generally improved under the condition of ensuring accuracy. F1_Score is a comprehensive evaluation of accuracy and recall rate. The higher the FI_Score, Classification performance is better. In summary, the classification performance of the algorithm proposed in this section is relatively better compared with the performance of the algorithm based on a single feature and the combination of different features.

4.2 Category Classification Experiment

Among them, eight kinds of medicinal materials were classified and identified: white mustard seed, yellow mustard seed, perilla seed, white sage seed, Nan Ting Li Zi, Bei Ting Li Zi, Xiaohua Sugar Mustard, and Fu Shi Zi. Among them, there are 220 micro-character pictures of each kind of medicinal material, a total of 1760 pictures, 1320 pictures are used as the training set, and the remaining 440 pictures are used as the test set. The size is uniformly normalized to 128*128*3. The specific results are shown in Fig. 2. (The same as the model indicator in the Figure above).

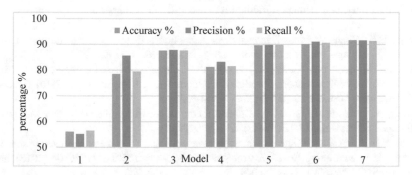

Fig. 2. Results of the eight-classification experiment

As shown in Fig. 2, the algorithm proposed in the article is significantly better than the algorithm based on a single feature and the algorithm of different feature combinations in the multi-class identification of Chinese medicinal materials. For example, 53 out of the chuen seeds were classified correctly, one was mistakenly classified as yellow mustard seeds, one was mistakenly classified as small sugar mustard, and another white mustard seeds and one yellow mustard seeds were incorrectly classified. It can be seen that the algorithm in the article has a better classification effect on Fructus fulvidraco, Bai Suzi, and Beiting Lizi, and relatively poor performance for White Mustard and Yellow Mustard. The recognition rates of White Mustard and Yellow Mustard are 87% and 84, respectively %, the main reason is that the mustard seeds have a high degree of similarity, and their color and shape are similar to those of the sesame seeds, which is difficult to distinguish.

5 Conclusions

Chinese medicinal materials are the most important material carrier of Chinese medicine in my country, and they have been responsible for the treatment and health maintenance of the Chinese nation for thousands of years. Even in today's western medical society, Chinese medicine is still an important medical treatment and health regimen for many Chinese and overseas Chinese. Authentic Chinese medicinal materials refer to those famous and authentic medicinal materials that have Chinese characteristics, are traditionally recognized, and come from specific production areas. However, due to the wide variety of medicinal materials and complex sources, many fake and inferior medicinal materials of authentic medicinal materials frequently appear in the medicinal material market, which leads to chaos in the medicinal material market and damages people's health and property. The Chinese medicine recognition based on the Internet of Things technology proposed in this paper has achieved good results under the test. The classification performance can reach more than 80%, and the recognition performance can reach a correct rate of more than 90%.

References

1. He, M., Shi, J.: Circulation Traceability System of Chinese Herbal Medicine Supply Chain based on Internet of Things Agricultural Sensor. Sustain. Comput. Inf. Syst. **30**(10), 100518 (2021)
2. Sun, E.H., Chen, Y.: Multi-layer wireless perception and recognition technology based on internet of things. Int. J. Online Biomed. Eng. (iJOE) **15**(1), 31 (2019)
3. Lu, J., Hu, K., Yang, X., Hu, C., Wang, T.: A cluster-tree-based energy-efficient routing protocol for wireless sensor networks with a mobile sink. J. Supercomput. **77**(6), 6078–6104 (2020). https://doi.org/10.1007/s11227-020-03501-w
4. Yang, Y., Peng, H., Li, L., et al.: General theory of security and a study case in internet of things. Internet Things J. IEEE **4**(2), 592–600 (2017)
5. Lee, S.H., Yang, C.S.: Fingernail analysis management system using microscopy sensor and blockchain technology. Int. J. Distrib. Sens. Netw. **14**(3), 155014771876704 (2018). https://doi.org/10.1177/1550147718767044
6. Zhang, X.B., Ge, X.G., Yan, J., et al.: Application of image recognition technology in census of national traditional Chinese medicine resources. Chin. J. Chin. Mater. Med. **42**(22), 4266 (2017)
7. Xiaojuan, D., Yuan, Z.: Application and implementation of intelligent community system based on the internet of things technology. Agro Food Ind. Hi Tech **28**(1), 1558–1562 (2017)
8. Chang, J., Zuo, X., Hou, B., et al.: Internet of things security detection technology based on grey association decision algorithm. Complexity **2021**(13), 1–12 (2021)
9. Wang, J.: Research on data fusion technology of mobile intelligent robot based on internet of things. IOP Conf. Ser. Earth Environ. Sci. **707**(1), 012007 (2021). https://doi.org/10.1088/1755-1315/707/1/012007
10. Siddiq, M.I., Wibawa, I.P.D., Kallista, M.: Integrated internet of things (IoT) technology device on smart home system with human posture recognition using kNN method. IOP Conf. Ser. Mater. Sci. Eng. **1098**(4), 042065 (2021). https://doi.org/10.1088/1757-899X/1098/4/042065
11. Shi, W.: Teaching reform of chinese international education major based on big data of internet of things. J. Phys. Conf. Ser. **1915**(4), 042036 (2021)
12. Wang, L.L., Li, B., Wang, Z.Y., et al.: Application of systematic biology technology in research of traditional Chinese medicine based on overall research. Chin. Tradit. Herbal Drugs **51**(19), 5053–5064 (2020)

The Application of Computer Multimedia Technology to the Informatization of Linguistic Content

He Chen(✉)

Northwest University of Political Science and Law, Xi'an 710122, Shaanxi, China

Abstract. With the advent of the era of information technology, multimedia technology has been used in the field of teaching, and achieved good results. Multimedia technology enriches the boring and difficult linguistic content. The use of multimedia technology, combined with pictures, voice and even images, makes the linguistic content gradually informationized. The application of multimedia technology in linguistics provides an effective language teaching and learning tool for learning linguistics. On the basis of linguistics, the content of multimedia technology is informationalized and interacted with learners. This paper mainly studies the use of computer multimedia technology (CMT) to make linguistic content information. By analyzing the current situation of multimedia assisted language teaching, this paper finds out that multimedia technology plays a very important role in assisted language teaching, but at the same time, it also has various shortcomings or disadvantages. This paper expounds the advantages of CMT in linguistic content informatization. The introduction of multimedia technology to assist linguistic content informatization can change the traditional teaching mode, change the traditional teaching methods and improve the traditional teaching efficiency. It also studies the computer multimedia linguistic content informatization technology. In order to study and understand the students' practical sense of using CMT to carry out linguistic content informatization, this paper adopts the method of questionnaire survey to carry out investigation and analysis. A total of 200 questionnaires have been issued in this survey, and all of them are valid questionnaires. The experimental results show that 73% of the students like the information-based language teaching method of CMT, and 21% of the students have a general sense of the information-based language teaching method of CMT, There are 6% of the students do not like this kind of CMT language teaching information means.

Keywords: Multimedia technology · Content information · Language teaching · Computer media

1 Introduction

With the continuous application and development of CMT in the field of teaching, multimedia language teaching has become the most popular language learning aids [1, 2]. The application of CMT makes up for the shortcomings of the traditional language learning methods, and changes the boring of traditional language teaching. Combined with image, video, PPT and other modes, the linguistic content is informationalized [3, 4]. With the help of CMT, language learning can provide learners with relaxed learning atmosphere, excellent language carrier, and mobilize their enthusiasm for language learning [5, 6]. On the other hand, the CMT has informationized the linguistic content, so that language learners can freely choose the language content they are interested in according to their own learning progress, not limited by time and space [7, 8]. Therefore, this paper studies the use of CMT to make linguistic content information, explores the connotation of CMT, studies and analyzes the characteristics of linguistic content information and multimedia technology, and summarizes the realization of multimedia technology in linguistic content learning [9, 10].

In the research of CMT language teaching, many domestic and foreign scholars have studied it and achieved some research results. Ji x points out that multimedia assisted teaching can stimulate students' interest in learning, change abstract into image, simplify the complexity, better help students break through the key and difficult points, and help students to actively learn in the teaching process. It has the advantages incomparable to traditional teaching forms [11]. Al seghayer k points out the characteristics of the kindness, gorgeous and easy interaction of multimedia works, which makes the traditional education model fresh, improves the single and boring educational means, increases the interest of the knowledge transfer process, enhances the interaction of the classroom, increases the information capacity of the classroom, effectively improves the students' interest in learning, thus arousing the initiative of students' learning, Improve learning effect [12]. These theoretical researches on the teaching of CMT and language provide theoretical support for the application of CMT in this paper to make the Information-based Research of linguistic content available.

This paper mainly studies the use of CMT to make linguistic content information. By analyzing the current situation of multimedia assisted language teaching, this paper finds out that multimedia technology has played a very important role in assisted language teaching, and it has indeed brought a lot of vitality and vigor to teaching. It has improved students' learning enthusiasm to a certain extent, but at the same time, it also has various shortcomings or disadvantages. This paper expounds the advantages of CMT in the informatization of linguistic content, and the introduction of multimedia technology to assist the informatization of linguistic content can change the traditional teaching mode, change the traditional teaching methods, and improve the efficiency of traditional teaching.

2   Using CMT to Make Linguistic Content Information

2.1   Current Situation of Multimedia Assisted Language Teaching

Multimedia technology with its unique advantages was first applied in the field of education, among which the United States, the European Union and other western countries took the lead in introducing multimedia technology, more than 10 years earlier than China. The characteristics of multimedia works, such as friendly, gorgeous and easy to interact, make the traditional education mode take on a new look, improve the single and boring means of education, increase the interest in the process of knowledge teaching, enhance the interaction in the classroom, increase the information capacity in the classroom, effectively improve the students' interest in learning, so as to arouse the students' learning initiative and improve the learning effect. At present, many colleges and universities in our country are using computer multimedia teaching mode, multimedia courseware also emerge in endlessly. However, multimedia technology can not cure all kinds of problems. For example, there are some problems in English vocabulary teaching at the present stage: first, most of the vocabulary courseware is a copy of the dictionary or a special training for a certain level of examination, which is out of touch with the textbook, and the pertinence is not strong. Most of the multimedia vocabulary courseware is a special training for the vocabulary of CET-4 or CET-6. Second, the classroom capacity is too large, too many words and usages are explained within the specified time, which violates the rule of memory and easily frustrates students' learning enthusiasm. Most vocabulary softwares emphasize the shorthand of words or memorize more words in a short time, which actually violates the rule of memory in learning and ignores the quality for the sake of quantity. Thirdly, it overemphasizes the spelling memory of words and ignores the role of words as morphemes in the language system. The basic unit of language of words is to carry information and can not be used independently without context. If we only pay attention to spelling and can't grasp the meaning and usage of words flexibly, the effect of vocabulary learning is bound to be poor. Through the above analysis, we can see that multimedia technology has played a very important role in assisting language teaching. It has indeed brought a lot of vitality and vigor to teaching and improved students' learning enthusiasm to a certain extent, but there are also various shortcomings or disadvantages.

2.2   CMT Makes the Information Advantage of Linguistic Content

(1)   Change the traditional teaching mode

We should change the traditional teaching mode and introduce multimedia education technology to make vocabulary teaching full of vitality. Due to the confinement of teaching equipment and teaching means, it has been in the form of teachers' dictation and students' notes. Vocabulary learning itself is very boring and less fun. Simply requiring students to memorize the arrangement and combination of those letters can not stimulate students' interest in learning. The dictation required by teachers is also in vain. Therefore, the use of modern means of education, making more exquisite teaching courseware, increasing students' participation and interaction, with interest to lead students to learn to think, vocabulary teaching will glow with new vitality.

(2)   Changing traditional teaching methods

In order to change the traditional teaching method, multimedia education technology is used to replace the teaching method of "blackboard and white characters", and words, pictures, pronunciation, even video and other media are introduced into vocabulary explanation and training, so that vocabulary learning can be changed from silent to sound. Multi sensory stimulation can improve the effect of memory and achieve the purpose of long-term memory. The ultimate goal of learning vocabulary is to understand and remember these words. The traditional way of memorizing words is reading books silently or copying, which mainly depends on visual stimulation. However, auditory stimulation is also an indispensable part of memory activities, and occupies a very important proportion. The purpose of long-term memory is to stimulate the memory activity of learners, enhance the response of sensory memory to input information, extend the storage time and achieve the purpose of long-term memory.

(3) Improve the efficiency of traditional teaching

The application of multimedia technology can greatly enrich the teaching content, change the teaching structure and enrich the teaching means. With the advantages of multimedia teaching, teachers collect and display pictures, audio, video and movie clips related to the teaching content to let students understand the cultural background knowledge, broaden students' knowledge, and better meet students' desire for knowledge. In addition, because teachers should prepare teaching materials, carefully design teaching activities and make multimedia courseware before class, they will save a lot of writing time on the blackboard and make classroom teaching more compact. Teachers can also grasp the key and difficult points in the teaching process, teach students in accordance with their aptitude, so as to achieve better teaching effect. At the same time, the multi-media courseware can provide a variety of multi-level exercises. Students can train purposefully according to their own conditions, so as to achieve the purpose of using the language skillfully.

2.3   Computer Multimedia Linguistics Content Information Technology

Today, with the rapid development of information technology, CMT can promote the informatization of linguistic content. Language learners can choose the language content they are interested in according to their own learning progress, not limited by time and space. The emergence of this kind of shared linguistic content will inevitably lead to illegal elements trying to invade and steal these information-based content. CMT not only promotes the informatization of linguistic content, but also guards against this phenomenon. In the process of content informatization, digital watermark anti-theft technology is adopted, the specific formula is as follows:

$$F(u, v) = A(u, v) + B(u, v)i + C(u, v)j + D(i, v)k \tag{1}$$

The analysis shows that the real part of the quaternion Fourier transform satisfies the following formula, where m and N are the size of the image, and the formula is as follows:

$$A(u, v) = -A(M - u, N - v) \tag{2}$$

If the corresponding frequency domain matrix of pure quaternion after QFT is not modified, the real part of quaternion obtained by inverse transformation of iqft is still zero. Because the real part of the color image after QFT has the characteristics of symmetry, if the real part is used as the embedding area, and the watermark information is embedded symmetrically according to the formula, the information carried by it is dispersed into the imaginary part of the original quaternion. It is proved that the real part of the new quaternion is zero. Therefore, the information embedded in the real part can not be lost after iqft, and the visual error caused by the embedded information is spread to the whole image, which is not easy to cause subjective detection. Therefore, it is more difficult for the processed linguistic information content transformed by CMT to be damaged by malicious theft.

3 Experimental Study

3.1 Subjects

In order to study and understand the students' practical sense of applying CMT to linguistic content informatization, this paper uses questionnaire survey to conduct investigation and analysis. CMT promotes the development of language teaching content in the direction of informatization. As the beneficiaries, students' experience sense is very important. A total of 200 questionnaires have been issued and all of them are valid.

3.2 Experimental Process Steps

This paper studies the students' practical sense of applying CMT to linguistic content informatization through questionnaire survey. It analyzes the current situation of multimedia assisted language teaching, clarifies the advantages of CMT in linguistic content informatization, and studies the technology of computer multimedia linguistic content informatization. In this paper, through the use of chart analysis method to analyze students' perception of CMT language teaching information means.

4 Using CMT to Make Linguistic Content Information Experimental Research and Analysis

4.1 Sensory Analysis of CMT Language Teaching Informatization

In order to study and understand the students' practical sense of using CMT to carry out linguistic content informatization, this paper uses questionnaire survey to carry out investigation and analysis. A total of 200 questionnaires were distributed in this survey, and all of them have been collected for statistical sorting. The sensory data results are shown in Table 1.

Table 1. Analysis of CMT language teaching information sensory

	Very like	Like	Commonly	Unlike	Very unlike
Number	68	78	42	12	0
Proportion	34	39	21	6	0

It can be seen from Fig. 1 that 73% of the students like this kind of information-based language teaching method of CMT, and 21% of the students have a general sense of using CMT for linguistic content information, There are 6% of the students do not like this kind of CMT language teaching information means.

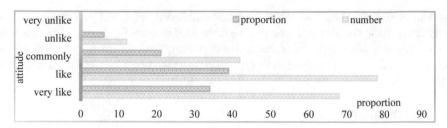

Fig. 1. Analysis of CMT language teaching information sensory

4.2   The Application of CMT in Language Teaching Informatization

The application of CMT in language teaching is mainly to promote the development of language teaching in the direction of informatization. There are many kinds of applications in it. This paper collects and sorts out the main application of CMT in language teaching informatization through data mining technology, and the results are shown in Table 2.

Table 2. Analysis of the application of CMT in language teaching informatization

	Rich pictures	Optimize courseware	Highlight content	Classroom interaction	Easy to operate	Others
Proportion	16.75	14.61	27.21	21.33	12.17	7.93

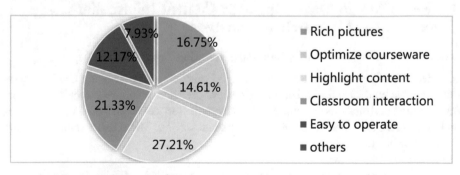

Fig. 2. Analysis of the application of CMT in language teaching informatization

As can be seen from Fig. 2, the application of CMT in language teaching, highlighting linguistic content accounted for 27.21%, enriching pictures accounted for 16.75%, optimizing courseware accounted for 14.61%, classroom interaction accounted for 21.33%, convenient operation accounted for 12.17%, and other functions accounted for 7.93%.

5   Conclusions

The application of CMT in the content teaching of linguistics is the product of the development of science and technology and the development of educational reform and improvement. CMT gives new vitality to linguistic content. The diversity and interaction of multimedia works help learners to learn linguistic content. Through the intervention of multimedia technology, learners' enthusiasm can be increased and their participation can be mobilized. The application of CMT, which promotes the informatization of linguistic content, is a major manifestation of the current progress of science and technology.

Acknowledgement. This research supported by The Research Fund for Foreign Languages Education, China, 2020 (ZGWYJJJJ10A016).

References

1. Tang, M.: Research on the Chinese language and literature teaching assisted by computer multimedia technology. IOP Conf. Ser. Mater. Sci. Eng. **569**, 052026 (2019). https://doi.org/10.1088/1757-899X/569/5/052026
2. Wang, Z., He, J.: On influence of computer multimedia technology in language learning. J. Phys. Conf. Ser. **1802**(3), 032092 (2021). https://doi.org/10.1088/1742-6596/1802/3/032092
3. Kabooha, R., Elyas, T.: The effects of youtube in multimedia instruction for vocabulary learning: perceptions of EFL students and teachers. Engl. Lang. Teach. **11**(2), 72 (2018)
4. Gonzalez-Acevedo, N.: Technology-enhanced-gadgets in the teaching of english as a foreign language to very young learners. ideas on implementation. Procedia Soc. Behav. Sci. **232**, 507–513 (2016). https://doi.org/10.1016/j.sbspro.2016.10.070
5. Wang, X.: Discussion on application of multimedia teaching in college english vocabulary teaching. Open J. Mod. Linguist. **6**(3), 177–181 (2016)
6. Zhao, Q.: Research on English teaching transition between rural primary schools and secondary schools in light of computer multimedia technology. J. Phys. Conf. Ser. **1578**, 012143 (2020). https://doi.org/10.1088/1742-6596/1578/1/012143
7. Bavaharji, M., Chetty, T.N., Ismail, Z.B., et al.: A comparison of the act and frequency of plagiarism between technical and non-technical programme undergraduates. Engl. Lang. Teach. **9**(4), 106 (2016)
8. Kjaergaard, H.W., Fougt, S.S.: But all we really wanted was a course! teacher professional development for innovative teaching with ICT. J. Educ. Multimedia Hypermedia **25**(4), 377–395 (2016)
9. Yawiloeng, R.: Second language vocabulary learning from viewing video in an EFL classroom. Engl. Lang. Teach. **13**(7), 76–83 (2020)
10. Jung, R., George, B.: Teachers' perceptions about teaching multimodal composition: the case study of Korean English teachers at secondary schools. Engl. Lang. Teach. **9**(6), 52 (2016)
11. Ji, X.: Community guidance model based on interactive multimedia system. Multimedia Tools Appl. **78**(4), 4723–4741 (2018). https://doi.org/10.1007/s11042-018-6856-x
12. Al-Seghayer, K.: Celebrating the story of my first contribution to call. Lang. Learn. Technol. **20**(2), 179–184 (2016)

Application of Network Security in Logistics Information Management Platform

Xu Liu[✉]

Institute of Logistics Science and Engineering, Shanghai Maritime University, Shanghai, China

Abstract. In recent years, with the adjustment and optimization of the industrial structure, the basis and core role of the logistics industry in urban construction has become more and more obvious. At the same time, ensuring the stable operation of the logistics information management platform and ensuring the safety of the information and data in the system have become issues that must be considered in the development of logistics information. This article focuses on the application of network security in the logistics information management platform. This article first summarizes the application of network security in the logistics information management platform, and studies the security system of the logistics information management platform from the aspects of logistics business data collection and transmission security, and the security of the external environment of the system. In this paper, the logistics information management platform architecture is planned and designed from the perspective of platform users, and platform functions are planned according to the needs of platform users, and the main functions of the logistics information management platform based on network security are designed. Finally, through the function and performance test of the logistics information management platform, the test results show that the system fully meets the actual application requirements. The test case response time is within 6 ms, and the operation success rate is 100%. Therefore, the system response speed fully meets the performance requirements of this research.

Keywords: Network security · Logistics information management platform · Demand analysis · Function design

1 Introduction

With the development of the logistics industry, the construction of a logistics information management platform has become an important foundation for promoting the development of modern logistics [1, 2]. The logistics information management platform under network security can integrate the logistics information resources of various links in the city and complete the inter-system data exchange and sharing, to achieve a high degree of integration and in-depth mining of information, to achieve the improvement of the soft environment for the development of information technology in the logistics industry, to meet the increasing diversified needs of different users, and to promote the rapid development of urban modern logistics [3, 4]. However, the existing logistics information management platform still has many network security problems. How to build a

J. Macintyre et al. (Eds.): SPIoT 2021, LNDECT 98, pp. 590–598, 2022.
https://doi.org/10.1007/978-3-030-89511-2_76

logistics information management platform that meets the network security according to the characteristics of the logistics information management platform has become a hot spot in the current logistics information security research [5, 6].

Regarding the research of network security and information management platforms, many scholars have conducted in-depth discussions on them. For example, Jayasinghe RS believes that the functional planning of the platform should be as comprehensive and in-depth as possible to avoid superficial and formal problems [7]; Adeitan DA proposed that the platform should establish a disaster recovery backup system, improve the disaster prevention capabilities of the logistics system, change the limitations of traditional data protection, and improve emergency response capabilities [8].

This article first summarizes the application of network security in the logistics information management platform, and studies the security system of the logistics information management platform from the aspects of logistics business data collection and transmission security, system external environment security, and network security. In this paper, the logistics information management platform architecture is planned and designed from the perspective of platform users, and platform functions are planned according to the needs of platform users, and the main functions of the logistics information management platform based on network security are designed. Finally, through the function and performance test of the logistics information management platform, the test results show that the system fully meets the actual application requirements.

2   Application of Network Security in Logistics Information Management Platform

2.1   Application of Network Security in Logistics Information Management Platform

(1) User identity authentication

System login is the first step for the user to enter the logistics information system. After confirming the legitimacy of the login user's identity, the user can add, modify, delete and other operations to the business system. Therefore, the user's identity authentication is very important in the access control security strategy of the entire system.

In order to ensure that the user identity authentication information needs to be encrypted and stored in the database, even if the information is leaked, the thief cannot immediately understand the true meaning of the ciphertext information [9, 10]. MD5 (Message Digest Algorithm) encryption algorithm is a one-way hash algorithm, input information of any length, after processing, all output as a 128-bit hash value.

1) Data padding: pad the original information into a number whose length is less than a multiple of 512 bit by only 64 bits, that is, $n * 512 + 448$. The filling method is to append a 1 after the message, followed by multiple 0s as required, until the message length is met. Then append a length of pre-padding information expressed in 64-bit binary. After the data is filled, the length of the information is exactly an integer multiple of 512.

2) Initialization variables: load four 32-bit initial link variables, they are: A = 01234567h, B = 89abcdefh, C = fedcba98h, D = 76543210h.

3) Processing packet data: the message is processed in units of 512-bit packets through 4 auxiliary functions. The 4 auxiliary functions are as follows:

$$F(X, Y, Z) = (X\&Y)|((\sim Y)\&Z) \tag{1}$$

$$G(X, Y, Z) = (X\&Y)|(Y\&\sim Z) \tag{2}$$

$$H(X, Y, Z) = X^\wedge Y^\wedge Z \tag{3}$$

$$I(X, Y, Z) = Y \sim (X|(\sim Z)) \tag{4}$$

Among them, & stands for AND operation, | stands for OR operation, ~ stands for non-operation, and ^ stands for XOR operation.

4) Output: After processing all 512-bit packets, the cascade of ABCD will be output as the result of MD5 hash.

Logistics business data collection and transmission security

Data collection in the logistics system is mainly through bar code recognition technology, wireless data transmission and wireless tag (RFID) technology to realize the collection of a large number of logistics data and improve the efficiency of logistics activities [10, 11]. Among them, the use cost of bar code technology is lower, and compared to RFID technology, it is more widely used in logistics systems. Through bar code technology, warehouse operations, transportation and other links are closely linked together to build a tracking system for the entire production process. As the carrier of logistics information, barcodes cannot be ignored in terms of security. By encrypting the QR code, the direct exposure of the information in the QR code is avoided.

The data transmission process is the link where data is most likely to be tampered with by illegal elements. In order to improve the security of client-server data transmission, this study uses a data transmission channel based on the HTTPS protocol. By adopting the HTTPS protocol, data is transmitted in cipher text during transmission, which effectively prevents the possibility of data being intercepted or tampered with, and ensures the security of data communication.

(4) The external environment of the system is safe

The security of the external environment of the system is mainly considered from the perspective of the information system, and its hidden dangers come from the users of the system on the one hand, and from the unknown natural environment on the other hand.

For users of the logistics information system, first, the user's identity needs to be verified when the user logs in. Secondly, it is necessary to strictly control user access rights. RBAC is currently the most common access control model. Based on RBAC, a dual authorization model of user-permission and role-permission is proposed, and the concept of user group is introduced to improve the flexibility of role assignment. At the same time, time constraints are added, which can increase the dynamic controllability of permissions and avoid users' unauthorized operations.

2.2 Functional Design of Logistics Information Management Platform Based on Network Security

(1) Functional design of the shipper's enterprise

1) Information release module

The information release module provides a unified way for shippers and enterprises to release transportation requirements, replacing the traditional release channels such as telephone, mail, and the Internet, and is more standardized and consistent, and the delivery of transportation information is more accurate. The shipper enterprise needs to maintain the delivery address and contact person of its own customer in the information release module so that it can be directly selected when releasing the information. If it is not convenient to carry out maintenance, you can also entrust the business personnel of the logistics company to do it on your behalf.

2) Order management module

The order management module provides a way for cargo owners to monitor the logistics business in real time. Different from the traditional logistics management system, it provides cargo owners with a channel to supervise the whole process of business operation, which will greatly enhance the customer experience. The shipper company can see the transportation status of the goods, check whether the issued transportation order status is received, dispatched, executed, delivered or completed, and can also view various details of the business, such as vehicles and the driver's certificate, photos of the vehicle's on-site operation, the location of the vehicle and the photo of the receipt. In addition, customers can also filter relevant data for viewing according to their own needs, and provide help for their own operational decisions.

3) Business settlement module

The business settlement module provides shippers with the management of checking freight, payables, settlement statements and settlement records. The shippers can check the settlement of freight, obtain the details of the settlement list, and facilitate settlement management, reducing the time and occurrence of reconciliation between the two parties.

(2) Functional design of logistics company

1) User data management module

The user data management module is mainly for the logistics company's information management of its customers and third-party carriers. The relevant business personnel upload and enter the customer and carrier information in the user data management module, including business license, address, account opening bank, contact information, etc., and upload the corresponding contract at the same time, enter the contract start date, and it will be reviewed after review call in the platform, remind to renew before the contract expires, and stop using the contract when it expires. This module can also have deeper applications, such as categorizing customers by keywords such as regions and customer industry categories, so as to provide a certain reference for market development.

2) Order management module

The order management module realizes the entire management and control of the logistics process, changing the situation of passively inquiring and understanding the logistics business situation through the phone in the past, can obtain first-hand information, improve the timeliness and transparency of logistics information acquisition, and increase the enterprise operating efficiency. Relevant personnel perform their duties according to their respective job responsibilities and provide data. Relevant users can quickly know which stage of the transportation business is based on the status of the order. At the same time, they can view on-site photos, vehicle positioning and receipt photos at any time, and achieve full control and control for all businesses. The process steps are traceable.

3) Inquiry module

The inquiry module can initiate an inquiry to a third-party carrier to arrange transportation tasks at the fastest speed and the best price when its own capacity cannot meet the transportation requirements.

4) Settlement Center Module

The settlement center module includes two parts: receivable and payable.

Receivable settlement means that after completing the customer's transportation business, the relevant personnel select the billable business in the settlement center, submit an invoicing application, and the corresponding management personnel can approve the invoicing. After the customer's freight is credited, the financial records are recorded in the settlement center.

Payable settlement means that after the carrier completes the transportation business, the relevant business personnel can select and pull the payment list in batches in the settlement center to make payment applications. At the same time, the amount of cash and oil card is allocated, and the corresponding management personnel will make payment after approval. Remitted to the carrier's account, the oil card automatically enters the oil card receiving and approval process, and is issued after the approval is approved. The third-party carrier's transportation invoice is received and the financial record is entered into the settlement center.

(3) Function design of third-party carrier

1) Bidding module

The bidding module is the main form for carriers to obtain business. It provides carriers with an opportunity to obtain first-hand business information. After the logistics company releases the supply information, the carrier will put forward a quotation based on its own vehicle conditions and cost accounting. After the bidding is successful Ready to operate.

2) Waybill management module

In the waybill management module, the carrier can view and manage business waybills. Without increasing development costs, they can obtain a free management system to view business information, schedule specific delivery vehicles, and view the status of the waybill.

3) Business settlement module

4) Job status report module

The job status report module can provide the real-time geographic location of the vehicle, which is the key to realizing the whole-process tracking of the logistics business. It improves the current situation of opaque cargo location in the logistics industry and provides support for improving the service level of the logistics industry.

3 System Test of Logistics Information Management Platform Based on Network Security

3.1 Test Methods and Means

The test method is a black box test method, and the method adopted is manual functional test one by one.

3.2 Test Environment

The test server is Alibaba Cloud Server ECS, the operating system is Centos, the database system is Mysql, and the Java programming language is used.

3.3 Performance Test

In order to ensure that the platform has a good user experience, the logistics management platform opens the second level cache at the back end, hoping to shorten the server response time and improve the system performance.

3.4 Test Items

(1) Review of outbound orders

On the outbound order management interface, click the outbound order to be reviewed to view the details of the outbound order, and click Save to complete the review.

(2) Order settlement

On the outbound order management interface, check the outbound order in the list of goods to be received, click Order Settlement, the order details page will pop up, select the payment method, enter the payment amount, and click OK to settle.

(3) Order adjustment

On the outbound order management interface, click the inventory order that has been reviewed, click the order adjustment, and the inventory result list will pop up, and then click the save button.

4 Data Analysis of Logistics Information Management Platform

4.1 Response Time

Send the same request to the server five times and record the time from the start of the request to the end of the response; after opening the cache, send the same user request as above five times to the server, and record the time from the start of the request to the end of the response time. The time comparison is shown in Table 1.

Table 1. Response time (ms)

	Add cache	No cache
1	100	200
2	74	176
3	59	178
4	56	154
5	61	203

As shown in Fig. 1, the five response times without cache are 200 ms, 176 ms, 178 ms, 154 ms, and 203 ms; the five response times for adding cache are 100 ms, 74 ms, 59 ms, 56 ms, and 61 ms. It can be seen that after the cache is turned on, the system response time has been greatly reduced, and the performance has been improved.

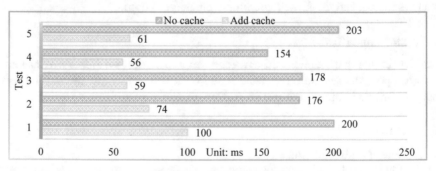

Fig. 1. Response time (ms)

4.2 Analysis of Some Performance Test Cases

Postma is a powerful API testing tool. The logistics information management platform studied in this paper uses this software to perform concurrent performance testing on all interfaces of the system. The concurrency number is set to 100. Due to space limitations, this article only tests some use cases 10 times, and averages the results as shown in Table 2.

Table 2. Part of the performance test case time

	Average response time (ms)	Success rate (%)
Log in	2.4	100
Outbound order review	3.1	100
Order settlement	2.5	100
Order adjustment	5.1	100

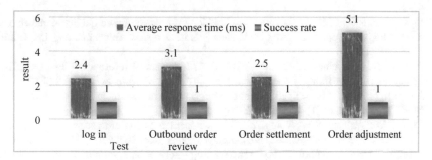

Fig. 2. Part of the performance test case time

It can be seen from Fig. 2 that the response time of the test cases is within 6 ms, and the operation success rate is 100%. Therefore, the system response speed fully meets the performance requirements of this research.

5 Conclusion

The construction of logistics informatization is one of the important ways to reduce costs and increase efficiency in the logistics industry. The application of network security in the logistics information management platform is the continuous update of information technology through the use of logistics to connect, analyze, mine, and flow data. As well as effective collaboration in an interconnected, open digital environment, and borrowing new information technologies such as big data and cloud computing to dig deeper into existing logistics information to meet the information needs of bilateral and multilateral markets, and to promote the logistics industry Fast, standardized and efficient development.

References

1. Lin, Y., Chen, A., Yin, Y., et al.: A framework for sustainable management of the platform service supply chain: an empirical study of the logistics sector in China. Int. J. Prod. Econ. **235**(3), 108112 (2021)

2. Guo, B.: Research on the construction of logistics information platform for coastal ports under the background of electronic commerce. J. Coast. Res. **115**(sp1), 120 (2020)
3. Zhang, N., Zheng, K.: Research and design of the architecture of the marine logistics information platform based on big data. J. Coast. Res. **106**(sp1), 628 (2020). https://doi.org/10.2112/SI106-142.1
4. Li, C., Jie, S., Liu, Y.: Study on architecture and implementation of port logistics information service platform based on cloud computing. New J. Chem. **27**(3), 1078–1083 (2015)
5. Agyabengmensah, Y., Ahenkorah, E., Korsah, G.: The mediating roles of supply chain quality integration and green logistics management between information technology and organisational performance. J. Supply Chain Manage. Syst. **8**(4), 1–17 (2019)
6. Zijm, H., Klumpp, M., Regattieri, A., Heragu, S. (eds.): Operations, Logistics and Supply Chain Management. LNL, Springer, Cham (2019). https://doi.org/10.1007/978-3-319-92447-2
7. Jayasinghe, R.S., Chileshe, N., Rameezdeen, R.: Information-based quality management in reverse logistics supply chain: a systematic literature review. Benchmarking Int. J. **26**(7), 2146–2187 (2019)
8. Adeitan, D.A., Aigbavboa, C., Bamisaye, O.S.: Influence of information flow on logistics management in the industry 4.0 era. Int. J. Supply Oper. Manage. **8**(1), 29–38 (2021)
9. Jovi, M., Schlierf, J.F., Heinen, B., et al.: Information management in reverse logistics. J. Marit. Transp. Sci. **58**(1), 155–167 (2020)
10. Choi, T.M., Wallace, S.W., Wang, Y.: Risk management and coordination in service supply chains: information, logistics and outsourcing. J. Oper. Res. Soc. **67**(2), 159–164 (2016)
11. Zhao, S.: Design and optimization of accounting information system under the background of internet logistics service and supply chain. Bol. Tec./Tech. Bull. **55**(7), 490–496 (2017)

Application of VR Virtual Simulation Technology in Medical Teaching Experiment

Yuan Zhou[1,2,3](✉), Jiejun Hou[4], Shanshan Wu[1], Nan Wang[1], Li Ma[1], Jianjun Guan[1], Yongchang Diwu[2,3,4], and Tao Wu[3,5]

[1] Department of Anatomy, Shaanxi University of Chinese Medicine, Basic Medical College, Xianyang, Shaanxi, China
[2] Shaanxi Key Laboratory of Chinese Medicine Encephalopathy, Shaanxi University of Chinese Medicine, Xianyang, Shaanxi, China
[3] Discipline Innovation Team of Shaanxi University of Chinese Medicine, Xianyang, Shaanxi, China
[4] Shaanxi University of Chinese Medicine, 1St Clinical Medical School, Xianyang, Shaanxi, China
[5] Shaanxi University of Chinese Medicine, Acupuncture and Tuina School, Xianyang, Shaanxi, China

Abstract. VR virtual simulation technology is a computer simulation system that can simulate three-dimensional dynamic space. It comprehensively uses real-time three-dimensional graphics technology, wide-angle three-dimensional display technology, sound, feeling, voice and other technologies to bring vision, hearing, and touch to users Such multi-sensory experience creates good participation and maneuverability. The application of virtual reality technology in medical education has very important practical significance. Some virtual reality systems used in basic medical experiments, clinical practice and research can solve many problems that restrict the development of practical teaching, and their advantages and effects are immeasurable. Medical education is different from other disciplines in teaching methods. It not only teaches learners medical knowledge, medical trends, understands the development trend of modern medicine, but also pays attention to the cultivation of practical skills. The teaching mode of traditional medicine has been far behind the speed of knowledge update. Many abstract teaching contents are difficult to express. Many practical teaching contents and emotional experience are difficult to realize in the teaching process. Moreover, experimental teaching resources are limited. Experiments are dangerous and there are not many opportunities for clinical practice. These contradictions seriously affect the quality and effectiveness of medical education. The purpose of this article is to study the application of VR virtual simulation technology in medical teaching experiments. This article mainly uses the questionnaire survey method to conduct VR virtual teaching experiments in a university in this province, and then conducts questionnaire surveys and analyzes the data of students. The experimental results show that virtual teaching is beneficial to students' knowledge, performance, and spatial ability.

Keywords: Virtual simulation · Virtual laboratory · Teaching experiment · Medical education

© The Author(s), under exclusive license to Springer Nature Switzerland AG 2022
J. Macintyre et al. (Eds.): SPIoT 2021, LNDECT 98, pp. 599–607, 2022.
https://doi.org/10.1007/978-3-030-89511-2_77

1 Introduction

The advancement of science and technology has greatly promoted the development of all walks of life, especially for the information network and computer industry, its booming trend far exceeds people's previous imagination [1, 2]. Education is no exception. In fact, technology plays an important role in education. It has changed the education and learning models at all levels from elementary school to university. Especially for medical education, science and technology entered the classroom and entered the medical experiment, which has revolutionized medical education [3, 4]. At present, virtual reality modeling and simulation technology has become an indispensable technical means for studying the physiological mechanism of the musculoskeletal system, formulating surgical plans, evaluating treatment effects, developing rehabilitation training systems, and optimizing the upgrading of medical equipment [5, 6].

Virtual reality technology originated in the United States, and its level of VR research technology is at the forefront of the world. The concept and practical application of VR technology was first completed by Ivan Sutherland in the mid-1960s [7]. The main concept of foreign training and teaching is student-oriented, focusing on cultivating students' abilities in actual teaching. However, due to factors such as hospitals or individual patients, there are few opportunities for hands-on practice on the patient's body in practical training [8]. The overall status of domestic virtual reality technology is that it started late and developed relatively late [9]. In clinical teaching, it still stays in the traditional teaching mode. Therefore, there is room for further expansion of clinical training teaching. There is a possibility of virtual reality application in medical vocational education training, such as arm puncture and cardiopulmonary resuscitation. Virtual reality technology can also be used on real people whose reality conditions are restricted [10]. The medical simulation laboratory and the medical simulation center are even more breakthroughs. Taking Shantou University Medical School as an example, in order to improve the level of virtual teaching, the Norwegian SimMan digital virtual person is used to explore the improvement of clinical teaching skills and further expand to the clinical application. Take Shanghai University of Traditional Chinese Medicine as an example. In 2003, it used virtual technology to create a modern simulated hospital and applied it to clinical teaching, which provided resources for clinical practice of medical students and expanded new ways [11, 12].

This paper conducts a questionnaire survey on students to understand the role of medical experiments based on VR virtual simulation technology in real teaching.

2 Application of VR Virtual Simulation Technology in Medical Teaching Experiment

2.1 Virtual Laboratory

The virtual LAB teaching platform and system are relatively mature, mainly using virtual reality technology and open Internet terminals to realize the digitization and virtualization of medical laboratory teaching and medical courses, so that students can learn and operate in a virtual experimental environment. On the one hand, it can effectively enhance students' interest in learning and protect their personal safety. On the other hand, it can also save corresponding expenses and reduce the loss of physical objects. More importantly, it can be recycled and entertaining.

2.2 Application of VR Virtual Simulation Technology in Medical Anatomy

How to enable students and residents to master the complex 3D anatomical structure, how to interpret the tomographic images of complex anatomy, and how to combine anatomy and images, has always been a difficult problem in medical education. The current form of education is the traditional 2D learning mode, such as textbooks, 2D wall charts, plastic models, 2D videos and lectures; the second is autopsy, which is expensive and difficult to obtain, and cannot be stored and reused for a long time. Complex, boring, and boring courses can hardly arouse students' enthusiasm for learning.

The 360-degree model experience effect of VR virtual simulation technology allows students to observe more clearly and stereoscopically those anatomical details that are usually overlooked. In clinical anatomy teaching, students can observe and operate the three-dimensional and vivid virtual anatomical model from multiple angles, repeated practice, and the high degree of classroom interest will undoubtedly increase the students' strong interest in learning.

2.3 Application of VR Virtual Simulation Technology in Surgical Operation Training

The quality of practical teaching links is even more important for learning this course. Medical students not only need rich theoretical knowledge, but also need a certain degree of hands-on ability. However, for a surgical resident, faced with the diversification of surgical approaches, the increasing updating of surgical instruments, and the minimally invasive surgical methods, the complex surgical approaches, the wide variety of instruments, and the limited surgical field of vision make medical Students and residents understand that it is becoming more and more difficult to master a surgical operation. At the same time, it is also increasingly unrealistic for the teaching teacher to clarify the operation technology while successfully completing the operation. However, in actual operation, in terms of the current medical environment and doctor-patient relationship in China, let an inexperienced beginner directly face the patient for invasive operation. Many patients will be more repelled and used as a teaching model. You may also feel uneasy because of your lack of self-confidence. VR virtual simulation technology can improve the surgical operation skills of untrained residents. Compared with the residents who received traditional training, the residents who use VR for colonoscopy and endoscopy training, when they actually treat the patient, the VR virtual simulation training group has higher accuracy and more errors.

2.4 Design Principles of Virtual Simulation Experiment

(1) Scientific principles

The scientific principle means that the experiment should be guided by scientific ideas and based on facts. This principle must be strictly observed whether it is traditional experiments or virtual simulation experiments. Therefore, it must be reflected in all aspects of the design of virtual simulation experiments.

For an experiment, the first thing is to clarify the purpose of the experiment, and to know why the experiment is done, so the purpose of the experiment must be scientific. The purpose of the experiment is the "baton" of the whole experiment, and all the experimental links are carried out around the purpose of the experiment. The purpose of the experiment is not set blindly, but clearly stipulated. According to the goal, the class will not deviate from the teaching.

In addition to the scientific purpose of the experiment, the subject of the experiment, that is, the content of the experiment must also be scientific. The design of the experiment must follow the laws of physics, and the experiment must not be designed against the facts, and there must be no mistakes in knowledge. This requires teachers to study textbooks carefully, consider experiments, and transform real experiments into virtual simulation experiments to assist teaching and improve teaching effects.

In addition, the experimental procedures must be scientific, and it must be clear what to do first and then what to do, so that the virtual simulation experiment can be carried out successfully, reducing unnecessary operations and avoiding wasting limited classroom time. For example, when conducting electrical experiments, pay attention to the connection sequence of experimental equipment, develop good experimental habits, and conduct scientific experimental research.

Finally, the scientific nature of the virtual simulation experiment is also reflected in the design of the experiment to pay attention to the development of students, the design of the experiment should take into account the students' own existing knowledge reserves and the law of the students' cognitive development of things.

(2) The principle of relevance

Appropriateness refers to what type of experimental teaching the virtual simulation experiment is suitable for assisting. As a product of modern information development, virtual simulation experiments often have two extreme attitudes in reality. The first is to fully accept, exaggerate the role of modern information technology, and use virtual experimental teaching instead of real experimental operations for any experiment. The second is to reject all, stick to the rules, unwilling to accept new things, and hold a negative attitude towards modern information technology. For virtual simulation experiments, we need to understand under what circumstances it is suitable to use it to assist teaching and improve teaching efficiency.

(3) Teaching principles

The pedagogical principle means that the virtual simulation experiment should serve education and teaching, and it should not only benefit the teaching of teachers, but also benefit the learning of students. Only by adhering to the principle of teaching and according to the law of students' cognitive development, can we design a virtual simulation experiment that meets the learning requirements of students. Virtual simulation experiment is not designed to directly impart knowledge to students. The original intention of virtual simulation experiment design is to enrich students' appearances and promote students' cognitive development. It should be based on students' cognitive development and enable students to experience the formation of physical models at close range. Observe those experimental processes that cannot be carried out through reality, so that students' cognition can be fully developed. As for how to design virtual simulation experiments to achieve the most optimized teaching, it requires constant communication and feedback between teachers and students, and continuous improvement of the virtual simulation experiments in order to achieve the best results and assist teaching.

(4) The principle of subjectivity

When designing the virtual simulation experiment, the students' factors should be considered. The principle of subjectivity is one of the important principles necessary for effective teaching. In the teaching process of virtual simulation experiment, teachers and students discuss together, and students feel the experiment process firsthand, summarize conclusions and gain knowledge. The knowledge acquired by students through self-exploration is more in-depth and meticulous than the traditional teaching-based teaching. The students' interest in learning and confidence in learning physics will also be improved, and the students will be fully developed while mastering the knowledge. As the main body of learning, students can learn independently through virtual simulation experiment software, even at home, they can also use virtual experiments to conduct "self-service experiments" for inquiry learning, so as to gain knowledge and exercise experimental ability.

3   Case Implementation

3.1   Experiment Object

In this experiment, the first-year medical students of a university in this city were selected as the experimental subjects. Class 1 was used as the experimental class to conduct medical experiment courses using virtual reality systems; Class 2 was the control class, using traditional multimedia courseware for teaching. Due to school conditions, the experimental class uses a first-level virtual reality system.

The experimental class and the control class are two parallel classes. The average scores of the two classes are basically the same, and the knowledge and intelligence are basically the same. The same teacher teaches, different teaching methods, the same teaching content. The teaching time is the same, all for one academic year. In teaching, the practice questions and unit test questions selected in the experimental class are the same as those in the control class, and the same indicators and test methods are used to test both the experimental class and the control class at the same time.

3.2   Data Processing

This article uses SPSS 22.0 software to count and analyze the results of the questionnaire, and conduct a t test. The t-test formula used in this article is as follows:

$$t = \frac{\overline{X} - \mu}{\frac{\sigma X}{\sqrt{n}}} \tag{1}$$

$$t = \frac{\overline{X_1} - \overline{X_2}}{\sqrt{\frac{(n_1-1)S_1^2+(n_2-1)S_2^2}{n_1+n_2-2}\left(\frac{1}{n_1} + \frac{1}{n_2}\right)}} \tag{2}$$

4 Effect Analysis

4.1 Likeness Analysis

The preference of the two classes of students for medical experiment courses was investigated, and the results are shown in Table 1.

Table 1. Preference survey results

Degree	Experiment group		Control group	
	Number	Proportion	Number	Proportion
Like very much	16	32%	8	16%
Like	18	36%	12	24%
General	9	18%	15	30%
Dislike	5	10%	9	18%
Dislike very much	2	4%	6	12%

It can be seen from Table 1 and Fig. 1 that the students in the experimental class obviously prefer the medical experimental courses, while the students in the control group are obviously not as fond of the organic courses as in the experimental group. There are 34 students in the experimental group who like (including very) medical experiments, more than half of the class, accounting for 68%. There are only 7 people who dislike and dislike very much, accounting for 14%. Attitudes indicate that there are generally 9 people, accounting for 18%. In the control group, there were 15 students who disliked and disliked medical experiments very much, which accounted for 30% of the class. There were 15 students who expressed general attitudes, which accounted for 30%. There are only 20 students who like and like medical experiments very much, accounting for 40%. It shows that medical experiment courses based on VR virtual simulation technology are more attractive to students and can increase students' interest in medical experiment courses.

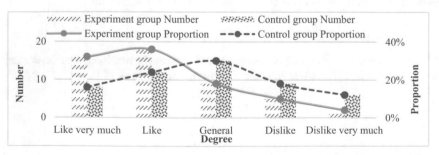

Fig. 1. Preference survey results

5 Exam Results

The first semester midterm exam is used as a pre-test, and the second semester final exam is used as a post-test.

Table 2. Test score

| | Experiment group | | Control group | |
	X1	X2	X1	X2
Minimum	49.00	54.00	50.00	49.00
Maximum	94.00	95.00	95.00	96.00
Mean	76.34	82.31	76.54	77.68
Std. Deviation	13.43	8.51	12.35	11.64

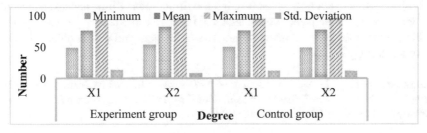

Fig. 2. Test score

As can be seen from Table 2 and Fig. 2, in the pre-test (X1), the mean test scores of the experimental and control classes were not significantly different (76.34, 76.54), and their passing rates were 85.71% and 88.24%, respectively. In the post-test (X2), the average academic performance of the experimental class increased by more than 5 (82.31–77.68) points compared to the control class, which is a significant teaching effect. In the pre-test, the difference between the two classes in all aspects of performance was not very large, the average score was almost the same pass rate excellence rate and so on was also similar, however, after one semester, the experimental class students' performance improved significantly, the average score, pass rate excellence rate were significantly higher than those of the control class, it can be considered that the virtual reality system plays a significant role in medical laboratory teaching.

Through interviews with students in the experimental class after the class, most of the students think that the virtual reality system is better than other means in medical experimental teaching. The advantages learned through the students are summarized. Most of the students said that using the virtual reality system for medical laboratory teaching can be immersive learning and more interested in medical laboratory learning; by processing the abstract medical laboratory things in image, it makes the medical laboratory knowledge easier to understand in image. In conclusion, through the experiments of virtual reality system-assisted medical teaching, it can be concluded that virtual reality system plays a significant role in medical experimental teaching, and is conducive to the mastery of students' knowledge, the improvement of their performance, and the enhancement of their spatial ability.

6 Conclusions

As a new type of teaching method, VR virtual simulation technology has attracted great attention from the education circle. Because VR virtual simulation technology will have a profound impact on the exploration and development of modern educational ideas, improving the level of educational technology, improving the experimental environment, optimizing the teaching process, and cultivating talents with innovative consciousness and innovative capabilities. Traditional teaching methods are difficult to express the abstract or practical teaching content in medical teaching intuitively and vividly. The introduction of virtual reality technology into medical teaching and the cofignstruction of medical virtual teaching can make up for these deficiencies. The medical virtual teaching system can not only break through the limitations of time and space, visualize abstract content, and help learners understand, but also can effectively convey practical teaching content through the interaction between learners and virtual objects in virtual reality technology.

Acknowledgements. This work was supported by Discipline Innovation Team of Shaanxi University of Chinese Medicine (2019-QN05) and Education and Teaching Reform Research Project of Shaanxi University of Chinese Medicine (21JG29).

References

1. Little, J.: Using virtual simulation to increase deep learning in radiography students. Radiol. Technol. **92**(4), 324–330 (2021)
2. Chang, W., et al.: Construction of virtual simulation teaching platform for elevator control. J. Phys.: Conf. Ser. **1848**(1), 7, 012120 (2021)
3. Li, M., Wang, L., He, Z.: Simulation application of VR technology in communication network. J. Phys.: Conf. Ser. **1881**(4), 7, 042004 (2021)
4. Xiao, X., Liu, X., Xiao, Z.: Construction and application of computer virtual simulation teaching platform for medical testing. J. Phys.: Conf. Ser. **1915**(4), 7, 042074 (2021)
5. Li, Z., Cao, Y., Luo, J.: Application of virtual simulation technology in chemistry teaching. E3S Web of Conf. **267**(9), 02067 (2021)
6. Sugimoto, M., et al.: Medical image processing in clinical extended reality (Virtual Reality, Augmented Reality, Mixed Reality) and automatic segmentation of the internal organs using artificial intelligence and deep learning. Med. Imag. Technol. **37**(1), 22–27 (2019)
7. Fang, M., You, F., Yao, R.: Application of virtual reality technology (VR) in practice teaching of sports rehabilitation major. J. Phys. Conf. Ser. **1852**(4), 042007 (2021)
8. Wang, Q., Zhe, Z., Xing, Y.: Application and research of VR technology in art design teaching. J. Phys.: Conf. Ser. **1345**(4), 4, 042026 (2019)
9. Zhou, Y., Hou, J., Liu, Q., et al.: VR/AR technology in human anatomy teaching and operation training. J. Healthcare Eng. **2021**(4), 1–13 (2021)

10. Ping, D., Li, A.: Simulation of physical education teaching based on FPGA and wearable VR equipment. Microprocess. Microsyst. **81**(3), 103773 (2020)
11. Jo, Y.J., Choi, J.S., Kim, J., et al.: Virtual reality (VR) simulation and augmented reality (AR) navigation in orthognathic surgery: a case report. Appl. Sci. **11**(12), 5673 (2021)
12. Lai, P., Zou, W.: The application of virtual reality technology in medical education and training. Glob. J. Inf. Technol. Emerg. Technol. **8**(1), 10–15 (2018)

Simulation and Simulation of Virus Transmission in Virtual Network Based on Time Series Network

Baisheng Zhong[✉]

Guangzhou College of Technology and Business, Guangzhou 510800, Guangdong, China

Abstract. While the Internet provides convenience for information transmission, it also provides a good breeding ground for the development of computer viruses. The function of computer virus self-replication and spread is very powerful, which brings a huge threat to the information security of everyone in modern society, and it is also regarded as the number one natural enemy of the advancement and development of mobile Internet technology. Since there are still many commonalities between computer viruses and biological viruses in their infection mechanisms, many scientists and scholars have conducted in-depth research with the help of theoretical ideas on the dynamics of infectious diseases, in order to better explore their internals for us. This paper aims to study the simulation and simulation of virus propagation in virtual networks based on time series networks. On the basis of analyzing the characteristics of computer viruses, classical computer virus propagation models and virus propagation dynamics research methods based on time series networks, a virus propagation dynamics model based on time series networks is constructed and simulated experiments are carried out. The experimental results show that the installation of firewalls and anti-virus software to check and kill the virus, that is, reduce the value of β, reduce the number of infected nodes in the system, thereby controlling the spread of the virus.

Keywords: Time series network · Network virus · Virus spread · SEIR model

1 Introduction

Since entering the Internet age, people have enjoyed various benefits given by the Internet, such as the improvement of life and the improvement of work efficiency [1, 2]. People are becoming more and more dependent on the Internet, and it is difficult to imagine what people's lives and work would be like without the Internet [3, 4].

The emergence of any technology is a double-edged sword, and the key lies in the people who develop and use the technology. While the Internet has brought great convenience to people, it has also opened up a shortcut for those individuals or organizations trying to obtain illegitimate benefits through the Internet. Various cyber security incidents have frequently erupted, posing a great threat to people's lives and property [5, 6]. A network security incident refers to an attack that is launched against a network or computer and can effectively damage the integrity, confidentiality, and availability

© The Author(s), under exclusive license to Springer Nature Switzerland AG 2022
J. Macintyre et al. (Eds.): SPIoT 2021, LNDECT 98, pp. 608–615, 2022.
https://doi.org/10.1007/978-3-030-89511-2_78

of data or systems inside the network. Computer virus attacks are the most frequent network security incidents that occur in our daily work and life [7, 8].

Based on the analysis of the characteristics of computer viruses, classic computer virus propagation models, and virus propagation dynamics research methods based on time series networks, this paper constructs a virus propagation dynamics model based on time series networks and conducts simulation experiments on them.

2   Simulation and Simulation of Virus Transmission in Virtual Network Based on Time Series Network

2.1   Characteristics of Computer Viruses

(1)   Enforceability

Computer viruses are essentially the same as ordinary programs in terms of format. However, compared with traditional programs, viruses are designed to perform destructive behaviors. In order to achieve this goal, the virus must be executed and only successfully executed on the infected host can it be destroyed [9, 10].

(2)   Infectious

Computer viruses usually have two transmission methods: on the one hand, viruses have strong self-replication ability and can spread; on the other hand, viruses spread through the transmission of host files (such as file replication and exchange). There are usually viruses in storage media such as hard disks, floppy disks, U disks, and CDs. With the continuous development of development technology, today's viruses are not limited to spreading to specific operating systems (such as Windows and Linux), but can also spread across platforms [11, 12].

(3)   Concealment

In order for computer viruses to perform destructive actions, they must do everything possible to prevent them from being detected before they break out. Once the virus successfully infects the host program, its behavior is the same as a regular program. Therefore, it can be spread quietly without the user's permission or knowledge. Virus concealment is mainly reflected in: first, there is a hidden format, the form and structure of the virus host program is not significantly different from the normal program, and it is difficult for users to find out whether the program is infected; second, the transmission behavior is concealed, before the virus breaks out, it will try to infect more files to cause greater damage, and it is difficult for users to know how many files are infected. Some viruses are very concealed and can even be modified so that the anti-virus software cannot control it and cannot do anything about it. Concealment is a prerequisite for computer viruses to hide from being discovered for a long time.

(4)   Latent

In order to cause greater harm, some computer viruses will not attack a specific host immediately after infecting them, but will exist on the host for a long time. The better the latency, the longer the virus stays on the host, the more files it infects, and the greater the damage caused during the outbreak. The incubation period of the virus is not fixed. It will not break out until "the time is right" for a few hours, days, and years. This "time" reflects the conditions that need to activate the virus outbreak.

(5) Derivative

Computer viruses are similar to biological viruses because they mutate. With the continuous advancement of anti-virus technology, anti-virus software can already detect some common viruses. In order to prevent the anti-virus software from being checked and killed, the computer virus designers used the idea of "biological virus mutations in response to the antibodies produced by the immune system" and modified and upgraded the original virus. The upgraded virus can also be used in the process of spreading. Other people make modifications, and after constant modification and upgrades, the final new virus form and structure are very complicated, and it is difficult to detect by anti-virus software. Therefore, the modified computer virus is more destructive.

2.2 Classical Computer Virus Transmission Model

(1) SIS model

The full name of the SIS model is the Susc eptible-Infected-Susceptible model. According to the various states of the system computer, the model divides the computers in the system into two states: susceptible (S) and infected (I). A vulnerable computer does not carry a virus, but when it comes into contact with a virus-infected computer, it is more likely to be infected. The infected computer is the spreader of the virus and can infect other nodes. When an infected person discovers that he or she is carrying the virus, he or she can take appropriate measures to become a susceptible person with a certain probability.

The differential equation expression of SIS is described as follows:

$$\begin{cases} \frac{dS}{dt} = -\beta SI + \gamma I \\ \frac{dI}{dt} = \beta SI - \gamma I \end{cases} \tag{1}$$

(2) SIR model

The full name of the SIR model is the Susceptible-Infected-Recovered model. Unlike the SIS model, the SIR model adds an immune storage R, that is, recovery. At present, all computers in the system are divided into different virus infection levels: (1) susceptible (S): computers that are vulnerable to viruses without protection capabilities, (2) infected (I): computers that have been infected with viruses, (3) Restorer (R): A computer with immune capabilities. The recovery program in the SIR model refers to a type of computer that is permanently immune to the infected computer and will no longer be infected with computer viruses.

The differential equation expression of SIR is described as follows:

$$\begin{cases} \frac{dS}{dt} = -\beta SI \\ \frac{dI}{dt} = \beta SI - \gamma I \\ \frac{dR}{dt} = \gamma I \end{cases} \tag{2}$$

(3) SEIR model

SEIR model is the abbreviation of Susceptible-Expos ed-Infected-Removed model. A new delay library based on the SIR model has been added to divide all system states into the following four libraries: (1) Vulnerable (S): computers that are susceptible to virus infection; (2) Latent (E): computers that are in the incubation period; (3) Infected person (I): A computer that has been infected with a virus; (4) Recovered (R): A computer that has been temporarily immunized. Among them, the computer in the latent state and the computer in the infected state both carry viruses, and the computer in the latent state refers to the computer that has been invaded by the virus code but does not show the characteristics of the virus and does not spread the virus, that is, the computer is in the latent state although the computer has a virus, but the virus has not yet broken out and will not damage other nodes.

The mathematical description of the SEIR model is as follows:

$$\begin{cases} \frac{dS}{dt} = -\beta SI + \mu I \\ \frac{dE}{dt} = \beta SI - \alpha E \\ \frac{dI}{dt} = \beta SI - (\gamma + \mu)I \\ \frac{dR}{dt} = \gamma I \end{cases} \tag{3}$$

2.3 Research Method of Virus Transmission Dynamics Based on Time Series Network

(1) Mean field method based on individual

The main idea of the average field method is to replace the sum of the individual effects with the average action effect, and collect the influence of the system environment on the object in a concentrated manner. Based on the individual average field method (IBMF), each node of the network is considered to be statistically independent of each other, that is, the state of each node is statistically independent of neighboring nodes. The solution of the IBMF method usually depends on the spectral characteristics of the adjacent matrix, especially its maximum eigenvalue. The prediction results are usually compatible with the numerical simulation results obtained using static networks. In addition, the method can also be extended using approximate methods.

(2) Mean field method based on degree

In addition to being a theoretical method, the average field method is also a way of thinking. Inspire researchers to study problems from an average and probabilistic perspective. The degree-based mean field method (DBMF) is the first theoretical method to analyze general dynamic processes in complex networks. Because it is suitable for analyzing dynamic processes on the Internet, it spreads well. In the DBMF method, the approximate method of the dynamic process in a complex network is to treat all nodes with degree k as statistically equivalent. That is, unlike the classification based on the node status in the IBMF method, the classification is based on the degree of the node in the DBMF method, and the nodes with the same degree are classified into one category.

3 Experiment

3.1 Model Establishment

By entering different computer entry and removal rates in the system, when an infected computer user detects that their computer is infected with a virus, it will move the computer out of the network as soon as possible. At this time, it is often faster to clear infected nodes than other nodes. Vulnerable nodes and latent nodes become immune system recovery nodes by regularly updating anti-virus software and fixing vulnerabilities. Considering the above factors, the SEIR model is established, which divides the system computer into the following four different states:

(1) Susceptible: A node that is not infected with the virus but does not have the ability to be immune. It is easy to be infected by an infected person and enter the incubation period to become a latent person.
(2) Exposed: A node that has been infected with a virus, but the virus has not been triggered, and is still in the incubation period. After an outbreak, it will become an infected node.
(3) Infected: A node where the virus has been triggered, and an infected node can infect a susceptible person.
(4) Recovered: A node that has been temporarily immunized in some way will become a susceptible node after losing immunity.

   The state of each node in the system is not only related to its own state, but also related to neighboring nodes and the network environment. This article defines the transformation rules of each state in the model as follows:

(1) In unit time, any computer that enters the system is a susceptible node, and the rate of entering the system is set to A.
(2) In unit time, the rate at which each node in the system moves out of the network due to computer damage, equipment aging, etc. is d; while the infected computer is to avoid becoming the source of infection, the rate to move out of the network is b. Combining the above two reasons, the rate at which an infected computer moves out of the network is b+d.
(3) Once the susceptible node contacts the infected node, it will be infected with the virus and enter the incubation period, and will be transformed into a latent person at a rate x per unit of time.
(4) In a unit of time, due to the installation of anti-virus software by susceptible persons, they are transformed into immune recoverers at a rate of δ.
(5) In unit time, due to regular updates of anti-virus software, reinstallation of the system, etc., the probability of a latent person and an infected person being transformed into an evictor is η and ε respectively.
(6) Once a computer virus is triggered in a unit of time, the latent node will be transformed into an infected node with a probability of β.

3.2 Basic Reproduction Number

The basic regeneration number is the threshold to determine whether the computer virus exists in the system, expressed as

$$x' = F_i(X) - V_i(x) \tag{4}$$

The Jacobian matrix of F(X) and V(X) at the non-toxic balance point P_0 is

$$F - DF_i(P_0) = \begin{bmatrix} 0 & \frac{\alpha A}{\delta+d} \\ 0 & 0 \end{bmatrix}, V - DV_i(P_0) = \begin{bmatrix} \eta + \beta + d & 0 \\ -\beta & \varepsilon + d + b \end{bmatrix} \tag{5}$$

After a simple calculation

$$R_0 = \frac{\alpha \beta A}{(\delta + d)(\eta + d\beta+)(d + b + \varepsilon)} \tag{6}$$

4 Discussion

Whether a computer virus can continue to exist in the network depends on the size of the basic number of regeneration R_0. Therefore, analyzing the key factors that affect the basic reproduction number R_0 is very important for controlling the spread of computer viruses. From the expression of the basic regeneration number R_0, we can see that the probability δ of a susceptible node transforming into a recovery node, the probability η that a latent node transforms into a recovery node, and the probability β that a latent node transforms into an infected node, these parameters are all related to the basic regeneration number R_0.

The evolution of the basic reproduction number R_0 with the parameter δ is shown in Fig. 1, and the system parameters of each line are shown in Table 1.

Figure 1 shows that the basic regeneration number R_0 decreases with the increase of the probability δ of the susceptible node transforming into a recovery node. Therefore, the number of computers of the infected person in the network will decrease with the increase of the parameter δ. Therefore, measures can be taken to increase the value of the parameter δ, such as improving the security awareness of netizens and installing anti-virus software on the computer.

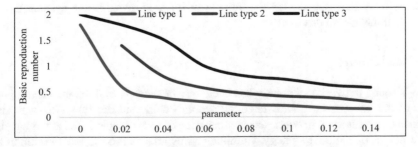

Fig. 1. Change of basic regeneration number with parameters δ

Table 1. System parameter value of each line

Line type	A	α	β	ε	η	d	b
1	0.6434	0.004	0.079	0.029	0.038	0.01	0.069
2	0.5026	0.021	0.037	0.1	0.038	0.007	0.02
3	0.35	0.496	0.038	0.46	0.059	0.3	0.069

Take the value of each parameter in the system as shown in Table 2, and get the evolution of the basic regeneration number R_0 with the parameter β as shown in Fig. 2.

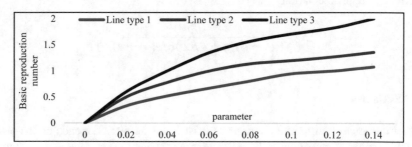

Fig. 2. Change of basic regeneration number with parameters β

Table 2. System parameter value of each line

Line type	A	α	β	ε	η	d	b
1	0.8	0.04	0.02	0.2	0.051	0.04	0.061
2	0.5026	0.021	0.038	0.1	0.04	0.007	0.02
3	0.35	0.48	0.059	0.51	0.057	0.03	0.068

Figure 2 depicts that the basic reproduction number R_0 increases with the increase of the probability β of the latent node transforming into an infected node, that is, the infected computer in the system decreases with the decrease of the parameter β. You can install firewalls and anti-virus software to check and kill viruses, thereby reducing the value of β, that is, reducing the number of infected nodes in the system, thereby controlling the spread of viruses.

5   Conclusions

Today's social network technology has been very developed, and the increasing popularity of the Internet has greatly changed people's work and lifestyle. However, since the birth of the Internet, network security has been an unavoidable topic, and it has risen to the height of national strategy. Computer virus attacks are the most common and typical network security incidents. Therefore, it is of practical significance to study the propagation law of computer viruses.

Acknowledgements. Research project of Guangzhou College of Technology and Businesss in 2018, Research and Design of Network Attack and Defense Platform Based on Virtual network (Project's number: KA201832).

References

1. Thimbleby, H., Anderson, S.O., Cairns, P.A.: A framework for modelling Trojans and computer virus infection. Comput. J. **4**, 324–325 (2018)
2. Yang, L., Yang, X., Wu, Y.: The impact of patch forwarding on the prevalence of computer virus: a theoretical assessment approach. Appl. Math. Model. **43**(Mar), 110–125 (2017)
3. Ansari, M.A., Arora, D., Ansari, S.P.: Chaos control and synchronization of fractional order delay-varying computer virus propagation model. Math. Methods Appl. Sci. **39**(5), 1197–1205 (2016)
4. El-Sayed, A., Arafa, A., Khali, M., et al.: A Mathematical model with memory for propagation of computer virus under human intervention. Progress Fract. Differ. Appl. **2**(2), 105–113 (2016)
5. Zdemir, N., Uar, S., Erolu, B.: Dynamical analysis of fractional order model for computer virus propagation with kill signals. Int. J. Nonlinear Sci. Numer. Simul. **21**(3–4), 239–247 (2020)
6. Tao, D., Wang, A., Liao, X.: Impact of discontinuous antivirus strategy in a computer virus model with the point to group. Appl. Math. Model. **40**(4), 3400–3409 (2016)
7. Tang, C., Wu, Y.: Global exponential stability of nonresident computer virus models. Nonlinear Anal. Real World Appl. **34**(Complete), 149–158 (2017)
8. Zhang, Z., Bi, D.: Dynamical analysis of a computer virus propagation model with delay and infectivity in latent period. Discrete Dyn. Nat. Soc. **2016**(10), 1–9 (2016)
9. Fatima, U., Ali, M., Ahmed, N., et al.: Numerical modeling of susceptible latent breaking-out quarantine computer virus epidemic dynamics. Heliyon, **4**(5), e00631 (2018)
10. Terzieva, T., Iliev, A., Rahnev, A., et al.: Comments on some modification of suja cumulative functions with applications to the theory of computer viruses propagation. VII. Int. J. Differ. Eqn. Appl. **19**(1), 83–95 (2020)
11. Dang, Q.A., Hoang, M.T.: Numerical dynamics of nonstandard finite difference schemes for a computer virus propagation model. Int. J. Dyn. Control **8**(3), 772–778 (2019). https://doi.org/10.1007/s40435-019-00604-y
12. Nguyen, V.T.: A computer virus detection method based on information from PE structure of files combined with deep learning models. Commun. Comput. Inf. Sci. **1306**(1), 120–129 (2020)

Information Management of Port Transportation Supply Chain Based on Computer Network

Yumei Zhang[✉]

Department of Shipping Management, Shanghai Maritime Academy, Shanghai 201204, China

Abstract. As the key hub of our country's logistics activities, the port is an important link in the international strategic layout, which is of unparalleled significance for our country's economic development and social construction. As the trend of economic globalization becomes more and more obvious, building an international comprehensive service port is an important way to promote economic development. Based on this, this paper conducts a research on port transportation supply chain information management based on computer network. This article first introduces the research status of port transportation supply chain, and then analyzes the advantages of computer network technology in the operation of commercial vehicles. Then, this article uses computer network technology to design an information sharing system about the port supply chain, and conduct related tests on the system. The test results show that the response time of the system is very fast, and the functional modules are also very reliable.

Keywords: Computer network · Port · Transportation supply chain · Information management

1 Introduction

Port supply chain refers to the establishment of an integrated and collaborative service system based on internet information technology, and then realizes the goal of cooperation, mutual benefit and mutual support between enterprises in the supply chain system [1, 2]. In the port logistics supply chain system, the port through the orderly management of all levels of the supply chain system to form a network topology service model with the port as the core enterprise and the auxiliary operations of related agencies. As the pivot point in the logistics supply chain, the port is more of a manager and commander. In the traditional port logistics model, the supplier, the shipper, and the receiver all act as independent entities to complete the entire business process [3, 4]. The port logistics supply chain is to coordinate the upstream and downstream enterprises in the supply chain from a holistic perspective, divide, reorganize, and allocate logistics business processes to related organizations with a better sequence to achieve resource sharing and information sharing, and then complete the efficient operation of the entire supply chain

© The Author(s), under exclusive license to Springer Nature Switzerland AG 2022
J. Macintyre et al. (Eds.): SPIoT 2021, LNDECT 98, pp. 616–623, 2022.
https://doi.org/10.1007/978-3-030-89511-2_79

system [5, 6]. Compared with traditional port logistics services, the modern port supply chain system has distinct characteristics: the core port logistics company is the core of the supply chain, the internal and external related industries are combined horizontally and vertically, the enterprise resource allocation plan is optimized, and each process is carried out.

Chinese scholar Li Jian believes that the rapid development of economy and the improvement of science and technology have enabled port logistics to be well developed. The design and analysis of port logistics supply chain information systems are of great significance. Based on this, starting from the design goal of the supply chain information system, the design content of the supply chain information system is analyzed, and its technical support and system architecture are explored [7]. Yang Kai proposed a cloud-based spatio-temporal geographic information sharing service platform architecture system, function and application model of the smart port. Taking a typical port in northern China as an example, the construction practice of a cloud-based spatio-temporal information geographic sharing service platform is introduced, which is of great significance for guiding the construction of smart ports in various regions [8]. Yu Qiulin pointed out that due to the continuous development of our country's economy, the development of logistics is also very rapid, in order to effectively promote the rapid development of our country's economy and reduce the impact of logistics on economic development. In terms of management, it is necessary to continuously optimize management methods and improve management systems. Using supply chain management, on the basis of meeting customer service requirements, reduce the input cost of the entire supply chain system as much as possible [9].

At present, the degree of informatization of port logistics is a major indicator to measure the status quo of a port's construction, and it is also a key factor to measure a country's logistics capabilities. The emergence of the port logistics supply chain system is in line with the trend of the times. Through the port logistics supply chain information system, various links in the originally loose supply chain are effectively and closely linked to form a stable network topology structure [10]. Ports are the core role in the logistics supply chain, supported by the port logistics supply chain information system, to complete the management, coordination and task assignment of related agencies. Through the port logistics supply chain information system, all links in the supply chain can be effectively coordinated and shared, and the optimal allocation of resources can maximize the utilization of resources, thereby reducing logistics costs and improving service quality [11, 12]. All in all, informatization is an indispensable part of the modernization of ports, and an informatized logistics platform will provide support for the interchange, integration and sharing of port information through new management methods. In addition, traditional handwritten documents will also be replaced by modern electronic documents, making the sharing of information in the port logistics supply chain more convenient, thereby reducing logistics operation costs and providing users with better logistics services.

2 Commercial Vehicle Business Optimization Based on Computer Network Technology

2.1 Related Computer Network Technology

Radio Frequency Identification (RFID) is a technology that realizes non-contact automatic identification of objects. RFID technology consists of three parts, namely, RFID tags, readers and information processing systems. The RFID tag is mainly used as a data carrier, it has the advantages of non-contact identification, long service life and strong signal penetration. The reader reads the data in the tag through radio frequency and transmits the read data To the information processing system. In the optimization of the commercial vehicle business process, RFID technology mainly realizes the marking of the vehicle and the storage of part of the vehicle information. Global Positioning System (GPS) is a technology used to determine the location of a target, which can realize real-time management of goods. For commodity vehicle logistics, the transportation process is a constantly moving process. If real-time management can be achieved, it will be of great help to the scheduling of cars and ships and the choice of transportation routes. In addition, the use of GPS technology can effectively combine the various links of the commodity vehicle transportation supply chain, so that the cargo owner can grasp the operation of the vehicle in real time, thereby improving the service level of the commodity vehicle transportation supply chain. Infrared technology is an automatic control device realized by infrared, which uses infrared transmitting and receiving devices to realize an infrared loop. When the infrared receiving route is blocked, the receiving device can immediately issue a warning. In this paper, infrared technology is used to realize the control of the entry of commercial cars and cars, and the anti-theft control of commercial cars.

2.2 Process Optimization Effect

Enterprises can use the information sharing platform to obtain the required business data, and prepare for operations in a timely manner, ensuring the smooth progress of business between enterprises. In addition, with the help of an information sharing platform, commodity vehicles can integrate the resources of the supply chain, provide vacant logistics resources to the required enterprises, reduce the empty-load situation of transportation vehicles, and reduce the cost of commodity logistics. By binding RFID tags, commodity vehicles in the commodity vehicle logistics supply chain can achieve "car card" correspondence, and the corresponding commodity vehicle information can be read through network information collection technology, which can reduce manual data collection and entry. On the one hand, this reduces the duplication of data entry and ensures the consistency of the data. On the other hand, this kind of information collection can achieve real-time data through regular collection of equipment. This provides a technical foundation for ports and other commodity vehicle logistics enterprises to realize real-time monitoring and precise management of commodity vehicles. By optimizing the business process of commercial vehicles, a set of information sharing system and network information specifications are used uniformly. The operation of enterprises on the port-centric commodity truck transportation supply chain will be more standardized, and the upstream and downstream supply chains can continue to advance. Collecting data by network information collection equipment can save human resource costs. After various data collection enters the information sharing platform, statistical analysis technology and data mining technology are used to provide assistance for corporate decision-making.

2.3 The Advantages of Computer Network Technology in the Operation of Commercial Vehicles

When a commodity car or a commodity car transporter passes through the gate of a port or enterprise, it reads the data in the RFID tag on the vehicle. Determine whether it is a vehicle entering the port, and automatically raise the gate, or prompt that it does not belong to the scope of entry and exit. The entire operation process does not require staff intervention, which speeds up the operation of the commercial vehicle and reduces labor costs. Combined with GPS technology and RFID and other monitoring and sensing technologies, real-time monitoring of commodity vehicles can be realized, including the transportation process and warehouse process of commodity vehicles. Through the "car card" binding, one car one card is realized, and the commodity car operator can automatically obtain the commodity car information through the data collection device. Errors in manual operations are avoided, the time for commercial vehicles to enter the port is reduced, work efficiency is improved, and the work pressure of operators is reduced. By laying antennas on the yard of commercial vehicles and setting up wireless data communication devices, the information of the commercial vehicles in the yard and the storage information of the commercial vehicles can be obtained in real time. The security of the warehouse is enhanced, and it is convenient for the staff to make reasonable arrangements for the warehouse. The port can stick the electronic ticket on the car, so as to realize the digitization of the car. When the car enters and exits the port, the automatic identification of the car transport can be realized, and the car transport can be remotely tracked, which improves the safety during the transportation of the commodity vehicle.

3   Design of Information Sharing System Based on Computer Network

3.1   System Design Overview

This article uses the.NET platform to build the B/S architecture, and the database uses SQL Server 2005, because these technologies have a strong backing as a technical guarantee. The problems that arise during development will be referred to, and the use of these mature technologies will also be of great help to the later maintenance of the system. Mature and advanced technology must be used to ensure the long-term use of the system. The Java development language used in this book guarantees the stability of the development system. The B/S three-tier mode selected in the architecture is the mainstream idea of the current system, which makes the operation of the system more convenient. It can ensure that the system has advanced technology, stable operation and high compatibility. From the perspective of the overall planning goal of the system, the system should have high information security protection to ensure that the system information data is not destroyed. The openness of the design system for the rapid development of the information age ensures that the system can keep up with the development. According to the needs of users, meet their suggestions and reflect the practicality of the system. By simplifying the operation steps of the system and the clear operation interface, the principle of simple operation of the system is maintained. In the process of system development and design, it should be selected and popularized as much as possible, with advanced and mature framework technology. In addition, the technology used will be relatively advanced for a long time to come and cannot be eliminated.

3.2   Overall System Architecture

The system architecture is the data information generated by the user accessing the server, and feedback to the user through the calculation of the server, so that the user can get the information they want. This process will be displayed at all levels of the system, and there is an interactive relationship between each level. Ensuring the security of data is the main principle of designing the system. The first consideration and key design in the design of the system is the safe construction of the logistics management system. The protection of hardware is the lowest layer of protection in the security architecture, but this layer is the most basic. The hardware device provides the carrier platform for the system. Without the hardware device, there is no system. Therefore, physical protection of hardware facilities is indispensable. When designing, pay attention to storage, and whether the top of the computer room of the hardware equipment is waterproof and leak-proof. It is necessary to set up beautiful facilities in the computer room to avoid fire. The computer room also has temperature-adjustable air conditioners, so that the temperature in the computer room can be guaranteed to make the equipment run better. There is also an anti-static floor in the machine room, which has many functions not only anti-static, but also to prevent overheating and heat preservation. Also consider the power supply problem of the computer room, set up dual power supply and UPS to ensure stable power consumption. In addition, a management system must be formulated for the computer room to ensure that the server can operate safely and stably.

3.3   System Function Module Design

The business management module mainly manages customer orders, and is a module for viewing logistics customer information, cargo information, and order information. Here you can see a lot of customer related information, generate transportation orders according to customer needs, and request to complete the assignment of transportation tools and personnel. You can also add new users, update departments, and modify user information for internal personnel in the enterprise, empower different roles, and manage external customer accounts and passwords. At the same time, information can be released through the system interface, which is convenient for customers to inquire about cargo information. The transportation management module is mainly responsible for the management of transportation tools and management and dispatching of enterprise vehicles based on customer business orders. According to the type and quantity of the goods ordered by the customer, formulate a transportation plan, specify the means of transportation, arrange transportation personnel, dispatch the existing transportation means of the enterprise, and arrange the transportation itinerary reasonably. The warehousing management module is mainly to record the information of the existing goods in the warehouse. It is mainly aimed at the logistics enterprise after receiving the customer's goods, during the transportation process or the customer's demand, the goods are transferred and directly stored in the warehouse. Evaluate the storage capacity of each warehouse according to the customer's goods, determine the storage location of the goods, and record it in the storage inventory. According to the customer's orders and requirements, the registration and review of the outbound and inbound goods are determined. The formulas used in the design of system function modules are:

$$T = \pi\alpha^3\omega^2\left(\frac{d}{2}\right)^2 \tag{1}$$

$$M = (\sigma_t + \gamma f_t k)W \tag{2}$$

4   System Test

4.1   System Response Time Test

Table 1. System response time test

Number of visitors	Response time (s)
500	0.15
1000	0.29
1500	0.52
2000	0.73
2500	0.88

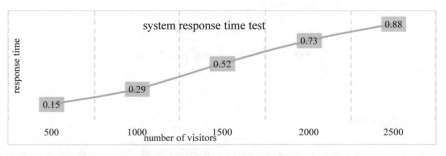

Fig. 1. System response time test

According to Table 1 and Fig. 1, it can be known that when the number of visitors is 500, the system response time is 0.15 s. When the number of visitors is 1000, the system response time is 0.29 s. When the number of visitors is 1500, the system response time is 0.52 s. When the number of visitors is 2000, the system response time is 0.73 s. When the number of visitors is 2500, the system response time is 0.88 s. The response time of the system is approximately linear with the number of visitors, and the response time is within one second. Therefore, the response time of the system is very fast.

4.2 System Function Module Test

Table 2. System function module test

Test results	Frequency	Percentage
Function correct	27	90%
Function error	1	3.3%
No response	2	6.7%

According to Table 2 and Fig. 2, we can know that the system function modules were tested 30 times, and the results showed that the function was normal 27 times, accounting for 90%. The second function does not respond, accounting for 6.7%. 1 function error, accounting for 3.3%. The cause of the function error is caused by a data entry error, and the cause of the unresponsive function is caused by network fluctuations. Therefore, excluding factors outside the system, the system function modules are still very reliable.

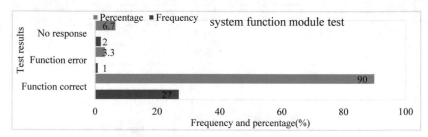

Fig. 2. System function module test

5 Conclusions

At present, the road of port information construction in my country has started and is still moving forward. Facing the increasingly competitive international market, it is an inevitable choice for the development of the port to enhance the core competitiveness of the port through port informatization. Facing the achievements already made, the logistics systems of major coastal ports have been established one after another. But we must guard against arrogance and rashness, and we must pay close attention to the problems that arise during development. Faced with the proposed strategic plan, we must implement it. On the basis of combining the existing level, continue to deepen the theoretical research of the port logistics supply chain information system, and actively learn from foreign advanced ideas and scientific management models. Starting from the actual situation of the port and constructing a suitable logistics supply chain information system according to the characteristics of the port, it will provide strong support for becoming an international comprehensive service port as soon as possible, and will also contribute to our country's economic development and social construction.

References

1. Cheng, Z., Wu, H.: The development model of smart port construction in the Beibu Gulf of Guangxi. China Ports **333**(3), 36–40 (2020)
2. Zong, Y.: Port "business and logistics integration" supply chain finance innovation. Ind. Innov. Res. **48**(19), 24–25 (2020)
3. Hu, L., Liu, Q.: Development strategy of port logistics in Lianyungang area of Jiangsu Free trade zone. Logist. Technol. v.39, **409**(10), 14–17+34 (2020)
4. Ribo, H.: Research on optimization of total quality management system of port vehicle logistics supply chain. Logist. Eng. Manag. **41**(4), 119–120 (2019)
5. Wang, R.: Accounting platform in the collection of port logistics supply chain fees. Port Sci. Technol. **177**(11), 11–17 (2020)
6. Yu, X.: Research on coordination and profit distribution of port logistics service supply chain. Knowl. Econ. **493**(9), 79–80 (2019)
7. Li, J., Wang, H.: Analysis and design of port logistics supply chain information system. Inf. Comput. **424**(6), 70–71 (2019)
8. Yang, K., Li, Z.: Research on the spatio-temporal geographic information sharing service platform of smart port based on cloud architecture. Waterways Ports **40**(1), 120–124 (2019)
9. Yu, Q., Zhang, Y.: Research on port logistics operation mode based on supply chain management. Metall. Ser. **5**(4), 275–276 (2020)
10. Ji, C., Yan, Y.: Research on the evolution and innovation trend of port enterprise supply chain finance mode under the background of the Internet. Soft Sci. v.33, **233**(5), 26–32 (2019)
11. Wang, S.: Port hinterland supply chain optimization based on multi-sector dynamic game. Logist. Technol. **38**(11), 93–99 (2019)
12. Han, B., Kang, J., Kuang, H.: Research on dynamic factors and related effects of green operation in port service supply Chain. Ind. Eng. Manag. 2020, v.25, **41**(2), 63–70

Development and Application of Intelligent Auxiliary Technology for Ships Maritime Foreign-Related Emergency Disposal

Jiajia Han$^{(\boxtimes)}$ and Ruixun Wang

Dalian Naval Academy, Dalian, Liaoning, China

Abstract. This paper not only introduces the development and application of AI, but also analyzes the demand of AI for maritime foreign-related emergency disposal, and innovatively puts forward various application forms of intelligent auxiliary technology in maritime foreign-related emergency disposal.

Keywords: Artificial intelligence · Maritime · Emergency disposal · Intellectualization

Within the increasing of national maritime trade and the continuous development of overseas interests, the situation of ships foreign-related emergency disposal has increased significantly. However, the maritime foreign-related affairs overseas are often in the forefront of international political and economic struggle, and each movement often involves knowledge and relevant skills in many fields such as world politics, culture, diplomacy, psychology, language, etc., that needs a strong political, overall situation and sensitivity, which puts forward high requirements for navigators. Artificial Intelligence (AI), as a science and technology to simulate human intelligence activities and expand human practical ability, can play the role of language recognition, modeling and imaging, information retrieval, auxiliary decision-making, robot service, etc. in maritime foreign-related emergency disposal, and effectively improve the maritime foreign-related emergency disposal ability of both military and civilian ships. Therefore, the application of artificial intelligence in the field of Maritime Foreign-related Emergency Disposal (MFED), undoubtedly the development of intelligent auxiliary system is a subject with great application value.

1 The Overview of Development and Application of Artificial Intelligence (AI)

With Artifcial Intelligence (AI) entering our lives in novel ways—both known and unknown to us—there is both the enhancement of existing ethical issues associated with AI as well as the rise of new ethical issues [1]. Artificial intelligence (AI) is an interdisciplinary and frontier science developed on the basis of computer science, information theory, cybernetics, neurophysiology, psychology, linguistics, mathematics, philosophy and other disciplines. As early as the 1955 World "Learning Machine Symposium",

J. Macintyre et al. (Eds.): SPIoT 2021, LNDECT 98, pp. 624–630, 2022.
https://doi.org/10.1007/978-3-030-89511-2_80

famous scientists Alan Newell and Oliver Seyfridge proposed the research of chess and computer pattern recognition respectively. At a conference held at Dartmouth University in 1956, the term "Artificial Intelligence" (AI) was formally used for the first time, and the initial development route and goals of artificial intelligence were discussed and determined. Nowadays, due to the proposal of the Hopfield multi-layer neural network model, the research and application of artificial neural network has been thriving. At the same time, artificial intelligence (AI) is widely used in Smart manufacturing, Self-driving cars, Consumer Intelligent Robots, Virtual assistants, Chatbots, Smart finance, Smart medical, Smart news writing, Smart lawyers, Smart cities and other fields, bringing significant economic and social benefits. Therefore, major countries in the world are taking a variety of measures to promote the development and application of artificial intelligence from top-level planning and project research and development. The world is therefore undergoing a leap development from "Internet + " to "Artificial Intelligence + " [2].

2 The Demand of Artificial Intelligence (AI) for Maritime Foreign-Related Emergency Disposal (MFED)

Ships are often isolated at sea, and maritime foreign-related affairs often involve a wide range of areas, strong political sensitivity and short response time [3]. If Artificial intelligence (AI) is applied to maritime foreign-related affairs, it will greatly alleviate the contradiction between the complexity of maritime foreign-related affairs and the lack of navigator ability to deal with maritime foreign-related affairs.

2.1 Solve Maritime Language Communication Obstacles

Maritime communication of ships is carried out by means of VHF. However, in the practice of dealing with maritime foreign-related affairs, it is often difficult for the navigators to recognize and understand the speech and semantics because of the big noise, unstable signal, strong accent, omission of continuous reading and swallowing, and even misunderstanding and misjudgment. Therefore, it is urgent to use Artificial intelligence (AI) technology to improve the accuracy of maritime communication, in order to solve the problem of navigator.

2.2 Search Information of Ship and Aircraft Fast

In the event of an emergency at sea, such as a pirate hijacking or a shipwreck, it is often necessary to seek help from the surrounding ships and aircrafts in time. The speed and accuracy of search and discrimination at sea directly affect the method and effect of disposal. However, due to the limitation of simple identification information of ship by navigation equipment and lack of information accumulation and disposal experience of navigator, it is difficult to fully grasp the information of foreign ships and aircrafts around. Therefore, it is urgent to use Artificial intelligence (AI) technology to improve the speed and efficiency of intelligent retrieval, so as to solve the problem that the information acquisition of foreign ships and aircrafts is not fast, accurate and complete [4].

2.3   Provide Expert Consultation in Various Fields

Maritime foreign-related emergency affairs often involve many factors such as international politics, economy, diplomacy, international laws and regulations, ship maneuvering, etc., and a little carelessness may have adverse effects on the overall situation. It requires navigator to make a correct judgment and rapid response in a short period of time by comprehensively considering many factors. However, due to the limitation of occupation and specialty, navigator can't have the professional quality and experiences of experts in all aspects at the same time [5]. Therefore, it is urgent to use Artificial intelligence (AI) technology to provide expert opinions and response plans for Maritime Foreign-related Emergency Disposal (MFED), so as to provide an important reference for on-the-spot decision-making.

2.4   Solve the Problem of Limited Vision

Ships at sea, many cases are beyond visual range, at night or in extreme weather, such as wind, waves, rain and snow, which greatly affects navigator's recognition and observation of passing commercial vessels, military ships and aircrafts and even pirates. If the image that has not been captured or extremely blurred can be imaged through intelligent filter and automatic modeling, it is as if us are in the dark and the enemy is in the light. It is bound to greatly improve navigator coping ability and achieve unexpected results. Therefore, it is urgent to use Artificial intelligence (AI) technology for 3D modeling and imaging to solve the problem of visual limitation in response to environmental impact.

2.5   Cooperate with Manual Disposal

Maritime emergency disposal has strong operability, covering communication, command, control and other practical operations. It is difficult for a single person to complete the task, and often requires the assistance from other personnel. However, the number of personnel on board is limited, especially in the command room. Therefore, it is urgent to use Artificial intelligence (AI) technology to provide coordination for on-the-spot commanders to solve the problem of single person operation.

3   Development and Application of Intelligent Auxiliary System in Maritime Foreign-Related Emergency Disposal (MFED)

The research and application fields of Artificial intelligence (AI) mainly include: natural language processing, automatic programming, expert system, robotics, pattern recognition, intelligent retrieval, problem solving, logical reasoning and theorem proving, machine learning, neural network, intelligent control, intelligent scheduling and command, distributed Artificial intelligence (AI) and agent, etc [6]. These researches and applications can cross and support each other, aiming at the needs of Maritime Foreign-related Emergency Disposal (MFED). This paper focuses on five aspects of application.

3.1 Maritime Foreign-Related Intelligent Speech Recognition and Translation System

NLP (natural language processing) is a technology that uses natural language to communicate with computers [7]. The key of processing natural language is to let the computer "understand" natural language, so it is also called computational linguistics, which is the intersection of language information processing and Artificial intelligence (AI). Natural language understanding can be divided into oral language understanding (speech recognition, analysis, etc.) and written language understanding (machine translation, automatic writing, etc.). In order to solve the obstacles of maritime communication, we can use natural language understanding technology and research results to establish "maritime foreign-related intelligent speech recognition and translation system". The system takes Artificial intelligence (AI) speech recognition and machine translation technology as the core, and comprehensively uses telecommunication signal intelligent noise reduction technology (including analog signal, digital signal, wired and wireless signal transmission and speech extraction, etc.), speech technology (including speech recognition technology, speech synthesis technology and various speech databases, etc.), data communication technology (including various database applications Text translation technology, various statistical analysis technology, etc.). Among them, natural language understanding is a bright spot in the application of Artificial intelligence (AI), especially in the two fields of speech recognition and machine translation, such as the widely used "automatic speech translation system for seven languages (English, Japanese, Italian, Korean, French, German and Chinese)", which effectively solves the communication problems of people from different speech countries [8]. The general process of the system is to transform the foreign voice signal into an electrical signal after being received by the microphone, which is used as the input of the speech recognition system, and then carry out intelligent noise reduction processing on the incoming signal through the oral recognition algorithm and speech model, filter the background noise such as wave, wind and rain, electromagnetic interference, etc., which often appear in the maritime environment, and extract the characteristic parameters. The feature parameters are compared with the original database, and the recognized language results are finally output and displayed in writing synchronously. According to the application needs of different maritime foreign-related situations, navigators can research and develop an application platform for maritime foreign-related intelligent speech recognition and translation, which can realize "man-to- machine dialogue" and communication between China and foreign countries fluently, so as to meet the actual needs of maritime foreign-related speech recognition and translation services.

3.2 Maritime Foreign-Related Intelligent Retrieval System

Intelligent retrieval based on natural language understanding technology, that means the system can understand the text in the linguistic sense [9]. When users inquire information, the intelligent retrieval system can also understand the sentences inquired from the information, and then carry out semantic concept matching the information. In order to realize information retrieving of foreign ships and aircrafts fast, accurately and comprehensively, we can use the intelligent retrieval, data mining and knowledge discovery technology and research results of Artificial intelligence (AI) to establish the "maritime foreign-related intelligent retrieval system". This system can build a friendly operation page, provide different paths for the navigator retrieve target to carry out the specified function retrieval, maintain the consistency between the retrieval interface theme and navigator retrieval target, and improve the efficiency of navigator target retrieval and browsing. The intelligent design of the retrieval module can also design the retrieval target model according to the frequency of navigator retrieval target. Through model matching and distributed retrieval technology, priority is given to push the information and key data about the subject, and effective filtering and intelligent filtering of document information are carried out based on navigator retrieval target model, so as to meet navigator retrieval needs. Using this system, navigator can quickly analyze their nationality, model, performance characteristics, route, crew natural conditions and other information through intelligent retrieval function after the navigation radar captures foreign ship signals, so as to help navigator to respond to maritime foreign-related affairs more scientifically and accurately.

3.3   Intelligent Auxiliary Decision-Making System for Maritime Foreign-Related Emergency Disposal (MFED)

Expert system is a kind of computer program system that simulates human experts to solve problems in some fields [10]. It has a lot of knowledge and experiences in a certain field of expert level, and can use the knowledge and problem-solving methods of human experts to solve problems in this field. The core of expert system is knowledge base and inference engine, which contains a lot of professional knowledge and experiences, and has the ability of storage, memory, reasoning, judgment and so on. According to the needs of Maritime Foreign-related Emergency Disposal (MFED), the expert system of Artificial intelligence (AI) technology can be used to establish the "Maritime Foreign-related Emergency Disposal (MFED) intelligent auxiliary decision system". The realization of the system needs to have the mechanism of experts in this field to solve practical problems, at the same time, it also needs to establish a perfect storage device, and be summarized and analyzed by special experts. Before sailing, it is necessary to store the information of foreign-related disposal cases, past performances responded, knowledge and experiences of relevant experts in the system database, and then use the reasoning proof of Artificial intelligence (AI), machine learning and other technologies to scientifically and comprehensively analyze and judge the international situation, ship performance, international regulations, psychological influence and other factors to quickly generate response plans, in order to provide navigator with expert level coping methods and solutions.

3.4   Intelligent 3D Modeling and Imaging System for Maritime Foreign-Related Emergency Disposal (MFED)

More than 80% of the external information received by human beings comes from vision, and about 10% comes from hearing. With the development of machine vision technology, computers have the ability to simulate human beings to receive external information, recognize and understand the surrounding environment. The intelligent

3D modeling and imaging system for Maritime Foreign-related Emergency Disposal (MFED) can be established by using intelligent machine vision technology. The system, with serial communication technology and image synthesis technology as the main research means, and 3D automatic imaging software as the control center, can connect the intelligent rotating platform and camera with the computer through the data line, pull the camera and intelligent rotating platform through 3D automatic imaging intelligent software to take multi-angle photos of the target, and quickly synthesize HML, Ls and other formats of product 3D panoramic display animation. The 3D panoramic display animation of imaging can enable navigator to see the details of pirate ships, foreign military ships and aircrafts and maritime environment from different angles, thus greatly improving the intuitive and realistic sense of foreign-related disposal. Compared with the traditional 2D animation display mode, 3D panoramic display animation can rotate 360 degrees automatically, drag left and right, enlarge HD viewing, which can play an important positive role in further optimizing the efficiency and accuracy of Maritime Foreign-related Emergency Disposal (MFED).

3.5 Multifunctional Intelligent Mobile Robot Assistant for Maritime Foreign-Related Emergency Disposal (MFED)

Robotics is another important branch of Artificial intelligence (AI). A kind of machine, that can simulate human behavior, robot integrates the research of environment perception and recognition, planning and decision-making, behavior and control, and can complete simple even complex tasks in some specific scenes. According to the needs of Maritime Foreign-related Emergency Disposal (MFED), we can use the robotics, robot arm, machine vision, speech recognition and other technologies of Artificial intelligence (AI) to develop "multi-functional intelligent mobile robot assistant for Maritime Foreign-related Emergency Disposal (MFED)". The robot assistant uses speech recognition technology and Bluetooth technology to realize man-machine connection, and realizes data acquisition, voice control, free movement, arm grasping and other assistance actions through various sensors and power devices. Navigator can remotely control the robot assistant through voice password to achieve preset action or real-time task action. The robot assistant is mainly composed of three parts, voice control terminal, mobile device and instrument handle. The mobile device takes STC12C5A60S2 as the control core, and also includes power module, Bluetooth wireless transmission module, flame module, motor drive module, temperature detection module, pan tilt actuator module, webcam and far infrared distance sensor module. The voice control terminal builds a man-computer interaction platform based on voice through a speech recognition module and a speech synthesis module, so that navigator can remotely control the robot assistant through voice. At the same time, the wireless module will receive the environmental monitoring information and video monitoring images sent by the robot assistant, and display them through the display module. Of course, robot assistant can also integrate intelligent speech recognition and translation system, intelligent retrieval system, intelligent assistant decision-making system, intelligent 3D modeling and imaging system, and even combine intelligent control technology, intelligent scheduling and command technology, robot arm technology, etc. in this way, the function of robot assistant will be greatly expanded.

References

1. Robbins, S.: AI and the path to envelopment: knowledge as a first step towards the responsible regulation and use of AI-powered machines". AI & Soc. **35**, 391–400 (2020)
2. Citron, D.K., Pasquale, F.A.: The scored society: due process for automated predictions. Wash Law Rev **89**, 1 (2014)
3. Hol: AI in the UK: Ready, Willing and Able? Select Committee on Artificial Intelligence. House of Lords (2018). https://publications.parliament.uk/pa/ld201719/ldselect/ldai/100/100.pdf
4. Robertson, L.J., Abbas, R., Alici, G., Munoz, A., Michael, K.: Engineering-based design methodology for embedding ethics in autonomous robots. Proc. IEEE **107**(3), 582–599 (2019)
5. Bourn, D.: Normalization equivalence, kernel equivalence and affine categories. In: Carboni, A., Pedicchio, M.C., Rosolini, G. (eds.) Category Theory. LNM, vol. 1488, pp. 43–62. Springer, Heidelberg (1991). https://doi.org/10.1007/BFb0084212
6. Csákány, B.: Primitive classes of algebras which are equivalent to classes of semi-modules and modules. Acta Sci. Math. (Szeged) **24**, 157–164 (1963)
7. Copeland, J.: The Church-Turing Thesis (2019). http://plato.stanford.edu/entries/church-turing/. Accessed 22 June 2019
8. Gilpin, L.H., Bau, D., Yuan, B.Z., et al.: Explaining explanations: an overview of interpretability of machine learning. In: 2018 IEEE 5th International Conference on Data Science and Advanced Analytics (DSAA), pp. 80–89 (2018)
9. Müller, V.C., Bostrom, N.: Future progress in artifcial intelligence: a survey of expert opinion. In: Müller, V.C. (ed.) Fundamental issues of artifcial intelligence, pp. 555–572. Springer, Switzerland (2016)
10. González, R.: Classical AI linguistic understanding and the insoluble Cartesian problem. AI & Soc., 441–450 (2019). "https://doi.org/10.1007/s00146-019-00906-x

Application of Computer Multimedia Technology in the Design of English Education Curriculum

Jingjing Du[✉]

Haojing College of Shaanxi University of Science and Technology, Xi'an, Shaanxi, China

Abstract. With the vigorous development of CMT (computer multimedia technology), it is widely used in teaching. At present, mankind is entering a new period of technological development. The emergence of CMT provides a new material foundation for the reform of English teaching methods and teaching models. The wide application of MT (multimedia technology) in teaching provides students with a brand-new learning form and promotes the innovation of traditional teachin. Many years of teaching method reforms have consumed a lot of financial and manpower, but the students' English proficiency has not been significantly improved. How to organically combine CM technology with traditional teaching methods to form teaching advantages is particularly important. The application of CM technology in the design of EE (English education) curriculum is the focus of this article. First, analyze the results of the two classes. The results show that the multimedia teaching mode can improve students' listening, reading, vocabulary, writing, and students' comprehensive English scores. The average score of the control group is 83.3. 12.4 points higher than the control group. Secondly, analyze the effect of CM technology on English courses. 41% of students believe that MT teaching can stimulate students' interest in learning English. 22 and 19 people respectively think that MT can strengthen the communication between teachers and students and enrich the teaching content. The remaining 18 people pointed out that this teaching model is conducive to cultivating teachers' quality, improving teaching structure and efficiency.

Keywords: Multimedia technology · English education · English performance · Applied research

1 Introduction

With the vigorous development of computer network technology and MT, new MT provides a variety of options for delivering the latest news and resources. New MT has promoted educational reform. Multimedia teaching can cultivate students' interest in learning English, and the MT teaching model is more novel than the traditional teaching model. This emerging technology can also strengthen the connection between teachers and students, so that students are no longer just individuals who passively accept knowledge. In the basic education stage, MT has promoted the reform of English teaching, so that more and more English teachers use CM in the classroom.

J. Macintyre et al. (Eds.): SPIoT 2021, LNDECT 98, pp. 631–637, 2022.
https://doi.org/10.1007/978-3-030-89511-2_81

There are not a few researches on the application of CM technology in EE courses. Gao L pointed out that in the context of world diversification, economic globalization, and education internationalization, the world is changing rapidly, and CM technology stands out from the crowd. With the vigorous advancement of education informatization, CM technology has been widely used in the teaching of English subjects [1]. Tosun E A believes that the teaching classroom using MT is different from the previous teaching methods, and this classroom model is more in line with students' cognitive laws [2]. Correia AP pointed out that with the strengthening of education management in our country, education methods are becoming increasingly diversified, and multimedia teaching represents the future development trend of English teaching [3].

This article mainly studies the application of CM technology in the design of EE courses. After expounding the basic concepts, characteristics and principles of multimedia technology, an experimental study was carried out. The research content mainly includes the following content: First, select two classes taught by the same teacher, and divide the two classes into experimental classes and control classes according to whether multimedia teaching is used or not. At last, analyze the results between two classes. Secondly, according to the results of the questionnaire feedback, the effect of CM technology on English courses is analyzed.

2   The Application of CM Technology in the Design of EE Courses

2.1   MT

Multimedia is the integration of multiple media [4]. MT is a technology based on digitization that uses computers to process text, pictures, sounds, animations and other information from a variety of media. The characteristics of MT are interactivity, integration, real-time, and intelligence. The intelligent design of MT has attracted the attention of students and is conducive to students' acceptance of teaching content [5]. There are four basic principles in the application of multimedia in modern teaching. Followed by educational principles, scientific principles, technical principles, artistic principles and economic principles. Among them, the scientific principle requires that MT be scientific. In the teaching process, the pictures and videos should not be highlighted blindly and the scientific principle should not be ignored [6]. The use of multimedia equipment by the teacher must be accurate and standardized, and the teaching content presented by the teacher in the classroom must be scientific and reasonable in terms of sound, image, color, etc. Taking into account the principle of economy, when making MT software, there must be a detailed and specific plan, a reasonable arrangement of manpower and material resources, and efforts to make the best use of everything [7].

2.2   Constructivist Learning Theory

Constructivist academic degree theory is one of the four learning theories. This article mainly uses the guiding ideology of constructing learning theory. Constructivism believes that intellectual development is a process of gradual establishment of cognitive structure in an individual way [8]. Constructivism believes that learning should be a mutual process, not just a process in which teachers impart knowledge unilaterally. In the teaching process, teachers should fully mobilize the initiative of students. Throughout the teaching process, teachers are no longer the indoctrinators of knowledge, but are transformed into helpers and guides in the construction of students' will. Teachers need to actively provide students with abundant information resources and support. The main point of constructivism is to guide students to build a new experience system from their original experience. In constructivism, learning is seen as a gradual process. Constructivism attaches importance to the establishment of conceptual tools, regards concepts as tools to solve unknown problems, and pursues new conceptual tools in solving unknown difficulties. This is the innovation of knowledge. Constructivist learning theory obviously provides very useful and specific guiding significance for educational methods and content [9, 10].

2.3   English Education Courses

Content, language, and learning strategies are the three dimensions of English teaching goals [11]. High-quality English teaching needs to consider these three dimensions at the same time when setting teaching goals, so as to make the teaching design clear and help students build a complete knowledge system. The first is the content goal. The English teacher selects teaching content suitable for students within the scope of the syllabus [12]. English is a subject and a language tool at the same time, so English learning is integrated with other subjects in content. Teachers should guide students to explore natural sciences objectively and accurately. The content goal of success is not limited to the knowledge in the textbook, but to combine knowledge and skills. Only for language knowledge is not a complete EE. The second is the language goal. The ultimate goal of language learning is the four practical skills of listening, speaking, reading, and writing. The language knowledge in the learning process consists of three parts: language function, professional vocabulary, and specific grammar. Among them, vocabulary accumulation and grammar knowledge should be mastered in contextual learning. Embedding vocabulary learning into grammar teaching in a meaningful and real context can greatly improve learning efficiency and improve language ability. In addition, teachers should also consciously guide students to pay attention to language application scenarios in the language target design, such as distinguishing written and social language, etc. Therefore, when setting language goals, teachers should not only consider the use of English, but also pay attention to cultivating students' English thinking. Learning strategy is a very important but often overlooked goal in English teaching. Teachers need to consciously design different learning strategies in different links. In each process of teaching new knowledge, teachers should provide students with enough time to discuss and practice repeatedly. Learning strategy objectives are divided into declarative knowledge and procedural knowledge training. In the teaching of declarative knowledge, in order to cultivate students' ability to think independently and learn independently, teachers should focus on cultivating students' autonomous cognitive awareness. In the learning of procedural knowledge, teachers need to impart specific skills to cultivate corresponding learning strategies, achieve the integration of teaching and learning, and reflect the value of "teaching" and the meaning of "learning" [13]. The following formulas are mainly used in evaluating the quality of English teaching:

$$E(Y) = E[g(\gamma)] = \sum_{k=1}^{\infty} g(x_k)p_k \tag{1}$$

$$\rho XY = \frac{\text{Cov}(X, Y)}{\sqrt{D(X)D(Y)}} \tag{2}$$

3 Experimental Research on the Application of CM Technology in EE Curriculum Design

3.1 Experimental Background

MT has the unmatched advantages of traditional teaching methods. This teaching mode is conducive to cultivating students' interest in learning and better helping students break through key and difficult points. MT is widely used in the field of education and teaching because of its comprehensive media performance capabilities and powerful interactive functions. Applying MT to English classroom teaching can prolong students' attention span and increase students' interest in English. The MT class transforms boring English knowledge into vivid, interesting, visual and audible knowledge, which is more conducive to students mastering knowledge, exercising English thinking, and improving their own comprehensive English skills. Then the application research of computer MT in EE curriculum design is the key content of this article.

3.2 Experimental Content

After understanding the concepts related to computer MT and EE, this article conducts the following experimental investigations based on the application of MT in EE courses: First, select two natural classes in a first grade in city A as the experimental objects. These two classes are taught by the same teacher, and there is no significant difference in average grades. The experiment time is 4 weeks, 4 class hours per week, a total of 16 class hours. The experimental class used MT in English during these 4 weeks, while the control class still used the traditional teaching method. After the experiment, test the four aspects of listening, vocabulary, reading and writing to compare and analyze the difference in performance of the two classes. Secondly, a questionnaire was distributed to a total of 100 students in these two classes. The content of the questionnaire was: How effective is the computer multimedia technology for English teaching?

4 Experimental Analysis of the Application of CM technology in EE Curriculum Design

4.1 Analysis of the English Scores of the Classes

Before using multimedia, the English scores of these two classes were tested. The results showed that the average score difference between classes was less than 1, In order to

Table 1. English performance of the experimental class and the control class

	Experimental class	Control class
Listening score	19.8	16.7
Vocabulary score	22.1	20.1
Reading results	19.9	16.3
Writing score	21.5	17.8
Average score	83.3	70.9

make the results of the two classes comparable. After the experiment, the results of the two classes are shown in Table 1:

It can be seen from Table 1 and Fig. 1 that the average English score of the experimental class is higher. The English score of the control group is 70.9 points, and the experimental group is 12.4 points higher than the control group. The listening score of the experimental group was 19.8 points, which was also higher than the 16.7 points of the control group. The average vocabulary score of students in the experimental group was 22.1, which was a full improvement of 2 points compared with the control. The average reading score of the experimental group was 3.6 points higher than that of the other group, while the reading score of the other group was only 16.3 points. The most important part is reading. The reading score of the experimental group is 3.7 points higher than the other group. After the experiment, the reading score of the experimental group is 21.5 points. MT can improve students' listening, vocabulary, reading and writing performance better than traditional teaching, and ultimately enhance students' comprehensive English ability.

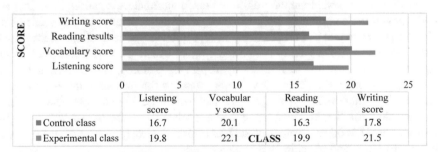

Fig. 1. English scores of two different classes

4.2 Analysis of the Effect of CM Technology on English Courses

MT can improve students' English performance, so what is its specific effect? The results of the questionnaire survey are summarized as shown in Table 2:

Table 2. The effect of computer multimedia technology on english course

	Number	Proportion
Stimulate interest in learning	41	41%
Enrich teaching content	19	19%
Promote teacher-student exchanges	22	22%
Improve teaching efficiency	13	13%
Other effects	5	5%

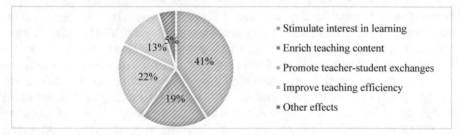

Fig. 2. The effect of computer multimedia technology on english course

MT is conducive to the development of EE. According to Table 2 and Fig. 2, 41% of students believe that MT can cultivate students' interest in learning English. The MT model is more novel than the traditional teaching model. 22 people think that MT can strengthen the communication between teachers and students, so that students are no longer just individuals who passively accept knowledge. 19% of the students also pointed out that MT can greatly enrich the content of EE. For example, teachers can convey a lot of information to students by displaying pictures, videos, etc., and enrich students' knowledge. Similarly, 13 students proposed that MT can improve teachers' literacy and provide teachers with space to display their teaching talents. The remaining 5% of people think that in addition to the above advantages, the use of MT in English classrooms also has other advantages, such as improving teaching efficiency, improving teaching structure, and enriching teaching methods.

5   Conclusions

MT has advantages that traditional teaching can't match. This teaching mode can increase students' interest in English. Multimedia-assisted English teaching can present the teaching content in a more intuitive form, explain knowledge points from multiple different angles, and improve teaching efficiency. The previous teaching mode is limited by time and space, so the amount of knowledge it can impart is very limited. Multimedia English teaching can enable students to get as much knowledge as possible. Therefore, the application research based on computer MT in EE curriculum design is of practical significance. It can make more teachers and students realize the advantages of MT mode and promote the application of this technology in English classrooms.

References

1. Gao, L., Yang, Y.: Research on the application of digital audio media based on computer in english teaching. J. Phys. Conf. Ser. **1648**, 022179 (2020)
2. Tosun, E.A., Zaydnl, B.: Teacher opinions on the applicability of the 2017 secondary education english curriculum based on the blended learning approach. J. Qualit. Res. Educ. **8**(2), 686–713 (2020)
3. Correia, A.P., Koehler, N., Thompson, A., et al.: The application of PhET simulation to teach gas behavior on the submicroscopic level: secondary school students' perceptions. Res. Sci. Technol. Educ. **37**(2), 193–217 (2019)
4. Ibrahim, Z., Alias, N., Nordin, A.B.: Need's analysis for graphic design learning module based on technology & learning styles of deaf students. Cogent Educ. **3**(1), 268–279 (2016)
5. Herakleioti, E., Pantidos, P.: The contribution of the human body in young children's explanations about shadow formation. Res. Sci. Educ. **46**(1), 21–42 (2016)
6. Williams, C., Beam, S.: Technology and writing: review of research. Comput. Educ. **128**, 227–242 (2019)
7. Na, W.: Research on the Korean language education curriculum reform based on multimedia platform application. J. Comput. Theor. Nanosci. **13**(12), 10326–10331 (2016)
8. Xu, B., Li, N.: Research on the application of computer technology in the construction and management of "double qualification" teachers in undergraduate colleges and universities under the background of combination of production and education. J. Phys. Conf. Ser. **1744**(3), 032044 (2021). (5pp)
9. Zhao, Q.: Research on the influence of computer information technology on rural preschool education. J. Phys. Conf. Ser. **1915**(3), 032066 (2021). (4pp)
10. Thohir, M.A., Sukarelawan, M.I., Jumadi, J., et al.: The effects of instructional design based web course on pre- service teachers' competencies. Int. J. Eval. Res. Educ. **10**(1), 230 236 (2021)
11. Zarei, A., Mohd-Yusof, K., Daud, M., et al.: A novel approach of multimedia instruction applications in engineering education. J. Theor. Appl. Inf. Technol. **93**(2), 472 (2016)
12. Shirahama, K., Grzegorzek, M.: Towards large-scale multimedia retrieval enriched by knowledge about human interpretation. Multimedia Tools Appl. **75**(1), 297–331 (2014). https://doi.org/10.1007/s11042-014-2292-8
13. Sani, M.S., Bello, H., Okegbile, S.A.: Development and validation of a multimedia package for teaching applied electrical and electronic component of basic technology. Jurnal Pendidikan Teknologi dan Kejuruan **25**(2), 165–176 (2019)

Experimental Information System Evaluation of Preparatory Students Based on Data Analysis

Ji Meng[⊠]

School of Education, Hubei Minzu University, Enshi, Hubei, China
virusking2020@tom.com

Abstract. In order to improve the information literacy of preparatory students, 53 preparatory students were selected as subjects, and a half-year evaluation experiment of information literacy cultivation was designed. Self-test scale of word processing, self-test scale of information-based learning ability and self-test scale of information-based scientific research level were selected to evaluate the subjects. Using SPSS software to analyze the experimental data, the results show that through the cultivation activities of combining theoretical teaching with practical teaching, the preparatory students' ability of word processing, information-based learning and information-based scientific research have been greatly improved, which is conducive to their better connection with information literacy education and learning at undergraduate level.

Keywords: Data analysis · Preparatory students · Cultivation of information literacy · Evaluation experiment

Minority preparatory students (referred to as "preparatory students") are some minority candidates that the government allows colleges and universities to lower their scores to a certain extent on the basis of the existing enrollment scores of ordinary college students in order to train more minority talents. After entering colleges and universities, preparatory students must have one-year special tutoring for basic cultural courses such as Chinese and mathematics, and only after passing the examination can they be transferred to the subsequent related study of ordinary college students. Information literacy is the core literacy in the 21st century [1]. As early as 1997, China's Ministry of Education incorporated basic computer teaching into the regular teaching system of colleges and universities, in order to cultivate college students' information literacy and form their basic ability to adapt to the information age. Compared with undergraduate students, although the teaching system of preparatory students is different, based on the needs of improving individual's own information literacy and connecting undergraduate computer basic education, colleges and universities generally carry out corresponding information literacy cultivation. In order to understand the degree to which the skill level and information processing ability related to individual information literacy of preparatory students have developed after a certain period of training, it is necessary to carry out targeted evaluation experiments.

© The Author(s), under exclusive license to Springer Nature Switzerland AG 2022
J. Macintyre et al. (Eds.): SPIoT 2021, LNDECT 98, pp. 638–644, 2022.
https://doi.org/10.1007/978-3-030-89511-2_82

1 Research Assumptions and Research Problems

1.1 Research Hypothesis

Starting from the information literacy level of preparatory students, adopting appropriate cultivation measures can enable students to master the basic computer operation skills required by information learning, and at the same time change, promote and enhance students' understanding and operation rules of relevant elements under the visual threshold of information education, so as to lay a solid foundation for them to receive information literacy education at undergraduate level in the future, and reduce the related burden for them to adapt to university information life [2].

1.2 Research Problems

(1) What are the main aspects of information literacy of preparatory students?
(2) How to improve the information literacy of preparatory students?

2 Research Process

2.1 Research Object

The subjects were selected from the preparatory students of Grade 2020 in Education College of Hubei University for Nationalities, and 53 students were selected as the subjects, including 25 girls and 28 boys, aged about 18 years old.

2.2 Research Tools

2.2.1 Information Literacy Level Self-test Scale

At present, the Likert scale widely adopted in China adopts a five-level scoring method, including five factors: self-judgment, word processing level, information learning ability level and information scientific research level. The higher the score of the subjects on the scale, the higher their overall information literacy level.

(1) Word processing self-test scale. Based on word, excel and PowerPoint, which is commonly, used in e-office, the more software application skills and proficiency the subjects have, the higher their scores will be [3].
(2) Information learning ability self-test scale. To investigate the students' familiarity with the usage of information learning platform and the acquisition of learning resources, the subjects scored according to their own situation [4].
(3) Self-test scale of informatization scientific research level. Investigate students' ability to carry out scientific research by means of information tools, including the awareness of building academic community, the query and acquisition of scientific research resources, and the information management ability of scientific research management platform [5].

2.2.2 Cultivation Activities

Cultivation activities consist of two parts. The first class is given priority to with regular classroom instruction, in which students master the brief history of information development and some applied theoretical knowledge through classroom theoretical study [6]. One is practical teaching in computer laboratory, in which teachers examine students' actual mastery level by arranging practical tasks and assessing students' completion.

2.2.3 Evaluation Process

(1) Pre-test. One week before carrying out cultivation activities, 53 subjects were tested for their information literacy level by using Word Processing Self-test Scale, Information Learning Ability Self-test Scale and Information Scientific Research Level Self-test Scale, so as to understand the students' relevant situation.
(2) Implementation. Cultivation activities are held once a week, each time including 4 h, of which 2 h are used for theoretical teaching and 2 h for practical teaching. A total of 16 weeks were cultivated. In the middle of the activity, a teacher-student exchange activity will be held, so that students can share their learning experiences and get first-hand information.
(3) Post-test. After 16 weeks of cultivation activities, the 53 subjects were tested again with the self-test scale of word processing, self-test scale of information learning ability and self-test scale of information scientific research level, and then the obtained data were compared with the pre-test data for analysis.

2.2.4 Data Processing

The research uses SPSS software to input and analyze data.

3 Research Results

3.1 Data Before and After the Word Processing Self-test Scale

In this study, 60 points is regarded as the qualification standard of word processing level. In this study, there are 50 people with a total score of less than 60 in the pre-test, and 94.3% of them have a low level of word processing; There were 5 people whose total score was below 60 in post-test, and 9.43% of them had lower level of word processing. The overall word processing ability has been greatly improved. The average score was 52.6 in the pre-test and 79.8 in the post-test, which greatly improved the average score.

Paired sample t test was carried out on the total average scores of the two tests. from Table 1, it can be seen that there is a significant correlation between the word processing ability of preparatory students and teaching activities, and the difference between the two total average scores is significant, namely $t = 28.4$ ($p < 0.05$), that is, through classroom teaching and practical teaching activities, the word processing ability of preparatory students has been significantly improved.

Table 1. T-test of paired samples before and after total averaging

	Pairing difference				
	t	Significance (double tail)	Mean value difference	95% confidence interval of difference	
				Lower limit	Upper limit
Total mean score before measurement-Post-test total mean score	−28.40	.000	−27.21	−26. 43	−27.65

3.2 Information Learning Ability Self-test Scale Before and After the Test Data Results

The demarcation value of the self-test standard score of information-based learning ability is 50 points. The pre-test time is poor (score below 50 points) in 13 people, poor (score 50–59) in 16 people, basically qualified (score 60–69) in 15 people, and good (score above 70 points) in 9 people. 5 people were poor (score below 50), 6 were poor (score 50–59), 25 were basically qualified (score 60–69), and 17 were better (score above 70). Among them, the number of poor students (score below 50) decreased by 8, and the number of students in good students (score above 70) increased by 8.

Through the paired sample test of the scores before and after the information learning ability, it can be seen that the average score of the post-test is obviously higher than that of the pre-test, that is, the information learning ability is stronger; There is a significant positive correlation between classroom teaching and practical teaching activities and level improvement, and the correlation coefficient is 0.895 ($P < 0.01$) (see Table 2).

Table 2. Correlation of paired samples of information learning ability

		N	Correlation	Significant
Pairing 1	Pre-test & Post-test	53	0.895	0.000

3.3 Results of Data Before and After the Self-test Scale of Informatization Scientific Research Level

The demarcation value of the standard score of the self-test scale for informatization scientific research level is 53 points. Through the pre-and post-test of preparatory students, there are 35 poor (score below 53), 6 poor (score 54–63), 7 basically qualified (score 64–73) and 5 good (score above 74); Eight people were poor (score below 53), five were poor (score 54–63), 23 were basically qualified (score 64–73), and 17 were better (score above 74). Among them, the number of poor students (score below 53) decreased by 27, and the number of good students increased by 12.

Through the paired sample test of the scores before and after the informatization scientific research level, it can be seen that the average value of the post-test scores is obviously higher than that of the pre-test, that is, the informatization scientific research ability is stronger (see Table 3); There is a significant positive correlation between classroom teaching and practical teaching activities and level improvement, and the correlation coefficient is 0.832 (P < 0.01).

Table 3. Matching sample statistics of informatization scientific research level

		Average value	N	Standard deviation
Pairing 1	Pre-test	42.56	53	9.59
	Post test	69.32	53	11.89

4 Research Conclusions

4.1 Is Conducive to Improving the Word Processing Ability of Preparatory Students

From the paired sample T test results of the data before and after the word processing self-test scale, the cultivation activities have a significant positive correlation effect on improving the students' ability to use common office software [7]. The possible reason is that in the practical teaching of cultivation activities, students can have enough practical situations to participate in the specific operation of related software; On the premise of completing the assignment of teachers' deployment, students can strengthen the specific usage, steps and precautions of tool use by communicating with each other and learning from the knowledge points of teaching materials. Meanwhile, they can deepen the application practice and form practical experience through half-semester experience sharing and summarization [8]. Under the guidance of teaching materials and teachers' teaching, students' learning process of word processing software is relatively systematic and comprehensive, with most knowledge points involved, and they have completed the establishment, upgrading and optimization of understanding of software applicability in practice. In addition to the above reasons, the systematic arrangement of teachers' experience sharing and summarization every half semester also encourages students to keep high concentration and pay attention to critical reflection in the whole learning process, which further promotes the improvement of students' word processing ability.

4.2 Is Conducive to Improving the Information Learning Ability of Preparatory Students

The research shows that the students' information learning ability has been greatly improved through cultivation activities. This may be due to the theoretical teaching link in cultivation activities, as well as its own unconscious related practice in the information-based learning environment. The development of information-based learning depends on the common instant communication tools in the information society, which may be related to information-based common sense, and may also require special learning of corresponding usage. In theoretical teaching, it generally includes the introduction of computer development history, information application history and common information tools, such as online education platform and conference mode in QQ software, which may provide basic support conditions for information learning, and their learning should be carried out in theoretical education; In addition, in recent 20 years, China has gradually established a relatively extensive information-based teaching ecology, such as the supply and acquisition of digital learning resources, the establishment of massive open online course education platform, etc., all of which provide a natural practice place for students to start information-based learning, and the practice process has continuously promoted the improvement of students' related learning ability [9].

4.3 Is Conducive to Enhancing the Information-Based Scientific Research Ability of Preparatory Students

It is a normal phenomenon that most of the subjects have poor information research ability. The reason is that our country has not put forward any requirements on scientific research for students in basic education stage, and naturally they also lack the external environment for the generation and growth of information-based scientific research ability [10]. Through cultivation activities, students have a basic understanding of the information chemical industry commonly used in scientific research. Even though they will not enter the substantive stage of information-based scientific research at present, their cognitive level has improved from scratch, which also lays a good foundation for them to carry out scientific research activities in combination with their own disciplines after entering the undergraduate study stage in the future. From another point of view, the cultivation process also makes them have the modernization consciousness of carrying out scientific research under the information ecology. For example, the establishment of academic community is conducive to enhancing the strength of scientific research teams, and can obtain more relevant academic resources beneficial to scientific research through information channels, which can be regarded as the cultivation of their basic information literacy in the initial stage of scientific research.

5   Conclusions

The half-year practical research has achieved good results, which has played a positive role in promoting the information literacy of preparatory students. The reason why the evaluation experiment can achieve such results is that the research team members have been engaged in the education, teaching and evaluation of students' information literacy for a long time, and are familiar with the actual needs of the development of information society and the design system of computer basic courses between different sections of universities, because the setting of evaluation standards is scientific to some extent. Secondly, the information literacy dimensions measured by the evaluation scale are various, which make the experimental results have higher objective reference value to a certain extent [11]. This evaluation experiment has laid a solid foundation for the systematic establishment of information literacy training curriculum system for preparatory students in the future, and can also be used for reference by other university peers.

References

1. Doyle, M., Foster, B., Yukhymenko-Lescroart, M.: Initial development of the perception of information literacy scale (PILS). Commun. Inf. Lit. **13**(2), 205–209 (2019)
2. Crist, E., Popa, D.: Information literacy and cultural context: chinese english language learners' understandings of information authority. Coll. Res. Libr. **81**(4), 646–661 (2020)
3. Gu, Y.: Enhancement of college english teachers' information literacy in information environment. Int. Educ. Stud. **13**(4), 106–112 (2020)
4. Matteson, S.M.: Chex mix data analysis activity. Coll. Teach. **69**(3), 121–125 (2021)
5. Apfeldorf, M.: Deepening student understanding of world war i-era attitudes with data analysis. Soc. Educ. **82**(2), 64–67 (2018)
6. Logan, T.: A practical, iterative framework for secondary data analysis in educational research. Aust. Educ. Res. **47**(1), 129–148 (2020)
7. Vogl, S.: Integrating and consolidating data in mixed methods data analysis: examples from focus group data with children. J. Mixed Methods Res. **13**(4), 536–554 (2019)
8. Howard, M.C.: Scale pretesting. Pract. Assess. Res. Eval. **23**(5), 289–301 (2018)
9. Glassman, M.: The internet as a context for participatory action research. Educ. Inf. Technol. **25**(3), 1891–1911 (2020)
10. Davies, P., Hughes, A.: Defining and assessing enterprise capability in schools. J. Educ. Work **28**(5), 507–527 (2015)
11. Dixon, H., Hill, M., Hawe, E.: Noticing and recognising AfL practice: challenges and their resolution when using an observation schedule. Assess. Matt. **14**(2), 42–62 (2020)

Application of Internet of Things Technology in Smart Home Interior Design

Ya Yang[✉]

Urban Vocational College of Sichuan, Chengdu, Sichuan, China

Abstract. With the progress of the times and the vigorous development of technologies such as sensors and the Internet, people's daily work and learning methods have been completely changed. The traditional home environment has begun to fade out of the stage. Comfort, convenience, intelligence, and high security are people's pursuit of future homes. The emergence of Internet technology has brought us a lot of convenience. Today's general understanding of SM (smart home) refers to the use of computer technology, network technology and other emerging technologies to integrate various IOT related devices in home life to achieve unified centralized management and remote monitoring. In recent years, with the vigorous development of cloud computing, it has also brought new flash points to the intelligent of homes and brought a brand new life experience to people. Therefore, the application of the IOT (Internet of Things) technology in the interior design of SM is the focus of this article. First, analyze the differences in the development of SM interior design at home and abroad. The results show that only 43% of newly renovated houses in China will have a certain degree of intelligent system, which is lower than that of developed countries such as the United States and South Korea. Secondly, it analyzes the application effect of the IOT technology in the design of SM buildings. The results show that 41% of professional designers believe that the use of IOT technology in residences will make SM installation more convenient, and 32% of designers believe that Internet technology will make SM maintenance more convenient. 7% and 10% of professional designers pointed out that the application of IOT technology in SM design will make SM easy to use and operate.

Keywords: IOT technology · Smart home · Interior design · Zigbee network technology

1 Introduction

As a kind of emerging technology, the IOT technology mainly analyzes and processes data from various sensors to provide users with services on specific occasions. The high technology related to the IOT is no longer ignored, and the comfort and convenience brought by technology can be felt in daily life, and SM have emerged. The SM system is a complex system with many functions. Traditional single-threaded, single-process programs of single-chip microcomputers can no longer handle such complex system tasks. The development of the IOT technology provides a perfect solution for the realization

© The Author(s), under exclusive license to Springer Nature Switzerland AG 2022
J. Macintyre et al. (Eds.): SPIoT 2021, LNDECT 98, pp. 645–651, 2022.
https://doi.org/10.1007/978-3-030-89511-2_83

of the home intelligent system. The gradual maturity of the IOT technology has greatly promoted the popularization of home intelligence. Therefore, the application of the IOT technology in the interior design of SM is a subject worthy of our in-depth study.

In the relevant research on the application of IOT technology in the interior design of SM. Many scholars at home and abroad have conducted various discussions on it. For example, Altaf E pointed out that smart interior design is a new fashion trend that aims to increase the value of interior design projects. The standard of interior design is to create a humanized and energy-saving good space. Improving functions and quality of life have always been two major components of interior design [1]. Lee S believes that people are paying more and more attention to their quality of life and living environment, traditional home design can no longer meet people's needs, SM interior design can have a new development trend [2]. Al-Ghaili AM proves that the development of information technology and Internet technology has made modern life intelligent development from the concept and architecture of the network, the application and prospects of the SM [3].

This article mainly studies the application of networking technology in the interior design of SM. Consult relevant information and combine the results of the questionnaire to carry out the following analysis in turn: First, analyze the development differences of SM interior design at home and abroad. Knowing that the development level of SM interior design in my country is still different from that of developed countries. Secondly, the effect of the IOT technology in the interior design of SM is analyzed, and the results point out that the IOT technology can make the installation, maintenance, and operation of SM more convenient, thereby greatly improving people's quality of life.

2 Theoretical Research on Smart Home Based on Internet of Things Technology

2.1 IOT Technology

The concept of networking has not only recently appeared in people's vision; it is an extension of the Internet [4]. The terminals of the IOT are not limited to computers, but also include intelligent devices such as embedded computer systems. Although the IOT was first proposed as early as twenty years ago, it was not until the last few years that it began to be truly transformed into tangible products to realize its own value [5]. The IOT technology abandons the previous traditional thinking. Reinforced concrete, cables, and infrastructure such as chips and broadband will be integrated into a unified whole. The IOT technology can greatly reduce costs while improving the economy, so it is widely used in all aspects of our lives, such as intelligent transportation, public safety, personal health and many other different fields [6]. We are familiar with Xiaomi Smart Home, ETC, etc. The application of the IOT makes people's lives more comfortable, safe and convenient. These have greatly improved our quality of life [7].

2.2 SM

SM is a concentrated manifestation of the IOT under the influence of the Internet. SM is safer and more convenient than traditional home [8]. The concept of SM was first proposed in the United States of the upper world. The concept of SM has gone through three different stages of development: The first stage is the budding period. At this stage, the entire industry is still in a stage of concept and product recognition, mainly for a single electrical appliance, no network is formed, and there is not much connection between electrical appliances. At this stage, there is also no professional smart home manufacturer. The second stage is the pioneering period. At this stage, most electrical appliances have formed their own simple networks. The third stage is the stage of home intelligence [9]. At this stage, the development of SM is faster than the previous two stages, and the market has a clear growth trend. SM are widely used in interior design: the first important application is smart lighting: the lighting function is an important part of the decoration design, which is different from the traditional lighting system, the smart home system is used to control the light switch or the color or brightness of the light in different scenes at any time [10]. The second is smart door locks: safety is a factor that cannot be ignored in interior design. With smart door locks, if an emergency occurs to the elderly and children at home, it can remotely alarm and remind to avoid accidents. Smart doors and windows: Smart doors and windows can automatically identify whether to open windows for ventilation through sensors and other smart devices based on the intensity of ultraviolet light and air quality on the day. The application of SM in interior design makes people's lives more comfortable and safer.

2.3 Technical Realization of SM Design System

The technical realization process of the intelligent home design system mainly relies on the Zigbee network and CGI. Zigbee and CGI are subdivisions of IoT technologies [11]. Among them, Zigbee network has three different node types: coordinator, router and terminal node. Zigbee supports star, tree and mesh network topologies. In the SM system, it is mainly the communication between the coordinator and the terminal node, so the star network architecture can fully meet the design requirements of the system [12]. The formation of Zigbee network is divided into two parts: network initialization and terminal node access to the network. Since only a full-function device can establish a network, the program first determines whether it is a coordinator device and whether it has the ability to build a network. Then determine whether the coordinator has joined the network created by other nodes.

CGI (Common Gateway Interface) is also called Common Gateway Interface, which is actually an interface specification or protocol. All programs developed in accordance with the CGI interface specification can be called CGI programs. Through CGI, the Web server can start different external programs and forward the request content to the CGI program according to different requests. Finally, after the program execution ends, the final execution result is returned to the client as a response.In addition, CGI is not portable, and CGI applications written for a specific platform can only run in this environment. For each request, a new process must be generated for processing. This method is very easy to implement, but it is very inefficient and difficult to expand. Because each process occupies a lot of server resources and time, it is very easy to cause the server to be unable to handle many concurrent requests at the same time.

3 Experiments on the Application of IOT Technology in the Interior Design of SM

3.1 Research Background

With the advancement of social science and technology and the rapid development of embedded technology of the IOT, people are paying more and more attention to people's quality of life and living environment. Traditional single-threaded, single-process programs of single-chip microcomputers can no longer handle such complex system tasks. The continuous development of the IOT technology makes it possible to make the home intelligent. How to make good use of the IOT technology is a very realistic and very important issue before us, and it is also a very urgent issue. Therefore, this article will conduct an in-depth study on the application of the Internet of Things technology in the interior design of smart homes.

3.2 Experimental Process Steps

In order to better study the application of the IOT technology in the interior design of SM, this paper uses the questionnaire star software and the printed paper questionnaire to conduct the investigation of this experiment. The advantage of using the questionnaire star software is that it saves time is convenient and fast, and the recovery rate is high; the advantage of using the paper version of the questionnaire is that after completing the questionnaire, you can have face-to-face communication with the respondents for in-depth understanding. At the same time, in order to improve the accuracy of the experiment, the subjects of the questionnaire survey are mainly professional designers who have been engaged in the home design industry for many years. In this survey, a total of 100 questionnaires were distributed, and 100 valid questionnaires were returned. The recovery rate was 100%. The main statistical information includes: differences in the development mode of SM design at home and abroad, and the analysis of the effect of the IOT technology in the interior design of SM.

4 Experimental Analysis of the Application of IOT Technology in the Interior Design of Smart Homes

After the third part of the experiment, in order to make the analysis results of this paper more credible, after consulting related literature and combining the survey results of the questionnaire and paper questionnaires, the development differences and material aspects of SM interior design at home and abroad are discussed. The effect of networking technology in the interior design of SM has been specifically analyzed. The results of the analysis are now shown as follows.

4.1 Differences in the Development of SM Interior Design at Home and Abroad

What is the current development status of domestic SM design? Approximately what proportion of newly renovated houses have a certain degree of intelligent functions?

After consulting domestic and foreign literature, the data results are sorted as shown in the Figure below:

As can be seen from Fig. 1, SM technology in developed countries has become mature. By 2020, 71% of newly renovated houses in the United States will have a certain degree of intelligent systems, such as intelligent lighting, intelligent doors and windows, and audio-visual entertainment. In South Korea and Singapore, 66% and 57% of newly renovated houses will take the concept of smart home into account when designing their interiors. In China, only 43% of newly renovated houses will have a certain degree of intelligent system, which is lower than that of developed countries such as the United States and South Korea, indicating that the level of SM development in the United States and other developed countries is at the world's leading level. The development level of SM interior design in our country is still different from that in developed countries.

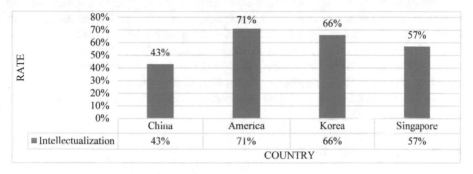

Fig. 1. SM development differences

4.2 The Effect of the IOT Technology in the Interior Design of SM

Before understanding the effect of the IOT technology in the interior design of SM, first introduce the application of the IOT technology in the interior design of SM. Mainly have the following characteristics: traditional manual control, retaining the original manual control switch, to meet the needs of the elderly at home for operation. Through a remote control or control app to realize the remote control of all home appliances; home equipment timing management, each kind of lamps and electrical appliances can be timed, fully meet the various specific requirements of the residents.

Table 1. The effect of IOT technology in smart home

	Easy to install	Easy to maintain	Easy to use	Easy to operate
number	41	32	17	10
proportion	41%	32%	17%	10%

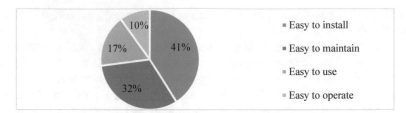

Fig. 2. The effect of IOT technology in smart home

The effective application of IOT technology plays a very important role in promoting the development of domestic SM interior design. It can be seen from Table 1 and Fig. 2 that 41% of professional designers believe that the use of IOT technology in residences will make smart home installation more convenient, eliminating many unnecessary and cumbersome installation procedures; 32 designers believe that unlike traditional smart home design, IOT technology will make SM maintenance more convenient; 17% and 10% of professional designers respectively pointed out that the application of IOT technology in SM design will make SM easy to use and operate. For example, smart wireless remote control makes home appliances more convenient to use, even when residents are outside. Home appliances can also be operated remotely, which greatly improves our quality of life.

5 Conclusion

SM interior design is a new fashion trend that can improve people's quality of life and living environment. The SM combined with the IOT technology will bring a lot of convenience to our lives and provide us with more convenient services. The use of the IOT technology in the residence will make the installation, maintenance, and operation of SM more convenient, and greatly improve the quality of human life. Although the combination of IOT technology and smart home has achieved certain results, it is learned through research and analysis that the development level of SM interior design in my country is still different from that of developed countries. Therefore, this article hopes that through the application of IOT technology in SM interior design, it can inspire more designers to explore and apply IOT technology in the future, thereby it contributes to promoting the development of the smart home interior design industry and reducing the difference in the development level of smart homes at home and abroad.

References

1. Altaf, E., Esther, M.S., Bijan, N.: Designing interiors to mitigate physical and cognitive deficits related to aging and to promote longevity in older adults: a review. Gerontology **64**, 1–11 (2018)
2. Lee, S.: Communication technology and application of Internet of Things (IoT) in smart home environment. Int. J. Control Autom. **10**(3), 397–404 (2017)
3. Al-Ghaili, A.M., Kasim, H., Othman, M., et al.: A review on building energy efficiency techniques. Int. J. Eng. Technol. **7**(4), 35–40 (2018)

4. Sinha, B., Sahay, S.S.: Role of augmented reality application in higher education learning. Aegaeum **8**(10), 926–937 (2020)
5. Fu, J., Hua, J., Wen, J., et al.: Optimization of energy consumption in the MEC-assisted multi-user FD-SWIPT system. IEEE Access **8**, 21345–21354 (2020)
6. Alimin, N.N.: DIY as interior design education "everybody can be designer." Int. J. Creative Arts Stud. **6**(1), 45–60 (2019)
7. Lv, X., Li, M.: Application and research of the intelligent management system based on Internet of Things technology in the era of big data. Mob. Inf. Syst. **2021**(16), 1–6 (2021)
8. Chen, H., Huang, J.: Research and application of the interactive english online teaching system based on the Internet of Things. Sci. Program. **2021**(S1), 1–10 (2021)
9. Li, T., Haibin, Y.: Application and research of new power supply in the construction of "Internet of Things Plus" smart city. J. Phys. Conf. Ser. **1852**(4), 042015 (2021). (7pp)
10. Yoon, S.J., Kim, J.B.: A Study on the user's value of the smart home service in the Internet of Things technology. Int. J. Future Gener. Commun. Networking **10**(6), 65–80 (2017)
11. Qiu, G.P., Sun, S.P., Wang, F.T., et al.: Architecture-oriented design method for smart home care Internet of Things system. Int. J. Adv. Comput. Technol. **8**(4), 8–19 (2016)
12. Vynohradova, M.: The citation of information technology attributes in visual communication design. Humanit. Sci. Curr. Issues, 54–59 (2021)

Analysis on the Application of Electronic Diagnosis Technology in the Components of Energy Meters in Intelligent Equipment

Zhengang Shi[✉], Chaofei Wu, Wenjie Fu, Peng Tao, Linhao Zhang, and Bo Gao

State Grid Hebei Electric Power Co., Ltd., Marketing Service Center, Shijiazhuang, Hebei, China

Abstract. The gradual intelligentization of EEM (electric energy meter) is inseparable from the development of electronic information technology. Due to some unavoidable factors, the EEM will inevitably fail. The failure of the EEM is not only related to the vital interests of every household. At the same time, it will also affect the reputation of EEM manufacturers. In order to prevent this from happening, this puts forward high requirements on the reliability of smart device EEM. To strengthen the diagnosis technology of the EEM of the smart device, it is far from enough to rely on the traditional backward fault diagnosis technology. It is necessary to combine the electronic diagnosis technology. Therefore, the application of ED (electronic diagnostic) technology in the components of smart equipment watt-hour meters is the focus of this article. First, analyze the health monitoring of electronic diagnostic technology in the circuit. The results show that the average method, ripple voltage method and wavelet method in electronic diagnostic technology can effectively monitor the health of the circuit. When using the mean value method, the measured early warning time was only 86 days, and the failure time was 101 days. Secondly, it analyzes the fault diagnosis process of electronic diagnosis technology in the circuit. The results show that using wavelet analysis combined with the electronic diagnosis technology of Mahalanobis distance, a higher fault diagnosis accuracy rate can be obtained in the circuit diagnosis process. When the inductance is reduced by 0.5 times, the R resistance is short-circuited and increased by 2 times, the diagnostic rate of wavelet analysis is as high as 100%.

Keywords: ED technology · Intelligent equipment · Electric energy meter · Components

1 Introduction

EEM will inevitably fail due to uncontrollable factors. Its reliability is not only related to the safe and reliable operation of the smart grid, but also related to the vital interests of every household. In the actual power grid, its operating state will always suffer from many kinds of disturbances. For example, lightning strikes cause short circuits in transmission lines; a series of unpredictable environmental changes such as heavy rainfall and hail will damage the lines and transformers. Once the energy meter fails, the data may be

J. Macintyre et al. (Eds.): SPIoT 2021, LNDECT 98, pp. 652–659, 2022.
https://doi.org/10.1007/978-3-030-89511-2_84

inaccurate and affect the self-interest of users. How to ensure the long-term safe and reliable operation of smart energy meters is a hot issue that the current energy meter industry pays attention to. In order to further reduce the failure rate of smart equipment EEM, in addition to strengthening quality control in the design and production links, ED technology can be applied to the fault diagnosis of intelligent equipment EEM. Therefore, the application of ED technology in smart equipment EEM components is a subject worthy of our in-depth study.

Many well-known professional scholars have made research on the application of ED technology in smart equipment EEM components. For example, Zhang L pointed out that in recent years, with the rapid development of science and technology, the EEM has gradually developed into a smart energy meter [1]. Qian X believes that the demand for electricity in production and people's lives in all walks of life is increasing, and electricity has become one of the important resources to promote social development. Smart EEM is professional equipment for collecting electricity information, which can improve the efficiency of information collection and shorten the overall collection time [2]. Rozenberg EN claims that the EEM is a very common electronic instrument in people's lives, and its automated function can bring convenience to the staff, reduce their workload and improve work efficiency [3].

This article mainly studies the application of electronic diagnosis technology in smart equipment EEM components. First, analyze the health monitoring of electronic diagnostic technology in the circuit, and understand that the average method, ripple voltage method and wavelet method in electronic diagnostic technology can effectively monitor the health of the circuit. Secondly, it analyzes the fault diagnosis process and effect of electronic diagnosis technology in the circuit. The results show that this electrical diagnosis analysis method using wavelet analysis combined with Mahalanobis distance can obtain a higher fault diagnosis accuracy rate in the circuit diagnosis process.

2 Research on the Application of ED Technology in Smart Equipment EEM Components

2.1 ED Technology

Due to the superiority and convenience of power electronic technology, this technology has become one of the indispensable key technologies in any high-tech system today [4]. When the power meter of the equipment fails, it is difficult to find the fault and deal with it in time with the efforts of the maintenance personnel alone. Power ED technology is much more complicated than ordinary diagnosis technology. This technology involves countless small parts [5]. It is precisely because of the high difficulty of this technology that the ED technology can quickly find the fault when the fault occurs, so that the experienced maintenance personnel can solve the problem in time. Therefore, it saves a lot of maintenance personnel's labor and shortens the failure time, greatly improve the accuracy of troubleshooting [6]. Therefore, timeliness and accuracy are the advantages of ED technology. New testing methods, preprocessing of fault information and fault identification are the key research contents of analog circuit fault diagnosis. At present, the most commonly used data preprocessing technologies are normalization processing, wavelet analysis, spectrum analysis, etc.Wavelet analysis, as the name implies, is a small waveform. The little finger here is debilitating, while wave refers to volatility. Wavelet analysis is a method used to analyze the frequency and time of a signal, which can show the local characteristics of the signal. At the same time, it can also greatly reduce the dimensionality of fault features. The distinguishing feature of the spectrum analysis method is that there are few measurement points and simple detection hardware, but the spectrum analysis method is not applicable to all occasions, and this method is not applicable in some occasions. The rough set method was first proposed by a Polish mathematician. It is a method of discovering and inferring knowledge from the data based on a large amount of displayed data and keeping the classification unchanged. There are also some difficulties in electronic fault diagnosis. In addition to the general circuit diagnosis difficulties, its online diagnosis requirements are high. And this new type of diagnosis has important differences with the general analog circuit fault diagnosis [7]. The duration of fault information is very short, and real-time dynamic monitoring is required, which often leads to an increase in cost [8].

2.2 EEM

The main function of the EEM is to measure, which can also be called an watt-hour meter [9]. When using the EEM, it should be noted that low voltage can be directly connected to the circuit for measurement, while high voltage cannot be directly connected to the circuit. It needs to be used with a voltage transformer. The working principle of the energy meter is the principle of magnetic induction. The intelligent EEM is composed of an EEM unit, a cost control unit, a communication unit, etc. In addition to the metering function of a traditional mechanical meter, it can also effectively prevent disputes caused by on-site meter reading, solve the problem of difficulty in collecting electricity, and systematically manage users' electricity purchase information to facilitate users to monitor their own electricity usage information. Smart EEM is a new type of EEM, which mainly adopts the design concept of electronic integrated circuit. Compared with inductive electric meters, smart electric meters have great advantages in performance and operating functions. First of all, the energy consumption of smart meters is lower than that of ordinary meters. According to experimental results, the power consumption of each smart meter is only about 0.7 W. Secondly, smart meters can improve diagnostic accuracy. The error range of the traditional EEM is large, and as the wear becomes more and more serious, the EEM may go slower and slower, which will eventually cause the error to become larger and larger. The EEM that adopts the electronic diagnosis technology can be connected to the computer through the relevant communication protocol to realize the control and management of the hardware. Therefore, in addition to its small size, smart meters also have functions such as remote control, multiple tariffs, identification of vicious loads, anti-stealing, and prepaid electricity consumption. According to different working methods, EEM can be divided into two types: induction type and electronic type. The electronic EEM uses an electronic circuit to drive the counting mechanism to count electric energy, while the induction EEM uses electromagnetic induction to generate torque to drive the counting mechanism to count electric energy. Induction watt-hour meters are widely used due to their low cost and simple structure [10].

2.3 Components

EC (Electronic components) include components and devices. Common EC often refer to certain parts of electrical appliances, radios, meters and other industries, such as capacitors, transistors, hairsprings, springs and other sub-devices. Electronic devices refer to finished products such as resistors, capacitors, and inductors that do not change the molecular structure and composition during production. Because the electronic device itself cannot produce electrons, it can't control and transform the voltage and current, so it is also called passive device. EC refer to finished products whose molecular structure and composition are changed during production and processing. Like transistors, electronic tubes, integrated circuits, etc. This EC itself can generate electrons, which can control and transform voltage and current. Generally speaking, electronic devices can be divided into 12 categories. The reliability of smart energy meter components is an important issue we need to consider. Reliability refers to the ability of a product to complete the specified functions under normal conditions. Under normal circumstances, the higher the reliability, the longer the trouble-free working time of the product. There is currently no unified standard for measuring the reliability of components. It can only be based on mathematical statistics and through the collection and analysis of a large number of samples to identify the model [11, 12]. The following formulas are mainly used in the identification process:

$$\mu_p = \sqrt{\frac{P(1-p)}{n}\left(1 - \frac{n}{N}\right)} \tag{1}$$

$$S_{yz} = \sqrt{\frac{\sum(y - \hat{y})^2}{n}} = \sqrt{\frac{\sum y^2 - a\sum y - b\sum xy}{n}} \tag{2}$$

3 Research on the Application of ED Technology in Smart Equipment EEM Components

3.1 Experimental Background

As the intelligent terminal of the smart grid, the smart energy meter is the link between the power company and the user. Whether smart meters can work reliably for a long time is not only an issue that electric power companies will pay attention to, but also attracts the attention of users. Because the failure of EEM equipment not only brings bad effects to the enterprise, but also brings unnecessary troubles to users. Once the quality of smart energy meters is not well controlled, it will inevitably affect the entire industry. How to minimize the damage caused by the failure of the EEM is very realistic and very important before us, and it is also a very urgent problem. Therefore, this article starts a research on the application of ED technology in smart equipment EEM components.

3.2 Experimental Process Steps

In this paper, the application of electronic diagnosis technology in the components of smart equipment watt-hour meters is studied. Taking the whole circuit as the research object, the whole experiment mainly includes the two core parts of circuit health monitoring and fault diagnosis.

(1) Health Monitoring

In the experiment, the health status of the monitoring circuit based on the Mahalanobis distance is mainly used, and the Mahalanobis distance is calculated based on the characteristic parameters of the healthy circuit and the characteristic parameters of the circuit to be tested, and then judge whether the circuit under test is healthy by analyzing the difference between the circuit under test and the standard circuit.

(2) Fault Diagnosis

In the actual electronic diagnosis technology, after detecting the hidden health hazard in the circuit, the circuit should be repaired in time to determine the fault of the circuit. This experiment is mainly based on wavelet analysis and Mahalanobis distance to study the electronic fault diagnosis method, to diagnose simple electronic circuits, and to determine the fault location and fault type.

4   Application Research and Analysis of ED Technology in Smart Equipment EEM Components

4.1   Research on Health Monitoring of Electronic Diagnosis Technology in Circuits

The deterioration of the health status of most electric energy meters is caused by the deterioration of the components in the circuit. According to relevant data, under the condition of a typical 25 °C, when the circuit fails, the probability of capacitor failure is 32%, the probability of inductive components failure is 8%, and the probability of resistor failure is 17%. The circuit can be divided into three states: health, warning and failure. Therefore, to determine the state of the circuit, two thresholds need to be set. The threshold is determined according to the normal distribution. The specific figures are shown in Table 1 below. The characteristic parameters extracted by different methods are different. When the characteristic parameters are extracted by the mean value method, the early warning threshold is 8.34 and the fault threshold is 15.7.

Table 1. Threshold distribution table

	Mean Method	Ripple voltage method	Wavelet method
Early warning	8.34	16.34	17.76
Failure	15.7	38.95	34.13

Comparing the Mahalanobis distance of a single original under three different methods with the fault threshold, we can get the time point of the circuit early warning and fault in the actual circuit degradation process. The specific data is shown in the following table:

Table 2. Time point of circuit state change

	Mean method	Ripple voltage method	Wavelet method
Early warning point	75	89	93
Failure point	101	117	111

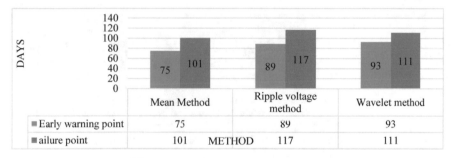

Fig. 1. Time point of circuit state change

The three methods have different monitoring effects on the circuit. The average method and the ripple voltage method have little effect on the monitoring results, and the effect is betteIt can be seen from Table 2 and Fig. 1 above that when using the mean value method, the measured early warning time is 86 days, and the failure time is 101 days. The pre-warning time obtained under the ripple voltage method is 89 days, and the failure time is 117 days. Although the early warning time and failure time measured by the wavelet method are different from the previous two methods, these three methods can effectively monitor the degradation of the circuit over time, and can judge the circuit based on the obtained Mahalanobis distance and its threshold. Health status.

4.2   Research on Fault Diagnosis of Electronic Diagnosis Technology in Circuits

In electronic circuit diagnosis methods, wavelet analysis is mainly used for circuit signal analysis to extract fault characteristic parameters, and Mahalanobis distance is used to judge the fault type of the circuit. When the Mahalanobis distance is the smallest, the fault is all the faults in the circuit. This experiment only considers the hard faults and soft faults of a single component. The single fault of the circuit and the correct rate of fault diagnosis are shown in Table 3 below.

It can be seen from Table 3 and Fig. 2 that the wavelet analysis combined with the Mahalanobis distance analysis method can obtain a higher fault diagnosis accuracy

Table 3. Diagnosis accuracy rate

Fault number	Failure mode	Diagnosis accuracy rate
1	R short circuit	100%
2	D open circuit	96.7%
3	L short circuit	94.3%
4	R increased by 2 times	100%
5	C is reduced to 0.5 times	98.2%
6	L reduced to 0.5 times	100%
Barrier-free	Barrier-free	97%

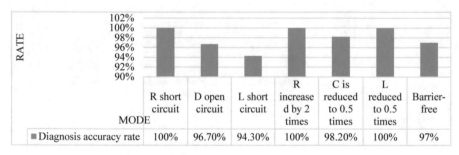

Fig. 2. Diagnosis accuracy rate

rate in the circuit diagnosis process. The parameter C in the above table is capacitance, R is resistance, L is inductance, and D is freewheeling diode. When the inductance is reduced by 0.5 times, the R resistance is short-circuited and increased by 2 times, the wavelet analysis diagnosis rate is as high as 100%. When the freewheeling diode is open and the capacitance is reduced to 0.5 times, the diagnostic accuracy of this analysis method exceeds 95%. Among them, when the capacitance is reduced by 0.5 times, the diagnostic accuracy rate reaches 98.2%. Only when the inductance is short-circuited, the diagnostic accuracy rate is 94.3%, which is slightly lower than 95%. On the whole, electronic diagnosis technology has a fast recognition speed and a higher fault diagnosis accuracy rate.

5   Conclusion

The rapid improvement of our country's economy is driving the vigorous development of the power industry. Due to some force factors, the operation of EEM is facing a huge test. Once the EEM fails, it will inevitably bring unnecessary troubles to the user, and it will also bring a bad influence to the electric power company and the EEM manufacturer. In addition to controlling the quality of the product at the factory, it is necessary to improve the diagnostic technology. Traditional fault diagnosis technology can no longer meet the current needs, and it is particularly important for ED technology to act in the components of smart EEM. Therefore, this paper studies the application of ED technology in the components of smart equipment watt-hour meters. It is hoped that through in-depth research on the application of ED technology in smart equipment energy meter components, in the future, more professionals can be inspired to explore and apply ED technology, gradually optimize the fault diagnosis process of EEM, and improve diagnosis efficiency.

Acknowledgment. Funded by the Science and Technology Project of State Grid Hebei Electric Power Co., Ltd. (B304DY200090).

Supported by the Science and Technology Project of State Grid Hebei Electric Power Co., Ltd. (B304DY200091).

References

1. Zhang, L., Wu, H.: Application of single chip technology in Internet of Things electronic products. J. Intell. Fuzzy Syst. **40**(2), 3223–3233 (2021)
2. Qiao, S., Qian, X., Zhou, M., et al.: Analysis on the maintainability of smart meter software for full lifetime. J. Phys. Conf. Ser. **1748**(4), 042025 (2021)
3. In'kov, Y.M., Rozenberg, E.N., Maron, A.I.: Simulation of the process of implementation of an intelligent electric power metering system. Russ. Electr. Eng. **91**(1), 65–68 (2020)
4. Garcia-Hernandez, J.J., Gomez-Flores, W., Rubio-Loyola, J.: Analysis of the impact of digital watermarking on computer-aided diagnosis in medical imaging. Comput. Biol. Med. **68**, 37–48 (2016)
5. Liao, F.F.: The design and application of electrical energy-saving in railway industry. J. Railway Eng. Soc. **34**(8), 98–103 and 108 (2017)
6. Shang, H., Liu, Y., Zhang, Q., et al.: Research on the quality data service platform architecture of intelligent electricity meters. J. Phys. Conf. Ser. **1693**(1), 012120 (2020). (7pp)
7. Li, T., Kou, Z., Wu, J., et al.: Multipoint optimal minimum entropy deconvolution adjusted for automatic fault diagnosis of hoist bearing. Shock. Vib. **2021**(5), 1–15 (2021)
8. Klimov, P.I., Samoylenko, V.M.: The concept of an automated educational system for the training of refueling complexes personnel. Civil Aviat. High Technol. **23**(4), 45–57 (2020)
9. Rosa, L., Almeida, C.M., Pereira, D., et al.: A systemic approach for assessment of advanced distribution automation functionalities. IEEE Trans. Power Deliv. **34**(5), 2008–2017 (2019)
10. Chen, X., Song, Z., Li, H., et al.: Research on fault early warning and the diagnosis of machine tools based on energy fault tree analysis. Proc. Instit. Mech. Eng. B J. Eng. Manuf. **233**(1), 095440541881684 (2018)
11. Ji, C., Ma, F., Wang, J., et al.: Real-time industrial process fault diagnosis based on time delayed mutual information analysis. Processes **9**(6), 1027 (2021)
12. Cellini, A., Blasioli, S., Biondi, E., et al.: Potential applications and limitations of electronic nose devices for plant disease diagnosis. Sensors **17**(11), 2596 (2017)

Application of BP Neural Network
in the Prediction of Population Aging

Xiaojun Xie[✉]

Guangzhou College of Technology and Business, Guangzhou, Guangdong, China

Abstract. In recent years, the total population of China has continued to increase and the aging process has accelerated, which has led to a continuous increase in the dependency ratio, increasing pressure on the social security system and public service system, and affecting the harmony of social intergenerational relations. Accurate prediction of population size can provide an important basis for formulating sustainable social and economic development plans. In this paper, BP neural network is applied to population prediction, and neural network prediction model is established using BP neural network toolbox function. Numerical test results show that the population prediction model based on artificial neural network is effective.

Keywords: BP nerve · Neural network · Population aging · Population prediction

1 Introduction

China is a populous country, and the population problem has always been one of the key factors restricting China's development. However, to determine a population development strategy, we must not only focus on population problems, but also handle the interrelationships between population and economic, social, resource and environment. Build a harmonious socialist society, and coordinate the resolution of population issues such as population size, quality, structure, and distribution. The existing literature has used differential equations, gray systems and curve fitting methods to study the population problem in China [1]. Based on some new characteristics of China's population development in recent years, as well as China's population sampling data in recent years and the existing national census data, this paper uses artificial neural network algorithms [2] to analyze and predict the Chinese population. Artificial neural network (ANN) can extract and approximate the non-linear relationship between input and output through learning. Therefore, prediction methods based on artificial neural networks have become a hot research topic in recent years. At present, BP neural network and local feedback neural network are mainly used. BP neural network is the most widely used algorithm in artificial neural network [1].

© The Author(s), under exclusive license to Springer Nature Switzerland AG 2022
J. Macintyre et al. (Eds.): SPIoT 2021, LNDECT 98, pp. 660–666, 2022.
https://doi.org/10.1007/978-3-030-89511-2_85

2 BP Neural Network Prediction Method

BP neural network is a one-way propagation multi-layer forward network [1], this paper selects a more typical three-layer structure network (that is, including input layer, hidden layer and output layer) for prediction, and its structure is shown in Fig. 1.

The modeling steps can be expressed as the following six steps:

The first step is to perform dimensionless processing and normalization processing on the data, so that the input data and the target data fall within the interval $[-1, 1]$;

In the second step, the number of neurons in the hidden layer is,

$$n = \frac{n_0 + n_1}{2} + a_n = \frac{n_0 + n_1}{2} + a$$

Among them, n_0, n_1 are the number of neurons in the input layer and output layer respectively, a is a constant and $a \in [1,10]$ $a \in [1,10]$. Of course, various parameters must be constantly adjusted in the actual process to strive for the best prediction effect [2].

The third step is to create a network. This article sets the transfer function of the input layer and the hidden layer to tansig, namely

$$f(x) = \frac{1 - e^{-x}}{1 + e^{-x}}$$

The transfer function of the output layer is set to purelin, the training function is set to trainlm, the weight learning function is set to learngdf, the performance function is set to mse, and the simulation function is set to sim.

The fourth step is to set the parameters. This article sets the display range of the training state to 50, the learning rate to 0.1, the momentum coefficient to 0.9, the number of training times to 1000, and the error accuracy to 0.001.

In the fifth step, in the process of training the network, the threshold and weight are repeatedly adjusted to reduce the value of the performance function mse, and the training is terminated when the preset error accuracy is reached [3].

The sixth step is to use BP neural network to perform simulation prediction and simulate output data.

3 The Establishment of Population Forecasting Model

3.1 Construction of BP Network

BP neural network is also called error back propagation neural network. It is a feedforward network composed of nonlinear transformation units. It can be used to adjust the weights of multilayer feedforward neural networks. From an institutional point of view, BP network is a typical hierarchical multi-layer network, which has an input layer, a hidden layer and an output layer [3]. The layers are mostly connected in a fully connected manner. There is no interconnection between units in the same layer, and the weights of each layer can be adjusted through learning.

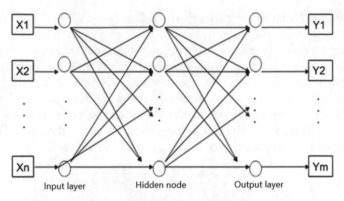

Fig. 1. BP neural network topology

3.2 Determination of the Number of Network Layers, the Number of Neurons, the Transfer Function and the Learning Algorithm

The fully learned three-layer BP neural network can approximate any function. The number of input layer nodes is generally equal to the dimension of the sample vector to be trained, which can be the dimension of the original data or the dimension of the extracted feature; the number of nodes in the output layer ultimately determines the input the number of layer nodes is 4, and the number of output layer nodes is 1. The choice of the number of hidden layer nodes is a very complicated issue. Too few hidden layer nodes, poor fault tolerance, and low ability to recognize and learn samples; too many hidden layer nodes will increase the network training time and reduce the number of samples [4]. The regular content of Central Africa is stored, and after repeated trials, it is determined that the number of hidden layer nodes is 9. The transfer function determines the connection mode of neurons, and it plays an important role in the neural network [4]. In BP network, there are many kinds of transfer functions involved, such as purelin, tansig, logsig and so on.

4   Application of Predictive Models

4.1   Sample Data Selection and Preprocessing

Based on the population of Hubei Province from 1978 to 2010, empirical analysis and testing are carried out. The population data over the years is a one-dimensional time series, but the BP neural network requires multiple sets of inputs to train the network, and the input range is generally set at [0, 1]. In the learning and training of the network, due to the output limitation of the transfer function, there are many orders of magnitude difference data in the learning process that need to be converted into an output value of 0 to 1 through certain calculations [5]. There are many ways to normalize the data. This article performs preprocessing, and the change rule is: $P'(t) = p(t)/10n$. n is the integer digits of the maximum value in all statistics, where $n = 4$. The preprocessed data sequence $p'(t)$ is divided into training samples and test samples according to the determined network structure, as follows: Input sample set $p = [p'(t - 4), p'(t - 3),$

$p'(t-2)$, $p'(t-1)$], output sample set $t = [p'(t)]$, where $t = 1978,1979,\ldots\ldots 2010$. At the same time, select the data from 1978 to 2007 as the learning sample set, and the data from 2008 to 2010 as the test sample set [5]. Use the learning sample set to train the network, and use the trained neural network to simulate.

4.2 Population Forecast of BP Network Based on Matlab

In Matlab, for the realization of the BP network, it provides three basic functions: newff, train and sim, which correspond to the three steps of new construction, training and simulation respectively.

1) Create newff function.

   To realize the training and simulation of a BP network, you first need to create a BP network, newff is to complete this operation [6]. The basic code for building a three-layer network model (including input layer, hidden layer and output layer) is as follows: net = newff(minmax(p), [NodeNum TypeNum], {'tansig','purelin'},'trainlm').

2) Train the train function.

   After the BP network is initialized, it can be trained. The function train uses batch processing to update the network connection weights and thresholds as follows, and proceed according to the set parameters (such as learning step size, error target, etc.): net.performFcn = 'mse';% means to use both square error to evaluate the network performance index; net.trainParam.show = 200;% means that the training interval is 200 times and the result is displayed once; net.trainParam.goal = le-8;% means the target value of the network mean square error; net. trainParam.epochs = 2000;% represents the maximum number of training cycles allowed; net = train(net, p, t);% starts training, where p and t are the input and output vectors respectively; the net obtained after the training is used for training The BP neural network model established by the sample establishes the mapping relationship between the input model and the output model [6].

3) Simulate the sim function.

   The simulation sim function is to use the established neural network for simulation. The calling form is: $X = \text{sim}(net, P)$, where net is the trained network object, P is the input vector or matrix, and X is the network output. If P is a vector, it is a single-point simulation, and P is a matrix, it is a multi-point simulation. Using the sim function, the input and output simulations can be performed before and after the network training, and compared, so that the network can be modified and evaluated. Finally, the test samples can be used to test the effect of the network model predictions obtained by training [7].

5 BP Nerve in the Prediction of Population Aging

For the BP neural network prediction, the input variable is the population aging index data. In this paper, the number of neurons in the input layer is 5, and the output layer is 1, and the number of neurons in the hidden layer is 4–12. After training, when the number of hidden layer neurons is 9, the error is the smallest [7]. At this time, the sum of squared errors of the in-sample prediction is 1.7641. It can be concluded that the combined prediction model is effective in both in-sample and out-of-sample predictions. Therefore, the model can be used to predict the level of population aging in China in the next few years [8]. In order to improve the accuracy of forecasting, we use the data from 1978 to 2016 as the in-sample data to construct a single model, and then obtain new weights, which are 0.2952, 0.3239, and 0.3809, respectively, and construct a combined forecasting model as follows:

$$x^t = 0.2952x^{1t} + 0.3239x^{2t} + 0.3809x^{3t}x^t$$

The level of aging in China from 2017 to 2022 is predicted, and the prediction results are shown in Table 1.

Table 1. Foercast of China's aging of population from 2017 to 2022

Year	2017	2018	2019	2020	2021	2022
Population aging level	10.964	11.176	11.548	11.794	12.056	12.478

Since China officially entered an aging country in 2000, although it has entered a late time but has developed too fast, the issue of China's aging has attracted the attention of many domestic and foreign authorities and has made relevant predictions. The United Nations predicts 2020 the aging level of China's population is around 11.5% in 2015. The Development Research Center of the State Council predicts that the aging level of China's population will be around 12% in 2020. It can be seen from Table 1 that the combined forecasting model is used to predict the aging level of China's population in 2020 at 11.7936%, which is between the forecasted values of the United Nations and the National Research Center, which is more realistic [8]. Therefore, the use of combined forecasting models has guiding and practical significance for forecasting the level of aging in China.

It can also be seen from Table 1 that the problem of population aging in China is still aggravating, and there is no slowing trend. To avoid the adverse impact of population aging on China's development, policy makers should make reasonable predictions on the level of aging and formulate relevant policy measures to avoid its adverse effects.

6   Forecast Validity Analysis

This paper selects the mean square forecast error (MSFE) of forecast as the evaluation index of forecast validity, and its calculation formula is as follows:

$$\text{MSFE} = \frac{1}{T}\sum T_t = (X_t - X^t)^2$$

In formula (10), T represents the total number of predicted samples, X_t represents the actual value, and X^t represents the predicted value. The smaller the MSFE, the smaller the difference between X_t and X^t, and the more effective the prediction of the model is [9].

6.1 Analysis of the Effectiveness of In-Sample Predictions

Using three single models and the established combination forecasting model to predict the population aging data from 1978 to 2010 in-sample, obtain the in-sample predicted values of the four models from 1978 to 2010, and then calculate the MSFE from the above formulas. The value of MSFE is shown in Table 2.

Table 2. The comparison of MSFE within the sample forecast in four forecasting model

Prediction method	MSFE	Forecast effectiveness ranking
Quadratic Exponential Smoothing Forecast	0.0825	3
BP neural network prediction	0.0535	2
Combined forecasting model	0.0467	1

It can be seen from Table 2 that, on the one hand, the MSFE of the combined prediction model constructed based on the three single models is the smallest, so its intra-sample prediction is the most effective; on the other hand, among the three single models, the intra-sample prediction of the BP neural network the most effective, followed by modified gray forecast, the worst in-sample forecasting ability is the quadratic exponential smoothing forecast [9].

6.2 Analysis of the Effectiveness of Out-of-Sample Predictions

It is difficult to reflect the prediction ability of the model in the future through in-sample prediction, while out-of-sample prediction can further analyze the effectiveness of model prediction. Although the combined prediction model has the highest prediction accuracy in the sample, it does not mean that it has the best prediction for the future [9]. Therefore, in order to select the optimal model to make the prediction of the future more accurate, this paper conducts an out-of-sample prediction of the population aging level from 2011 to 2016, and the obtained MSFE values are shown in Table 3.

Table 3. The comparison of MSFE outside the sample forecast in four forecasting model

Prediction method	MSFE	Forecast effectiveness ranking
Quadratic Exponential Smoothing Forecast	0.0884	3
BP neural network prediction	0.0594	2
Combined forecasting model	0.0470	1

From Tables 2 and 3, the following conclusions can be drawn: First, whether it is in-sample or out-of-sample prediction, the MSFE of the combined prediction model is the smallest among the four prediction models [10]. Therefore, the combined prediction model is in terms of predicting population aging. It has certain validity; secondly, among the three single models, whether it is in-sample or out-of-sample prediction, the prediction accuracy of BP neural network is the highest, followed by modified gray prediction, and the worst prediction ability is the quadratic index smooth prediction; third, the research on the effectiveness of model prediction is mainly to study the predictive ability of the model in the future. Although in this example the prediction accuracy rankings of the in-sample and out-of-sample are the same, in fact, the model with high in-sample prediction accuracy the accuracy of out-of-sample prediction is uncertain [10]. Therefore, when analyzing the problem of prediction validity, more attention should be paid to the effectiveness of out-of-sample prediction under the premise of reasonable in-sample prediction accuracy.

7   Conclusion

The BP network prediction results show that it is closer to the actual value than other mathematical methods. This is because China's population aging is affected by many factors, and the population growth is nonlinear. The advantage of the BP network lies in its Non-linear approach and generalization ability. In this paper, the method of combining adaptive learning rate and additional momentum method is much better than applying a single method. The BP network has its limitations. The design of the hidden layer is artificially set, which can easily cause errors. Therefore, how to better combine the BP network with other neural networks and apply it to the population aging prediction system is the direction of our future efforts.

Acknowledgments. Funding Project: 2021 school-level scientific research project of Guangzhou College of Technology and Business (KA202132).

References

1. Xie, J.F., Li, Y.M.: Beijing city population forecast based on BP artificial neural network. Agric. Network Inf. **11**(7), 86–89 (2010)
2. Luo, R.G., Huang, M.M.: Research on population prediction of the Yangtze river basin based on BP neural network. J. Wuhan Univ. Technol. **12**(10), 90–92 (2014)
3. Wang, B.T., Ma, X.Y.: Population prediction of Zhejiang Province based on BP neural network. Sci. Technol. Inf. **10**(06), 114–115 (2011)
4. Wang, X.L., Yang, G., Zhang, B.H.: The application of gray and its improved model in population prediction. World Sci. Technol. Res. Dev. **31**(04), 757–759 (2019)
5. Jiang, C., Yang, L., Fu, M.: The mathematical model of population prediction in China. J. Neijiang Norm. Univ. **3**(12), 33–35 (2018)
6. Xing, L.Y., Huang, D.W.: Application of BP neural network in china's population forecasting. Basic Sci. J. Text. Univ. **23**(13), 386–388 (2019)
7. Chen, H., Tian, B.L.: Analysis of characteristics of population aging in Jiangsu Province based on Leslie equation prediction. Mod. Electr. Technol. **28**(13), 91–92 (2015)
8. Liu, R.J., Yang, B.: Analysis of the influencing factors of China's population based on grey correlation analysis. Guangdong Sci. Technol. **12**(21), 65–68 (2012)
9. Yuan, S.: Chinese population prediction based on BP neural network. Sci. Technol. Plaza **9**(10), 98–101 (2014)
10. Chen, Y.H., Li, Y., Su, C., Sun, F.H.: Application of radial basis function neural network model in the prediction of population aging. Econ. Geogr. **32**(4), 32–37 (2017)

Analysis of Social Network Data Analysis Technology in the Evolutionary Law of Scientific Research Teams

Wei Zhao and Haoxiang Xia[✉]

School of Economics and Management, Dalian University of Technology, Dalian 116024,
Liaoning, China

Abstract. One of the purposes of the scientific research team is to undertake national theoretical foundations or applied research projects. It has clear tasks and directions. Most research teams have their own unique academic leaders and have achieved certain scientific research results. Since the evolution of the scientific research team, there is a certain evolutionary law, and the analysis of the evolutionary law of the scientific research team has certain practical value. Therefore, through the analysis of social network data technology, this article studies the application of social network data technology in the evolution of scientific research teams. First, this article outlines the basic content of social networks, and then describes and examines the basic framework of social network technology. The final test results show that the reliability and validity of the technology are basically close to 1, indicating that the technology can effectively promote the research of the evolutionary law of scientific research teams.

Keywords: Social network · Data analysis · Analysis technology · Scientific research team

1 Introduction

With the rapid development of science and technology, the era of knowledge economy has arrived. The traditional development model that determines the success or failure of economic development by resources and capital has gradually lost its effectiveness, and knowledge has become an internal core factor of economic growth. As a basic resource, knowledge will become an important driving force for social development and organizational innovation [1, 2].

The earliest origin of social network analysis theory was in the 1930s when a British anthropologist first applied the theory to community research. And the book that studies the beginning of the relationship between actors is none other than the book "Division of Labor in Society". For the first time, foreign scholars put forward the concept of "community map", which directly transforms the social structure hidden behind the phenomenon into a schema, and then analyzes it. The growth stage of social network analysis was in the 1930s and 1960s. At this stage, more and more scholars began to apply the concept of "social structure" to the research fields of sociology, anthropology, and statistics, and began to explore the "network structure" in life [3, 4].

The ultimate goal of theoretical research is to provide guidance and reference to reality. The research purpose of this paper is to summarize the problems existing in the evolutionary law of the university's scientific research team through the performance of social network data analysis technology and the overall network and individual network analysis, and to put forward the research conclusions of this article. It makes the research results of this paper play a realistic role in the actual knowledge management of the scientific research team in colleges and universities. It improves the efficiency and level of knowledge sharing of the scientific research team of the university, and ultimately increase the amount of scientific research results of the university, so as to better serve all aspects of society [5, 6].

2   An Overview of Social Network Data Analysis Technology in the Evolution of Scientific Research Teams

2.1   Basic Concepts of Social Networks

(1)   Social network
     The term social network refers to a group of participants and the connections between them. There are two basic elements in social networks, "nodes" and "connections". "nodes" are participants in social networks, and "connections" are the relationships between participants. Nodes can be different types of social connections. The contact method can be directional, such as the connection between a company's management and managers and employees, or it can be non-directional, such as the connection between colleagues and colleagues. Contact information can be expressed in different ways. Contact with friends, contact with employees, contact with scientific research cooperation, contact with national sales, etc. Social network is a perspective based on sociological organization structure, and it is also a series of analysis of sociological organization structure. Therefore, people are called social network analysis, which represents the complementarity and indispensability in research [7, 8].

(2)   Condensed subgroup
     The cohesive subgroup is a subset of actors that meet the following conditions. The participants in the set have relatively strong, direct, close, frequent, or active social networks, which is a particularly useful concept in social networks [9, 10].

(3)   Individual network
     Individual network analysis research focuses on self-centered networks between individuals, and defines social networks from the perspective of individuals. The basic concepts include range, density, and strong and weak network connections [11, 12].

(4)   Overall network
     The overall global network analysis method focuses on the overall network structure or the relationship structure of each role in the group. We are currently studying the internal relationships of small groups, changes in network structure over time, and the contact information of network members.

(5) Network density
The density of a social network is the ratio between the number of rows and the maximum possible number of rows. It represents the strength of connections between participants in a social network. The higher the density, the closer the relationship between network members. Density is the most commonly used metric in social network analysis, representing the cohesion level of a graph of social networks. The more the links between points on a fixed scale in the binary matrix are not aligned, the greater the density of the graph. The density of a binary graph is the ratio of the actual number of links in the graph to the maximum possible number of rows, expressed as:

$$p = 1/[n(n - 1)/2] \tag{1}$$

Among them, 1 represents the actual number of rows in the figure, and the value of p ranges from 0 to 1. In the analysis tool of UCINET social network, the network density of the directed value-added graph is expressed as:

$$q = k/n(n - 1) \tag{2}$$

Where k is the total value of the lines in the figure. Since each line has a direction, the sum is the sum of the values of the two lines between the two points. If the strength of the relationship is s, its value is counted as S. Because of the directed graph, each edge is bidirectional, and n (n−1) represents the total number of relations (edges). The density depends on two network structure parameters, the inclusiveness of the social network graph and the sum of the degrees of each point on the graph. Inclusion is the total number of key points minus any outliers.

(6) Cut point
The cut-off point means that after removing this point in the social network diagram, other originally connected points will be split into two unconnected social sub-network points.

(7) Length of characteristic path
The average length of the shortest path connecting two points in a relational network is an index to measure the overall nature of the network.

2.2 Concept of Scientific Research Team

The term team originated in Japan, but prevailed in the United States. However, due to differences in cognition, there are certain differences in the definition of the concept of team. The management master Drucker believes that the so-called team refers to those groups that are together for specific goals in the organization on the basis of complementary knowledge and skills, where the team has dedication, specific tasks, and common goals. Among them, scientific research teams generally exist in major universities, research institutes, and research centers, and the number is generally between 10 and 20 and is spontaneous. The scientific research team is a special group in the team. The members of this special group are composed of highly educated personnel with different disciplinary backgrounds and professional skills to jointly complete corresponding academic research or scientific research projects. These groups are the main force of scientific research in our country, and they play a strong role in promoting the innovation of our country's knowledge, the progress of society, and the development of science and technology. The members of the scientific research team are generally divided into the following categories:

(1) Team academic leader. This type of person is not only the leader of the scientific research project, but also the manager of the entire team. Most scientific research tasks in the scientific research team require the guidance and supervision of the academic leader.
(2) Core scientific research personnel, these personnel are the core force of scientific research tasks, mainly to complete corresponding academic research under the guidance of academic leaders.
(3) General scientific research personnel. These personnel are mainly auxiliary personnel in the scientific research team. The main task is to assist the scientific research team in daily administrative work, and is the right-hand assistant of the team's academic leader and core scientific research personnel. The scientific research team is a knowledge-intensive organization. This kind of team is different from other team organizations and has its own unique characteristics. Based on the above analysis of the concept of the scientific research team, the characteristics of the scientific research team are further summarized:
(1) Clear scientific research goals, each member of the scientific research team is working together around certain scientific research tasks and goals.
(2) Implicit scientific research team rules. In scientific research teams, unlike other ordinary teams, they have their own team rules. However, the scientific research team does not have clear rules, but has accumulated and formed its own unwritten implicit team rules for a long time.
(3) Relying on key laboratories, each scientific research team has its own academic research environment and experimental equipment, which is also a prerequisite for academic research.
(4) The sharing of team knowledge. The sharing of knowledge in the scientific research team can make up for the shortcomings between members, release the creativity of team members, and promote the development of the team.

2.3 The Impact of the Evolutionary Law of Scientific Research Teams on Social Networks

Social network analysis was originally mainly used to analyze the connections between people in the social environment. At the same time, I also think that when drawing a community map, we should pay attention to the roles of different network nodes. Some nodes are closely connected as an important role, while some nodes are isolated secondary roles. This is the first time that a community map is used. The way to intuitively express the complex social network relationship. Later in the 1960s, scholars widely used social network analysis methods to study social structure and social network structure. With the advent of the Internet era, social network analysis methods have been widely used in the fields of sociology, communication, and management. Foreign countries have begun to use this software to conduct centrality, location analysis algorithms, and faction analysis on related research issues. At the same time, as the software is updated and upgraded, relevant statistical analysis can also be carried out, such as factor analysis, multiple regression analysis, cluster analysis, etc. Based on the above advantages of the software, this paper selects UCINET software as the data processing tool for the empirical analysis of the evolution law of a scientific research team, and displays the exchange diagram of the evolution law of the scientific research team in a more intuitive way, and uses the software to perform calculations. The result is a concrete analysis of the evolutionary law of the scientific research team.

3   The Establishment of Models of Social Network Data Analysis Technology in the Evolution of Scientific Research Teams

3.1   The Framework of Social Networks in the Evolution of Scientific Research Teams

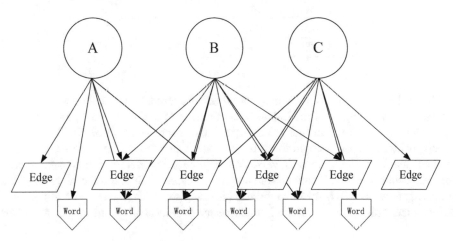

Fig. 1. Social network data structure diagram

As can be seen from Fig. 1, data points A, B and C of the social network represent three members, and the connection between members is represented by an edge. Some of them have multiple connections with each other, and each member also has internal attributes. These form a data framework for social networks that can be used to analyze the evolution of research teams.

3.2   Model Parameter Setting

(1) The number of Gibbs sampling iterations is 1000, the hyperparameter a is set to $k-1$, and k is the number of communities; the hyperparameter β is set to 0.01. The result of each experiment is the average result of ten repetitions.
(2) Cloud platform experimental environment: 2 PCs are used as the main node, and 10 PCs are used as the child nodes of the cloud platform.

(3) The configuration of each node: Huawei RH2285.2U server, Intel (R) Xeon (R) CPU E5530 @ 2.40 GHz 2 dual-channel 4-core 16G memory 4T hard disk.

4 Data Analysis of Social Network Data Analysis Technology in the Evolution of Scientific Research Teams

4.1 The Degree of Relevance Between Social Networks and Scientific Research Teams

Table 1. Relevdegree

	Response	Identification value	Influence factor
7	86	34	0.774
18	106	96	0.921
19	92	84	0.841

It can be seen from Table 1 that not all active members are academic backbones. Through the calculation of the influence coefficients of members No. 7, 18 and 19, we found that although the influence coefficient and response value of No. 7 player are relatively high, the low recognition value indicates that although he is more active in the research team, he has fewer followers. The calculation coefficient of member 18 shows that his response value, identity value and influence coefficient are the highest, so it is judged that member 18 is the main force in the scientific research team. Therefore, social network technology has a profound impact on the evolution of scientific research teams.

4.2 Reliability and Validity Test

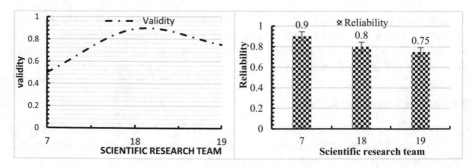

Fig. 2. Reliability and validity test

It can be seen from Fig. 2 that the reliability of this technology in the evolutionary law of the scientific research team is basically close to 1, indicating that the reliability of the technology is relatively high; in addition, the validity of the technology in the evolutionary law of the scientific research team is basically above 0.5. Effectiveness is close to 1.

5 Conclusion

In the study of the evolutionary law of scientific research teams, data analysis techniques based on social networks can provide guidance to scientific researchers. This article mainly discusses this field at home and abroad from the following aspects. First, it introduces related concepts, basic theories, and current research results; then, it explains the current status and existing problems in this area at home and abroad; finally, it puts forward corresponding countermeasures and suggestions and looks forward to future development trends, hoping to provide reference and reference significance and value for academic research in universities in our country. In this article, it was based on the social network data analysis technology and combined with the evolution law of the scientific research team to study its evolution mechanism.

References

1. Alberto, R.T., et al.: A bibliometric and network analysis of lean and clean(er) production research (1990/2017). Sci. Total Environ, **653**, 765–775 (2018)
2. Taddeo, R., Simboli, A., Vincenzo, F.D., et al.: A bibliometric and network analysis of lean and clean(er) production research (1990/2017). Sci. Total Environ. **653**(FEB. 25), 765–775 (2019)
3. Najmi, A., Rashidi, T.H., Abbasi, A., Travis Waller, S.: Reviewing the transport domain: an evolutionary bibliometrics and network analysis. Scientometrics **110**(2), 843–865 (2016). https://doi.org/10.1007/s11192-016-2171-3
4. Tan, S., Yang, Y.N.: Research on the status quo of data visualization technology: based on the perspective of social network analysis. Value Eng. **037**(036), 215–216 (2018)
5. Jianhua, L., Hui, Y.: A sociological analysis of the alienation of network technology in the "Micro-era"%Liu Jianhua Yang Hui. Forward Pos. **000**(001), 72–78 (2019)
6. Wang, G., Kong, Q., et al.: The dilemmas of scientific research cooperation and their resolution from the perspective of evolutionary psychology. Front. Psychol. **10**, 2561–2561 (2019)
7. Gupta, Y.K., Mittal, T.: Empirical aspects to analyze population of India using Apache Pig in evolutionary of big data environment. Int. J. Sci. Technol. Res. **9**(1), 238–242 (2021)
8. Bashar, F.R., Zijoud, S.M.H., Khaghanizade, M., et al.: Service-based learning in nursing: a Rodgersian evolutionary concept analysis (review study). J. Military Med. **19**(6), 532–543 (2018)
9. Grossmann, S., Moura, M., Matias, M., et al.: The use of social networks in scientific research with questionnaires. Brazilian J. Oral Sci. **17**, 1–8 (2018)
10. Linlin, Z.: Analysis of the information processing technology of university libraries in the big data era. Agro Food Ind. Hi Tech **28**(1), 2036–2040 (2017)

11. Fernando, C., Barbosa, M.W., Gomes, R.R.: Measuring participation in distance education online discussion forums using social network analysis. J. Am. Soc. Inform. Sci. Technol. **70**(2), 140–150 (2019)
12. Fengting, W., Hongyu, P., Chuchu, W., et al.: Analysis of the problems and countermeasures in college students' scientific research innovation ability training. Mod. Educ. Equip. Chin. **000**(023), 72–74 (2017)

Optimization of Computer Aided Mandarin Proficiency Test System

Bingjie Han[✉]

Shandong Vocational College of Light Industry, Zibo, Shandong, China
Bingjie_Han81@haoxueshu.com

Abstract. With the in-depth development of computer technology, it has brought earth-shaking changes to our lives and studies. More and more schools use computers to assist in teaching in various subjects. The requirements for the equipment of computer teaching facilities and teachers to master computer technology methods are becoming more and more demanding. In particular, the use of computers for Mandarin proficiency test not only brings different experiences to teachers, but also presents many advantages, reducing the human capital investment and capital investment in the process of Mandarin proficiency test, but it is inevitable in the process of using it. There are also some drawbacks. This article proposes a computer-assisted Mandarin proficiency test system to optimize some of the existing drawbacks.

Keywords: Computer · Assisted · Mandarin · Proficiency test · System optimization

1 Introduction

With the widespread popularity of Mandarin in Chinese teaching classrooms, more and more teachers and parents have paid attention to the importance of Mandarin, but how to determine the level of Mandarin has always been an alarming headache. The traditional method is to arrange test centers and interview teachers, which generally include three test stages of listening, reading, and writing. Students who need to take a Mandarin proficiency test go to designated places and test centers. This not only wastes a lot of human capital investment, but also increases the economy for the candidates. burden [1].

2 The Development Process of Computer Adaptive Mandarin Proficiency Test at Home and Abroad

Computer adaptability testing is an ideal testing method for computer-assisted language testing, and its theoretical basis is project response theory. The application of computer-assisted teaching has become more and more in-depth, which has affected language testing. Language teaching and language testing are inseparable, and scientific language testing can better test language teaching and get positive teaching feedback. This shows that the role of language testing cannot be underestimated. The fitness test is a computer-based exam [1]. Adaptive testing means that the test software will automatically adapt to the specific situation of the testee.

2.1   The Development of Adaptive Language Proficiency Testing in the United States

Each adaptability test should include at least five basic steps: question bank establishment, initial project, project selection, ability calculation and end point criteria [2]. First of all, the establishment of the question bank is very important. The success or failure of adaptive testing depends to a large extent on the quality of the test question bank. However, there are huge problems in the invention and establishment of the question bank. The establishment of a question bank is by no means an easy task that can be completed in a day or two. It is a long-term problem that requires a large investment in manpower and material resources and takes at least a few years to put it into use [2].

2.2   The Development of Adaptive Testing in China

However, computer adaptability testing has not yet been adopted in China. For example, the current college students' English Band 4 and 6 exams still use the traditional paper-and-pencil answering format. The process consumes a lot of time and manpower, and the accuracy of the test is not satisfactory [3]. Computerized testing, but adaptive testing can reduce students' worries about the test; Han Shaojie and others designed a comparative experiment between adaptive test and conventional test.

3   Analysis of Advantages in the Process of Using Computer-Aided Mandarin Proficiency Test

Computer-assisted Mandarin proficiency test is not only an advanced testing technology, but also an organic combination of modern information technology, teaching theory and testing system. This kind of auxiliary test emphasizes the design, production and application of Mandarin proficiency test programs and systems with a computer platform as the core [3].

3.1   Break Through the Unity of Traditional Test Methods and Rich Test Methods

First, the computer equipment can accurately display the contents of the test, such as watching, listening, and speaking, and can also accurately record the test candidates' test conditions, avoiding the unfairness of the test. Computer-assisted Mandarin proficiency test is mainly for students logging in to the computer language test system, interacting and communicating with the examiner through computer equipment [3]. During the test, language materials with pictures, text, sound and images are presented, which not only reduces the student's test the pressure has also fully mobilized the examinees' various organs such as vision and hearing, and devoted themselves to the test, which can greatly stimulate the students' potential and give full play to everything, as shown in Fig. 1. Second, the use of computers to create a relaxed and pleasant test environment for test takers can accurately grasp the true ability of the test takers, effectively avoiding the unfair factors of human examiners [4].

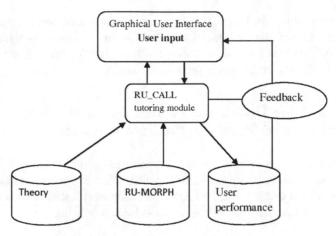

Fig. 1. Computer-aided testing architecture

3.2 Large Amount of Stored Information, Effectively Improving the Efficiency of Mandarin Proficiency Test

The traditional Mandarin proficiency test stays on the recorder and paper. Due to various factors of time and space, it wastes time and fails to achieve the expected results. The overall process can only be done step by step, which greatly reduces the efficiency of the Mandarin proficiency test [4]. On the one hand, the use of computer multimedia equipment to assist the Mandarin proficiency test accurately conveys a large amount of information of the candidates on a vivid, intuitive and scientific basis. It not only enriches the content of the language test, but also breaks through the prefabrication of traditional tests, which not only saves manpower Capital investment and scheduling time, and compact and methodical presentation of the computer-assisted testing procedures, have a comprehensive understanding of the candidates' thinking and imagination abilities, through students' reflections, you can see the candidates' response abilities and the content they have mastered comprehensiveness [5].

3.3 Effectively Stimulate the Candidates' Thinking Potential

Traditional language testing needs to be organized to deal with, clarify the design and arrangement of the testing process and steps, and the application of specific testing standards and methods need to be discussed. Computer-aided Mandarin proficiency testing creates a human-computer interaction environment of "candidates + computers". In the corresponding test, specific scenarios can be created to enhance the contextualization and pertinence of the test [6]. First, it can promote the candidates to adapt to changes, not being solidified by the fixed test mode, which is conducive to stimulating the candidates' ability to adapt to changes and innovative thinking. English learning itself is a practical function, which is a combination of knowledge and skills. If the test of students' innovation and processing ability is neglected during the Mandarin proficiency test, the test taker will not be able to use English proficiently in actual English communication situations. Affected the application of the acquired knowledge. Second, using computer information technology, the language testing system can be designed into various forms of question or situation testing, which can fully stimulate the divergent thinking of candidates, not only for the purpose of testing the level, but for the purpose of training, using computer assistance Mandarin proficiency test [6].

4 Analysis of the Drawbacks in the Process of Using Computer-Aided Mandarin Proficiency Testing

Everything has both sides, and the use of computer-assisted Mandarin proficiency test is no exception. The limitations shown on the basis of its advantages are also obvious. The advancement of computer technology is not only manifested in human-computer interaction [7]. Computers cannot be as active as human thinking.

4.1 Loss of the candidate's Main Role and Incomplete Testing of the candidate's Mandarin Proficiency

Relying entirely on computer-assisted Mandarin proficiency test, practice has proved that not only the leading role of candidates is lost, but also the test of Mandarin proficiency of candidates is not comprehensive. On the one hand, in the process of computer-assisted Mandarin proficiency test, candidates still play the main role. Through the design and production of Mandarin proficiency test software, it reflects the outline and plan of Mandarin proficiency test. It is impossible to accurately find individual differences among students and problems encountered by candidates there are many differences, and some functions of the Mandarin proficiency test in the traditional test center are not replaced by computer-aided equipment [7].

4.2 Easy to Be Interfered by External Factors and Fail to Achieve Expected Results

During the Mandarin proficiency test, all the test material questions and pictures are input into the computer through the corresponding program, but in the specific production process, unavoidable factors are often encountered [8]. If the computer-aided equipment and test requirements are not met, the examinee is extremely susceptible to interference from external factors and cannot judge the examinee's knowledge ability based on unexpected circumstances. The second is that candidates are more random in the process of testing. Other candidates may need a quiet environment, which will have a greater impact on other candidates, or the sound recorded by the computer may be noisy and indistinguishable, failing to achieve the expected Mandarin proficiency test results [8].

5 Optimization of the Mandarin Proficiency Test System

Computer-assisted Mandarin proficiency test has certain advantages and disadvantages. The use of computer-assisted Mandarin proficiency test has the disadvantages of losing the main role of test takers, incomplete testing of test takers' Mandarin proficiency ability, easy to be interfered by external factors, and failing to achieve the expected results, and it has a breakthrough in traditional test methods [9]. The unity of the test, the rich test methods, the large amount of stored information, effectively improve the efficiency of the Mandarin proficiency test, and effectively stimulate the candidates' thinking potential, as shown in Fig. 2.

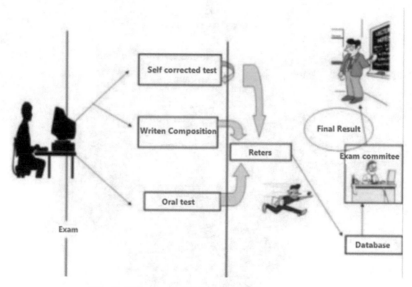

Fig. 2. Computer-aided language testing system

The first thing to do: judge whether the pressure test method and tools (lr, ab, multi-threaded client written by myself) are correct or not.

The second thing is to use the resource monitoring commands of the operating system (see the linux command), cpu utilization, load, memory usage (cache, swap, application occupancy), context switching, io wait, network data volume, packet loss, file handle configuration, etc. Judging from these indicators, when user concurrency increases, which indicators have changed drastically, and have even become a bottleneck [9]. If you are a java application, then look at the gc of jvm, thread dump to see if there is a large number of locks, or a single thread eats a fixed cpu and so on.

The third thing is to begin to locate what caused these basic resources to become a bottleneck. First of all, we must eliminate the problem of dependent systems. The so-called dependent systems, such as web containers, centralized caches, db, etc., usually you have no way to debug these systems [8]. The most important thing is to check their logs. Then start to disassemble each module of your application, mock method to ensure no consumption interface dependency, and then repeatedly test and locate the problem. When a certain module has an impact, you must pay attention to it. If you have written a language with pointers, you will understand it, and the problem often occurs at a non-burst point. At the same time, it should be reminded that many times when you really judge a bottleneck point, A and B two modules, A is the bottleneck tps of 30, B The tps is 35. At this time, you optimize A to a tps of 50, but for B, the pressure is obviously increased [9]. At this time, B may start to decrease due to resource pressure, ay also drop), as shown in Fig. 3.

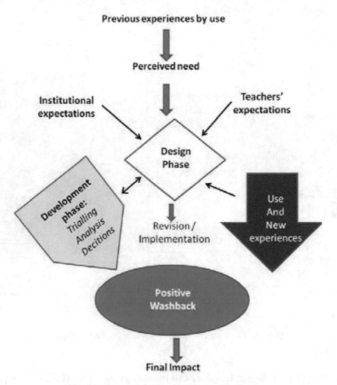

Fig. 3. Schematic diagram of optimization

1) Reduce the total RT of critical business paths. In many cases, entangled with the so-called code level saving, it is better to optimize the business directly. (Except the basic system)
2) Bottleneck resources, such as all the indicator resources mentioned in the second step and system-dependent resources (such as the number of threads of the web container, etc.) [10]. Two types of treatment are performed on bottleneck resources: a. Efficient use. (For example, batch processing of db, batch acquisition of cache, batch flushing of disk) b. Use less. Some local caching can be done if the data consistency requirements are not high. c. Quick release. Students who have written event-driven code should have a deep understanding that after the business processing is cut and refined, the original hold resources will also be used in fragments. (In fact, we can see this popular truth when we study the evolution of computer cpu), rapid release means that the same resource can serve more requesters [10]. d. Exchange resources. Disk for memory, multi-core CPU processing power for storage.

3) Changing dependencies, including web containers, centralized caches, disks, etc., changing them is not necessarily bad for them, but it may require multiple operations (centralized caching) due to unsupported data structures, and there are more changes between multiple systems a good private interaction protocol (between reverse proxy and web container) has itself been revolutionized (solid state drive) [10].

6 Conclusion

With the trend of computer-assisted Mandarin proficiency testing, it is necessary to design a computer-assisted Mandarin proficiency testing system to improve the reliability and validity of language testing. It is necessary for designers to understand and grasp the guiding ideology and theoretical basis of language testing, and strictly control scoring. Only by ensuring the reliability and validity of the language test, can it truly reflect the test takers' real English proficiency and truly promote the development of Mandarin teaching.

References

1. Chen, S.: Talking about the application of computers in the Mandarin proficiency test. Campus English **11**(16), 30–34 (2017)
2. Gu, T., Zhang, H.L.: Application of computer multimedia technology in english listening test. Shanghai Int. Stud. Univ. **33**(01), 53–56 (2016)
3. Xiong, X., Sun, X.Y.: Several factors affecting the validity and reliability of reading tests. Sci. Educ. Wenhui (first half month) **9**(12), 1898–1900 (2015)
4. Huang, J., Duan, S.: A study on the Validity of English reading tests. J. Jianghan Univ. (Humanities Edn.) **48**(3), 7–10 (2017)
5. Zhao, Q.J., Zhao, Q.: Reliability and validity in college English achievement tests. J. Jishou Univ. (Soc. Sci. Edn.) **32**(6), 983–988 (2017)
6. Liu, C.P.: General principles of use cases for learning Chinese as a foreign language dictionary research, **52**(3), 110–118 (2013)
7. Wang, K.: On the oral English test of non-English majors. J. Jilin Normal Univ. Eng. Technol. **22**(5), 346–352 (2016)
8. Zhang, Y.L.: A preliminary study on the reliability and validity of the higher education self-taught test (Shanghai) oral English test (English). Teach. Eng. China **65**(9), 72–74 (2018)
9. Wang, X.T.: The application of reliability and validity measurements in college English tests. Mod. Bus. Trade Ind. **11**(04), 264–268 (2018)
10. Zhou, Z.: Reliability and validity of second language testing. Chin. J. Liter. **10**(11), 214–217 (2018)

Intelligent Integrated Software Development Based on Neural Network Fuzzy Algorithm

Yizhi Wu[✉]

Guangzhou College of Technology and Business, Guangzhou, Guangdong, China

Abstract. Because there are a large number of data collection points and different standards in some large projects, there are many problems in data processing and maintenance. Therefore, the "intelligent integrated platform" came into being in this case and has become an effective means to solve the problem of information processing. This paper studies the intelligent integrated software development based on neural network fuzzy algorithm, introduces the neural network fuzzy algorithm into the intelligent integrated software development and application, and tests the developed software. The test results show that, according to the phenomenon that the smaller the variance is, the better the data fusion effect is, it is known that the software developed in this paper has a good effect in the data fusion, the minimum variance reaches 0.04, the maximum variance reaches 0.3, and the value of the variance is small in general. Then, in terms of the usage evaluation of the software, the overall usage evaluation of the software developed in this paper is good, about 75% of the software content is good, and about 69% of the software function is good, but the software has some shortcomings, mainly the human-machine interface design of the software is not very good.

Keywords: Neural network fuzzy algorithm · Intelligent integration · Software development · Data fusion

1 Introductions

Today, there are between some of the engineering system data independence, there are technical barriers between different business systems, low level of information, can not be cross-sectoral, multi-platform, multi-disciplinary function [1, 2]. After the development of intelligent integrated software, the management system can be integrated management and integrated access in interdisciplinary, multi-platform, and cross-professional applications, so that management is "intelligently integrated", and different data systems can be decomposed and provided appropriately information services [3, 4]. Then it can complete data integration between business systems at all levels, effective solution to the problems between applications and data from bad information, provide technical support [5, 6] for the inter-application integration and data exchange of different, so for the intelligent integration of software development there are also more and more researches [7, 8].

Development of software for intelligent integration, there were researchers for the integrated system of intelligent manufacturing industry research, the integration of intelligent software framework, the framework is a multi-agent system language service-oriented architecture combined with manufacturing systems, and related the function is explained, and a manufacturing system that can be automatically controlled is developed [9]. Other researchers have conducted research on smart home life and developed an intelligent integrated furniture control software to integrate and control the smart furniture at home, which brings convenience to life. There is no need to equip each smart furniture with a remote control app [10]. There are researchers in the intelligent switch for home study, conducted by research on the issue of integration of intelligent switches, developed an intelligent integration of intelligent switch control software, you can achieve intelligent control, it has brought to people's lives convenient [11]. Researchers have also conducted research on the problem of information and data isolation in smart substations, and developed an intelligent integrated information platform to solve the problem of data isolation. The platform processes, integrates, and shares related system data in smart substations to improve the performance of smart substations work efficiency [12]. In summary, it can be seen that there are many researches on the development of intelligent integrated software, but few researches on the application of neural network fuzzy algorithms.

This paper studies the development of intelligent integrated software based on neural network fuzzy algorithm, summarizes the significance of intelligent integrated development on the basis of relevant literature, mainly understands the problems that intelligent integrated software can solve, and then selects the direction of software development. This paper selects the intelligent integrated software development of smart grid, tests the software developed in this paper, and draws relevant conclusions through the test.

2   Research on Intelligent Integrated Software Development

2.1   The Significance of Intelligent Integrated Software Development

(1)   Today, the project's video surveillance, intelligent inspection, security, and fire protection systems are still independent of each other, unable to enter the integrated information platform and become an information island. The analysis and processing of comprehensive information is helpful for intelligent management, the development of advanced integrated information practical applications, and the full use of information mixing platform for system information integration, which can effectively solve the problem of information islands. Most of the information in the system is mainly based on unstructured data such as tables, images, and videos. Real-time monitoring will provide a large amount of historical data. Conventional data analysis and processing methods cannot meet actual needs, and useful data analysis can be obtained quickly and efficiently. Integrated information software development is its foundation.

(2)   With the continuous improvement of digitization and informatization of engineering systems and the continuous installation and operation of various systems, intelligent integrated software development standardizes business and process management and improves personnel work efficiency. However, there is also a problem that can

not be ignored, that is, all data of each system are stored in different data areas, and devices such as firewall and partition wall are also developed, which makes it difficult for all data between different operating systems to interact and simple data management and maintenance. The data between different business systems is independent of each other, resulting in multiple isolated data groups, which makes the data uniformity and completeness relatively poor. The data in the grid format is stored in multiple business systems, models and their interfaces they are not uniform, and the specific data names are also inconsistent with the coding standards. The same object can be defined with different names in different systems. Therefore, it is necessary to build a complete data sharing platform for users to integrate and centrally manage scattered and inconsistent data information to realize the free exchange and sharing of data between different systems, while reducing workload and improving work efficiency.

2.2 Technical Requirements for Intelligent Integrated Software Development

(1) Unification and digitization of data requirements
One of the characteristics of smart substations is the comprehensive application of various manufacturers and different types of measurement and control devices. Integration refers to the use of unified communication protocols and interfaces for the transmission of data within the station to maintain the integrity and accuracy of the data; the data is first converted to a pure digital mode through the intelligent terminal, and then transmitted through the optical fiber.
(2) Data integration
On the basis of data standardization and digitization, the data is integrated, processed and modeled, and various data fusions are completed.
(3) Bidirectional transmission of information flow
It should have a single access interface to facilitate the interaction between the final information of the monitoring station and related information such as scheduling, so as to realize the two-way flow of information, while ensuring the accuracy and safety of the information.

2.3 Neural Network Fuzzy Algorithm

In the cooperative fuzzy neural network, the fuzzy system and the artificial neural network operate independently of each other. Both artificial neural networks and fuzzy systems can use them as input to another system. In this type of system, according to its different characteristics and uses, the fuzzy system and artificial neural network are integrated to build a new type of system. The artificial neural network learns the fuzzification method from the training data:

The first layer: fuzzy operation is performed on the input variables according to the membership function, and the membership of the variable pair and a certain fuzzy concept are defined. Assuming that the participation function of x and y is $\mu_a(x)$, then the output of the first layer is:

$$\mu_a(x) = \frac{1}{1 + [(\frac{x-c}{a})^2]^b} \tag{1}$$

Where {a, b, c} is the parameter set of the function, which is called the premise parameter.

The second layer: stimulation intensity layer. Represents the excitation intensity under each fuzzy rule, which is the superposition of the participation degree of the fuzzy set. Generally speaking, the superposition method adopts the form of algebraic product, but it can also adopt the form of small, finite product and strong product. The nodes at this level are fixed and cannot be trained.

Level 3: function group level. Each node in this layer corresponds to a function f, that is to say, under certain fuzzy rules, use this function to approximate the actual output. f is generally a linear function. As the number of inputs increases, the number of function parameters also increases.

$$O_3 = \overline{\varpi}(px + qy + r) \tag{2}$$

$\overline{\varpi}$ is the normalized excitation intensity from the third layer, so the number of conclusion parameters {p, q, r} in this layer is 3. The parameters of this layer are adaptive and can be obtained through training based on input data.

3 Development of Intelligent Integrated Software Based on Neural Network Fuzzy Algorithm

3.1 Intelligent Integrated Software Analysis

This paper develops intelligent integrated software based on neural network fuzzy algorithm. After consulting a large number of documents, it is found that the current research hotspot is the intelligent integrated research in the direction of smart grid, so this paper chooses the information integration based on smart grid platform software development.

3.2 Data Collection

Through the grid transportation object coding standard technology and model construction technology, a panoramic model of "transmission, conversion, distribution" grid operation is constructed, and an intelligent data center network is built for the network operation integrated in the main distribution. The grid operation information is combined with the main network to realize the distribution integrated power grid applications: user data and models, power grid fault analysis, user power path analysis, early warning of major risks for users and other customers-user-oriented service support.

3.3 Data Fusion

In the process of data collection, due to unstable environment, such as voltage instability, temperature affecting equipment sensitivity, equipment failure, etc., it often leads to abnormal collection. The data and resistance data clearly revolve around the actual results. This is called a small error. In the data aggregation process, if the original data collection equipment is used directly, the reliability and accuracy of the data aggregation result will be reduced due to its large errors. Therefore, it is necessary to determine the amount of serious errors in the data collected by each device group and delete them. However, if you judge whether there are a lot of errors in a particular set of data, you cannot judge at will. The data must be analyzed and judged according to the corresponding error scale, so the data must be preprocessed before data fusion.

3.4 Data Security

Intrusion detection data usually contains multiple components. Immediate classification of wear characteristics can easily lead to the continuation of the disaster. Reducing the size of data is an effective way to solve such problems. Reducing the size means that the initial data in the large-size blank is mapped to a smaller-size space while retaining its core characteristics. The main purpose of dimensionality reduction is to select the most representative features from the multi-dimensional shape, to remove the correlation and redundancy between the original long-dimensional data, and to provide a concise data structure for the subsequent arrangement and mining operations.

3.5 Data Sharing

Intelligent data exchange activities emphasize the basic and logical capabilities of the information platform to distribute aggregated data within the site. First of all, the platform should be able to change the order of data execution according to data characteristics, and adjust the database structure in real time according to the priority of the data. The higher the priority is, the higher the storage and retrieval capabilities are; secondly, the platform must be able to process data in advance to ensure the convenience of information exchange. In addition, the platform should also be able to provide guidance data and information in accordance with the information request instructions submitted by the requesting department, and have guidance and identification functions.

4   Software Detection

4.1   Data Fusion Test

After preprocessing the collected data using t-test criteria, 5 data are randomly selected for fusion, and the variance of the data is used to characterize the data fusion effect. The relevant data results are shown in Table 1:

Table 1. Data fusion test

	Variance	Weight
1	0.04	0.14
2	0.30	0.03
3	0.14	0.04
4	0.05	0.08
5	0.04	0.17

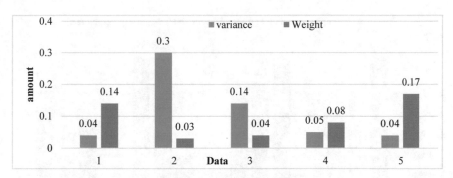

Fig. 1. Data fusion test

It can be seen from Fig. 1 that the calculated variance value is still relatively small. According to the smaller the variance value, the better the data fusion effect, indicating that the data fusion effect in this paper is better, with the minimum variance value reaching 0.04.

4.2 Software Use Evaluation

In this paper, the developed software will be evaluated by related experts. A total of 35 people have been recruited to evaluate the content, functions and human-computer interaction of the software development. The relevant data results are shown in Table 2:

Table 2. Software use evaluation

	Developed content	Software function	Human-computer interaction
It is good	37%	38%	17%
Better	34%	35%	24%
Generally	10%	16%	34%
Have no idea	19%	11%	22%

It can be seen from Fig. 2 that the overall evaluation of the software is relatively good. The content of the software is well evaluated in more than 75%, and the software function evaluation is good in more than 69%, but the interface evaluation of the software is not very good, may not consider the operation of human-computer interaction.

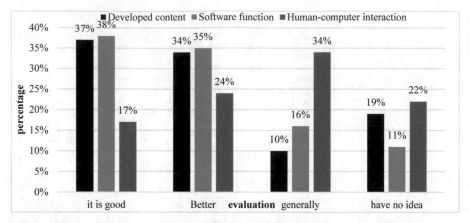

Fig. 2. Software use evaluation

5 Conclusions

This paper studies the intelligent integrated software development based on neural network fuzzy algorithm. On the basis of consulting a large number of literature, this paper selects the intelligent integrated software development in smart grid, mainly to study the problem of data isolation in smart grid system. Then the relevant functional modules are designed according to the problems, and the developed software is tested. The results show that the software designed in this paper has a good effect on data fusion, but there are still some deficiencies in the research process of this paper, mainly because there is no experimental verification of the algorithm, which will be improved in the future research work.

References

1. Li, Z., Dong, Y., Fu, L., et al.: Integrated research on power distribution intelligent switching equipment. Int. Core J. Eng. **6**(1), 48–54 (2020)
2. Rybina, G.V., Blokhin, Y.M.: Methods and software implementation of intelligent planning for integrated expert system design. Sci. Tech. Inf. Process. **46**(6), 434–445 (2019)
3. Yang, X., Wang, Y., Rao, D., et al.: Design and application of IED in integrated monitoring intelligent component of transformer. Dianli Xitong Baohu yu Kongzhi/Power Syst. Prot. Control **45**(16), 130–135 (2017)
4. Mizukami, Y., Own, H., et al.: Applying integrated R&D process in process innovation research:estimating the impact of a process change in automotive ECU development on organizational flexibility and product quality. Int. J. Jpn Assoc. Manage. Syst. **9**(1), 7–17 (2017)
5. Rybina, G.V.: Intelligent technology for construction of tutoring integrated expert systems: new aspects. Open Educ. **4**, 43–57 (2017)
6. Jin, H., Yao, X., Chen, Y.: Correlation-aware QoS modeling and manufacturing cloud service composition. J. Intell. Manuf. **28**(8), 1947–1960 (2015)
7. Liu, H., Mao, S., Li, M., et al.: A GIS based unsteady network model and system applications for intelligent mine ventilation. Discret. Dyn. Nat. Soc. **2020**(5), 1–8 (2020)

8. Santos, J., Rodrigues, J., Casal, J., et al.: Intelligent personal assistants based on Internet of Things approaches. IEEE Syst. J. **12**(2), 1793–1802 (2018)
9. Kulkarni, R.H., Padmanabham, P.: Integration of artificial intelligence activities in software development processes and measuring effectiveness of integration. IET Software **11**(1), 18–26 (2017)
10. Li, D., Zhang, C., Shao, X., Lin, W.: A multi-objective TLBO algorithm for balancing two-sided assembly line with multiple constraints. J. Intell. Manuf. **27**(4), 725–739 (2014)
11. Chen, S.-H., Perng, D.-B.: Automatic optical inspection system for IC molding surface. J. Intell. Manuf. **27**(5), 915–926 (2014)
12. Uzam, M., Li, Z., Gelen, G., Zakariyya, R.S.: A divide-and-conquer-method for the synthesis of liveness enforcing supervisors for flexible manufacturing systems. J. Intell. Manuf. **27**(5), 1111–1129 (2014)

Application of Computer VR Technology in Campus Landscape Design

Kunming Luo[✉] and Li Zhang

Jiangxi Tourism and Commerce Vocational College, Nanchang, Jiangxi Province, China

Abstract. VR technology is a high-tech type based on computer technology. Its main characteristics are interactivity, imagination and immersion. It is mainly the application of the leading role of humans in VR. Campus landscape design is a field that combines art, technology and innovation. Through the application of VR technology, it can fully improve the efficiency of campus landscape design, reduce design costs, and meet the actual needs of campus landscape design. This article analyzes and discusses the application of VR technology in campus landscape design.

Keywords: Computer · VR technology · Campus landscape · Landscape design

1 Introduction

VR technology is a new interactive method based on computers. It uses a computer platform to create three-dimensional images, and combines a series of perceptions such as vision, hearing, and touch to form a new interactive experience, which provides a realistic reference for campus landscape design. Value, the generational and visual characteristics of VR will give people a comprehensive and detailed experience, feel the emotional details changes brought about by changes in the campus landscape, provide a realistic basis for campus landscape design, avoid blind campus landscape development and utilization, and improve campus landscape the practicality and aesthetics of the design [1].

2 The Meaning of VR Technology

The essence of VR is to merge virtual and reality. Analyzing from a theoretical perspective, VR technology is a computer simulation system that can create and experience a virtual world. It uses computer technology to form a simulation environment and allows users to integrate into this environment. VR technology uses various data in real life, combines electronic signals with output devices through computer technology, and transforms them into phenomena that everyone can feel. These phenomena can be real in daily life, or they can be three-dimensional. The model presents substances that everyone cannot see [1]. The real world simulated by computer technology is called VR. In the development of modern society, VR technology has been widely favored by everyone.

J. Macintyre et al. (Eds.): SPIoT 2021, LNDECT 98, pp. 690–696, 2022.
https://doi.org/10.1007/978-3-030-89511-2_89

Users get more real feelings in the VR world. The simulated environment has a certain degree of authenticity, which brings everyone an immersive feeling. Reality has all the perceptual functions that everyone has, such as hearing, vision, touch, taste, smell and other perception systems [2]. In addition, VR technology has strong simulation capabilities, which helps to achieve human-computer interaction and obtain the most realistic feedback during actual operations.

3 Features of VR Technology

3.1 Perceptual

The most basic "feeling" of VR technology is the key content that everyone pays attention to. The earliest visual experience can no longer meet the actual needs of customers. This requires innovation from the technical level and the introduction of more auxiliary equipment to achieve the unity of the five senses. Bring a full range of experiences such as hearing, vision, touch, taste, etc., so that you can experience the feeling of reality [2]. For example, VR roller coaster, VR dinosaur park, etc., through glasses and gloves, bring users a real feeling, interact in actual scenes, and bring a comprehensive visual experience. In the process of art design teaching, students can obtain various scenes in different situations. The actual teaching mode requires a lot of interaction costs. VR technology can not only save a lot of costs, but also enable students to get more real feelings.

3.2 Interactivity

Interactivity refers to the fact that students use VR technology to truly feedback the operability of things in the world, which is embodied in the following two aspects, namely behavior and visual interaction. In the process of art design, the interactivity of VR technology helps students to truly perceive the changing process of art design [3]. In the development of the new era, more and more VR-related auxiliary equipment, such as stereo glasses can observe the dynamics of three-dimensional space; VR gloves can complete grasping, moving, and manipulation; head-mounted displays can realize free walking and movement in space, through rotation watching the VR environment highlights the feature of interactivity to a large extent [3].

3.3 Specificity

The specificity of VR technology is that the design effect can be seen intuitively in the design process, the application effect and actual phenomenon of the work can be displayed concretely, vividly and vividly, and the visual evaluation of the work before the design is completed, thereby improving the rationality of the design plan performance and feasibility to meet the anticipated design requirements to the maximum extent [3].

3.4 Conceptual

VR technology has the characteristic of conception and is also called autonomy. This characteristic is mainly reflected in the rich imagination space of VR technology, which effectively expands the scope of knowledge and vision of everyone. Autonomy can truly feedback the virtual environment, and feel things through touch, feedback the material and toughness, so that the virtual scene can bring you a real experience and immerse yourself in it [4].

4   Principles of Contemporary Campus Landscape Design

4.1   The Principle of the Integration of Technology and Art

With the rapid development of modern science and technology, the abstract beauty in many works of art could not be conveyed before, and now it can be intuitively expressed with the support of modern science and technology [4]. The use of modern technology in garden campus landscape design can visually present the artistic beauty of campus landscape design, thereby guiding more innovative ideas for campus landscape design and reflecting the contemporary value of campus landscape design [5]. In campus landscape design, science and technology are combined with traditional art.

4.2   Functional Principle

Functionality reflects the extensive material needs of modern people. The functional principles followed in the landscape design of garden campus are designed to highlight the value of campus landscape design. At the same time, its function setting must insist on people in campus landscape design. The text, everything is based on the needs of people, to provide humanized services based on people, to meet people's functional needs for the garden campus landscape [5]. Therefore, in the design stage of the garden campus landscape, it is necessary to conceive the overall function of the campus landscape and the expected goals, and combine the basic needs of the garden as a place for public leisure and entertainment, and realize all the functions that meet the needs of the public.

4.3   The Principle of Simplicity and Practicality

In today's garden campus landscape design, more people have an essential pursuit of returning to nature. Therefore, this need must be considered in the design stage, following the principle of simplicity and practicality, combining the characteristics of the environment of the garden campus landscape itself, and adopting the most concise. The design ideas and methods of the school present the characteristics of simplicity and practicality of the campus landscape [6]. Specifically, it is necessary to ensure that the original style of the campus landscape is not destroyed to the greatest extent during the construction.

4.4   Relatively Unified Principle

Garden campus landscape design is a key component of urban construction planning. Combining with the needs of garden campus landscape design, construction needs to be carried out in accordance with unified planning requirements, while referring to regional characteristics to ensure that the final presentation of the garden campus landscape effect is in harmony with the surrounding environment [6]. In addition, in the process of garden campus landscape design, the development of modern technology has also prompted innovations in design theories and methods, but relatively uniform principles still need to be firmly followed to ensure the rationality of campus landscape design.

4.5   Ecological Application Principles

It is also necessary to pay attention to the issue of ecological balance in garden campus landscape design. Only by comprehensively considering the biological characteristics of various types of vegetation can the internal stability of the garden be ensured. The reasonable arrangement of cover vegetation can effectively prevent vicious competition among plants [7].

5   The Concrete Application of VR Technology in Campus Landscape Design

5.1   Construct a Virtual Three-Dimensional Environment Space

If you want to use VR technology to construct a virtual environment space, you must first be able to collect data and information related to the building, such as the floor plan, elevation, and cross-sectional view of the building. The aspect content is presented more realistically in the form of images, making the spatial relationship and spatial reference of the building more intuitive. In the process of modeling, it is necessary to reduce the number of points and segments of the model as much as possible based on meeting the design requirements [7]. When designing its vision, it does not need to have extremely high precision, to avoid the reduction of work efficiency. Light is also extremely important for the 3D virtual environment. Without light, the transparency and shape of the building cannot be rendered authentic. Therefore, for designers, it is necessary to make a clear distinction between real light and virtual light, grasp the influence of light changes on the scene, and use 3DS-MAX software to make virtual light truly appear, as shown in Fig. 1. Therefore, in the process of material design, it is necessary to consider various factors that may affect the visual effect.

Fig. 1. Three-dimensional environment space constructed by virtual technology

5.2 Interactive Optimization Design and Program Selection

Applying VR technology in the process of campus landscape design can enhance the interaction of campus landscape design. In the process of designing it, the relevant designers should use the interactive system functions of VR technology to fully apply the software advantages of the computer system, and at the same time combine with natural conditions, economic conditions and social background [8]. According to the actual needs of customers for design, after formulating a variety of design plans, through effective comparison and analysis, timely adjustments to relevant data, and some specific parts of the revision work, the revised plan and the previous plan Make a comparison to ensure that it can meet the actual needs of customers [9]. Secondly, in the process of continuously deepening the design, through the application of VR technology, it is possible to avoid large design conflicts between related disciplines and optimize the space design work, as shown in Fig. 2. In addition, the application of VR technology in the campus landscape design process can also realize the application of network paths, strengthen communication with the same industry, learn advanced experience and technology, and ensure the scientific improvement of the design plan [8].

Fig. 2. Virtual technology is used for optimization in design

5.3 Fully Display the Overall Information of the Building

At present, in the process of campus landscape design, many design methods cannot fully present the building information. However, in the process of campus landscape design for VR technology, because it relies on computer technology, through the application of computers, virtual buildings with strong authenticity can be displayed in front of people, so that people have a kind of exposure [9]. With the use of VR technology, everyone can know the design content of the architectural model, which makes the communication between users and designers more convenient, and users can more effectively participate in campus landscape design work, as shown in Fig. 3.

Fig. 3. Virtual technology is used to show the whole building

5.4 Effective Demonstration of Building Construction Technology

When developing a construction project, it often requires the participation of various units, such as design units, construction units, and construction units [10]. As the primary link of the project, at the beginning of the campus landscape design work, designers need to fully consider the relevant content of the design work, such as the actual needs of the construction unit for the building, and whether the construction technology selected by the construction unit is relatively high. Strong feasibility, as shown in Fig. 4. Therefore, in the campus landscape design link, it is also necessary to effectively demonstrate the feasibility of the construction technology. The application of VR technology can verify various problems that may be caused by the construction technology during the construction process through the simulation of dynamics, so as to adjust the construction technology selection of the construction unit and ensure the effectiveness of the construction technology selection [10].

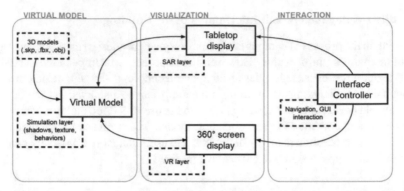

Fig. 4. Demonstration of virtual technology used in construction technology

6 Conclusion

In summary, as a virtual environment, VR can not only create but also embody. In the VR environment, people are just like in the real environment. The application of VR technology in campus landscape design can change the traditional plane campus landscape design, which is of great significance to the improvement of campus landscape design efficiency. Therefore, in order to fully improve the quality and level of campus landscape design, in-depth research and application of virtual display technology are also needed. In the context of the continuous development of computer technology, the important role of VR technology will also be brought into full play.

References

1. He, B.Z.: Research on the application of VR technology in environmental art design. Popular Literature **11**(07), 95–96 (2012)
2. Wang, Z.: Research on the application of VR technology in environmental art design. J. Chifeng Univ. Nat. Sci. Ed. **31**(24), 49–51 (2015)
3. Wang, Z.: Research on the application of VR technology in environmental art design. Shandong Ind. Technol. **12**(02), 275–277 (2016)
4. Wang, R.H.: Application of VR technology in environmental art design. Art Educ. **7**(02), 212–215 (2016)
5. Li, X.C.: Research on the application of VR technology in environmental art design. Art Educ. Res. **6**(17), 68–71 (2014)
6. Shan, J.: Application of VR technology in environmental art design. Hous. Real Estate **8**(03), 296–299 (2017)
7. Tan, Y.: Research on the application of VR technology in environmental art design. Kunming Univ. Sci. Technol. **3**(10), 33–37 (2016)
8. He, D.G.: Research on the application of VR technology in environmental art design. Popular Literature **4**(15), 95–98 (2018)
9. Ding, J.L.: The application of VR technology in garden landscape design. Inform. Comput. (Theor. Ed.) **12**(21), 78–81 (2019)
10. Fan, L.C., Li, L.L.: Application analysis of VR technology in landscape architecture design. City Weekly **10**(30), 13–15 (2018)

Construction and Application of 2D Product Classification Hybrid Model Based on Internet of Things

Cangying Chen[✉]

School of Business Central, South University, Changsha 410083, China

Abstract. Since Internet technology has been widely used in daily life, more and more attention has been received to the Relevant technologies of E-commerce industry. The emergence of machine learning technology has made product classification into the ranks of artificial intelligence for automatic classification. This paper proposes a Naive Bayesian product classification hybrid model based on word2vec, using the word2vec model for word vector training and Laplace smoothing coefficient to modify the Naive Bayesian classification model, then forming a hybrid product classification model. Finally, the product text information of Jingdong Mall is crawled through the Houyi collector as experimental data, and the evaluation index obtained by the experiment is improved compared with the other two models which verifies the feasibility and effectiveness of this model.

Keywords: Product classification · Naive Bayesian · word2vec

1 Introduction

In recent years, with the rapid development of science and technology, economic level, and the widespread application of the Internet, e-commerce has gradually become an important part of people's daily life by virtue of its advantages of a wide range of product categories, convenient purchase channels and affordable prices. In the process of e-commerce transactions, a large amount of product data information will be generated. In order to reduce information overload, it is difficult for users to find the required information. Therefore, these product data information need to be classified reasonably and effectively. Product classification can make e-commerce websites better Serving suppliers and consumers locally is also one of the important technical means of e-commerce platforms.

The classification of commodity data information is essentially a text classification of commodity information. Commodity classification is to abstract commodity data information through machine learning, and extract key commodity text information as the text features and attributes of the classification, so as to perform commodity classification. Nowadays, commodity classification technology combines search engines, related recommendations, data mining and other technologies, so that e-commerce and consumers can better control commodity information, take advantage of commodity information big

J. Macintyre et al. (Eds.): SPIoT 2021, LNDECT 98, pp. 697–705, 2022.
https://doi.org/10.1007/978-3-030-89511-2_90

data, and reduce information overload. Therefore, researching and improving commodity classification algorithms is a powerful boost for the better development of e-commerce and has certain practical significance.

Nowadays, popular machine learning classification methods include vector space algorithm, naive Bayes algorithm, neural network algorithm, etc. On the basis of traditional classification algorithms, scholars have made many improvements to better apply them to practical problems. Theoretically, compared with other classification algorithms, the Naive Bayes algorithm has the smallest error, but due to the characteristics of the Bayes formula and the assumption that each feature of Naive Bayes is independent of each other, the theoretical effect cannot be achieved in practice. Therefore, scholars have made improvements to Naive Bayes from all aspects in order to achieve the desired effect.

In the Bayesian formula, the phenomenon of probability 0 has an obvious influence on the Bayesian classification model. Therefore, scholars have proposed many smoothing methods, including Laplace smoothing parameters, M-based estimation smoothing parameters and so on. The smoothing parameters based on M estimation perform better on the multi-relational naive Bayes classifier when performing attribute filtering based on extended mutual information [9]. Many literatures can see that the performance of the naive Bayes classifier has been improved after the smoothing parameter is added, but the gap between the smoothing parameter and the classifier is not large, and compared with the other smoothing parameters, the Laplace smoothing parameter has The calculation is simple and easy to understand.

In the processing of the input word vector of the naive Bayes model, there are vector space model, TF-IDF, one-hot, word2vec, etc. The vector space model is the most common and mature text representation model, and different weights can be added to the features [1]. TF-IDF is often used to evaluate the importance of a word to a single document in a text library. The more the word appears in the text, the stronger the importance of the word, but as the frequency of other texts in the text database increases, the importance of the word gradually decreases [2]. The one-hot word vector is a vector of N*1. The dimension of N is the size of the vocabulary, which is generally 0. Only one dimension has a value of 1. This dimension represents the current word, which is easy to cause dimensional disasters [8]. Word2vec can quickly and effectively express a word into a vector form through an optimized training model according to a given corpus, combining the advantages of neural networks, and is a new tool for application research in the field of natural language processing[6, 7] [12].

In summary, this paper combines the advantages of neural networks and naive Bayes models in machine learning, and uses Laplace for smoothing, and proposes a naive Bayesian product classification hybrid model based on word2vec.

2   The Key Technology of Commodity Classification

2.1   Word2vec Word Vector Training Model

Word2vec is an algorithm model proposed by Google in 2013 to represent words as real-valued vectors, that is, word vector representation using Distributed Representation, which is characterized by simplicity and efficiency. Word2vec contains two training models, namely the CBOW model (Continuous Bag-of-Word Model) and the Skip-gram model (SG). As shown in Fig. 1, the former predicts the current word based on the context, and the latter predicts the current Word prediction context. In order to improve the efficiency of word vector training, simplify the calculation, and reduce the computational complexity of the output layer, word2vec also provides two optimization methods, namely Hierachy Softmax (HS) and Negative Sampling (NS). Thus, combining the training model and the optimization method can get four combinations, as shown in Table 1. In view of the fast training speed of the CBOW model and its suitability for big data processing, this paper selects the CBOW model and uses the Hierachy Softmax method for optimization.

Table 1. Word2vec combination method.

Model	CBOW	Skip-gram
Hierachy softmax	CBOW + HS	SG + HS
Negative sampling	CBOW + NS	SG + NS

As shown in Fig. 1, the CBOW model consists of an input layer, a projection layer and an output layer. Its working principle is to predict the probability of the current word by knowing the words ω_{t-1}, ω_{t-2}, ω_{t+1}, ω_{t+2} before and after the current word ω_t. The objective function uses the log-likelihood function commonly used in neural networks, as shown in formula (1).

$$L = \sum_{\omega \in c} \log p(\omega | Context(\omega))\ \ \ \ \ \ \ \ \ (1)$$

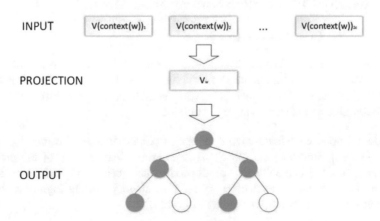

Fig. 1. The training framework of CBOW model based on Hierachy Softmax

2.2 Evaluation Indicators

After the product classification model is trained, it needs to be evaluated. Because it is a classification model, classification accuracy is used as the evaluation index. This article uses three indicators: accuracy, recall, and F1.

The accuracy rate, also known as the precision rate, refers to the ratio of the number of product categories $Classify(j, i)$ a correctly classified by the product classification model to the total number of all product categories $\sum_{j=1}^{n} Classify(j, i)$. The expression is shown in formula (2).

$$precision = \frac{Classify(i, i)}{\sum_{j=1}^{n} Classify(j, i)} \tag{2}$$

The recall rate is the ratio of the product text correctly classified by the product classification model $Classify[i][i]$ to the total number of documents actually belonging to the classified product $\sum_{j=1}^{n} Classify[i][j]$, as shown in formula (3).

$$recall = \frac{Classify[i][i]}{\sum_{j=1}^{n} Classify[i][j]} \tag{3}$$

The recall rate is the ratio of the product text correctly classified by the product classification model to the total number of documents actually belonging to the classified product, as shown in formula (3).

$$F_1 = \frac{2precision \times recall}{precision + recall} \tag{4}$$

3 Commodity Classification Optimization Model

3.1 Naive Bayes Classification Algorithm

The premise of the naive Bayes classification algorithm is to assume that all the calculated feature words are independent of each other, and its theoretical basis is the full probability formula of Bayes' theorem. After deduction, the probability of a certain product belonging to a certain category is obtained.

Definition 1 Suppose the feature word set of a certain product data information is $W_i = \{\omega_1, \omega_2, \cdots, \omega_n\}$, where ω_j $(1 \leq j \leq n)$ is a certain feature word of the product i. $C = \{c_1, c_2, \cdots, c_m\}$ is a collection of all product categories, and find the probability that the product i belongs to each category c_k. The category with the highest probability is the category of the product, as shown in formula (5).

$$P(ci \, max\{P(c_1|W_i), P(c_2|W_i), P(c_3|W_i), \cdots, P(c_m|W_i)\}_{max} \tag{5}$$

The probability of a product belonging to each category can be calculated by Bayesian formula, as shown in formula (6).

$$P(c_k|W_i) = \frac{P(W_i|c_k)}{P(W_i)} = \frac{P(\omega_1\omega_2\cdots\omega_n)P(c_k)}{P(W_i)} = \frac{P(\omega_1|c_k)P(\omega_2|c_k)\cdots P(\omega_n|c_k)P(c_k)}{P(W_i)} = \frac{P(c_k)\prod_{j=1}^{n}P(\omega_j|c_k)}{P(W_i)} \quad (6)$$

Among them, $P(W_i) = \sum_{k=1}^{m}P(W_i|c_k)P(c_k)$, and from formula (6) we can see that each category $P(W_i)$ is the same, so when comparing the probabilities, you only need to compare the numerators. The calculation method of the numerators is shown in formulas (7) and (8), Suppose that there are M numbers characteristic words contained in all commodity categories, and ω_l ($0 < l < M$) represents a characteristic word of a certain commodity category.

$$P(c_k) = \frac{\text{Number of feature words of commodity category } c_k}{\text{Total number of feature words in all product categories}} = \frac{Count(\omega_l, c_k)}{Count(\omega_l, C)} \quad (7)$$

$$P(\omega_l|c_k) = \frac{\text{The number of times the feature word } \omega_l \text{ appears in the product category } c_k}{\text{Total number of feature words in commodity category } c_k}$$
$$= \frac{Count(\omega_l, c_k)}{\sum_{l=1}^{M}Count(\omega_l, c_k)} \quad (8)$$

3.2 Smoothing Method

Since the premise of the naive Bayes classification model is that each feature item is independent of each other, it is often difficult to achieve in actual situations, so it will have some impact on the classification result, but the impact is limited and within an acceptable range. In addition, it can be seen from the Bayesian formula that if a certain conditional probability is 0, the error will be too large, so the naive Bayes classification model needs to be smoothed.

The Laplace smoothing method is an effective method to solve the problem of zero probability and overfitting by modifying the posterior probability, also known as the plus one smoothing method. It can be seen from Sect. 2.1 that the determinants of the posterior probability in the Naive Bayes formula are the prior probability $P(c_k)$ and the conditional probability $P(\omega_j|c_k)$, so the Laplace smoothing method is used to modify the conditional probability, as shown in formula (9).

$$P(\omega_l|c_k) = \log \frac{Count(\omega_l, c_k) + \lambda}{\sum_{l=1}^{M}Count(\omega_l, c_k) + T\lambda} \quad (9)$$

Among them, $\lambda \geq 0$, when it is equal to 0, it is the maximum likelihood estimation, and the value is generally 1, which is the number of all feature words contained in the product data information.

3.3 Commodity Classification Hybrid Model

The commodity classification hybrid model is divided into four steps.

Step 1: Data pre-processing. First, use the jieba word segmentation technology to pre-process the product data information. The jieba word segmentation system is a type of word segmentation technology based on knowledge understanding. The text of the product information is usually relatively streamlined and has a certain structure, with entries, keywords and features for each product category The descriptions are all short texts, so the word segmentation technology used in this article is the jieba word segmentation system for data pre-processing, removing stop words and symbols and other irrelevant content before text segmentation.

Step 2: word2vec word vector training. Input the processed product text information into the word2vec model for word vector training, and each word will get a 100-dimensional word vector.

Step 3: Naive Bayesian classification model based on Laplace smoothing. Input the trained word vector into the Naive Bayes classification model to obtain the final classification result, and then calculate the model evaluation index for comparative analysis.

4   Experiment and Analysis

4.1 The Choice of Corpus

This article first uses the Houyi collector to crawl the merchandise category information of Jingdong Mall, and selects two categories of women clothing and luggage leather goods as the experimental data set. The women clothing category grabs 1800 product information from each bottom category, and the luggage leather goods category From each of the bottom-level categories, we grabbed 3,200 product information, grabbing a total of 115,200 product-related information, and the training set and test set were randomly divided by 1:9. In the middle-level classification, two types of products, namely, high-quality men bags and trendy women bags, are selected for the second model verification. There are a total of 38,400 product information, and the training set and the test set are still randomly divided by 1:9. In the bottom layer classification, each selects a total of 6,400 backpacks and messenger bags in trendy women bags, and randomly divides the training set and the test set according to 1:9.

Data pre-processing This article selects the jieba word segmentation system, and all the product information obtained is cleaned first, and only Chinese and English are retained, and symbols and stop words are removed. Then customize the product subcategory into the jieba word segmentation thesaurus, so that the Chinese word segmentation through the jieba word segmentation system is more accurate. Then generate a Chinese corpus, and then enter the word2vec bag-of-words model as an input file for Chinese word vector training, and obtain each word vector of 100 dimensions. Finally, the Naive Bayesian commodity classification model is smoothed by Laplace smoothing method, and the smoothing parameters are directly taken.

4.2   Experiment and Result Analysis

4.2.1   Word2vec Parameter Selection

This article runs the python program on the Anaconda software platform to train and test the classification model. The word2vec model in Python has many parameters. Choosing the CBOW model and using the Hierachy Softmax method for optimization can be determined by the choice of parameters. The word2vec used in this article The parameters are shown in Table 2. The specific parameters are set as $sg = 0$, $hs = 1$, $size = 100$, $window = 5$, $min\_count = 2$, $workers = 4$, $alpha = 0.025$.

Table 2.   word2vec training parameters.

Parameters	Description
sg	Whether to use the Skip-gram model, when sg = 1, use the CBOW model when sg = 0
hs	Whether to use hierarchical softmax method, when hs = 1, use NS when hs = 0
Window	Set the maximum distance between the current word and the predicted word in a sentence
min_count	After setting, words whose word frequency is less than min_count will be discarded. The default value is 5
works	The parameter controls the number of parallel training
Size	Set the dimension of the feature vector, the default is 100
alpha	Learning rate

4.2.2   Comparative Analysis of Bayesian Model Performance

In this experiment, we will train the naive Bayesian product classification model based on the word2vec model, the naive Bayes classification model based on TF-IDF, and the naive Bayes classification model based on one-hot. After the training, input the test set and output Classification results, calculate accuracy, recall and F1 value for performance evaluation.

Finally, the naive Bayes product classification model based on the word2vec model, the naive Bayes classification model based on TF-IDF, and the naive Bayes classification model based on one-hot are tested and evaluated in the upper, middle, and bottom categories. The results are shown in Fig. 2.

It can be seen intuitively and clearly through the histogram that when the upper category is classified, the accuracy of the two products is relatively obvious, so the accuracy is relatively high, and the difference is weakened. Compared with the other two models, the naive Bayesian product classification model based on word2vec has improved in all aspects of evaluation indicators. In the middle-level category and the bottom-level category, the feature similarity is higher, and the advantages of this model are more prominent, which is significantly higher than the other two models.

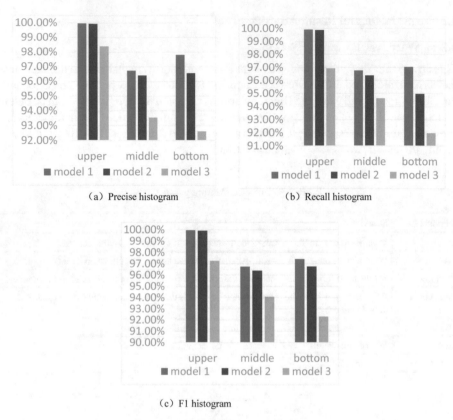

(a) Precise histogram (b) Recall histogram

(c) F1 histogram

Fig. 2. The evaluation result histogram

5 Conclusion

This paper starts from the actual problems of commodity classification on actual e-commerce websites, proposes a naive Bayesian hybrid commodity classification model based on word2vec, and uses actual data to verify its feasibility. Among them, the word bag model of word2vec is used to optimize the input vector of the naive Bayes classification model, and the Laplace smoothing method is used for smoothing on the naive Bayes classification model to improve the performance of the traditional naive Bayes classification model. performance. Finally, the actual product classification data on the e-commerce platform of Jingdong Mall is used to evaluate the performance of the model, which verifies the feasibility and effectiveness of the hybrid product classification model.

References

1. Mo, S.Y., Cui, X.Y.: The Research of Commodity Classification Algorithm Based on Double-layer Bayesian Model. Chinese Science paper Online, Beijing. http://www.paper.edu.cn/releas epaper/content/201512-584

2. Zhang, B., Huang, X.F.: Convolutional neural network for news text classification optimization based on TF-IDF. J. Southwest Univ. Sci. Technol. **35**(01), 64–69 (2020)
3. Dilmen, E., Beyhan, S.: Novel online LS-SVM approach for regression and classification. IFAC-PapersOnline **2017**(50–1), 8642–8647 (2017)
4. Xi, Y., Mengxuan, T.: An improved Naive Bayesian classification model based on attribute weighting. J. Stat. Mech. Theory Exp. **1550**(2), 022017–18 (2020)
5. Cheng, Y.K., Jiang, Y., Cheng, G.: Subject of website classification based on word2vec. Comput. Digital Eng. **47**(01), 169–173 (2019)
6. Jia, R., Zhou, L.: Cleaning off the English obstacles in literature retrieval course in higher vocational college. Tech. Info. Dev. **25**(02), 145–148 (2015)
7. Xue, J.C., Jiang, D., Wu, J.D.: Research on automatic patent text classification based on word2vec. Comput. Technol. **44**(02), 73–77 (2020)
8. Lai, W.H.: Research on Chinese Text Classification Based on Deep Learning Theory, pp. 9–29. South China University of Technology (2019)
9. Xu, G.M., et al.: Improving multi-relational naive bayesian classifier using smoothing methods. Comput. Eng. Appl. **53**(5), 69–72 (2017)

Multi Evaluation System of Quality with Cloud Computing

Jianrong Li[✉], Xiaoyu Wang, and Xiaowei Feng

Heihe University, Heihe 164300, Heilongjiang, China

Abstract. The popularization, informatization and networking of higher education put forward new requirements for College English teaching. The multiple evaluation system of College English teaching quality refers to a series of methods that use a variety of effective technical means and evaluation methods to evaluate students' learning status and learning effect, teachers' teaching process and teaching effect in unstructured teaching situation. This is not only the change of teaching mode, but also the change of evaluation concept, evaluation method and evaluation implementation process. At present, an outstanding problem in College English curriculum is the unreasonable, imperfect and unscientific rating system. Therefore, this paper will apply the multi intelligence theory in teaching practice based on cloud computing, which provides a new perspective for the reform of education and teaching. Practice has proved that the construction of the evaluation system based on the theory of multiple intelligences has a positive practical significance to improve the teaching effect and students' ability in all aspects.

Keywords: Cloud computing · College English course · Multi evaluation system of teaching quality · Reform in education

1 Introduction

The promulgation of the new college English teaching requirements marks the beginning of a new round of College English teaching reform. While emphasizing the transformation of teaching methods and activities, it also points out the shortcomings of the existing college English teaching evaluation system, which puts the scientific teaching evaluation at an unprecedented height and calls for the transformation from test teaching to evaluation teaching. In the actual teaching, it has become a difficult point in the current education reform to emphasize both educational equity and educational efficiency. From the perspective of education evaluation, Multiple Intelligences Evaluation provides us with a feasible entry point, that is, to really improve students' abilities in all aspects as the fundamental goal of education and training, and to adapt to the new requirements of social development for teaching evaluation. The evaluation view based on multiple intelligences theory provides a new idea for teaching evaluation.

The reform of evaluation methods and evaluation system is an important part of the whole college English teaching reform «College English teaching requirements (Department of higher education [1], Ministry of education, 2007:7–8) emphasizes the reform

© The Author(s), under exclusive license to Springer Nature Switzerland AG 2022
J. Macintyre et al. (Eds.): SPIoT 2021, LNDECT 98, pp. 706–711, 2022.
https://doi.org/10.1007/978-3-030-89511-2_91

of curriculum, teaching mode, teaching content and teaching means, and points out that "teaching evaluation is an important part of College English teaching, which is an important basis for teachers to obtain teaching feedback, improve teaching management and ensure teaching quality, It is also an effective means for students to adjust learning strategies, improve learning methods and improve learning efficiency. However, the traditional teaching evaluation method is mainly summative evaluation, which focuses on Teachers' evaluation of students and ignores students' self-evaluation and mutual evaluation; Emphasis on students' evaluation of teaching, ignoring other subjects' evaluation of teaching; Emphasis on teaching evaluation, ignoring teaching management evaluation research. The network environment provides a very favorable condition for the construction of multiple intelligence evaluation model.

2 Multiple Intelligences Evaluation View

In 1983, Howard Gardner, a famous psychologist of Harvard University, put forward a new theory of human intelligence structure multi – intelligences after years of research. He thinks that everyone has at least eight kinds of intelligence in his mind, and everyone has his own superior intelligence. These intelligences have unique and diverse forms of expression. Only by adopting multiple evaluation standards can we truly and accurately reflect the intelligence level of students, provide opportunities for each student to show his or her own style, and let each student experience the joy of success. At present, there is no definite definition of multiple evaluation. Some scholars believe that "multiple evaluation is based on the values of innovative education, with the fundamental task of developing the potential of students and teachers and promoting the innovative quality of students and teachers, and the essence of encouraging students and teachers to constantly surpass themselves. Education evaluation marked by multiple goals, multiple standards and diverse methods. Generally speaking, the main content of teaching evaluation is to measure quantitatively and qualitatively the performance of teachers in the whole teaching process, the accuracy of teaching content, teaching atmosphere, teacher-student interaction, etc., and finally point out the problems existing in classroom teaching, so as to help teachers effectively improve teaching strategies. As an important part of teaching reform, teaching evaluation can show the teaching effect in a specific way, so as to help teachers understand the students' learning situation in the first time, find out the main problems existing in the learning process, and provide corresponding guidance for the improvement of teachers' teaching strategies, so as to ensure the effective allocation of teaching resources and the absolute improvement of teaching quality.

3 Construction of Multiple Evaluation System Under Cloud Computing

There is no doubt that teaching evaluation can promote teaching. Based on the theory of multi – intelligences, we set up a multi – Evaluation System of College English under the network environment to measure the teaching effect comprehensively and effectively step by step. It mainly includes: formative evaluation of students' learning process, hierarchical evaluation according to students' intelligence level, sub project evaluation of students' listening, speaking, reading, writing and translation abilities, and developmental evaluation of students' language progress. One of the following is shown in Fig. 1, with three evaluations.

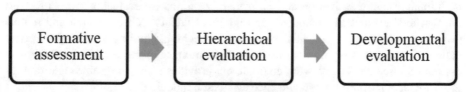

Fig. 1. Multiple evaluation system

1. formative evaluation.

In the content of evaluation, teachers first identify the students' intelligence advantages through questionnaire and other ways, and open up to the level of M 6 martial arts in the field. In teaching, interpersonal father orientation, eye-seeking cognition and practical intelligence are integrated into English classroom teaching. The task of teaching experience of Ling Sha is in the process of intelligent teaching and is based on the students' knowledge of various tasks [2].

The completion of the students' learning process is comprehensively evaluated. The pre class preview, classroom participation, task completion after class, online self-study records and performance in various teaching activities are included in the evaluation scope, and their learning process is recorded and feedback is made. In teaching, we should pay attention to the students' advantage intelligence and encourage them appropriately, help them to establish their self-confidence in autonomous learning, and thus generate the motivation of language learning. In the evaluation subject, the evaluation of the ten thousand journal changes the teacher as the main role of the evaluator, the right of evaluation will be transferred to the students' own hands, so as to give full play to the main role of students, reduce the pressure on the teaching evaluation, make students become the evaluation participants and introspectives. In the evaluation way, the method of combining student self-evaluation, mutual evaluation and teacher evaluation will be implemented to enhance the enthusiasm of students' evaluation and promote learning by evaluation.

2. evaluation by levels.

Teachers group students according to their actual English level, carry out layered teaching in classroom teaching, and arrange different tasks and assignments for different levels of students online. The actual level of language intelligence is fully considered, and the students are taught according to their aptitude. The students of different levels are given different difficult experiential teaching tasks. After teaching, the evaluation paper of corresponding difficulty is designed according to the intelligence level of each group of students.

3. developmental evaluation.

The paper collates the results of the process test that students have obtained in the course of English learning in each semester, and compares the final results of each semester and their achievements in CET-4 and CET6, and evaluates the progress of language learning in different learning stages.

4 Construction of Multiple Evaluation System for College English Curriculum

The multiple evaluation system of College English teaching quality refers to a series of methods that use a variety of effective technical means and evaluation methods to evaluate students' learning status and learning effectiveness, teachers' teaching process and teaching effectiveness in unstructured teaching situation. Its essence is to evaluate students and teachers comprehensively and truly, so as to provide feedback information for improving teaching, So as to promote the all-round development of students and teachers. The multiple evaluation system of College English teaching quality of Chongqing University of science and technology takes the development of students and teachers as the fundamental goal, takes the teaching process as the main line, and takes "teaching evaluation" and "learning evaluation" as the dual core to build a new college English teaching quality evaluation system. It is based on modern quality management theory, constructivism theory and multiple intelligence theory. The evaluation of teachers is usually at the end of the semester. After a certain period of class and face-to-face communication with teachers, students can have a more comprehensive and objective evaluation of teachers. Finally, the evaluation results and statistical analysis of all parties are fed back to teachers and school leaders for reference (as shown in Fig. 2).

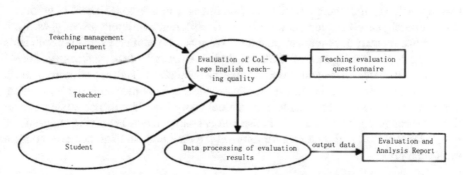

Fig. 2. Data flow chart of multiple evaluation of College English teaching quality

The new multiple evaluation system of College English teaching quality has many advantages. First, the indicators of multiple evaluation not only cover the "overall orientation of the curriculum (Teaching Philosophy), teaching content, teaching methods, teachers' basic literacy, teaching efficiency" (Liu Limei, 2011), but also highlight the student-centered evaluation. Second, the results of students' evaluation, teaching management department (experts/leaders/supervision group) evaluation and peer evaluation provide important reference and basis for teachers' Reflection on teaching behavior, adjustment of teaching strategies, improvement of teaching methods, optimization of teaching content, adjustment of teaching schedule and selection of teaching means. Thirdly, diversified evaluation standards and methods, which distinguish young teachers from middle-aged and old teachers, change the traditional disadvantages of using the same standard to evaluate all teachers. It is conducive to mobilizing the enthusiasm of young teachers, realizing the fairness of teaching evaluation, and promoting the growth and personality development of young teachers [3]. Diversified evaluation standards and methods, different evaluation standards for students with different learning objectives (– General requirements, higher requirements and higher requirements), truly realize the people-oriented evaluation, so that students with different English levels can experience English, experience happiness and experience success. Fourthly, the diversified classroom teaching evaluation system pays more attention to students' monitoring and evaluation of language learning process, which is conducive to students' correct learning attitude, confidence, interest and motivation, and the cultivation of students' personalized learning and autonomous learning ability. Fourthly, the diversified curriculum teaching quality evaluation system pays attention to students' classroom performance and group activities, It is conducive to the cultivation of students' team consciousness, collective concept, cooperation ability, innovative spirit and inquiry ability.

5   Significance of Building Multiple Evaluation System in Cloud Computing Environment

1. Use the incentive mechanism of multiple evaluation system to fully mobilize the enthusiasm of students.

Gardner believes that there are multiple intelligences in everyone. There is no difference in intelligence between students, only the difference in intelligence types. Each student has its own development potential. We make diversified evaluation standards according to the intelligence of students' different strengths and weaknesses. Such a standard will enable all students to experience the happiness of success, so as to establish self-confidence. At the same time, according to the incentive mechanism of evaluation, teachers always use an acceptable and non defensive tone to enhance students' confidence through positive feedback, praise and encouragement; By giving suggestions, the students realize that they need to improve the direction, so as to achieve the ideal evaluation effect.

2. The content of teaching evaluation is more substantial, which greatly develops students' personality.

The theory of multiple intelligences tells us that each kind of intelligence plays a unique role in our daily life and work. We attach importance to the influence of multiple intelligences and put it into the evaluation system of students. This way of evaluation is a good supplement to the traditional evaluation system, so that the evaluation of students will be more comprehensive. The evaluation of students is only based on their understanding and application of basic concepts and knowledge, which is not comprehensive.

We should also pay attention to their ability, attitude, emotion and sense of value in communication, competition and cooperation.

3. Attach importance to the role of self-evaluation and mutual evaluation to form a multi evaluation subject.

The theory of multiple intelligences emphasizes people-oriented, the effectiveness of evaluation and the promotion of development. Guided by the theory of multiple intelligences [4], teachers and students can understand and trust each other. On this basis, we can form a multiple evaluation system with teachers and students as the main body. At the same time, we combine students' self-evaluation with group members' mutual evaluation. In this way, the interaction between all kinds of evaluation subjects is enhanced, and the evaluation information sources are richer and the evaluation results are more authentic, It also helps to promote the development of students' cooperative ability.

6   Conclusion

Evaluation system is an indispensable part of College English teaching. As a new evaluation model, the multi-element evaluation system in the network environment shows more and more advantages and attractiveness. Classroom teaching is still the main link of teaching, and the quality of teaching directly affects the quality of talent training in Colleges and universities. Therefore, the evaluation of classroom teaching quality plays a very important role in the cultivation of talents and the development of colleges and universities. However, the selection of evaluation object, evaluation method and evaluation standard directly affects the construction of evaluation system. But it is not a matter of a day to build a perfect evaluation system of College English curriculum. It will be improved with the development of English teaching and the renewal of teachers' ideas, so that it can truly achieve "promoting teaching by evaluation and learning by evaluation".

References

1. Yang, Z., et al.: 2000 New Edition of ISO 9000 International Standard Terminology Manual, p. 7. National Defense Industry Press, Beijing (2004)
2. Liu, L.: New discussion on the classroom teaching evaluation system of newly established local universities. J. Hebei Normal Univ. (Educ. Sci. Ed.) **10**, 79 (2011)
3. Zhang, L.: Research on the Formation Mechanism, Effective Teaching Evaluation and Quality Management System of Higher Education. Nanjing University of Technology (2008)
4. Zhao, Y.: New development of learning theory and Its enlightenment to modern foreign language teaching. China Adult Educ. **3**, 171 (2007)

Evolutionary Game Analysis of Environmental Accounting Information Disclosure Based on SD Model

Xiaoping Chen and Ying Song[✉]

Liuzhou Institute of Technology, Liuzhou 545000, Guangxi, China

Abstract. The worsening of social environment is a major problem facing the society today. The disclosure of environmental accounting information by listed companies can show the use of resources and pollution control in the production process to the society, which is helpful to the governance of outstanding environmental problems. Through the evolutionary game analysis between listed companies and regulators, SD model is established by system dynamics to simulate the game. It can directly see the evolution state of game between the two sides, and put forward suggestions from three aspects: reducing the harm to society, increasing the punishment of illegal enterprises and reducing the supervision cost, so as to promote the sustainable development of ecological environment.

Keywords: Environmental accounting · Information is exposed · Government supervision · Evolutionary game analysis

1 Introduction

In the early 1990s, China began to introduce the theory of environmental accounting. The convening of the 19th National Congress has made the environmental accounting issue a hot spot again, and has been paid attention by the whole society. The final result of environmental accounting work of listed companies is to disclose environmental accounting information to the public, which is the most important part of the environmental accounting system. Nowadays, the ecological environment is deteriorating day by day. The disclosure of environmental accounting information by listed companies has great help to the severe environmental problems.

The listed companies can not only timely grasp their own environmental related behaviors, but also objectively and accurately account for their business performance. At the same time, it is beneficial to the fundamental transformation of the business model to strengthen the internal management of the company and establish a good and responsible corporate image. In this process, listed companies can timely understand their environmental responsibility and whether they are in place, strengthen the environmental pollution control, and make rational investment decisions according to the relevant situation. However, the disclosure of environmental accounting information by listed companies is costly. Most companies are reluctant to disclose. Only when

J. Macintyre et al. (Eds.): SPIoT 2021, LNDECT 98, pp. 712–718, 2022.
https://doi.org/10.1007/978-3-030-89511-2_93

the government attaches importance to the work of environmental accounting information disclosure, can the supervision be strengthened, the attention of the society will be increased, and the listed companies can bear their environmental protection responsibilities and Disclosure Responsibilities better under the pressure of public opinion. Therefore, as one of the influencing factors, regulators have a great influence on whether the listed companies disclose environmental accounting information.

In game theory, the participants adjust their own countermeasures in the equal situation based on the existing information according to the changes of the other party's strategies to obtain the optimal strategy. From this point of view, environmental accounting information disclosure is a game, and the two main bodies of the game are regulators and listed companies, and there is a game relationship between the two parties. Through evolutionary game, SD model is established by system dynamics to simulate, which reflects the dynamic evolution process of game. It can study the relationship between environmental accounting information disclosure cost, violation cost and government supervision cost more objectively and clearly, analyze the dynamic strategy change and trend between them, and propose suggestions for both parties to maintain sustainable development together.

2 Research Review

2.1 The Status of Disclosure of Enterprise Environmental Accounting Information

Environmental problems are related to the vital interests of citizens, and the environmental information disclosure of enterprises in China is facing a series of constraints. Compared with other enterprises' information disclosure, the disclosure of environmental accounting information in China started late and developed slowly. The reasons include the imperfection of relevant laws and regulations and accounting system in environmental accounting, and the failure of the environmental accounting system to get a good growth in China. The current situation is that the legal and legal system of the enterprise environmental accounting report is lack, the public has not exercised the supervision power of environmental information, the power of enterprise resources to provide environmental accounting report is insufficient, and the current disclosure mode of environmental accounting information is not perfect [1]. The specific problems include the lack of accounting, which causes the difficulty of information sources, and the pollutant emission and environmental policies can only be described in narrative form; At present, China has not released the guidance of environmental information disclosure, which makes the decision-making of enterprise information disclosure less relevant; The government can not keep up with the environment situation, the management efficiency is low, and the enterprises are exposed to the negative treatment of environmental information due to the low efficiency of environmental information utilization; In the process of operation, information is scattered, exposed less and the proportion of quantification is less, there is no unified form, and the comparability is poor! 31; The scope of disclosure is inconsistent, formalized and obvious industry.

In recent years, scholars in relevant fields of China have done a lot of research on the methods, forms and contents of environmental accounting information disclosure.

The report of environmental accounting information disclosure has gradually become a prototype, but the enterprise is not motivated and negatively cooperated, which makes it difficult to be further improved in practice, and the enterprise environmental accounting information disclosure system is difficult to form.

2.2 Incentive Measures for Disclosure of Enterprise Environmental Accounting Information

There are two ways to encourage the disclosure of environmental accounting information. First, although the Disclosure Measures for plutonium state information (Trial) have some requirements, there are still no standardized and systematic measures such as accounting, environmental disclosure contents, which can be classified as top-down path; Second, it can be classified as the bottom-up path by encouraging enterprises with an effective mechanism and means to make them actively and actively disclose environmental accounting information. It can be divided into short-term benefits and long-term benefits in terms of time: short term cost reduction and income increase, including tax preference, special awards, patent granting and so on for environmental accounting information disclosure of enterprises, increasing non operating income and reducing taxes payable; In terms of long-term cost reduction and income increase, the main reason lies in the active implementation of social responsibility and credit. The affirmation of the brand and reputation of the society will improve the reputation of the enterprise and increase the intangible assets.

Through the improvement of relevant mechanism, enterprises can gradually disclose environmental information, and make the economy move forward in the direction of sustainable development [2]. Among the factors that encourage the disclosure of environmental accounting information, the government policy and the degree of attention are positively related to the disclosure of environmental accounting information and the increase of market value of enterprises, and then further encourage the disclosure of environmental accounting information. The disclosure level of environmental accounting information in different industries is different, and the disclosure level of environmental accounting information in heavy pollution industries is relatively high. However, in fact, the incentive results of environmental accounting information disclosure in China are not optimistic. Some listed companies hold negative attitude in the aspect of environmental information disclosure, and the quality of environmental information is not high, incomplete and comparable, and the data is rarely quantitative. Even if the environmental accounting information is disclosed actively, it is only limited to actively disclose good news which is conducive to the corporate image", which is only distinguished from other enterprises with punitive and reported bad news by the media. There are not a few enterprises that are negative to the disclosure of environmental accounting information or only to disclose their own content. Enterprises worry about the negative effect of environmental accounting information on the image of enterprises. It shows that positive reporting on environmental information can really shape a good image of enterprises. From the perspective of the public, the enterprises that pay attention to environmental protection embody social responsibility, which is more popular with consumers. The enterprise undertaking environmental protection responsibility is one of the important embodiment of the social responsibility of the enterprise. The fulfill-ment of social responsibility by enterprises can help reduce transaction costs, improve economic performance, bring indirect economic benefits and form a social benefit chain.

3 Theoretical Analysis and Research Hypothesis

(1) Theoretical analysis

Evolutionary game theory is based on the limited rationality, taking participants as the research object, and based on the basic concept of evolutionary stability strategy, emphasizes the dynamic process of game, and explores which equilibrium state to reach in various equilibrium under different initial conditions and evolution paths. Because of the uncertainty and risk in the strategic emerging industrial clusters, the innovation behavior of cluster subject is a slow evolution process. Therefore, it can be analyzed by the dynamic mechanism model of biological evolution. In the process of specific innovation game, each member chooses to adjust their own strategy according to their relative adaptability in the group. The cluster subject with limited information adjusts its strategy continuously on the margin according to the vested interests, in order to pursue the improvement of its own interests and finally achieve dynamic balance [3].

(2) Research hypothesis

There are peer and homogeneous imitation behaviors in the disclosure of enterprise environmental information. The enterprises in the same industry learn from each other to imitate the behavior of compiling and disclosing environmental accounting reports, which provides an important premise and basis for the evolution of game in this paper.

Under the premise of cost-benefit assumption, this paper analyzes the conditions that cost-benefit needs to meet, and finally gets the variables that cost income meets, which can make the dynamic balance point of the enterprise approach the new report which can improve the disclosure of enterprise environmental accounting information. The paper focuses on the role of media publicity and communication on the cost and benefit in the process of enterprise evolution.

Hypothesis 1: in the market economy environment, fair competition, enterprises pursue the maximum profit. This is the basic requirement that the evolutionary game of enterprise environmental accounting information disclosure can carry out. Because the establishment of this model mainly considers the important role of media, in order to eliminate interference, the market environment is regarded as the control variable, and the fair market competition environment is taken as the background of the game model. The cases and special cases under unfair competition are not the scope of this model. At the same time, the essence of enterprises in market economy is to maximize the pursuit of interests, balance and consider more cost-saving costs and bring more income and profit to the production and operation.

Hypothesis 2: symmetrical game between a and B enterprises with two scales and equivalent strength in the same industry. Because there are imitation behaviors between enterprises in the same industry, the paper makes a conclusion that the evolutionary game must have the universal imitation behavior, and this paper carries out the conclusion that the universal imitation behavior exists, so the scope of this model is only controlled between the enterprises in the same industry.

Hypothesis 3: both enterprises adopt the new report and the old report. The state rewards the enterprises that use the new report of environmental accounting information disclosure. The new report is the environmental accounting information disclosure report prepared by the authoritative accounting institution or the third-party enterprise of the industry. The content is more complete and easy to be supervised by the public. The countries using the report will give certain rewards, but the enterprises using the report will need to improve the technology and pay a certain cost; The old report is the report used by Party A and enterprise itself. When using, the existing technology is used to collect the environmental information of the enterprise, which is low cost (assumed as 0 here), but the state will not give rewards.

Hypothesis 4: different strategic returns are selected, and when both parties choose new reports, the income is R; When one party adopts the new report and the other party adopts the old report, the party adopting the new report will get the income of 2R, and the income of the former report party is 0; When both parties adopt the old statements, the income is 0. The total cost of equipment technology improvement for the enterprises adopting new report is C in order to collect the environmental information of the enterprise and to make the report content meet the environmental protection standards [4].

The possibility of adopting new statements is p when the two enterprises disclose, and the possibility of using old statements is $(L-P)$. Under the symmetrical model, the risk cost and benefit faced by both enterprises are symmetrical, and the possibility faced by the two enterprises is symmetrical.

The details are as follows (Table 1):

Table 1. An analysis of symmetric game between a and B enterprises

		Enterprise B	
		New report (P)	Old report $(1-p)$
Enterprise a	New report (P)	$r-e$, $r-\in$	$2r-e$, 0
	Old report $(1-p)$	0, $2r-e$	0,0

4   Model Setting and Analysis Description

This model is a symmetric game model. Choosing either Party A or Party B will not affect the results. Therefore, we may as well choose the game strategy choice and income of enterprise a to analyze the results.

On the basis of the above variable design and assumptions of evolutionary game, we get the expected return of enterprise a using the new statements for disclosure π_N is:

$$\pi_N = (r-e)p + (2r-e)(1-p) \tag{1}$$

Income of enterprise a's environmental accounting information disclosure by using the old statements π_0 is:

$$\pi_0 = 0 \tag{2}$$

Therefore, the weighted average expected return is obtained by taking into account the possibility P factor of different strategies adopted by enterprise a π For:

$$\pi = p\pi_N + (1-p)\pi_0 \tag{3}$$

According to the expected income of the new statement adopted by enterprise a π_N. Expected income from using old statements π_0, and weighted average expected return π, The replication dynamic equation is obtained

$$p = p(\pi_N - \pi) = p(1-p)(\pi_N - \pi_0) \tag{4}$$

The fixed point is obtained by copying the dynamic equation above

$$0, 1, \frac{2r-c}{r} \tag{5}$$

When r > c, $\frac{2r-c}{r}$ > 1, the dynamic equation is always greater than 0. The results of evolutionary game simulation in the initial state are shown in Fig. 1. The probabilities of both sides tend to be stable after varying degrees of floating. Because this model is a symmetric model, the expected return, weighted average expected return and evolutionary game conclusion of the new and old statements faced by enterprise B are consistent with that of enterprise a.

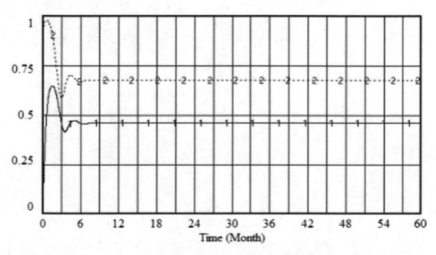

Fig. 1. Game evolution process of mixed strategies under dynamic penalty strategy

5   Conclusion

Environmental issues are the whole human society, should be concerned by all aspects, listed companies and regulators should cooperate with each other to stop the rapid deterioration of the ecological environment, while pursuing the maximization of economic benefits, and jointly undertake the responsibility of environmental protection, and maintain the sustainable development of ecology.

References

1. Ting, W.: Problems and Countermeasures of Environmental Accounting Report in China. Jiangxi University of Finance and economics, Nanchang (2010)
2. Shixing, Z.: Research on Enterprise Environmental Information Disclosure Based on environmental performance evaluation. Ocean University of China, Qingdao (2009)
3. Li, X., Li, C., Yang, L.: Empirical study on disclosure of enterprise environmental information. Traffic Account. 9 (2003)
4. Wu, D., Tang, G.: Research on environmental accounting and corporate social responsibility. Account. Res. 1 (2012)

Reliability Evaluation Algorithm of Power Communication Network Based on Service Risk Balance Degree

Yinian Gao[✉] and Guoying Liu

Shenzhen Power Supply Co, Ltd, Shenzhen 518001, Guangdong, China

Abstract. Based on the service characteristics of intelligent grid and the typical structure of power communication network, the service of power communication network is analyzed, the concept of service importance is proposed, and the measurement of service importance is determined. The concepts of risk degree of channel segment service and risk balance of whole network service are proposed, and the reliability evaluation benchmark, evaluation model and solution of power communication network based on risk balance of whole network service are established. It can provide reference for the implementation of power communication network reliability assessment, optimization of business network, scientific arrangement of business channel and operation mode of communication network from the business level.

Keywords: Electric power communication network · Reliability · Communication business: risk balance

1 Introduction

As the construction of smart grid has accelerated over the past few years, the development of computer technology and communications technology has become more and more rapid for the development of telecommunication networks. The safety stable control system and the management automation system constitute three columns of the safe and stable operation of the modern smart grid as a special network of the power system. The power system consists of five parts: power generation, substation, power transmission, distribution, and power. All systems are operated under higher automated conditions. The concept of a power communication network has been submitted to ensure secure and stable operation of each module. The power grid is the basis of modern smart grids and power systems. The important way of transmission is the three modernization of the power system (the operation of the power network, the management automation of the power network, and the modernization of the power network management, the basis of the modernization of the power network is the construction of the smart grid, and the power system operation increasingly depends on the power network.) The fault of the power grid becomes more and more affected by the power system, and the reliability of the power network becomes higher.

© The Author(s), under exclusive license to Springer Nature Switzerland AG 2022
J. Macintyre et al. (Eds.): SPIoT 2021, LNDECT 98, pp. 719–724, 2022.
https://doi.org/10.1007/978-3-030-89511-2_94

The analysis of electric power communication business shows that the importance of business is based on the balance of business risk and the reliability criteria of the telecommunication network business, and forms the model and method. The risk criterion and risk evaluation problem of all network power communication tasks were solved.

2   Establishment of Business Importance Evaluation System for Power Communication Network

2.1   Business Overview of Power Communication Network

In such an environment, in order to adapt to the development speed of power system, power system private network has been produced. As a special communication network of power system, together with security and stability control system and dispatching automation system, modern intelligent grid has three pillars of security, stability and economic operation. The formation of power network has two reasons. On the other hand, the public network is complex and lack of security, which can not meet the communication needs of characteristic services of power system. Compared with private network, public network is developing slowly and can not meet the basic communication needs. With the development of intelligent grid, its role in the power system of power network is improving.

2.2   Business Types

According to the business attributes, The service types of power communication network can be divided into the following categories. 1) voice business, including administrative telephone business and dispatching telephone business; 2) Data services, including dispatching data network services, integrated data network services, special line data services, enterprise information services, etc.; 3) Video business, video conference business, video monitoring business, consultation video business, etc.

According to the service object, It can be divided into the following types. 1) production and scheduling services, including protection services, automation services, telecontrol services, security and stability control services, scheduling telephone, scheduling data network services, etc.; 2) Enterprise business, including marketing service business (95588), user electricity information collection business, etc.; 3) Enterprise management and information business, including data disaster recovery center business, sg-erp business and so on [2].

2.3   Establishment of Evaluation Index System

This paper uses consistency based fuzzy analytic hierarchy process to establish and analyze such problems. Similar to AHP, FAHP is a top-down analysis method. According to the above analysis, firstly, the index is hierarchical. The complex multi-objective decision-making problem is divided into three parts: objective layer, middle layer and scheme layer. The first layer is the target layer, which is the final problem to be solved. The target layer has only one element. The second layer is the middle layer, which is also called the criterion layer. According to the complexity of the problem, the criterion layer can be divided in detail according to the sub criteria under each criterion. The third layer is the scheme layer, which is an alternative to achieve the goal. The main characteristics of this method are: from top to bottom, except for the first level, other levels are directly related to the upper level. Generally, there is no dominant relationship between elements of the same level. In the whole hierarchy, the number of layers is not limited. The calculation process starts from the second layer and calculates the weight value of each layer relative to the upper layer. Then the relative weight of each sub criteria layer to the target layer is calculated by the comprehensive evaluation method. According to the different "requirements" of each sub criteria layer, the total weight of each alternative after weighting is calculated by 1 ~ 9 scale method, and the higher the weight is the optimal solution of the multi-objective decision problem [3].

According to the factors involved in completing a task in a specific environment, a three-level in Fig. 1.

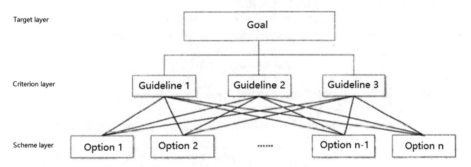

Fig. 1. Evaluation index system

3 Measurement Index and Algorithm of Business Risk Balance Evaluation

3.1 Business Risk

Business risk refers to the measurement that affects the safe and stable operation of power network when the business is completely interrupted or the business reliability is reduced. For a single business on the channel of power communication network, the risk level is as follows.

$$R(E_S(i,j)_{s_i}) = I_{s_i}(l - P_{E_S(i,j)}) \tag{1}$$

Where: $R(E_S(i,j)_{s_i})$ represents the service risk value of the s_i service carried by the $E_S(i,j)$ side in the network; I_{s_i} represents the importance of the ith service on the $E_S(i,j)$ side of the network. $P_{E_S(i,j)}$ denotes the reliability value of the $E_S(i,j)$ th side carrying the i-th service in the network.

The comprehensive service risk degree of channel segment refers to the comprehensive value of the risk degree of various services carried by a certain channel segment in the network, that is, the corresponding edge in the network.

3.2 Business Risk Balance

1) The whole network business risk degree.

The whole network service risk degree $R(G_s)$ is defined as the comprehensive measurement of the whole network service risk of network $G_s = (N_s, E_s, S)$. Considering the bi-directional nature of the service, $R_{Es\,(i,j)}$ and $R_{Es\,(j,i)}$ represent the service risk value of the same network side, then $R(G_s)$ can be expressed as

$$R(G_s) = \left[\sum_{i,j} R(E_s(i,j)) \right] / 2 \tag{2}$$

There are two steps to solve the whole network business risk: firstly, sum up the risk value of the business carried by the $Es\,(i,j)$ side of the network; Secondly, the sum of all the risk values of all side services belongs to business network G_s.

2) Average business risk of the whole network.

The average risk of network traffic across the network is based on the degree of risk across the network and the variables in the network topology.

$$R(G_s)_{average} = R(G_s)/N \tag{3}$$

Where N is the number of edges in the whole network.

3) Network business risk equilibrium.

The risk balance index of network service is calculated by the following formula:

$$B_{R(G_s)} = \frac{\sqrt{\sum_{i,j(i<j)} (R(E_s(i,j)) - R(E_s)_{aveage})^2}}{N} \tag{4}$$

Using the whole network business risk degree equilibrium degree $B_{R(Gs)}$, we can measure the overall network business allocation equilibrium from two aspects of network business risk degree and risk distribution equilibrium degree. If the index value is too high, it means that the services carried by the service channels in the network are unevenly distributed, individual links are overloaded, or individual links are lightly loaded, or important services (high service importance) on individual links are too concentrated; If the index value tends to 0, it means that the business arrangement risk of the whole network is balanced and the network operation risk is small.

4 Reliability Analysis of Power Communication Network

4.1 Basis of Network Reliability

The definition of communication network reliability is as follows:

(1) The ability of the whole network to maintain connectivity under specified conditions and within specified time.

(2) Communication network reliability refers to the ability of communication network to meet the normal communication needs of users in the actual continuous operation process.

(3) According to the acceptable communication service quality standards and business requirements, the power communication system continuously provides the measurement of communication connection capability to the power system.

The definition (1) points out that the research purpose of network reliability is to ensure network connectivity, and the definition (2) points out that the reliability of communication network is to "complete the normal communication needs of users" in the continuous operation process. The purpose of the definition is from the previous reliability of maintaining connectivity to the current "normal communication needs", It fully reflects that the research of communication network reliability takes object-oriented as the research center, which also marks the phased progress of communication network reliability. Definition (3) refers to the minimum requirement that the communication network can provide business services to meet the needs of users when it fails naturally or is attacked [4]. From the above three definitions, the purpose of reliability analysis of communication network is to meet the business needs of users.

4.2 Service Based Reliability Analysis of Power Communication Network

Based on the reliability research of power communication network based on traditional network topology, this paper studies the importance of service. As an important parameter to study service loss, it is added. And by reducing the importance of the whole network business and adjusting the network business flow, the power communication network business is in a state of high reliability and high stability. The network topology model and attack model based on service importance are established in the network reliability model, and the network topology model based on service importance calculates and analyzes service importance as important network parameters. The attack model calculates and arranges the edge betweenness of each edge in the network topology to select the order of attacking the network topology edges when deliberately attacking. At the same time, the concept of polymorphic network is added to the reliability model, and the reliability of the network is quantitatively analyzed by the loss of business importance. At the same time, it analyzes the side effect of reducing network reliability when attacking. The power grid evaluation index introduces and calculates the "network risk value" and the above data, comprehensively analyzes and improves the business process, relatively reduces the business loss in the same attack of the power grid, reduces the vulnerability of the power grid and improves the reliability of the power grid. Firstly, a service-based network topology model is established.

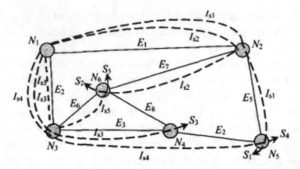

Fig. 2. Reliability evaluation model of power communication network based on business risk equilibrium

5 Conclusion

The operational risk balancing index proposed in this paper is important to evaluate the safety and reliability of the power communication network and to enhance the reliability of the telecommunication network. The evaluation standard and method is the existing network. It has broad application prospects in the evaluation of business comprehensive risk measurement, reasonable security service channels, organization and operation mode, etc.

References

1. Zhao, Z., Chen, X., Liu, J.: Discussion on the establishment of reliability management system of power system communication network. Power Syst. Commun. **27**(168), 58–61 (2006)
2. Gao, Y., Yang, J., Han, J.: Structural reliability analysis of UHV multi loop tower. Power Grid Technol. **34**(9), 181–184 (2010)
3. Li, Y., Zheng, N., Xiao, N., Hong, W., et al.: Method for determining design wind speed of UHV transmission line tower based on reliability theory. Power grid Technol. **34**(7), 5–8 (2010)
4. Ding, L., Tan, M., Xiao, W.: Simulation analysis of invulnerability of complex communication network based on service. J. Nanhua Univ. Nat. Sci. Ed. **23**(1), 81–84 (2009)

Design and Development of Network Education Platform Based on Clustering Algorithm

Fuyang He[✉]

Xichang University, Xichang 615000, Sichuan, China

Abstract. With the development of computer technology, in the field of education in our country, there is a discussion on the new education model represented by educational technology, Among them, distance education is a very prominent part. The physical education in the University for nationalities is a bilateral educational activity with high practicality and strong participation. In physical education, the demonstration and imitation of actions are very suitable to be expressed in the form of multimedia content. The construction of physical education platform of Western University for nationalities is based on mature campus network and network technology, computer technology, multimedia technology and modern communication technology.

Keywords: Platform design · Network education · Sports

1 Introduction

After thousands of years of development, the reason why the national traditional sports can last forever is that it contains the excellent cultural gene of our country. Under the restriction and influence of political, economic and cultural development, the National Traditional Sports highlight the overall level of material civilization and spiritual civilization of the times. There are many kinds of national traditional sports. We can set up some national traditional sports events in college sports fitness projects to stimulate students' interest. It can not only strengthen the body, but also understand the national culture and promote national unity.

At present, with the rapid development of world science and technology, the rapid development of information technology and the wide application of computer related technologies, human society has entered the era of knowledge and network. The knowledge economy with high technology as the core has driven the development of the world economy. A country's comprehensive national strength and international competitiveness are more and more educated. With the development of science and technology and the level of knowledge innovation, The rapid development of modern science and technology needs a large number of innovative talents. This new development makes the traditional education model unable to meet the needs of higher education. requires great leap forward development in the field of education, but also provides unprecedented opportunities for the modernization of educational means. At present, the current situation of education in China is: the scale of education is expanding, the structure

© The Author(s), under exclusive license to Springer Nature Switzerland AG 2022
J. Macintyre et al. (Eds.): SPIoT 2021, LNDECT 98, pp. 725–730, 2022.
https://doi.org/10.1007/978-3-030-89511-2_95

of education is constantly optimized, the concept of education is constantly updated, and the forms of education are increasingly diverse. More and more people want to go back to school. Colleges and universities also provide many learning opportunities for these students, but at the same time, they also bring pressure to the traditional education. The popularization of distance education, especially network education, is not only an effective way to solve this problem, but also an important driving force to innovate the traditional education mode. Modern distance education provides a broad space for the leap forward development of China's education. All kinds of new network universities, virtual universities and network education bases are gradually formed. The formation and development of modern distance education has become an important part of modern education, but also reflects the development of modern science and technology and its application in the field of education [1].

This paper mainly studies the design and implementation of sports network platform in University for nationalities. Firstly, it introduces the concept and characteristics of modern distance education, focusing on the development status of sports network education at home and abroad and the development trend of modern sports network education. I propose the construction and implementation of the platform. Based on the actual situation and specific needs of school physical education, the overall design, detailed design and system structure of the system are analyzed, basic principles and characteristics of the network platform of physical education are discussed. It also analyzes and discusses the network learning mode of physical education, video stream transmission, website page design and information organization technology of physical education teaching information resource database. In the overall design of the platform, it introduces the basic structure, functional modules, software and hardware configuration of the platform; In the detailed design of the platform, the work flow and the security of the system are described in detail.

2 Overview of National Sports Network Education

2.1 Overview of Online Education

Distance education is a kind of education form that students and teachers, students and educational institutions mainly use a variety of media means to communicate and teach systematically. Two way interactive video, one-way video broadcasting, video on demand and data sharing are commonly used in distance education. Physical education network education mainly meets the needs of school physical education, promotes and supplements the education of teachers and students, breaks through the traditional school physical education mode, gets rid of the limitations of time and space, and constructs classroom education, radio education, television education and diversified physical education system. This paper mainly studies the distance education of physical education network based on Internet and telecommunication network.The so-called distance network education is a new stage of the development of distance education, which is different from traditional education and full-time education. The modern network distance education, which emerged in the late 1990s, is carried out in the form of receiving courseware through computer network or directly transmitting courseware through network communication technology, so that students can choose by themselves. As shown in Fig. 1 below.

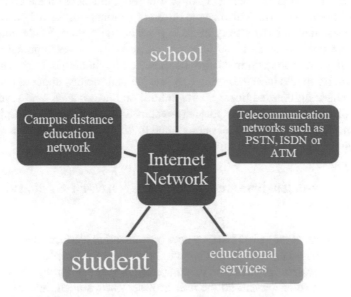

Fig. 1. Model diagram of network teaching platform based on Internet component

Distance education platform is an overall network learning solution, which generally includes: online learning and management system, courseware making system, online synchronous audio and video classroom system, recording room, supporting network equipment and server, etc. it can complete online learning courses, learning management, resource management, courseware making, online synchronous audio and video classroom, course recording and other functions. Its biggest feature is remote and interactive. The distance teaching mode will become the most widely developed teaching mode [2].

2.2 National Traditional Sports

National traditional sports is a comprehensive cultural form of competitive entertainment and education created by all ethnic groups in order to obtain the skills to enhance their physique. It is a complex of a dynamic process in which all ethnic groups take physical exercise as the basic way. It is a traditional cultural way of life formed by the progress of national civilization and has unique characteristics of all ethnic groups.

Due to the differences in geographical environment, social production, lifestyle, cultural level and religious customs, the traditional sports of ethnic minorities in China have formed their own distinctive characteristics. These characteristics also reflect the main forms of the development of our national culture from different angles and levels, mainly including nationality and modernity.

First of all, the national traditional sports should have distinct nationality. China is a multi-ethnic country. Affected by the factors of region, environment, living habits and attitude, there are certain differences in the cultural characteristics of different nation-alities. Different cultural characteristics have different influences on material, spiritual, life and social relations, which also create different nationalities, namely nationality.

The Chinese nation has a history of 5000 years of civilization. While inheriting the ancient civilization, we should also face the future. With the development of society, national traditional sports will inevitably be impacted by modern culture. The national traditional sports should have the courage to face the real society, undertake the respon-sibility of inheriting the traditional sports culture and advantages, and give full play to the role of national traditional sports in modern life, so as to occupy a place in the development of world sports and let our national traditional sports culture take root and germinate in the world sports culture.

3   Development Environment of Sports Network Education Platform

The design purpose of the sports distance network education platform is to provide users with an Internet-based network learning environment. All the functions of the platform are: the sports network education platform system is realized through the interaction between the application server and the user browser. Save student information, sports photos, sports technology demonstration videos, sports teaching materials, student inter-action information, sports network test scores and other information and data. These data are saved in different forms in the database server, which is organized and maintained by the database server. As shown in Fig. 2, it is a sketch of the application mode of sports distance network education platform. [3].

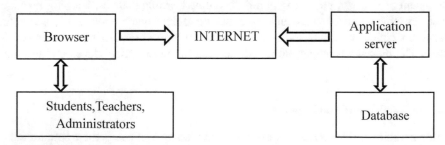

Fig. 2. The teaching mode of sports distance network education based on Internet

Firstly, the working mode reflected by the image must be connected with the net-work of users (students, teachers and managers) to participate in sports network edu-cation. Each browser sends a request to the application server. The application server corresponds to the user's request, retrieves the data of the database server as needed, and sends a request to the user through the browser. Feedback service results such as response and retry.

In the above application server and database server, especially in this study, the software platform of my SQL database server and Tomcat web server comprehensively analyzes the needs of sports distance network education platform. On the basis of studying the advantages and disadvantages of other courses network education system, the gymnasium is studied according to the platform development standard. The composition of distance network education platform is divided into five levels. That is, my SQL database server, Tomcat network server, campus network and its ancillary facilities, education platform management system and client education system.

The function of the database is to provide a course plan interface for teachers to view and edit their own courses, so its data table should include the required teacher information and course information. Here, in order to save space, we use the basic teacher information table and the specific course plan table, Extract the course number from all courses in the teacher basic information table as a keyword, and then compare it with the form name in the course plan library to find the required form, such as course (n) here. In fact, it may be course (C001) or course (matho01), and then operate it. The specific SQL statements for creating two table frame structures are as follows:

Create teacher information in the format "wxdata".

(SIR ID varrhar (10) null default value".

"Sir name" var (20) is an empty default value.

"Sir pass" variable (20) is an empty default value.

"Email" var (50) idle base value"

"Scx" variable (10) is an empty base value.

I don't think the year of "birth" (4) is "0000".

"Pro" variable (20) null basic value"

Alternate "varrhar (20) null base value".

"Title" varrhar (20) null default ".

"Procedure" varrhar (200) empty base value"

"Empty base value for process varrhar (200)"

Format: mysom).

4   The Advantages of Teaching Application in Physical Education Network

The main functional advantages of physical education distance network education in Universities for nationalities are reflected in the leading role of teachers and the dominant position of students. Giving full play to the complementary role inside and outside the classroom is conducive to cultivating students' learning autonomy and innovation ability, and is conducive to the conservation and efficiency of Physical Education resources. Promoting the use of resources has a significant effect on the learning mode of the whole education. The sports distance network education platform uses programming technology for the object. The overall detailed design takes the object as the core, which has good openness and expansibility. Lost. It is a good platform for follow-up development. I believe that using this platform will get good economic and social effects. At the same time, other related systems also have certain reference and practical value [4].

The platform has a high satisfaction of the students' needs in public sports courses, but the targeted function design for professional athletes and sports specialty students is less. This is a serious blank in the design of this platform. Therefore, in the process of system test, we should pay attention to the basic attitude of students participating in public sports and international distance education on online sports courses, students' subjective evaluation of learning effect, and teachers pay special attention to statistics and analysis of students' assessment results. The technical requirements of demonstration and explanation of action technology in sports teaching video resources should be standardized and clear.

5   Conclusion

The physical education technology teaching platform in national colleges is very important in the school physical education network education system. It deals with users directly. Whether the platform can be used conveniently and correctly directly affects the effect of distance learning of sports. It can be said that the success of the design of teaching support system directly affects the performance of the whole platform. Through the research of this paper, the design and Realization of the supporting system of physical education and daily management in Ethnic Colleges and universities can be reduced to some extent, and some valuable reference can be provided for the overall design of the sports network education platform for a wider range of objects.

Acknowledgements. Liangshan Prefecture Ethnic Culture Research Fund Planning Project: Study on the Sports Culture Path of Promoting Liangshan Ethnic Minorities to Adapt to Modern Social Civilization from the Perspective of Cultural Integration—20201102.

References

1. Shuiwen, L., Ran, L.: The value and promotion strategy of folk dance Qingjiang dance. J. Cent. S. Univ. Nationalities [Hum. Soc. Sci. Ed. 1., **31**(4), 166–68 (2011)
2. Lu, B.: Introduction to Chinese Traditional Sports Culture, pp. 108–123. People's Education Press, Beijing (2005)
3. Li, S.: Feasibility analysis of introducing martial arts film and television into martial arts teaching. Front. Econ. Cult. 61158–61160 (2010)
4. Li, W.: Introduction to Chinese Wushu, pp. 526–528. People's Sports Publishing House, Beijing (2005)

Application of Internet of Things in Online Teaching Platform

Rong Wang[⊠]

Gannan Medical College, Ganzhou, Jiangxi, China

Abstract. In the era of knowledge economy in the 21st century, modern science and technology has penetrated into the field of education, which has brought great changes to China's education. Artificial intelligence has been gradually applied to education, greatly innovating educational methods, and has become an important trend in the development of educational informatization in China in the future. Machine learning is an important part of artificial intelligence. With the development of educational big data, machine learning is an educational application. It can effectively analyze and predict a large number of educational data and better realize intelligent education, starting from the significance of artificial intelligence. This paper analyzes the necessity of independent research and development of women's university network education platform, and introduces the architecture, design idea and function of the system.

Keywords: Online teaching platform · Artificial intelligence · Higher education

1 Introduction

The pace of scientific and technological innovation is irresistible. With its continuous development in modern society, artificial intelligence technology is constantly maturing. At present. Artificial intelligence has gradually begun to penetrate into all areas of society, The application in the field of education has brought new major educational changes, and the development of information technology and big data technology has produced intelligent education. China's exploration of wisdom education has also entered a stage. Machine learning is an important part of artificial intelligence. The application in education provides more possibilities for intelligent education. Let students become more personality. In order to prove the effectiveness of education, exploring the educational application and innovative exploration of machine learning from the perspective of artificial intelligence has important practical significance for improving the educational level of modern education.

How to promote the innovation and development of informatization through knowledge education and make the innovation and development of education become the inevitable trend of the information age. Personalized learning is the core element of intelligent education. How to better support the development of individualized learning through technology? According to the needs of the research field of intelligent education, the practical application of personalized learning is mainly autonomous learning.

J. Macintyre et al. (Eds.): SPIoT 2021, LNDECT 98, pp. 731–735, 2022.
https://doi.org/10.1007/978-3-030-89511-2_96

Although we attach great importance to personalized guidance and the adaptation of learning content, there is no differentiated learning service in the whole learning. The main reason is that the development of technology can not fully meet the needs of personality learning [1].

Machine learning is the core and most popular technology in the field of artificial intelligence. It is a large number of data automatic recognition patterns. According to the discovery law, predict students' academic performance and meet the needs of intellectual education and personality learning. At present, there is no systematic research on the educational application of mechanical learning at home and abroad. By comprehensively sorting out the potential, development and challenges, it provides theoretical and practical basis for researchers and educators to implement intelligent education and personalized learning.

2   The Concept of Machine Learning

2.1   Interaction Between Machine Learning and Intelligent Education

Machine learning is the core of artificial intelligence. It mainly collects the knowledge of multiple disciplines and realizes intelligent machine learning through research and analysis of computer simulation and human learning behavior. The application of machine learning in intelligent education can provide students with continuous and effective feedback. Provide services to improve students' learning efficiency. Now in practical wisdom education, interactive expression mainly includes the following aspects.

First, teacher-student interaction. The best effect of interaction between teachers and students is that teachers can make targeted teaching and counseling programs according to the different personality characteristics of students. However, due to the limited energy of teachers, it is difficult to make personalized programs for all students.

Second, the interaction between students. The best effect of this kind of interaction is effective and in-depth knowledge exchange between students. But the actual situation is that the quality of students' spontaneous interaction is difficult to control, and it is very difficult to achieve the best interaction.

Third, the interaction between students and the system content, that is, students learn through the effective, accurate and massive knowledge provided by the computer system. But this kind of interaction is the computer to convey knowledge to students, but it does not have strong pertinence. The computer system is like an excellent teacher, in class for thousands of students, personalized teaching service is difficult to achieve, of course, the quality of teaching can not be guaranteed [2].

The interaction between machine learning and intelligent education is mainly to make the interaction between students and system content intelligent. In short, it means that in the process of learning, students can guide what and how to learn through intelligent computers, deeply analyze the characteristics of students, and provide one-to-one personalized teaching services for students. In addition, in the application of artificial intelligence in the field of education, fully consider the complex characteristics of different students' learning process. In the process of education, comprehensively consider students' learning situation and learning level, and explore what, how and how to let students learn. Confirm whether the course is completed and provide students with the most appropriate learning content.

2.2 Relevance of Machine Learning and Intelligent Education

Smart education is mainly through the use of information technology, so that teachers can innovate efficient teaching methods, provide better teaching services for students, improve the quality of teaching, and promote the development of students. In the intelligent education environment, all aspects of students' data information can be collected completely, forming a huge education big data. Intelligent machine learning can fully mine and analyze these data information, and can make reasonable prediction, which provides more advantageous teaching content and teaching mode for the development of intelligent education. Therefore, using machine learning to improve the quality of smart education is the inevitable way of its development. Through machine learning, the effective information in education big data can be clearly provided to educators, so as to promote the improvement of teaching quality. At the same time, through the intelligent teaching platform, the analysis of students' individual information data can be strengthened, So that teachers can provide students with personalized teaching programs, more conducive to students' learning.

3 Design of Online Teaching Platform for Higher Education

In the architecture of large-scale distributed application system, the separation of front-end and back-end has become the industry standard way of Internet project development. The front-end uses JavaScript language to write web program, uses react library to build user interface, the back-end uses. Net core technology to develop distributed API service, uses Visual Studio code as program development tool, and the database uses my SQI, one of the most popular relational database management systems.

The platform is designed hierarchically from four aspects: data acquisition, communication, business and application. The mobile terminal and PC are used as hardware terminals for data acquisition; The communication layer is based on wired/wireless LAN for communication between internal servers, and Internet for communication between external acquisition terminal and application server; The business layer includes data platform and function platform. The data platform is divided into data acquisition and storage platform and data processing and sharing platform. The two platforms jointly realize the functions of data acquisition, storage, analysis and sharing. The function platform includes three main parts: online teaching platform, online education and background management; The comprehensive application layer provides users with specific software applications, which can be divided into classroom evaluation, classroom record, teacher portrait, student portrait and so on. The overall architecture of the system is shown in Fig. 1.

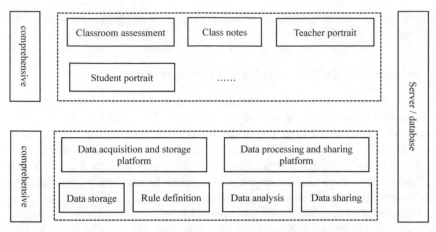

Fig. 1. Overall architecture of online teaching platform system

To achieve the above goals [3], teachers need to first adjust the teaching design, adjust the teaching objectives of knowledge transfer, and turn to help students establish a correct cognitive model, cultivate their learning ability and thinking ability as the curriculum goal. Teaching design should be closely around the "student-centered", and classroom activities should mobilize the enthusiasm of students. Secondly, teachers should update the teaching materials, including adjusting the original courseware, deleting unnecessary contents and highlighting the key points; The introduction of rich teaching cases, especially those closely related to the reality of life, causes students to think, explore and pay attention to life; We should increase the frontier of subjects or introduce new technologies and methods to provide students with a broader vision.

4 Application of Online Teaching Platform

The construction of online teaching platform of higher education in the perspective of artificial intelligence can record meaningful teaching activities in class and upload it to the "flipped classroom" module, which can connect students and construct big data in class, which provides strong support for process assessment. Students upload the team results to the "results display" module to show the results of the student team cooperation.

Flipped classroom, as a new teaching method to promote the reform of classroom teaching mode, has been widely explored and practiced by the majority of educators. Chen Zhaoxi and others carried out the reform and practice of flipped classroom teaching mode supported by micro class; Li Wenwen and others studied the current situation of Hybrid Teaching Mode in domestic medical education, and introduced the hybrid teaching mode based on flipped classroom in detail" The online teaching platform, which focuses on students, fully highlights the students' learning subject status, and emphasizes that students apply knowledge into skills in practical projects. Teachers arrange courses with certain challenges and difficulties in the course. Organize students to complete the tasks in groups, and submit the learning results to the "achievement display" module. Teachers encourage and propose suggestions for improvement of students' works and star rating for the groups with good performance. Through this innovative group results, the interaction and feedback between teachers and students is more active and effective. Teachers understand the personalized thinking content of students in listening, students recognize the shortcomings and improvement direction in the works in the teacher comments, which is conducive to further improving the thinking ability and further optimizing the works [4].

Teachers can flexibly arrange the class time limit tasks or homework after class, and multi-dimensional detection and consolidation of students' learning effect, and promote the accumulation of students' ability" Online teaching platform "is closely around the whole process of teaching function design, which solidifies the teaching activities involved in pre class, in class and after class into software, pays attention to the distribution of materials and tasks before class, emphasizes group discussion and classroom test in class, and sets up homework after class to promote the review of curriculum content and knowledge application. Through the arrangement of homework, the review, extension, expansion and deepening of key knowledge can be realized, so that students can master the contents of the course and cultivate the intelligence factors of students. Students in the process of completing homework, must pass a certain effort to complete, to a certain extent, conducive to the cultivation of students perseverance, fear of difficulties and other non intellectual factors. Moreover, homework has a good feedback function. Besides the way of classroom questioning and student demonstration, teachers can also check and fill in the missing, optimize the teaching content and improve the teaching method according to the students' homework completion, and improve the teaching effect.

5   Conclusion

Robot learning in the vision of artificial intelligence has attracted more and more attention in the current social background, and has very important scientific research value. Through the effective integration of robot learning and education, and more support is given to the scene technology innovation. In the era of big data, the application innovation of robot learning will certainly provide more assistance for learning and education.

Acknowledgements. Deep learning oriented development of college students' computational thinking ability and reformation of innovation practice—take Basic computer courses for instance in Jiangxi province in 2017 with the project number JXJG-17-13-19.

References

1. Zhu, Z.: Leading the innovation and development of education informatization with wisdom education. China Educ. Inform. (9), 4–8 (2014)
2. Multi dimensional perspective of Wang Yun: Hua Fu's wisdom education in Weinian: also on the future development of wisdom education. Curr. Russ. Educ. Technol. (2), 21–27 (2020)
3. Zheng, Y,, Fu. Q., Zhao, Y.: Analysis on the research status and theme structure of artificial intelligence education in China. Digit. Educ. (1), 21–26 (2020)
4. Wang, Y., Tian, D., Liu, X., et al.: Ten key issues in the research of Educational Informatization – Based on the salon records and related literature of "face to face with the editor in chief". Mod. Dist. Educ. Res. (1), 133 (2020)

Construction of 1 + X Specialty Group Based on Personalized Optimization Recommendation Algorithm

Qifeng Han[✉], Yanchun Xu, and Chuanye Wang

Shandong University of Engineering and Technology, Shandong, China

Abstract. Specialty group construction is an important cornerstone of characteristic development and connotation promotion of higher vocational colleges. In practice, there are path dependence in concept, system, positioning, teaching, evaluation and so on, which brings a series of problems. In order to realize the high-quality development of specialty group, it is necessary to deal with the philosophical relationship between big and small, far and near, main and secondary, internal and external, gain and loss, and realize "five changes" in the implementation strategy.

Keywords: 1 + X · Professional group construction · Path dependence · High quality development · Strategy

1 Introduction

Specialty group refers to a collection of related specialties with a strong comprehensive strength as the core, which is composed of several specialties with "similar professional foundation, similar technical field, related occupation and post, and shared teaching resources". In 2006, the Ministry of education and the Ministry of Finance jointly issued the "opinions on implementing the construction plan of National Model Higher Vocational Colleges and accelerating the reform and development of Higher Vocational Education", which put forward the concept of specialty group in the policy document for the first time. Then, on the basis of national guidance, provincial coordination and school implementation, after two rounds of national demonstration (backbone) higher vocational college construction projects, the construction of specialty group in higher vocational colleges has achieved a series of results. However, the overall level of specialty group construction in higher vocational colleges is not high, and the cluster effect is not obvious. Fundamentally, due to the long-term dependence on the traditional concept of specialty construction and management mode, it has formed a strong path dependence, and shows the trend of continuous strengthening, which leads to many problems in the method and mode innovation of specialty group construction, and the effect is not obvious. At present, China's economy has changed from a high-speed growth stage to a high-quality development stage. With the acceleration of economic restructuring and industrial transformation and upgrading, higher requirements are put forward

J. Macintyre et al. (Eds.): SPIoT 2021, LNDECT 98, pp. 736–742, 2022.
https://doi.org/10.1007/978-3-030-89511-2_97

for the high-quality development of higher vocational education and the cultivation of high-quality technical talents. The national vocational education reform implementation plan proposes to build 150 backbone professional groups as an important lever to promote resource integration, realize agglomeration effect, promote structural optimization, improve service function and promote high-quality development of higher vocational education. Therefore, breaking the path dependence of specialty group construction in higher vocational colleges has become an urgent practical problem to be solved [1].

2   Connotation and Challenge of 1 + X Certificate System

2.1   The Connotation of 1 + X Certificate System

"1 + X certificate" refers to one academic certificate and several vocational skill level certificates. In 2019, the notice of the State Council on printing and distributing the implementation plan of the national vocational education reform clearly pointed out that it is necessary to "deepen the reform of the training mode of compound technical and skilled talents, learn from the general practice of international vocational education and training, formulate the work plan and specific management measures, and start the pilot work of the 1 + X certificate system". Under the organization of local governments, colleges, enterprises and training institutions, in accordance with the requirements of the program, strive to "develop both academic education and training education", develop all kinds of vocational skills and grade certificate standards, and actively invest in talent training, so as to use various types of vocational skill certificates to broaden students' employment and entrepreneurship channels, and promote the pilot work of the 1 + X certificate system launched by the State Council.

2.2   Challenges of 1 + X Certificate System

The 1 + X certificate system is not only the innovation of China's vocational education, but also the supplement to China's vocational education system. In order to realize the large-scale and high-quality development of vocational education and meet the "three education reform" of vocational education, the quality and skills of higher vocational students meet the requirements of employers, and the implementation of 1 + X certificate system has a long way to go. As of July 2020, the national vocational skill level certificate information management service platform has announced that 72 training and evaluation organizations have participated in the development of vocational skill level certificates. At present, there are 92 kinds of vocational skill level certificates. Enterprises, institutions and colleges across the country have actively responded to the development of certificates, which also reflects the urgent requirements of various industries for employees with negative vocational skill levels, However, there are also challenges in the process of certificate development and implementation [2]. From the perspective of "X" certificate development process, in order to enhance its influence in the industry and seize the commanding height, the training and evaluation organizations did not conduct sufficient investigation and evidence collection, and started to apply for certificate development when the software and hardware measures did not meet the standards, which would undoubtedly reduce the quality of certificates. From the perspective of the certificate itself, the 1 + X certificate system has not completely replaced the dual certificate system. Under the coexistence of the two systems, how to organize and transform the curriculum content and form, how to realize the credit exchange of vocational qualification certificate and academic certificate, and whether the relevant education and teaching policies and management systems are in place are the urgent problems to be solved in the implementation of the 1 + X certificate system. Judging from the recent implementation of the 1 + X certificate system, some colleges, enterprises and training evaluation organizations do not have a good understanding of the relevant policies, the designation of the primary, middle and Advanced Certificate levels is not strict enough, the training process is rough, the implementation standards of the pilot colleges are not strict, and the "1 + X" teacher construction and related training resources are lack, The lag of new skilled personnel training has become an obstacle to the development of the 1 + X certificate system, which will eventually affect the training of compound skilled personnel.

3   The Main Performance of the Current Transition Path Dependence of Colleges and Universities

The path dependence theory holds that once a system with normal feedback mechanism is adopted by the system under the influence of external accidental events, it will develop along the path defined by -. And it is difficult to replace it with other potential or even better systems. It has three effects, such as path locking, invalid innovation and track tenacity. The path dependence of the current university transformation is mainly manifested in the following three aspects.

First, the transformation goal depends on. There are some misunderstandings in the direction of transformation goal and value pursuit. Many colleges ignore the characteristics and advantages of long-term running, and put forward the goal of transformation and upgrading. For example, some normal colleges, agricultural forestry and art colleges propose to build multi-disciplinary colleges into comprehensive colleges, while some single science colleges have sought to develop into multi-disciplinary colleges. Some higher vocational colleges with good application-oriented skills training characteristics or colleges take upgrading to their own colleges as the goal of career development. Some newly-built undergraduate colleges with relatively general conditions and comprehensive strength still need to be improved actively seek to expand from the simple development of undergraduate education to both undergraduate and graduate education, and from teaching university to teaching research university. Some universities with graduate education experience have put forward the transformation from teaching research university to research-based university or research university in the development planning. One sided "seeking high, big, complete and new" has become the "bright mark" and "value orientation" of the upsurge of university transformation.

Second, the transformation mode depends on. Some universities still continue the traditional extensive growth, extension development and investment driven mode in the process of transformation. Based on the concentration of school running funds, investment, introduction of high-level discipline leaders and discipline teams, greatly improve the level of experimental facilities construction, expand the scale of capital construction, pull up various school running indicators in a short time, and achieve the "quality improvement and upgrading" of school running level and level. In the process of rapid development, some schools ignore the comprehensive balance of the structure, function, quality and benefit of running schools. The introduction of discipline leaders and innovation teams and the construction of talent platform are in a gap. The integration and fusion effect between new teams and original teachers needs to be further demonstrated; The linkage mechanism of discipline construction, teaching reform and talent training mode innovation needs to be further improved; The construction of campus hardware and the shaping of school spirit in the new era.

Third, the transformation measures depend on. The dependence of transformation measures is mainly reflected in the alienation of transformation objectives and transformation measures, the innovation and growth of transformation means. Some schools have put forward "distinctive characteristics", "have a high visibility", and lack of specific reference standards and connotation definitions for the running objectives such as "first-class school running level" and "high level". The educational ideas of "running schools by layers", "characteristic running", "connotation development" and so on lack of targeted response in practice. Different schools have some convergence in the development measures of teaching reform, talent training mode innovation, scientific research, production, learning and research cooperation, social service, etc. the different development measures of the same school in different stages of development need to be further clarified. Generally speaking, the new school running idea and goal established by the transformation of colleges and universities in the new period need to be better penetrated into the innovation of running mode and mechanism, further refining and concentrating the characteristics and advantages of running schools, and realizing the "God shape and dual possession" of the transformation.

4   Path Dependence and Solution Strategy of 1 + X College Specialty Group Construction Under Optimization Algorithm

The path dependence of the construction of professional groups affects the effect of professional agglomeration and the benefit of resource allocation. To solve this path dependence, five "changes" should be done.

4.1   The Source of Construction Movement Changes from "External Drive" to "Endogenous"

The external evaluation of professional qualification assessment and level evaluation is an important grasp of the management department to promote professional construction, and also the driving force leading the direction of professional construction and determining the allocation of professional resources. The original purpose of the construction of professional group is to produce the benefit of "1 + 1 > 2" through the construction and integration of professional agglomeration. To build a high-level professional group is no longer the construction of some aspects and certain indicators, and the driving force should not rely on external evaluation, but rather the reasonable allocation and mutual "chemical action" of various elements such as environment, technology, equipment, talents, system and culture. The formation of this internal factor endogenous driving force requires that the construction of professional group team be put first, the sense of belonging and responsibility of team members should be strengthened, the creativity and enthusiasm of the team should be activated, the professional group should be built into a community scientifically allocated by teaching resources, a carrier of school enterprise culture integration, a collection of innovative thought and practice, and the cultural value system of the professional group should be built, Release the collaborative innovation vitality of various elements needed for the construction of professional groups, promote the continuous improvement of talent training quality and the breeding and output of high-level achievements, finally realize the accumulation of cultural internal forces, and stimulate the deep motivation of high-level construction.

4.2 The Architecture Changes from "Combination" to "Integration"

Professional group construction is an important way to meet the requirements of industrial transformation and upgrading, technology renewal iteration and post group of production front line, and to train talents with complex and cross-border technical skills. Compared with the construction of a single specialty, the development of professional collection and group is conducive to integrating the resources of running schools, reducing the cost of running schools, improving the efficiency of running schools and enhancing the overall competitiveness. At the same time, compared with the loose professional system, the technology (service) field of professional group docking industrial chain is a cluster professional structure formed by new and old, high-level and low-level similar fields, the combination of different specialties and orderly collection. From "professional" to "professional group", it is the transformation from "individual" to "collective", a transition from "combination" to "integration", a change from "closed" to "open", a transformation from one-way transfer of professional construction elements to comprehensive sharing, and a key problem to promote the construction of high-level professional groups. Among them, the integration of teaching organization and curriculum, the co construction and sharing of teaching resources, and the improvement of cooperative development mechanism are the symbols of professional deep integration, and the key to the construction of high-level professional groups.

4.3 The School Running System Changes from "Dual" to "Pluralistic"

Open and inclusive is the basic requirement of Higher Vocational Education in the new era. After a long period of development, many professional groups of higher vocational colleges have carried out the "dual" schools in the form of school school cooperation, school land cooperation and school enterprise cooperation, which has realized the mode innovation and characteristic development, and improved the quality of talent training. The basic premise is to build a professional group leading reform, supporting development, Chinese characteristics and world level, and high-quality education and teaching. Specifically, the construction of high-level professional groups should adhere to the quality line, innovation and reform power, explore the "multi" cooperation mode of government, industry, enterprises and schools, promote the sharing and integration of resources, information, technology, capital, talents, equipment and other factors, forming a good situation for school and enterprise sports personnel with dual main players. At the same time, we should fully explore the service function of professional groups, explore the realization ways of professional research achievements to feed professional teaching, and form the internal cycle of innovation of talent training mode; To explore the operation mechanism of the marketization of professional scientific research achievements, and form an external circulation of innovation in the application and promotion mode of technical achievements; Finally, the construction effect of professional group can be transformed into talent support and intellectual support for economic and social development and industrial transformation and upgrading.

4.4 The Operation Mechanism is Transformed from "Management" to "Governance"

The construction of professional groups requires not only the administrative level from top to bottom institutional arrangement, but also the interest expression of stakeholders from the bottom-up. It is necessary to break through the path dependence of traditional management, handle the relationship between academic and administrative, and strive to change to modern governance. To achieve the high-quality development of professional groups, we should also highlight the "teacher management" and "professional management". The so-called "teacher management" means to play a decisive role in education and teaching. Teachers should be encouraged to ask and teach by words and deeds. By adhering to academic freedom and academic norms, we should innovate and develop teaching objectives and teaching scheme design, constantly regulate and influence students' academic behavior, pay attention to society and industry, and integrate industrial elements into teaching Teaching materials and classes. The so-called "professional management" is to separate the administration from the academic. The administrative affairs are undertaken by the special administrative personnel, reduce the interference of administrative factors on academic affairs, avoid short-term profit seeking behavior, and also eliminate the acquisition of academic resources by administrative means to ensure the sustainable and healthy development of professional groups.

4.5 Talent Training Changes from "Tool" to "Value"

Mr. Gu believes that the essence of education is to improve the quality of life and to improve the value of life. However, in the practice of education and teaching, influenced by the traditional concept of "teachers, top-down, lower effects", the educational relationship is often defined as the upper and lower relations, and the concrete implementation of the relationship has evolved into "Inculcation". In teaching, students are only emphasized to "do things right", but ignore the guiding students to "do the right things", and there are instrumentalism, pragmatism and utilitarianism. In the construction of professional group, we should position the talent training in the cultivation of "human", follow the internal growth and development law of the educated, not only the mastery of students' knowledge and skills, but also the development of students' values and thinking, that is, the transformation from instrumental thinking to value thinking.

5   Conclusion

"1 + X certificate system" is a major reform to meet the needs of the national industry, improve the quality of vocational education and students' employability. It is a fit practice of the new reform of teaching mode in Higher Vocational Colleges and the employment demand after the transformation of enterprises in the new era. The solution to the path dependence of university transformation is a social system engineering. Its promotion also depends on the inheritance and development of advanced school running culture. The utilitarian goal of running a school and the extensive growth mode are essentially a distortion of the mission and culture of running a school. The transformation of colleges and universities should inherit and carry forward the advanced cultural tradition of running a school, follow the law of educational development, promote the continuous improvement of the strength of running a school of higher education with a scientific outlook on educational development, a broad educational vision and a strong sense of educational mission, and run a higher education satisfactory to the people, so as to serve the building of a moderately prosperous society in an all-round way.

Acknowledgements. 1. Construction of Applied Software Technology Specialty Group.
2. A Study on the Certificate System of 1 X Vocational Education.

References

1. Liu, H.: Discussion on the practice of 1 + X certificate system for rail transit majors. J. Nantong Vocat. Univ. (6), 32–33 (2019)
2. Li, Z.: Vocational Education 1 + X certificate system: background, positioning and pilot strategy I. Interpretation of national vocational education reform implementation plan. Vocat. Educ. News (3), 30–36 (2019)
3. Lu, X.: Discussion on the construction mode of "loose" specialty group in Higher Vocational Colleges based on the concept of collaborative innovation. Res. Contin. Educ. (1), 51–52 (2014)
4. Shi, X., Geng, G., Li, H.: New explanation of "path dependence" theory. Economist (6), 53–64 (2014)

Network Public Opinion Crisis Early Warning Model of Mobile Social Platform

Yi Li[✉]

Henan University of Animal Husbandry and Economy, Zhengzhou 450046, Henan, China

Abstract. With the continuous development of social economy in China, the sudden problems caused by social conflicts and conflicts of interest in the process of economic development often occur, which leads to the outbreak of social public opinion, which leads to the network public opinion becoming the main way for the public to choose to express their own views, and the possibility of the crisis of network public opinion is constantly improved. The sudden public crisis has a negative impact on the public, society and even the state. Therefore, this paper designs a crisis early warning model, classifies it by using support vector machine algorithm and naive Bayesian algorithm, and combines AHP and fuzzy comprehensive evaluation to carry out the network public opinion crisis warning.

Keywords: Mobile social platform · Network public opinion · Crisis warning

1 Introduction

The development of mobile information technology under the background of big data promotes the efficient integration of Internet and information industry, which makes network media gradually become an important way for people to communicate information and emotion. Mobile social platform can not only provide instant messaging services for Internet users, but also provide Internet news reporting function for Internet users, Netizens express their views on the crisis events through the Internet, and the network public opinion is expanded to meet the needs of netizens' comments and information forwarding. The mobile social platform has become a new channel of public opinion. Mobile social platform accelerates the information transmission of crisis events, which is easy to cause qualitative change of public opinion on mobile social network. It makes the research on the crisis of network public opinion on the mobile social platform have certain application value. Based on the real public opinion, it studies the network public opinion formed on the mobile social platform, provides certain reference for the development and solution of the crisis, strengthens the government and relevant departments' attention to the crisis management of network public opinion on the mobile social platform, and improves the government's credibility, It is of practical significance to reduce the degree of public questioning and polarization of crisis events. Through the mobile social platform, the government and network public opinion managers can timely grasp the public opinion, and provide reference for the correct decision-making to eliminate false news and rumors, reduce the social harm of crisis events, and maintain the order of the network environment and the real environment [1].

J. Macintyre et al. (Eds.): SPIoT 2021, LNDECT 98, pp. 743–749, 2022.
https://doi.org/10.1007/978-3-030-89511-2_98

2 Msnp Network Public Opinion Crisis Influencing Factors and Early Warning Process Analysis

2.1 Msnp Network Public Opinion Crisis Influencing Factors

(1) Main factors. It mainly includes netizens (platform users) and the government. The government carries out its rights as a "big" netizen, so it belongs to the category of netizens. Netizens belong to the public and have a certain perceptual and rational psychology. In response to sudden crisis events, some people release false information on the Internet in order to satisfy their curiosity. It is very easy to arouse netizens to make divergent comments on the gap between the rich and the poor, the social status gap, and the pertinence of events, which aggravates netizens' anxiety, Increasing the correlation between similar events and their own experience, set off a new climax of public opinion. The key factors include netizens' psychology, attention, opinion orientation, opinion leaders' role, netizens' quality, government credibility, government cohesion, etc.

(2) Objective factors. It mainly refers to public emergencies, which usually refers to the events that can cause serious social harm without any omen or early performance is not obvious. The consequences of such incidents are uncertain, which can easily cause netizens to express their views from multiple perspectives. The object of network public opinion is the reason and direction of the development of public opinion, and it is also the direct subject factor that easily leads to public opinion crisis. The key factors include social influence and event sensitivity.

(3) Subject and object behavior factors. This paper describes the behavior set of subject and object factors in the formation and development of network public opinion on mobile social platform, which is the concrete embodiment of public opinion. The common behaviors include the number of comments published on the platform, the amount of forwarding on the platform, the interaction and expression intensity of opinions on the platform, as well as the social attention and discussion heat on the platform. If the number of speeches with group polarization reaches more than 50% of the overall level of public opinion, it is very easy to promote the trend of the whole public opinion. If it is again affected by the network push, the wrong control of the interaction and expression intensity of opinions and opinions can also cause the crisis of public opinion.

(4) Environmental factors. Generally, the environment of public opinion on mobile social platform is divided into network and reality, which is divided into social, economic, natural, technological and cultural environment. If the external environment changes greatly, it will change the development trend of events. For example, social environment can enhance the interaction between stakeholders of public opinion. The economic environment reflects the emotional changes of economic stakeholders caused by the event. Technology environment simply shows that the development of events and the spread of public opinion are controlled by technology [2].

2.2 Msnp Network Public Opinion Crisis Early Warning Process Analysis

The basic principle of early warning of public opinion crisis on mobile social platform network is to find the alarm source of public opinion crisis on mobile social platform network through the determination of the purpose, method and model of early warning. On this basis, the obtained information of the alarm source is analyzed, so as to clarify the stage of the alarm, and the crisis early warning model is used to determine the warning of this stage, Judge the alarm level of crisis early warning, according to the given alarm limit of similar events, give the scope of public opinion crisis early warning of mobile social platform network according to the alarm limit, that is, give the corresponding forecast results, so as to provide the forecast results for the government and relevant managers, eliminate the possible crisis (that is, the police), improve the current crisis early warning system, etc., as shown in Fig. 1.

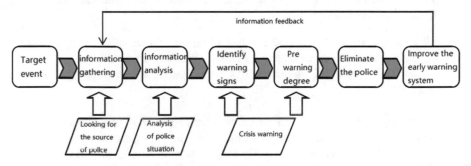

Fig. 1. Network public opinion crisis warning process

3 Msnp Network Public Opinion Crisis Information Dissemination Mode and Path

3.1 Msnp Network Public Opinion Crisis Information Dissemination Mode Analysis

In this study, the media is divided into media and online media. Traditional media report the actual information with its inherent authority and effectively control the crisis information, but online media increase the attention of Internet users by exaggerating events, and use big data technology to count and disseminate the crisis reflection and government public relations of Internet users; As a new media, mobile social platform can improve the spread speed of crisis information; As an important node of information processing and control of public opinion crisis, the government can understand the actual situation and enhance the openness and transparency of crisis events by communicating with event stakeholders. The information dissemination mode of network public opinion crisis is shown in Fig. 2.

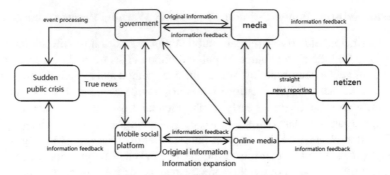

Fig. 2. Network public opinion crisis information dissemination mode

3.2 The Analysis of Msnp Network Public Opinion Crisis Information Dissemination Path

According to the six degree segmentation theory, this paper subdivides the participants, explains the functions and roles of the participants from the perspective of multi-agent, and describes the specific process of network public opinion from generation to extinction.

(1) Event triggers information generation. Sudden public crisis is the "fuse" of public opinion crisis on mobile social platform. If the parties, organizations or enterprises who deal with the incident improperly, they will easily become the "soul of the pen" of the network pusher. In order to timely and effectively control the information dissemination of network public opinion crisis, the event itself should timely control the social influence, reduce the degree of harm to society, and timely and properly handle.

(2) Media and online media control. Information report with the implementation of the regulations on information disclosure by the government, both the government and the media report the information related to the crisis according to the regulations. The media reports the news according to the routine, while the online media is vulnerable to network attacks, forming the phenomenon of false news dissemination. It is an important way to quickly identify the crisis, find the signs of the crisis, and put forward a reasonable and effective crisis warning scheme. In the process of information dissemination, the media gradually changes from information protection to information orientation.

(3) Internet users participate in information dissemination. As the core subject of public opinion crisis information dissemination, Internet users usually face crisis events with a variety of psychological characteristics. If curiosity, conformity and polarization play a leading role, it is easy to cause "secondary processing" of crisis information. If Internet users view crisis events with rational psychology, it is easy to make the news silent.

(4) Mobile social platforms spread information. With the rapid formation of mobile social platform and social network, netizens can express their views on crisis events with pictures, videos and audio anytime and anywhere. The emergence of new

media has both good and bad, which makes it convenient for netizens to express their feelings and opinions, and also provides convenience for Internet pushers, making rumors and false news spread on the Internet earlier than official news, causing netizens' panic and making people lose confidence in the government and relevant departments in handling the incident [3].

(5) Government oriented information dissemination. The dissemination of crisis information is the standard to verify the government's handling of crisis events. Generally, the government reports the real information generated by crisis events with the media. In order to avoid causing social panic, it usually chooses to reduce the transparency of information disclosure. However, it is affected by the rapid communication channels of mobile social platforms, which makes the relevant stakeholders, onlookers, the public, the media, the media, etc. People who know the truth have accelerated the development of the situation through the mobile social platform, resulting in the rapid growth and dissemination of network public opinion crisis information.

4 Construction of Msnp Network Public Opinion Crisis Early Warning Model

4.1 Msnp Network Public Opinion Crisis Information Analysis

In the context of big data, in order to eliminate the fragmented and useless "dirty data" in mobile social networks and accurately analyze the emotional tendency of network public opinion, it is necessary to preprocess the data before data mining, that is, to process the original data through data preparation, cleaning, conversion and reduction.

(1) Data preparation. Select the original data from mobile social platforms, such as Sina Weibo, wechat, Zhihu, etc., extract the information related to the specified crisis events, and use machine learning method to obtain some difficult to understand texts, and build a corpus suitable for manual analysis.

(2) Data cleaning. By determining missing values and deleting outliers to clean the original data, the purpose is to standardize the format of public opinion information on mobile social platform, remove abnormal data, correct errors and remove duplicate data. Data cleaning mainly uses metadata to determine missing values, which can test the attributes and missing values of original data by data omission, data interpolation and data sampling weighting. The cluster analysis method is to separate the value deviation, centralize the attribute values, and eliminate the values with large dispersion.

(3) Data conversion data conversion. The main application of the probability of data, standardization and other ways to achieve the transformation of data form. It usually detects the error of the original data, and then redefines and modifies the error in a way suitable for emotional orientation analysis. This operation abstracts and analyzes the data of different dimension areas by means of data generalization, which can improve the efficiency of information mining on the mobile social platform. Data normalization is the process of re dividing and re organizing the fuzzy information in the original data, and processing it in a standardized way.

(4) Data reduction. Due to the huge amount of original data, the time of data mining and analysis is too long. Data reduction can realize the operation on the premise of keeping the integrity of the original data. Dimension reduction can eliminate the noise data and redundant data in the original data, improve the operation rate of data mining through dimension reduction, and realize attribute reconstruction based on attribute changes [4].

4.2 Msnp Network Public Opinion Crisis Early Warning Model Design

Combining naive Bayes reasoning technology with a posteriori probability, this paper analyzes the text and characteristic words of public opinion crisis information on mobile social platform network on the basis of emotion dictionary, uses triangular fuzzy number to deal with the fuzziness of experts' opinions, de fuzzifies and averages the statistical data according to the information data processing rules, and determines the variable triangular fuzzy number evaluation value, The fuzzy comprehensive evaluation method is used to weight it. Triangular fuzzy number contains three possible values, namely lower possible value, possible value and higher possible value. The membership function of triangular fuzzy number is determined according to the probability distribution function. Given a random triangular fuzzy number $\tilde{A} = (A^l, A^m, A^u)$, and $0 \leq A^l \leq A^m \leq A^u$, then the membership formula of variable b belongs to $\tilde{A}$ is as follows:

$$\mu_{\tilde{A}}(b) = \begin{cases} \frac{b-A^l}{A^m-A^l}, & b \in [A^l, A^m] \\ \frac{A^u-b}{A^u-A^m}, & b \in [A^m, A^u] \\ 0, & b \in (-\infty, A^l] \cup [A^u, +\infty) \end{cases} \tag{1}$$

Because the performance of the network public opinion crisis is very fuzzy, the process of early warning of a network public opinion crisis in an emergency must be judged according to the experience of experts, so as to determine the basis of Bayesian network crisis assessment.

In order to effectively warn the public opinion crisis of mobile social network, this paper uses the fuzzy analytic hierarchy process to construct the crisis warning index system, and uses the triangle fuzzy number to determine the index weight, so as to determine the risk index of the early warning index of the network public opinion crisis. The network public opinion crisis warning index system is shown in Fig. 3.

From the perspective of six main factors, the popularity of online public opinion is mainly affected by the media, online media and mobile social platforms; As for the attitude of Internet public opinion, it is mainly influenced by the government and Internet users; As for the behavior of online public opinion, it is mainly affected by events, media, online media and mobile social platforms. Therefore, for different indicators, it corresponds to different main influencing factors. This study will give the overall effect of each main factor and its internal influencing factors for online public opinion crisis early warning in the simulation of crisis early warning effect. For the subjective influence level, the triangular fuzzy evaluation value given by experts is used to determine the importance of the index, and the evaluation judgment matrix and the fuzzy positive and negative value matrix are determined.

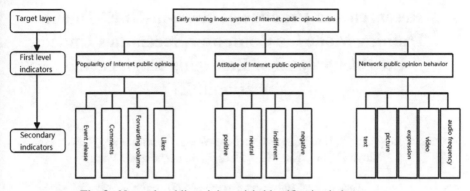

Fig. 3. Network public opinion crisis identification index system

5 Conclusion

With the rapid development of information technology under the background of big data, the efficient use of mobile social network can not only enhance the communication ability between the government and the public, but also avoid the network impact caused by emergencies to a certain extent, give full play to the quickness and convenience of the network, overcome the complexity of the network, and make the network public opinion emotional and emotional caused by emergencies Therefore, the research on the early warning of public opinion crisis on mobile social platform network has certain political and economic value to the real society. This study provides a new research perspective for our government, network public opinion managers and other public opinion management, and can provide reference for their research on the formation degree of network public opinion crisis caused by unexpected events and the treatment of relevant main influencing factors.

References

1. Tang, K.: Analysis of the impact of Internet public opinion on national security. China Soft Sci. (6), 56–62 (2008)
2. Huang, Y., Yu, F., Wang, X.: Reasons for the formation of Internet public opinion in the period of social transformation in China. J. Cent. China Normal Univ. (Humanit. Soc. Sci). **49**(3), 49–57 (2010)
3. Xia, L.: Analysis of the network public opinion transmission mechanism of emergencies. Editor. J. (1), 104–107 (2016)
4. Liu, H., Ren, D.Y.: Communication and control mechanism of Internet public opinion. Inf. Sci. **33**(4), 20–24 (2015)

Research and Practice of "Five-in-One" Talent Training Mode for Computer Specialties Under the Background of Emerging Engineering Education (3E)

Xianchao Wang[✉], Hao Wang, Xianchuan Wang, Huiling Wang, Dayou Hou, Xiuyou Wang, Yan Chao, Heng Li, and Zhongxin Wang

Fuyang Normal University, Fuyang 236037, Anhui, China

Abstract. Under the background of Emerging Engineering Education (3E), cultivating innovative talents with sustainable competitiveness is the new mission of higher education. This paper put forward the "Five-in-One" talent training mode based on the educational concept of "student-centered, outcome-oriented, and continuous improvement", and then focused on its connotation, methods and mechanism. At the same time, the outcomes of this mode after several years of implementation are given.

Keywords: Student-centered · Output-oriented · Continuous promotion · Sustainable competitiveness

1 Introduction

A new round of technological and industrial revolution is reshaping the global innovation landscape and reshaping the global economic structure. China is implementing major development strategies such as "Made in China 2025", "Artificial Intelligence 2.0", "Internet plus" and innovation-driven development. The rapid development of new economy urgently needs the support of new engineering talents with thick scientific foundation, high comprehensive quality and strong engineering ability, so as to promote the vigorous development of new economy characterized by new technology, new business form, new mode and new industry. Therefore, the reform and development of engineering education in the new period need us to set up new concept about comprehensive, innovative, and full cycle engineering education, to build new structure of engineering combined with a traditional engineering discipline specialized, to explore new mode for the implementation of engineering education personnel training, to build new quality of internationally competitive engineering education, to establish and perfect new system of engineering education with Chinese characteristics.

However, in the context of accelerating the development and construction of Emerging Engineering Education (3E), there are few research reports on how to develop and build 3E in local universities.

J. Macintyre et al. (Eds.): SPIoT 2021, LNDECT 98, pp. 750–755, 2022.
https://doi.org/10.1007/978-3-030-89511-2_99

The rest of the paper is organized as follows. Section 2 introduces thenew requirements for talent training of 3E. Section 3 focus on the connotation,method and mechanism of the "Five-in-One" talent training mode. Outcomes of the talent training mode are briefly listed in Sect. 4. Finally, we conclude and give future work in Sect. 5.

2 New Requirements for Talent Training of 3E

The concept of 3E was introduced in 2016. In 2017, the Ministry of Education in China launched the research work of 3E. On February 18, 2017, 30 universities jointly discussed the connotation and characteristics of 3E, as well as the path selection of 3E construction and development, which was called the "Fudan Consensus" [1]. On April 8, 2017, more than 60 universities agreed that training a large number of diverse, innovative excellence engineering science and technology talents was urgent and long-termed, formulated the new target of engineering construction forming an action route of 3E (Tianjin University action) [2]. On June 9, 2017, more than 30 experts deliberated and approved *the Guidelines for 3E Research and Practice Projects*, put forward the guidelines for 3E construction, forming the Beijing Guidelines [3]. They constitute the "trilogy" of 3E construction, which plays the main melody of talent training and opens up a new path of engineering education reform. Therefore, 3E is the reform direction of engineering education in our country, based on the new demand of national strategic development, the new situation of international competition, the new requirements for fostering virtue through education [4].

As new technologies continue to emerge, the engineers of the future will need to apply technologies that don't exist today to solve problems that don't exist yet. At the same time, the innovation cycle of the new economy is getting shorter and shorter; the boundary of technology development and industrialization is becoming increasingly blurred; technology updating and achievement transformation are more rapid; and industrial updating is accelerating. All these require engineering and technical personnel to have the consciousness and ability of innovation and entrepreneurship.

As the core of a new round of technological and industrial revolution, the Internet and Artificial intelligence both have the particularly strong cross-border penetration ability. Their industrial innovation mode requires engineering science and technology talents to have not only knowledge reserve but also interdisciplinary and cross-industry ability [5]. Therefore, under the background of the country layout of 3E for the future, speeding up the development and construction of 3E, to explore more diversified and personalized training mode for computer specialties of local undergraduate universities, to train creative ability and cross-border integration ability of engineering science and technology talents, to enhance the students' sustainable competitiveness have important practical and long-term strategic significances.

3 "Five-in-One" Talent Training Mode

3.1 Connotation of the Talent Training Mode

Under the background of 3E, cultivating innovative talents with sustainable competitiveness is the new mission of higher education. Therefore, the transformation from "teaching-centered" to "learning-centered" has become an urgent problem to be solved in the connotative development of higher education. In the long-term practice of application-oriented talent training, it is gradually convinced that the key to realize the transformation is to focus on the cultivation and promotion of students' ability with sustainable competitiveness as the core, to reform the teaching methods and service management mechanism and to optimize the talent training mode. We created "Five-in-One" talent training mode shown in Fig. 1 based on some provincial subject research, and practiced it for several years. In the mode, we are guided by value guidance, social demand and specialty objective; the education and teaching philosophy is student-centered, output-oriented and continuous improvement; moral education, teaching, competition, management and service are all valued and integrated into one; Six dimensional evaluation, including the curriculum evaluation, teaching evaluation, learning evaluation, outcome evaluation, specialty evaluation and social evaluation, is applied to check the achievement of goal. Thus, the "3356" talent training mode has been formed, the innovation of talent training mode has been achieved, and the coupling problem of "Five-in-One" has been successfully solved.

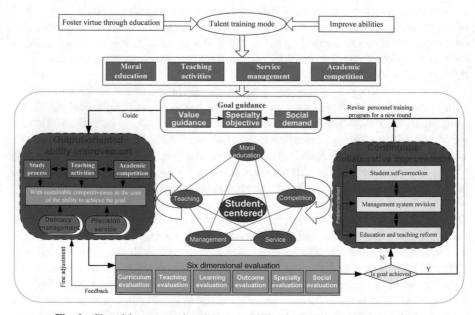

Fig. 1. Closed-loop operating diagram of "Five-in-One" talent training mode

The student-centered "Five-in-One" is tightly coupled and coordinated. In moral education, we have constructed an infiltrating value building system to implement ideological guidance by combining course education, league and student activities, academic competitions, and culture edification. In terms of teaching, we have promoted the achievement of students' abilities with sustainable competitiveness as the core by focusing on the achievement of specialty certification and building a competence-oriented curriculum system. In the aspect of academic competition, we have gotten through the first and the second classes to improve the comprehensive quality of students in an all-round way by constructing an output-oriented project training system with full staff coverage to promote teaching and learning in competition. In terms of management, we have constructed a closed loop continuous improvement mechanism by joining problem orientation, improvement measures and effect evaluation with the guidance of refinement. In terms of service, the student-centered service culture has been formed by implementing precise assistance to students. Together, they provide synergy for talent development.

Obviously, the talent training mode is for the full development of each student. And it is output-oriented and promotes the transformation from teaching-centered to learning-centered in an all-round and multi-channel manner by use of the formative evaluation and continuous and effective improvement of students' ability with sustainable competitiveness as the core.

3.2 Method and Mechanism of the Talent Training Mode

Teaching, youth league and college work together to create a positive and progressive cultural atmosphere for the students. And an integrated and infiltrating value shaping system is built by combining teaching, activities, competitions, culture and bases. We implement the ideological and political reform of curriculum by combining project seminar, special lecture and personal experience. We have combined the education of Marxist standpoint, viewpoint and method with the cultivation of scientific spirit in specialized courses to improve students' ability to correctly understand, analyze and solve problems. At the same time, we have place emphasis on strengthening the education of engineering ethics among students, cultivating students' craftsmanship spirit of striving for perfection, and inspiring students' national feelings and mission of serving the country with science and technology. In addition, with the help of "Internet+", special activities such as three rural areas, and other competitions, we have explored local traditional culture and red culture resources, and carried out characteristic activities such as red search and voluntary teaching.

Teaching and competition go hand in hand. We have constructed systematically the output-oriented teaching and competition activities which center on the cultivation and promotion of students' ability with sustainable competitiveness as the core, and carried out multi-dimensional evaluation to promote the achievement of the goal. Specific measures are as follows.

- To meet the needs of industry and society, the university and enterprise jointly optimize the talent training program, and carry out the specialty construction of 3E according to the international general certification standards.
- We carefully sort through the supporting relationship between courses and students' ability to achieve, and reconstruct all the specialized courses.
- We implement project-based academic and skill competition system, which strengthens the cultivation and promotion of sustainable competitiveness including innovation and entrepreneurship ability and cross-border integration ability.
- Finally, we implement the innovation of teaching evaluation, reform the evaluation system from the simple teaching evaluation to value, knowledge, ability and accomplishment of students to achieve the goal of multi-dimensional evaluation. At the

same time, based on the evaluation results, we continuously improve the teaching, management and service related to personnel training.

To implement the cooperation of the "Five-in-One", we have optimized the management system, improved the operation mechanism, and created a people-oriented service culture. The party and government jointly manage, build a collaborative mechanism, and provide students with all-round services from teaching, learning, competition, activities and other aspects. In the refinement-oriented management, we construct the closed-loop continuous improvement mechanism by linking problem oriented, improvement measures and effect evaluation. Moreover, we implement the academic tutor system to provide students with accurate services, such as precise help in learning, precise help in competition and precise help in creation.

4   Outcome of the Talent Training Mode

We has obtained rich theoretical research results. About 30 research papers on education and teaching have been published in journals including "*Computer Education*", "*Journal of Fuyang Normal University (Natural Science)*", "*Journal of Anqing Normal University (Natural Science)*" and "*Journal of Langfang Normal University (Natural Science)*".

The team has won about 20 provincial-level quality engineering projects and teaching research projects, including one major teaching research project, one first-class online and offline mixed course, two teaching talents, one ideological and political demonstration course and one online teaching excellent class.

We have won 10 provincial teaching achievement awards, including four first prizes and eight other teaching awards. Sustainable competitiveness of students significantly increases. About 30 students have won the national innovative entrepreneurial training projects in recent years. Students have obtained academic competition at or above the provincial level more than 200 items, including the college students' mathematical modeling competition, smart car competition, national software and information technology professionals competition and computer design competition for Chinese college students. Students have applied for or participated in the application of 12 patents, including 6 national invention patents applied in the past three years.

Therefore, the quality of talent training has improved significantly, and students' satisfaction has increased year by year. The "Five-in-One" talent training mode systematically solve the problem of the effectiveness of talent training and has theoretical significance and practical application value.

5   Conclusions and Future Work

This paper focus on discussing the connotation, method and mechanism of "Five-in-One" talent training mode. In the future, we will continue to take 3E concept as the guidance, engineering education certification as an effective means, implement the education concept of "student-centered, outcome-oriented, continuous improvement", further deepen the application of the talent training mode, and cultivate more innovative engineering application talents.

Acknowledgements. This work was supported in part by Anhui Province Major Teaching Research Project under Grant 2018jyxm0507 and in part by Anhui Province General Teaching Research Project under Grant 2020jyxm1384.

References

1. The consensus of Fudan University on the construction of emerging engineering education. Res. High. Educ. Eng. (1), 10–11 (2017)
2. The construction action route of emerging engineering education ("Tianjin University Action"). Res. High. Educ. Eng. (4), 24–25 (2017)
3. The construction guide of emerging engineering education ("Beijing Guide"). Res. High. Educ. Eng. (4), 20–21 (2017)
4. Zhong, D.H.: The Connotation and action of emerging engineering education construction. Res. High. Educ. Eng. (5), 1–6 (2017)
5. Liu, D.F., Wang, X.C., Wang, H., et al.: Cultivation and improvement of the sustainable competitiveness of students majoring in computer under the background of 3E. J. Fuyang Normal Univ. (Nat. Sci.) **37**(2), 119–122 (2020)

Multi-dimensional Analysis of University Library Borrowing Based on Big Data and Statistical Analysis

Xiu Dong[1], Lu Cai[2], Hongliang Fan[1], and Xianchao Wang[1(✉)]

[1] Fuyang Normal University, Fuyang 236037, Anhui, China
[2] No. 31 Middle School, Bengbu 233000, Anhui, China

Abstract. The interaction between readers and books produces the borrowing data that is unique to libraries. This paper focuses on the borrowing data of our university library from 2018 to 2020, takes the number of annually borrowed books (NABB) as indicators, and uses statistical analysis method to mine the hidden law of reader borrowing behavior, and studies the change trend of the numbers of monthly borrowed books (NMBBs) with time. The results show that undergraduates were the main body of readers in our university; literary books were the most popular; the average of borrowed books in liberal arts colleges was generally higher than that in science colleges. NABBs of classes showed a long-tailed distribution. The NMBBs had a periodicity, and NMBBs of spring semesters had a trend of decreasing, while the NMBB of autumn semesters had a trend of increasing first and then decreasing.

Keywords: Long-tail distribution · Number of borrowing books · Paper book

1 Introduction

With the development of information technology and automation technology, university libraries have gradually entered the era of large circulation. In order to realize the maximum circulation of books, the information of readers and each behavior node are collected. Thus a large amount of circulation data is generated. It contains the reader's borrowing tendency and the change of the library service. Both the borrowing quantity of paper books and the borrowing situation of readers are reflected in the circulation data. In the era of big data, statistical analysis of borrowing data can reveal the rules of readers' behavior, explore the changing trend of book reading, and then understand the potential needs of readers. It is conducive to the library to provide accurate services for readers. Therefore, the multi-dimensional statistics and analysis of borrowing data is particularly important. This paper intends to integrate and make a comparative analysis on the borrowing data of our library, summarize the current situation of borrowing, analyze the reasons and put forward corresponding solutions.

Analyzing the reading situation of paper books can mine the value of borrowing data. Xu et al. built a mathematics model based on the analysis of time series to analyze the

J. Macintyre et al. (Eds.): SPIoT 2021, LNDECT 98, pp. 756–762, 2022.
https://doi.org/10.1007/978-3-030-89511-2_100

changing rule of borrowing data using the borrowing data of Jiangnan University [1]. Li et al. analyzed the big data of reader behavior and summarized the accurate services that could be realized based on the big data of reader behavior, such as optimization of collection layout, accurate information service, scientific decision-making of electronic resources procurement [2]. Bian et al. established regression equation model based on the borrowing data of Nanjing Tech University from 2012 to 2017 and obtained the results that total annual borrowing books shows the descending trend with the increase of the year, and the borrowing rates in the spring and autumn semesters all show a trend of decline [3]. Wang et al. combined the cluster analysis and Apriori to analyze the big data of reader borrowing behavior of Tianjin Commerce University, excavated the borrowing behavior characteristics of readers of different grades and majors, and applied the results to the transformation of collection space, the innovation of service and the extension of service content [4]. Lee et al. analyzed the borrowing data of university libraries and found that the borrowing behavior of readers is periodic and power-law [5]. Based on time-driven activity-based costing (TDABC), Kissa et al. studied the book borrowing situation of Macedonian University to evaluate the borrowing and return process of the library [6]. Tsuji et al. used support vector machine to recommend books based on borrowing records [7]. It can be seen that they mainly used different models and data analysis methods to study the borrowing data. This paper will conduct a statistical analysis on the borrowing data of our university from 2018 to 2020 to find the borrowing rules of readers.

The rest of the paper is organized as follows. Section 2 focuses on statistical analysis of borrowing behavior data. Section 3 analyzes the relationship between the number of monthly borrowed books (NMBB) and borrowing time. Section 4 puts forward some advice. Finally, we conclude in Sect. 5.

2 Statistical Analysis of Borrowing Behavior Data

2.1 Analysis of the NABBs According to Reader Identity

The readers mainly included undergraduate students, graduate students and the staff. From 2018 to 2020, the numbers of annually borrowed books (NABBs) were 151671, 130886 and 60900 respectively. The reason that the NABB in 2020 reduced greatly compared with those in 2018 and 2019 was that students did not return to our campus in the first half of 2020 because of COVID-19 pandemic. Meanwhile, the NABBs of un undergraduate students, graduate students and the staff were (146035, 123238, 54543), (332, 2531, 3115) and (5304, 5117, 3242), respectively. And the annual percentage of undergraduate students, graduate students and the staff were (96.28, 94.16, 89.56), (3.50, 3.91, 5.32) and (0.22, 1.93, 5.11), respectively. It can be seen that with the increase of graduate majors in our university, the number of graduate students and the NABB of them were also increasing and the NABB and annual percentage of undergraduate readers were in the dominant position.

2.2 Analysis of the NABBs by Subject Category

The borrowed books are classified according to 22 subject categories such as economy (F), literature (I), general theory of natural science (N), mathematical science and chemistry (O), language and characters (H), industrial technology (T), culture, science, education, sports (G). And the NABBs by subject categories are shown in Fig. 1.

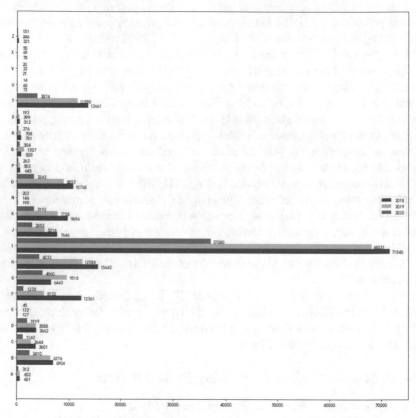

Fig. 1. The NABBs by different subject categories in 2018–2020

It can be seen that the books borrowed of different categories were very unbalanced: some categories reach tens of thousands, others only reach dozens. The top five categories of books borrowed from 2018 to 2020 are shown in Table 1.

Table 1. The top five categories of books borrowed from 2018 to 2020

2018		2019		2020	
Category	NBB	Category	NBB	Category	NBB
I	71540	I	68031	I	37080
H	15642	H	12584	G	4860
T	13661	T	11590	H	4232
F	12361	G	9518	T	3874
O	10768	O	8853	O	3243

From 2018 to 2020, the top five categories accounted for 80.48%, 76.02% and 78.29%, respectively. The reason why these books were widely borrowed is that they were mostly tool books which could provide great help for readers in different professional fields in daily study, scientific research, postgraduate entrance examination and certification examination. At the same time, the literary books were mainly novels which only needed shallow level, fast food reading and had fast circulation speed.

2.3 Analysis of the NABBs Per Capita by the School

The NABB is an important indicator of the book circulation. However, it is not enough to reflect the real situation of the borrowed books in each school because the number of readers in each school is very different. Therefore, we use the NABB per capita to show the borrowing situation of each school. There are 17 second-level schools in our university. The NABB per capita of each school is shown in Fig. 2.

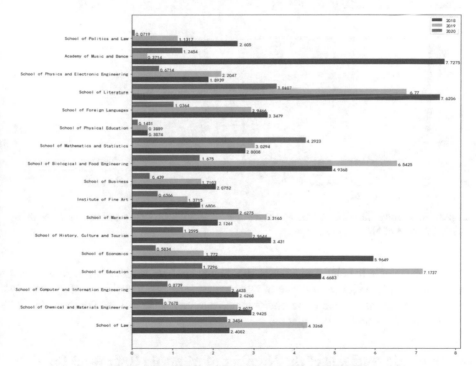

Fig. 2. The NABB per capita by school in 2018–2020

It can be seen that the maximums of NABB per capita of each school in 2018–2020 were 7.7275, 7.1737 and 4.2923, respectively. The NABBs per capita in 2018–2020 were 3.5260, 3.0525 and 1.4208, respectively. In addition, generally speaking, liberal arts schools had higher NABB per capita than science schools.

2.4 Analysis of the Borrowing Situation by Class

The NABB of a class can reflect the climate for learning of the class to a certain extent. Now we focus on investigating the borrowing situation of the undergraduate classes by considering the NABBs of the classes and the NABBs per capita of the classes. In 2018, there were 553 undergraduate classes, the maximum of NABBs by class was 1999 and the median was 122. In 2019, there were 571 undergraduate classes, the maximum was 2700, and the median was 102. There are 619 undergraduate classes in 2020, the maximum was 1504, and the median was 48. The minimums in these three years were all one. From 2018 to 2020, the maximums of NABBs per capita by class were 35.97, 48.62 and 27.00, respectively; the medians were 3.13, 2.75 and 1.30, respectively; and the minimums were 0.03, 0.03 and 0.02, respectively. The borrowing situation of undergraduate classes from 2018 to 2020 is shown in Fig. 3. As can be seen from Fig. 3, both the distributions of NABBs and NABBs per capita by class are long-tailed. In other words, the classes with high NABB or NABB per capita are small, while those with low NABB or NABB per capita accounted for the vast majority.

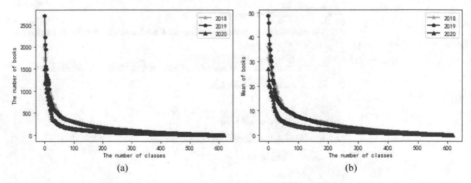

Fig. 3. Borrowing situation of undergraduate classes in 2018–2020 (a) distribution of NABBs, (b) distribution of NABBs per capita

It is reasonable to believe that the NABB and NABB per capita of each class have a great relationship with the guidance of teachers, especially counselors. If they actively guide their students to read more books inside and outside the class, the indicators of the class led by them will be higher.

3 Dynamic Analysis of the Number of Monthly Borrowed Books

The changing trend of NMBBs can reflect the common decision of readers under the condition of natural selection. The NMBBs of readers by semester in 2018–2020 are shown in Fig. 4.

As you can see, if it is not affected by the epidemic in the first half of 2020, the NMBBs of the three years have significant time consistency convergence trend. In the first semester i.e. the spring semester beginning in March, the NMBBs show decline trend. In other words, at the beginning of the semester, the NMBBs reached the maximum in March (21500 in 2018 and 16684 in 2019), decreased gradually with the progress of the semester, and reached the minimum (175 in 2018 and 4024 in 2019) in August. The dominating reason is that about 25% of the seniors and senior graduate students were engaged in the experiments and writing of their thesis, and gradually left in May and June. CET examinations in June and final examinations are also important factors to affect the NMBBs. The NMBBs in second semester starting in September first increased and then decreased. That is, the NMBBs gradually increased from September increased reaching the maximum (41157 in 2018, 17897 in 2019 and 28109 in 2020) in November, and then gradually decreased, reaching the minimum (3390 in 2018, 6602 in 2019) in February. The main reasons for the increasing trend from September to November are as follows. First of all, in September, about 25% of freshmen and a few graduate students enrolled, and the absolute increase in the number of students leaded to the increase of the NMBBs. Secondly, senior students and senior graduate students participated in the internship, which required them to borrow more books to learn more professional knowledge for guiding their own internship. At the same time, they had more leisure time to read their favorite books. Thirdly, quite a number of students spent a lot of money on shopping during the Double 11 every year. Thus, they couldn't go out to play or participate in other related activities, borrowing books to amuse themselves or study hard. And the main reason for the decreasing trend from November is that the CET-4 and CET-6 exams in December and the subsequent preparation for the final exams made them have no spare time to borrow books. In short, the NMBBs had a clear periodicity in some degree. In addition, the NMBBs of spring semester showed a decreasing trend while the ones of autumn semester increased first and then decreased.

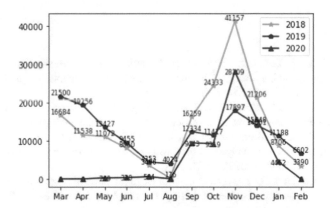

Fig. 4. Change trend of the NMBBs during 2018–2020

4   Advice

First of all, upgrading of library services needs speeding up. With the high informatization and intelligent development of library circulation business, the related work has changed gradually from resource center to service center. The transformation and upgrading of library services need precise and intelligent services to meet the personalized needs of modern readers. Therefore, the library should actively carry out the analysis and mining of the reader behavior data to effectively use the value of the reader behavior data, making the service more convenient and more humanized.

Secondly, more attentions need paying to cultivating student habit of reading. The NABBs of many classes are very low, which shows that the students have not developed the habit of reading. The main reason is short of correct reading guidance. Therefore, the teachers especially counsellors should often encourage students to go to the library to read books especially professional books. Meanwhile, teachers pay more attentions to the combination of the recommendation of professional books and the actual teaching process so that more students can benefit from in-depth professional reading.

5   Conclusions

To further improve the service level of the library, this paper focuses on the borrowing data of our university library from 2018 to 2020, and analyzes the data by statistical analysis. The results show that undergraduates were the main body of readers in our university, literature books had the largest borrowed volume, the NABBs by class showed a long-tailed distribution, the NMBBs had a periodicity.

Acknowledgements. This work was supported by the innovation team from Fuyang Normal University under Grant XDHXTD201703.

References

1. Xu, Z.R., Chen, Q., Guo, L.X.: Research on the book loan of university library based on the time series theory: taking library of Jiangnan University as an example. J. Libr. Inf. Sci. Agric. **30**(10), 107–110 (2018). (in Chinese)
2. Li, R., Wang, Q.S., Li, S., Dong, C.W., Song, J., Wang, L.: Research on library precision service and management innovation based on readers' behavior big data: taking the library of Dongbei University of Finance and Economics as an example. Libr. Res. **3**, 19–23 (2018). (in Chinese)
3. Bian, G.F., Li, T., Wu, L.: Analysis on borrowing behavior in university library based on data driven. Inf. Res. **7**, 79–90 (2019). (in Chinese)
4. Wang, L., Gao, X., Ma, B.: The path analysis of the application of borrowing behavior big data to the service innovation of university library. J. Acad. Libr. Inf. Sci. **38**(6), 107–120 (2020). (in Chinese)
5. Lee, T.H., Lee, J.W.: Self-organized human behavioral patterns in book loans from a library. Physica A **563**, 125473 (2021)
6. Kissa, B., Stavropoulos, A., Karagiorgou, D., Tsanaktsidou, E.: Using time-driven activity-based costing to improve the managerial activities of academic libraries. J. Acad. Librariansh. **45**,102055 (2019)
7. Tsuji, K., et al.: Book recommendation based on library loan records and bibliographic information. Procedia – Soc. Behav. Sci. **147**, 478–486 (2014)

Multi-scale Convolutional Neural Networks-Based Severity Prediction

Guangliang Liu, Huaming Liu, Fei Ding, and Zhonghua Yao[✉]

Fuyang Normal University, Fuyang 236037, Anhui, China

Abstract. The software defect report is an important part of the software repair process, and its severity attribute is very important. According to the description of the severity attribute of the software defect report, it can assist the developer to solve the defects described in the important defect report within a limited time. But manually assessing the severity is a tedious task. In this paper, we use a multi-scale convolutional neural network algorithm to automatically predict the severity attribute of defect reports. First, we use natural language processing technology to preprocess the defect report. Then create a vector for the defect report. Finally, the constructed vector input classifier based on multi-scale convolutional neural network is used to predict the severity. By comparing with other severity prediction algorithms, it can be found that our method improves the accuracy of the severity prediction of defect reports.

Keywords: Severity prediction · Convolutional neural network · Defect report

1 Introduction

Software defect repair is the key content of software version updates and software services, so the control of the software defect repair process has become the main problem faced by open source software [1]. The software defect tracking system can record defect information reported by software users. Through the analysis of the related attribute information of defect report, the process of the generation and repair of software defects can be visualized [2]. Therefore, the evaluation of severity attributes has become an important research direction in the field of software repair. The severity attribute is one of the many attributes of the software defect report, and it runs through the life cycle of the defect report. An accurate description of severity attribute can determine the defect's repair rate, resolution time and holder, etc. [3]. At the same time, this attribute will also affect the priority classification of software defects.

The severity attribute is mainly divided into six categories. There are blockers, critical, major, and normal with the higher severity attribute level, and minor, trivial, and enhancement with the lower level. For the severity classification of software defect reports, evaluations are usually made based on characteristics such as resolution time, repair rate, and holders.

Software defect reports can extract relevant attributes through text processing. According to these extracted attributes, the severity attribute of software defect report is

J. Macintyre et al. (Eds.): SPIoT 2021, LNDECT 98, pp. 763–769, 2022.
https://doi.org/10.1007/978-3-030-89511-2_101

classified. Some researchers have already done some work in this field. Zhang et al. [4] used the improved CP algorithm and the improved REP algorithm to classify severity and achieved good experimental results. Lamkanfi et al. [5] proposed the use of natural language processing to analyze the text description of the bug and achieved good results. However, in their analysis, they only divided the severity of the bug into two categories, serious and non-serious, and did not further divide it. Lamkanfi et al. [6] also studied the effects of four different algorithms on the prediction of bug severity. These four methods are Naive Bayes polynomial algorithm, KNN, Naive Bayes and SVN, and compare their severity prediction accuracy rates. The classification of software defect reports is researched through text mining methods, and the severity attributes of the defect reports are analyzed. Yang et al. [7] used a topic-based model to predict the severity of bugs. This method is first classified according to the product category and component category to which the bug belongs. Then the descriptive sentences in the bug report are segmented, and finally the severity is predicted by the topic model. Menzies et al. [8] proposed an automated method SEVERIS to help testers judge the severity of a bug. SEVERIS is based on standard text mining and machine learning techniques, and only uses the descriptive text of the bug as the original data. The method proposed by Antoniol et al. [9] is also based on text mining and improved on the feature selection method. Tian et al. [10] used a method of information retrieval. The innovation is the calculation of document similarity based on the BM25 algorithm, with the purpose of automatically predicting the severity of defect reports.

Many methods are not fine in the granularity of the division of the severity of defect reports, which is not very helpful for the priority arrangement of bug fix. In this paper, when classifying the severity attribute categories, we have taken into account the severity of the 6 types in a more fine-grained manner. Previous research algorithms generally only focused on Bayesian algorithm, KNN and other algorithms. Such algorithms have limited effects in feature selection and require manual selection operations. In this paper, the neural network model used eliminates the complicated steps of feature selection. It can select features by itself, which is more scientific and accurate without wasting manpower.

2   The Proposed Method

In this paper, the data set used is generated by the Eclipse software project, and its severity attributes have 7 categories. In the process of analyzing the severity attributes, it is found that defect reports with a severity level of enhancement have little impact on software functions and generally will not be assigned and processed. Therefore, this paper only selects the defect reports with the severity of blocker, critical, major, normal, minor, and trivial as the data set for severity prediction. Because it is a multi-classification task, it is different from the general two-class classification. In binary classification problems, the Logistic function is generally used, but its polynomial regression, that is, the softmax function, can solve multi-classification problems. If the probability of being recorded in a certain defect severity category is the greatest, it means that the severity of the bug report is in this category.

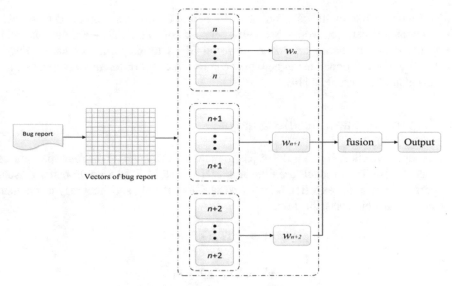

Fig. 1. The framework of MCNN-SP

Figure 1 shows our method MCNN-SP. The process of this method to predict the severity attribute is as follows:

First, we use natural language processing technology to preprocess the defect report.

We tokenize each defect report, correct spelling, remove stop words, etc., to remove irrelevant text descriptions in the defect report. We use the Natural Language Toolkit (NLTK) [12] to preprocess the defect report. The defect report includes many attributes. These attributes are basically described in words, so there are meaningless symbols in the description. These symbols do not help predict the severity attributes, so the relevant words need to be tokenized and filtered out meaningless symbols. The presence of stop words in the defect report may increase the amount of data, thereby increasing time for data processing, and may reduce the efficiency of the classification algorithm. We remove stop words from the preprocessed defect report. Word inflection converts some plural forms of words into singular forms, and some comparative and superlative words are converted into basic words through lemmatization. In order to avoid reusing words with the same basic terminology, we will carry out word inflection and lemmatization at the same time. Finally, we convert all preprocessed words to lowercase.

Second, each defect report needs to be converted into a vector. Through Word2vec proposed by Mikolov et al. [11], we can convert the defect report into a fixed-length vector. This vector can extract the relationship between the words in the defect report. The model built to generate this vector is essentially a neural network that predicts context words based on the central word. Words can be converted into vectors through a well-trained network.

Finally, as shown in Fig. 1, the vector obtained by word2vec is input to a classifier based on a multi-scale neural network to predict the severity of the defect report. The reasons for predicting the severity of defect reports using convolutional neural networks are as follows: First, multi-scale convolutional neural networks can obtain deep semantic relationships between words. Secondly, because convolutional neural networks can perform parallel computing on modern powerful GPUs, training time is greatly reduced [13]. Third, convolutional neural networks can use filters of different sizes to avoid the explosion gradient problem [14].

3   Experiment and Results Analysis

In this paper, we chose the Eclipse project to include defect reports for four open source products. These four products are Platform, PDE, CDT, JDT. In order to better present the performance of the experiment, we use Precision, Recall, and Accuracy to evaluate the severity attribute classification.

$$\text{Precision} = \text{TP}/(\text{TP} + \text{FP}) \tag{1}$$

$$\text{Recall} = \text{TP}/(\text{TP} + \text{FN}) \tag{2}$$

$$\text{Accuracy} = (\text{TP} + \text{TN})/(\text{TP} + \text{TN} + \text{FP} + \text{FN}) \tag{3}$$

TP (True Positive) means that a positive sample is judged as a positive sample.
TN (True Negative) means that a negative sample is judged as a negative sample.
FP (False Positive) means that a negative sample is judged as a positive sample.
FN (False Negative) means that a positive sample is judged as a negative sample.

In order to study the performance of the severity attribute prediction method, we compared the proposed method with Naïve Beyes and LDA.

3.1   Precision Comparison

Figure 2 shows the precision comparison between the Naive Bayes method, the topic model-based method LDA, and our proposed multi-scale convolutional neural network method on the classification of four datasets. The calculation method is the same as formula (1). In the four datasets, the precision of the naive Bayes method is about between 0.58 and 0.62, and the topic model-based method is slightly better than the Naive Bayes method, and the precision is about between 0.68 and 0.72. However, the precision of MCNN-SP is between 0.75 and 0.79. Obviously, our method is superior to the Naive Bayes method and topic model-based method LDA.

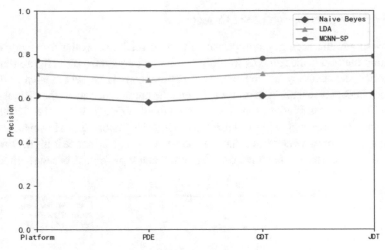

Fig. 2. Precision comparison

3.2 Recall Comparison

Figure 3 shows the recall comparison of the three methods on the four datasets, and the calculation method is the same as formula (2). In the four datasets, the recall of the naive Bayes method is about 0.57–0.63, and the topic model-based method is slightly better than the Naive Bayes method, and the recall is about 0.72–0.75. However, the recall of our proposed method is between 0.75 and 0.79. Obviously, our method is superior to the Naive Bayes method and topic model-based method LDA.

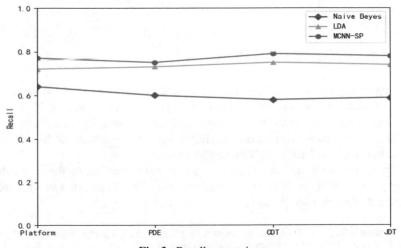

Fig. 3. Recall comparison

3.3 Accuracy Comparison in Six Categories

Figure 4 shows the accuracy comparison of three methods on the 6 categories. The calculation method is the same as formula (3). It can be seen from the figure that in the 6 categories, the accuracy is uneven and the difference is large, which is caused by the difference in data distribution. In these datasets, the amount of data in the "Blocker" and "Trivial" categories is relatively small compared to other categories, so the classification performance of three methods on these two categories is not high. However, it can be seen from the other four categories that the classification performance is high. Obviously, MCNN-SP is superior to the Naive Bayes method and topic model-based method LDA.

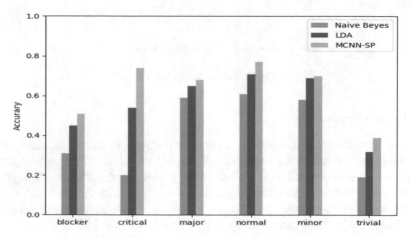

Fig. 4. Accuracy comparison in six categories

4   Conclusion and Future Work

In this paper, we proposed a severity prediction method named MCNN-SP. First, natural language processing techniques are used to preprocess defect reports. Then we use Word2Vec to extract the vector of defect report. Finally, the severity attribute is predicted based on the multi-scale convolutional neural network. The results show that MCNN-SP has relatively high severity prediction performance.

In the future, we would like to apply more machine learning methods to software severity prediction to better serve software repairs. We will also extend our severity prediction method to other projects.

Acknowledgments. This work was supported by PhD research startup foundation of Fuyang Normal University (No. 2020KYQD0037), the Basic and Frontier Technology Research Project of Henan Province (Nos. 132300410433, 122300410302 and 122300410384), and the natural science fund project of colleges and universities in Anhui province (No. KJ2020ZD46).

References

1. Ni, C., Liu, W.-S., Chen, X., Gu, Q., Chen, D.-X., Huang, Q.-G.: A cluster based feature selection method for cross-project software defect prediction. J. Comput. Sci. Technol. **32**(6), 1090–1107 (2017). https://doi.org/10.1007/s11390-017-1785-0
2. Ryu, D., Jang, J.-I., Baik, J.: A hybrid instance selection using nearest-neighbor for cross-project defect prediction. J. Comput. Sci. Technol. **30**(5), 969–980 (2015). https://doi.org/10.1007/s11390-015-1575-5
3. Strate, J.D., Laplante, P.A.: A literature review of research in software defect reporting. IEEE Trans. Reliab. **62**(2), 444–454 (2013)
4. Zhang, T., Chen, J., Geunseok, Y., Byungjeong, L., Luo, X.: Towards more accurate severity prediction and fixer recommendation of software bugs. J. Syst. Softw. **117**, 166–184 (2016)
5. Ahmed, L., Serge, D., Emanuel, G., Bart, G.: Predicting the severity of a reported bug. In: Proceedings of the 7th International Working Conference on Mining Software Repositories, pp. 1–10.IEEE, Cape Town (2010)
6. Lamkanfi, A., Demeyer, S., Soetens, Q.: Comparing mining algorithmsfor predicting the severity of a reported bug. In: European Conference on Software Maintenance and Reengineering, pp. 249–258. IEEE, Oldenburg (2011)
7. Yang, G., Zhang, T., Lee, B.: Towards semi-automatic bug triage and severity prediction based on topic model and multi-feature of bug reports. In:Computer Software and Applications Conference, pp. 97–106. IEEE, Vasteras (2014)
8. Menzies, T., Marcus, A.: Automated severity assessment of software defect reports. In: IEEE International Conference on Software Maintenance, pp. 346–355. IEEE, Beijing (2008)
9. Antoniol, G., Kamel, A., Massimiliano, D., Foutse, K., Yann-Gaël, G.:Is it a bug or an enhancement?: a text-based approach to classify change requests. In: Proceedings of the 28th Annual International Conference on Computer Science and Software Engineering, pp. 2–16. ACM, Ontario (2018)
10. Tian, Y., Lo, D., Sun, C.: Information retrieval based nearest neighbor classification for fine-grained bug severity prediction reverse engineering. In: 19th Working Conference on Reverse Engineering, pp. 215–224. IEEE, Ontario (2012)
11. Mikolov, T., Sutskever, I., Chen, K., Corrado, G., Dean, J.: Distributed representations of words and phrases and their compositionality. In: Advances in Neural Information Processing Systems 26, Nevada, USA, pp. 3111–3119 (2013)
12. Christopher, D., Surdeanu, M., Bauer, J., Jenny, R., Bethard, S., McClosky, D.: The Stanford CoreNLP natural language processing toolkit. In: Proceedings of the 52nd Annual Meeting of the Association for Computational Linguistics, MD, USA, pp. 55–60 (2014)
13. Ren, S., He, K., Girshick, R., Sun, J.: Faster R-CNN: towards realtime object detection with region proposal networks. IEEE Trans. Pattern Anal. Mach. Intell. **39**(6), 1137–1149 (2017)
14. Geoffrey, E., Hinton, Nitish, S., Alex, K., Ilya, S., Ruslan, S.: Improving neural networks by preventing co-adaptation of feature detectors. CoRR abs/1207.0580 (2012)

Research on UAV Scheduling Optimization in the Forest Fire

Qihang Yu, Huan He, Mingyang Li, Dayou Hou[✉], Jie Zhang[✉], and Xianchao Wang[✉]

Fuyang Normal University, Fuyang 236037, Anhui, China
houdy@fynu.edu.cn

Abstract. Wildfire is a kind of fire that occurs in the natural ecosystem, which will affect a series of processes such as ecosystem succession, carbon cycle and natural climate change. In recent years, wildfires continue to occur around the world, and the losses caused by wildfires are increasing. In this paper, we will take New South Wales and East Victoria in Australia as the research object, establish the corresponding model, and give the best UAV combination method to solve the wildfire extinguishing action. We use the UAV scheduling algorithm to establish a mathematical model about UAV combination, and then consider the threat factors such as mountain terrain synthetically, and further simplify the model. Also, we use data visualization to show the relevant information more concretely.

Keywords: UAV · Scheduling algorithm · Wildfire

1 Introduction

Wildfire, as a kind of fire occurring in natural ecosystem, affects a series of processes such as ecosystem succession, carbon cycle and natural climate change. Furthermore, because wildfires are prone to severe drought and persistent high temperatures, the intensification of global warming also seriously affects the of wildfires in various regions [1, 2].

Over the years, Australia's fire season has caused devastating wildfires in every state, and firefighters have been using drones to monitor fires and situational awareness. The related equipment carried by surveillance and situational awareness-unmanned aerial vehicle (SSA-UAV) can better the monitoring of the environment and achieve maximum security [3, 4]. For it to play its maximum role, the optimal number of combinations of SSA drones and related equipment should be determined, which is currently missing. Therefore, we will carry out relevant data analysis and model building to determine the optimal number and combined of SSA-UAV and its related equipment [5]. In this paper, we will take New South Wales and East Victoria in Australia as the research object, and use the unmanned aerial vehicle (UAV) combination model to give the best combination mode of SSA-UAV and related equipment to solve the problem of extinguishing wildfire.

The composition of the rest of the paper is as follows. In second 2, we introduce the UAV combination model, the third section focuses on the construction process of the model, and the fourth section is the combination results and its correlation analysis. Finally, we summarize the model in section five and generalize it.

J. Macintyre et al. (Eds.): SPIoT 2021, LNDECT 98, pp. 770–777, 2022.
https://doi.org/10.1007/978-3-030-89511-2_102

2 UAV Scheduling Algorithm

We can assume that the total number of UAV pairs is M, each UAV moves in different environments, and select the next position according to the state transfer rules. Assuming that the UAV is k in position at the moment of t_i, then the probability of UAV choosing the next position is

$$P_{ij}^k(t) = \frac{\tau_{ij}^\alpha(t)\eta_{ij}^\beta(t)}{\sum \tau_{iv}^\alpha(t)\eta_{iv}^\beta(t)} \quad j \in V, v \in V \tag{1}$$

The V indicates that the UAV-k can choose the next position set, α is the position information heuristic factor, the larger the α, the more the UAV-k tends to choose the path of most UAVs, and the β indicates the expected heuristic factor, which reflects the influence of visibility information on the UAV's next position.

In the Eq. (1), V denotes the set of locations where UAV-k can choose the next, α is the location information heuristic factor, and a larger α indicates that UAV k tends to choose the path taken by the majority of UAVs. β denotes the expectation heuristic factor, which reflects the magnitude of the role of visibility information on the UAV's selection of the next position, and a larger β value indicates that the UAV k tends to select the position close to the target point. $T_{ij}(t)$ represents the position information on the t time path (i, j); $\eta_{ij}(t)$ represents the heuristic information on the t time path (i, j), which is defined as

$$\eta_{ij}(t) = \frac{1}{d_{(j,g)}} \tag{2}$$

Where g is the target location, and $d_{(j,g)}$ represents the distance between j and g.

$$d_{(j,g)} = \sqrt{(x_j - x_g)^2 - (y_j - y_g)^2} \tag{3}$$

Because the UAV group will leave the position information when passing through a certain position, and the position information will change continuously over time, so after each UAV walks a position, the local position information should be updated. The updated formula is as follows

$$\tau_{ij}(t+1) = (1 - \rho)\tau_{ij}(t) + \Delta\tau_{ij}^k \tag{4}$$

$$\Delta\tau_{ij}^k = \begin{cases} Q/L & 1 \\ 0 & 2 \end{cases} \tag{5}$$

In the Eq. (4), where ρ $(0 < \rho < 1)$ is the degree of update of local location information, the degree of decrease in the amount over time Q is a constant, representing the information left by each UAV after searching the entire path once, L_{se} is the path length of the UAV. After obtaining the position information of a UAV passing through a certain position point, the residual value and value of the path information of all UAV's passing through the position point are the pheromone values of the position at that time. Then calculate the transfer probability and select the next position point.

The Eq. (5) represents 1 means that the drone selects j as the next position 2 means that the drone does not select j as the next position.

The core part of the algorithm is to simulate the transfer probability selection behavior of the ant colony by approximating the pheromone of the ant colony with the position information of the UAV and using the heuristic function value for transfer probability calculation, in which the inverse of the distance between the node and the target point is used as the heuristic information in the process of UAV state transfer is not conducive to the pre-avoidance of obstacles, and in the complex pair path planning environment, the algorithm search in a large space, the location information density on the path at the early stage of optimization is small, and the positive feedback information is not obvious, especially the "blind search" in the process of random solution generation generates a large number of local cross paths, which reduces the operation efficiency of the algorithm and is easy to fall into the local optimum, and after the search proceeds to a certain degree, it is easy to After a certain level of search, the stagnation phenomenon occurs, and all individuals find the same solution, so that no further search can be performed, which is not conducive to finding a better solution.

Based on the UAV scheduling algorithm, a UAV combination model is established. The problem can be expressed as under the condition that the center of the combination and the location of the task point to be executed are known, the UAV will be arranged to execute the task from the center of the combination to the designated point. In the combined scheme, a mission point can only be docked by one UAV, but each UAV can go to multiple mission points. In addition, the flight distance of UAV can not exceed its maximum limit, and various factors such as task priority, time constraint and terrain difference should be considered in the execution process. The ultimate goal is to maximize the benefits of the combined scheme at the minimum cost.

3   UAV Combined Model

By using the UAV combination model to solve task 1, the goal is to determine the optimal planning combination of UAV when the task is completed, so that the corresponding objective function can reach the maximum. We express the objective table function J, as follows

$$max\, J = \omega_1 \times R - \omega_2 \times C_1 - \omega_3 \times C_2 \qquad (6)$$

Among them, ω_1, ω_2, ω_3 represent the weight of income, the weight of distance cost and the weight of risk cost, R represent benefits for completing the task portfolio model, C_1 is the distance cost to complete the task combination model, C_2 is the risk cost to complete the task combination model.

We define income s priority x timeliness x distance, from the following formula, you can get income efficiency

$$R = \frac{\sum_{i=1}^{m} \sum_{j=1}^{n} x_{ij} \times r_j \times P_j \times l_j \times V}{T} \qquad (7)$$

Among them, x_{ij} represent decision-making factors, indicates whether the mission j is completed by the UAV i, r_j is the priority of the task j; P_j is the timeliness of task completion; l_j is the distance of the task; V is the unit normalized constant.

The constraints of the model are as follows

- A mission can only be carried out by one UAV, which can be expressed as

$$\sum_{i=1}^{m} x_{ij} = 1(j = 1, 2, \ldots, n) \tag{8}$$

- The start time of the task is not less than the execution start limit time, and the end time of the task is not greater than the execution termination limit time, which can be expressed as

$$t_s \leq s_i$$
$$e_i \leq t_e \tag{9}$$

- Total range not exceeding its maximum flight distance

$$\sum_{i,j=1}^{n} \sum x_{ijr} d_{ij} \leq L_{,max} \tag{10}$$

- Total flight time can not be greater than its maximum flight time

$$\sum_{i,j=1}^{n} \sum x_{ijr} t_{ij} + \sum_{i-1}^{n} \sum x_{ir} t_i \leq T_{max} \tag{11}$$

4 Presentation and Analysis

4.1 Results and Analysis of Data Visualization

We need an intuitive, interactive and responsive visual environment to help us do further research. Data visualization is a theory, method and technology that uses computer graphics and image processing technology to transform data into graphics or images to be displayed on the screen and to process interactively. Its essence is to communicate and communicate information clearly and effectively using graphical, so that the content expressed through data is easier to understand.

As the core of data visualization, it is particularly important to find the right chart for the right data. To make the form of data visualization effective and clear, according to the characteristics of the latitude, longitude, brightness, scan, track, acq_time, satellite, confidence, bright_t31, frp, daynight, type, we determine the characteristics of the graph as distribution class, proportion class and interval class, select histogram, pie chart and other visual charts, select four variables brightness, confidence, bright_t31, frp, to visualize the data in different ways.

We obtained the terrain data of Australia through NASA, and drew the topographic map of Australia by Airgis software in Fig. 1.

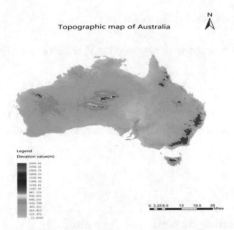

Fig. 1. Topographic map of Australia

Then, we performed a series of analyses by Kaggle [6] to obtain data from the MODIS C6 and VIIRS 375m of NASA satellite instruments from August 1, 2019 to January 11, 2020 in Australia.

By using a map API data visualization platform, we import the coordinates of fire one by one and get Fig. 2. According to Fig. 2, we carry determine the dense occurrence of fire spots.

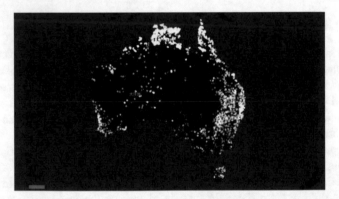

Fig. 2. Distribution of wildfire coordinates

Using the obtained fire-related data and using day and night as the dividing points, we made the proportion map of the number of wildfires occurring during day and night, shown in Fig. 3.

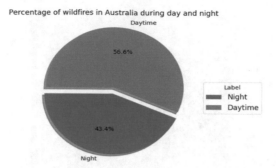

Fig. 3. Proportion of wildfires occurring during day and night

We divided the number of wildfire fires from August 2020 to January 2021, and made the frequency distribution histogram every ten days, shown in Fig. 4.

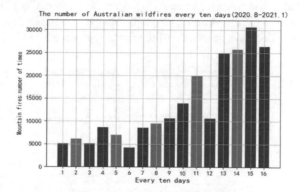

Fig. 4. Histogram of frequency of wildfire

According to the figure, we can know that wildfires occur very frequently in autumn and winter, which should be caused b they dry climate. And it needs to attract our attention to autumn and winter.

According to the figure, we can know that the proportion of wildfires occurring during the day is higher than that of wildfires at night. We speculate that this may be related to sunlight, so we also need to strengthen the monitoring. Night monitoring should not be taken lightly.

Using the obtained fire-related data, the longitude, latitude and time of the wildfire are formed into a coordinate system, and then combined with the brightness of the flame, the data visualization diagram is obtained, shown in Fig. 5.

Fig. 5. Distribution of flame brightness in a coordinate system

4.2 Results and Analysis of the Model Solution

In order to simplify the model, we use a circle to represent the region that is threatened by the UAV, the radius of the circle represents the range of influence of the threat, the center of the circle represents the threat point, the farther away from the center, the smaller the threat degree of the UAV. We get a map of the threat regions, shown in Fig. 6.

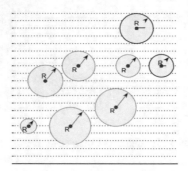

Fig. 6. Regional threat model map

By calculation, the final result is found that the SSA-UAV needs 26, while the Radio Repeater drones only needs 3, so that the cost can be minimized and the utilization efficiency can be maximized on the premise of completing the required tasks.

5 Conclusions and Future Work

The UAV uses information technology such as mobile Internet, Beidou positioning system and Internet of Things to provide intelligent decision-making and information services for flyersThe problem of intelligent deployment of UAVs selects the area to be operated according to the selected search range and working time window, and after a comprehensive analysis of the operating cost of UAVs and the estimated operating revenue, the goal is to minimize the operating deployment cost and maximize the operating revenue. The capture of location information can be effectively improved by loading a small number of UAVs with spectrum devices, which greatly saves costs and ensures that UAVs can cover the target area in a more balanced manner; the UAV scheduling algorithm model oriented to minimizing differences can enable rapid convergence of reconstruction results and achieve high accuracy requirements.

Acknowledgements. This research was supported in part by the Project of National Natural Science Foundation of China under Grant 61672006, the Key Project of Natural Science Research in Anhui under Grants KJ2017A340, KJ2019A0533 and KJ2019A0535, the Quality Engineering Project in Anhui under Grant 2018JYXM0507, 2020jyxm1384, and 2020xsxxkc327, the innovation team from Fuyang Normal University under Grants XDHXTD201703 and XDHXTD201709, the Natural Science Research Key Project of Fuyang Normal University under Grant 2019FSKJ06ZD, and the Fuyang Normal University Young Talents Key Project under Grant rcxm202004. And we would like to thank the reviewers for their beneficial comments and suggestions, which improves the paper.

References

1. Oris, F., Asselin, H., Ali, A.A., Finsinger, W., Bergeron, Y.: Effect of increased fire activity on global warming in the boreal forest. Environ. Rev. **22**(3), 206–219 (2014)
2. IPCC's Special Report on Global Warming of 1.5°C.https://www.cseindia.org/ipcc-s-special report-on-global-warming-of-1-5-c-9056
3. Chen, Y.H.: Forest fire detection of UAV based on convolutional neural network. Xi'an University of Technology, Xi'an (2019)
4. Shen, F.Q.: Automatic Planning Method of UAV Route Based on 3D Surface Model. Southwest Jiaotong University, Chongqing (2015)
5. Chen, X., Zhang, Y.: Simulation and analysis of mathematical model for global optimal scheduling of UAV channel. Microelectron. Comput. **34**(4), 135–139 (2017)
6. https://www.kaggle.com/carlosparadis/fires-from-space-australia-and-new-zeland

Short Paper Session

Internet Finance Development Opportunities and Challenges Based on Text Analysis Methods

Megan Wang[✉]

Dulwich College Beijing, Beijing 101300, China

Abstract. Internet finance is the product of the integrative development of internet technology and the financial industry. The businesses such as online business, payment and settlement business, and so on derived from internet finance have tremendously changed people's consumption habits and directly or indirectly affected the real business form. However, with the rapid development of internet finance, the existing space for traditional finance is further occupied, which brings both opportunities and challenges to traditional finance development. Based on the analysis of the development status and trend of internet finance, this paper conducts analysis and research on how traditional finance development is compatible with internet finance's opportunities and challenges for renewal development.

Keywords: Internet finance · Traditional finance · Opportunities and challenges · The development strategy

1 Introduction

The representative product of internet technology penetrating the financial field is the emergence of internet finance. The emergence and development of internet finance signify the emergence of e-commerce, investment and financing, third-party payment, and other brand-new functions in the financial field. As for traditional finance, the operation for internet finance business is characterized by low investment, high return, and comprehensive coverage, which is a double-edged sword. To survive in the financial industry, internet finance's opportunities and challenges should be fully understood in the traditional finance industry and take the initiative to seek new development under the circumstance of mastering all the conditions.

2 Analysis on the Current Development Status and Trend of Domestic Internet Finance

2.1 The Current Development of Internet Finance in China

Internet finance can be regarded as a new form of traditional financial mode supported by internet technology. Internet finance is no longer constrained by traditional finance ideas, which significantly improves the participation of many business links, thus greatly reducing the cost input required for the transaction. New types of internet finance, such as P2P network loans, have emerged, given the current development situation [1]. Moreover, the state has also introduced relevant laws and regulations to control all kinds of existing or new internet finance businesses reasonably based on the stability, safety, and rapid development of internet finance [2]. Simultaneously, the online service and transaction mode brought by the development of internet finance plays an irreplaceable role in the improvement of user experience and work efficiency, which is also the main reason why consumers increasingly rely on internet financial products and their businesses.

2.2   Analysis on the Future Development of Domestic Internet Finance

Nowadays, the deep integration of internet technology and various social life fields has become an irreversible development trend. On this occasion, the cutting-edge competition among internet finance industries will be intensified with the deep integration of internet technology and the financial industry [3]. Many departments of traditional finance, relying on their original advantages in offline funds, have gradually begun to explore personal financial services, which means that the degree of competition in the domestic internet financial market will be further intensified in the future.

To this end, the risk management that is closely connected with the development of internet financial will further develop toward high efficiency and precision. Internet finance is a brand-new product after internet technology is fully applied in the financial sector, while internet finance is still at a preliminary stage for exploration and development, which is true to the supervision. The lack of a targeted supervision system will negatively impact the in-depth development of the internet finance industry.

Under such circumstance, the further implementation of rising based supervision and effective control for internet finance can ensure its further development within a reasonable range. Simultaneously, in the process of the in-depth development of the internet finance industry, its matching degree with consumers' consumption behavior will also be improved accordingly [4]. All operation business platforms in the internet industry, especially the most popular e-commerce platforms, are gradually developing a customized product line for individual users' demands backed by all the online shopping technologies under the premise that related internet technologies' financial operation conditions are constantly improved and developed.

3   Opportunities and Challenges Brought by the Development of Internet Finance to the Re-development of Traditional Finance

3.1   The Opportunities Brought by Development of Internet Finance to Traditional Finance

From a particular standpoint, the emergence of internet finance is an inevitable product of the development in an era. The deep integration of internet technology and various social life fields is an important technical and social support for its emergence. The advantages of a significant reduction of transaction costs and remarkable improvement of coverage area can ensure the further improvement of financial institutions' economic benefits and improve customers' satisfaction for all kinds of financial behaviors involved in purchasing financial products.

To some extent, internet finance development has effectively alleviated the original imbalance between supply and demand in China's financial market. Moreover, as an embodiment of various characteristics of internet technology, internet finance provides many jobs for the whole society. While enriching the types of talents needed for social development greatly alleviates the domestic job market's incredible pressure [5]. It is also worth mentioning that the development of internet finance has also made China's original financial, ecological environment healthier and more efficient. These factors also provide a new financial ecological environment and opportunity for the acceleration of the implementation of the transformation of traditional financial strategy.

3.2 Challenges Brought by Development of Internet Finance to the Renewal of Traditional Finance

The development of internet finance is vital for the survival and development of traditional finance, and internet finance, on itself, is a double-edged sword. From one perspective, the business model adopted by various institutions involved in traditional finance is difficult to effectively and comprehensively meet customers' diversified demands in the current financial, ecological environment [6]. Against the background of the rapid, safe, and stable development of internet finance, all kinds of small business customers and young consumer groups need more personalized and differentiated financial services in the context of financial consumption.

Besides, the service experience requirements of various financial products and financial behaviors are also on the rise. However, various operating institutions involved in traditional finance still follow a very traditional service model in the current situation. Such a rigid model of products and service experience makes it difficult to fully and effectively meet financial customers' diversified and personalized needs.

From another perspective, the accelerated development of financial disintermediation has dramatically influenced the development and innovation of the traditional financial institutions' internal business, which is gradually permeated into all fields of credit financing of small and micro businesses in the business development and innovation. Besides, the relevant institutions with the third party payment have gained significant momentum in development. This kind of behavior will lead to a considerable impact on the traditional businesses of traditional financial institutions such as credit services. As one of the traditional representatives on finance, it can be easily discovered that the Commercial Bank is stereotyped in services with low flexibility and relatively high transaction cost compared with internet finance [7]. The existence of these problems puts traditional financial in the inferior position in competition with internet finance [8].

4 Traditional Finance Seeks a New Strategy Under the Development of Internet Finance

4.1 The Accelerated Implementation of Traditional Financial Transformation Strategy

By observing the change of people's living habits and policies issued by the state on the development of internet finance, it is easy to find out that the integrated development of internet technology and the finance industry is an irreversible developing trend, which will lead to profound and massive influence on financial products and businesses. In the development process, the financial industry has been supported by internet technology, which can improve the overall service quality and product innovation efficiency on the premise of further expanding the degree of openness of the financial industry to the outside world, and thus build a multi-layered financial service system [9]. To achieve re-development in the new ecological financial market environment, traditional finance must effectively deal with internet finance opportunities and challenges. While in this process, the strategic layout for traditional finance needs to be further changed, which is also a critical point in the development of traditional finance. This work's practical implementation plays a significant role in improving the traditional financial market's competitive strength. The change of ideology should be the leading factor for the implementation of any action. A high level of awareness of potential risks in case of unexpected incidents should be cultivated with the premise that the development trend of the times should be thoroughly mastered in the development of traditional finance. The concept for the development of internet finance is applied in the development strategy of traditional finance. As a result, innovation in traditional financial services and demands is further achieved to truly realize the improvement of profitability. Based on the development status of the internet and the traditional finance, it is not hard to see that countries are gradually relaxing the level of financial regulation, which was the key factor to achieve improvement in the openness and freedom of the financial market, which, however, also makes the competition that is already fierce between financial markets further intensified [10]. In this case, the traditional financial institutions need to implement the transformation strategy as soon as possible and provides quality, personalized, and diversified financial services and products based on the accumulated funds in the process of financial development and backed by the internet technology with the premise that the innovative development strategy.

4.2 Further Development of Interactive Cooperation

Under the background of the current social and economic development era is the primary driver of information technology. From this perspective, the relationship between the internet and traditional financial enterprises is not a zero-sum game. A win-win scenario can be achieved only by both sides' joint cooperation by drawing upon each others' advantages. The advantages of internet enterprises and financial enterprises are apparent. The former undoubtedly possesses the most high-end information technology in the development process of the times [11], while the latter possesses the advantages of financing accumulated in the process of financial development. Traditional financial need to cooperate with internet enterprises in an all-round manner for re-development. The software system should be developed with its own characteristics based on the business operation and enterprise development strategy to truly improve the financial service quality and innovation efficiency of products [12]. The traditional enterprises, in the process for further development, should invest more funds in the product innovation and cultivate talents for the intellectual output to build a talented team with high professional quality in order to enrich further and improve the knowledge structure and working competence of employees in ideology, which laid the foundation for the sustainable development of traditional finance.

4.3 The Continuous Improvement of Risk Control Mechanism

The internet itself is characterized by high risks, which means that in the development of traditional finance, it is necessary to further improve the internal control system under the premise of fully complying with the internet financial regulation policies and legal system formulated by the state. To be specific, we can ensure operation and management safety by constructing a security system and optimizing initial business processes.

Besides, it is also necessary to establish a corresponding security risk supporting system and then build the real-time monitoring and early warning platform with internet technology's help and work out the corresponding security emergency treatment mechanism on the premise of comprehensively and effectively predicting various financial risks that may occur [13]. Moreover, in this process, a scientific and rational risk measurement model should be established on the premise that traditional finance should be comprehensively exploited, and the business characteristics are followed. Backed by big data technology, the various clients should be evaluated on multiple dimensions to further improve the risk management system because the critical risks in the financial industry's sound development are sorted out.

5   Conclusion

The rapid and stable development of internet finance brings opportunities such as an ideal environment for traditional finance development. As a rising star, internet finance also makes the original market environment of the financial industry more complex and brings new challenges for further developing traditional finance. Traditional financial can genuinely realize the goal of re-development for development based on the opportunities and challenges brought by internet finance by implementing strategic plans and improving risk control, and deepening cooperation with internet enterprises.

References

1. Metawa, N., Elhoseny, M., Hassan, M.K., Hassanien, A.E.: Loan portfolio optimization using genetic algorithm: a case of credit constraints. In: Proceedings of 12th International Computer Engineering Conference, ICENCO 2016: Boundless Smart Societies 7856446, pp. 59–64 (2016)

 2. Yang, Y.: The development of traditional finance industry from the perspective of internet finance. China J. Commer. **33**, 48–49 (2018)
 3. Chen, R., Chengwei, G.E., Jun, Y.: The transformation and development of traditional commercial banks from the perspective of internet finance. Mod. Manag. Sci. **09**, 97–99 (2017)
 4. Zhou, J.: Challenges and opportunities of internet finance development to traditional banks in China from the perspective of function. Technol. Econ. Guide **07**, 226 (2017)
 5. Zou, Y., Tao, S.: Interactive development between internet finance and traditional finance from the perspective of functional. Cred. Ref. **12**, 80–86 (2016)
 6. Yan, D.: Development status of internet finance from a new perspective and countermeasures and suggestions of traditional banks. Stat. Inner Mongolia **02**, 8–9 (2016)
 7. Eassa, A.M., Elhoseny, M., El-Bakry, H.M., Salama, A.S.: NoSQL injection attack detection in web applications using RESTful service. Program. Comput. Softw. **44**(6), 435–444 (2019)
 8. Mu, N., Wang, H.: Research on the development opportunities and challenges of traditional commercial banks in the context of internet finance. E-commerce **2016**(02), 36+92 (2016)
 9. Zhou, Q.: Analysis of big data application mode and value based on internet finance. Heilongjiang Sci. **11**(18), 156–157 (2020)
10. Han, Y.: Preliminary exploration on the reform direction of internet finance supervision in China. Times Financ. **26**, 14–15 (2020)
11. Shankar, K., Elhoseny, M.: Trust based cluster head election of secure message transmission in MANET using multi secure protocol with TDES. J. Univ. Comput. Sci. **25**(10), 1221–1239 (2019)
12. Duan, Y.: Development of crowdfunding and risk management. Times Financ. **26**, 113–114 (2020)
13. Yin, X.: Application of Big data technology in internet finance in the information age. Fortune Today (Intellect. Prop. Rights China) **10**, 31–32 (2020)

The Influence of Blockchain on the Innovation and Development of the Securities Industry Under the Background of Information Technology

Haiyun Zhao[⊠]

College of Business Administration, Zhejiang University of Finance and Economics,
Hangzhou 310018, Zhejiang, China

Abstract. The rapid development of block chain technology has attracted extensive attention of the securities industry, which is of great value to the cost reduction and risk aversion of the entire securities market. At present, block chain technology poses severe challenges to the innovation, supervision, risk and other systems of China's securities market, which requires us to attach great importance to the innovative development of block chain technology in the securities industry, and we should adopt a cautious and gradual path strategy. This paper expounds the application value, challenges and innovation path of block chain in the securities industry in order to promote the standardization and innovative development of block chain in China's securities industry.

Keywords: Block chain · The securities industry · Innovative development

1 The Application Value of Block Chain in Securities Industry

The rapid development of block chain technology has attracted the attention of the entire securities market. The application of block chain technology in the securities industry has caused extensive discussions, and the block chain technology may have a far-reaching impact on the original traditional business of the securities industry. The application value of block chain to the securities industry is mainly reflected in the following aspects:

1.1 It is Beneficial to Reduce the Cost of Securities Industry

The innovative application of block chain technology in the securities industry is conducive to helping the securities industry reduce costs, such as transaction costs and review costs [1]. The application of block chain in the securities industry can greatly reduce the processing of intermediate links in the securities industry, and the original complicated process was done by computer technology, turning it into an automated process. For example, the application of blockchain technology of distributed synchronous ledger can change the simple standardized cooperation among securities industry into a

J. Macintyre et al. (Eds.): SPIoT 2021, LNDECT 98, pp. 787–791, 2022.
https://doi.org/10.1007/978-3-030-89511-2_104

concise real-time modification operation of common ledger, thus reducing the collaboration cost between institutions to a certain extent [2]. The information sharing brought by the application of block chain technology can improve the work efficiency of the securities industry. More importantly, the information storage technology applied by the blockchain technology is not easy to be changed maliciously, which is conducive to the confidentiality of information and provides convenience for the supervision and audit work of the securities industry, thus reducing the review cost. Therefore, the application of block chain technology in the securities industry has brought practical significance to its development to a certain extent, and promoted the securities industry to continue to develop in a good trend.

1.2   It is Conducive to Reducing the Risks Faced by the Financial Infrastructure of the Securities Industry

The application of block chain technology to the securities industry can effectively avoid transaction risks, credit risks and liquidity risks faced by institutions such as securities settlement system and CCP, and reduce the default risks among counterparties to a large extent [3]. The application of blockchain technology to the securities industry can fully guarantee the security of the entire trading system in securities industry and reduce the risk of a single point of failure of a central counterparty. In addition, the blockchain technology has advantages in confidentiality and the recorded data is not easy to be tampered with, and liquidation and settlement can be carried out immediately after the transaction, so as to ensure the reduction of default risk, which in turn promotes the reduction of other risks, especially in settlement risk [4].

1.3   It is Conducive to the Improvement of Work Efficiency of the Securities Industry

Blockchain de-intermediation technology can greatly promote the work efficiency of the securities industry, and is conducive to stimulating the innovation vitality of the securities industry. The use of the "common ledger" in blockchain technology can help the securities market realize real-time settlement across locations, thus greatly improving the efficiency of trading links. More importantly, the application of blockchain technology can make breakthroughs and innovations on the original model of the securities market, and the "blockchain" is integrated in product innovation. With the emergence of big data and 5G, the securities industry can use more and more technologies and carry out more inclusive innovation [5].

2   Challenges Faced by the Securities Industry in Innovative Development of Block Chain Technology

2.1   The Risk Regulation of the Securities Industry is Confronted with the Impact of Blockchain

The application of blockchain technology to the securities industry poses severe challenges to the original risk supervision system and brings new risks. It is necessary to readjust the supervision system according to the application degree of the blockchain technology and the securities market. First of all, it is necessary to consider whether the existing risk regulatory framework can effectively supervise and integrate blockchain technology. When the existing regulatory system cannot effectively match the blockchain technology, it is necessary for construction from the aspects of technical risk, market risk, liquidity risk, etc. [6]. Secondly, based on the original risk regulation and block chain technology, new regulatory priorities are selected to improve the risk regulation of the securities market, and loopholes are improved according to the impact and challenges brought to the securities industry by the use of block chain technology. In the future development of the securities industry, the risk supervision system will be gradually improved.

2.2   The Stability and Security of the Securities Industry is Challenged by the Block Chain

Block chain technology has unique advantages of disintermediation. While improving work efficiency, it can also improve the transparency of data in the securities industry. However, great risks also occurred, such as money laundering and fraud. In addition, many securities institutions conduct illegal fund-raising activities, illegal issuance of securities and other illegal and criminal activities under the guise of block chain, which pose a challenge to the security and stability of the entire securities market. In the face of such uncertain risks, it is necessary to bring the blockchain technology into the scope of supervision of the securities industry in a timely manner, standardize the development of various projects of the blockchain. Besides, a red line for risk supervision of the securities industry should be established to prevent illegal and criminal behaviors hidden under the blockchain technology, and thus guaranteeing the stability and security of the securities industry [7].

2.3   The Legal System of the Securities Industry is Challenged by the Block Chain

The application of blockchain technology will pose a severe challenge to the legal system of the securities industry. In addition to the securities law, there are a series of other legal systems. In the securities market, the legal restrictions on blockchain technology are not formulated, which is likely to result in the illegal qualification of subjects. For example, traditional securities institutions conduct trade and operation through coding rules when applying smart contracts, but the division of the responsibility subjects and legal responsibility subjects of such securities institutions is unclear. The application of block chain technology is made without the participation of intermediaries, which solves information asymmetry and moral hazard to a certain extent. However, the block chain technology still lacks the support of judicial subjects and it poses a challenge to the existing legal system of China's securities market [8].

3   The Path and Strategy of the Innovative Development of Block Chain in the Securities Industry

Despite the rapid development of blockchain technology, there are still many defects in risks, laws and other aspects. In view of the use of blockchain technology in the securities industry, a gradual process should be observed and the core technology of blockchain technology should be explored in the early stage of application to understand the suitable part of the securities industry, which can guarantee the safety of the data, reduce the cost of trading in settlement and liquidation. By fully understanding the superiority of blockchain technology, the block chain technology, infrastructure and the core competitiveness of the securities industry can be developed. From the perspective of the innovative development path of the securities industry in China, the integrated innovation of the securities industry and the block chain can be carried out through over-the-counter issuance, ABS and equity financing.

3.1   OTC Issuance Transactions

There are many problems such as high trading risk, lack of standardization of products, and miscellaneous procedures of equity registration, which have been affecting the conduct of OTC transactions. While the blockchain technology can solve some problems to a certain extent through low cost and achieving "real-time settlement", which is conducive to reduce the risk of transactions. The private key of Block chain can be a proof of equity transactors. Such activity can be completed without the participation of the third party under such technological means, as a result, the intermediate links are reduces, efficiency is improved and and cost is reduced to achieve rapid change of equity. Therefore, OTC transactions are smoothly implemented and the data security is ensured without falsification. The two parties can fullfill their contract under the requirements of designed contract [9]. With the help of decentralized blockchain technology, OTC transactions can be carried out smoothly, which will promote OTC transactions.

3.2   Asset Securitization (ABS)

Asset securitization occupies a broad space in the securities market, and the application of block chain technology can redesign the issuance, trading and other links in the securities marketization, so as to improve the efficiency of asset securitization and product innovation. To some extent, ABS can be effectively integrated with block chain technology, which can be adopted in all links of the securities market, such as loan initiation, securities issuance, reviewing rating and other links. The application of block chain technology in ABS can make public the transaction records and risk changes of its underlying assets, which is conducive to the supervision of the underlying assets and timely warning of the risks generated, thus effectively avoiding the risks [10]. From the perspective of the current practical application of blockchain technology, the most suitable blockchain technology in the field of asset securitization is the creditor's right assets. As the blockchain technology becomes more mature, it will be applied in more fields of asset securitization in the future.

3.3   Internet Based Equity Financing

Compared with the traditional equity financing, equity financing is endowed with the following characteristics: The small amount, mass transfer and demand, etc., the frequent

Internet based equity transfer causes excessive transaction costs and risk, and the lack of regulation in the securities market will result in many illegal and frauds. The equity registration, management and tracking and other links can be standardized with the application of blockchain technology so as to reduce the deal risk of the securities market in equity financing and there will be fewer illegal and fraudulent conducts. Block chain technology has a broad application prospect in Internet based equity financing, which will provide practical ideas for the innovative development of China's securities industry market.

References

1. Zhao, X.: Application prospect and analysis on business development path of the securities industry based on block chain technology. J. Shanghai Lixin Univ. Account. Financ. (06), 94–103 (2018)
2. Zhao, X.: Challenges posed by blockchain technology to the development of the securities industry and analysis on path strategy. Wuhan Financ. Monthly (12), 4–9 (2018)
3. Wang, Z., Li, D.: Influence of blockchain technology on financial industry and countermeasures. Econ. Res. Guid. (36), 70–71 (2019)
4. Wu, Y., Chen, D.: A brief analysis of the application of block chain technology in securities industry. Mod. Commer. (27), 124–125 (2019)
5. Li, W., Xia, M., Hua, M.: Blockchain technology: reform and ethical challenges in the field of finance. J. Shanghai Lixin Univ. Account. Financ. (03), 17–29 (2019)
6. Li. B., Zheng, B., Guo, Z., Wang, H.: Development and prospect of the application of blockchain technology in finance. J. Appl. Sci. **37**(02), 151–163 (**2019**)
7. Tang, W.: Research status and innovation trend analysis of blockchain technology in the financial field. Financ. Econ. (04), 136–137 (2019)
8. Yu, W.: Legal supervision path of the application of block chain technology to the securities market. Fujian Financ. (07), 17–21 (2017)
9. Liu, Y., Zhou, S.: Application and exploration, problems and challenges and regulatory countermeasures of securities blockchain. Financ. Regulat. Res. (04), 89–109 (2017)
10. Cao, F., Song, T.: Exploration on application of block chain technology in securities Market. Tsinghua Financ. Rev. (04), 42–45 (2017)

The Application of Artificial Intelligence in Architectural Design and Planning

Lili Peng[(⊠)] and Yingjuan Ye

Chongqing College of Architecture and Technology, Chongqing, China

Abstract. Artificial intelligence (AI) technology has been initially used in site design and architectural design, and has gradually played a role in actual project planning. It can simplify the design process and improve design efficiency. The use of AI technology to complete architectural design work will become the development trend of the construction industry in the future. This article analyzes and studies the application of AI in architectural design and planning.

Keywords: Artificial intelligence · Architecture · Design · Architectural planning

1 Introduction

AI technology is an emerging technology, which adapts to the development and needs of today's society, and is suitable for many fields. As for the application of AI technology, different people hold different opinions. Many people have a positive attitude towards AI technology, believing that it will make human production and life more convenient. At the same time, many people think that AI is widely used. Maybe it will bring some practical problems [1]. The use of AI technology still has more advantages than disadvantages. For now, AI can greatly improve their operational efficiency in some areas. In recent years, the construction industry has developed very fast, and with the continuous improvement of AI technology, the construction field has continued to develop towards intelligence [1].

2 Introduction to the Application of AI to Architectural Design

2.1 Application of AI in Building Structure Design

The application prospects of AI technology in the field of building structures have been widely recognized. Many people have changed their attitudes because of the convenience brought by AI, from the initial questioning to the current dependence. The application of AI technology in the design of building structures has also become the main trend in the development of intelligent building structures [2]. More and more people are optimistic about the combination of AI technology and the realm of life. Nowadays, AI technology has also launched some applications in building structures. In the design of building structures, the choice of materials is very important. With the advancement of science

J. Macintyre et al. (Eds.): SPIoT 2021, LNDECT 98, pp. 792–797, 2022.
https://doi.org/10.1007/978-3-030-89511-2_105

and technology, more and more economical and effective new materials are gradually applied to life, such as intelligent building materials. Intelligent building materials are used in civil engineering. Very large application potential. As far as civil engineering is concerned, the rise and development of smart materials not only means the increase of structural functions, the optimization of structural design forms, and more importantly, the updating of traditional design concepts [3]. In use, smart materials can monitor the state of the building structure in real time to ensure the safety and reliability of the facility and reduce its maintenance costs.

2.2 The Impact of AI on Architectural Design

Due to the rapid development of China's economy, the construction industry as the foundation of human society has also achieved rapid development. The construction of the building is first to carry out the exterior design of the building, and then to the architectural design of the building. Architectural design is mainly divided into three stages: architectural plan stage, architectural design stage and construction drawing design stage. Traditional architectural design relies on construction engineers to calculate the rationality, comfort, and safety of the building, and use appropriate materials to construct comfortable, beautiful, and economical buildings. Architectural design tends to be more technical and more concrete, unlike architectural exterior design, which tends to be more abstract. The traditional design relies on people's logical computing capabilities [4]. Obviously, computers have higher computing capabilities. When AI is combined with architecture, it is easier to obtain reasonable design results based on human thinking and the powerful computing power of computers, which greatly facilitates human life. With the further development of science and technology in the future, AI technology will become more and more mature, and its impact on the architectural design industry will increase. At the same time, it will inevitably also bring some social problems.

3   Advantages of AI Building Design

Fig. 1. Application of AI in building construction

From the evaluation of land value to the stage of forced arranging of residential buildings, there are many repeated revisions in traditional architectural design methods [3]. The use of intelligent design methods supported by AI technology will greatly reduce working hours and design costs, and improve design efficiency, as shown in Fig. 1. The architect only needs to input the site location, floor area ratio and other parameters into the software interface, and recommend the best result to the architect through his own calculation result evaluation plan, which greatly reduces the time and labor cost of the initial investment [5]. In addition, the intelligent design method can also generate multiple schemes in a short time. The creation and implementation of a plan requires the government to meet the rigid indicators (such as building density, floor area ratio, etc.) that the government proposes to the developer. For the developer, the profit of the building needs to be maximized, and each step needs to repeat the plan. Adjustment, and the expression and presentation of the plan takes more time. Using the intelligent design method of AI technology to quickly create profit-maximizing solutions under the constraints of various indicators, greatly changing the status of the industry, and getting rid of the constraints of rigid indicators on the conception and realization of ideas. AI combined with big data can accurately reflect the needs and laws of people's behavior in the venue and surrounding environment. Arrangement and analysis based on massive data can reflect the economic value of the site more objectively and reliably [4]. At the same time, the modification of technical indicators can also be reflected in the plan in time, and the analysis results of the plan can be output to the architect in a more intuitive way [6].

4   Application of AI in Architectural Design and Planning

The application prospects of AI technology in the field of construction have been widely recognized. Many people have changed their attitudes because of the convenience brought by AI, from the initial questioning to the current dependence [7]. The application of AI technology in architectural design has also become the main trend in the development of intelligent buildings. More and more people are optimistic about the combination of AI technology and the realm of life. Nowadays, AI technology has also launched some applications in architecture. In architectural design, the choice of materials is very important. With the advancement of science and technology, more and more economical and effective new materials are gradually applied to life, such as intelligent building materials, which are very important in the field of civil engineering [7]. As far as civil engineering is concerned, the rise and development of smart materials not only means the increase of building functions, the optimization of architectural design forms, and more importantly, the updating of traditional design concepts. In use, smart materials can monitor the state of the building in real time to ensure the safety and reliability of the facility and reduce its maintenance costs.

4.1   Design Better Through Generative Design

Architects have always used AutoCAD for drawing work, and a lot of time was spent in drawing without reflecting the real architectural design. The architectural art, ideas and innovative inspirations cannot be better displayed. With the in-depth research on AI technology in the architectural design industry, at present, all advanced concepts and technologies that accompany the entire design of the two-dimensional graphic description and three-dimensional space performance are adopted, and the system architecture and operation mode reflect the characteristics of architectural design [8]. AI software is a CAD system designed to provide a design environment for the architectural profession. It integrates digitization, humanization, parameterization, intelligence, and visualization. It is built on an AI platform and adopts advanced core technology for custom objects, with architectural components as the basic design. Unit, multi-view technology realizes the synchronization of two-dimensional graphics and three-dimensional models, as shown in Fig. 2. The application of AI technology in architectural design allows architects to bid farewell to the past drawing labor and experience design and creation in the true sense. Engineering drawing files are no longer simple line stacking, but are composed of building components that contain digital technology [8]. The system interface is humanized, component creation is parameterized, component association is intelligent, and the design process is visualized.

Fig. 2. AI forms a building information model

4.2 Prevent Cost Overruns

Robotics, AI and the Internet of Things can reduce construction costs by as much as 20%. Engineers can put on virtual reality goggles and send mini robots into buildings under construction. These robots use cameras to track work progress. Real-time interaction of on-site construction personnel, machines and objects, and reminds managers of potential safety issues, construction errors, and productivity issues [9]. AI can also be widely used to plan the wiring of electrical and plumbing systems in modern buildings, and use AI to develop safety systems on construction sites.

4.3 Risk Mitigation

Construction projects have various forms of risks such as quality, safety, time and cost. The more complex the project, the greater the risk, because there are multiple processes for construction, and multiple worksites are used to handle different industries [9]. AI can help optimize the process and monitor and remind the workplace. Risks can be identified in time and prioritized, so the project team can focus their limited time and resources on the biggest risk factors. Some process AI that cannot avoid risks can be performed instead, as shown in Fig. 3. AI is used for automatic reminders and alarms, so construction managers can work closely with high-risk teams to reduce risks [10].

Fig. 3. AI application and construction design

4.4 Project Planning

With the acceleration of urbanization, the scale and number of buildings are continuously increasing, and the corresponding building land area is constantly shrinking, which to a large extent has brought greater difficulties to engineering construction. How to plan the construction site three-dimensionally and scientifically is a prominent problem that needs to be solved in the current construction safety management, which is to reasonably divide the construction area, office area, living area, etc. according to the actual situation. While maintaining construction safety, we must also strive to maximize space utilization in order to obtain more ideal economic benefits [8]. Applying AI technology to the site planning link can measure the location parameters and red line range of the building space. When building a three-dimensional model, the above information needs to be integrated so that technicians can do a three-dimensional spatial analysis under visual conditions. Such an approach can not only optimize the site planning, but also identify potential safety hazards and risks in a timely manner, and effectively formulate prevention and emergency plans [10]. In particular, it can be based on the three-dimensional space to ensure the determination of the safety distance, avoiding frequent material handling and unreasonable space applications during the cross-construction process.

5   Conclusion

The development of the times and human needs for materials have changed the problems that need to be solved in the field of architecture. The current contradictions in the field of architectural design are the increasing demands of urban construction for the design industry on the one hand, and on the other hand, the traditional design methods are complicated in process and slow in efficiency. Design methods have been unable to meet the higher demands of modern people for architecture. At present, AI technology has already possessed application value in the forced platoon phase of real estate plans, and has provided a reference basis for later design decisions. Today, architects should begin to understand and accept the energy-related software of AI in the field of architecture, and use it in the process of assisting plan design, free from repeated plan modification and re-optimization, so that architects can be more focused based on the program design itself, it creates architectural value and promotes the healthy development of the industry.

References

1. Dang, J.Y.: Research on the ethical and legal risks of artificial intelligence. Inf. Secur. Res. **11**(12), 1080–1083 (2017)
2. He, Y.Y.: The application of artificial intelligence technology in intelligent buildings. Commun. World **11**(06), 256–257 (2017)
3. Wu, J.: Research on the application of artificial neural networks in architecture. Eng. Constr. Des. **10**(05), 16–19 (2015)
4. He, W.Y., Yang, X.D.: Artificial intelligence design, from research to practice. Times Archit. **9**(01), 38–43 (2018)
5. Liu, D.M., Zhang, R.N.: Discussion on the "form-finding" method of large-span buildings based on AI technology. Urban Archit. **10**(8), 10–13 (2018)
6. Xu, S., Miao, Y.F.: The second opportunity for artificial intelligence in architecture and urban design. Times Archit. **11**(05), 32–37 (2018)
7. Jin, Z.: Artificial intelligence is the next application arena in architecture. Intell. Build. **12**, 22–24 (2017)
8. Wang, B.N., Jiang, P.H., Qing, X.X.: Artificial intelligence technology and its application in the construction industry. Microcomput. Appl. **8**(08), 4–7 (2014)
9. Huang, K.: The application of intelligent control technology in intelligent buildings. Low Carbon World **6**(18), 164–165 (2016)
10. Chu, R.Z., Zhou, X.N., Li, S.G.: Artificial neural network and its application in building construction. Industr. Constr. **7**(11), 962–964 (2016)

Speech Perception Model Based on Artificial Intelligence Technology

Jingtai Li and Zijian Tang[✉]

School of Foreign Languages, Jiaying University, Meizhou 514015, Guangdong, China

Abstract. This article analyzes various factors that affect English phonological perception, points out that artificial intelligence technology can help improve English learners' English phonological perception, and puts forward some suggestions on specific teaching modes. The teaching of English phonological perception should follow the development of artificial intelligence, promote its own teaching reform and innovation, build an education system to realize personalized learning and lifelong development.

Keywords: Artificial intelligence · Improvement · English · Phonological perception

1 Introduction

There are many factors that affect listening level, but the process of perceiving the original encoding of sound is the first stage of listening comprehension. The starting point of the process of hearing comprehension is phonological perception. At present, English phonetics teaching mainly focuses on correcting students' pronunciation errors and training from the aspect of phonetic production. Due to the limitation of class time and the number of students, the effect of phonetics teaching is often not satisfactory [1]. The new round of technological change of education will take intelligent education as the guide, inject new ideas into education and teaching, provide new methods and tools, drive the fundamental transformation of education and teaching mode, and promote the quality improvement of teaching effect. The teaching of English phonological perception should follow the development of artificial intelligence, promote its own teaching reform and innovation, build an education system to realize personalized learning and lifelong development, and promote the transformation of education from low-level, extensive to high-level and accurate.

2 Concept of English Phonological Perception Teaching Mode by Artificial Intelligence

The new generation of learners are changing from passive acceptance learning to active discovery and exploration, from large class teaching to personalized teaching, from knowledge consumers to knowledge creators, and their demand for intelligent learning

J. Macintyre et al. (Eds.): SPIoT 2021, LNDECT 98, pp. 798–802, 2022.
https://doi.org/10.1007/978-3-030-89511-2_106

environment, ubiquitous learning resources and personalized teaching is also increasing [2]. The key of contemporary teaching reform is to meet the new needs of learners for teaching and learning methods, explore the possible effect of artificial intelligence technology in boosting education and teaching innovation, and solve the problems in the current education development. Intelligent personal assistant is a kind of artificial intelligence, which constructs natural human-computer interaction between human and machine [3]. Intelligent assistant can be applied in a variety of educational scenarios, such as companion learning, language learning and so on. Intelligent teaching assistant is not to replace teachers, but to explore how to connect teachers, students and technology to improve the efficiency of teaching and learning. By embedding technology into learners' daily life and supporting students' culture, practice and goals, teaching robots can promote learners' English phonological perception.

3 English Phonological Perception Teaching Mode by Artificial Intelligence

3.1 Change the Course of Learning

Digital learning is of great significance to the development of human learning, leading human learning into the era of networking, digitalization and globalization. Digital learning refers to the process of learning in a digital learning environment with the help of digital learning resources in a digital way [4]. It contains three basic elements: digital learning environment, digital learning resources and digital learning methods. The digital learning environment is mainly constructed by multimedia equipment, interactive electronic whiteboard, computer and Internet. Digital learning resources have the characteristics of diversity and richness, which can realize a wide range of open sharing and meet the diversified learning needs of learners [5]. The support of digital learning resources and learning environment provides conditions for diversified learning methods and helps to promote the comprehensive development of learners.

3.2 Correct Understanding of Self Deficiency and Self Advantage

A correct understanding of self deficiency and self advantage is the basis for learners to use appropriate methods to improve themselves. In the process of intelligent learning, learners can obtain adaptive learning resources, get timely feedback through intelligent evaluation tools, and find their own cognitive characteristics, learning preferences, advantages and disadvantages [6]. Intelligent learning can make learners know their own learning goals, position their own development direction, recognize their own value, tap their own potential, and realize personalized growth.

3.3 The Improvement of Comprehensive Quality and Innovation Ability

The ultimate goal of intelligent learning is to improve learners' practical ability, innovation ability and lifelong learning ability. Intelligent learning emphasizes situational awareness, acquires knowledge in the situation, applies knowledge in practice, inspires learners' sense of innovation, and constantly stimulates learners' thirst for knowledge, so that learners can improve their comprehensive quality and innovative ability in the process of exploring knowledge [7].

3.4 Intelligent Interactive Learning

Students interact with the external environment during the learning process, which helps to gradually build up their own cognitive structure, thereby effectively improving learning efficiency [8]. Researchers in the field of artificial intelligence have also begun to explore various new technical interaction methods, such as natural language processing, pattern recognition, etc., which can be used to improve the performance of educational artificial intelligence applications. Besides, human-computer interaction is an important part of the research in the field of artificial intelligence. Human-computer interaction can reconstruct the learning experience, provide more interactive teaching, and even influence people's cognition from the visual, auditory, and tactile senses [9].

4   Problems of English Phonological Perception Teaching Mode in My Country

First of all, traditional learning mainly relies on textbooks. Students perform the learning process of memorizing and recitation. Learning is only for the improvement of knowledge, which only examines the degree of knowledge mastery of students, and ignores the cultivation of comprehensive qualities and abilities [10]. As a result, students only pay attention to test scores and form an exam-oriented thinking, restricting students' creative initiative.

Secondly, in the traditional teaching process, students are relatively passive in learning, with consistent learning content, learning tools, and learning activities, and lack of personality characteristics [11]. Standardized learning makes it easy for learners to follow the trend and it is difficult to truly understand their own shortcomings and advantages.

Thirdly, in the course of traditional classroom teaching, there is a lack of effective interaction, and most students are in a passive learning position. Teachers generally only focus on students with better or worse grades. Frequently there are cold scenes where students deliberately bow their heads and are unwilling to raise their hands to answer.

5   The Strategies for the Improvement of English Phonological Perception by Artificial Intelligence in My Country's Colleges and Universities

5.1   Intelligent Detection of Blind Spots in English Phonological Perception

In the past, students often need to do a lot of exercises before teachers can discover where the learners are lacking in knowledge [12]. However, blind learning often results in a waste of time. Artificial intelligence can achieve refined matching for learners, intelligently detect the blind spots and repetition rate of learners, so as to guide or help people reduce the time of repeated learning and improve learning efficiency. For teachers, who have a full set of learners' learning trajectory data, which will improve the efficiency of providing teaching services.

5.2   Interest Driven, Guided Learning

Artificial intelligence learning companions should provide learners with personalized learning resources such as text, video, and audio according to their learning interests and knowledge mastery level, and automatically adjust the difficulty and depth according to the learners' learning progress [13]. The artificial intelligence learning partner gives thumb up to learners when they complete the learning tasks, and gives supervision and encouragement when they don't complete the tasks, so that learners can feel humanistic care and actively complete the learning tasks, without the need to learn passively under the pressure and requirements of teachers and parents [14]. The process of autonomous learning establishes the learner's dominant position. The learner sets the learning goals and learning process by himself, and launches learning activities independently.

5.3   Real-Time Interaction, Inspiration and Guidance

As an encyclopedia, intelligent robots can accompany learners, answer questions for learners at any time, and guide learners through interactive inspiration, so that learners can think by themselves, and provide learners with space for thinking and imagination. After the learner completes the pronunciation exercises assigned by the teacher, the artificial intelligence learning partner can automatically correct the learner's pronunciation [15]. On the one hand, it helps the learner to correct the pronunciation and make up the weak link in the pronunciation knowledge. On the other hand, it can find the shining points of learners, fully exploring the advantages of learners, and stimulating the learning interest of English phonological perception.

6   Conclusion

In the era of rapid change of knowledge and information, if learners only learn passively by "waiting and relying", they will eventually be eliminated by the society. Artificial intelligence learning partners can guide learners in autonomous learning and help learners master autonomous learning methods, because learning methods are far more important than learning content [16]. Learners should dare to innovate, have the belief that they can surpass teachers, and take the initiative to research and explore. Artificial intelligence technology is not only limited to facilitating learners to learn specific and structured knowledge and skills, but also to help learners acquire advanced abilities such as complex problem solving, critical thinking and deep learning, so as to improve learners' English phonological perception level.

References

1. Abramson, L., Garber, J., Seligman, M.E.P.: Learned helplessness in humans: an attributional analysis. In: Garber, J., Seligman, M.E.P. (eds.) Human Helplessness. Academic Press, New York (1980)
2. Bachman, L.F.: Fundamental Consideration in Language Testing. Oxford University Press, Oxford (1990)

3. Barnes, T., Boyer, K., Sharon, I., et al.: Preface for the special issue on AI-supported education in computer science. Int. J. Artif. Intell. Educ. (1) (2017)
4. Black, J.W., Hast, M.H.: Speech reception with altering signal. J. Speech Lang. Hear. Res. 5(1), 70–75 (1962)
5. Carroll, J.: Tools for Teaching in an Educationally Mobile World. Routledge, Abingdon (2015)
6. Davies, J., Brember, I.: The closing gap in attitudes between boys and girls: a five year longitudinal study. Educ. Psychol. 21, 103–115 (2001)
7. Goksel-Canbek, N., Mutlu, M.E.: On the track of artificial intelligence: learning with intelligent personal assistants. Int. J. Hum. Sci. (1), 593–601 (2016)
8. Holotescu, C.: MOOCBuddy: a chatbot for personalized learning with MOOCs. In: Iftene, A., Vanderdonckt, J. (eds.) Proceedings of the International Conference on Human-Computer Interaction - Ro CHI 2016. Matrix Rom, Bucharest (2016)
9. Lecumberri, M.L.G., Cooke, M., Cutler, A.: Non-native speech perception in adverse conditions: a review. Speech Commun. 52(11), 864–886 (2010)
10. Mc Arthur, D., Lewis, M., Bishary, M.: The roles of artificial intelligence in education: current progress and future prospects. J. Educ. Technol. 4, 42–80 (2005)
11. Pinkwart, N.: Another 25 years of AIED? Challenges and opportunities for intelligent educational technologies of the future. Int. J. Artif. Intell. Educ. 2, 771–783 (2016)
12. Roll, I., Wylie, R.: Evolution and revolution in artificial intelligence in education. Int. J. Artif. Intell. Educ. 2, 582–599 (2016)
13. Timms, M.J.: Letting artificial intelligence in education out of the box: educational cobots and smart classrooms. Int. J. Artif. Intell. Educ. 2, 701–710 (2016)
14. Vail, A.K., Grafsgaard, J.F., Boyer, K.E., Wiebe, E.N., Lester, J.C.: Predicting learning from student affective response to tutor questions. In: Micarelli, A., Stamper, J., Panourgia, K. (eds.) ITS 2016. LNCS, vol. 9684, pp. 154–164. Springer, Cham (2016). https://doi.org/10.1007/978-3-319-39583-8_15
15. Woolf, B.P., Lane, H.C., Chaudhri, V.K., et al.: AI grand challenges for education. AI Mag. 4, 61–84 (2013)
16. Zimmerman, B.J., Risemberg, E.: Self-regulatory dimensions of academic learning and motivation. In: Phye, G.D. (ed.) Handbook of Academic Learning, pp. 105–126. Academic Press, London (1997)

Exploration of Intelligent Grammar Cohesion Based on Deep Learning Technology

Jingtai Li and Xiao Zhong[✉]

School of Foreign Languages, Jiaying University, Meizhou 514015, Guangdong, China

Abstract. In the intelligent age, English writing teaching should not only focus on the transmission of knowledge, but also focus on cultivating students' [autonomous learning ability and deep learning awareness]. In classroom teaching, teachers can make use of artificial intelligence to develop personalized teaching methods, use the rich teaching resources of artificial intelligence to select unit theme teaching content, use the diversity of artificial intelligence evaluation to carry out continuous evaluation, use the efficiency of artificial intelligence teaching to make hierarchical teaching plan, improve the development technology of artificial intelligence, and boost students' deep learning.

Keywords: Artificial intelligence · Improvement · English · Grammatical cohesion

1 Introduction

In the age of intelligence, classroom teaching should not only be limited to imparting knowledge, but also focus on cultivating students' self-learning ability, deep learning consciousness and the ability to flexibly use the knowledge they have learned. English writing teaching is one of the ways to cultivate students' higher-order thinking. Problems such as low interest in English writing and low writing level are still common in English writing teaching [1]. Therefore, this article starts with the analysis of the characteristics of English writing teaching in the artificial intelligence-assisted teaching mode, and explores the teaching path of English writing teaching in the intelligent age to promote students' grammatical cohesion in English Writing.

2 Concept of AI-Assisted Teaching Mode for Grammatical Cohesion in English Writing

With the application of artificial intelligence in English teaching, education has entered a new stage of development [2]. The combination of artificial intelligence and education and teaching makes it easier for teachers to individualize teaching. The rich teaching resources and diversified evaluation methods in teaching are more conducive to improving teaching efficiency. Individualized teaching fully takes into account the differences of each learner. For the development of each student and the all-round development

© The Author(s), under exclusive license to Springer Nature Switzerland AG 2022
J. Macintyre et al. (Eds.): SPIoT 2021, LNDECT 98, pp. 803–807, 2022.
https://doi.org/10.1007/978-3-030-89511-2_107

of students, teachers can create more meaningful learning environment in the teaching process. Students are no longer passive learners, but are immersed in learning, actively exploring and devoting themselves to learning [3]. Artificial intelligence can record the entire learning process of students. Through the analysis and processing of these data, a "digital portrait" of each student is finally generated. Teachers and parents can learn about students through the "digital portrait", and schools can use these data to set the most suitable training program for students.

In English teaching, teachers should consciously use the artificial intelligence platform to strengthen the application of grammatical cohesion means in English writing through practical exercises, so that students can use various grammatical means proficiently, improving the ability of discourse construction [4]. Many students are not able to accurately express the coherence of English texts in English writing due to the influence of Chinese thinking. Therefore, teachers can use artificial intelligence resources to focus on the similarities and differences in the grammatical cohesion methods of English and Chinese texts, and enable students to grasp the differences in grammatical cohession mechanisms between English and Chinese in translation practice. Teachers should pay attention to the cultivation of students' English discourse construction ability, especially through the artificial intelligence assisted teaching technology to enable students to master and use English grammatical cohesion devices, enhancing students' sense of discourse and discourse coherence, and improving their English writing level [5].

3   AI-Assisted Teaching Mode

3.1   Develop Individualized Teaching Methods

In order to realize the individualized development of students and promote students' deep learning in the teaching of English writing, the learners should first be guided to master individualized learning methods [6]. Using the artificial intelligence platform, teachers can flexibly use the network platform to guide students in writing, and can also understand the learning situation of each student through the system, and then guide students to practice English writing in a targeted manner. Teachers can also analyze the learning situation of different learners in the system and arrange reasonable learning plans for learners. The learning system can also track the learning situation of students and give detailed and real-time feedback, which is conducive to learners' deep learning step by step [7].

3.2   Choose the Teaching Content of Unit Theme

English writing can be taught by unit theme teaching content, which refers to the integration of core knowledge related to the subject into a unit according to the curriculum standards [8]. This unit can fully reflect the development of curriculum knowledge, the depth of subject content and the law of subject content learning. Unit theme teaching content can stimulate students' learning motivation, and systematic knowledge system can improve learners' learning efficiency and promote students' deep learning [9]. In the era of artificial intelligence, teachers can design teaching activities as a whole by summarizing rich network teaching resources.

3.3   Carry Out Continuous Evaluation

The continuous evaluation in the teaching of English writing needs to formulate a continuous evaluation plan. The teacher should let every learner understand the evaluation criteria, including: the dimensions of the evaluation, the content of the evaluation, the method of evaluation, etc. [10]. Students can carry out self-evaluation against the evaluation standards, and the teacher's evaluation can be more targeted and reliable. Artificial intelligence can provide teachers with various evaluation methods through online mobile terminals. Teachers can better follow up on students' learning through mobile app evaluation and implement continuous evaluation.

3.4   Develop a Hierarchical Learning Plan

Hierarchical teaching means that teachers scientifically divide students into several groups according to learners' knowledge and ability levels, and teachers adopt appropriate hierarchical teaching strategies according to different groups, so that each learner can play his maximum potential [11]. In the intelligent age, in addition to dividing students into groups with different foundations, teachers can also consider hierarchical teaching content. Teachers can also use a hierarchical approach when assigning English writing assignments. For learners with a good foundation and excellent grades, teachers can assign essays on more difficult topics; For learners with intermediate grades, teachers can assign moderately difficult topics; For learners with a poor foundation and low interest in learning, teachers can ask them to recite good words and sentences from the articles they have learned, and to imitate or rewrite them [12]. With the help of the artificial intelligence platform, each assignment can help learners review the content learned in the class, so that students with different foundations can slowly achieve the purpose of deep learning in the process of continuous thinking and practice.

4   Problems of English Writing Teaching Mode in My Country

First of all, under the conditions of traditional education, teachers need to explain to dozens of students in a fixed place and space, which makes differentiated teaching particularly difficult.

Secondly, traditional English writing teaching focuses on summative evaluation, and pays less attention to process evaluation and continuous evaluation. On the other hand, continuous evaluation takes the development of learners as the core and takes the development of the core literacy of students as the orientation, which can effectively evaluate the comprehensive abilities of students [13].

Thirdly, due to limited resources, it is difficult for teachers to implement hierarchical teaching, making it difficult for learners to realize their full potential. Most of the students already have the most basic vocabulary and grammar foundation, but even the students who have a good command of English vocabulary and grammar, there are many problems such as poor writing, loose structure, unclear level, improper cohesion, lack of coherence and so on [14].

5   The Strategies for Adopting AI-Assisted Teaching Mode for Grammatical Cohesion in English Writing in My Country's Colleges and Universities

In the field of artificial intelligence writing, intelligent writing can be regarded as creative writing in the digital age. It is the process and form of human-like writing that combines existing algorithms, automation technology, basic conditions and laws of writing, and formalization of language symbols.

5.1   Further Understanding of the Theory of Writing Creation

As a kind of human-like intelligence, artificial intelligence is much stronger than ordinary people in many aspects, such as the collection, memory and processing of materials. The fact that intelligent robots can write reinforces the idea that ordinary people can write. Judging from the status quo of intelligent writing, some writing software has appeared on the market. As an artificial intelligence-assisted writing form, they provide ordinary people with writing formats and content templates to assist human authors in faster creation or templated batch creation. Therefore, with the participation and encouragement of intelligent writing robots, more and more people will increase their confidence in writing and make writing a necessary form of daily life just like using mobile phones and private cars. The era of national writing envisioned by creative writing will really come.

5.2   "Workshop System" Activities as a Comprehensive Discipline

Today's artificial intelligence writing machine is the result of collective creation. Its research and development process is a collection of multidisciplinary collaborations such as literature and art, writing, linguistics, semiotics, mathematics, computers, automation, psychology, neuroscience, and brain science. The crystallization of multi-disciplinary teamwork is difficult for one person to complete independently. Artificial intelligence writing also fully proves its possibility and rationality as creative writing in terms of its imitative and scientific nature. More importantly, as a new form of creative writing in the digital age, artificial intelligence writing has promoted the realm and connotation of creative writing.

5.3   Rising Digital Cultural and Creative Industries

The development of artificial intelligence writing is not to compete with human writing, nor is it to accompany human writers in writing games in the laboratory, but as a new form of creative writing that better serves the current language education [15]. At present, artificial intelligence writing can be divided into two categories: artificial intelligence-assisted writing and robot creation. Artificial intelligence assisted writing provides writing format and content templates, which can help human authors to create more quickly or create in batches. This kind of auxiliary writing can provide a large amount of cultural creativity and software services, which is favored by creators because it can greatly reduce the burden on creators and improve creative efficiency. The vitality of artificial intelligence works also lies in the fact that as a cultural consumer product, it will face the education market and obtain commercial aesthetic value.

6 Conclusion

In short, as a new form of creative writing in the digital era, artificial intelligence writing is not only an inevitable manifestation of the rapid development of digital technology in the field of writing, but also the result of people's huge drive for a better life in the creative era. It is an era of the integration of technological productivity and educational art. Teachers should make full use of the artificial intelligence platform, pay attention to the integration of English grammar cohesive methods into the teaching of English writing, teaching them as the key and difficult content, so as to help students effectively improve their English writing ability.

References

1. Abramson, L., Garber, J., Seligman, M.E.P.: Learned helplessness in humans: an attributional. Analysis. In: Garber, J., Seligman, M.E.P. (eds.) Human Helplessness. Academic Press, New York (1980)
2. Bachman, L.F.: Fundamental Consideration in Language Testing. Oxford University Press, Oxford (1990)
3. Barnes, T., Boyer, K., Sharon, I., et al.: Preface for the special issue on AI-supported education in computer science. Int. J. Artif. Intell. Educ. (1) (2017)
4. Bayne, S.: Teacherbot: interventions in automated teaching. Teach. High. Educ. (4) (2015)
5. Crowhurt, M.: Interrelationships between reading and writing persuasive discourse. Res. Teach. Engl. 23(4), 110–122 (1991)
6. Davies, J., Brember, I.: The closing gap in attitudes between boys and girls: a five year longitudinal study. Educ. Psychol. 21, 103–115 (2001)
7. Goksel-Canbek, N., Mutlu, M.E.: On the track of artificial intelligence: learning with intelligent personal assistants. Int. J. Hum. Sci. (1), 593–601 (2016)
8. Holotescu, C.: MOOCBuddy: a chatbot for personalized learning with MOOCs. In: Iftene, A., Vanderdonckt, J. (eds.) Proceedings of the International Conference on Human-Computer Interaction - Ro CHI 2016. Matrix Rom, Bucharest (2016)
9. Leontiev, A.: A Psycholosy and the Language Learning Process. Oxford Pergamon Press, London (1990)
10. Mc Arthur, D., Lewis, M., Bishary, M.: The roles of artificial intelligence in education: current progress and future prospects. J. Educ. Technol. 4, 42–80 (2005)
11. Pinkwart, N.: Another 25 years of AIED? Challenges and opportunities for intelligent educational technologies of the future. Int. J. Artif. Intell. Educ. 2, 771–783 (2016)
12. Roll, I., Wylie, R.: Evolution and revolution in artificial intelligence in education. Int. J. Artif. Intell. Educ. 2, 582–599 (2016)
13. Timms, M.J.: Letting artificial intelligence in education out of the box: educational cobots and smart classrooms. Int. J. Artif. Intell. Educ. 2, 701–710 (2016)
14. Vail, A.K., Grafsgaard, J.F., Boyer, K.E., Wiebe, E.N., Lester, J.C.: Predicting learning from student affective response to tutor questions. In: Micarelli, A., Stamper, J., Panourgia, K. (eds.) ITS 2016. LNCS, vol. 9684, pp. 154–164. Springer, Cham (2016). https://doi.org/10.1007/978-3-319-39583-8_15
15. Woolf, B.P., Lane, H.C., Chaudhri, V.K., et al.: AI grand challenges for education. AI Mag. 4, 61–84 (2013)

Analysis of Economic Growth and Environmental Pollution of VAR Model Under Big Data Technology

Haiyun Chen[✉]

Industrial Economics, Shanghai University, Shanghai, China

Abstract. Due to the continuous progress of China's economic growth and social development, the economic development has made remarkable achievements, but at the same time, this extensive development has brought serious pollution problems to the environment. As an irreplaceable resource, it is particularly important to better understand and deal with the contradiction in the pollution and economic growth. As a super first tier city and economic center in China, Shanghai, as a representative of rapid economic growth, has become increasingly serious in environmental pollution. Based on the data of GDP and industrial pollution in Shanghai from 2000 to 2017, this paper uses a VAR model to makes an empirical analysis on the relationship between economy and industrial pollution. It is concluded that there is a two-way mechanism between economy and industrial pollution in Shanghai, and whether the "inverted U" curve between the industrial pollution and economic growth is existed depends on the data of the region and different ways to measure the level of the industrial pollution.

Keywords: Economic growth · Industrial pollution · VAR model

1 Introduction

Due to the continuous development and progress, the economy has achieved great and rapid progress. The average of annual economic growth rate reaches 9.8%. However, this extensive economic growth in China, especially the economic growth driven by factor input and sacrificing environment in the late 1990s, has brought serious environmental problems. Therefore, in recent years, the sustainability of China's economic development has attracted much attention. Environment is an irreplaceable resource. Many data show that the cost of resources and environment for China's rapid economic growth is heavy, and economic growth will also affect the environment. In the future economic development, how to deal with the contradiction between economy and governance in the environment has become a very urgent and practical problem.

2 Literature Review

The empirical research on the relationship between economy and pollution began in the 1990s. In 1995, Grossman and Kreuger [1] proposed the Environmental Kuznets

© The Author(s), under exclusive license to Springer Nature Switzerland AG 2022
J. Macintyre et al. (Eds.): SPIoT 2021, LNDECT 98, pp. 808–813, 2022.
https://doi.org/10.1007/978-3-030-89511-2_108

curve after analyzing the changes of environmental pollutant emissions in many countries, that is, the relationship between the pollution of environment and the trend of per capita income in most countries is an "inverted U" curve. Subsequently, more and more researches on environmental pollution and economic growth appeared. However, egli analyzed the environmental data of Germany in 2002 and denied the existence of "inverted U" curve of Environmental Kuznets. In China, Zhang (1998) established a time series based on China's data from 1985 to 1995. After analysis, the relationship between per capita GDP and the level of environmental pollution in China is not obvious "inverted U-shaped" weak environmental Kuznets curve. Chen et al. [2] (2004) found that the Environmental Kuznets curve hypothesis is valid for most indicators through the analysis of environmental indicators in Shanghai for 11 years. Bao et al. [3] (2005) investigated the panel data of China from 1996 to 2002, and concluded that the selection of estimation method and pollution index affected the inverted "U-shaped" curve. Wang et al. [4] (2015) found that a "U-shaped" curve relevance between urban income and urban air pollution condition was existed in China. Many literatures focus on the relation between pollution and economy, and the research on this kind of relationship is mainly to verify whether the Environmental Kuznets curve of the impact of economic growth on the environment exists. This problem reflects a destructive relationship between economic growth and environmental quality, which is a one-way causal relationship (Apergis and Payne [5], 2010). Dinda [6] (2004) thinks that most of the literature on the Environmental Kuznets ignores the bidirectional influence mechanism between economic growth and environmental pollution. Yang and Ren [7] (2011) pointed out that environmental quality and economic growth quality complement each other, and there is a linear relationship with different slopes between them. The slope depends on the contribution of environmental quality to the quality of economic growth. Adhering to the green development with high efficiency of economy and low pollution of the environment is the only way for the coordinated development of economic growth quality and environmental quality. Li and Zhou [8] (2013) established VAR model by selecting per capita GDP and environmental pollution index data of Shandong Province, and concluded that whether the "inverted U" curve exists depends on the regional data and the indicators of environmental pollution. Li [9] tested whether there is the inverted "U" curve relation between pollution of environment and industrial structure. Avik Sinha and Bhattacharya [10] (2017) used the data of Indian cities to test the environmental Kuznets curve for SO2 emission.

3   Model Setting and Data Analysis

3.1   Model Setting

In this paper, the VAR method is used to research the two-way interaction mechanism in economic growth and pollution indicators in Shanghai. VAR model uses every endogenous variable of the system as a function of all the endogenous variables' lag value to construct VAR model, thus extending the single one to the VAR model formation of multiple time series variables. The basic form of the model is:

$$Y_t = \alpha_1 Y_{t-1} + \alpha_2 Y_{t-2} + \alpha_3 Y_{t-3} + \cdots + \alpha_p Y_{p-1} + \beta X_t + \varepsilon_t \tag{1}$$

Yt is an variable in the K dimension, XT is an variable in the D dimension, and a 1, P is the K × k-dimensional coefficient matrix which will be estimated, B is the K × d-dimensional coefficient matrix to be estimated; $\varepsilon t \sim i.i.d$ $(0, \sum)$(where $\sum$ is the variance matrix of the k-dimensional vector εT); εt can be synchronously correlated, but it is usually not related to the lag value or the variable on the right side; P is the lag order of the model. In this paper, the per capita GDP (last year = 100) is used to represent the economic growth of Shanghai, and the pollutant emission index is used to measure the pollution. The pollution emissions can be further divided to three parts: gas pollution, liquid pollution and solid pollution. Since the reform and opening up, the main source of environmental pollution in Shanghai is industry. To ensure the availability and continuity of the statistical data, this paper selects three indicators: industrial waste water, waste gas and solid waste discharge. The time series data of industrial pollution indicators and economic growth indicators are selected from 2000 to 2017. In order to eliminate the possible heteroscedasticity and obtain the stationary time series, all the data are processed by logarithm and first-order difference, as shown in Table 1 below.

Table 1. The variables

The variable	The definition	The variable	The definition	The variable	The definition
gdp	Per capita gdp	lngdp	gdp after logarithm	dlngdp	Lngdp after difference
water	Total industrial waste water discharge	lnwater	Water after logarithm	dlnwater	Lnwater after difference
gas	Total industrial waste gas discharge	lngas	Gas after logarithm	dlngas	Lngas after difference
solid	Total industrial waste solid discharge	lnsolid	Solid after logarithm	dlnsolid	Lnsolid after difference

3.2 Stability Test of VAR Model

In this paper, the first-order difference of GDP, industrial waste water, waste gas and solid waste emission in Shanghai is taken as variables to construct VAR model. The AR root estimation will be used to measure whether the model is stable.

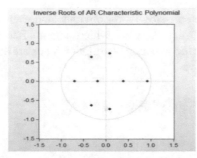

Fig. 1. The AR root estimation result

Through the unit root test in Fig. 1, it could be found that the reciprocal of all roots is in this unit circle. The results indicate that the model is stable enough and the results are effective to go on the study. On this basis, the impulse response of the generalized VAR model is applied to analyze the response between GDP and various pollution, and the dynamic relationship among variables is depicted. Here, the impulse response model with 20 lag periods is selected.

3.3 Generalized Impulse Response Analysis Based on VAR Model

It can be found in Fig. 2 that after the impact of dlngdp in this period, it can be observed that the total amount of industrial waste gas emissions presents an upward trend, then reaches the maximum value in the third period, then falls back in the fourth period, and then gradually tends to be stable, showing a certain degree of Environmental Kuznets "inverted U-shaped" curve, This indicates that the waste gas has an important impact on the environment of Shanghai. On the contrary, the "inverted U" curve of waste water

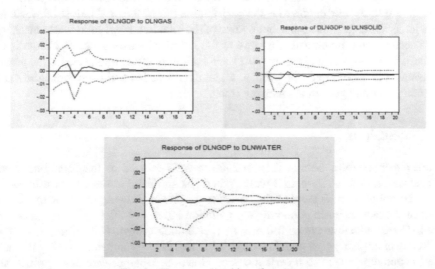

Fig. 2. The generalized impulse response

emission and solid waste emission is not obvious, and even the total industrial solid waste discharge presents "micro U type", which supports Bao et al. (2005) to think that the estimation method and the selection of pollution index are important factors affecting the inverted U-shaped curve.

3.4 The Analysis of Variance Decomposition

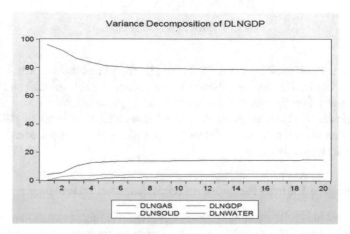

Fig. 3. The Variance decomposition result

Variance decomposition is to evaluate the importance of diverse structural shocks by analyzing the contribution of the structural shock to the change of different variables. It can be observed in Fig. 3 that in the error decomposition of economic growth prediction, from the average contribution rate, the contribution rate of the total industrial waste gas emission is the largest, the contribution rate of the total industrial solid waste discharge is the second, and the contribution rate of the industrial waste water discharge is the smallest. This shows that at present, the main inhibitory effect on the economic growth of Shanghai is the emission of waste gas, and the impact of solid and industrial waste water on economic growth is relatively small.

4   Conclusions

In this paper, GDP of Shanghai from 2000 to 2017 is used as the economic development index, and the industrial waste water, gas and solid emission are used as the industrial pollution indicators of the environment. Through using the VAR model, the interaction mechanism in economy and pollution index variables in Shanghai is analyzed. The conclusions are as follows: first, economic growth will affect the level of environmental quality to a certain extent. The level of environmental quality will also affect economic growth to a certain extent. However, whether economic growth and environmental pollution present an environmental Kuznets "inverted U" curve or not depends on the selection of data from different countries or regions, the selection of different indicators to measure environmental pollution and the choice of models. Second, in the study of the relation of the economy and pollution in Shanghai, it is found that the main problem of environmental pollution with the economic growth of Shanghai is too large amount of industrial waste gas emissions, while the industrial waste water and solid waste emission is relatively small.

References

1. Grossman, G.M., Krueger, A.B.: Economic growth and the environment. Q. J. Econ. **110**(2) (1995)
2. Chen, H., Liu, K.: Economic growth and environmental quality: an empirical analysis of the environmental Kuznets curve. Fudan J. (Soc. Sci. Ed,) **02**, 87–94 (2004). in Chinese
3. Bao, Q., Peng, S., Yang, X.: Is there an environmental Kuznets inverted U curve? – an empirical study based on six kinds of pollution indicators. Shanghai Econ. Res. (12), 3–13 (2005). (in Chinese)
4. Wang, M., Huang, Y.: Environmental pollution and economic growth in China. Economics (Quarterly) **14**(02), 557–578 (2015). in Chinese
5. Apergis, N., Dincer, O.C., Payne, J.E.: The relationship between corruption and income inequality in U.S. states: evidence from a panel cointegration and error correction model. Publ. Choice **145**(1–2), 125–135 (2010)
6. Dinda, S.: Environmental Kuznets curve hypothesis: a survey, **49**(4) (2004). (Elsevier B.V.)
7. Yang, F., Ren, B.: Quality of China's economic growth: evaluation from the perspective of carbon emissions. Soft Sci. **25**(11), 89–93 (2011). in Chinese
8. Li, Z., Zhou, D.: Empirical analysis of the relationship between economic growth and environmental pollution based on VAR model: a case study of Shandong Province. Enterp. Econ. **32**(08), 11–16 (2013). in Chinese
9. Li, P.: Testing the inverted "U" curve relationship between industrial structure and environmental pollution – an analysis based on the joint influence of industrial structure adjustment and economic growth rate. Econ. Issues (10), 21–26+109 (2016). in Chinese
10. Sinha, A., Bhattacharya, J.: Estimation of environmental Kuznets curve for SO2 emission: a case of Indian cities, **72**, 881–894 (2017)

Treatment Robot for Children with Autism Based on Deep Learning

Suzhen Qiu and Zhixiang Lu[✉]

College of Information Engineering, Nanning University,
Guangzhou 511458, Guangdong, China

Abstract. Autism is a condition of social communication disorder and repetitive sensory use. In recent years, the incidence of autism has gradually increased, and the number of children with autism has also increased. According to the reality of relevant data investigation, the time to diagnose autism is directly proportional to the effect of intervention treatment. At present, more and more experts have put forward insights on the diagnosis and treatment of children with autism, incorporating the technology of therapeutic robots into it. The application scenario of machine as the main technology builds a system for treating children with autism. By introducing classroom learning videos of children with autism, the current emotions of students can be judged in time, and students' performance can be evaluated and evaluated in a deep learning environment. Tracking effectively reduces the pressure on teachers.

Keywords: Deep learning · Children with autism · Therapeutic robot

In the study of treatment robots for children with autism based on deep learning, many scientists have discovered related performances under the same external stimulus among children with autism from different cultural backgrounds. At the level of the treatment of children with autism, it is very important to lead children with autism to interact efficiently with the outside world. The ability to detect the emotional response of children with autism in time is of great significance for the deep learning research of children with autism.

1 Research Technology of Treatment Robot for Children with Autism Based on Deep Learning

1.1 Early Diagnosis

Robotic technology has the advantages of interactivity, predictability, and simple operation. According to relevant data to investigate the reality, robotics technology has achieved good results in the treatment of children with autism, including early diagnosis, emotion recognition, social skills, and detection. Technology and early diagnosis, etc. [1]. Between the ages of 3 and 12, autism is prone to typical emotional expression behaviors, which can be displayed in social smiles, moodiness and other emotions. I know that some children will have these problems at an earlier time. If early diagnosis can be treated in the best time. Traditional autism diagnosis technology, doctor Pinggu, analyzes children's personality characteristics and behaviors, and is highly subjective. Robot technology can lie about the problem, make more accurate judgments, and use interactive methods to measure children's society. Time conditions ensure that the doctor's diagnosis method is more efficient. When the robot interacts with children, social signals can be automatically measured and recorded, without the need for doctors to operate on the spot, and effective data can be quickly obtained. Different sound clips can be played in time through the button device, and the time, date and auditory information can be automatically recorded, to obtain relevant diagnostic value [2]. Robots can have the advantage of sensors in perception system vendors, and can record and interpret children's socially relevant information in a timely manner without interaction. For example, in the process of developing a vision system, a robot integrates with a computer to determine the focus of attention and direction of the child's line of sight [3]. During the children's concentration machine, the difference between normal children and autistic children can be detected. The stereo camera in the visual system can detect the specific position movement of the child, and simply perform data statistics, including children and adults. Keep your distance, spend time together, how you will react to adults, etc.

1.2 Monitoring, Tracking and Recording, Analysis

Robotic research for children with autism is a relatively long process, requiring staff to practice records regularly, monitor and record the behavior of children with autism, and adjust the treatment plan in time during comparative analysis. The entire process will consume the treatment staff's time. And patience. In the current environment, the application of robotic technology in the treatment of children with autism presents the characteristics of interactivity, accuracy and objectivity. The robot system can detect in time the needs of children with autism in the interactive environment, and complete real-time transmission at the same time. Data storage and behavior analysis. The treatment robot system covers auxiliary systems, personal digital auxiliary equipment, remote servers, exchange centers, cameras, etc. The observation data can be input through PDA devices. The wireless channel data can also be transmitted to the remote server, with video storage, audio information storage, etc. Ok [4]. The robot system can observe the reaction of autistic children in time, and transmit data information to staff, doctors, parents, etc., to set a more complete dislike. The integration of robotics and other technologies is to synchronously capture the movements and emotional states of children with autism. The Kinect device is introduced into the robotic system to capture different joint movements of humans.

1.3 Emotion Recognition

Emotion theory is an individual's ability to perceive other people's emotions. Effective emotions belong to the ability to interpret situations, gestures, facial expressions, and sounds. In the current environment, many intervention methods can interpret the emotions of children with autism. However, due to the complex emotional language of human beings, and the rapid speed of body movements and facial expressions, children with autism are prone to enter a state of sensory overload. Robot technology has many advantages in emotion recognition and comprehension training for children with autism. For example, robot emotion detection with multiple emotion settings can understand each other's body movements and facial expressions [5]. Scientists use music to perform auditory stimulation to guide autistic children and robots to recognize emotions in social interaction. By combining music with robots, a complete game interaction plan can be set up, and an objective evaluation of the emotional effects of human-computer interaction can be put forward. Robot technology helps doctors understand the emotional state of children with autism more clearly. In professional emotional assessment and detection, they can understand and track the emotional reactions of children with autism, such as anger, disgust, surprise, and happiness, and identify their emotional motives in time. Can realize multi-dimensional treatment.

1.4 Social Skills Training

Children with autism have social barriers and rarely interact with society. They emphasize individualism and neglect their connection with the outside world. Social barriers are not conducive to children's social relationships, healthy mental state, and peer acceptance. According to early data research, it is found that in behavior predictions made by peers, children with autism are more likely to experience anxiety. Robotic technology has many advantages in social intervention, including human attachment to things and overcoming fear of the outside world. Enhance the confidence of social interaction, etc., which is conducive to the development of social skills of autistic children in the social environment [6]. When a child with autism triggers the sensor, the machine will respond with sound, vibration, and light. The therapist will help promote the interaction between the child with autism and the people around him during the process of controlling the instrument. The interconnection of other interactive sensors can obtain timely information on the internal physiological state of autistic children, including stress level, internal temperature, heart rate, etc., interactive robot system, and timely assess the response level of autistic children, and promote children's social interaction Skill development.

1.5 Behavior Modification

Children with autism are Changhu who show repetitive and stereotyped behaviors. Executive dysfunction can effectively explain this problem, including organizational planning, working memory, impulse control, and attention maintenance. Children with autism have difficulty in controlling personal attention and suitability, and have difficulties in planning and coordination. Applied behavior analysis is the current advanced behavioral intervention method, but there is a problem of abundant resources in treatment. Continuous training needs to be controlled within three years. Most children need to complete professional service guidance, and robots will not have fatigue problems. Only need to complete a lot of repetitive work, no time limit. According to research findings, robotics can help children with autism complete adaptive behaviors and have achieved certain results in promoting healthy lifestyles [7]. The current robotics technology mainly uses the ABA intervention principle to strengthen behavioral intervention in the stimulus response. For example, the robot-assisted behavioral intervention system meets the requirements of the ABA unit teaching method. The system currently has the following functions: to detect the eye detection of children with autism; the system recognizes the corresponding frontal tactile response when the child is in contact with the sensor; during behavior control, the system is used to grasp the behavior of children with autism and decide whether to Turn on the treatment, encouragement and pause mode. When the child with autism maintains the treatment mode and implements it, the robot will respond positively; when the robot does not pay attention to the training scene, the system will turn on the "encouragement" mode, through dance, music, and fun Positive reactions such as actions; when a child with autism feels "tired", the robot will start the "pause" mode.

1.6 Guide Game Interaction

The diagnostic criteria for children with autism released in the United States emphasized that children with autism have difficulty understanding imaginative play activities, and it is difficult for children with autism to engage in play activities with children except for machine games. The robot has the basic function of a toy and can communicate with it. For people, robots get along well with children with autism. Robot technology can create a simple interactive environment, relying on the communication and interaction of children with autism as a companion, and can improve children's social skills, including language expression, social interaction, language imitation, etc. There are many remote control game modes for robots, including hat games, posture transfer and so on. During the game interaction, the robot can send out interesting actions to attract the attention of autistic children, and at the same time use language to praise [8]. Robots will also require children with autism and their game partners to use social language, while paying attention to the politeness of greetings. According to relevant data surveys, it is found that children with autism can show social communication skills just like normal children. Concerned that game cooperation in real life requires frequent exchanges and communication, scientists have designed a series of game content to simulate real scenes, and set up game cooperation content with the help of attractive videos to encourage children with autism in organized The team and the robot complete the cooperation goal together. In experiments, it was found that children with autism can complete good game interaction or game collaboration with robots.

2 A Typical Representative of a Treatment Robot for Children with Autism based on Deep Learning

The important logic of deep learning is to essentially affect the degree of knowledge mastery of children after learning. In the study of treatment robots for children with autism based on deep learning, the appearance design of the robot is a key channel for the training of the emotional and cognitive abilities of children with autism, and children with autism have a relatively longer gaze for non-social structure subjects. The appearance design of autism intervention robots mainly includes the following categories: robots, animal robots, and other modeling robots. Because robots are similar in appearance to humans, they are easy to be recognized by children with autism. However, some robots do not need to be real. In order to strengthen their social cues and avoid distraction of children with autism, animal-type robots and simulated animal-type robots can be used. It is a robot pet and has attracted the attention of children with autism. According to relevant data surveys, children with autism prefer animal robots. In the current environment, many types of robots have been developed on the market, and they are equipped with sensors to ensure that the robots have the ability to react and perceive. The installed sensors cover light sensors and acceleration sensors, which can well control the movement trajectory while avoiding obstacles in front. Under normal circumstances, the common functions of these robots include: emotion recognition and functional expression, automatically recording the expressions and voices of children with autism, and recognizing their emotional state. The robot uses facial expressions and body language to express its emotions [9]. In order to enhance the intuitiveness of robot emotions, the researchers also set up color buttons on the eyebrows of some robots. In communication, robots can maintain basic frontal communication with autistic children. Some robots that lack language capabilities can use body language to express themselves. Robots with language capabilities can speak many languages [10]. In the intelligent entertainment interaction with autistic children, the robot effectively improved the exhaustion of the metal rods of the autistic children. During a large number of exercises, the game interacted with the autistic children, which improved the children's social recognition. Knowing ability.

3   Conclusion

In summary, with the help of assistive tools, children with autism can be helped in time. The development and application of robotics technology effectively solves the research direction of the treatment of children with autism. According to relevant data research, it is found that robotic technology is used in the symptoms of children with autism. The upper treatment is remarkable. Compared with the traditional method, it has achieved a positive effect. At present, robotic technology has played an active role in the treatment of children with autism. Robot technology is an auxiliary tool and cannot directly replace the therapist. There are active individual differences in children with autism. Robots need to solve many problems in the intervention of children with autism. This article discusses the research of treatment robots for children with autism based on deep learning. It is hoped that robot technology will become intelligent and adaptive. Direction development.

Acknowledgements. This work was supported by Nanning University 2021 Professor Cultivation Project (Level: School Level), Project Number: 2021JSGC03, Project Title: Research on the Application of Deep Learning-based Expression Recognition Algorithm in the Treatment of Autism Spectrum Disorder.

References

1. Yang, S., Gan, Y., Chen, W., et al.: Research on the effect of humanoid robots in assisted rehabilitation of children with autism. Mod. Diagn. Treat. **9**, 1345–1347 (2020)

2. Liang, Y., Li, T., Tang, L., et al.: Research on early risk prediction and realization path of child autism based on affective computing. China Sci. Technol. Educ. (4), 26–27 (2020)
3. Zhai, Q., Feng, L., Zhang, G., et al.: The application of artificial intelligence in the field of mental health. Zhejiang Med. **42**(10), 96–102+109 (2020)
4. Jiang, Y.: Artificial intelligence to improve the social ability training of children with autism spectrum disorder, **000**(006), 212–213 (2020)
5. Ma, B., Zeng, X., Gong, R., et al.: Research progress in artificial intelligence-assisted diagnosis of autism spectrum disorder and rehabilitation training abroad. J. Rehabil. **029**(006), 65–69 (2019)
6. Deng, R., Lin, J., Wang, J.: A rehabilitation training auxiliary system for children with autism spectrum disorder. CN111489808A (2020)
7. Yang, Z., Han, R., Wang, B.: A review of the application of expert systems to children with autism. J. Suihua Univ. **10**, 46–52 (2019)
8. Ji, W.: A review of intervention research on autism spectrum disorder. Seek. Knowl. Guide **000**(019), 17–18 (2019)
9. Xu, Y., Zhang, W., Xu, H., et al.: Integration of educational robots and language rehabilitation training for children with autism. Adv. Psychol. **10**(8), 10 (2020)
10. Ma, B., Zeng, X., Gong, R.: Research on multimodal discourse and intelligent-assisted diagnosis and intervention of autistic people in foreign countries. Lang. Strateg. Res. **5**(26(02)), 53–62 (2020)

Risk Control Strategy of Internet Finance Based on Financial Big Data Background

Jianjun Xiao[✉]

Xiamen Institute of Technology, Xiamen 361000, Fujian, China

Abstract. In the process of continuous development of the network, big data has been widely used and has received increasing attention and importance. Financial big data belongs to a brand-new field involved in the continuous development of Internet finance. In this context, internet finance presents a more obvious openness and diversity, which also makes it more dynamic in its development. But some risks will also be magnified by this. Therefore, in terms of the internet finance, it is necessary to focus on the implementation of risk control.

Keywords: Financial big data · Internet finance · Risk control

The overall development of the current financial services industry is relatively rapid, and its development is mainly investment-driven as the main content; the return will be relatively low, though some of the presented returns will be relatively high, the risk is relatively large. In the era of big data, internet finance faces both opportunities and challenges; the use of the internet to reduce financial investment risks and to achieve effective operation are the key issue to be considered in the development of Internet finance [1].

1 Basic Concept of Internet Financial Risk

The financial risks in terms of the Internet are mainly the risks that occur in the financial industry in the traditional mode, but also in terms of Internet technology. Compared internet finance to traditional finance, it is cheaper in terms of operating costs, has a wider overall coverage, and is actually more efficient [2]. However, compared with the regulatory system implemented in traditional finance, the system does not occupy an advantage in terms of legitimacy and system standardization, and has more deficiencies; the influence of the above-mentioned problems will lead to more uncertain risks in internet finance, so it is necessary to conduct a detailed analysis of the risks that exist in internet finance.

2 The Main Forms of Internet Financial Risk

2.1 P2P

In internet finance, P2P belongs to a common form in financing, and it serves small businesses and individuals. However, the financial transactions implemented in P2P are with unfamiliar merchants and unfamiliar individuals [3]. In this case, there are greater security risks, which are mainly due to the difficulty of getting detailed and comprehensive information about the real lender. In the process of financial transactions, it is difficult for creditors to obtain effective guarantees of their rights and interests, and credit risk is also an obvious issue. Since P2P lending is carried out online and highly mobile, it is difficult to gain a comprehensive understanding of customers in each region and to propose a more targeted solution based on regional characteristics, and in this case, it is necessary to ensure maximum circulation of formal certificates of interest and expand the scope of actual transactions. However, in this process, it is difficult to avoid market risks. For creditors, whose main purpose is to obtain economic benefits, the interests of creditors may also change due to the changing market interest rates. In terms of the current social development status quo, the lack of perfection of laws related to internet finance, in the event of issues, creditors in the interests of the legal guarantee will be difficult to obtain, so internet finance in the development will face the legal aspects risk. In recent years, P2P platforms have shown a significant downward trend in terms of number, as shown in Table 1.

Table 1. Number of P2P platforms and year-on-year growth over the years

	2016	2017	2018	2019
Platforms number	2488	2356	1162	343
Year-on-year growth	−30.26%	−8.45%	−58.46%	−68.09%

2.2 Crowd-Funding

The uniqueness of crowd-funding is more obvious compared to traditional financing. Some people have difficulty in expressing their creative ideas and creative perspectives due to financial constraints. Crowd-funding can be used to obtain the necessary funds and projects will not be difficult to implement due to lack of funds. However, the use of crowd-funding can be risky, and some people may use crowd-funding to raise funds illegally [4]. In order to avoid these situations, crowd-funding often requires the approval of the relevant authorities for permission. In case of illegal crowd-funding or suspicion of term debt, the relevant authorities will launch an investigation against the crowd-funder. The technical issues are also the key and difficult problems, and to make the process smooth, it is necessary to start the relevant monitoring and to punish the relevant persons in case of illegal crowd-funding.

2.3 Third-Party Payment

Currently, third-party payment has gained wide popularity and application, and many people are more attracted to this method. It is a payment tool that brings together many aspects of purchasing, fund raising and settlement, providing convenience to people; as long as the network exists, this payment method will work [5]. However, third-party payment can also carry certain risks, as payment may be made by anonymous transfers, in this case, they can be prone to underhanded problems. It can also make it difficult to obtain adequate assurance of information security involved in the transaction process. In the era of big data, when conducting online transactions, everyone may leave personal information online, which may lead to information security risks. In the process of using third-party for payment, information maintenance needs to be strengthened to avoid the problem of fund loss under the influence of information exposure.

3   The Effective Measures for the Implementation of Risk Control of Internet Finance in the Context of Financial Big Data

3.1   Pay Attention to the Establishment and Improvement of Risk Prevention and Control System

For Internet finance, credit risk belongs to its main type of risk, which will affect the sustainable development of Internet finance, and the overall regulation and control should be strengthened for the emergence of this risk. The specific implementation process can be carried out in these aspects: first, to promote the continuous improvement of the risk control system and to strengthen the control and assessment of user risk; second, for private loans, customer information needs to be reviewed before the formal loan is made, promoting the implementation of risk control-related procedures and operations, paying attention to the timely repayment of funds, ensuring the overall safety of funds, and avoiding damage to the corporate image under the influence of customer qualification problems and credit problems, which in turn makes it difficult for creditors to obtain adequate guarantees in terms of legitimate rights and interests [6]; finally, Internet finance needs to comprehensively collect customer information through legal channels, raise the threshold of loans appropriately, and set the loan amount reasonably based on the actual needs of loan customers and their specific conditions, so that the risks arising from Internet finance can be effectively controlled [7].

4   Promote the Continuous Improvement of Laws and Regulations

Internet finance in the development process, the continuous improvement of relevant laws and regulations can, to a certain extent, ensure the effective implementation of supervision. In this process, the relevant departments need to fully consider the type of Internet finance risks, risk characteristics and the actual operating characteristics of enterprises, and then use a targeted approach to formulate relevant laws and regulations, so that the existence of risk omissions in Internet finance can be effectively compensated; and the operation mode of Internet enterprises in the business is to carry out supervision, to ensure that the operation can be carried out in the scope of legal provisions, to prevent the Internet financial-related enterprises prevented from operating in violation of the law, and the legitimate rights and interests of social groups prevented from being damaged. At the same time, government agencies should focus on the role of supervision and guidance to promote the continuous improvement of financial management in the system, and also need to set up professional law enforcement agencies to effectively regulate the behavior of individual law enforcement officers [8]. When new Internet finance companies enter the market, government agencies need to make comprehensive use of a variety of ways to investigate and assess their qualifications, so as to avoid the entry of illegal enterprises in the Internet finance-related market and avoid the negative impact on the stable and sustainable development of the Internet finance industry.

5 Pay Attention to the Use of Professional Technology

In the development of enterprises in this area of Internet finance, there are often risks of user default, so it is necessary to strengthen the prevention of such risks. In practice, the information data related to users need to be carefully analyzed and then judged holistically for users. In the process of realizing the above measures, advanced technology is needed as a support. For Internet finance companies, they can strengthen the use of big data; information analysis is based on big data to ensure that the information analysis and information processing in the scientific and rational implementation; the user in the case of low qualifications should decrease lending, or to avoid lending relationship; after borrowing if they fail to repay on time, it will affect the development of the platform to a greater extent and will also hinder the platform's smooth progress of the work. At the same time, in the process of risk control, Internet finance-related enterprises need to focus on the use of big data analysis-related systems to ensure the use of scientific and reasonable nature; and for the investment project information, it needs to carry out systematic analysis, and make a possible negative impact security analysis based on multiple dimensions and levels for the project in the process of operation of risk factors; the use of reasonable methods control the risk within the tolerable range, so that the enterprise and the company's operation and development are fully guaranteed [9].

6 Strengthen the Overall Control of Payments

In the big data environment, Internet finance needs to ensure the clarity of the payment platform for the flow of funds in order to achieve sustainable development, so that risk management can obtain good conditions in the implementation process. The specific implementation process can be carried out in the following aspects: first, the account needs to use the real-name system to manage the registration of customer information, to ensure that customer information can be entered in a complete state; secondly, according to the specific payment function of Internet finance, the corresponding payment mode should be set, and the business form should be clearly different from that of banking institutions to ensure the clear identity of enterprises in the payment system.; at the same time, enterprises in Internet finance need to make scientific planning for the transaction process and strengthen communication and cooperation with banks, which will transmit relevant information to the payment system based on payment instructions, and then the settlement work will be completed successfully; finally, to promote the continuous improvement of the market supervision system, especially the need to strengthen the effective supervision of the operation mode of Internet finance, we need to pay full attention to the supervision and post-event supervision [10]. Through the use of big data, it is possible to realize the whole process of supervision and fully utilize the advantages of the industry database, which is a more effective and direct way to control risks.

7 Conclusion

In conclusion, Internet finance has obvious randomness as well as complexity; in the context of the era of big data, the industry development faces more conditions, and we should focus on the overall control in terms of financial risk, which is an important way for Internet finance to continuously enhance its competitiveness and enable enterprises to achieve in the development of profit and avoid harm. In this process, it is necessary to actively build a risk control-related system, promote the continuous improvement of laws and regulations, strengthen the use of new technologies, achieve effective communication and cooperation between enterprises, and constantly summarize relevant experience, thus enable Internet finance to achieve a better development.

References

1. Ren, W.: Research on risk management in the internet finance industry in the era of big data. Knowl. Econ. **5**, 33–34 (2020)
2. Bian, Y.: Exploring the risk of internet consumer finance from the perspective of big data. Small Medium-Sized Enterp. Manag. Technol. **11**, 98–99 (2020)
3. Diao, Y.: Risk control of internet finance in the context of big data. Invest. Entrep. **2**, 125–126 (2020)
4. Zheng, S.Y.: The risk of internet finance in the context of big data and how to response. Mark. Res. **1**, 42–43 (2020)
5. Chen, J.Y.: Exploring the risks and preventive measures of internet finance in the context of big data. Mall Mod. **20**, 139–141 (2020)
6. Zhang, B.: Big data technology and its application in the field of internet financial risk monitoring. Financ. Technol. Era **5**, 43–49 (2020)
7. Li, W.: Discussion on risk control strategy of internet finance under the background of financial big data .Consum. Guide **000**(004), 228, 247 (2020)
8. Feng, P.: Discussion on risk prevention and control methods of internet finance in the era of big dfigata. Mod. Mark. (Next Issue) **08**, 161–162 (2017)
9. Zhang, Y.: Analysis of internet financial innovation and risk prevention and control strategy in the era of big data. Consum. Guide **000**(019), 190–191 (2019)
10. Chen, S.: Risk and prevention of Internet finance in the context of big data. J. Enterp. Sci. Dev. **8**, 264–265 (2018)

Simulation of a Visualization Model Wind Tunnel

Zhou Li[1(✉)] and Liu Qin[1,2]

[1] Hubei Minzu University, Enshi 445000, Hubei, China
[2] Ningbo GMF New Material Technology Co., Ltd., Cixi 315300, Zhejiang, China

Abstract. This article mainly introduced the design of a simple low-speed wind tunnel, and analyzed the impact of axial fans rotation flow on flow field quality. The results show that the wind tunnel flow field is uniform and stable and verify the feasibility of axial fan in the wind tunnel design.

Keywords: Wind tunnel · Axial fans · Fans analysis

1 Introduction

Wind tunnel is a specially designed air experiment device generating velocity-adjustable air current in the test section through power plant to conduct various researches on aerodynamics [1]. According to the theory of flow relativity and similarity, when air flow through the model with a certain velocity, air forces acting on the prototype is obtained by measuring the parameters in flow field and aerodynamic response of the model. The wind tunnel with direct current low velocity is mainly composed of stable section (honeycomb is installed in stable section), contraction section, test section, diffusing section and power section, etc. [2].

Axial flow ventilator is often chosen as power end when design low-speed miniature wind tunnel but people ignore the impact of the fan rotating flow field on the wind tunnel flow field quality [3]. The paper detailed analysis of wind tunnel simulation model with fan rotation, validated the wind tunnel design.

2 Wind Tunnel Design

Test section: model test section is the place where the air flow field experiment is conducted. It is an important component of the wind tunnel [4]. The air flow of test section should be stable, the distribution of speed and direction should be uniform, original turbulence, noise intensity, and static pressure gradient should be low [5, 6]. The characteristics of the test section airflow decide flow field quality. Test section size is determined by the size of the model. Figure 1 Schematic diagram of wind tunnel Test model ratio is 1:18, the test section is set at $180 \times 180 \times 1190$ mm. Material is transparent organic glass, its thickness is 6 mm.

J. Macintyre et al. (Eds.): SPIoT 2021, LNDECT 98, pp. 825–830, 2022.
https://doi.org/10.1007/978-3-030-89511-2_111

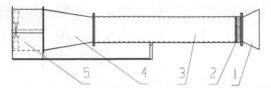

1-contraction section 2-stable section 3-test section 4-diffusing section 5-power
section

Fig. 1. Schematic diagram of wind tunnel

Power section: compared with centrifugal fans, axial fans will produce vortex, but axial fans' installation is easier. Using the same amount of power, axial fans can produce a larger amount of wind. Figure 2 Wind tunnel model Test section requires wind speed is 20 m/s, according to the test section cross-sectional area axial fans type is SF3-2R(air volume 3000 m^3/h, speed 2800 rpm, power 0.37 KW, adjustable speed).

Contraction section: contraction section uniformly accelerates airflow from the stable section, and improves the quality of the flow field of the test section. The design of contraction section will meet following requirements: When air flows through the contraction section, monotonically increasing airflow to prevent flow separation occurring on the wall; flow velocity distribution at the outlet of contraction section is uniform, the direction of air flow should be straight and steady. The main factors affecting the contraction section are contraction ratio and contraction curve. This paper chose linear shrinkage curve, contraction ratio 0.42, contraction section length 130 mm.

Stable section: Honeycomb panel is used to improve of the quality of air flow when air flowed past the honeycomb panel. In the term of the flow resistance, the minimum loss is hexagonal, followed by a square, round the worst. Longer the frame length of the rectification grid can reduce the lateral air flow, but the length of the frame increases, the friction loss of air flow increases. This article chooses thickness of 45 mm hexagonal honeycomb core as a rectifier grid, mesh side length 5 mm, aluminum thickness 0.1 mm.

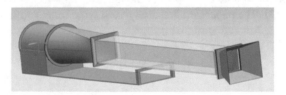

Fig. 2. Wind tunnel model

3   Simulation

The simulation of the wind tunnel whose power section is axial fans ignores impact of axial fan rotation flow on quality of the wind tunnel test section flow field. This paper introduces the fan rotation simulation model to improve the flow field simulation authenticity Fig. 3 Model grid.

Fig. 3. Model grid

1. Pretreatment: In order to simplify the model to increase computing efficiency, establish a 1:2 model in UG and import it to Gambit for pretreatment. A volume which is just wrapped around the blade is considered as the rotating fluid. The overall model subtracted the rotating fluid. Mesh the model. Set outlet as PRESSURE-OUTLET, set inlet as PRESSURE-INLET, set blades as WALL named "blade", set rotating fluid as FLUIDE named "rotate". Finally, exported mesh file.
2. Solver and turbulent flow model: Import the mesh file into Fluent, check mesh, setting mesh scale, set the units and define materials. Chose RNG $k - \varepsilon$ option in Viscous Model. Compared with RNG $k - \varepsilon$, standard $k - \varepsilon$ model is more general and easier to convergence, but RNG $k - \varepsilon$ can simulate eddy more accurately. [7, 8] Chose Standard Wall Function option in Near-wall Treatment.
3. Boundary conditions and Cell zone conditions: Set inlet as pressure-inlet, set outlet as pressure-outlet, Gauge Total Pressure is 0,chose Intensity and Hydraulic Diameter option in Specification Method, Turbulent Intensity was 5%, Hydraulic Diameter of Inlet and outlet would be obtained in Eq. (1). Define "rotate" as rotation zone, Rotational Velocity is 2800 rpm, mode of motion is moving reference frame (MRF) [9]. According to the right-hand rule to determine the direction, in this article the rotation-axis is X, so fill 1 in X of Rotation-Axis Direction, fill 0 in Y and Z. Set "blade" as moving wall, its velocity is as the same as rotation fluid, speed of relative to adjacent cell zone was 0 Fig. 4 Boundary conditions and Cell zone.

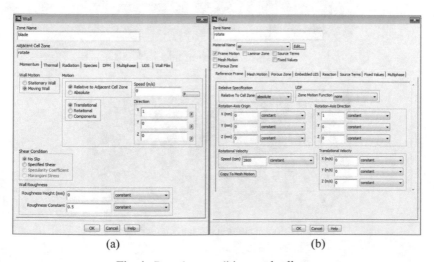

(a)   (b)

Fig. 4. Boundary conditions and cell zone

$$d_c = 4 \times \frac{S}{L} \tag{1}$$

Where S is flow cross acreage, and L is wetted perimeter.

4. Setting residual absolute criteria: Fill 0.0001 in absolute criteria, other numeric keep the default value. After solution initialization, run calculation.
5. Result: About 600 iterations, the air flow tended to be stable and solution tended to be converged Fig. 5 is pathlines of inlet, Fig. 6 is velocity contours of longitudinal section. According to the two pictures, the effect of the contraction section is not so good, the shrinking tendency of linear shrinkage curve was not gradual enough, and velocity variation rate was high [10]. But after air flow through the stable section, velocity distribution was uniform, the velocity near wall is low, the velocity of cross section center could reach 25 m/s, the vortex from power section don't impact the quality of test section flow field, airflow of power section, diffusing section and the place near diffusing section was turbulent, so putting test-board in first half of test section would be reasonable Fig. 7 was static pressure contours of walls [11]. The static pressure of test section maintains a stable value, −400Pa, which meet the quality requirements of the flow field.

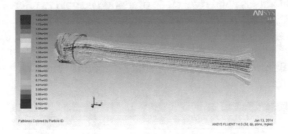

Fig. 5. Pathlinesof Inlet

Fig. 6. Velocity contours of longitudinal section

Fig. 7. Static pressure contours

Fig. 8. Static pressure contours in conventional simulation methods

Fig. 9. Velocity contours in conventional simulation methods

Figure 8 is the static pressure contours in conventional simulation methods, Fig. 9 Velocity contours in conventional simulation methods. it doesn't show the rotation of the fans. Though static pressure of test section is similar with the result of this article, pressure of contraction section and power section isn't accurate without rotation flow of the fans. The velocity result without rotation flow of the fans doesn't have a reference value in the wind tunnel with an axial fans as its powerplant. According to the simulation results, the model is appropriately modified and processed Fig. 10 was wind tunnel which was made on the basis of the model in this article.

4 Conclusion

This paper describes the selection of the parameters of various parts, and the fans analysis application in wind tunnel simulation. In this method simulation result with rotate flow of fans is more accurate. According to simulation results, there is no negative impact of axial

Fig. 10. Wind tunnel

fans rotation flow on flow field quality, the wind tunnel reached the basic requirements with basic conditions for further experiments.

References

1. Yang, W., Peng, W., Ding, Y.: A visual miniature wind tunnel. CN102023080 (2011)
2. Gu, Z.: Automotive Aerodynamics. Communication Press, Beijing (2005)
3. Fares, E.: Unsteady flow simulation of the Ahmed reference body using a lattice Boltzmann approach. Comput. Fluids **35**(8–9), 940–950 (2006)
4. Makowski, F.T., Kim, S.E.: Advances in external-aero simulation of ground vechicles using the steady RANS equation. SAE Paper 2000-01-048
5. Longmireek, E.: Structure of a particle- laden round jet. J Fluid Mech. **236**, 217–257 (1992)
6. Fang, F.M., Chen, J.C.: Experimental and analytical evaluation of flow in a square-to- square wind tunnel contraction. J. Wind Eng. Ind. Aerodyn. **89**, 247–262 (2001)
7. Fluent Inc.: FLUENT User's Guide.Fluent Inc (2003)
8. Luca, V., Pier, P.: A CFD-FEM Approach to Study Wing Aerodynamics under Deformation. SAE Paper 2004-01-04
9. Ma, Z., Wang, Z.: Application of airfoil test and numerical simulation in transonic vind tunnel. Exp. Technol. Manag. **37**(5), 128–136 (2020). (in Chinese)
10. Ren, B., Zhu, P., Wei, S.M.: A numerical wind tunnel software system. J. North West. Poly Tech. Univ. **18**(1), 44–47 (2000). (in Chinese)
11. Qi, Y., Li, J., Rao, J.Y.: A portable smoke wind tunnel for flow visualization. Exp. Technol. Manag. **23**(4), 40–43 (2006). (in Chinese)

Application of Virtual Simulation Technology in Building Construction Network Learning

Lili Xiao[✉]

Xi'an FanYi University, Xi'an 710105, Shaanxi, China

Abstract. Practice teaching is an indispensable key link in the undergraduate teaching of environmental design. Combining with the current needs and shortcomings of practice teaching of Environmental Design Specialty, this paper constructs a virtual simulation practice teaching platform with multi-module combination, to realize the functions of online demonstration teaching, training, practical guidance and examination feedback, and to form a practical teaching system integrating production, learning and research. The platform is convenient for students to study independently outside class, enrich practical teaching projects, realize fine teaching quality control, and develop students' innovative thinking and comprehensive application ability, it can be used as a reference for other similar courses or specialties that require higher practical operation skills to carry out network teaching and construct practical teaching platform.

Keywords: Virtual simulation technology · Practical teaching · Structure and construction of architectural decoration

1 Introduction

Virtual simulation experiment is a new kind of experiment mode based on virtual reality technology and Internet sharing technology. By simulating the real instrument, equipment, material and experiment environment on computer, breaking through the limitation of time, space and nature, students can carry out experimental operation through virtual simulation platform, which provides ample experimental opportunities for students. Because of its immersion, verisimilitude and interactivity, virtual simulation technology is widely used in practice teaching in colleges and universities, which is beneficial to the sharing of experimental teaching resources and can stimulate students' learning enthusiasm and interest in experiments, improving the quality of practical teaching.

In recent years, with the development of virtual simulation technology, virtual simulation experiment teaching has developed rapidly and become one of the important means of classroom teaching and traditional experiment teaching reform. This teaching method has obvious advantages in large-scale or comprehensive experiment teaching with complicated environment, irreversible operation, high danger and high cost. Many colleges and universities are planning or constructing virtual simulation laboratories with subject characteristics, in order to realize the teaching and experimental functions that the traditional laboratories do not have, and to cultivate students' practical operating ability

J. Macintyre et al. (Eds.): SPIoT 2021, LNDECT 98, pp. 831–835, 2022.
https://doi.org/10.1007/978-3-030-89511-2_112

and consciousness of innovative development, Some scholars have studied the virtual simulation experiment teaching in different fields or specialties. For example, Wu Juan and others have built a multi-level and modular virtual simulation course system [1], Ma wending and others have studied the virtual simulation experiment teaching platform for Mining Engineering [2]; Zhu Shengxue and others have discussed the construction experience of the virtual simulation experiment teaching platform for transportation [3]; Wang Yong and others have discussed the construction experience of the virtual simulation experiment teaching platform for economics and management majors [4], the framework of virtual simulation experiment teaching platform of professional cluster is put forward. These researches have accumulated a great deal of experience for the construction of virtual simulation experiment teaching system and laboratory.

2 Requirements for Practical Teaching of Architectural Decoration Structure and Construction Course

The construction and construction of architectural decoration is a comprehensive and practical course, and the practical teaching is an important teaching link. In the traditional teaching method, the teacher can only demonstrate the construction technology by means of multimedia courseware and video materials, and the students can only learn the construction technology by watching, but cannot participate in it. At the same time, students passively accept knowledge, face complex construction drawings, students space imagination is not enough, do not understand the drawings, slowly lose interest in the course. So it is very important to increase the proportion of practical teaching in teaching.

It is necessary to construct a construction training room, a large number of different engineering situations and various construction scenes in the course of architectural decoration and construction. In the construction teaching, each time the curriculum consumes the massive material, the component is the very big expense, the normal teaching is difficult to support. At the same time, in each class may also produce a large number of irreversible damage to the training base, such as drilling holes, such as the need for regular maintenance, the resulting construction waste is not easy to remove, such as stone wet hanging method in the grouting of mortar, the adhesive mortar of metope pavement, after dry hard can only look for special worker to clear, to routine teaching is very difficult to do. This part of experimental project has the characteristic of high cost, high consumption, the actual operation difficulty is big, the serious restriction experiment teaching effect, causes the student ability and the quality raise insufficiency, has become in the specialized experiment teaching the universal common problem.

3 The Application of Virtual Simulation Technology in the Course of a Architectural Decoration and Construction

Virtual simulation technology is an essential key technology in digital and intelligent design, control and manufacturing. In recent years, modern science and technology, represented by Information Science and network technology, have penetrated into every field of social and economic life, and information technologies such as virtual reality, multimedia, human-computer interaction, database, etc. have developed vigorously and become more and more mature, it brings a new opportunity for improving the quality and efficiency of the experimental teaching of the course of architectural decoration, construction and construction. In the 《Outline of the national plan for medium and long-term educational reform and Development (2010–2020)》 accelerate the process of educational informatization is specially expounded as a part of it [5]. The Ministry of Education has also strengthened its macro-guidance on experimental teaching and the informatization of experimental teaching in colleges and universities, The 10-year development plan for Education Informatization (2011–2020) was formulated, and the development plan and objectives were defined, namely, to promote the deep integration of information technology and higher education, and to innovate the training mode of talents. The Ministry of Education " on the overall improvement of higher education quality of a number of opinions & quot; also clearly put forward to strengthen the practice of educating people, improve the level of experimental teaching.

According to the training objective of environmental design specialty and the characteristics of architectural decoration construction and construction course, the guiding ideology of quot scientific planning, highlighting key points, improving efficiency and sustainable development quot; is adhered to, based on the basic principle of quot; the combination of the virtual and the real, the mutual complement, the ability and the real quot; with the sharing of high-quality experimental teaching resources as the core, and the construction of information-based experimental teaching resources as the focus, the construction task is to realize the teaching function which the real experiment does not have or is difficult to complete, and it covers the virtual simulation practice teaching of many module projects such as ground engineering, top engineering, wall engineering, partition wall and so on, to achieve the goal of professional training.

Taking the dry hanging stone in the wall engineering module as an example, the project provides an experiential experimental operation platform for students majoring in environmental design and interior design to help them master the construction principle, construction process and technical points of dry hanging stone in the virtual building interior decoration construction process. In each construction process stage, there are different materials, sizes, practices and other options, so as to arouse students' thinking and discussion, and train students' ability to solve problems. This experiment adopts a variety of experimental methods, such as autonomous learning, engineering survey analysis, experimental operation, machine judgment, self-evaluation, self diagnosis and so on [6].

In the self-learning stage, that is, at the beginning of the experiment, the students first enter the warming up stage for self-learning, and learn about the reserved knowledge of the material performance, basic type, basic treatment method, construction method, etc. involved in the experiment in advance. The lack of relevant theoretical knowledge, especially the lack of building material performance knowledge, limits the students' understanding of the construction principle and the correct choice of construction scheme [7]. Therefore, it is necessary and effective to reserve relevant building material theory knowledge in advance to improve students' deep understanding of the principle.

In the stage of project overview analysis, there are three kinds of common building wall base: concrete base, brick base and hollow masonry base. Students choose different scenes. Different scenes have different actual engineering situations. Enter the simulation system to directly observe different project profiles, analyze different wall base, and choose different stone dry hanging schemes. At the same time, key information is captured and recorded to prepare for the following tasks.

In the machine judgment stage, select one of the scenes and start the construction operation process. The system will prompt the task, and the students will respond according to the task requirements. The system judges students according to their responses. Each point will have a corresponding score. If the score is enough, the next step will be taken; If the score is not enough, the system prompts to try again until the score is enough [8].

In the self-evaluation stage, at the end of the experiment, the students evaluate themselves according to the experimental objectives and feelings. The evaluation content is divided into three dimensions: quality objectives, knowledge objectives and ability objectives. Through the self-evaluation, students can understand their own experimental standards and learning progress.

In the self diagnosis stage, after self-evaluation, the system will automatically pop up the personal performance window, and then enter the self diagnosis link. In this session, students learn about their scores in each round [9].

Through the selection and discussion in the whole construction process, students can understand the practice of dry hanging stone in principle, and choose the correct construction method under different building base conditions. In the construction process, the operation can meet the specification requirements.

4 Conclusion

To sum up, it is a new requirement of our country for the development of higher education to construct a highly simulated virtual experimental environment and experimental objects, and to develop and realize the informationization of experimental teaching, it is also an inevitable way to deepen the reform of professional experimental teaching mode. Through the development of a high-level virtual simulation experiment teaching platform, a complete professional practice teaching system characterized by the organic combination of real construction training and virtual simulation construction experiment is formed, it is expected to enhance the depth and breadth of experimental teaching in this specialty, expand the space and time of practical teaching, and enhance students' autonomous learning ability, engineering practice ability and innovative consciousness, in order to adapt to the country, the society to this specialized talented person's high quality compound type request [10]. Therefore, in combination with the training objectives of environmental design and the requirements of the teaching program of architectural decoration, construction and construction, we should actively develop a reliable, economical, safe, content-rich and benefit-wide virtual simulation experiment teaching mode, it is of great practical significance to form a practical teaching mode combining virtual simulation experiment with real training, which can effectively overcome many restrictions in real practice environment and stimulate students' initiative of autonomous learning.

References

1. Wu, J., Sun, Y., Wei, L.: Southeast university virtual simulation experiment, teaching center construction planning ideas and progress. Exp. Tech. Manag. **31**(10), 5–9 (2014)
2. Ma, W., Zuowu, W., Wan, Z.: Virtual simulation experiment teaching, system construction and practice of mining engineering. Exp. Technol. Manag. **31**(9), 14–21 (2014)
3. Zhu, S., Bao, X., Xia, J.: Transport virtual simulation experiment teaching platform, platform construction and practice. Logist. Eng. Manag. **40**(7), 161–162, 127 (2018)
4. Wang, Y., Pei, L., Mao, Z.W.: An empirical study on the construction of cross-specialty virtual simulation teaching platform of economics and management. Ind. Inf. -Based Educ. **11**, 74–79 (2017)
5. Li, L., Zhao, Y., Li, Z.: Virtual simulation experiment teaching of materials science and engineering, construction of center. Exp. Technol. Manag. **31**(2), 5–8 (2014)
6. Jiang, X., Deng, J., Zhan, Y.: Virtual simulation experimental teaching design for forging process and die design. Die Ind. **41**(12), 63–66 (2015)
7. Li, P., Mao, C., Xun, J.: Developing the construction of national virtual simulation experimental teaching center to improve the informatization level of experimental teaching in colleges and universities. Lab. Res. Exp. **32**(11), 5–8 (2013)
8. Yuelin, D., Huang, G., Wang, F.: Building virtual simulation experimental platform and exploring innovative talent training mode. Exp. Technol. Manag. **32**(12), 26–29 (2015)
9. Li, P., Mao, C., Jin, X.: Developing the construction of national virtual simulation experimental teaching center to improve the informatization level of experimental teaching in Colleges and universities. Lab. Res. Explor. **32**(11), 5–8 (2013)
10. Xie, D., Lin, C.: Virtual simulation psychological experiment center promotes the construction of innovative talent training system. Lab. Res. Explor. **36**(6), 165–168 (2017)

New Sales Mode of Agricultural Products Based on Digital Economy

Song Li, Xin Zhang, and Qiong Wu[✉]

School of Economics and Management, Shenyang Institute of Technology, Shenyang, Liaoning, China

Abstract. China's digital economy has maintained rapid growth. Driven by new information and communication technologies and new formats, traditional industries have radiated new vitality and become one of the powerful engines of national economic growth. Sales of agricultural products is an important industry related to people's livelihood in traditional industries. Although the computer science and Internet have promoted the production, processing, circulation and sales of agricultural products to a certain extent, the sales of agricultural products in today's market is still in the situation of low intelligence and chaotic management system. This paper puts forward a solution on how to increase value of data and integrate with agricultural products sales to drive agricultural products sales to informatization, intelligence and unified standardization.

Keywords: Digital economy · Agricultural products traceability · Agricultural products sales · Big data

Since 2019, the contribution rate of digital economy to GDP growth has reached 67.7% [1], which has become one of the powerful engines of national economic growth. Digital industrialization and industry digitization are the core of the development of digital economy. With the continuous evolution and upgrading of digital industrialization and the comprehensive integration and development of service industry, it is expanding the scope of real economy. Industry digitization has evolved from single point application to industry collaborative development. It uses digital technology to reduce costs and increase efficiency in an all-round and whole chain way, and promotes high-quality development, which has become an increasingly important growth power of digital economy [2]. The combination of data economy and traditional industries is the only way for the reform and development of traditional industries. However, there are still many problems in the sales of agricultural products industry, so it is urgent to explore the sales mode to adapt to the information age.

1 Overview of Agricultural Products Sales

The sale of agricultural products plays a very important role in the development of the whole national economy, which is related to the basic life of the people and the three rural issues [3]. To help agricultural products establish efficient new sales mode, not only

J. Macintyre et al. (Eds.): SPIoT 2021, LNDECT 98, pp. 836–841, 2022.
https://doi.org/10.1007/978-3-030-89511-2_113

to meet the needs of market consumers for the quality and level of agricultural products, but also to increase the income of farmers and enterprises, and promote rural farmers to become rich. At the present stage, the market of agricultural products in our country is at an early stage of development. A small number of enterprises can obtain better profits by registering product brands for processing. Most of the suppliers of agricultural products are individual farmers, lack of production tools and means of production, and no access to information, low level of management, poor risk resistance. The proportion of China's digital economy scale in GDP and Internal Structure of China's Digital Economy are shown in Fig. 1 and Fig. 2.

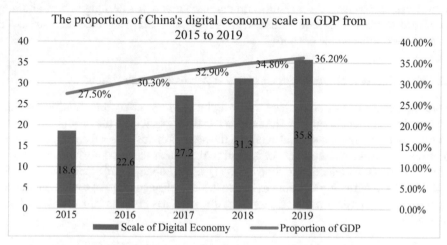

Fig. 1. The proportion of China's digital economy scale in GDP from 2015 to 2019

Fig. 2. Internal structure of China's digital economy from 2015–2019

1.1 Low-Income Level of the Farmers'

China is a large agricultural country, and the sales of agricultural products are related to the income of farmers and economic development. At present, there are about 600 million farmers in China [4]. Although the distribution provinces are different and the planting products are different, the income of farmers in most areas is low, which is lower than the international level. For example, although Northeast China is a major grain producing area, due to the lack of access to market information and lag, the income depends on nature, has no planning adjust to the changing of the market, which they just no changing years by years. The management of the production is lack of the ability to control market changes. As a result, farmers' enthusiasm for farming is low, and they are forced to manage and difficult to manage the land. Farmers' income restricts the development of the whole economy.

1.2 Food Safety and Quality Assurance of Agricultural Products

From the perspective of consumers, more and more food safety and quality problems of agricultural products appear. The serious excess of pesticides, the residue of harmful substances, or the use of chemical agents in the process of business, all endanger the vital interests of consumers. The safety of agricultural products has become the current focus, more consumers pay attention to food safety. Quality problems also emerge in endlessly. In the market of agricultural products, due to the lack of standards and evaluation means, the phenomenon of poor products or selling dog meat has disturbed the normal order of the market.

2 Sales Mode of Agricultural Products Based on Digital Economy

The amount of information in the sales process of agricultural products is low, the information that merchants can master and share is very less, and the information that consumers can get even less. At the moment of digital economy, it should be able to help businesses and consumers get enough information of whole market, which brings to improve efficiency, new value and more benefits for increasing.

2.1 Establish Traceability System for Agricultural Products

In the aspect of agricultural products traceability, many scholars and experts have provided feasible solutions, such as information sharing of APP platform, and information collection can be set up in every link of agricultural products from planting, growing, harvesting, processing to circulation [5]. Another example is the application of agricultural product supply chain traceability based on blockchain and multi chain. Due to the large amount of data form and storage, block information traceability is adopted. Multiple accounts are allocated to each link block in the supply chain to adapt to the complex interaction between different enterprises or manufacturers in each link of the supply chain [6], effectively avoiding the problems of low security of traceability data and high storage pressure in the block chain; Combined with the actual demand of the traceability system of agricultural products supply chain, the traceability architecture is designed, and the intelligent contract is written to ensure the security and reliability of all parties' transactions in the agricultural products supply chain [7].

2.2 Establish an Information Platform for Demand and Supply

An important reason for the low income of farmers is that they could not grasp the market information. To establish an information sharing platform for agricultural products market, farmers can understand the market demand information. The platform can collect and process the integrated sales information of agricultural products in the past years under the background of big data, understand when and where consumers buy, and understand the changes in the market. It can help farmers to make production plans, reduce the blindness of planting, reduce the planting risk caused by not knowing the supply and demand of the market, and increase the income of farmers. The analysis model after big data processing can better understand the preference characteristics of market demanders for agricultural products, guide farmers to effectively meet the special needs of consumers, and increase the accessory value of products.

2.3 Establish Order Before Production Mode

Internet sales provides a network sales platform for traditional agricultural products sales, through which more customers can be contacted. Now the network has become a sales platform for agricultural products retail. But this could not guide agricultural producers to guide production [8]. Under the background of big data information, the new sales mode can use modern information processing to establish data order distribution, help merchants place orders first, and then grow, reduce unsalable and waste, and reduce the production risk of producers.

3 Safeguard Measures

In order to realize the sharing of resources and information, several basic conditions are the important basic conditions to ensure the new sales mode of agricultural products. The government needs to introduce a unified quantitative standard for agricultural product inspection, provide a platform for information sharing such as traceability, and ensure that users can use and can use information technology.

3.1 Unified Inspection Standards for Agricultural Products Launched by the Government

How to identify the safety standards and quality levels of agricultural products, the government should develop and open the indicators and standards for different agricultural products. Consumers and businesses can log in app or web page, and directly provide unified inspection standards for the products they search. Not because they don't know the business, don't understand the standard and purchase difficult or deceived things development [9]. For different levels of the same product, reference data should also be provided to help the market establish corresponding standards to measure the quality level of products. Such as shape, weight, sweetness, month and so on, the reasonable reference level is given. Consumers are no longer difficult to deal with how to choose goods because they have less common sense of life.

3.2   Establish and Improve the Information Sharing Platform

Because of the product tracking platform and regional tracking module, the information is isolated and modular. How to improve the barrier for businesses to coordinate and share information [10], the government and enterprises must solve how to connect and establish the information sharing platform. Only information sharing can give full play to the maximum value of information and provide more effective and high-value information resources for farmers, consumers and the government.

3.3   Help Farmers Understand and Use Digital Information

Most of the farmers have low educational background and are located in remote mountainous areas away from the city. Their concept and consciousness are relatively backward and they do not know much about the use and value of data information and Internet. At present, although most areas of our country have covered the Internet, hardware facilities such as smart phones and computers can be popularized. However, there is still a lack of understanding about the understanding of digital information and how to use it reasonably. The government can train and publicize to increase farmers' concept and awareness, and enterprises can also go deeply into rural areas to promote intelligent information collection and tracking products [11].

4   Summary

Accelerating the integration of agricultural products sales industry and data information technology is the internal driving force to promote the rapid development of the industry. Using digital information to increase the management level and improve the efficiency of supply chain. Through the establishment of agricultural products traceability program and the establishment of information sharing platform, help merchants to have orders first and grasp the demand characteristics of the new mode of application. The new marketing mode of agricultural products can increase farmers' understanding of the market information, guide production plans, and increase income [12]. At the same time, it can enhance consumers' mastery of food safety and product quality information, and protect the benefits of consumers'.

References

1. Liu, Y.: Efficiency analysis and countermeasures of agricultural digital transformation. Econ. Horiz. (07) (2020). https://doi.org/10.16528/j.cnki.22-1054/f.202007106
2. Sun, J.: Guidance and policy suggestions for high quality agricultural development in China. J. Manag. (06) (2019)

3. Luo, Y.: On the application of we media in agricultural and rural development. Gansu Agric. (12) (2019)
4. Xinling, Xiaoning: Evaluation system construction and measurement analysis of China's agricultural high quality development. Econ. Aspect (05) (2019)
5. Zhang, C., Liu, Q.: Research on the evaluation and path of agricultural high quality development under the background of rural revitalization strategy. Econ. Forum (04) (2019)
6. Liu, H.: Driving agricultural modernization with precision agriculture to accelerate the digital transformation of modern agriculture. China's Agric. Resour. Reg. (01) (2019)
7. Gao, Y., Xiangwen, L.V., Yuan, L., Li, M.: Research on application of trusted traceability of agricultural products safety based on blockchain. Comput. Appl. Softw. (07) (2020)
8. Lu, G., Xie, L., Li, X.: Comparative study on consensus algorithms of blockchain. Comput. Sci. (S1) (2020)
9. Zhai, H.: Research on the application of blockchain technology in agricultural product safety information traceability. Internet Things Technol. (01) (2020)
10. Li, G.: Development path of modern agricultural industry under industrial Internet mode. Discuss. Mod. Econ. (07) (2015)
11. Xia, H., Chen, L.V.: Empirical study on e-commerce CSFs of 78 listed textile and garment enterprises. E-commerce (07) (2013)
12. Qi, L.: Characteristics and advantages of agricultural intelligent measurement and control system based on Internet of things and its application in greenhouse. Anhui Agric. Sci. (30) (2011)

Classroom Teaching Assessment in College English Under Information Environment

Xiaoshu Xu[✉]

Foreign Languages College, Dalian Neusoft University of Information, Dalian, Liaoning, China

Abstract. With the continuous development and integration of information technology and higher education, the advantages of a multiple assessment system based on Internet thinking have gradually become prominent. The assessment system with the main characteristics of subject diversification, method diversification and standard diversification is becoming more and more perfect. The classroom teaching assessment strategy, method of tool's compiling under the information environment and the framework of classroom teaching assessment system from the angles of classroom observation and classroom quiz are demonstrated in this paper. Based on the research results and teaching practice in the field, this paper demonstrates the application value of the assessment system constructed through information technology in college English teaching.

Keywords: Classroom assessment · College English · Information environment

1 Introduction

Classroom assessment refers to the general term of evaluation forms such as classroom observations, classroom exercises, homework and classroom tests, whose essence is a formative assessing activity to improve teaching and provide feedback for learning [4]. The research of college English teaching assessment under the information environment is the crucial and difficult point of in-depth study of classroom teaching effect.

According to the demands of college English curriculum reform, new college English classroom assessment ought to focus on the diagnosis and improvement, while the traditional assessment pays more attention to the evaluation purpose itself [5]. The new assessment will put forward that classroom evaluation is mainly based on teacher self-evaluation, while traditional assessment mostly implements evaluation by others [2]. The new assessment advocates formative evaluation, while the traditional one still pays too much attention to summative assessment [1]. Compared with the traditional teaching evaluation whose drawbacks is emphasizing examinations and neglecting practicality, the new assessment focuses on how to breaking the barriers of time and space, and how to fit the educational concepts of modern college English [3].

J. Macintyre et al. (Eds.): SPIoT 2021, LNDECT 98, pp. 842–846, 2022.
https://doi.org/10.1007/978-3-030-89511-2_114

2 Assessment of CE Classroom Teaching Under the Informationized Environment

The goal of the classroom assessment is to conduct real-time, accurate, and multi-angle analysis of informationized classrooms [7]. On one hand, it could provide teachers with formative and summative evaluations, so as to adjust and improve teaching activities in time; on the other hand, in an information technology environment, it will provide the scientific and automatic evaluation of classroom teaching.

Under the informationized environment, in particular, the effective research tools and methods are insufficient for the assessment of College English(CE) classroom teaching, which includes the quality of teaching interaction, teaching mode analysis, and real-time feedback processing etc. [6]. Therefore, it is significant to explore the strategies and methods for college English classroom teaching assessment in an informationized environment.

2.1 Application of CE Classroom Assessment

Research approach which based on classroom observation includes classroom teaching model analysis and the quality analysis of teaching interactive. In the selection of research objects, a sample of 185 students and 3 teachers in a College English course for non-English majors in a college in Dalian was selected. In the sample, boys accounted for 72% and girls accounted for 28%. The class teachers involve 1 titles of associate professor and 2 lecturers, and their working years range from 7 to 10 years.

The internet-based College English learning in this university has been implemented for nearly five years, and the teachers are very familiar with the purpose of internet-based learning, teaching methods, and classroom evaluation. According to the teaching theme of each unit, the related teaching are designed by the course group, 4 h a week, lasting 16 weeks. The study uses two research methods, qualitative analysis and quantitative analysis, to evaluate the research results.

According to the experimental test, the three-level evaluation standard is determined, which is made into a scale suitable for classroom assessment, forming a relatively complete comprehensive evaluation scale for the integration of information technology and college English classroom teaching. The assessment index system established in this study can reflect the characteristics of college English classroom teaching, and can objectively reflect the quality of college English teacher classroom teaching in the environment of information technology integration from both qualitative and quantitative aspects [8].

The assessment scale for the integration of information technology and college English teaching includes four first-level indicators: teaching design, teaching implementation, teaching effect and teaching environment. The author here uses the factor analysis method to analyze all the index items to examine whether they are consistent with the four first-level indexes, and at the same time try to isolate the representative index items [7]. The questionnaire is tested whether factor analysis can be performed. In the test result, the KMO test coefficient is 0.691, which is greater than 0.5, indicating that it is suitable for factor analysis. The chi-square value of Bartlett's sphericity test is 5138.411, and the concomitant probability value is $0.000 < 0.01$, reaching the significance level, which also shows that it is suitable for factor analysis.

Table 1. Factor analysis results of first-level indicators

Dimension	Factor 1	Commonness	Contribution rate	Cumulative contribution rate
Teaching design	.989	.982	94.38%	94.38%
Teaching implementation	.981	.965		
Teaching effect	.924	.878		
Teaching environment	.978	.973		

The factor analysis results of the 4 first-level indicators are in line with the theoretical conception of the scale (see Table 1), indicating that the scale has good structural validity in determining the dimensions of the 4 first-level indicators. Through the above work, we have obtained a more reasonable "Information Technology and College English Teaching Integration Evaluation Questionnaire". In the verification of the index system, we increase the reliability and validity of the test methods to carry out verification research, in order to present a more scientific, reasonable and effective classroom teaching evaluation index system [10].

2.2 Case Study of Two CE Classrooms Teaching

Using the scale to test the classroom teaching of two college English teachers, it showed that there is a slight gap between the scores of Teacher A and Teacher B. According to the further analysis of the four first-level indicators of teaching design, teaching implementation, teaching effect and teaching environment, it is found that: A and B teachers are both excellent in teaching design. Teacher B's ability in teaching implementation is relatively lacking; Teachers A and B have similar performance in teaching effect and teaching environment. The following is a detailed analysis of the two teachers' scoring results on these four aspects.

First, in terms of teaching design, through the observation of the two teachers in the classroom, it is found that both teachers are aware of the necessity of analyzing student's groups. They have clear understanding of the importance for the application of information technology and students' practical experience. In the "design of teaching objectives", teacher A not only clarifies the goal of cultivating students' cultural awareness in the teaching plan, but also has designed more practices in the teaching process. Both teachers have the teaching design concept based on students and teaching materials, and have designed practical teaching programs that are easy for students to practice and understand.

Secondly,in terms of teaching implementation, in the two indicators of "interactive activities between students and information technology" and "interactive activities between teachers and students through information technology", both teachers received good evaluations. In terms of the two indicators of "integration of subject knowledge and information technology" and "evaluation of the effectiveness of teaching methods", Teacher A is better at linking students' lives with reality and actively encourages students to imagine with the assistance of information technology means such as videos and animations.

What's more, in the dimension of teaching effect, both teachers have been well received, and both have successfully completed the teaching tasks of this class and achieved the teaching goals of this class. In terms of "students' mastery of knowledge", both teachers can make full use of information technology to draw inferences from one another and help students understand knowledge outside of their subject. In the index of "to achieve the students' advanced skills", it can be seen that the two teachers attach great importance to the cultivation of students' independent problem-solving ability and the cooperative inquiry learning method.

The last but not least, as for the indicator of "evaluation of the teaching atmosphere", although teacher B's classroom atmosphere is not particularly strong compared to teacher A, and the students' enthusiasm for learning is not so high, when answering questions students' performance is still relatively positive. The simultaneous use or interspersed use of information technology makes the entire teaching process more humane and more convenient for students to conduct exploratory learning [9]. In the index item of "evaluation of teaching media equipment", teacher B appears to be more familiar, which reflects that the new generation of teachers are more proficient in the using, developing and maintaining of new things such as modern information technology.

3 Conclusions

The data analysis of the above scale shows that the "College English Teaching Classroom Assessment Scale" has good validity and reliability, and can more comprehensively evaluate the college English project classroom environment. And each dimension has a good correlation with the learning effect, and formative and summative evaluation are closely related.

To conclude, the integration and development of information technology and curriculum evaluation have received great attention from the international education community since the 1990s [9]. As a brand-new way to apply to education and teaching, it has gradually been recognized as a relatively effective teaching evaluation method applied to the teaching process. Teachers conduct quantitative and qualitative evaluation by classroom observation, which actively response to the demands of the new curriculum reform. Consequently, teachers analyze the real-time test results based on the classroom teaching to effectively guide the teaching and promote the mutual development between teachers and students.

References

1. Alan, C.K.C., Robert, E.S.: The effectiveness of educational technology applications for enhancing mathematics achievement in K-12 classrooms: a meta-analysis. Educ. Res. Rev. **22**(9), 88–113 (2013)

2. Cronbach, L.J.: Coefficient alpha and the internal structure of tests. Psychometrika **28**(6), 297–334 (1982). https://doi.org/10.1007/BF02310555
3. Carol, M.C.: The ISI classroom observation system: examining the literacy instruction provided to individual students. Educ. Res. **26**(7), 85–99 (2009)
4. Douglas, E.: Developing indicators for a classroom observation tool on pedagogy and technology integration: a Delphi study. Comput. Sch. **36**(9), 27–36 (2012)
5. Jody, B., Jerrel, C.: The technology integration assessment instrument: understanding planned use of technology of classroom teachers. Comput. Sch. **27**(9), 49–61 (2008)
6. Kearney, P., James, C.: Relationships among teacher communication style, trait and state communication apprehension and teacher effectiveness. Commun. Yearb. **4**, 19–23 (1980)
7. Randall, D.: Understanding technology literacy: a framework for evaluating educational technology integration. TechTrends **5**, 45–47 (2011). https://doi.org/10.1007/s11528-011-0527-3
8. She, H., Fisher, D.: The development of a questionnaire to describe science teacher communication behavior in Taiwan and Australia. Sci. Teach. Commun. **24**, 58–65 (1999)
9. Walberg, H.J., Anderson, G.J.: Classroom climate and individual learning. J. Educ. Psychol. **59**(6), 414–419 (1968)
10. Waldrip, B., Fisher, D.: Identifying exemplary science teachers through their classroom interactions with students. Learn. Environ. Res.: Int. J. **16**, 157–174 (2003)

The Training of Logistics Talents in Higher Vocational Colleges Under the Background of Intelligent Logistics

Ke Wang(⊠)

School of Business, Chongqing Vocational College of Transportation, Chongqing 402247, China

Abstract. Wisdom logistics under the background of higher vocational logistics personnel training research in this paper, the future demand for wisdom logistics under the background of logistics talents are analyzed, and puts forward the three major aspects of future logistics demand wisdom, wisdom is a skill type pipe type logistics talents, the second is data management and analysis logistics talents, the third is commerce application management wisdom logistics talents, In view of these three types of talent needs, the future reform direction of vocational intelligent logistics talent training is put forward. First, the new technology-driven innovation training mode is highlighted; second, the school-enterprise cooperation is strengthened to optimize the curriculum setting; third, the channel characteristics are given full play to cultivate talents with characteristics.

Keywords: Intelligent logistics · Higher vocational · Talents training

1 Introduction

With the development of new technology and business model, China's logistics industry is in a critical period from traditional logistics to smart logistics., artificial intelligence, cloud computing, big data, block chain, digital twin technology, quantum technology in-depth application in the field of high and new technology in logistics are constantly changing the operation of the logistics industry and the management pattern, the traditional personnel intensive jobs is to develop in the direction of unmanned, operation mode, management mode and the positions and functions change then there is the demand of change. In the future, with the continuous development of intelligent logistics in the logistics industry, the demand for intelligent logistics talents corresponding to its development will also become an important issue to be urgently solved. This paper will take the development of intelligent logistics as the background to deeply discuss how to train talents for the logistics major of higher vocational colleges according to the characteristics of the demand for intelligent logistics talents [1–6].

2 Demand Characteristics of Intelligent Logistics Talents in Higher Vocational Colleges

According to the implementation opinions of relevant documents on the national development of smart logistics, smart logistics will become an important support for the

J. Macintyre et al. (Eds.): SPIoT 2021, LNDECT 98, pp. 847–852, 2022.
https://doi.org/10.1007/978-3-030-89511-2_115

national strategic development and a new economic growth point. Under the background of the further deepening of the national strategy, logistics enterprises will complete the transformation and development from traditional logistics to intelligent logistics enterprises in the next 5 to 10 years. At the same time, the traditional logistics jobs will be gradually replaced by type wisdom logistics jobs, with wisdom logistics for the needs of a large number of new jobs will be caused, such as: intelligent logistics equipment operation and maintenance management, big wisdom logistics data analysis, logistics financial management, cross-border electricity wisdom logistics management, AI digital storage and distribution operation management wisdom logistics new jobs, and these jobs compared with traditional logistics jobs demand for talent is more prominent performance in can understand technology, management, business intelligence applied logistics talents. In the next 5 to 10 years, there will be a huge gap in intelligent logistics talents for Chinese logistics enterprises. Based on the analysis of national strategies, relevant logistics enterprises and the development status of intelligent logistics, the following three types of intelligent logistics talents will be most needed in the future development of Chinese logistics enterprises: Skilled management of intelligent logistics talents, data management and analysis of intelligent logistics talents, business application management of intelligent logistics talents [7–10].

2.1 Skilled Operation and Management of Intelligent Logistics Talent Demand

In Beijing, Shanghai, guangzhou, shenzhen and other logistics enterprises developed economic regions, the survey found over the next five to 10 years intelligent logistics facilities will gradually replace the traditional logistics facilities, in particular, some large logistics enterprise logistics facilities of intelligent than to reach more than 90% from the current 40%, traditional logistics jobs will be replaced by intelligent logistics facilities, Such as sorting, distribution, circulation processing and other traditional personnel intensive positions will gradually realize unmanned under the replacement of intelligent logistics equipment; But at the same time, we can see as smart logistics equipment to replace the traditional logistics jobs, at the same time also appear with the new position, such as smart logistics equipment cooperative operating post, intelligent logistics equipment operation and maintenance management positions for smart logistics equipment maintenance will operation, good management, to understand the smart skills type pipe position, and at present, most vocational colleges have yet to this aspect of talent cultivation, which makes this type of talent will become one of the most shortage of talent in the future.

2.2 Data Management and Analytical Intelligent Logistics Talents Need Urgently

A significant characteristic of intelligent logistics is digitalization, which is often referred to as digital logistics. Through a large number of data analysis, enterprises can accurately control the whole link of logistics to achieve the optimization of logistics management, so as to maximize profits. Most logistics enterprises in China form trade networks with major logistics nodes in China or the world through railway, air, road, water and other transportation modes. As a result, a large amount of logistics data, especially cross-border logistics data, is produced every day. Due to the variety, huge amount of data, rapid change, internationalization, system differences and other characteristics of these data, a large number of logistics data management and analytical intelligent logistics talents are needed to manage and analyze these data. From the data found in the growth of logistics efficiency to maximize economic efficiency of logistics enterprises, especially with the further development of Chinese economy in the next five years, the demand of talents will be into explosive growth, but at present, most vocational colleges have yet to this aspect of talent cultivation, which makes this type of talent will become one of the most shortage of talent in the future.

2.3 Business and Trade Application Management of Human Intelligence Logistics Talent Demand is Urgent

Chinese logistics enterprises through railway, air, road, water and other modes of transportation with the major domestic or world logistics nodes to form a trade network, for "One Belt And One Road" international trade services. Making every day a large number of commercial activities need logistics companies to complete, especially international trade logistics, because the international logistics has variety of information, large difference of system environment, general business, high degree of internationalization, etc., which requires a lot of business application management wisdom logistics personnel to manage these cross-border logistics activities in the international trade, In order to ensure the efficient and smooth logistics operation in international business activities, so that the quality of logistics service can be further effectively improved, especially in the next five years, with the further development of China's economy, the growth rate of international logistics will be significantly increased, and the demand for talents in this area will be explosive growth. At present, Most of the logistics major in higher vocational colleges have not carried out the professional curriculum setting or the formulation of talent training program, which makes this kind of talents will become one of the most urgently needed talents in the future.

3 Training Mode of Intelligent Logistics Talents in Higher Vocational Colleges

Based on the above analysis of the characteristics of the demand for intelligent logistics talents and the shortcomings of the current training methods for higher vocational logistics talents, this paper puts forward three training methods for intelligent logistics talents in higher vocational colleges based on the development of intelligent logistics.

3.1 Highlight New Technology Traction Innovation Training Mode

The core of intelligent logistics is the continuous application of new technology in logistics, so as to change the operation mode, management mode and business mode of the whole logistics industry. Can be thought of as new technology is the core of the whole logistics industry development and operation of traction, the new technology of traction with the development of logistics industry, and the training of logistics personnel in the service of the development of logistics industry, as a result, the new technology is the core of higher vocational logistics personnel training of traction, this requests us in the training of logistics personnel highlight the new technology to the traction and traction innovation personnel training mode in the new technology; Integrate the learning, application and development of new technologies into our curriculum, the formulation of talent training programs and practical training; Change the previous training mode of emphasizing management over technology, increase the proportion of new technology in personnel training, and highlight the study and application of new technology, especially the proportion of new technology in practical training. At present, artificial intelligence, big data, block chain, quantum technology, digital twin technology, Internet of things and cloud computing will be the key contents to be integrated into the personnel training of higher vocational logistics majors in the direction of new technology in the future.

3.2 Strengthen School-Enterprise Cooperation to Optimize Curriculum Provision

Higher vocational talents training goal is to carry the talent they need for enterprise development, enterprise's real demand is the fundamental requirement of higher vocational logistics personnel training, enterprise need what kind of person, what would need to develop talents in higher vocational colleges, and enterprises demand only through effective cooperation between colleges can truly effective real-time transmission to the school, the school can according to enterprise's talents needs to constantly adjust the talent training scheme, optimizing the curriculum in order to adapt to real enterprise demand for talents of different levels of ability; At present, due to various reasons university-enterprise cooperation in research and consulting enterprises, employment cooperation such as simple level, enterprise real dominated the school personnel training and the curriculum model has not yet formed, the enterprise development, talent demand, post Settings, and so on as the basic requirement of higher vocational logistics professional curriculum, make the enterprise post demand directly into the curriculum, to achieve a post a class or more than one duty class curriculum mode, in the process of enterprise development continued demand for post personnel changes in the real-time optimization of higher vocational curriculum truly and talent post one to one correspondence courses, the real-time optimization. At present, smart logistics equipment operation and maintenance management, smart logistics big data analysis, smart logistics financial management and green smart logistics management will be the key contents to be integrated into the curriculum of higher vocational logistics specialty in the future.

3.3 Cultivate Talents with Regional Characteristics

There are many regional economic zones in China, each of which has its own regional characteristics and development patterns. For example, the coastal areas are mainly export-oriented, and the Cheng-Chongqing areas are mainly manufacturing-oriented. This requires that each region should develop higher vocational intelligent logistics talents that adapt to its regional economic characteristics according to its own regional characteristics; Its main cultivation direction; First, focus on training intelligent logistics personnel with skills such as operation and maintenance management of intelligent logistics equipment and operation personnel of storage and distribution logistics equipment of intelligent logistics hub; Second, focus on the training of e-commerce logistics big data analysis talents, smart logistics financial management talents and other data management and analysis smart logistics talents; Third, talents in operation of international business intelligent logistics management platform, cross-border e-commerce intelligent logistics operation and management, and other business application management intelligent logistics talents.

4 Conclusion

In this paper, the demand for intelligent logistics talents in the future is analyzed. The demand for talents is mainly shown in the following categories: skilled management intelligent logistics talents, data management and analysis intelligent logistics talents and business application management intelligent logistics talents. The demand for these three types of talents will be the most urgent for intelligent logistics enterprises in the future. In view of these three types of talent needs, the paper puts forward the talent education and training reform in the future vocational intelligent logistics talent training from three aspects of highlighting the innovative training mode driven by new technology, strengthening the school-enterprise cooperation and optimizing the curriculum setting, and giving play to the characteristics of the channel to cultivate characteristic talents. The research of this paper aims to provide some ideas and methods to promote the training of higher vocational talents under the intelligent logistics. At the same time, the suggestions put forward will further change with the emergence of new technologies and new business models in the future, which is also the direction of the future research of this paper.

Acknowledgments. This work was supported by Chongqing Vocational College of Transportation science and technology project (CJKJ201902), 2018 Chongqing education science "13th Five Year Plan" project (2018-GX-099), Science and Technology Research Project of Chongqing Education Commission (KJ1500521), Chongqing Education Commission's 2021 Higher Vocational Education Teaching Reform Key Project (Z212068).

References

1. Wang, K.: Logistics management model of fresh agricultural products in the new international land-sea trade corridor based on information network. Revista de la Facultad de Agronomia de la Universidad del Zulia **36**(4), 1078–1085 (2019)
2. Wang, K.: Research on the application of VR technology in logistics equipment application and management course teaching. In: ICEIEM 2019, Penang, Malaysia, vol. 2, pp. 77–79, August 2019
3. Zhou, M.: Pragmatically promote the construction of new land and sea corridors in the west. Guangxi Econ. **31**(11), 46–47 (2019). (in Chinese)
4. Wang, K.: Study on the structure of waste material reverse logistics system based on Cainiaoyizhan. Logist. Technol. **38**(6), 145–147 (2019). (in Chinese)
5. Wang, K.: Research on training talents of intelligent logistics in higher vocational colleges under the background of New Western Land-Sea Corridor. In: SSEER2020, Penang, Malaysia, vol. 455, pp. 77–79, July 2020. [SSEER2020, Singapore, pp. 152–155, 2020]

6. Dong, J.: Research on the demand of smart logistics talents based on the construction of Hainan free trade zone. Value Eng. **38**(6), 1–3 (2019). (in Chinese)
7. Li, X.: Discussion on the training mode of applied logistics talents under the background of intelligent logistics Cainiaoyizhan. Logist. Eng. Manage. **40**(3), 130–132 (2018). (in Chinese)
8. Zhang, L.: Research on the teaching reform value of "blockchain+higher vocational tourism management specialty"—Huzhou vocational and technical college as an example. Teach. Innov. **2**(3), 53–55 (2020). (in Chinese)
9. Grather, W., Kolvenbach, S., Ruland, R., et al.: Blockchain for education: lifelong learning passport. Eur. Soc. Soc. Embed. Technol. **2**, 11–21 (2018)
10. Jirgensons, M., Kapenieks, J.: Blockchain and the future of digital learning credential assessment and management. J. Teach. Educ. Sustain. **1**, 145–156 (2018)

Basketball Data Analysis Based on Spark Framework and K-means Algorithm

Ning Zhu[1] and Qiongjie Dai[2(⊠)]

[1] Basic Teaching Department, Ordos Vocational College, Ordos, Inner Mongolia, China
[2] School of Mathematics and Computer Engineering, Ordos Institute of Technology, Ordos, Inner Mongolia, China

Abstract. With the extensive development of basketball, traditional data processing and analysis methods can no longer satisfy the optimization of basketball training indicators and the evaluation of training status. This severely restricts the digitization and management of basketball information data in our country. At present, my country's basketball big data analysis still faces the following problems: (1) At present, there is still a lack of research on the platform and frame application of data calculation and analysis in the field of basketball sports. As the modern computer data storage and collection facilities, the information processing equipment and functions of sports data are becoming more and more perfect, the workload of information storage and calculation of basketball sports data continues to expand. The existing sports data analysis service platform is far from being able to meet the needs of modern basketball big data analysis technology. (2) Because people's understanding of deep learning modeling and machine learning algorithms is not deep enough, it is limited to the query of individual action data and the statistical description of the competitiveness level, and the data mining and analysis are not deep enough. And it is unable to provide effective support for my country's basketball sports information prediction and strategy implementation [1].

Keywords: Spark framework · K-means algorithm · Basketball · Data analysis

With the development of information technology, the continuous progress of the times, and the arrival of the 5G era, the development trend of big data is unstoppable. The essence of big data is a combination of all data. Under the requirements of new technology, the collection of big data is no longer the simple way before, but more advanced technology. There are so many types of big data. If the use of big data is strengthened in production and life, it will inevitably have an irreplaceable effect on the accuracy of work and the quality of life and production. Therefore, in the future development, strengthen the use of big data, promote the progress of life, social development and economic construction, and cultivate professional talents to promote the correct and reasonable use of big data.

J. Macintyre et al. (Eds.): SPIoT 2021, LNDECT 98, pp. 853–857, 2022.
https://doi.org/10.1007/978-3-030-89511-2_116

1 Build a New Open Source Network Big Data Analysis and Calculation Platform

Hadoop mapreduce technology has been considered a sphere, and it has been widely used in the basket of big data platforms. Because the mapreduce system needs to write the intermediate results of a task directly back to a disk, it needs to re-copy these data from every node in the entire network, and spend most of the time in the disk storage environment in the entire network [2]. Due to the technical limitations of hadoop mapreduce, it affects the real-time analysis and calculation of basketball sports big data. In order to adapt to the computational speed and scope limitations of hadoop mapreduce, this subject research builds a parallel computing platform based on big data on the basis of an open source general parallel computing framework proposed by amplab of the University of California at Berkeley. Smack is a general-purpose big data computing service platform dedicated to supporting multiple application scenarios. It can solve the core problems of batch processing, interactive query and stream computing in the process of big data calculation. The data storage in this production environment is realized by the hadoop distributed file system hdfs. Although smark has many advantages of hadoop mapreduce, it is different from mapreduce in that the intermediate output and results of spark can be directly stored in the content. So there is no need to read and write hdfs complicatedly. Therefore, spark is more suitable for data mining and machine learning for those who need to replace algorithms [3].

2 Players Based on the Grabbing Framework and Ajax Technology

Based on web capture technology, it automatically collects relevant statistics and data of this technology, and uses it as a sample of machine learning to construct a new type of deep learning network model. It uses machine learning algorithms to conduct in-depth data mining and analysis, which provides decision-making support for the development of sports trainers. Scrapy is a framework technology that implements software crawling function through python language. It is mainly used to capture and scrape the data obtained from the website and through these pages. It can be widely used in data development, monitoring and control. [4] ajax (asynchronous javascript and xml). That is to say, js language and xml are used on the server at the same time, that is, asynchronous interaction between the two languages on the server through asynchronous javascript. The ajax technology is mainly implemented based on the json module in python. The basic data type in python is that the sequence is converted into a string in a standard format by methods such as json. dump, and then it can be directly stored or transmitted via the Internet. The ajax technology is mainly used to obtain web pages related to the player, and then obtain the player data and export the CSV file according to the network data captured by the scrapy framework [5].

3 Analyze the Mobilization Situation of the Stadium Based on the Algorithm

Analyze the person who is most likely to get a ball on the basketball court through data analysis and visualization (pointguards) based on the k-means clustering algorithm (pointguards). Usually considered to be an offensive organizer of the basketball team. This paper selects the point displacement as an actual case, and uses the k-means clustering algorithm to analyze and visualize the data. This study divides point guards into five categories, namely assist percentage and average points per shot. We can also use more information about players to cluster them. After we obtain the cluster information, we can obtain the information of all participants in each cluster. Teams and players of the same level in the group stage. If a player in a cluster has received great appreciation and attention, pay great attention, because they belong to a high-level group of players. One of the greatest advantages of the k-means algorithm is its ease of implementation. When clustering, the distance from each point to the center of mass should be saved, [6] and this point should be understood as an error. Then, the lower the value of ssh, the closer the data points will be to their centroids, and the better the clustering effect. Since the error point is a square, it is more important to find a point farther from the center. Its purpose is to obtain a better simulation of actual basketball team information. The more detailed the division of the clusters, the smaller the sum of squares of errors [7]. The relationship between the number of clusters and the error sum of squares in this actual case is shown in Fig. 1. In addition, due to the different centroids of random initialization, the final clusters are likely to be different. Therefore it may not be the optimal solution. Therefore, it is necessary to perform multiple similarity measures, by sequentially filtering the elements D10, D9, D8, etc. in the training set to measure the parameters of the results in the test set with the k-means clustering algorithm. The specific process is shown in Fig. 2 [3].

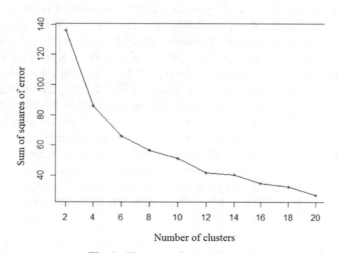

Fig. 1. The sum of squared errors

Cross validation is a method of dividing data set D into k mutually exclusive subsets of similar size, where each subset maintains the consistency of the data distribution as much as possible, that is, through the data set D Obtained by means of stratified sampling [4].

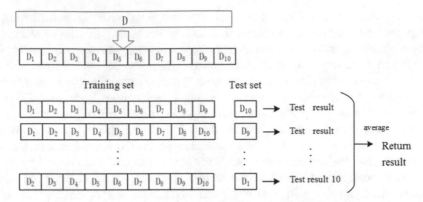

Fig. 2. K-value cross-validation to determine training parameters

4 Conclusion

In this era of big data, the ability of basketball training and game data mining not only directly affects the statistics and analysis mode of basketball game data, but also directly affects the future progress and development of basketball training and games in our country. The main research objectives and focus of this article are: (1) Based on the Spark framework and cloud computing technology, it integrates the existing scattered and isolated basketball sports information and data resources. Construct a large-scale cluster sports competition data analysis and calculation platform based on a clustered data computing environment. This can promote and promote the informatization process of basketball sports game information data analysis. (2) Based on a large number of scientific and information technology statistics and learning samples, a set of network models based on deep learning are constructed. Using machine learning algorithms to conduct in-depth data mining and comprehensive analysis of basketball players. Intuitively and accurately describe the competitions and decisions that various basketball players participate in. Improved the intelligence of his basketball knowledge and decision-making. In the following research and work, on the basis of the Spark framework, two technical methods of caching and radio broadcasting are used to optimize and adjust the parallel computing program, thereby further improving and increasing the efficiency of computing. At the same time, through the previous preprocessing, the data fully meets the needs of the classification algorithm, reducing the complexity of the later preprocessing [5].

References

1. Song, W.: Research on basketball sports data analysis based on Spark framework and K-means. Chinese Sports Science Association. In: Collection of Abstracts of the 11th National Sports Science Conference. Chinese Sports Science Association: Chinese Sports Science Society, p. 3 (2019)
2. Zhao, Z., Hang, L.: Bibliometric analysis of Chinese basketball sports research literature——based on CNKI (1979–2019) data research. Shandong Sports Sci. Technol. **41**(05), 29–33 (2019)

3. Lan, N.: Feasibility Study of the Chinese Men's Basketball Team in the 31st Olympic Games. Wuhan Institute of Physical Education (2015)
4. He, Z.: A comparative analysis of the technical and tactical data of the 2014 NBL Jiangsu Tongxi Team Finals. Youth Years **4**(07), 260 (2015)
5. Zhang, H., Li, G.: Data analysis in basketball. Xueyuan **4**(08), 188–189 (2015)
6. Miao, X.: Research on Exercise Load of High-Level Competitive Basketball. Beijing Sport University Press. 201305.177
7. Zhao, Z., Hang, L.: Bibliometric analysis of Chinese basketball research literature-based on CNKI (1979–2019) data research. Shandong Sports Sci. Technol. **41**(05), 29–33 (2019)

Automatic Analysis of RPA in Digital Service Testing of Grid System

Linhong Chen[✉], Fei Lou, and Tianqi Fu

State Grid Zhejiang Electric Power Zhoushan Company, Zhoushan 310015, Zhejiang, China

Abstract. As science and technology in China has continuously developed, more and more programs and technologies are moving towards automation in power companies, especially in power digital service systems. The use of automated programs such as artificial intelligence can effectively improve the service capabilities of grid companies and meet the daily needs of enterprises and provide better services to users. The function of RPA is important in the digital service of grid system. With the use of function of RPA, the digital power service can be fully realized, grid enterprises can be effectively assisted in power system maintenance, power dispatching, real-time power grid monitoring, fault alarm processing and digital service functions It has completely changed the traditional power grid service system by combining with grid system and is an important reform of grid system. Therefore, this paper mainly elaborates the concept, development status and development prospects of RPA, and optimizes the difficulties of RPA during the testing to make it run stably.

Keywords: Artificial intelligence · Grid system · Digital service system · Automatic analysis

1 Introduction

In artificial intelligence systems, the use of RPA platforms can effectively realize digital service. Especially in the information age, more and more users rely on smart devices and systems, and more and more promotion of intelligent service platforms make people's lives more convenient in daily life. Especially in the grid system, in order to effectively meet the work and life needs of grid companies and users, the use of intelligent digital platforms in power dispatching, power supply services and grid operations can effectively improve the efficiency of grid companies and make the life of grid users more convenient. Among them, in the test of power digital service system, the introduction of RPA platform can effectively provide smart service system for grid companies, and the use of RPA can improve the operating efficiency of enterprises, enhance the value of manpower of enterprises, and effectively accelerate the process of digital reform in grid enterprises [1]. Therefore, the introduction of RPA in grid companies can assist them in creating their own software robots to make their service system and business processes intelligent and automated. As a result, the efficiency and service capabilities of enterprises can be effectively improved.

© The Author(s), under exclusive license to Springer Nature Switzerland AG 2022
J. Macintyre et al. (Eds.): SPIoT 2021, LNDECT 98, pp. 858–863, 2022.
https://doi.org/10.1007/978-3-030-89511-2_117

2   RPA Platform

2.1   Concept of RPA Platform

The RPA platform refers to the software for Robotic Process Automation, which is an artificial intelligence virtual system adopting the latest technology. It improves users' work efficiency by setting related automation programs. With the continuous development of information technology, RPA has been developed, especially in recent years; it has been widely used in various industries in China, improving the process automation ability of various industries. Especially in the digital service system of grid system, with the use of RPA, the service ability of grid system can be effectively upgraded and transformed, which makes it stride forward to the field of automation. When using RPA in the service system of grid system, its main work core is to transform human resources into automatic service procedures through the use of automation technology and intelligent technology, to bring human resources through the establishment of relevant digital service system, to effectively reduce the labor cost and reduce unnecessary repetition in the service system. In addition, the service system of RPA is not interfered with by artificial decision-making, which can effectively improve the service efficiency of power system and reduce the occurrence of unnecessary errors [2]. In addition, the basic architecture of RPA platform mainly includes developers, cloud expansion editors, cloud expansion consoles, scheduling systems, and monitoring systems. The specific structure is shown in Fig. 1.

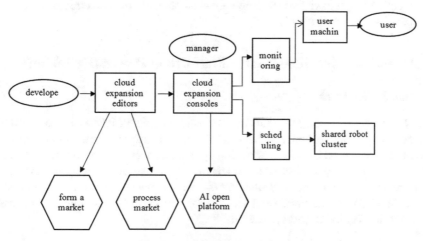

Fig. 1. Basic structure of RPA

2.2   Functions of RPA Platform

The main functions of RPA platform include collecting data from users, integrating and processing user information according to cloud computing, and executing corresponding instructions according to user needs through professional calculation, so as to provide users with convenient and efficient services. And the use of RPA platform can also effectively improve the work efficiency of grid enterprises. It can also effectively carry out remote control, statistics, fault handling, equipment upgrade, safety management and real-time database management of power system. By adopting the intelligent RPA platform, it can ensure the normal operation of the service system and operating system of enterprises. In the intelligent RPA platform, it has a sound authority management function, which can effectively query user information and service processes, and carry out power dispatching. RPA software can fully realize enterprises' automation capabilities; realize dispatch automation of power, and operation automation and digital service automation of power grid. It has characteristics of reliability, safety and real-time performance [3].

2.3 Application Status and Development Trend of Intelligent RPA Software

With the continuous development of information technology and science and technology in China, intelligent RPA software has been effectively used in power operation systems and service systems, and its development prospects are better. At the current stage, automation technology is adopted in the application of intelligent RPA software in power system. By adopting the cloud service function of intelligent RPA software, it can effectively connect the national power dispatching system and power service system. The development trend of RPA system mainly includes information digitization, communication digitization, decision communication digitization, management digitization, etc. Through its digital development and application, it can effectively carry out full operation in the market [4].

3 Problems of Intelligent RPA Platform in Power Digital Service

3.1 Not Widely Used

In power system of China, the start time of using intelligent RPA platform for digital power service is relatively late, so that it is still unable to achieve the integrated operation and detailed operation of power operation system and service system. Due to the lack of technical coverage, there are certain technical vulnerabilities during its application, and there are deviations in the collection of relevant power operation data and power service data by relevant staff. As a result, RPA cannot fully automate the process of digital service testing of grid system [5].

3.2 Unstable Data Processing and Analysis Capabilities

In the process of RPA power digital testing, due to the late start of RPA technology, there is still a large development space for its dispatch technology in power dispatching and operating systems, and work efficiency and transfer efficiency still need to be improved, especially the information storage function and capacity of cloud platform need to be expanded. Because RPA technology is not mature and perfect enough, there are still some details in the process of testing power digital service system, especially in the process of processing user information and service information, there is a problem of insufficient data processing and analysis capabilities, and its working ability is not stable, causing certain power service problems [6].

3.3 Loopholes in Network Security

RPA is a kind of software based on information technology, which needs to be connected with network system in the process of power digital service. However, in the process of RPA development, due to the incomplete understanding of power service system by RPA developers, there are some loopholes in the design of RPA program, which leads to certain network security risks in the process of RPA digital power service testing. In the process of network operation, it is easy to be interfered by network viruses and hackers, and problems of network security cannot be guaranteed [7].

3.4 Lack of Talents for RPA Technology Development

In the development and design of RPA program, the lack of talents for RPA technology development in power enterprises leads to certain defects in the development of their own digital service system. Because many RPA technical personnel do not know the power service system and process of power enterprises very well, there is a lack of pertinence in the design of relevant power digital service system, and a lack of pertinence in the design of relevant service functions, resulting in the lack of pertinence in RPA program. In addition, due to the lack of RPA technology management talents, it faces certain difficulties when developing and maintaining RPA system, which is not conducive to the maintenance and upgrading of RPA system [8].

4 Optimization Strategy for Automation of RPA System

4.1 To Improve the Coverage of RPA Power Digital Service System

In order to optimize and upgrade intelligent RPA technology, it is necessary to comprehensively improve the coverage of RPA power digital service system. By upgrading the nationwide power service system and applying intelligent RPA technology to realize the function of power digital service, a full coverage picture will be formed nationwide, and the ability of integrated operation and comprehensive operation of power operation system and service system will be improved, as well as that of meticulous operation. And it can also improve the loopholes of RPA technology in power digital service system, strengthen the cooperation ability with the international, and improve the ability of power digital service [9].

4.2 To Improve the Comprehensive Performance of RPA Technology

Combined with the current situation of unstable data processing and analysis capability of RPA technology, the staff effectively improve its functions and technologies according to the actual needs and conditions of power operation system and service system. The power digital service system runs stably by strengthening the research ability of automation technology of RPA technology, improving the digital service ability of RPA technology, and upgrading the cloud computing ability and storage ability of RPA technology. As a result, the data processing and analysis capabilities of RPA technology can be improved. The comprehensive performance of RPA technology is improved from different angles by combining with various efficient automation systems, and RPA technology is improved [10].

4.3 To Improve RPA Network Security

In order to effectively improve RPA network security, a scientific and comprehensive network security protection system needs to be established in the RPA program. Through the use of stand-alone defense and network defense systems, the ability to prevent network viruses is strengthened. In the process of network security investigation, the relevant service data and instructions is checked by installing relevant prevention system to prevent the infection of network viruses. At the same time, the setting up of firewall in the RPA program can avoid hacking, and effectively block the problem of illegal data packet implantation. In addition, by establishing a system database security precaution system, user information and related programs are encrypted to avoid problems such as data leakage [11].

4.4 To Introduce Key Talents of RPA Technology

In order to improve the development, maintenance, operation and upgrade capabilities of RPA technology, it is necessary to introduce key talents of RPA technology, conduct regular training for the technical personnel, and conduct targeted training for the power service system, so that they can reasonably design an efficient and reasonable digital power service system according to the needs of power service system when developing and using related technologies, improving the service capacity of grid enterprises [12].

5 Conclusion

In summary, in the process of using RPA function for power digital service, due to the imperfect technology promotion, it cannot cover the integrated power digital service system. The function of RPA needs to be upgraded and improved, and there are certain network security and related talents, leading to certain difficulties in the use process. In order to effectively strengthen the application ability of RPA technology, it can effectively upgrade and optimize RPA technology by improving the coverage of RPA technology digital power service system, improving the comprehensive performance of RPA technology, improving RPA network security, establishing network security protection system and introducing key talents of RPA technology, so that RPA technology can better serve the power digital service system, and improve the intelligence and automation capabilities of power companies.

References

1. Sheng, N.: Discussion and function analysis of process automation robot. Sci. Technol. (11), 15 (2020)
2. He, X., Wang, R., Wu, J.: Information characteristics of power electronic transformation and information control technology from discrete digital to intelligent power energy. Chin. J. Electr. Eng. **40**(05), 1579–1587 (2020)
3. Fu, C., Xiao, Z.: Research and application of digital transformation technology in power enterprises. China New Commun. **22**(04), 43–45 (2020)
4. Quan, W.: Application of RPA technology in financial intelligence of electric power enterprises. Telecommun. Sci. **36**(01), 139–143 (2020)
5. Quan, W.: Practice of digital financial transformation based on RPA technology in power enterprises. Account. Learn. (30), 15–16 (2019)
6. Yang, F., Sun, Y., Yuan, W.: Application of digital technology in power marketing. Integr. Circuit Appl. **36**(11), 90–91 (2019)
7. Siemens builds digital substation for glitreenerginett to support the application of power Internet of Things. Electr. Appl. **38**(10), 2 (2019)
8. Quan, W.: Practice of RPA Technology Based on Financial Intelligence in Power Enterprises. Editorial Department of Telecommunications Science, p. 414. People's Post and Telecommunications Press (2019)
9. Yu, D.: Building a digital labor force and cultivating a new kinetic energy of digital economy: introduction to RPA technology and application. Mil. Civ. Integr. Internet Inf. Technol. (05), 24–27 (2019)
10. Quan, W.: RPA technology helps financial internal control management of power enterprise. Account. Learn. (15), 245–247 (2019)
11. Daphne, G., Shi, W.: Robot process automation. Shanghai Qual. (02), 15–18 (2019)
12. Song, P., Li, Y., Wang, P.: Digital realization method of power electronic nonlinear load. J. Electr. Mach. Control **20**(05), 7–13 (2016)

Exploration and Practice of Network Training and Teaching Platform Based on Virtual Simulation Technology

Jinliang Li[1], Yang Yang[2(✉)], and Xiaohai Qu[2]

[1] College of Materials Science and Engineering, Jilin University, Changchun, Jilin, China
[2] Engineering Training Center, Jilin University, Changchun, Jilin, China
yyljl@jlu.edu.cn

Abstract. Engineering training, as a compulsory course for colleges and universities to carry out engineering practice teaching, plays a huge role in learning basic technological knowledge, creating innovative thinking, and cultivating engineering practice and innovation capabilities. Combining the characteristics of engineering training practice teaching in colleges and universities, and aiming at the safety education problem in the engineering training practice teaching process, it is proposed to build a virtual simulation training teaching platform for engineering training safety. The construction of this platform can realize online and LAN single-machine learning, online and single-machine learning Examination access, systematizes and standardizes engineering training safety education, displays dangerous safety accidents through vivid virtual resources to achieve the purpose of warning education, assists the safety education link in engineering training teaching, and forms teachers and students Online exams can enter the safety management mode of practical training. The construction of this platform fills the gaps in the relevant syllabus for engineering training of Jilin University and provides important reference for engineering training safety education in other colleges and universities.

Keywords: Virtual reality · Engineering training · Safety education

1   Introduction

With the development of new era industries such as Made in China 2025, Internet+, Internet of Things, Industry 4.0, and artificial intelligence, emerging engineering and scientific talents are required to have solid engineering practice capabilities, outstanding innovation capabilities, broad international vision, and the ability to discover and solve problems. Ability to complex engineering problems.

Engineering training, as a compulsory course for colleges and universities to carry out engineering practice teaching, plays a huge role in learning basic technological knowledge, creating innovative thinking, and cultivating engineering practice and innovation capabilities.

J. Macintyre et al. (Eds.): SPIoT 2021, LNDECT 98, pp. 864–869, 2022.
https://doi.org/10.1007/978-3-030-89511-2_118

2 Safety Issues in the Process of Engineering Training Practice Teaching

In order to better cultivate the ability of college students to innovate and solve complex engineering problems, most of the engineering training centers are as close to the actual production environment of the factory as possible in the hardware construction and practical teaching implementation process, and students' own safety awareness and prevention Ability is relatively weak, such as: not understanding the training dress requirements, unfamiliar with safe walking routes, there are potential safety hazards; different equipment operation methods are different, students are not familiar with the working principle and performance of the equipment, and do not operate the equipment in accordance with the safety operating procedures or cannot be in time If problems with the equipment in operation are found, it is easy to have safety hazards in the process of practical teaching, which can cause teaching accidents and endanger the lives of students [1].

Under the new situation, with the continuous development of engineering training, the engineering training curriculum system of colleges and universities is constantly reforming and innovating. In order to cultivate students' awareness of "big engineering", engineering training has changed from traditional practical teaching modules such as ordinary turning processing, Fitter, general milling, grinding, casting, welding, etc. have gradually expanded into modern and advanced practical teaching modules such as CNC turning processing, machining centers, 3D printing, laser processing, industrial measurement, industrial robots, intelligent manufacturing, artificial intelligence, etc., strengthen The practice teaching content of new materials, new technology and new technology is adopted, and the integration of multiple disciplines is emphasized. The full openness of practical teaching resources in space and time has been realized to meet the independent creation and processing of students. But at the same time, the probability of safety accidents in engineering training has also greatly increased. Once a safety accident occurs, the negative impact on students, teachers, engineering training centers and even schools is particularly obvious [2].

Therefore, it is imperative to reform the safety practice teaching in the engineering training process.

3 Constructing an Engineering Training Safety Virtual Simulation Training Teaching Platform

Aiming at the engineering training safety practice teaching link, a practical teaching method of "complementing reality with virtual reality" is proposed, that is, the development of engineering training safety virtual simulation training platform, which presents safety accidents that are difficult to show by traditional education methods to engineering training safety virtual simulation training In the platform, examples of typical safety accidents are explained in the form of animations, pictures, and videos, providing students and training instructors with an "autonomous", "interactive", and "open" engineering training safety training teaching environment to enable students Have a more vivid and true understanding of the practical operations to be faced, and actively abide by the safety operating procedures, consciously and consciously avoid or even eliminate some potential safety hazards [3, 4].

Teachers and students of the platform can learn about the safety operation procedures of the relevant equipment used in engineering training design through online or stand-alone learning, understand the hazards of safety accidents, and pass relevant exams to achieve continuous engineering training practice teaching the goal of. Failed to pass the exam, can not continue training and learning, this kind of learning and passing the exam can enter the access teaching system of the training link, and ultimately will achieve a deep understanding of the content of the training safety syllabus, greatly reducing and avoiding safety accidents happened.

3.1 The Overall Construction of the Safety Virtual Simulation Training Platform for Engineering Training

The engineering training safety virtual simulation training platform includes the teacher end and the student end. The teacher end is mainly divided into resource management system, data analysis system, information maintenance system, examination system, etc.; the student end is mainly divided into learning system and examination system.

3.2 Construction of Safety Digital Teaching Resources for Engineering Training

Engineering training safety digital teaching resources are mainly for training safety general principles, equipment safety operation regulations in each training teaching module, safety cases prone to teaching accidents, etc. The training safety teaching resource library includes electronic courseware library, two-dimensional/three-dimensional animation Library, video library.

3.3 Construction of a Virtual Simulation Interactive Platform for Engineering Training Safety

Using computer virtual reality technology to create a virtual interactive safety practice operating environment, the operator can complete the safety operating procedures of each module, the safety cases prone to teaching accidents, etc., as if in the real environment, practical training, such as engineering training practice The dress code for training is shown in Fig. 1, and prohibit measuring the rotating workpiece in Fig. 2. The use of network technology to realize the network virtual simulation training platform, truly realize an open engineering training training environment, break the time and geographical constraints and be fully open to students, and expand the dimension of the engineering practice teaching platform [5, 6].

Fig. 1. Dress code for engineering training

Fig. 2. Prohibit measuring the rotating workpiece

3.4 Construction of Engineering Training Safety Assessment System

Set up engineering training safety assessment mechanism, so that students deeply understand the importance of safety training, so as to firmly establish safety awareness. The examination system contains modules such as test question maintenance, test subject maintenance, and test paper selection. Teachers can import and export training safety test questions in batches, or edit the test questions online, and can choose to manually or randomly group papers. The system interface is based on All subjects are managed in a tree structure, and the content of the examination is a sub-category work test, and finally a statistical analysis is carried out.

4   The Effect of the Engineering Training Safety Virtual Simulation Training Teaching Platform

4.1   Enriching the Practical Teaching Content of Engineering Training

Practical teaching is the basic task of the engineering training center of colleges and universities, and ensuring the safety of students' practical training is the top priority of practical teaching. Safety accidents in engineering training are difficult to show using traditional educational methods. For example, female classmates have long hair involved in a lathe and all their scalp is removed; students wear gloves to operate a milling machine, and their hands are involved in injuries caused by rotating tools. Adopting a teaching mode that combines virtual simulation and actual training teaching, breaking the limitations of the original engineering training safety practice process, providing teachers and students with a more intuitive teaching method, so that students can learn what they cannot understand in traditional training safety education Content [7, 8].

4.2   Mobilize Students' Enthusiasm for Active Learning and Training Safety

Under the conditions of virtual simulation training, engineering training safety practice teaching is no longer limited to the classroom, nor is it limited to simple safety hazard pictures or videos [9]. The space for practical teaching activities has been expanded; at the same time, for students, The mastery of safety knowledge is no longer limited to the classroom, you can use your spare time to learn, and students' study time has also been expanded, which helps to deepen the understanding and mastery of safety knowledge. In addition, through the setting of the software, teachers can count the students' learning time, whether the operation is completed, whether there are illegal operations or non-operations, etc., and can more intuitively understand the time for students to learn each safety knowledge point, and improve the quality of practical teaching.

4.3   Improve the Safety of Engineering Training Practice Teaching

There are many practical training content for engineering training, including 23 training teaching modules such as machining center, intelligent manufacturing, CNC lathe, CNC wire cutting, lathe, milling machine, welding, forging, heat treatment, fitter, etc. Students not only need to process the above teaching content The learning of craftsmanship and processing methods requires the operation training of practical training equipment. When students are unfamiliar with the performance of the machine tool and their operating proficiency is not high, there is bound to be the possibility of safety accidents due to operating errors. Therefore, in the process of practical training and teaching, you can use the engineering training safety virtual simulation platform to familiarize students with the safe operation procedures of the equipment, train students to operate the equipment in a standardized way, and enter the engineering training center for internships after passing the assessment, which is conducive to enhancing students Safety awareness, standardize safety behaviors, and improve the safety of training [10].

5   Conclusions

Through the construction of the engineering training safety virtual simulation training platform, the engineering training safety education is systematized and standardized, and a safety management mode is formed in which teachers and students can enter the training after passing the online exam. The construction of this platform helps to enhance students' Safety awareness, standardize safety behaviors, and improve the safety of engineering training practice teaching, facing Jilin University Engineering Training A, Engineering Training B, and Engineering Training C, so that there are 37 majors in the school's mechanical and near-mechanical departments and 11 colleges. 3,700 students per year benefit. With the opening of courses such as the integration of medicine and engineering and the integration of art and engineering, the scope of benefits will be further expanded in the future, and at the same time, it will provide important reference for engineering training safety education in other colleges and universities.

References

1. Ye, S., Mao, X.: Construction and exploration on security system of engineering practice teaching. Exp. Technol. Manage. **26**(12), 139–140 (2009)
2. Zou, W.: Safety education in engineering training center. Res. Explor. Lab. **29**(2), 181–183 (2010)
3. Wang, R., Zhang, X., Zhang, M., Zhu, G., Zhou, S.: Construction and exploration of internet security virtual simulation practice teaching for new engineering. Exp. Technol. Manage (2019)
4. Bi H., Wang, T., Zheng, H., Ma, Y., Shi, Z.: Research and practice on engineering training security guarantee system in colleges. Lab. Sci. **22**(6), 203–205 (2019)
5. Tie, F., Hong, S.-D., Shu, Q.-M.: Exploration and ponderation on safety guarantee problem of engineering training. Exp. Technol. Manage. **30**(4), 208–210 (2013)
6. Zhang, G.: Thoughts and practices on reinforcing safety education of engineering training for intelligent manufacturing. Exp. Sci. Technol. **16**(6), 132–135 (2018)
7. Bi, H., Cheng, Y., Wang, T., Ma, Y., Zheng, H., Shi, Z., Zhang, Y.: Research and practice on safety culture system of engineering training under the new situation. Lab. Sci. **21**(2), 209–212 (2018)
8. Shi, J.T., Li, X.K., Zhang, H.L.: Construction of information content security experimental course based on virtual simulation cloud platform. Exp. Technol. Manage. **34**(4), 9–13 (2017)
9. Wang, A., Shi, D.: Security and protection in the training of numerical control practicing. Res. Explor. Lab. **36**(12), 290–292 (2017)
10. Wang, Y.-L., He, Y.-X., Zhu, Q.: Research and practice on safety culture system of engineering training under the new situation. Lab. Sci. **21**(2), 210–212 (2018)

Application of Network Database Security Technology Based on Big Data Technology

Lang Liu, Zilong Xu[✉], and Dapeng Zhou

Gongqing College of Nanchang University, Gongqingcheng 332020, Jiangxi, China

Abstract. A network database system is an information warehouse in an open environment. Once data is lost, damaged, illegally altered or compromised, it will cause great losses. Nowadays, the research on the security mechanism of network databases is continuously deepening, and the security management of network databases has become one of the hotspots of research by experts and scholars. The purpose of this article is to study network database security technology based on big data technology. This article first summarizes the basic theory of big data, and then extends the core technology of big data. Combined with the current research status of network database security, big data technology is used to study its network database security. This article systematically elaborates the database design and security strategy analysis, and designs and implements the database security model system. This article uses comparative analysis and observation methods to study the subject of this article. Experimental research shows that the average detection rate of network database security systems based on big data technology is 0.138 μs, and the average detection rate of traditional quota network database security detection systems is 0.565 μs, which fully reflects the excellent performance of the system studied in this article.

Keywords: Big data technology · Database · Security analysis · Application research

1 Introduction

The network database application system is a system that combines database technology and network technology for information processing. Network technology is one of the most rapidly developing technologies in the world today, and as an extremely effective data processing tool, database has developed relatively mature technology from early hierarchical databases, mesh databases to relational databases [1, 2].

In terms of network security, the key is the study of network reliability. The content involves the security and confidentiality of the network system. At present, there are various corresponding programs to deal with the security threats from various aspects of the network, and at the same time, new network security issues are constantly raised [3, 4]. Some of these key technologies include analysis and research of security protocols, virtual network technology, firewall technology to prevent network vulnerabilities, etc. [5, 6].

J. Macintyre et al. (Eds.): SPIoT 2021, LNDECT 98, pp. 870–875, 2022.
https://doi.org/10.1007/978-3-030-89511-2_119

The purpose of this article is to improve network data security, and to study network database security technology based on big data technology. Through the demand analysis of the database, the establishment and implementation of data models, and the description of the secure database management system, this article Research the theme.

2 Research on the Application of Network Database Security Technology Based on Big Data Technology

2.1 Database Design and Security Strategy Analysis

(1) Content composition

The database security information system studied in this paper is composed of several modules: supply and demand hotline, data query, expert introduction, product information, business directory, member registration and friendship link [7, 8].

(2) User analysis

The users of the system are mainly corporate users. According to the requirements of system business operation, a membership system is adopted [9, 10]. Therefore, when designing the database, users are first divided into three categories, general users, member users and super users. General users are unpaid users and can only get general shared information of the system. Member users are paying users and have established a long-term cooperative relationship with the system. It can obtain high-level technical information and materials [11, 12]. The super user is the system administrator, responsible for the entire database management, and has the right to modify the database.

(3) Data model establishment and realization

In the database system of this article, information and services are provided to different users through modules such as supply and demand hotline, data query, expert introduction, product information, exhibition information, and membership registration. Therefore, it is necessary to highly abstract the user table and mark its identity characteristics, and the control requirements can be realized based on the mark. At the same time, the above information sometimes overlaps in content. In order to reduce the redundancy of the data, it is also necessary to abstract the content of the above sections and establish a data model.

2.2 The Overall Design of the Network Database Security Access System

The network database security access system is divided into three parts: interface layer (presentation layer), logic layer (data access layer) and data layer, each of which performs different functions. Role-based access control has different implementations in each part. At the same time, it combines the security strategy provided by the development environment and the corresponding components and class libraries to meet the needs of system security access control. The presentation layer provides an interactive interface for users to implement form authentication and SQL injection prevention procedures. The access layer is located between the user and the data layer, and implements the control logic, including the realization of the authentication program and the encryption and decryption of data. The data layer realizes the storage of basic data and encapsulates the stored procedures for call. The program logic of the network database security access system is located on the authentication server.

2.3 Smart Card Authentication Scheme Based on Hash Function

The scheme is put forward based on analyzing the characteristics of the one-way hash function. It is easy to use, requires simple equipment, and can be implemented on a personal computer.

(1) Smart card initialization phase

When user U submits a registration application to the smart card authentication server, he first submits the password PW he chose. The smart card authentication server assigns an ID to the user, and calculates the user's card personal information X.

$$X = h(ID, S) \oplus PW \tag{1}$$

The user card personal information X, DI and hash algorithm h() are stored in the smart card.

(2) Login phase

1) When the user logs in, he inserts a smart card into the terminal and enters his own PW.

2) Smart card identification C calculation

$$C = h(X \oplus T \oplus PW) \tag{2}$$

Where T is the current time of the system.

3) Form a login request R = (ID, C, T), and send the request to the system.

(3) Identification stage

Assuming that the time when the system receives R is T, the request is verified through the following process:

1) Verify whether the function is correct, if it is not correct, the login will be rejected.

2) Calculate $T' - T, T' - T > \Delta T$ if (ΔT is the maximum time interval allowed), it means timeout, and this login is rejected

(4) Analysis of the correctness of the algorithm

Proof of the scheme algorithm: If the whole process is carried out legally, the agreement is reliable.

From formula (1) (2), we can know:

$$C = h(X \oplus T \oplus PW) = h(h(ID, S) \oplus PW \oplus T \oplus PW) = h(h(ID, S) \oplus T) \tag{3}$$

Therefore, when the user is a legitimate user and the submitted password is correct, the agreement is correct.

3 Experimental Research on Network Database Security Technology Based on Big Data Technology

3.1 Experimental Protocol

In order to make this experiment more scientific and effective, this experiment compares the network database security system based on big data technology studied in this article with the traditional network database security system. This experiment uses the same data for the two systems Set for virus detection and run on simulation software. In order to make the experiment more scientific, a total of 5 groups were set up for comparison in this experiment, and the results obtained were calculated and analyzed using mathematical statistics.

3.2 Research Methods

(1) Comparative analysis method

This experiment compares and analyzes the network database security system based on big data technology and the traditional network database security system studied in this article and records the data, and organizes and counts the recorded data. These data not only provide a theoretical reference for the topic selection of this article, but also provide data support for the final research results of this article.

(2) Observation method

This study compares and analyzes the detection rates of the two systems on the simulation software and obtains data. These data provide a reliable reference for the final research results of this article.

(3) Mathematical Statistics

Use relevant software to make statistics and analysis on the research results of this article.

4 Experimental Analysis of Network Database Security Technology Based on Big Data Technology

4.1 Analysis of the Detection Speed of the Network Database Security System

In order to make this experiment more scientific and effective, this experiment will analyze the detection speed on the simulation software, and the data obtained are shown in Table 1.

Table 1. Analysis of detection speed of network database security system

	Test1	Test2	Test3	Test4	Test5
Big data	0.132	0.137	0.142	0.135	0.130
Traditional	0.561	0.554	0.564	0.563	0.569

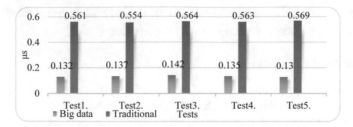

Fig. 1. Analysis of detection speed of network database security system

It can be seen from Fig. 1 that the average detection rate of the network database security system based on big data technology researched in this paper is 0.138 μs, and the average detection rate of the traditional quota network database security detection system is 0.5655 μs, which fully reflects the excellent performance of the system studied in this paper.

4.2 Analysis of Detection Accuracy of Network Database Security System

In order to further research and analyze this experiment, this experiment conducted an experiment on the detection accuracy of the network database security system, and the data obtained is shown in Table 2.

Table 2. Analysis of detection accuracy of network database security system

	Test1	Test2	Test3	Test4	Test5
Big data	72.3%	76.8%	70.7%	76.1%	77.2%
Traditional	64.7%	66.2%	65.1%	63.4%	66.7%

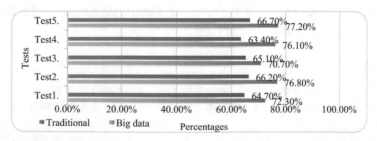

Fig. 2. Analysis of detection accuracy of network database security system

It can be seen from Fig. 2 that, compared with the traditional network database security system, the network database security system based on big data technology studied in this article has an accuracy higher than 70% in multiple tests, which is far beyond the traditional data security system, which fully reflects the feasibility of this research.

5   Conclusion

Network database security has always been a hot issue in recent years. According to different aspects of network database, many security strategies have been proposed. There are many technologies such as user-oriented security access control, network-oriented secure transmission and back-end database-oriented data encryption and audit trails. Among them, user-oriented security access control technology can not only solve the traditional illegal operation problem of illegal users, but also Moreover, legal users can be controlled hierarchically, and its access control model is flexible and easy to expand, so it has attracted much attention.

References

1. Allard, M.W., Strain, E., Melka, D., et al.: The PRACTICAL value of food pathogen traceability through BUILDING a whole-genome sequencing network and database. J. Clin. Microbiol. **54**(8), 1975 (2016)
2. Khatami, R., Luca, G., Baumann, C.R., et al.: The European Narcolepsy Network (EU-NN) database. J. Sleep Res. **25**(3), 356–364 (2016)
3. Anwar, A., Mahmood, A.N.: Anomaly detection in electric network database of smart grid: graph matching approach. Electr. Power Syst. Res. **133**(7), 51–62 (2016)
4. Fei, Y., Wang, R., Li, H., et al.: DPMIND: degradome-based plant miRNA–target interaction and network database. Bioinformatics **9**, 9 (2018)
5. Kalathur, R., Pinto, J.P., Sahoo, B., et al.: HDNetDB: a molecular interaction database for network-oriented investigations into Huntington's disease. Sci. Rep. **7**(1), 5216 (2017)
6. Solli, A., Helsper, C.W., Ader, R.J.M., et al.: Primary care management of women with breast cancer-related concerns-a dynamic cohort study using a network database. Eur. J. Cancer Care **25**(6), 1005–1014 (2016)
7. Pinto, J.A., Kumar, P., Alonso, M.F., et al.: Coupled models using radar network database to assess vehicular emissions in current and future scenarios. Sci. Total Environ. **761**(1), 1207 (2020)
8. Pinto, J.A., Kumar, P., Alonso, M.F., et al.: Kriging method application and traffic behavior profiles from local radar network database: a proposal to support traffic solutions and air pollution control strategies. Sustain. Cities Soc. **56**(1), 102 (2020)
9. Chandan, J.S., Thomas, T., Gokhale, K.M., et al.: The burden of mental ill health associated with childhood maltreatment in the UK, using The Health Improvement Network database: a population-based retrospective cohort study. Lancet Psychiatry **6**(11), 145 (2019)
10. Oliveri, M.E., Mcculla, L.: Using the occupational network database to assess and improve English language communication for the workplace. ETS Res. Rep. Ser. **2019**(1), 15–17 (2019)
11. Erdong, C., Martha, S., Hongyan, L., Ashley, B., Curtis, T., Elisabeth, W., et al.: Demographic and clinical correlates of seizure frequency: findings from the managing epilepsy well network database. J. Clin. Neurol. **14**(2), 206 (2018)
12. Komiyama, A.K., Nakamura, F.M., Tanabe, B.K., et al.: In-hospital mortality analysis of Japanese patients with acute coronary syndrome using the Tokyo CCU network database: applicability of the GRACE risk score. J. Cardiol. **71**(3), 251–258 (2018)

Application of Intelligent Control Technology in Tennis Special Training

Yongchao Chu, Guoquan Wang, Dongzhe Shi, Jianmin Ding, and Chang Chen[✉]

Department of Sports Science and Physical Education, Guangzhou XinHua University, Guangzhou, Guangdong, China

Abstract. With the deepening of the nationwide fitness strategy and the further development of professional competitions, school sports have developed rapidly. As an indispensable part of school sports, college sports have played a role in promoting the development of school sports. As a new sport, tennis is popular among students and has a strong momentum of development. Tennis special training, as a commonly used training method in physical education and training, has played an active role in sports teaching and training. The use of intelligent control technology in tennis special training can not only enrich the training methods and methods of college tennis teams, and can arouse the enthusiasm of students and improve the efficiency of student training.

Keywords: Intelligent control · Tennis training · Special training · Applied research

1 Introduction

Since tennis was listed as an official competition event in my country, more and more provinces, municipalities, and autonomous regions in China have gradually established a systematic training system, which has enabled tennis players to continuously improve their technical level and accelerate their professionalization process [1, 2]. Tennis is a sport dominated by foot movement. Without a solid and stable footwork movement and multi-directional movement ability, this sport is equivalent to losing its foundation [3, 4].

There are many methods of special tennis training. In China, there are two main types of tennis training in my country: one is the training of sports schools at all levels for the purpose of achieving competition results, and the tennis coaches go to the subordinate regional schools to select talents [5, 6]. The second is that clubs and training institutions focus on cultivating interest, and recruit students through advertising and other methods [7, 8]. As time goes by, a large number of high-level provincial, municipal, and national team athletes are directly engaged in tennis training after retirement. Most of these athletes were selected by coaches from a very young age, and gradually changed from half training (half of the time for cultural lessons) to full training (tennis training all day) [9, 10].

This article uses intelligent control technology to study the automatic ball machine, and then tests and compares it, confirming the machine's merits.

© The Author(s), under exclusive license to Springer Nature Switzerland AG 2022
J. Macintyre et al. (Eds.): SPIoT 2021, LNDECT 98, pp. 876–882, 2022.
https://doi.org/10.1007/978-3-030-89511-2_120

2   Application of Intelligent Control Technology in Tennis Training

2.1   Development Principle of the Intelligently Controlled Tennis Ball Machine

The assumptions are as follows:

(1)   During the entire operation of tennis, the nutation corner is assumed.
(2)   The tennis ball is an axisymmetric body.
        According to these two hypotheses, tennis can be regarded as a movement in which the entire mass of tennis is concentrated in the center of mass and one mass point.
(3)   The ground surface is flat;
(4)   The magnitude of the gravitational acceleration remains unchanged and the direction is always straight downward.
(5)   Coriolis acceleration is zero. That is, the influence on the rotation of the earth only considers the part of inertial centrifugal force contained in gravity.
(6)   The weather conditions are standard and there is no wind or rain.

2.2   Describe the Calculation Formula for the Motion of the Center of Mass of Tennis

Establish a tennis ball center of mass motion equation system with time as the independent variable. The equation of motion of the center of mass of tennis ball in the natural coordinate system is:

$$*\frac{dv}{dt} = -c\pi(y)F(v_\tau) - g\sin\theta \tag{1}$$

$$*\frac{d\theta}{dt} = -\frac{g\cos\theta}{v} \tag{2}$$

$$*\frac{dv}{dt} = v\sin\theta \tag{3}$$

$$\frac{dx}{dt} = v\cos\theta \tag{4}$$

$$v_\tau = v\sqrt{\tau_{on/\tau}} \tag{5}$$

2.3   Tennis Technical Training

Tennis skills include forehand and backhand, serve, receive, volley before the net, high-pressure ball, chipping and volley. Since most students have not been exposed to the sport of tennis before entering colleges and universities, tennis skills are relatively poor and lack of systematic training. Diverse skills and lack of venues restrict the further development of student athletes, making some athletes lack targeted guidance in terms of technical structure and special qualities. The video feedback training method can make each athlete's technique into a separate video, and then make a specific analysis based on the problems in the video. Forehand and backhand racquet problems, service continuity problems, and service connection problems can all be passed Video feedback is effectively resolved. The application of video feedback training method not only achieves the training effect, but also enriches the training methods of coaches, thereby mobilizing the enthusiasm of athletes in training and improving the training efficiency of athletes [11, 12].

2.4 Tennis Tactical Training

Tennis tactics play an important role in the game. Singles' bottom-line stalemate, defensive counterattack, and serve-to-net tactics are the most common tactics in the game, and they are also widely used. The key to singles tactics lies in the athlete's ability to read the game on the court. How to grasp the opponent's weakness in confrontation and give the opponent a fatal blow is an urgent problem for athletes to solve. The doubles event puts forward higher requirements on tactics. Scientific and reasonable tactics are the key factors for winning the game; serving tactics, receiving tactics, high-ball tactics, double bottom line tactics, double net tactics and the choice of standing position all require athletes to choose thousands of exercises can be used flexibly. The video feedback training method effectively solves the tactical problems of training. The movement and adjustment of singles tactics, the use of doubles tactics and the choice of position can all be clearly fed back in the video, so that different tactics can be selected for targeted and improved training. Therefore, purposeful training of tennis singles and doubles tactics can effectively improve athletes' athletic skills and competition results.

2.5 Tennis Line Training

In tennis training, line training is an important basis for technical and tactical launching. In the usual training, it is necessary to strengthen the practice of the line to cultivate the athletes' clear line awareness and clear technical and tactical intentions. Due to the different characteristics of singles and doubles, the training of lines is also different; singles focuses on changes in lines to mobilize opponents, and doubles focuses on avoiding players in front of the net and the ability to respond to changes in lines. The video feedback training method can train different lines according to the needs of technical and tactical training, and can timely feedback the problems in the training to the athletes, and the athletes can adjust the training plan and training content according to the feedback information. The video feedback training method clarified the content of the line training, formed a mature tactical thinking, and put forward higher requirements for the athletes' ball control skills. Due to the diversity of line changes in the game, athletes will try their best to return the ball to the target area, which has important practical significance for the improvement of students' technical and tactical level.

2.6 Tennis Mobile Training

Movement is the basic factor for tennis players to participate in the game. The characteristics of tennis movement are the variability of the direction of movement and the flexibility of turning back movement, which is obviously different from pure speed movement. Athletes tend to ignore the importance of movement in competition or training, which is also the main problem faced by most amateur athletes. In fact, a series of problems will often occur when the movement is not in place during training. The imbalance of the movement structure, the deviation of the hitting point, and the decline in the ability to control the ball will directly affect the performance of the mobilization technique and tactics. Therefore, athletes can make adjustments with the help of video feedback training method. The front and back turning, left and right movement, diagonal acceleration and four-corner transformation all require the players to move in place to achieve the training effect. The video feedback training method plays an important role in the mobile training of tennis, laying a solid foundation for athletes to play their skills and tactics and improve their competitive level.

3   Intelligent Control Tennis Ball Machine System Test

3.1   Machine Assembly

According to the detailed design of the mechanical system and electrical system of the ball machine, the engineering drawings of the mechanical system components are drawn and finally processed, the selected components required for the electrical system arc purchased, and then the machine is assembled, and finally the whole machine Wire and test.

3.2   Machine Debugging

After the ball machine is assembled, the debugging of the machine is carried out in two parts. First carry out hardware debugging, mainly including the correctness debugging of the forward and reverse wiring of each motor, the debugging of the detection distance of the photoelectric switch, and the consistency debugging of the direction of the speed read by the encoder and the steering of the brushless motor. Wiring; secondly, software debugging, including coding to realize the control strategy of the transmitter and its incremental PID control algorithm, to find out the appropriate PID parameters, and then coding to realize the automatic control of multi-sensor control and continuous serving process, and finally, the coding realizes the program flow control of the whole machine and online debugging to ensure that the ball machine can serve according to the set function.

4   Comprehensive Testing and Analysis

4.1   Accuracy of the Ball Machine's Drop Point

First of all, test the accuracy of the drop point of the ball machine, and compare it with the existing ball machines on the market. The test results of the ball speed at 60 km/h and 100 km/h, and the tennis drop point at an elevation angle of 30° are compared. The result is shown in Fig. 1.

It can be seen from Fig. 1 that the focus of the intelligent control server's drop point is better, and the ball drop point is within a circular area with a radius of 350 mm (ignoring the situation where the ball collides and bounces far); while the existing ball machine predicts that the drop point will be in a circular area with a radius of 550 mm based on the court line. In contrast, the ball machine in this article has been greatly improved to achieve the function of focusing the ball drop point.

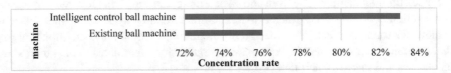

Fig. 1. Drop point accuracy

4.2 Basic Functions on the Upper Computer APP are Repeatedly Tested

The results of the twenty repeated tests are shown in Table 1:

Table 1. Basic function test results

Test name	Number of testing	Number of successes	Number of failures
Bluetooth connection	20	20	0
Bluetooth disconnected	20	20	0
Bluetooth disconnection alarm	20	20	0
Boot up	20	17	3
Shut down	20	18	2

It can be seen from Fig. 2 that the Bluetooth connection and disconnection and disconnection alarm functions are all successfully tested, and the switch machine mainly tests the success rate of the communication response between the upper computer and the lower computer. This is a multiple debugging, and an appropriate response delay is added. The result with a higher success rate can be seen for multiple response failures, and the reason cannot be found temporarily, and further research is needed later. For normal use, the command can be sent repeatedly until it succeeds. After the transmitter warms up successfully, the serving mode can be controlled without response, so it does not affect the normal use of the serving machine.

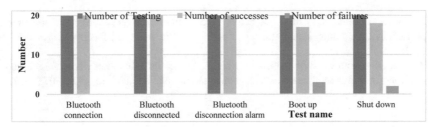

Fig. 2. Basic function test results

5   Conclusions

Special training is a convenient and efficient training method that has played an active role in tennis training. The coach's use of the steering training method not only improves the efficiency of training and mobilization, but also enriches the training mode and saves training costs and training time. In view of the easy operation, good effect and high efficiency of special training based on intelligent control technology, coaches should actively promote it in teaching and training to further promote the sustainable development of tennis.

References

1. Balkanli, D.T., Ahan, A., Erman, K.A.: Examine the effect of resistance band training applied concurrent with tennis training on strength, speed, agility and targeting performances. Turkiye Klinikleri J. Sports Sci. **12**(3), 313–321 (2020)
2. Peng, Y., Zhang, T., Fu, Q.: Research on real-time evaluation algorithm of human movement in tennis training robot. Microprocess. Microsyst. **81**(1), 103683 (2021)
3. Kilit, B., Arslan, E.: Effects of high-intensity interval training vs. on-court tennis training in young tennis players. J. Strength Cond. Res. **33**(1), 188–196 (2019)
4. Cho, E.-S., Ahn, et al.: Effects of table tennis training on salivary cortisol concentration and physical fitness in male adolescents. Korean J. Sports Sci. **26**(3), 1099–1107 (2017)
5. Tolgahan, F.T.: Effect of 12 weeks basic tennis training on tennis skill, strength and simple reaction time in children. Int. J. Soc. Humanit. Sci. Res. (JSHSR) **6**(47), 4319–4326 (2019)
6. Vélez, A.J.G., López-Miarro, P.N.: Acute effect of tennis training on sagittal spinal curvatures in standing. Cultura Ciencia y Deporte **13**(39), 243–252 (2018)
7. Peng, D., Li, R., Yan, X.: Implementation of tennis training management system based on data mining sharing platform. IPPTA Q. J. Indian Pulp Pap. Tech. Assoc. **30**(7), 595–601 (2018)
8. Chen, Q.: Color gamut method for computer-based tennis training. Agro Food Ind. Hi Tech **28**(1), 1935–1939 (2017)
9. Moisescu, P.C., Buctaru, R.M.: Physical training of beginner tennis players at the age of 8–10. Ser. IX Sci. Hum. Kinet. **13**(62)(1), 67–72 (2020)
10. Colibanu, F.C., Branite, G.: Effectiveness of the application of tactical training models of tennis players in the trading shots phase and completion of the point. Ser. IX Sci. Hum. Kinet. **13**(62)(2), 35–42 (2020)

11. Fortes, L., Almeida, S.S., Junior, J., et al.: Effect of motor imagery training on tennis service performance in young tennis athletes. Revista de Psicologia del Deporte **28**(1), 157–168 (2019)
12. Zhu, L.: Research on skill movement training model for table tennis beginners based on induction training method. Revista de la Facultad de Ingenieria **32**(15), 803–807 (2017)

Data Preprocessing Technology of Internet Opinion in Big Data Environment Based on MapReduce

Zhe Li[1], Hongwei Zhao[2], Jun Li[3(✉)], and Yongchang Ren[4]

[1] College of Literature, Bohai University, Jinzhou, Liaoning, China
[2] Beijing Academy of Science and Technology, Beijing, China
[3] School of Physical Education, Bohai University, Jinzhou, Liaoning, China
[4] College of Information Science and Technology, Bohai University, Jinzhou, Liaoning, China

Abstract. Data pre-processing is to process the collected original data and prepare for data analysis. In this paper, based on MapReduce technology, MapReduce calculation is effectively decomposed into Map and Reduce calculation process to maximize the degree of parallelism. The framework and workflow of MapReduce are analysed. The core work consists of three parts. One is text pre-processing, which divides the text data into several data blocks with appropriate size, and then processes them by Map function and Reduce function respectively. Second, feature selection, using Bayesian classifier, is divided into preparation stage, training stage and application stage. The third is text vectorization, which uses TF-IDF algorithm to represent text as a series of vectors that can express text semantics.

Keywords: MapReduce · Big data environment · Internet opinion · Data preprocessing

1 Introduction

Under the big data environment, the internet opinion is generated rapidly. The data volume is large, and it is extremely complex. The internet opinion is facing unprecedented challenges, and it also brings new opportunities for the guidance of internet opinion. Therefore, we should actively establish the concept of big data and deeply explore and reasonably use the value of big data in the guidance of internet opinion. At the same time, we also need to innovate the guidance thinking of internet opinion, grasp the essential characteristics of internet opinion, and explore the internal law of the evolution of internet opinion in order to establish the guidance mechanism of internet opinion to adapt to the big data environment [1]. Big data analysis provides a new means for the guidance of internet opinion. It is also a value-added process of internet opinion information to excavate the hidden valuable information, discover the potential hidden dangers of events as soon as possible, which can strangle the crisis in the cradle or make decision preparation for coping with the crisis [2]. Most of the data collected in the real world are not detailed, inconsistent or dirty data, which cannot be analyzed or explored.

J. Macintyre et al. (Eds.): SPIoT 2021, LNDECT 98, pp. 883–888, 2022.
https://doi.org/10.1007/978-3-030-89511-2_121

It needs to go through the data processing process, such as cleaning, supplementing, smoothing, merging and standardization of the initial data collected so as to meet the needs of big data analysis.

2 Frame Structure and Working Process of MapReduce

MapReduce is a distributed parallel computing model proposed by Google, which is mainly used to process and analyze huge data sets. MapReduce was initially designed to give priority to system availability and scalability, aiming to achieve parallel processing of big data through large, inexpensive server clusters. MapReduce is easy to program, suitable for offline processing of massive data above Pb level, sequential processing of data, avoiding random access, hiding system layer details for application developers, and dedicated to the algorithm design of application computing problems. MapReduce is easy to program, which is suitable for offline processing of massive data above Pb level, sequential processing of data, and avoiding random access. It can hide system layer details for application developers, and dedicate to the algorithm design of application computing problems. The framework structure of MapReduce is shown in Fig. 1.

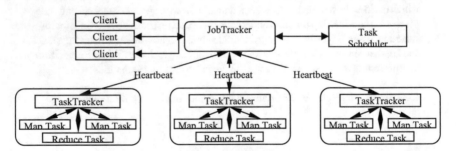

Fig. 1. MapReduce frame structure

MapReduce consists of five parts [3], namely Client, Job Tracker, Task scheduler, Task tracker and Task. The system automatically divides a job into multiple data blocks, and each computing task processes one data block. When a task is finished and resources are idle, Task scheduler will select the appropriate task to execute and coordinate the execution of the whole job. Task tracker starts an independent Java virtual machine for each task. At the same time, it uses the operating system to achieve resource isolation. In the process of running, it sends the running state of each task to Job tracker through heartbeat mechanism at regular intervals. MapReduce workflow is shown in Fig. 2 [4].

The workflow is divided into four stages. First, Input. It refers to input data and preprocess, verify the input format, and then divide the input file into several logical Input split instances. Second, Map Task. The number of maps is determined by the number of partitions. Different map tasks do not communicate with each other. Key value pairs are processed according to requirements and mapped into new key value pairs. Third, Reduce Task. There is no information exchange between different reduce tasks. Each processing in a reduce task is to reduce the data with the same key. Fourth,

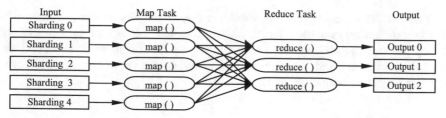

Fig. 2. MapReduce working process

Output. It checks whether the output result type is correct and verifies whether the output directory exists. If both conditions are met, it will output the reduce result to the file system.

3 Text Preprocessing

Text preprocessing is a very tedious work, involving many details, which has a great impact on the follow-up work. Removing stop words refers to remove some meaningless words to save storage space and improve search efficiency. Word frequency statistics is used to evaluate the repetition of a word for a file or a domain file set in a corpus. Generate inverted index is used to find records according to the value of the attribute in practical application. Using MapReduce programming model to construct appropriate Map function and Reduce function realizes distributed parallel processing of text mining preprocessing process, which can improve the efficiency of text preprocessing. The specific implementation process is shown in Fig. 3.

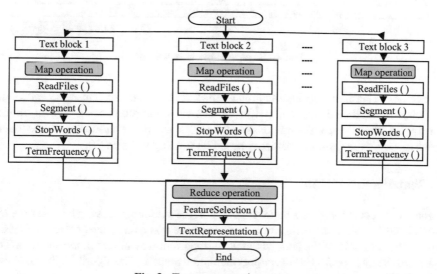

Fig. 3. Text preprocessing process

The text preprocessing process shown in Fig. 3 consists of three processes [5]. One is text data segmentation, which divides text data into several data blocks with appropriate size. Second, the Map function outputs the list of words generated after word segmentation. Third, the reduce function outputs the vector representation of the processed document name and text.

4   Feature Selection

Feature selection is an important data preprocessing process. In the filtering method, feature selection is performed on the data first, and then the classifier is trained. The wrapping method takes the performance of the final learner as the criterion of subset evaluation. In the embedded method, the learner selects features automatically in the training process. Feature selection plays an important role in classification. On the one hand, in the case of limited samples, it is not appropriate to train the classifier with a large number of features. This topic usesBayesian classifier. The flow is shown in Fig. 4 [6].

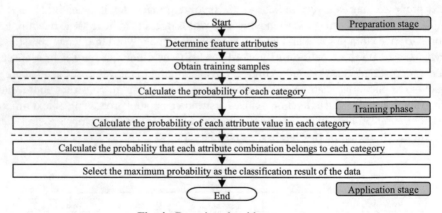

Fig. 4. Bayesian algorithm process

Bayesian classifier is a classification method of statistics, which uses the knowledge of probability and statistics to classify, and can solve the problem of "known training instance set, classifying unknown class instances". Naive Bayesian is a generative model, which models every possible category based on training samples [7].

5   Text Vectorization

A good text vector can give a better mapping in the vector space, which makes the text computable [8]. Text vectorization is to express the text as a series of vectors that can express the semantic of the text, which is an important way of text representation. This topic uses the improved TF-IDF weight calculation method. The algorithm flow is shown in Fig. 5 [9].

TF-IDF (Term Frequency - Inverse Document Frequency) is a common weighting technique for information retrieval and text mining. TF is the frequency of words in

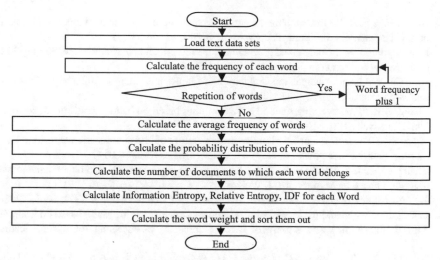

Fig. 5. TF-IDF algorithm process

a document, and IDF is the importance of a given word. A feature item may appear more frequently in a certain type of documents, but it can be ignored in other types of documents, which indicates that this feature item has good ability to distinguish categories and should be given higher weight. TF-IDF combined the word frequency with the reverse text frequency. The importance of a word increased in direct proportion with the frequency of it appearing in the file, and also decreased in reverse with the frequency of it appearing in the file. It met the requirement of showing the weight of a feature point [10].

6 Conclusion

The data preprocessing is to deal with the collected original data and prepare for data analysis. There are many data processing techniques in the big data environment. Based on MapReduce technology, this paper reasonably decomposes the whole job into multiple MapReduce tasks. It also effectively decomposes MapReduce calculation into two function calculation processes of Map and Reduce so as to maximize the degree of parallelism and meet the needs of internet opinion analysis in big data environment.

Acknowledgments. This work is supported by 2018 annual natural science foundation in Liaoning province (20180550541): Research on Analysis Technology and System for Internet Public Opinion of Emergency in Big Data Environment.

References

1. Li, H.X.: Innovate online public opinion guidance in the era of big data. Guangming Dly. (11) (2017)

2. Chu, J.W., Zhu, L.L.: Research on the network public opinion warning of emergencies based on big data analysis. Inf. Stud. Theory Appl. **40**(86), 61–66 (2017)
3. Blog Garden: Deeply understand the architecture and principle of MapReduce, 30 May 2021. https://www.cnblogs.com/claireyuancy/p/7281806.html
4. Blog Garden: Basic principle and application of MapReduce, 30 May 2021. https://www.cnblogs.com/lixiansheng/p/8942370.html
5. Zhang, A.K.: Preprocessing method of text mining based on hadoop platform. J. Shanghai Univ. Eng. Sci. **31**(2), 115–119 (2017)
6. Xing, Y.Y.: The research of MapReduce implementing of text classification algorithm based on mass data, Master's thesis of Harbin Engineering University (2014)
7. Zhang, C.Y., Liu, L.Z., Deng, K.W., Liu, J.: Text classification method of naive Bayes algorithm based on MapReduce. J. Wuhan Inst. Technol. **43**(1), 102–105 (2021)
8. Ji, Y.X.: Summary and analysis of text vectorization. Electron. World **40**(22), 10–12 (2018)
9. Cheng, L.: Design and implementation of information extraction system based on improved TF-IDF algorithm, Master's thesis of Beijing University of Posts and Telecommunications (2019)
10. Xu, D.D., Xie, T.Y., Wan, Z.H., Huang, B.H.: Detection of SQL injection based on TF-IDF text vectorization. J. Guangxi Univ. (Nat. Sci. Edit.) **43**(5), 1818–1826 (2018)

The Application of Smart Phone APP Mobile Platform in College English Online and Offline Learning

Lu Dai[(✉)]

Liaoning Institute of Science and Engineering, Jinzhou, Liaoning, China

Abstract. With the development of information technology, mobile terminals such as smart phones and tablet computers have been integrated into people's lives and studies, bringing students the good visual and auditory experience, and becoming the important path to enhance students' interest. In order to meet the needs of online and offline mixed teaching of college English, this paper is based on the mobile learning platform of the smart phone App. First, recommended to develop the architecture of the smart phone App mobile learning platform; then, design the smart phone App mobile learning platform software architecture to improve the quality and efficiency of software development; finally, purpose specific application strategies: improve the usability of smart phone App mobile learning platform, improve the interactivity of the smart phone App mobile learning platform, innovate teachers' deep participation in the new model of English teaching based on smart phone APP, and improve students' autonomous learning ability based on smart phone APP.

Keywords: Smart phone · APP · Mobile learning platform · College English · Online and offline mixed teaching

1 Introduction

In the era of mobile information, students can use mobile phones and tablet computers to obtain massive learning resources anytime and anywhere, implement autonomous learning, and make up for the lack of classroom teaching [1]. App (Application) usually refers to application software installed on a smart phone or tablet computer. Mobile APP refers to an application software installed and running on a smart phone. The mobile APP in this paper mainly refers to education APP. At present, there are mainly three kinds of mobile apps, which are Andriod system, IOS system and Windows phone system. Educator Herbart believes that interest is the direct purpose of teaching activities, and Dewey regards interest as the focus of teaching activities, which plays a key role in improving student performance. The smart phone App is rich in content, integrating pictures, sounds, and animations. It can bring students good visual and auditory experiences. It highlights the characteristics of personalized settings, convenient operation, rich resources, and anytime, anywhere to become the source of interest. Smart phone

J. Macintyre et al. (Eds.): SPIoT 2021, LNDECT 98, pp. 889–894, 2022.
https://doi.org/10.1007/978-3-030-89511-2_122

App mobile learning is the product of the development of multimedia technology and wireless network technology. Smart phone apps will complement and develop together with traditional classroom teaching.

Mixed teaching is changing from the traditional teacher-centered classroom face-to-face teaching model to student-centered online and offline mixed model, combining the advantages of traditional classroom teaching with the advantages of online learning, and using the advantages of information technology to make up for the shortcomings of traditional classroom teaching realize the expansion of space and the reconstruction of teaching, so as to obtain the best teaching process and the best learning effect [2]. The online and offline mixed teaching of college English supported by the smart phone App mobile learning platform has become the future development direction. At present, there are still many problems in the online and offline mixed teaching of college English based on the smart phone App [3], and there is a large gap between the quality of the smart phone App and the needs of students, which cannot fully meet the requirements of students' learning and teachers' teaching; the intelligentization of smart phone apps reduces students' interest in English classrooms, making students doubt about the significance of classroom education activities; lack of use of smart phone apps, scientific guidance of English teaching has hindered the functioning of smart phone apps. Through the research of this project, these problems can be effectively solved and the quality of English course teaching can be improved [4].

2 PhoneGap Architecture

It is recommended to use PhoneGap for the development of smart phone App mobile learning platform. The architecture is shown in Fig. 1.

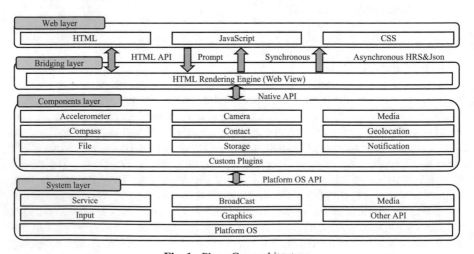

Fig. 1. PhoneGap architecture

PhoneGap is the open source development framework, developed and maintained by Adobe in the United States. It aims to allow developers to use Web APIs such as

HTML, Javascript and CSS to develop cross-platform mobile applications [5]. Phone-Gap supports multiple mobile platforms, including mainstream iOS, Android, Windows Phone and Symbian, etc., next to mobile web applications. PhoneGap can easily use the hardware functions of the mobile terminal. In appearance, PhoneGap does not use native UI elements, but uses CSS to imitate the native interface, which looks more similar to a mobile web application. In terms of processing speed, applications developed by Phone-Gap are fast and user interaction is also very smooth [6]. Compared with traditional mobile Web programs, PhoneGap can directly access the hardware interfaces of mobile devices, including accelerators, cameras, compasses, GPS, and file access. It can easily call system hardware resources and make up for the shortcomings of traditional mobile Web. The PhoneGap framework provides developers with a series of APIs, which can be called in JavaScript code; in the actual development process, through these API calls, the PhoneGap program can interact with the mobile platform SDK API used by the developer. After the development is completed, html, css and JavaScript files written using Web technology can be packaged in the form of mirror files and published to the mobile platform where the user is located.

3 Software Architecture of Smart Phone APP Mobile Learning Platform

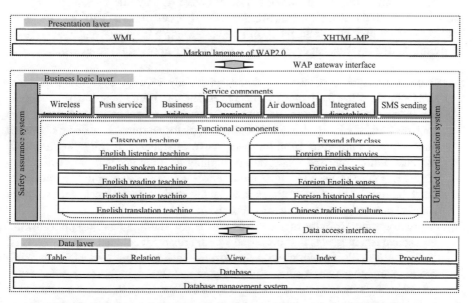

Fig. 2. Software architecture of smart phone APP mobile learning platform

The software architecture is the collection of structural elements with a certain form. The three-tier structure is a more flexible architecture that separates the display logic from the business logic and improves the quality and efficiency of software development.

The software architecture of the smart phone App mobile learning platform studied in this paper is shown in Fig. 2.

The software architecture of the smart phone App mobile learning platform shown in Fig. 2 is composed of the presentation layer, the business logic layer, and the data layer. Among them, the presentation layer is the data display and operation interactive interface provided for users. It is implemented based on the WAP2.0 markup language. The same technology can be used to develop Web and WAP sites, and any Web browser can also be used to access WAP2.0 applications with backwards compatible. The business logic layer is supported by the security assurance system and the unified certification system, and is composed of service components and functional components. The service components are used to complete the related functions of mobile services, the functional components are used to implement the related functions of the system, and the business logic layer is in the middle layer. Maintaining relative independence from the back-end system is conducive to system expansion. The data layer is used to store data and interact with the business logic layer through the data access interface. This platform uses the relational database system to store data and provide data access services.

4 Application of Smart Phone APP Mobile Learning Platform in Online and Offline Mixed Teaching of College English

In order to promote the application of the smart phone App mobile learning platform in the online and offline mixed teaching of college English, cultivate students' good learning quality, improve their comprehensive ability to use English language, and enhance their core English literacy, this paper proposes the following application strategies:

(1) Improve the usability of the smart phone App mobile learning platform [7]. Information technology has facilitated people's lives, and new things appear to make people enjoy "laziness". If smart phone apps can show ease of use and convenience, then students will be willing to use smart phone apps for learning. Developers of smart phone apps should think from the perspective of students, highlighting the basis of transcending time and space, and combining learning needs to improve convenience and ease of use. The operation interface should be as simple as possible, make full use of big data methods, understand student preferences, and carry out targeted promotion and information provision.

(2) Improve the interactivity of the smart phone App mobile learning platform [8]. Interactive teaching was first proposed by American educational psychologists A. L. Brown & A. S. Palincsar, with the purpose of improving students' communication, comprehension, reading and autonomous learning abilities. Interactivity is conducive to achieving the teaching goals of college English, and is conducive to making up for the shortcomings of "assisted classroom" and "autonomous learning". The interactive methods of the smart phone App mobile learning platform include "the interaction between the learner and the smart phone APP platform, the interaction between the learner and the teaching resources, the interaction between the learners, the interaction between the teacher and the learner", etc.

(3) Innovative teachers deeply participate in the new model of English teaching based on smart phone APP. English teachers must have the ability to teach informatization, have the full understanding of the content and format of popular English education apps, and be able to combine English teaching experience to provide guidance for students to choose apps scientifically. English teachers are good at using English education APP to assist classroom teaching, use U campus to publish learning tasks, students use APP to complete the mastery of relevant knowledge before class, and teachers use APP to sign in or test during class to enhance the interest of the classroom [9]. Use the interactive function of APP to provide personalized guidance to students.

(4) Improve students' autonomous learning ability based on smart phone APP. With the advent of the Internet age, smart phone apps for English learning have exploded, and it has become a common way to use smart phone apps to assist in English learning. Students make learning plans, find learning resources, complete knowledge learning, publish learning results, and carry out learning evaluations through the smart phone APP [10]. Develop the habit of using smart phone APP to learn English step by step, make full use of fragmented time, improve self-discipline, optimize the means and content of autonomous English learning, and meet the needs of personal growth and career development.

5 Conclusion

Mobile learning is a new teaching model and a new focus of teaching reform in colleges and universities. The online and offline mixed teaching model of college English based on smart phone APP realizes the effective integration of online and offline resources, breaks through the limitations of traditional space and time, implements modern student-centered teaching theories, and makes full use of fragment time to learn. Students can choose the resources they like on their own, and have the characteristics of personalized learning; related subjects can communicate better, and they also have the characteristics of interactivity. Teachers should start from the three aspects of pre-class preparation, classroom teaching and after-class feedback, insist on classroom teaching, use smart phone APP as an auxiliary tool to ensure the main status of classroom teaching, promote the transformation and upgrading of English teaching model, and improve students' English core literacy and cross-cultural communication skills.

Acknowledgments. This work was financially supported by Teaching reform project in Liaoning Institute of Science and Engineering in 2021 (LGB2021005): Research on the teaching reform of English major courses: Take "comprehensive English" as an example.

References

1. Miao, X., Zhang, X.N.: Research on Japanese teaching model in vocational colleges based on mobile APP learning platform. Educ. Teach. Forum **12**(40), 377–378 (2020)
2. Xu, Y.Y., Liu, C.Q.: Discussion on the problems of online and offline blended teaching mode. Sci. Educ. Artic. Collects **15**(5), 48–49 (2021)

3. Li, X.Q.: Research on the construction of a new model of college English teaching by mobile APP. J. Heilongjiang Inst. Teach. Dev. **39**(8), 142–144 (2020)
4. Lian, R.: Practical research on improving english autonomous learning ability of higher vocational college students by online and offline mixed teaching. Overseas Engl. **22**(3), 102–103 (2021)
5. Cui, L.M., Xue, F.F.: The discussion of cross-platform mobile application development base on PhoneGap method. Comput. Knowl. Technol. **12**(15), 30–33 (2016)
6. Wang, J., Yang, L.J.: Cross-platform mobile application development based on PhoneGap. J. N. China Inst. Aerosp. Eng. **27**(2), 19–21 (2017)
7. Hu, J.: A case on the current situation of college students' learning based on mobile APP: a case study of a college on undergraduate students in H Province, Master's thesis of Hunan Normal University (2019)
8. Dou, S.H.: The application study of smart phone APP on the mode network interactive teaching Chinese as a second language, Master's thesis of Inner Mongolia Normal University (2016)
9. Wu, M., Wen, J., Cheng, Y.: Research on the application of english education APPs in college students' language learning. Sci. Technol. Vis. **10**(20), 14–15 (2020)
10. Zhang, Y., Liu, J.J., Hu, Y.Z.: Study on the influence of mobile APP on college students' independent English learning in the new era. Overseas Engl. **21**(16), 191–192 (2020)

The Design of the Mental Health Test System for College Students Based on Android Platform Technology

Di Li(✉)

College of Physical Education, Bohai University, Jinzhou, Liaoning, China

Abstract. Using the mental health test system makes it easier for college students to accept the feeling of communicating with the computer, which is more conducive to protecting privacy and more truly reflects their psychological state. In order to meet the needs of college students' mental health testing in the mobile information age, this paper is based on Android research and designed the layered architecture consisting of "user use layer, business logic layer, data access layer, data storage layer", which is in line with the basic ideas of software engineering, to improve the efficiency of system development and maintainability. In response to the mental health of college students, interventions are proposed: reduce the impact of online games on college students' mental health, carry out extensive mental health consultation work for college students, construct the rapid response mechanism for mental health, and guide college students to actively participate in physical exercises.

Keywords: Android platform · Mental health testing system · Hierarchical architecture · Interventions

1 Introduction

Mental health refers to the ability of individuals to adapt to changes in the environment, have the characteristics of self-improvement, and have the ability to actively regulate [1]. Modern society is the era of diverse values, ideological and cultural turmoil, and rapid social changes, and it is prone to psychological crises and even a series of mental health problems [2]. Mental health and physical health are inseparable, and mental health is conducive to physical health. Psychological emotion is also a kind of mental health. If the level of mental health is not good, it will have an impact on physical health. The pace of modern society is getting faster and faster, receiving more and more information, and competition is becoming more and more fierce. College students are facing the dual pressure of entering the higher education and employment, and mental health is very important. Mental health is a necessary condition and foundation for a person's all-round development. Excellent psychological quality plays an important role in improving the overall quality of college.

Mental health education, as an important way for college students to develop in an all-round way, helps to shape a healthy personality. Unbalanced mental health may lead

J. Macintyre et al. (Eds.): SPIoT 2021, LNDECT 98, pp. 895–900, 2022.
https://doi.org/10.1007/978-3-030-89511-2_123

to abnormal life and learning behaviors, and long-term unbalance will form misconceptions, which will not only harm the college students themselves, but also harm others, families and society; a balanced mental health state is conducive to the full spirit of college students, put the state into life and study, and enhance the sense of happiness [3]. In the age of network information, mental health testing has become an important means of assisting college students' mental health education. Mental health testing is based on certain psychological theories and using certain operating procedures to determine a quantitative value for people's behavior. The effective tool for understanding and identifying people's psychological characteristics. Facing the increasing psychological problems of college students and the severe reality that the quality of college students needs to be improved, in-depth exploration of the existing problems of college students' mental health education and corresponding countermeasures have important theoretical and practical significance for enhancing the effectiveness of college students' mental health education.

2   Android Platform Parsing

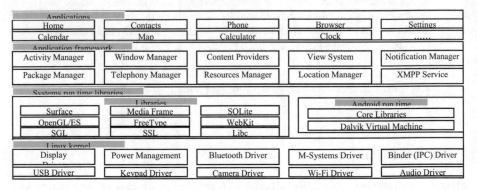

Fig. 1. Android platform architecture diagram

Mobile phones have become the most important communication tool in people's daily lives, and the popularizing rate of mobile phones among college students has reached 100%. Android is the smart phone operating system with the highest market share. The development of a mental health test system for college students based on the Android platform can be quickly and generally applied to college students to improve the effect of mental health education. The Android platform architecture is shown in Fig. 1.

3   Analysis of College Students' Mental Health Problems

According to survey data [4], 56% of college students said that their emotions are easily affected, 23% of college students' choice behaviors are easily affected, and 20% of college students are not affected by the surrounding environment. More than half of college students are emotionally vulnerable to environmental infections. Girls are significantly higher than boys in the three factors of depression, anxiety and fear; boys are higher than girls in factors such as paranoia, sleep and diet. Rural college students have higher mental health problems than urban ones, and are particularly prominent in four aspects: obsessive-compulsive disorder, interpersonal sensitivity, depression and fear. At present, the number of college students who suspend or drop out of school due to psychological problems that cannot study and live normally is increasing year by year. Based on the survey and research data, referring to relevant literature, the problems of college students' mental health are summarized into four aspects.

3.1 Online Game

Many college students get a sense of accomplishment from virtual online games in order to escape the pressure [5]. The content of "attack, fighting, violence, pornography" in the game corrodes the mental health of college students, seeking excitement and sense of gain in the virtual online game world, making it difficult for college students to define the distance between reality and virtuality, and their ability to adapt to life is getting worse. The body's immune function is reduced and various diseases are induced.

3.2 Interpersonal Communication

Many contemporary college students are the only child. They grow up in the excessive care of their families. They are accustomed to self-centeredness and cannot accommodate and understand the living habits of others. Some college students adapted quickly, and some eventually chose to forbear. There are also a few college students who choose not to accommodate and tolerate, and finally express it in an extreme way, and even cause serious consequences to the family, society and others.

3.3 Personal Reason

As college students "grow up" rapidly in physiology, the balance between physiology and psychology is greatly broken, and the balance between various psychological factors has been broken. Personal factors are one of the main reasons for mental health problems. Many college students are mentally fragile, emotionally extreme, negative emotions, jealous, self-destructive, and complaining about the public, all of which may have serious mental health problems.

3.4 Employment Pressure

The employment problem has become a problem for many people in recent years. Although the country has issued a series of active employment policies, the problem of difficult employment for college students has not fundamentally changed, and they are still facing a more severe and complex employment environment. Many college students bump into walls everywhere when looking for a job, feeling that the world is gloomy, and even suspicious of life, and many students break their contracts because of the gap between their ideals and reality.

4   Hierarchical Architecture of College Students' Mental Health Testing System

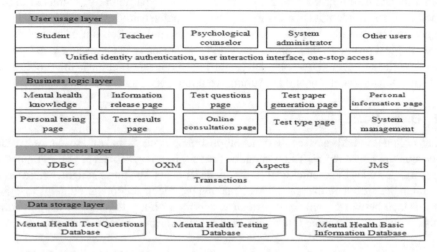

Fig. 2. Hierarchical architecture of college students' mental health testing system

The development of the college student mental health test system uses the layered architecture, as shown in Fig. 2.

5   Interventions of College Students' Mental Health

From the perspective of becoming talents in college, a strong physique is as important as a healthy mind. Sound personality and good psychological quality have an important impact on the future life and career development of college students [6]. Under the training of test-oriented education, college students have low awareness of mental health and lack the ability of self-adjustment [7]. Therefore, it is very important to strengthen the mental health education of college students, improve their self-cognition ability, and carry out effective psychological intervention through a variety of means.

5.1   Reduce the Impact of Online Games on College Students' Mental Health

While educating students to use online games for entertainment, they should be alert to the harm to their physical and mental health, and increase punishment for students who are addicted to online games [8]. Carry out a variety of campus cultural activities, cultivate college students' hobbies, and distract attention from online games. Strengthen administrative supervision, and never allow online games with negative content such as violence, pornography, and reactionary colors to be spread in the online environment. Parents should also play a supervisory role, control game time, and reduce the possibility of online game addiction.

5.2   Carry Out Extensive Mental Health Consultation Work for College Students

Colleges should set up mental health counseling centers as a base for quality education for college students, providing free counseling services for students with pressure, growth or adaptation problems, and assisting students to improve their adaptability and develop their personal potential. Through the development of psychological counseling work, create a platform for college students to vent their emotions; through the continuous guidance of counselors, lay the foundation for college students to regain their confidence; through the construction of a comprehensive professional psychological counseling and counseling mechanism, provide college students with professional psychological counseling services [9].

5.3   Construct the Rapid Response Mechanism for Mental Health

Once a college student's mental health incident occurs, the relevant departments will respond immediately and quickly implement decisive intervention measures to help students get out of their predicament to the greatest extent and prevent tragic incidents from occurring. Strengthen the coordination and unity of mental health processing related units, realize the interconnection and interaction of heterogeneous systems, and respond to mental health crises flexibly and quickly [10]. Strengthen the tracking and feedback of intervention effects, evaluate the results of interventions, find the best solutions, and adjust intervention methods at any time to minimize the impact of mental health.

5.4   Guide College Students to Actively Participate in Physical Exercises

Human body and mind are closely related. A healthy mind resides in a healthy body, and an unhealthy mind will lead to physical abnormalities. Physical exercise can not only enhance students' physical fitness, but also has a positive effect on promoting students' mental health. Infiltrate mental health knowledge in physical exercises, subtly help college students overcome physical and psychological obstacles [11], cultivate a good mental state while improving physical literacy, regulate depression, depression and tension and other emotions, and cultivate willpower to overcome difficulties and perseverance spirit.

6   Conclusion

Mental health is the foundation for college students' academic achievement, career success, and happiness in life. College students are not yet mature in their psychology, but their physiological growth rate is relatively fast. When there is an imbalance between the psychological and physical development, friction, disputes or contradictions are likely to occur. Using the mental health test system, college students are more likely to accept the feeling of communicating with the computer, and can more truly reflect their mental state. This paper studies the key technologies of the college students' mental health test system, and proposes intervention measures for college students' mental health problems, helping to cultivate more new era college students with good psychological quality, healthy social mentality, optimism, open-mindedness and full of vigor.

Acknowledgments. This work was financially supported by Humanities and social sciences research project of department of education of Liaoning province in 2020 (sz2020005): Research on the formation of humanistic quality and behavior cultivation from the perspective of multi culture.

References

1. Liu, Z.N., Cheng, G., Li, Y.F.: Study on mental health problems and coping measures of college students in the new era. Occupation **28**(3), 34–35 (2021)
2. Guan, M.Y., Zhou, B.M.: Mental health problems and countermeasures of contemporary college students. Neijiang Technol. **41**(12), 32–33 (2020)
3. Liu, M.D., Xue, Y.Q.: Problems and countermeasures of mental health education for college students in the new era. J. Liaoning Educ. Adm. Inst. **38**(1), 35–40 (2021)
4. Zhang, R.R.: Mental health problems and educational strategies of college students in the new era. China J. Multimedia Netw. Teach. **20**(1), 128–130 (2021)
5. Li, L.X.: The problems and solutions of mental health education of college students in the new era. Psychologies **16**(9), 213–214 (2021)
6. Zhang, Y.M.: Problems and countermeasures of mental health intervention for college students. PR Mag. **16**(8), 194–195 (2020)
7. Liu, L., Chen, S.G., Gui, Y.Y., et al.: Investigation and intervention study on mental health awareness of college students. Hum. Resour. Dev. **28**(20), 38–40 (2020)
8. Zou, C.L., Yu, Z.Q.: The influence of online games on college students' mental health and countermeasures. Sci. Educ. Artic. Collects **15**(5), 44–45 (2021)
9. Guo, D.Y.: College students mental health problems and crisis intervention exploration. Comp. Study Cult. Innov. **3**(9), 184–185 (2019)
10. Li, Y.: Discussion on intervention measures for mental health crisis of college students. Think Tank Era **4**(12), 102–103 (2020)
11. Niu, Y.G.: Intervention of sports activities on college students' mental health. Bull. Sport Sci. Technol. **27**(11), 168–170 (2019)

Effect of Superstar Learning Link APP on Academic Emotion and Self-regulated Learning Ability

Zhihai Hu$^{(\boxtimes)}$ and Mengmeng Wang

School of Educational Science, Huangshan University, Huangshan, Anhui, China

Abstract. Objective: To investigate the influence of SuperStar Learning Link on academic mood and self-regulated learning ability of college students. Methods: The Distance Learner Emotional Scale compiled by Li Xiaoying and the Research Scale for Student self-regulated Learning Ability compiled by Li Xikun were used. A random sampling method was used to select 205 undergraduates in the school to participate in the questionnaire survey. Independent sample t-test and analysis of variance were performed on the data. Results: The face-to-face and live-stream classes were in pleasant ($P < 0.05$), helpless/lonely ($P < 0.05$), depressed/inferiority ($P < 0.05$), and nervous/anxious ($P < 0.01$) and other significant emotional differences, especially in tension/anxiety. The students who regularly use Superstar Learning Link are significantly better than other groups of students in making autonomous learning plans ($P < 0.01$). The time each week students use Superstar Learning Link is negatively related to their feelings of helplessness/loneliness, boredom, and boredom ($P < 0.05$). Conclusion: Long-term use of Superstar Learning Link can help college students adjust their academic mood and improve their self-regulated learning ability.

Keywords: Superstar learning link · Online learning · Academic emotions · Self-regulated learning ability · College students

1 Introduction

The network teaching platform is an important foundation of education and teaching informationization, and plays a very important supporting role in education and teaching activities. Open educational resources can help students to enhance their ability in self-regulated learning [1]. Superstar Learning Link is a professional mobile learning platform developed by Superstar for mobile terminals such as smart phones and tablets. It is promoted and used by many universities, and universities are the main users of the software. According to the official data of the Superstar Group, there are currently more than 30 million registered users of the Superstar Learning Link APP. At present, the research on Superstar Learning Link focuses on the category of mixed teaching mode based on it.

People's various behaviors are deeply affected by emotions. Positive emotions can strengthen self-efficacy and accomplish academic goals [2]. Academic emotions refer

J. Macintyre et al. (Eds.): SPIoT 2021, LNDECT 98, pp. 901–906, 2022.
https://doi.org/10.1007/978-3-030-89511-2_124

to various emotional experiences related to students' academic activities during teaching or learning. From the perspective of students' growth and development, academic emotions affect their cognitive processing and academic performance, as well as their psychological and physical health. Pekrun (2002) put forward the concept of academic emotions for the first time, considering it to be directly related to academic learning, classroom teaching, and academic achievement [3]. Researchers found that students with Reading difficulties had lower hope and higher anxiety toward reading than those without RD [4]. For the study of students' academic emotions, most studies focused on primary and secondary school students [5, 6]. As for the dimensions of academic sentiment, some scholars divide it with a dimension of pleasure; other scholars divide it into positive and negative dimensions (such as the DES, PANAS scale); and some scholars divide it into positive and neutral emotion and negative emotions [7]. With the development and popularization of mobile intelligence, the learning environment and learning methods of learners have undergone great changes, which has also led to more complex and diverse sources of learners' emotions and different impacts on learners [8]. Holec (1981) first proposed the concept of self-regulated learning, that is, the ability to take charge of self-learning. Self-regulated learning is a modern learning method that takes students as the main body of learning and corresponds to traditional acceptance learning [9]. It takes students as the main body of learning, and achieves learning goals through independent analysis, exploration, practice, questioning, and creativity.

It can be seen that network teaching has a very important impact on learners. As the network becomes more and more connected with us, the study of learner autonomy based on the context of network teaching has become a focus [10]. Superstar Learning Link is one of the main channels for college students to conduct online learning. This study aims to explore the influence of SuperStar Learning Link on college students' academic mood and self-regulated learning ability.

2   Research Methods

2.1   Research Objects and Processes

The research objects are undergraduates in a certain university. In this study, a questionnaire survey method was used. 230 printed questionnaires were sent to students, and a total of 230 were collected. 25 invalid questionnaires (incomplete questionnaires and obvious regularity of answers) were eliminated, and 205 valid questionnaires were finally recovered. The recovery rate was 89%. There are 86 boys and 119 girls; 60, 74 and 71 are freshmen, sophomores and juniors respectively. In the process of testing, it is required to actively inform the subjects of the research purpose of the study, and at the same time to obtain the personal wishes of the subjects.

2.2   Research Tools

In this study, the "Self-evaluation Scale for Distance Learners' Emotion" compiled by Li Xiaoying (2011) was used. The scale consists of 15 questions, including two dimensions of positive emotion and negative emotion. Positive emotions such as happiness, excitement; negative emotions include helplessness, boredom,etc.. The Cronbach's α standard reliability coefficient of the questionnaire was 0.732, and the Cronbach's α standard reliability coefficients of the positive emotion dimension and negative emotion dimension were 0.856 and 0.911.Using the "self-regulated Learning Ability Scale" compiled by Li Xikun (2017), the scale consists of 66 questions, including four dimensions: learning motivation, planning, learning environment, and self-management. Among them, the Cronbach's α coefficient of the questionnaire is greater than 0.80, and it has high internal consistency and high validity.

According to the different courses used by Superstar Learning Links recently, the subjects were divided into face-to-face classes (face-to-face lectures with teachers in the classroom), live lessons (watch live lectures from teachers using the Internet), and combined lessons (both face-to-face lectures) Live broadcast lesson).

3   Results and Analysis

3.1   Impact on Academic Mood

According to the statistics of data description, the average value of face-to-face classes is higher than that of live courses. The independent sample T test results showed that the different curriculum methods were pleasant ($P = 0.048 < 0.05$), helpless/lonely ($P = 0.025 < 0.05$), depressed/inferiority ($P = 0.034 < 0.05$), and nervous/anxious ($P = 0.001 < 0.01$) There are significant differences in emotions, especially in tension/anxiety. The results are shown in Table 1.

Table 1.   Differences of college students' academic mood in different courses

Academic mood	Course style	N	Mean	t	p
Happy	Face-to-face	121	2.661	1.98	.048
	Live lesson	169	2.437		
Helpless/lonely	Face-to-face	121	1.933	.549	.025
	Live lesson	169	1.668		
Depression/inferiority	Face-to-face	121	1.727	.091	.034
	Live lesson	169	1.508		
Nervousness/anxiety	Face-to-face	121	1.834	.122	.001
	Live lesson	169	1.497		

3.2   Impact on Self-regulated Learning Ability

The analysis of variance on the data showed that the subjects had significant differences in the ability to make plans for different courses ($P < 0.01$). The results are shown in Table 2.

Table 2. Differences of college students' self-regulated learning ability in different courses

	Course	N	Mean	SD	F	P
Motivation	Face-to-face	36	2.733	.549	1.003	.369
	Live lesson	84	2.745	.424		
	Blended teaching	85	2.832	.443		
Making plans	Face-to-face	36	2.805	.590	9.369	.000
	Live lesson	84	2.957	.467		
	Blended teaching	85	3.178	.410		
Learning environment	Face-to-face	36	2.755	.328	.080	.923
	Live lesson	84	2.783	.350		
	Blended teaching	85	2.771	.367		
Self-management	Face-to-face	36	3.278	.531	.400	.671
	Live lesson	84	3.363	.517		
	Blended teaching	85	3.346	.418		

3.3 Impact of the Use of Superstar Learning Pass Weekly on College Students' Academic Mood

College students use the Superstar Learning Link each week to divide into high and low groups (27% of both ends of the sample). After independent sample t-test analysis, the results show that the difference between the helplessness/loneliness of college students who use Superstar Learning Pass every week and the helplessness/loneliness of college students who use Superstar Learning Pass per week Significantly at the 0.05 level. As shown in Table 3.

Table 3. A comparison of the effects of using time on academic mood

	M ± SD	t	P
Happy	2.45 ± .917	.665	.507
	2.55 ± .996		
Helpless/lonely	1.63 ± .949	2.456	.015
	2.01 ± .993		
Boring	2.00 ± 1.000	2.059	.041
	2.32 ± 1.000		
Depression/inferiority	1.51 ± .748	.632	.105
	1.73 ± .901		

4 Discussion

The differences in the curriculum methods affect the academic emotions of students to a certain extent. Online learning is that teachers and students are not face-to-face, and there is no comparison between other students. Therefore, the generation of happy emotions is not as strong as face-to-face lectures. When the learner develops "tension/anxiety" during the learning process, due to the asynchronous nature of online learning, the teacher and the student are not face to face, and the teaching activities do not need to be carried out immediately. At this time, the learner has a "gasp" time to regulate your emotions. In the face-to-face class, when the learner develops "tension/anxiety", because the teacher and the student face to face, and the teaching activities need to continue, this bad mood cannot be mediated and will continue to deepen and bring adverse effects. Teachers should actively use the online teaching platform in teaching activities, pay attention to the emotional state of the learners, and give the learners a timely response and affirmation.

College students' ability to make plans in combined classes is higher than their ability to make plans in face-to-face classes and live classes. In the combined lessons, teachers can help learners to clarify goals and formulate corresponding learning plans. At the same time, teachers can use online teaching platforms to help and supervise learners to implement learning plans, and continue to help learners develop and improve their ability to make plans. Motivation and self-management are internal factors of the learner. In this regard, teachers need to stimulate students' interest in learning, cultivate self-learning ability. Superstar Learning Link must also continuously improve learning resources, optimize operations, and continue to innovate, arousing the learning interest of learners.

The longer the college students use Superstar Learning Link each week, the lower the feelings of helplessness/loneliness, boredom/boredom. When learners do not know how to find the information they need in the webpage, they will feel "helpless", and the learning tasks they are performing will be confused and unable to start. They will become "disappointed", "helpless", "boring" "To wait for emotions and shorten the time of learning activities. When the learner continues to carry out learning activities, it indicates that the learner enjoys the learning process. The longer the use, the higher the level of investment, and the lower the "boring" and "helpless" mood. The use of Superstar Learning Link for too short a time will cause the learners to fail to find the fun of the learning content or fail to find the content they want, resulting in the learners being "boring" or "helpless". When using Superstar Learning Link for learning activities, learners must choose the appropriate learning content and enjoy learning, and cannot easily stop learning. At the same time, we should pay attention to the negative emotions of "boring" and "helplessness", and take appropriate measures in time to regulate such emotions and promote the learning activities.

Acknowledgements. This work was supported by Anhui Provincial Teaching Team "Teaching team of Applied Psychology" (2019JXTD100) & Horizontal research projects of Huangshan University.

References

1. Wong, T.-L., et al.: How to facilitate self-regulated learning? A case study on open educational resources. J. Comput. Educ. **7**(1), 51–77 (2019). https://doi.org/10.1007/s40692-019-00138-4

2. Burr, J., Beck Dallaghan, G.L.: The relationship of emotions and burnout to medical students' academic performance. Teach. Learn. Med. **31**(5), 479–486 (2019)

3. Pekrun, R., et al.: Academic emotions in students' self-regulated learning and achievement: a program of qualitative and quantitative research. Educ. Psychol. **37**(2), 91–105 (2002)

4. Sainio, P.J.: The role of learning difficulties in adolescents' academic emotions and academic achievement. J. Learn. Disabil. **52**(4), 287–298 (2019)

5. Janke, K.K., Wilby, K.J., Zavod, R.: Academic writing as a journey through "chutes and ladders": how well are you managing your emotions? Curr. Pharm. Teach. Learn. **12**(2), 103–111 (2020)

6. Riikka, H., et al.: The role of academic buoyancy and emotions in students' learning-related expectations and behaviours in primary school. Br. J. Educ. Psychol. **90**(4), 77–83 (2019)

7. Rogaten, J., Moneta, G.B.: Use of creative cognition and positive affect in studying: evidence of a reciprocal relationship. Creat. Res. J. **27**(2), 225–231 (2015)

8. Alshahrani, S., Ahmed, E., Ward, R.: The influence of online resources on student–lecturer relationship in higher education: a comparison study. J. Comput. Educ. **4**(2), 87–106 (2017). https://doi.org/10.1007/s40692-017-0083-8

9. Lee, Y., Choi, J.: A review of online course dropout research: implications for practice and future research. Educ. Tech. Res. Dev. **59**(5), 593–618 (2011)

10. Stefanos, G., Loizos, S.: The influence of instructional design and instructional material on learners' motivation and completion rates of a MOOC course. Open J. Soc. Sci. **08**(11), 190–206 (2020)

A News Application Based on the Computer Language Application Framework

Zhiheng Zhang and Qinghai Wu(✉)

Software Engineering Major, Jilin Agricultural Science and Technology University, Jilin, Jilin, China

Abstract. At present, no news app is mainly oriented to college students, which makes it difficult for college students with great demand for news to obtain the latest realtime information. The "look around" app is committed to making a model that passes the real name authentication of college students in the front end and stores it in the background database, provide accurate information for college students, and confirm the identity of the news parties and witnesses through their authentication information. Use Dubbo, layer UI and react JS to make and improve the system. This development needs to produce native app and webapp respectively, and the required languages for both are Java and HTML.

Keywords: News app · Application framework · Computer language

1 Introduction

News, is a kind of style to record society, spread information and reflect the times. The development of modern news industry is relatively stable, among which college students have a high degree of acceptance of social news and a wide range of contact, but there is no news system for college students at present, and the college students who need news most can not keep pace with the society due to the lack of industry.

As a news platform, "Look Around" is not only about the quality and quantity of the news content, but more importantly, we can lock the identity of the parties and bystanders of the news events through the college students' files provided by the school, and top the content sent by the news channel, which directly saves the tedious steps of reporter survey and questionnaire.

"We are not the initiator of making news for popularity, we are just the porter of the right news."

1.1 App Introduction

"Look Around" combines important current political news with university anecdotes.

Through the frontend college students to fill in the information of real name authentication to determine the information of the parties and witnesses, in the case of its permission, the content of its release will be pushed.

For college students of different majors, colleges and universities realize cross school platform connection and frequent academic exchanges.

© The Author(s), under exclusive license to Springer Nature Switzerland AG 2022
J. Macintyre et al. (Eds.): SPIoT 2021, LNDECT 98, pp. 907–912, 2022.
https://doi.org/10.1007/978-3-030-89511-2_125

1.2 Framework and Advantages

We use a variety of different frameworks to complete the program design.

The cross platform reuse code is used to solve the cross platform problem of Android and IOS systems.

Use Layer UI framework to design background program. Compatible with all mainstream browsers including IE6. The framework shows more robust functions with less code as much as possible, the outputs of Dict Layer are more discriminative and class specific than that of the traditional FC Layer. With less code as much as possible. The outputs of Dict Layer are more discriminative and class specific than that of the traditional FC Layer [1].

React JS framework to solve the front page and comment intelligent recognition system. Reactjs is a non MVC framework developed by Facebook. The framework allows the creation of a reusable UI component. The user interfaces of Facebook and instagram are developed with reactjs. We mainly use a standard HTML element to connect with the background. React may reduce up to 50% the effort required to build an application from scratch [2].

We implement JMeter website concurrency test. Apache JMeter is a Java based stress testing tool developed by Apache. It was originally designed for web application testing, but later extended to other testing areas. Team members test the project according to the preliminary planning. The search time of this system was shorter and the search result accuracy was higher [3].

1.3 Development Approach and Technology

We use two different development approaches and technologies. Our experimental results demonstrate that our service is effective and efficient in automatically transforming Web content into semanticretained formats suitable to be displayed on handheld devices [5].

Webapp development (Web Development).With the application of HTML development technology, although the "Look Around" app tries its best to make a news app, it still needs a download page on the web side. Html is the most commonly used language in web design. With its rich media compatibility, it has won the favor of many developers. See Table 1. With the feature of "one development, multiple compatibility", it can design and create more advanced and complex applications with the minimum development cost, and is not limited to devices. Web Quality Assurance is a very important part in Web engineering [6].

Native app development (software development).Using java language, Java language has the characteristics of objectoriented, platform independent, simple, execution interpretation, multithreading, security and so on. Compared with C language, most other languages can only compile specific CPU chips, which easily leads to errors and affects the process. Java is different in that it can be designed as a program to understand, interpret and execute, translate and execute every sentence, so as to effectively avoid the problem of lowefficiency. The Java programming language has been widely accepted as a generalpurpose language for developing portable applications, toolkits, and applets [7].

Table 1. Development approach and technology [4]

Number	Applied technology	Explain	Application purpose
1	Dubbo	A high performance and excellent service framework. Resource scheduling and management center management tools	As the middleware of service layer, it is used to solve the issue of publishing and calling
2	Layer UI	Compatible with all mainstream browsers including IE6. The framework shows more robust functions with less code as much as possible	As a background system, it mainly solves the problem of background program design
3	React JS	Reactjs is a non MVC framework developed by Facebook. The framework allows the creation of a reusable UI component. The user interfaces of Facebook and instagram are developed with reactjs. We mainly use a standard HTML element to connect with the background	As a frontend system, it mainly solves the problem of front page and comment intelligent recognition
4	Solr	Full text retrieval server	To solve the problem of search engine

2 Construction of App

2.1 Overall Structure of the Project

Information on the Web, which are conglomeration of heterogeneous data [2]. The project uses serviceoriented architecture, which can deploy, combine and use the loosely coupled coarsegrained application components through the network according to the requirements. The service layer is the foundation of SOA, which can be directly invoked by applications, so as to effectively control the human dependence of interaction with software agents in the system. SOA provides better flexibility to build applications and business processes in an agile way by combining existing applications to generate new services. See Fig. 1:

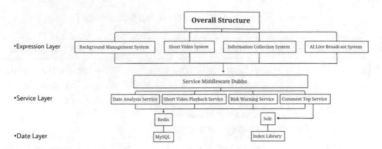

Fig. 1. Overall structure of the project

The presentation layer includes background management system, short video broadcast system, information collection system and AI live broadcast system; The service layer includes short video content service, data analysis service, danger warning service and comment top service The background system mainly uses layer UI, which is compatible with all mainstream browsers including IE6. The framework shows more robust functions with less code as much as possible.

The frontend system mainly uses react JS. Reactjs is a non MVC framework developed by Facebook. The framework allows the creation of a reusable UI component. The user interfaces of Facebook and instagram are developed with reactjs. We mainly use a standard HTML element to connect with the background.

2.2 Website Process Construction

Users can log in after real name registration to publish works and comments. If real name users are news witnesses or parties, they can apply for comment top for other users' reference. We will determine their identity through positioning and college students' files.

Other users can also apply to log in as tourists, but tourists can only watch news content, not post comments. See Fig. 2

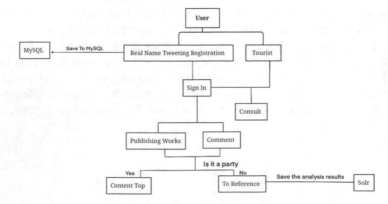

Fig. 2. Website process

3 Discussion

In order to build a complete and perfect "look around" app, the team used java and HTML to make mobile app version and web version respectively. During the construction process, we used a variety of different frameworks to give us great inspiration and efficiency. In addition to studying the construction of database and the application of frontend system, we also focused on users. Consider the user's psychological and physiological reactions when interacting with the app.

As a news app, its main operation is not only practical, but also how to fully consider the user experience in every aspect of mobile app interface design. We are the first to establish the identity of the participating members in the news to determine whether the content is top set. With the comments and blessings of witnesses and main participants, it is convenient for other users to understand the news more objectively and truly.

This requires a huge database, so we use a variety of different frameworks to improve the backend database, and judge the identity of frontend users in the news through the information they fill in. We build a management culture based on business through professional oriented editing team management. This solution involves a large number of news and a large number of users' real name registration, so it should be supported by a powerful database, which requires developers to master the development principle of relational data model and practical operation.Let programmers tap into each language's features as they develop Web and businss applications [8].

As the plethora of mobile apps continues to grow, app publishers are increasingly challenged on the appropriate strategies to adopt in order to improve app success [9].

As a news app, if its objectoriented is all mankind, then it is doomed not to be the most accurate one. On the contrary, "look around" only aims at college students. Although the user's browsing volume is slightly reduced by contrast, its pertinence and popularity have more significant social benefits. Consumers are generally watching news with the mentality of gossip. We think that it can enhance the value of news, thus forming a sharp contrast with TV news and paper news, so that users can still get the growth of news knowledge in the happy process of watching mobile phones [10].

4 Results

The background database matching with the application is created; A fully formed cross platform (Android and IOS) news app was produced; Complete the running and testing of app and software copyright.

Acknowledgements. Student innovation project of JiLin Agricultural Science and Technology University in 2021 [S20211149047].

References

1. Chen, Y., Su, J.: IEEE/CVF Conference on Computer Vision and Pattern Recognition Workshops (CVPRW) Dict Layer: A Structured Dictionary Layer: 2018, vol. 1, pp. 535–539 (2018). https://doi.org/10.1109/CVPRW.2018.00083

2. Lampropoulos, A., Ampatzoglou, A., Bibi, S., Chatzigeorgiou, A., Stamelos, I.: 11th International Conference on the Quality of Information and Communications Technology (QUATIC) REACT A Process for Improving OpenSource Software Reuse: 2018, vol. 1, pp. 251–254 (2018). https://doi.org/10.1109/QUATIC.2018.00044

3. Dong, Y., Wu, Y.: 11th International Conference on Measuring Technology and Mechatronics. Automation. Shopping Website Search System Based on Solr: 2019, vol. 1, pp. 708–711 (2019). https://doi.org/10.1109/CMTMA.2019.00162

4. ARENA: Asynchronous Reconfigurable Accelerator Ring to Enable DataCentric Parallel Computing, December 2021, vol. 32, pp. 2880–2892 (2021). https://doi.org/10.1109/TPDS.2021.308107

5. Yang, S.J.H., Zhang, J., Tsai, S.T.C.: An automatic semantic segment detection service for HTML documents. In: IEEE International Conference on Service for Computing, vol. 1, pp. 210–217. https://doi.org/10.1109/2008.155

6. Hu, R., Wang, Z., Xu, J.: IITA International Conference on Services Science, Management and Engineering (SSME), Web Quality of Agile Web Development: 2009, vol. 1, pp. 426–429 (2009). https://doi.org/10.1109/SSME.2009.112

7. Java a surge of wireless and mobile platform standardization activities led by Sun Microsystems

8. Ortega, B.: staff editorIT ProfessionalNews Briefs, March/April 2009, vol. 11, p. 69 (2009). https://doi.org/10.1109/MITP.2009.34

9. Dibia, V., Wagner, C.: 48th Hawaii International Conference on System Sciences (HICSS), Success within App Distribution Platforms: The Contribution of App Diversity and App Cohesivity: 2015, vol. 1, pp. 4304–4313 (2015). https://doi.org/10.1109/HICSS.2015.515

10. Peisachovich, E.: 9th International Conference on Information, Intelligence, Systems and Applications (IISA) SafeHome App: 2018, vol. 1, p. 13 (2018). https://doi.org/10.1109/IISA.2018.8633634

Innovation of Collaborative Education Evaluation System Based on Information Technology

Chunli Ma[✉]

Yunnan Technology and Business University, Kunming, Yunnan, China

Abstract. With the introduction of Application-oriented University, the multi-party collaborative education model advocated by the Ministry of Education has become an important direction for the reform of education management in universities across the country. However, it is only for some developed regions and some information technology majors. The other regions where the depth and breadth of the integration of industry and education are insufficient, the evaluation system is lacking, and there is no long-term mechanism. It is difficult to achieve the desired effect. The situation is even worse for Yunnan province, which is located in the southwestern border of the China, and the economy is backward. This project conducts in-depth research on the current situation and existing problems of the integration of education with industry in private colleges and universities, and the multi-party collaborative education mechanism. This paper is taking YTBU as an example, and aiming at the outdated model of multi-party collaborative education, actively explore new models of multi-party collaborative education.

Keywords: Integration of education with industry · Collaborative education · Operating mechanism

1 Introduction

The integration of education with industry and multi-party collaborative education are an important direction of the country's education reform, and the country has issued a lot of policies, for example, Decision of the State Council on Accelerating the Development of Modern Vocational Education, Modern Vocational Education System Construction Plan (2014–2020), Opinions on Comprehensively Improving the Quality of Higher Education and adopted by the Third Plenary Session of the 18th CPC Central Committee, The strategic deployment of Deepening the Comprehensive Reform in the Field of education, etc., all pointed out the important implementation path of the talent training goal of Application-oriented University - the Integration of education with industry, and the collaborative education of multiple parties [1].

2 Present Situation in Domestic and Abroad

Under the recent situation, the integration of industry and education is further deepened. Multi-party collaborative education is an effective way to promote the reform and

J. Macintyre et al. (Eds.): SPIoT 2021, LNDECT 98, pp. 913–918, 2022.
https://doi.org/10.1007/978-3-030-89511-2_126

development of higher education and enhance the ability to serve economic and social development. Many countries and local governments have actively promoted the industry through the introduction of policies, optimizing the environment, and improving systems and mechanisms.

The representative universities and regions for the development of multi-party collaborative education in the United States include Stanford University, Harvard University, Massachusetts Institute of Technology and other universities, which have made great contributions to the development of economy, technology, and society in Silicon Valley, Boston, and the entire US.

The representative case of multi-party collaborative education in the United Kingdom is the integration and development of Oxford University and Oxfordshire. At present, graduates of Oxford University found 80% of high-tech enterprises in Oxfordshire [2]. Technological innovation and the transformation of technological achievements have promoted the rapid development of Oxford.

In recent years, the multi-party collaborative education model advocated by the Ministry of Education has become an important direction for the reform of education management in colleges and universities in various regions.

A typical case of multi-party collaborative education development in Beijing is the construction and operation of Zhongguancun. At present, Zhongguancun has gathered nearly 20,000 high-tech enterprises, such as Lenovo, forming a high-tech industrial cluster represented by mobile Internet and other advantageous industrial clusters and integrated circuit and other industrial clusters, as well as high-end modern service industries. The development pattern of industrial clusters in a district has its own characteristics, and has become a high-end industrial function zone across administrative regions in the capital. The economic and social development of universities, Zhongguancun and even Beijing has achieved deep integration, symbiotic and coordinated development.

Shanghai has 67 colleges and universities, and colleges and universities provide strong talents and scientific technological support for Shanghai's economic and social development. The strength of economic and advanced higher education development policies also provide a good guarantee for the development of colleges and universities in Shanghai [3].

Shenzhen also practice the multi-party collaborative education development. Shenzhen's economic development ranks among the top in the country, but the scale of higher education, school level and its contribution to Shenzhen's economic and social development are not obvious, and the gap between the requirements of Shenzhen's social, economic, technological, and cultural development is obvious [4]. Collaborative education has become the soft power for Shenzhen to further promote the integrated development of industry and education.

3 The Existing Problems

With the introduction of Application-oriented University, it is necessary to integrate resources from all sources to cultivate talents in order to achieve the goal of training technology talents. This points out a new direction for private universities, taking the path of integration of education with industry, and multi-party collaborative education.

Domestic and foreign research on the integration of industry and education and multi-party collaborative education has last for many years, but it is only aimed at some developed regions and information technology majors. The other regions have insufficient depth and breadth of industry and education integration, and it is difficult to reach the desired effect.

Many colleges and universities have introduced related companies for the integration of education with industry and collaborative education, but many projects have the following problems:

a. Lack of evaluation system.
b. Lack of supervision and management.
c. No effective operating mechanism has been formed.
d. The multi-party collaborative education model was obsolescence.

In response to the above-mentioned problems, this paper conducts in-depth research on the practice of multi-party collaborative education based on the original research, aiming to focus on the outdated model of multi-party collaborative education in the research process within the YTBU as an example, and actively explore multi-party collaborative education. The result of the research will eventually be widely promoted to implement multi-party collaborative education. All colleges and universities across the country actively responded to deepen education reform, improve teaching quality, establish school-enterprise partnerships with multiple companies, improve teaching quality, strengthen professional construction, optimize curriculum structure, and improve cultivation system in the model of integrate production and education, and multi-party collaborative education [5].

4   Solutions and Suggestions

In response to the above-mentioned problems, this article combines the actual situation in order to achieve the connection between the talent-training goal and the needs of the industry and enterprises. The general curriculum standards will be deeply integrated with the industry standards, and industry enterprises will deeply participate in the training process, in order to strengthen the application ability and core competitiveness of students [6]. It has gradually formed the characteristics of the education model of integration of industry and education, school-enterprise collaboration, and actively explored a new model to achieve multi-party collaborative education.

YTBU actively explores the practice and innovation mode of the multi-party collaborative education mechanism, and has been continuously verified in the teaching process. It has achieved outstanding results, which provide basic and relevant information for the innovation and practice of the government, industry, and university.

4.1   Work-Learning Alternation and Integration of Education with Industry

Through continuous practice, YTBU has gradually formed the two goals of cultivating students' Professional Quality and Professional Ability, with School and Enterprise cooperation as the double subject, and Professional Internship and Graduation Comprehensive Internship as the carrier divided into two stages. The work-learning alternation and integration of education with industry model that combines learning and work internship training has been implemented [7], that is, Two Goals, Two Subjects, Two Stages, and combination of learning and training.

4.2 Construction of Simulation Practice Base on Campus

The on-campus simulation practical teaching experiment center has been established in many colleges and universities in Yunnan Province and the whole country. The establishment of this practical teaching system can not only help improve the teaching system, but also help the application of similar practical courses [8]. With the increase of data, more realistic results of big data support practical teaching. Moreover, it analyzes and summarizes theory and practice to provide a strong theoretical and practical basis for follow-up accumulated research.

4.3 Construction of Off-Campus Practice Bases

In order to strengthen students' practical skills and adapt to the job requirements as soon as possible, YTBU has established off-campus internship practice bases with 16 companies including Zhonglian Group Education Technology Co., Ltd. and Yunnan Pingyun Certified Public Accountants, which accept a total of 205 student internships from 2017 to 2020.

4.4 School-Enterprise Joint Training of Talents

The college and Zhongbo Education jointly established the International School of Finance and Economics, as well as Bangwei Technology Co., Ltd. And established an asset evaluation professional major with Zhonglian Assets. YTBU and enterprises jointly formulate talent training programs, jointly cultivate application-oriented talents, continuously improve the relevance of majors to enterprises and industries, improve students' practical ability and overall quality, improve the quality of application-oriented talent training, and enhance employers' employers' satisfaction with schools and students [9].

4.5 School-Enterprise Co-construction of Teachers

The school focus on the training of teachers with the ability to cultivate application-oriented talents and the ability to develop collaboration between enterprises, universities and research institutes. During summer vacation, teachers return to enterprise to acquire the practical knowledge. the combination of professional teachers and professional in the industries has been realized, and more than 130 teachers have been trained. The purpose is to have the Double-type Teacher faculty team that meets the needs of application-oriented talent training and integration of schools and enterprises.

4.6 Interdisciplinary School-Enterprise Cooperation

Follow the guiding ideology of Market-oriented, Professional-led, Project-based, and Alliance-based and the construction goals of Cooperation and Exchange, Resource Sharing, Mutual Benefit, and Win-Win for all parties, and implement the Service-oriented Approach, Employment-oriented Education policy, establish a long-term and stable connection and linkage mechanism between schools and enterprises, colleges and colleges, majors and majors. And realize the close connection between talent training goals and the needs of industries and enterprises, also the courses general standards and industry standards are deeply integrated. The School of Accounting has cooperated with Kingdee, SAP and other companies to build cross-professional training classrooms, offering cross-professional training courses for accounting, computer science and technology, e-commerce, business management, marketing and other majors.

4.7 Open Classes

The relevant content of courses can take open classes, take advantage of the corporate resources, and take students to the enterprises to learn on the ground. The corporate tutor will be the main lecturer, and the role of the schoolteacher will be changed to a teaching assistant, or the corporate tutor can be invited to the classroom according to the adjustment of the course content. In-school tutors and corporate tutors are jointly trained to reduce the gap between corporate needs and training in the school [10].

4.8 Win-Win Cooperation Between the Campus and Enterprises

Taking advantage of the policies and resources of the industrial cluster and the enterprises, the school can further strengthen professional construction and internship training, continuously explore new talent training models, consolidate the teachers and students, and deliver and train more suitable and practical employees for enterprises. The schools and the professions will work hard to improve the level of running a school, and work together to draw a new blueprint for school-government-enterprise cooperation.

4.9 Education with Industry Organizations

In order to further strengthen the cooperation among universities, government, and enterprises, help college students to Double Innovation and achieve win-win cooperation, and increase the publicity of the Belt and Road tax policy, YTBU and China Registered Taxation Association signed a strategic cooperation agreement in 2019. The two parties will further deepen cooperation in the fields of tax accountant reserve talent training, tax accountant industry continuing education and teaching, applied tax-related service subject research, dual-teacher talent training, and joint construction of student internship training bases.

References

1. He, Y.B.: Theoretical model of industry-university-research collaborative innovation. Sci. Res. **02**, 165–174 (2012)
2. Hu, Y.F.: Research on the social practice mode of college students in western regions. Ideol. Front. **37**, 428–430 (2011)
3. Jie, P.Y.: Thoughts on the practical education model of multi-party joint construction based on collaborative innovation
4. Liu, C.S.: Research on the effective mechanism of practical education in colleges and universities. J. Ideol. Theor. Educ. **12** (2016)
5. Li, W.H., Chen, C.: Application-oriented talent training model under the framework of school-enterprise cooperation-taking chemistry and technology as an example. Heilongjiang Educ. (Theory Pract.) **10**, 21–23 (2018)
6. Liu, Y.: Innovation of college students' social practice mode. Beijing City Univ. Stud. **3** (2016)
7. Qu, C.H., Wang, Y.F.: Research on diversified and collaborative cultivation of college students' professional literacy. Educ. Theory Pract. **35**(12), 3–5 (2015)
8. Zhao, X.: Restriction and realization of college students' social practice education advantages. Educ. Theory Pract. **24**, 09–10 (2012)
9. China Internet Network Information Center: The 47th China Statistical Report on the Internet Development, February 2021. http://www.cac.gov.cn/2021-02/03/c_1613923423079314.htm
10. China Mosaic: Online shopping boosts enthusiasm of Chinese consumers, May 2020. https://mp.weixin.qq.com/s/K8afC0iubzgB7dscp1WKIg

Location of Electric Vehicle Charging Station Based on Particle Swarm Optimization

Ding Chen[1], Min Zhou[1], Yutong Cui[2], Weijun Mao[2], Dawei Zhu[2], and Ying Wang[2(✉)]

[1] Jiaxing Power Supply Company of State Grid Zhejiang Electric Power Co., Ltd., Jiaxing, Zhejiang, China
[2] China Ji Liang University, Hangzhou, Zhejiang, China

Abstract. The distance of the electric vehicle driving to the charging position is one of them, and reducing the loss is the primary research purpose. Query data to get road node distance, traffic flow and other information, charging demand can be obtained by analyzing the above data. After getting the necessary information, the ordinary particle algorithm is used to simulate the model. It is found that the ordinary particle algorithm has some shortcomings in the optimization, which may lead to the result is not the optimal solution. In order to obtain a more reasonable solution, the catfish particle algorithm and mutation particle algorithm are used to solve the location model. In order to compare the advantages and disadvantages of the three algorithms in the results, software simulation is adopted, The simulation software is MATLAB; Choose the charging station distribution coverage model of mutation particle algorithm with better optimization effect, and get the final model by integrating other conditions. In this paper, the feasibility and resource saving are considered to make the experimental results more reasonable.

Keywords: Electric vehicle · Charging station · Particle swarm optimization

1 Introduction

This paper mainly aims at owners and reasonable planning to reduce the energy consumption of builders and users in the process of charging stations to meet the environmental protection requirements. Taking the road conditions and actual traffic flow into account, the actual requirements of construction and several different optimization methods, the optimal location model is finally obtained through the simulation results.

In this article, we first put forward the research purpose, the user charge road losses to a minimum, the objective function, and the algorithm for the optimization results, the algorithm to get the actual charge demand information and the actual traffic information, the distance. After the analysis of the algorithm results, some particles are transformed to improve the optimization results [1, 2].

2 Location of Electric Vehicle Charging Station Based on Mutated Particle Swarm Optimization

2.1 Electric Vehicle Charging Station Location Model

In order to solve the problem of how to shorten the distance between users and charging stations to the minimum and use it as a solution, the formula is as (1):

$$\min Z = \frac{\sum D_a X_{ab}}{g} p \times 265 + \frac{\sum D_a X_{ab}}{v} k \times 365 \tag{1}$$

In the formula: $\sum D_a$ represents the sum of the driving distance of all test vehicles to No. a charging station; g represents the distance the vehicle can travel per unit electric quantity; p represents the price of charging; v is the speed of the car; k is the owner's time value; X_{ab} is a random variable [3–5].

2.2 Constraint Condition

When assuming the location of the charging station in the experimental area, the requirements of all nodes in the test area should be met, so that the charging distance of all test points is relatively minimum, so as to achieve full coverage of users and meet the charging needs of all car owners. Therefore, the following constraints should be met in the operation process:

$$\sum_{a=1,b=1}^{N_{ch}} X_{ab} = 1 \tag{2}$$

$$\sum_{a=1}^{N_{pq}} I_i = N_{ch} \tag{3}$$

$$X_{ab} \leq I_i \tag{4}$$

$$X_{ab}, I_i \in \{0, 1\} \tag{5}$$

Equation (2) guarantees that the charging station set can cover all test points guarantees that the charging station set can cover all test points X_{ab}. When X_{ab} is 1, it means that the charging station set at b can meet the charging demand of test point a. When X_{ab} is 0, it means that the charging station n at b cannot meet the charging demand of test point a. Equation (3) standardizes the number of charging stations in the region to avoid the construction of too many charging stations. Equation (4) indicates that the charging demand at a certain point can only be solved when the charging station is completed. Equation (5) restricts X_{ab} to either 1 or 0.

2.3 Particle Swarm Optimization

Particle swarm optimization (PSO) is an advanced algorithm derived by simulating the collection activities of birds. After sharing different information of experimental entities with birds, the overall activities become regular and can be followed.

Particle Swarm Optimization (PSO) is first introduced by the problem of birds looking for food sources. Suppose a group of birds are searching for a fixed food in an area, the exact location of the food is unknown but the distance of each bird to the food is known then the best way to find the food is to search the area around the bird with the shortest distance to the food. It is then assumed that the solutions to the problem to be optimized are all a bird, a "particle", all of which are initialized by the function at the beginning of the algorithm. All random particles are then iterated to find the optimal solution. Iterative engineering involves a process of self-renewal of particles. Where Pbest is the optimal solution of the particle itself and Gbest is the optimal solution of all particles.

Particle swarm optimization algorithm is widely used in solving some specific problems, such as combinatorial problems, optimization problems with constraint properties, nonlinear continuous problems and neural network training. In the process of iteration, the particle swarm optimization algorithm has fewer times and faster operation speed. In the study of charging station installation, it is necessary to integrate several factors, establish a hypothesis model, and take the minimum construction cost as the optimal solution. When solving the model, different constraints such as construction cost, whether the construction meets the needs of users and charging power should also be considered. Particle Swarm Optimization (PSO) is the main solution for solving complex problems of upper class.

2.4 Catfish Particle Swarm Optimization Algorithm

We have discovered a new law of animal movement. After analyzing the catfish effect, relevant experts have improved the particle swarm optimization algorithm and proposed an improved optimization algorithm – catfish particle swarm optimization algorithm. The catfish particle swarm algorithm is to replace the sardine which has insufficient searching ability through the competition between the catfish swarm and the sardine swarm. The improved algorithm can make the calculation result get a better solution from the local optimum, improve the accuracy and stability of the algorithm, and the result is reasonable.

The current position of the particle needs to be redefined, its physical properties defined by its previous position and velocity. In this algorithm, the catfish particle is added as a disturbance, and the optimization speed of the particle is mutated to achieve the goal of optimization.

2.5 Variation Particle Swarm Optimization Algorithm

The Particle Swarm Optimization (PSO) algorithm usually used lacks the best global search function. There is no guarantee that the value found is globally optimal. To solve this problem, we use a method to add mutated particles to the existing particle algorithm to make the experimental results more consistent with the requirements. The particle whose fitness is higher than the average is selected, and the particle is mutated to become a mutated particle, so that the particle has higher searching ability than other particles, and the probability of finding the optimal result is improved.

The maximum optimization speed of the experimental particle is set as V_{max}. If the particle's speed exceeds the maximum speed in the optimization process, the particle's speed is reset to V_{max}, and then the optimization operation is continued. According to this operation, the minimum speed of local optimization is set as V_{min}. If the particle optimization rate is too low in this process, the applied particle rate V_{min} is determined again and then the optimization process is continued. The position information of the particle is the solution of the problem, and the local solution can be obtained by substituting the instantaneous position of the particle [6–9].

3   Case Study

3.1   The Experimental Model

The experimental model used is analyzed, and the area of the experimental area reaches. There are 48 test points in the model, and 110 road connections connect to these nodes. The road specification diagram for this example is shown in Table 1, and the numbers in the table represent the corresponding nodes.

Table 1. Road structure code diagram

X/km	Y/km							
	1	2	3	4	5	6	7	8
1	1	2	3	4	5	6	7	8
2	16	15	14	13	12	11	10	9
3	17	18	19	20	21	22	23	24
4	32	31	30	29	28	27	26	25
5	33	34	35	36	37	38	39	40
6	48	47	46	45	44	43	42	41

3.2   Simulation Experiment Parameters

In this experiment, the convergence trend and location model of the loss caused by charging distance under different algorithms are obtained by dynamic simulation through MATLAB. The specific values of parameters selected in MATLAB simulation are shown in Table 2.

3.3   Analysis of Experimental Results

Figure 1 shows the convergence trend of energy consumption caused by charging under the standard particle algorithm. Figure 2 shows the convergence trend of distance energy consumption caused by charging under the catfish particle algorithm. Figure 3 shows

Table 2. The algorithm is substituted with parameters

Parameter	Value	Unit
α	10%	–
μ	15%	–
C_v	50	kW
p	0.8	$/kW·h
p^*	0.56	$/kW·h
v	20	km/h
k	17	-

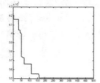 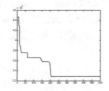

Fig. 1. Standard particle swarm optimization simulation convergence graph	**Fig. 2.** Catfish particle swarm optimization simulation convergence graph	**Fig. 3.** Variation particle swarm optimization convergence graph

the convergence trend of energy consumption caused by charging under the mutation particle algorithm.

According to the graph convergence trend, the optimization result of elementary particle swarm converges to 3.5 × 105, and the optimization operation is completed when the iteration is about 150 times. The optimized catfish PSO and mutated PSO converge to 3.35 × 105 and 2.79 × 105, respectively. The catfish particle swarm optimization algorithm is smaller than the ordinary particle swarm optimization algorithm, and the mutation particle swarm optimization algorithm is smaller than the catfish optimization algorithm. The catfish algorithm takes about 200 iterations to complete the iterative operation, while the mutation particle swarm optimization algorithm takes 300 iterations to complete the iterative operation. The comprehensive simulation information shows that the final consumption convergence value decreases obviously with the improvement of the algorithm, but the number of iterations increases slightly. Compared with the previous two algorithms, the mutation particle swarm optimization algorithm has disadvantages in terms of the number of iterations, but the final optimization effect has obvious advantages.

4   Conclusions

(1) Although the particle swarm optimization algorithm selected at first has fewer iteration times and fast solving speed, it is obviously trapped in local optimal condition, which leads to the optimization result is not optimal. The improved catfish particle

swarm optimization algorithm has the variation of the optimization speed, and the optimization ability has been improved obviously. In order to get a model with less travel consumption, the mutation particle swarm optimization algorithm is used to vary the speed and position, which can ensure that the mutation particle can constantly break through the local optimal value and find the whole global optimal value when searching for the optimal value, and can also ensure that the result is more accurate.

(2) It can be clearly seen from the simulation data that, with the improvement of optimization, although the number of iterations is significantly increased, the convergence trend of the journey loss is significantly reduced, and the optimization effect is improved with the improvement of the algorithm. By comparing the simulation results, it can be concluded that the site selection model based on the mutation particle swarm optimization algorithm can achieve better results in saving resources than the previous methods. Therefore, the coverage information of charging stations under the mutation particle swarm optimization algorithm is selected as the final site selection model. The model can save the cost of both the consumer and the builder.

References

1. Bayati, M., Abedi, M., Gharehpetian, G.B., Farahmandrad, M.: Short-term interaction between electric vehicles and microgrid in decentralized vehicle-to-grid control methods. Protect. Contr. Mod. Pow. Syst. **4**(1), 1–11 (2019). https://doi.org/10.1186/s41601-019-0118-4
2. Huang, Z., Fang, B., Deng, J.: Multi-objective optimization strategy for distribution network considering V2G enabled electric vehicles in building integrated energy system. Protect. Contr. Mod. Pow. Syst. **5**(1), 48–55 (2020)
3. Schutte, J.F., Reinbolt, J.A., Regly, B.J.F., et al.: Parallel global optimization with the particle swarm algorithm. Int. J. Numer. Methods Eng. **61**(13), 2296–2315 (2004)
4. Yang, L.: A fast and elitist multi-objective particle swarm algorithm: NSPSO. In: The 2008 IEEE International Conference on Granular Computing, GrC 2008, Hangzhou, China, 26–28 August 2008. IEEE (2008)
5. Das, S., Abraham, A., Konar, A.: Automatic kernel clustering with a multi-elitist particle swarm optimization algorithm. Pattern Recogn. Lett. **29**(5), 688–699 (2008)
6. Alsghaier, H., Akour, M.: Software fault prediction using particle swarm algorithm with genetic algorithm and support vector machine classifier. Softw. Pract. Exp. **2020**, 50 (2020)
7. Chen, G., Chacón, L., Nguyen, T.B.: An unsupervised machine-learning checkpoint-restart algorithm using Gaussian mixtures for particle-in-cell simulations. J. Comput. Phys. **436**(1), 110185 (2021)
8. Kashyap, N., Kumari, A.C., Chhikara, R.: Service composition in IoT using genetic algorithm and particle swarm optimization. Open Comput. Sci. **10**(1), 56–64 (2020)
9. Chhabra, S., Singh, H.: Optimizing design of fuzzy model for software cost estimation using particle swarm optimization algorithm. Int. J. Comput. Intell. Appl. **19**(1), 2050005 (2020)

Design and Development of Big Data Cloud Platform for Cattle and Sheep IoT Breeding Based on SaaS

Xiaoli Zhang[✉] and Yimin Zhang

Department of Information Management, Dalian Neusoft University of Information, Dalian, Liaoning, China

Abstract. The Chinese government attaches great importance to animal husbandry, adhere to giving the high priority development of agriculture and rural areas, and accelerate the Rural Revitalization Strategy. In recent years, animal husbandry has developed continuously, infrastructure has been improved, and the concepts of resource conservation, quality safety and high-quality ecology have been strengthened. With the increasing acceptance of new things and new technologies by farmers and the continuous development of modern animal husbandry, farmers increasingly need a low-cost, high-efficiency and easy-to-operate cattle and sheep breeding system, which can efficiently record livestock breeding records and assist farmers in digital management of livestock. This paper introduces the design and development of big data cloud platform for cattle and sheep breeding based on SaaS mode, and describes the feasibility and operability of this technology in detail, which can help the transformation and upgrading of animal husbandry, and has good feasibility and prospects.

Keywords: SaaS · Cattle and sheep breeding · Internet of things · Big data

1 Introduction

During China's 14th Five Year Plan period, Chinese animal husbandry is facing great development opportunities. The Chinese government attaches great importance to animal husbandry and make major arrangements. The general office of the State Council issued a series of policies and measures on promoting the high-quality development of animal husbandry, which highlights the strategic position of animal husbandry. The Party Central Committee adheres to the "three rural" work as the top priority, speeds up the strategy of rural revitalization, and constructs a new development pattern, so as to provide significant benefits for the transformation and upgrading of animal husbandry and improving the quality and efficiency. With the implementation of the policy of strengthening agriculture and benefiting agriculture, the pace of scale, standardization, industrialization and regionalization of animal husbandry has accelerated. In recent years, the informatization of animal husbandry has been continuously upgraded, but there are few livestock raising and management platforms for farmers. On the one hand, there is a lack of talents in agricultural informatization, and farmers' education level is not high, informatization

J. Macintyre et al. (Eds.): SPIoT 2021, LNDECT 98, pp. 925–930, 2022.
https://doi.org/10.1007/978-3-030-89511-2_128

consciousness is not strong, and informatization work level is low; On the other hand, the infrastructure of agricultural informatization is not perfect, the network facilities are weak, there is no computer, no laptop, and it is impossible to deploy high-definition cameras [1].

With the construction of 5G infrastructure in China, the popularity of smart phones and the wide application of IoTs technology, farmers can connect to the Internet online at any time through their own smart phones. Therefore, the cattle and sheep breeding cloud platform that follows the interface interaction design principles of "easy to understand" and "less to move" can be accepted and used by more and more farmers.

2 Introduction of Related Technologies

2.1 SaaS

SaaS (software as a service) is a new completely innovative software application mode rising in the 21st century, since the development of Internet technology and the maturity of application software. The traditional sales method of software is mainly to sell licenses, that is, the software is deployed to the terminal, and the fee is charged according to the number of publishing terminals. SaaS defines a brand new way delivery method, so it makes the core of software sales return to service, which is the essence of software production. SaaS providers provide enterprises with a network information service platform. Enterprises no longer need to build network infrastructure and purchase software and hardware operation platforms. At the same time, SaaS providers will also be responsible for all design and implementation of the information system, the maintenance and other services. Enterprises can use it through the Internet without purchasing software and hardware, building computer rooms and recruiting IT personnel. Just like turning on the light switch can use electricity, enterprises rent different versions of software services from SaaS providers according to their actual needs.

The price of a SaaS application is usually inclusive of all services, including not only the cost of using the software, but also the cost of operating and maintaining the software, software technical support services, and so on. Users only need to pay for the duration of use. The most suitable users of SaaS are small and medium-sized enterprises, and it is the best way for them to realize informatization by using advanced technology. For providing software services, SaaS has improved the traditional mode. It avoided a large amount inverstment on local deployment, further highlighting the nature of information software as a service. It has become the mainstream delivery model in today's software market.

2.2 Internet of Things

The Internet of things (IOT) is a concept that describes a huge and growing number of digital devices operating between networks with a global scale [2]. Internet of things is to point to by various kinds of equipment and technology such as information sensor (e.g. radio frequency identification technology, global positioning system, infrared sensor, laser scanner, etc.). It will be anything connected to the network for information exchange and communication, and to collect sound, light, heat, electricity, mechanics, chemistry, biology, position required for all kinds of information, such as through a variety of possible Internet access, To realize the ubiquitous connection between objects and objects, objects and people, and to realize the intelligent identification and management of objects and processes. The Internet of Things takes the Internet and the traditional telecommunications network as the carrier. It enables all the ordinary physical objects that can be addressed independently to form the Internet, so as to realize the connection between objects and objects and between objects and people [3]. The Internet of Things (IoT) is a cutting-edge technology that combines machines and devices into a global network that enables them to interact with each other [4].

2.3 Big Data

Big data refers to the data sets that cannot be captured, managed and processed by traditional software tools within a certain time range. Massive, high growth rate and diversified information asset requires new processing mode with insight ability, stronger decision-making power, and process optimization ability. The strategic significance of big data technology is to professionally process these meaningful data, not only to master huge data information. In other words, the key to big data making the industry profitable is to improve the "processing capacity" of data and realize the "value-added" of data through "processing". Big data can be structured (business data; basic data; input data; structured sensor data, etc.), unstructured data (traffic photos and map data; network pictures and social network data; meteorological and hydrological survey data; seismic observation data; audio and video data, etc.) [5].

3 Cloud Platform Based on SaaS Mode

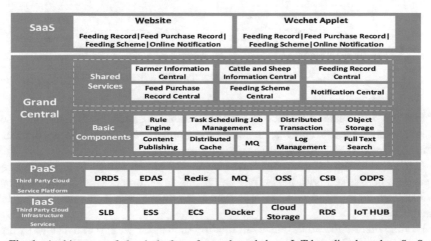

Fig. 1. Architecture of cloud platform for cattle and sheep IoT breeding based on SaaS

The development of animal husbandry industry in China still faces some major problems, such as small scale of farms, low degree of standardization, high labor cost and low content of science and technology [6]. The scale of domestic cattle and sheep farmers is more than ten, dozens of small farmers. Saas mode of precision agriculture provides agile demand service of precision agriculture software development. Based on SaaS cloud platform function standardization, it is characterized by low cost, fast online speed and high quality, which is more suitable for small farmers' daily feeding information needs. Farmers can apply online, and the platform side can simply set up. Medium sized farmers can meet the needs through the bottom configuration, while large farms can be customized based on some functions of this framework. Figure 1 is the architecture of big data cloud platform for cattle and sheep IoT breeding based on SaaS. It can improve management efficiency in the process of poultry breeding [7].

4   Core Functions of Cattle and Sheep Breeding Cloud Platform Based on Internet of Things

The platform has six core business modules: farmer information, cattle and sheep information, feeding record, feed purchase record, feeding scheme and online notification (see Fig. 2). So as to form a closed loop of cattle and sheep from entry, feeding, epidemic prevention to exit, and realize the digitization and traceability of cattle and sheep breeding. Based on this platform, other functional modules can be expanded in the future, such as online transaction, livestock insurance, health monitoring, online consultation, intelligent reminder, analysis and statistics, online micro video of scientific epidemic prevention feeding, etc.

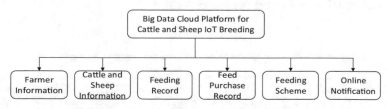

Fig. 2. System function module structure diagram

Farmers' information: Farmers' personal information, such as name, mobile phone number, village, specific address, breeding type, number of cattle and sheep, number of cattle and sheep in advance and other basic information.

Cattle and sheep information: Based on the IoTs technology, one livestock and one code, and marked with the entry time, listing time, breed, type (calf/cow), status (on hand/pre listing) and other information. Each animal has a unique code for later management [8].

Feeding record: record the basic information such as the weight of livestock by times or by cycle, as well as the feed ratio and drinking water situation of livestock feeding. At present, most of them have adopted mixed feeding, so recording the feeding process on time is convenient for analyzing the relationship between livestock growth and feed, adjusting the feeding scheme in time and feeding accurately.

Feed purchase record: record the commodity name, brand, supplier and kilogram of each feed purchase. On the one hand, it is convenient to evaluate the feed quality and feeding effect; On the other hand, it can scientifically estimate the next purchase time and quantity, and try to control reasonable inventory to ease the capital pressure.

Breeding scheme: the platform combines industry experts and professional companies to provide breeding scheme. After sorting and processing, it provides relevant information to farmers in the principle of simplicity and accuracy.

Online notice: including animal epidemic prevention notice, the latest policy and other consulting information.

From the aspects of big data collection and integration, big data analysis methods and resource shortage, there are some problems in the application of big data in animal husbandry. However, it is the basis of big data platform to collect core data related to livestock breeding through farmers. In recent years, more and more researchers are committed to the research in this field, such as poultry breeding technology [9–11].

5 Conclusions

Based on SaaS, the big data cloud platform for cattle and sheep IoT breeding can effectively promote the informatization of animal husbandry. Farmers can realize scientific breeding through digitization of basic information of cattle and sheep, scientific feeding scheme, visualization of feeding process, and using Internet of things and big data technology. The application and promotion of the platform is conducive to realizing of the digitization of agricultural and animal husbandry production, reducing agricultural production costs, improving agricultural production efficiency and increasing the economic benefits of agriculture.

References

1. Xu, M., Zhang, J.: Analysis on the development of agricultural informatization in China. Comput. Knowl. Technol. **15**(27), 291+294 (2019)
2. Chou, S.-C.: Controlling information flows during software development. Int. J. Secur. Privacy Trust Manag. **5**(3), 1–7 (2016)
3. Liu, C., Jing, X., Dong, G.: Technical characteristics and wide application of internet of things. Sci. Consult. **9**, 86 (2011)
4. Vardan, M., Alexey, F., Vsevolod, C., et al.: Big data and internet of things (IoT) technologies' influence on higher education: current state and future prospects. Int. J. Web-Based Learn. Teach. Technol. **16**(5), 137–157 (2021)
5. Arifin, F., Hariadi, M., et al.: Extracting value and data analytic from social networks: big data approach. Adv. Sci. Lett. **23**(6), 5286–5288 ((2017))
6. Lian, J., Li, H.: Informatization empowers the upgrading of animal husbandry in China. Poult. Sci. **4**(08), 9–10 (2019)
7. Zheng, H., Zhang, T., Fang, C., Zeng, J., Yang, X.: Design and implementation of poultry farming information management system based on cloud database. Animals **11**, 900–914 (2021)

8. Challoo, R., Oladeinde, A., Yilmazer, N., Ozcelik, S., Challoo, L.: An overview and assessment of wireless technologies and co-existence of ZigBee, Bluetooth and Wi-Fi devices. Procedia Comput. Sci. **12**, 386–391 (2012)

9. Rowe, E., Dawkins, M.S., Gebhardt-Henrich, S.G.: A systematic review of precision livestock farming in the poultry sector: is technology focused on improving bird welfare? Animals **9**, 614–618 (2019)

10. Norton, T., Chen, C., Larsen, M.L.V., Berckmans, D.: Review: precision livestock farming: building 'digital representations' to bring the animals closer to the farmer. Animals **13**, 3009–3017 (2019)

11. Astill, J., Dara, R.A., Fraser, E.D.G., Roberts, B., Sharif, S.: Smart poultry management: smart sensors, big data, and the internet of things. Comput. Electron. Agric. **17**, 105–109+291 (2020)

Font Design Based on Visual Interaction Technology

Yuan Wang[✉]

School of Art and Design, Xi'an Eurasia University, Xi'an, Shaanxi, China

Abstract. This paper discusses the font design based on visual interaction technology, in order to provide reference for the future direction of font design. With the development of science and technology, the way people read and receive information is also changing. The new media makes the font design from static to dynamic, breaks through the two-dimensional limitations, and realizes the three-dimensional multi form space effect. Dynamic font changes the traditional design mode and extends the meaning of font communication. At the end of last century, the emergence of multimedia has brought three-dimensional animation design. The application of animation and hypertext technology has added more dimensions to font design, mainly reflected in movement and time. From the initial animation technology to the later artistic expression, the appearance of three-dimensional dynamic font design completely changed the thinking mode of font design. With the development of new media technology, the information exchange under the network environment goes beyond space and time, and new interactive design means appear in font design.

Keywords: Visualization · Interactive technology · Font design

1 Introduction

The retinal cells of human eyes are saturated. When observing the external images, the observed objects will produce a dynamic stimulation to the retinal cells. At the end of the observation, the stimulation of observing the external image will gradually disappear, the retinal cells will gradually reach equilibrium saturation, and the just observed image will gradually blur or even disappear. The change of dynamic design, dynamic graphics or dynamic font will make up for the visual decline caused by static stimulation of traditional graphic design, and can maintain a comfortable and natural visual observation for a long time. Moving objects can attract people's attention more than static objects [1]. Making good use of the communication principle of dynamic design and human perception theory to carry out dynamic design can improve the audience's attention and information reception in the process of communication.

Dynamic font design is on the rise in the early stage, full of the whole design market, designers should strengthen the learning of related skills. However, we pay too much attention to the formal change of dynamic font, often ignoring the phenomenon of dynamic effect accumulation on the whole, and losing the functional significance of

J. Macintyre et al. (Eds.): SPIoT 2021, LNDECT 98, pp. 931–936, 2022.
https://doi.org/10.1007/978-3-030-89511-2_129

information cognition, which may lead to the ambiguity of information transmission in the overall logical performance. Designers pay too little attention to the information transmission of dynamic font content subject, and pay more attention to some external visual forms [2].

2 Dynamic Font Design

The visualization of new media technology changes the thinking mode of font design through real-time, mobile and interactive visual experience. Among them, graphics software and digital audio processing technology of new media digital information processing and generation technology give birth to the emergence of dynamic font [3]. Dynamic font has the characteristics of time and movement, as well as the interaction of visual video image and network, which promotes people to participate in the experience, including the experience of vision, hearing, touch and smell. Dynamic font design has the characteristics of luminous, rotating and dynamic effects, which can stimulate and attract eyes, increase the virtual and interactive of font space-time, and become the direction of continuous exploration of font design in the future.

In the real communication process of dynamic font design, the information that viewers want to get in the picture is not received, which also increases a lot of reading burden. From the perspective of dynamic communication designers, we must explore effective ways of visual communication through systematic perceptual research, and practice in our own works, accurately grasp the information theme of the works, and highlight the information we want to express.

Spread to the audience under the positioning, so that the information can accurately cause the aesthetic and emotional resonance of the docking vertical audience. By analyzing the audience's different levels of aesthetic cognitive concepts and processes, dynamic information can be more effective and purposeful interpretation of information content, so as to attract the attention of different levels of viewers and improve the transmission of dynamic font information. Making the font dynamic is not only a sense of form, but also a real complement to the font, which can better convey information, make it dynamic and better serve the market and the public [4].

3 Research on Related Concepts of Font Design

The font comes from "typography", which originally means printing and arrangement. Dynamic font design is defined as dynamic typography in English, which also translates into dynamic typesetting design. It refers to the design of the arrangement and integration of design elements such as glyphs in a constantly moving and changing state. Visual experience shows that people's senses and visual perception are more easily attracted by moving objects. The more dynamic it is in a static background, the more so it is. For example, cattle and horses in the grassland, people crossing the road at traffic lights [5]. Dynamic design is easier to attract the audience's senses, which is much faster than the static design of information transmission speed. It is also the case that visual designers should make good use of this point to do a good job in dynamic visual order, and reasonable dynamic effect and visual guidance can transmit the information to the audience.

Information is not an actual object, but the unity of symbol and meaning. It is an objective substance based on facts and not transferred by human will. Symbol is the material carrier of information. Font is also a symbol in a sense. Text is a kind of "visible, readable and understandable" symbol agreed after creation. Dynamic font design is not only to arrange the characters in a different way, but also to endow the characters with emotion and emotion in the dynamic effect, so that the transmission content presents the personality and content that static design cannot transmit. Its advantages of fast transmission, rich vision, interesting interaction and low cost attract the audience, There are great differences between traditional static Chinese characters and dynamic Chinese characters in various aspects. Table 1 shows the difference between dynamic font design and traditional font design [6].

Table 1. Differences between dynamic font design and traditional font design

	Dynamic font design	Traditional font design
Design principles	Graphic design rule + dynamic design rule	Graphic design principles
Format	Changeable, typesetting is not limited by time, changes in the picture	There can only be one fixed style font design in a graphic work
Color	Apply to digital screen, no printing, reduce error	Generally printed on paper, is monochrome or a combination of colors, can gradient
Body	As time goes by, the same font can show countless changes	We can only see the visual effect of our current display
Information focus	Stimulate the sense of the audience from the visual and auditory aspects to avoid the interference of Irrelevant information	The feeling is relatively single
Sensory organs	Vision, touch, hearing	Vision
Audience	Receive information actively	Receiving information passively
Form of expression	Three dimensional, four-dimensional media, multimedia forms of expression	Paper, computer and other traditional media

4   Font Interaction Design in the Context of New Media

In the context of new media, Chinese font design mostly uses the combination of shape and meaning, shape and character, shape and color to increase the image effect of font. The font design is based on a certain space, which is composed of two-dimensional space and three-dimensional space. When the font appears on the screen as a three-dimensional virtual space, compared with the two-dimensional hand-painted design on paper media, it becomes a dynamic space form limited by multi-dimensional direction and time. The direction is the track of motion, and time describes the occurrence and end of things [7].

Dynamic font design can control the movement state, rhythm and speed of font design through key frame technology. The visual psychological feeling caused by the track of dynamic font design and the speed of movement is different. The fast rhythm has tension and vitality, while the slow font design image movement state appears quiet and calm. The dynamic font presented by the screen, in addition to the use of key frames for rhythm changes, can also use 3D animation software to design and produce perspective, gradient, light and shadow, transparent, virtual and real, material background. Dynamic fonts are presented in visual video media, either single or multi font combination, or different motion changes in the stroke structure of the font in the combination, which spread in the way of dynamic image presentation.

Another characteristic of new media is the function of interaction, which can realize the communication between artists and audiences. The realization of virtual reality art, it uses computer graphics, simulation technology, multimedia technology, artificial intelligence technology, computer network technology, sensor technology. The basic characteristics of virtual reality art are multi perception, existence, interaction and conception. The traditional hand-painted font design is a two-dimensional creative art, while the new media uses virtual reality technology to guide the audience into the virtual interactive font game through the screen, assuming the font design at the beginning of the film, the font design of web advertising, and the font art form of virtual conception. In particular, virtual font design is a kind of non-existent, but can use virtual reality technology to create fonts. The form of font design is no longer just static, and can even make the font produce new images with the user's operation, or the combination of graphics and strokes, or the flow of three-dimensional font [8].

5 Interactive Design in Visualization

Through visual interaction design, we can find patterns, relationships and exceptions in the samples. Excellent interaction design in visualization can make users understand and analyze data better. The interactive design of visualization mainly includes data source connection, user clustering, event analysis, data modeling, data visualization, data collaboration, index management, user behavior analysis and other functions. Design is carried out around the user, purpose, scene, media and behavior. This paper mainly analyzes the visual interactive design of font from the perspective of interactive task [9].

Dimension in product refers to the angle from which people analyze things. To analyze the new users of a product, we can look at it from the geographical perspective, from the time perspective of the past three days and seven years, or combine these two dimensions. Dimensions must have enumerable values. From the perspective of time, new users have year, month, week and day. Interactively, they can filter dimensions. For dimensions with hierarchical relationship, dimensions can be changed interactively, such as the analysis dimension for new users, from the province of the region to the specific city, county and township. Indicators refer to the dimension elements that can

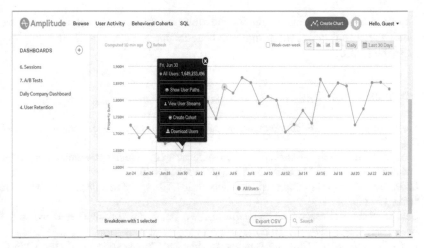

Fig. 1. The trend chart provided by amplitude

be quantitatively analyzed by numerical values. Generally, they are expressed by specific values or ratios. Composite indicators are established by operation among multiple indicators [10].

Database records are involved in user clustering. User clustering is the process of selecting user attributes, user data is actually a table, selecting the enumeration value I want to analyze for discrete data, selecting the range I want to analyze for continuous data matching, and reducing the original data table. The exploratory analysis of charts and dashboards is similar to professional visualization software. Users choose dimensions and indicators for exploratory analysis. Reconfiguration provides different perspectives to observe the data. For example, amplify provides a line chart to see the trend. The comparison can compare different time and space ranges, customize dimensions, and compare the overall data with specific data. Figure 1 shows the trend chart provided by amplitude.

6   Conclusion

The modeling of font design has changed from static combination to dynamic, virtual and interactive new features, which also shows that font design is constantly changing and adapting to new requirements, and it is necessary to keep a new idea space. Dynamic font design expands the space and connotation of font art. The visual language of dynamic font shortens the distance between information coding and decoding, so that visual information can be transmitted in a more efficient and effective way. As an information symbol, the ultimate goal of font design is to transmit information. Dynamic font design integrates sound, text, graphics, animation and other media, extending the connotation and significance of font design. The font design based on visual interaction technology has visual hearing, movement track, virtual reality and interactivity, which expands the new ideas of font design. People can not only interact with the machine, but also scatter and graft the strokes of the font structure into new font images. Stories and games can make the interaction between designers and viewers resonate. The interaction of works formed by the virtuality of font design will become one of the challenging topics of font design in the future.

References

1. Yu, X.: Application of interactive teaching mode in font design course. Design **34**(05), 97–99 (2021)
2. Cheng, Y.: Research on the application of packaging visual communication design in interactive experience. Popular Lit. Art **4**(15), 53–54 (2020)
3. Li, D.: On the composition of visual art elements in UI design of mobile games. Drama House **4**(14), 129 (2020)
4. Gan, Y., Wang, T.: Research on font performance of public guide system based on new media. Arch. Cult. **4**(12), 192–193 (2019)
5. Wang, D.: Research on Chinese character structure and Chinese style font design. Chin. Char. Cult. **4**(S1), 71–72 (2019)
6. Qin, X., Zhang, X.: Analysis of the influence of digital media technology on font design. Art Educ. **4**(10), 205–206 (2019)
7. Tian, P.: Analysis of the interaction of font design in the new media environment. Drama House **4**(16), 129–130 (2019)
8. Ge, M.: Teaching practice innovation of font design in the context of new media. J. Beijing Univ. Print. **25**(06), 55–57 (2017)
9. Lu, L., Tang, D.: Research on visual language of dynamic font. Design Art Res. **7**(04), 91–96+103 (2017)
10. Wang, M., Fan, T.: Research on the development prospect of font design in new media. Beauty Times (Part 1) **4**(07), 55–57 (2017)

Grid Asset Health Evaluation Model Based on Smart Grid Big Data Technology

Jianbin Wu[✉], Jinxi Dong, Zhiwei Liu, Huiwen Qi, Zhenbo Xu, and Xiangyu Zhang

Economics and Technology Institute of State Grid Shanxi Electric Power Company, Taiyuan, Shanxi, China

Abstract. With the development of information technology, the traditional power system is evolving into a smart grid. All the data generated during the operation period of the power system can provide effective value for the grid development through big data technical support. In this paper, we take the development of big data technology of smart grid as the background to build the evaluation model of grid asset health level and trace tracking model, so as to provide a thinking path for the sustainable development of smart grid and promote the optimization and upgrading of smart grid system.

Keywords: Smart grid · Big data · Grid asset management

1 Introduction

Assets are the fundamental basis for grid development, and there have been scholars who have conducted studies on asset management issues such as asset size accounting [1, 2] and asset use efficiency [3, 4], and some scholars have proposed problems facing grid asset management in the new environment of transmission and distribution price reform [5], but asset health is always the basic goal of asset management [6]. Further, grid asset health is directly related to the state of equipment operation. Only by scanning the status of devices, networks and systems in real time, establishing their health profiles, implementing dynamic tracking, and building corresponding databases, can the state assessment of grid assets be realized [7].

In the context of smart grid big data development, by recording and analyzing grid equipment grid equipment operation parameters and data, the reliability indexes and change laws of power equipment in each stage can be analyzed in depth, thus analyzing its operation status and risk level more objectively and accurately; it is also easier to identify the design defects of equipment. The level of visualization of the operational status of power assets and equipment has been continuously enhanced [8], and with the joint efforts of scholars and researchers, the ways of detecting and evaluating the status of power grid assets have been gradually diversified [9]. In addition, there are also scholars from the new demand-side operation mode as the entry point for research on the impact of grid asset status [10].

In a comprehensive view, smart grid big data technology provides a reliable basis for power asset health evaluation, while the results of power asset health evaluation can be

J. Macintyre et al. (Eds.): SPIoT 2021, LNDECT 98, pp. 937–942, 2022.
https://doi.org/10.1007/978-3-030-89511-2_130

effectively fed back to the power system, thus providing an effective auxiliary decision on operation and maintenance strategy and system operation mode. Therefore, the evaluation of power grid equipment health level is of great significance to the sustainable development of power grid.

2 Grid Asset Health Evaluation Model

Relying on smart grid big data technology to monitor the status of grid assets in real time, the grid asset health measurement results are shown as a comprehensive assessment of the results of the value, the greater the value is the healthier, the value is below 70 points, you need to find solutions to effectively improve the asset health.

2.1 Grid Asset Health Evaluation Index System

The evaluation indexes should reflect the operation of grid assets comprehensively and scientifically. Considering the important responsibility of State Grid as a state-owned mega backbone enterprise, this paper constructs a grid asset health assessment index system by considering three aspects: physical, economic and social aspects of grid operation. The details are as follows (Table 1):

Table 1. Grid asset health assessment index system

Grid asset health evaluation	
Tier 1 indicators	Tier 2 indicators
Physical v_1	Comprehensive line loss rate v_{11}
	Load capacity ratio v_{12}
	Annual load ratio v_{13}
	N-2 passing rate of double-return line with the same tower v_{14}
	Average life of retired equipment of 220 kV and above transmission and substation assets v_{15}
Economic v_2	Revenue from electricity sales per unit of assets v_{21}
	Asset debt ratio v_{22}
	O&M cost per 10,000 grid assets v_{23}
	EBITDA margin v_{24}
Social v_3	10 kV grid line interconnection rate v_{31}
	Reliability rate of power supply v_{32}
	Number of annual safety accidents v_{33}
	Percentage of renewable energy access v_{34}
	Social investment as a proportion of fixed asset investment v_{35}

2.2   Standardization of Evaluation Indicators

Due to the complex data types of the evaluation indexes of the health level of the grid assets, different indicators also have the problem of inconsistent objectives, so before starting the evaluation, the evaluation indexes need to be standardized. In this paper, according to the nature of indicators, indicators are standardized according to four categories: trapezoidal function, left trapezoidal function, right trapezoidal function and monotonic increasing function.

2.3   Design of Evaluation Index Weights and Health Level Measurement

After standardization, a hierarchical analysis was used to assign weights to the asset health evaluation indicators, and the steps were as follows.

(1)   Establishing a hierarchical model
    The factors affecting each relevant factor are decomposed into several levels, and the factors on the same level are subordinate to or have influence on the factors on the higher level, and at the same time dominate the factors on the lower level or are influenced by the factors on the lower level. The hierarchical model here corresponds to the asset health level evaluation index system.
(2)   Construction of judgment matrix
    Starting from the second level of the hierarchical model, a judgment matrix is constructed by the two-by-two comparison method for factors of the same level that belong to (or influence) each factor of the higher level, up to the last level. In addition, the expert scoring method is used to judge the importance of the inter-indicators, and the judgment matrix is finally obtained as follows.

$$A_{ij} = \begin{bmatrix} a_{11} & a_{12} & \cdots & a_{1m} \\ a_{21} & a_{22} & \cdots & a_{2m} \\ \cdots & \cdots & \cdots & \cdots \\ a_{m1} & a_{m2} & \cdots & a_{mm} \end{bmatrix} \tag{1}$$

Where: a_{ij} denotes the mean of the score results obtained from factor i compared to factor j.
(3)   Calculation of the weight vector
    The specific process of weight calculation is as follows.

1)   Sum the data in each column, $b_j = \sum_{i=1}^{m} a_{ij}$, to obtain the sum vector, $B_j = [b_1, b_2, ..., b_m]$;
2)   Calculate the normalized vector matrix, C_{ij}, where, $c_{ij} = \frac{a_{ij}}{b_j}$;
3)   Calculate the weight vector ω_i

$$\omega_i = \frac{\sum_{j=1}^{m} c_{ij}}{\sum_{i=1}^{m} \sum_{j=1}^{m} c_{ij}} \tag{2}$$

4) Calculation results of weights

According to the structural model of hierarchical analysis method, the judgment matrix is constructed by expert scoring method, and the consistency test is conducted for the weight judgment matrix of criterion layer and indicator layer respectively, and if it does not pass, the weights are reassigned until the judgment matrix can all pass the consistency test.

5) Calculation of comprehensive score

The comprehensive score A of the evaluation object is calculated by the sum of the products of the standardized values of each index and the corresponding weights:

$$A = \sum_{i=1}^{m} \omega_i u_i \tag{3}$$

2.4 Trace Tracking Analysis Model

The ultimate purpose of the grid asset health evaluation model is to guide the sustainable development of the grid; therefore, after the measurement results are derived, further analysis of the results is required. In this paper, a trace model is used to analyze the evaluation results and identify the problematic attributes and specific indicators. The specific steps are as follows.

(1) Determine the impact of each attribute on the healthiness of grid assets

The TOPSIS method is used to rank the comprehensive evaluation value of each attribute, determine the strength of the effect of each attribute on the negative impact of the health level of grid assets, and perform deviation rate analysis on all indicators.

(2) Calculate the deviation rate of basic indicators.

Based on the standardized basic data, all indicators are analyzed for deviation from the optimal indicators, and the deviation rate of each basic indicator is calculated $v_{lj}(l = 1, 2, ..., L, j = 1, 2, ..., n, j \neq k)$:

$$v_{lj} = \frac{y_{lk} - y_{lj}}{y_{lk}} \tag{4}$$

(3) Calculate the deviation contribution and contribution rate of basic indicators

Combining the combined weight $\gamma_l(l = 1, 2, ..., L)$ of each indicator and the deviation rate of the indicator, the deviation contribution of the basic indicator and the contribution rate $\theta_{lj}(j \neq k)$ are calculated.

$$\theta_{lj} = \frac{v_{lj}}{\sum_{l=1}^{L} v_{lj}} \tag{5}$$

(4) Calculation of indicator deviation contribution ranking

The deviation contribution of indicators is ranked to find the basic indicators that affect the health of grid assets in turn, as a way to promote the sustainable development of the grid.

3   Smart Grid Big Data Technology-Based Asset Health Assessment of L Provincial Grid

3.1   L Province Grid Asset Data

Based on the grid asset health level assessment index system constructed in the previous section, this paper obtains the following data by referring to the relevant contents released by L Provincial Power Company and development reports, etc.:

$$v_1 = \{ 5.67 \ 1.94 \ 90 \ 100 \ 15.7 \}$$
$$v_2 = \{ 0.5 \ 55.44 \ 351.17 \ 9.98 \}$$
$$v_3 = \{ 63.19 \ 99.87 \ 17 \ 12.48 \ 9.25 \}$$

(6)

3.2   Grid Asset Health Evaluation Index System

(1) Calculation results of evaluation index weights

The results of the obtained model weights for evaluating the health level of grid assets are as follows.

$$\omega = \{ 0.64 \ 0.12 \ 0.24 \}$$
$$\omega_1 = \{ 0.16 \ 0.07 \ 0.27 \ 0.43 \ 0.07 \}$$
$$\omega_2 = \{ 0.17 \ 0.44 \ 0.24 \ 0.15 \}$$
$$\omega_3 = \{ 0.21 \ 0.31 \ 0.35 \ 0.05 \ 0.08 \}$$

(7)

(2) Asset health level evaluation results

Combining the grid asset health level evaluation model and the grid data of L province, the following results can be obtained.

$$u = \{ 89.26 \ 86.68 \ 88.36 \}$$
$$A = \{ 88.73 \}$$

(8)

From the results, it can be seen that the assessment results of L Province 2020 asset data are all above 80 points, i.e., the asset health level of L Province grid is good, however, L Province grid still needs to further improve the overall level of asset health while maintaining its current asset health status.

Further, measurements based on the trace trace analysis model show that annual load factor, average life of retired equipment of transmission and substation assets, EBITDA margin, and the number of annual safety incidents are the significant causes affecting the asset health level of L Provincial Grid.

4   Conclusions

This paper constructs a grid asset health level assessment model based on the hierarchical analysis method, and in order to clarify the influence factors of grid asset health level, this paper further proposes a post-evaluation model-trace tracking analysis model for the assessment model. Finally, an example analysis is conducted based on the actual data of L province power grid in 2020. The analysis results show that the overall asset health level of power grid in L province is good, in addition, the asset intelligence level can be continuously improved with the help of smart grid big data technology in order to further improve the asset health level of power grid.

Acknowledgements. This work was supported by the science and technology project "Evaluation of the Efficiency of Grid Investment Project Under Cost Regulation" of State Grid Corporation of China (SGSXJY00JYWT1800028).

References

1. Peng, J., Li, R., Xu, Y.: Determination of reasonable power grid asset scale adapting to electric power reform. Environ. Sustain. Dev. (2017)
2. Cai, D., Feng, D., Ding, H., Zhang, Z., Ma, L., Shen, N.: Research on real-time monitoring model of power grid assets, based on operation-finance integrated value management. Mod. Manag. **09**(05) (2019)
3. Lei, Y., Zeng, Y., Zhao, C.: Study on the calculation of power grid asset utilization index. Adv. Mater. Res. 2912–2915 (2014)
4. Navigant Research; Revenue for power grid asset management, condition monitoring devices and solutions is expected to reach $6.5 billion in 2025, according to navigant research. J. Eng. (4–5) (2014)
5. Shao, J., Meng, X., Zhang, J., Wang, Y.: Research on asset management of power grid enterprises under the situation of transmission and distribution price reform. Chin. Ind. Econ. (16), 123–124 (2020)
6. Wei, P., Li, X., Liu, Q.: Construction of a comprehensive evaluation index system for analyzing physical assets of power grids. J. Green Sci. Technol. **4**(24), 42–44 (2020)
7. Bu, R., Zhang, H., Liang, W.: Construction of identity code for power grid asset based on "internet of things+". Adv. Sci. Technol. Appl. Res. Center (2018)
8. Sun, Q., Zhu, B., Li, Q., Liu, Y., Yang, T.: An algorithm of analytic hierarchy process model based on electric power big data of smart grid. In: Proceedings of 2016 3rd International Conference on Engineering Technology and Application, pp. 491–497 (2016)
9. Su, Z., Xia, X., Wang, Y., Pang, Z.: Application of security label in power grid asset management Smart Grid **8**(5) (2018)
10. Mao, D., Gao, Z., Wang, J.: An integrated algorithm for evaluating plug-in electric vehicle's impact on the state of power grid assets. Int. J. Electr. Pow. Energy Syst. **105**(5) (2019)

Power Load Peak-Valley Time Division Based on Data Mining Method

Huiwen Qi[✉], Jianbin Wu, Jinxi Dong, Zhenbo Xu, Xiangyu Zhang, and He Huang

Economics and Technology Institute of State Grid Shanxi Electric Power Company, Taiyuan, Shanxi, China

Abstract. With the exploitation of wholesale desultory distributed power sources, the user's demand response mechanism has become an essential resource for the steady manipulation of grid. The user-side peak-valley time-of-use price as the core measure of demand reaction projects has gradually been fully popularized and promoted in the country, showing diversity. The realization effect of peak-valley time-of-use price depends on whether its design can effectively stimulate the demand response potential of power users. The scientific and reasonable division of peak-valley time periods is the foundation for the pricing of peak-valley time-of-use electricity prices. However, due to the limitation of measurement conditions, the time interval of traditional time-of-use meters and its meter reading system is difficult to adjust flexibly. Since the realization of time-of-day tariff in some areas, the time interval settings have remained unchanged for many years. This paper uses data mining methods, uses fuzzy semi-gradient membership functions, and takes into account the distribution characteristics of the load curve to divide time periods, which can be used for the time period in areas where time-of-day tariff are implemented for the first time.

Keywords: Data mining · Fuzzy membership · Time-of-day tariff

1 Introduction

For purpose of further the power supply dependability, power quality and load rate of the power grid, and guide users' power consumption, various demand-side management policies and systems have been introduced one after another [1–4]. The peak-valley time-of-use electricity price is a valid demand-side governance method that has developed accordingly [5]. It sets different electricity prices for different power consumption periods according to the difference in the peak and valley power demand of users, so as to reduce the peak and fill the valley, and alleviate the power consumption [6, 7]. It has played a very good role in stressing, increasing the load rate, improving the security and economy of grid, and enhancing the operating efficiency and stability of the system, and has achieved very significant economic and social benefits [6–8].

The scientific and reasonable division of peak-valley time periods is the elements for the pricing of time-of-use electricity prices [9]. Considering the necessity response to optimize the time-of-use electricity price ratio research has received a lot of attention,

© The Author(s), under exclusive license to Springer Nature Switzerland AG 2022
J. Macintyre et al. (Eds.): SPIoT 2021, LNDECT 98, pp. 943–947, 2022.
https://doi.org/10.1007/978-3-030-89511-2_131

and accumulated certain experience and results. However, due to the limitation of measurement conditions, the time interval of traditional time-of-use meters and its meter reading system is difficult to adjust flexibly. Since the realization of time-of-day tariff in some areas, the time interval settings have remained unchanged for many years. At the same time, there is little accumulation of theoretical research on time segmentation of time-of-use electricity prices [10]. This paper uses data mining methods, uses fuzzy semi-gradient membership functions, and takes into account the distribution characteristics of the load curve to divide time periods, which can be used for the time period in areas where time-of-use electricity prices are implemented for the first time. Divided design; divide the time separation on the foundation of the characteristics of supply cost changes, and realize the price function to image the diversity in power provision cost to consumer buyer.

2 Peak-Level Valley Period Division Model

The scientific and rational segmentation of time periods is the foundation for the pricing of time-of-use electricity prices. For purpose of enable the division of periods to correctly reflect the changes in the actual power load of users, this part uses the semi-trapezoidal membership function, threshold optimization theory and the principle of maximum days to divide the peak and valley periods of users, and constructs the height and trough period division based on threshold optimization. Data mining model, the technical route of model construction is shown in Fig. 1:

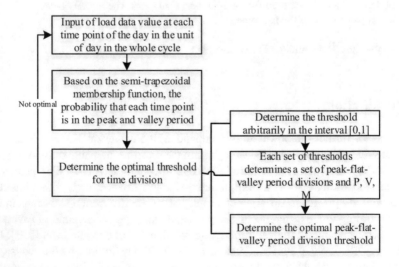

Fig. 1. Peak and valley time division steps

2.1 Load Curve Distribution Analysis

Do with the vague semi-trapezoidal qualification obligation method, the peak-valley interval is divided according to the possibility that each core on the load give is in the height period and the trough period, so as to determine the possibility. The membership function follows the following rules:

(1) The feasibility that the maximum point on the load give is in the height period is 100%, and the feasibility that the lowest point is in the height period is 0;
(2) The feasibility that the lowest point on the load give is in the trough period is 100%, and the feasibility that the highest point is in the trough period is 0;
(3) The feasibility of other points on the load give being in the height or trough period is decided by the qualification function in fuzzy mathematics.

For the convenience of calculation, only the longitudinal role of the qualification function is used for analysis: suppose the set of time cores on the i-th day load give is $T_j = \{t_1, t_2, t_3, \cdots t_{24}\}$, j represents the unit day time point, and the load value set corresponding to each time point is $Q_i = \{q_{i1}, q_{i2}, q_{i3}, \cdots q_{i24}\}$, suppose The lowest valley value and highest peak value on the load curve of a certain day are a_i and b_i, respectively, and the probability $A(q_{ij})$ and $B(q_{ij})$ that the load at a certain time point t_j is in the height period and the trough period is expressed as:

$$A(q_{ij}) = \frac{q_{ij} - a_i}{b_i - a_i} \tag{1}$$

$$B(q_{ij}) = \frac{b_i - q_{ij}}{b_i - a_i} \tag{2}$$

2.2 Threshold Optimization

In order to reduce the subjectivity of artificially setting height and trough thresholds, on the elements of determining the possibility that each period is in the height period and the trough period, the set classification is used to determine the peak and valley period optimal thresholds λ_1 and λ_2.

On the basis of knowing the possibility that respective time core of the i-th day is in the height period and the trough period, let λ_{i1} be the threshold value of the peak period of the i-th day, the number of time points corresponding to the peak period is n_1, and each time point in the trough period corresponds to the charge set of is P, $P = \{p_1, p_2, \cdots p_{n_1}\}$. Let λ_{i2} be the threshold of the i-th day valley period, the number of time points corresponding to the valley period is n_2, and the load set corresponding to each time point in the valley period interval is V, $V = \{v_1, v_2, \cdots v_{n_2}\}$. Suppose the number of time points in the flat time period is n_3, and the load set corresponding to each time point in the flat time interval is M = $\{m_1, m_2, \cdots m_{n_3}\}$.

Define a peak-valley time division function $f_i(x)$ on Q_i:

$$f_i(\lambda_{i1}, \lambda_{i2}, Q_i) = \begin{cases} P, A(q_{ij}) > \lambda_{i1} \\ M, other \\ V, B(q_{ij}) > \lambda_{i2} \end{cases} \tag{3}$$

Define the set Z as a feasible region:

$$Z = \{(\lambda_{i1}, \lambda_{i2}) | 0 < \lambda_{i1} < 1; 0 < \lambda_{i2} < 1\} \tag{4}$$

For any $(\lambda_{i1}, \lambda_{i2}) \in Z$, a peak-valley period division result $f_i(\lambda_{i1}, \lambda_{i2}, Q_i)$ on Q_i can be obtained. Corresponding to the division result, there should be a real-valued function $S(f_i(\lambda_{i1}, \lambda_{i2}, Q_i))$ as the evaluation index of its division effect:

$$S = \frac{(P_{min} - M_{max}) + (M_{min} - V_{max})}{(P_{max} - P_{min}) + (V_{max} - V_{min}) + (M_{max} - M_{min})} \tag{5}$$

In the formula: $P_{max}, P_{min}, V_{max}, V_{min}, M_{max}, M_{min}$ separately conduct oneself the maximum and minimum load importances in the peak, valley, and flat time intervals.

The physical meaning expressed by the function S is: if the difference between the load set P in the height hour and the load set M in the normal hour is more obvious, the difference between the load set M in the normal hour and the load set V in the valley hour is more obvious, And the three conditions that the more ambiguous the sample differences in the sets P, M, and V are, the larger the function value S will be; the larger the S value, the more obvious the economic meaning of the peak-valley time division effect.

Therefore, the threshold optimization problem can be transformed into the following optimization model:

$$\begin{aligned} max \ & S(f_i(\lambda_{i1}, \lambda_{i2}, Q_i)) \\ s.t \ & (\lambda_{i1}, \lambda_{i2}) \in Z \end{aligned} \tag{6}$$

Solving Eq. (6), we can get the optimal threshold $opt_i(\lambda_{i1}, \lambda_{i2})$ for dividing peak, flat, and trough periods on Q_i. Incorporating it into Eq. (3), the peak-valley time division result under the optimal threshold on the i-th day is obtained.

3   Conclusions

With the development of smart grids, user interaction and response capabilities have developed rapidly. It is necessary to establish a normal adjustment mechanism for the division of time intervals to promote good economic benefits of height shaving and trough filling. Therefore, in view of how to take into account the existing user responses in many areas where time-of-use electricity prices have been implemented at present, and adjust the division of time periods reasonably, this article starts from the response of power users' electricity demand and discusses the time segmentation method of time-of-use electricity prices ground on user responses. The fuzzy semi-gradient qualification function method is invested to distribute the time period in consideration of the distribution characteristics of the give, which can be used to distribute the time period in areas where the time-of-use electricity price is implemented for the first time.

Acknowledgements. This work was supported by the project "Evaluation of the Efficiency of Grid Investment Project Under Cost Regulation" of State Grid Corporation of China (SGSXJY00JYWT1800028).

References

1. Zhang, L.: Mechanism construction and method exploration in China's transmission and distribution price reform. Price: Theory Pract. (02), 29–31 (2016). (in Chinese)
2. Hu, Z., Chen, T., Ji, H., et al.: Benefits analysis on application of demand-side management (DSM) in Beijing. Autom. Electric Pow. Syst. 21(7), 22–25 (1997). (in Chinese)
3. Bialek, J.: Topological generation and load distribution factors for supplement charge allocation in transmission open access. IEEE Trans. Pow. Syst. 12(3) (1997)
4. Lin, M., Zhu, Y., Hu, B.: Peak and valley periods division and time-of-use price based on power generation cost. East China Electric Pow. 33(12), 90–91 (2005). (in Chinese)
5. El Ghaoui, L., Oks, M., Oustry, F.: Worst-case value-at-risk and robust portfolio optimization: a conic programming approach. Oper. Res. 51(4), 543–556 (2003)
6. Zhu, S.S., Fukushima, M.: Worst-case conditional value-at-risk with application to robust portfolio management. Oper. Res. 57(5), 1155–1168 (2009)
7. Huang, D.S., Zhu, S.S., Fabozzi, F.J., et al.: Portfolio selection under distributional uncertainty: a relative robust CVaR approach. Eur. J. Oper. Res. 203(1), 185–194 (2010). (in Chinese)
8. Zhao, J., Tan, Z., Li, Q.: Analysis of time-of-use (TOU) power price in China. Mod. Electric Pow. 22(2), 82–85 (2005). (in Chinese)
9. David, A.K., Li, Y.Z.: Effect of inter-temporal factors on the real time pricing of electricity. IEEE Trans. Pow. Syst. 8(3), 44–52 (1993)
10. Kirschen, D.S., Strbac, G., Cumperayot, P., et al.: Factoring the elasticity of demand in electricity prices. IEEE Trans. Pow. Syst. 15(2), 612–617 (2000)

Application of SAP BI in Financial Data Analysis of L Company

Shuang Gao[✉], Xiaoyan Hou, and Yanling Li

Information Management and Information System, Dalian Neusoft University of Information, Dalian, Liaoning, China

Abstract. SAP BI (Business Intelligence) is a popular business intelligence solution, which has won the favor of enterprises with its mature data extraction and data warehouse technology. This paper takes SAP BI as the main research object, uses SAP BW, SAP Hana and other core technologies, and takes the implementation of SAP BI project of L company as the background, carries out the model design, model establishment, data extraction and report design and display of financial data. Based on SAP BI, through the relatively flexible and scalable data warehouse and data mining technology, it can meet the needs of L company for financial data analysis and decision-making.

Keywords: SAP BI · BW · HANA

1 Introduction

L company is a large-scale anthracite production base. After years of development, it has already become a diversified enterprise group with coal and coal chemical industries as the leading industries. The current ERP system for L company provides an integrated system for the enterprise, which promotes the informatization level of the enterprise to a great step forward. However, the ERP system focuses on transaction processing, which is long way from reaching the decision support function expected by the enterprise [1]. At the same time, because of the increasing amount of data, the demand for the data analysis is more and more urgent, and the degree of specialization is also stronger. How to do a good job in financial data management has always been the focus of the company's quality management. In order to ensure the future sustainable development, the comprehensive promotion of management capacity and information construction, L company chooses to implement SAP BI solution [2]. This paper will also focus on SAP BI financial data analysis.

2 BI System Architecture Design

According to the data integration and improvement requirements of L company's business systems and the needs of realizing a unified platform for future access, management and decision analysis, this paper considers the design of BI system architecture scheme, as shown in Fig. 1 [3].

© The Author(s), under exclusive license to Springer Nature Switzerland AG 2022
J. Macintyre et al. (Eds.): SPIoT 2021, LNDECT 98, pp. 948–952, 2022.
https://doi.org/10.1007/978-3-030-89511-2_132

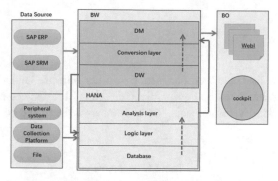

Fig. 1. System architecture design

The technical architecture of BI reflects the improvement direction of data flow and data processing. According to the latest technical characteristics of BW on HANA, the performance between database and BW application system can be improved to the greatest extent. BI system model includes extraction layer, transformation layer, analysis layer and display layer [4].

Extraction layer (HANA base layer, BW first layer extraction layer): Extract the data source data 1:1 without any processing. At this time, the data source includes the manually imported data source and the system data source, and the system standard data source is extracted into the BW database one by one. When designing, try to include all the fields of the data source as much as possible. At the same time, in Hana, it is also necessary to establish the basic model layer corresponding to all fields corresponding to the extracted data, so as to prepare for the later Hana logic layer.

Conversion layer (Hana logic layer, BW second layer conversion layer): the main purpose of this layer comes to realize business logic processing in combination with the requirements and the basic model layer of Hana. And establish the Hana logic layer. This layer is divided into sub layers according to business segments and the segment is divided into sub layers according to business themes [5].

Mart layer (HANA analysis layer): This layer is a logical processing layer based on Hana, which is established according to the theme of business segments (such as group coal, chemical industry, aluminum and electricity). It can conduct comprehensive analysis by theme (such as finance and materials) from the perspective of business segment. At the same time, the business theme is subdivided into management cockpit and fixed report sub layer.

Presentation layer (enterprise portal): The presentation layer uses the enterprise portal studied by itself as the presentation tool for the management cockpit, BOE as the tool for the presentation of fixed multidimensional reports, which is integrated into the portal. The management cockpit and reports are respectively set up folders to store reports according to each theme. Group-level reports and management cockpits are stored in separate directories.

3 Modeling

3.1 InfoObject

InfoArea is like a folder, which can group various objects. Under the InfoArea, it can include other InfoArea and InfoObject directory. InfoObject catalog is a catalog dedicated to classify InfoObjects. Table 1 shows a list of the company's financial InfoArea.

Table 1. InfoArea

List of InfoArea		
1	Financial management	RBI_TF_R_FICO
2	Financial bazaar	RBI_DM_FICO
3	Personal budget plan	OPSM_PBP
4	General ledger plan	OFIN_GL_MCA_EHP6_GL
5	Financial management extraction layer	RBI_DW_FICO
6	Financial master data	RBI_MD_FICO
7	Historical budget	OPSM_GM

The InfoObject is the simulation of the business evaluation object in real life in the SAP BI system. It is the smallest unit of data storage in SAP BI and the basis for creating a data model. InfoObjects are divided into 2 types: characteristics and key figures [6].

3.2 Advanced Data Storage Objects

The advanced data storage object is called ADSO, which is an upgraded version of DSO. It combines InfoCube and DSO, and the functions are also combined together. It can store detailed data and summarized data, and the data can support the implementation of the total and coverage. You can also analyze reports and create reports.

3.3 Data Source

A data source is a place where data is stored, also known as PSA. The data of the source system first enters the PSA through the sap BI extraction tool. The extracted data includes master data and transaction data. The master data also includes attributes and text. Before creating a data source, it needs to create the source system and application components. BW can connect many different types of source systems, including SAP systems and non-SAP systems.

3.4   Create Transformation

Through data transformation, the data from the source system can provide meaningful data analysis. The transformation definition of SAP BI is the rule for processing data. When data is loaded from one BI object to another BI object, the data is transformed, connecting two related objects. There are mainly the following types of transformations: transformation from business data sources to DSO, transformation from master data attributes to InfoObjects, transformation from master data text to InfoObjects, and transformation from DSO to InfoCubes.

3.5   DTP

The data transfer process (DTP) is mainly used for data loading inside the BW system, and it always appears in pairs with transformation. DTP is triggered by execution data. Also, DTP will have a unique ID after the setup is complete and will not be enabled until it is activated.

3.6   HANA Modeling

HANA has complete functions and can quickly build models. HANA views are divided into attribute view, analytical view, and calculation attribute. These three views are all non-materialized views. By establishing the HANA calculation view, and then connecting with the WebI report, a reasonable fixed report can be designed [7].

4   Data Extraction

After the model is established, it is necessary to prepare to start data extraction related work, mainly through ETL for data extraction, loading and transformation, mainly including ADSO data extraction, running DTP, will load the data into ADSO, and then activate the data to complete the effective data storage [8].

5   Report Development and Presentation

L company's report development needs mainly include income analysis table, cost analysis table, accounts receivable table, period expense table and other reports. According to the report model provided by the customer, the report is drawn up and connected with the HANA model data. Through a series of operations in WebI, the report is completed as required, and diverse information is fully presented. Then connect the cockpit, and finally display it. From the perspective of customers and users, using WebI to make reports can make the reports more convenient and intuitive, and achieve the best results. When displaying a graph, the default graph display corresponding to the histogram or the line graph is performed according to the dimension most frequently viewed by the user; free single-value, multi-value and range filtering. Users can perform free single-value, multi-value, range of filtering and other functions for different dimensions of information according to the filter [9].

The financial cockpit displays some of the types that appear in the report through revenue, profit, cost, expense, inventory, cash flow, and accounts receivable. According to the input data, the current month value, as well as the budget increase and decrease, the beginning of the year, and the same period is displayed. This interface is the financial cockpit diagram.

6   Conclusion

In the process of enterprise development and expansion, long-term considerations are required. The overall architecture design not only responds to the needs of the current project, but also needs to meet the needs of future sustainable development [10]; The flexible expansion requirements are met through module and segment design. This paper uses a new and future oriented technical scheme based on Hana and BW to integrate the financial data of L company, scientifically and reasonably display and analyze the financial information of L company by using advanced modeling tools and analysis methods, fully tap the value of information assets, improve enterprise management efficiency, support scientific decision-making, and lay the data foundation for fine management, It not only meets the data integration across business lines, but also supports the future development needs of enterprises. I hope this paper can provide reference for the design and implementation of BI system in other enterprises.

References

1. Zhou, Y., Yang, X.: Deepening application of SAP BW data warehouse technology in supply chain management. Electric Pow. Inf. Technol. (10), 76–80 (2010). (in Chinese)
2. Zeng, C.: Analysis on best practices of BW data warehouse based on SAP. China Sci. Technol. Inf. (16): 30–31 (2016). (in Chinese)
3. Cody, W.F., Kreulen, J.T., Krishna, V., et al.: The integration of business intelligence and knowledge management. IBM Syst. J. **41**(4), 697–713 (2002)
4. Brunel, R., Finis, J., Franz, G., et al.: Supporting hierarchical data in SAP HANA. In: 2015 IEEE 31st International Conference on Data Engineering (ICDE). IEEE (2015)
5. Plattner, H., Leukert, B.: The In-Memory Revolution: How SAP HANA Enables Business of the Future. Springer, Cham (2015). https://doi.org/10.1007/978-3-319-16673-5
6. Bai, V.S., Karthikeyan, P., Murugesan, S., et al.: Business application with SAP BI. In: Foundation of Computer Science (FCS) (2012)
7. He, Z., Li, L., Wang, C., et al.: Research and application of SAP report performance improvement method based on Hana. East China Sci. Technol. Acad. Ed. (9), 386–387 (2017). (in Chinese)
8. Liu, X., Wang, Y.: SAP business intelligence data incremental extraction mechanism and application. Sci. Technol. Wind (14), 88 (2011). (in Chinese)
9. Sun, J.F.: Research on report development in SAP system. Electric Pow. Inf. Technol. (2012)
10. Larson, D., Chang, V.: A review and future direction of agile, business intelligence, analytics and data science. Int. J. Inf. Manag. **36**(5), 700–710 (2016)

Application of Genetic Algorithm in Enterprise Marketing

LinHan Wang(✉)

School of Sport Science, Zhuhai College, Jilin University, Zhuhai 519000, Guangdong, China
yaoyao1037@sina.com

Abstract. According to the self-determination theory, in physical education, the self support environment created by teachers can meet the three psychological needs of autonomy, ability and belonging, thus promoting the formation of autonomous motivation. Therefore, based on the self-determination theory and its empirical research results, this paper explores how to create a self-supporting teaching environment in physical education. Based on the analysis of the theoretical basis of autonomous support environment, Reeve and his colleagues put forward five kinds of autonomous support teaching behaviors: cultivating intrinsic motivation resources; providing explanation; using information and non controlling language; admitting and accepting negative emotions; showing patience, and illustrating their application in physical education.

Keywords: Physical education · Self-determination theory · Independent support

1 Introduction

Sports are a valuable experience for children and youth. Regular participation in sports provides students with the skills, knowledge and values to achieve and maintain a healthy lifestyle. However, not all students enjoy playing sports. As adolescents enter adulthood, they become less motivated to learn in the classroom. Students who are less motivated, or whose motivation does not match the desired outcome of the course, are more likely to become dissatisfied with their learning experience and leave the class. For these students, physical education cannot meet their potential [1].

In physical education, the lack of students' motivation is not the latest problem. Many studies are trying to find a solution. Research based on autonomous learning theory shows that student motivation is influenced by the level of autonomy support in the classroom. Autonomy support is a powerful mechanism for student motivation because it promotes student autonomy and has a more lasting effect than other behavioral modifiers. Autonomous learners are motivated to learn because they are intrinsically motivated to learn, rather than because they feel pressure from external sources such as guilt, threats, or material rewards. External stimulation may be effective in the short term, but it fails when it dries up or when the goal is to maintain the behavior in the long term. Since the overall goal of physical education is to train students to participate

J. Macintyre et al. (Eds.): SPIoT 2021, LNDECT 98, pp. 953–956, 2022.
https://doi.org/10.1007/978-3-030-89511-2_133

in sports activities for life, teachers must understand the autonomous motivation and explore teaching strategies to promote autonomous learning.

2   Self Determination Theory: The Basis of Self Supporting Environment

Self determination is a kind of motivation view, which refers to the individual to engage in the behavior that they think is important or meaningful to personal development to improve their own wishes [2]. For example, in physical education classes, independent students work on learning tasks related to long-term physical behavior because they believe physical activity is important. Self determination theory provides a comprehensive explanation for students' learning motivation. Dec and Ryan described self-determination theory and students' motivation through three-phase linear process, including social environment, psychological needs and motivation level. Social environment refers to the development and implementation of supporting the psychological needs of students. The social environment is strongly influenced by the authoritative individuals (such as teachers), which can be divided into autonomous support or control. The emphasis of autonomous support environment is to support the teaching of students' will, control and sense of choice. On the contrary, the control environment only focuses on achieving teaching objectives through external measures (such as reward, guilt).

When contacting and participating in a certain social environment, students will perceive the satisfaction degree of three psychological needs: autonomy, ability and belonging. Autonomy refers to the individual's choice and control of a particular situation. Competence is a person's belief that he or she is capable of some level of learned behavior or action and is competent in that activity. Needs are related to relationships with others and belonging to a group, i.e., care, understanding, and support from the environment and others. In other words, a social environment that satisfies these three psychological needs will contribute to the internalization of extrinsic motivation.

3   Creation of Self Supporting Environment Based on Self Determination Theory

At first glance, it seems impractical to support students' autonomous learning, because the school environment is based on controlling principles such as students' behavior rules and students' expectation of complying with teachers' authority. However, it is important to realize that a structured learning environment can foster autonomy by providing opportunities for experiences that support student autonomy, such as choice, initiative and problem solving. Students are also more likely to accept guidance from their teachers if it aligns with their core values and what is required of them. Autonomy is not the same as independence, as children and adolescents tend to automatically rely on adults for guidance and decision-making [3].

The "meaningful teaching method" mentioned in this study should take Rogers and Ausubel's meaningful learning theory as its core concept. Meaningful learning means that learners can be responsible for learning, self-control, choose learning objectives and self-evaluation; be enthusiastic about learning and willing to continue learning; know how to transform knowledge and creatively solve practical problems; be good at collaborative learning or work. It is a kind of learning mode with learners as the main body and teachers as the leading. Only learners can really carry out meaningful learning, which is conducive to the development of higher-order ability, especially higher-order thinking ability. According to Jonassen, David h, meaningful learning has five characteristics: active, intentional, constructive, authentic and cooperative (Fig. 1).

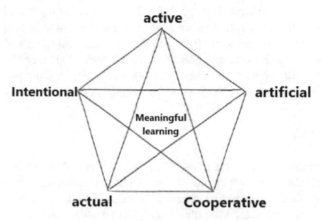

Fig. 1. Five characteristics of meaningful learning

The pricing law of the task is related to the distance from the task point to a certain center. We assign the random clustering center k = 20. For each task point, we find the nearest cluster center by calculating the distance, and put it into the class. At the same time, when the task points are put into the nearest cluster according to the distance, the cluster center will deviate, which will have a great impact on the clustering of the subsequent task points, so it is necessary to correct the cluster after each task point clustering. Therefore, the cluster center is defined as the member center, the member is the main body to complete the task, and the member is the object [4]. Through the average geographic location of the members, it is calculated as follows (1).

$$\bar{\phi} = \frac{\sum_{i=n}^{n} \phi_i}{n} \tag{1}$$

$$\bar{\lambda} = \frac{\sum_{i=n}^{n} \phi_i}{n} \tag{2}$$

4   Acknowledge and Accept Negative Emotions

In any class, the behavior of students may determine the teaching style and communication mode of teachers. Teachers try to use rules and regulations to create a safe and appropriate environment for teaching, but sometimes students do not follow, understand, or appreciate the process or demands of the lesson, leading to the expression of negative emotions. Therefore, it is important for teachers to find ways to deal with negative emotions. For example, showing negative body language (e.g., head bowing, depressed facial expressions) or failing to make an effort during classroom activities. A controlling teacher may use aggressive language or threats to motivate students or simply ignore the reasons why students try to avoid the learning process. Teachers who support autonomy, on the other hand, investigate students' negative behaviors. The teacher will approach and ask the students, "what's the matter? Let the students express their feelings. Teachers should be sensitive to students' feelings (e.g., "I understand how you feel") and, where appropriate, receptive to their feelings (e.g., "Let me think about what I can do to change things"). It is normal for students to have different feelings, and it is a natural part of the learning process. It is important that teachers do not prevent students from expressing negative feelings, but support them in expressing them through more positive channels (e.g., in class discussions or in regular short conversations with students).

5 Conclusion

At home and abroad, the core qualities of basic education and the goals of the physical education curriculum focus on the holistic growth of the learner and on "holistic education". There are three levels to understand the holistic growth of students: macro level, meso level and micro level. At the micro level, the physical education and health curriculum can promote the holistic development of students in two ways. General subject literacy Quality of physical education teaching.

Teaching methods in physical education refer to clear and effective ways for teachers and students to engage in learning activities in order to achieve specific learning objectives. There are three levels of teaching methods in physical education: macro, meso, and micro. The elements of physical education teaching mode include the relationship between teachers and students and various teaching elements reflecting teaching behavior, such as teacher-student language, teaching strategies, organizational forms, teaching methods, teaching means and teaching evaluation.

References

1. Fan, L.: Physical Education Teaching Theory. People's Sports Press, Beijing (2002)
2. Feng, J.: New Thinking of Teaching Methods: "Nine Character Universal" Model and Compatible Teaching Method. Guangxi People's Publishing House, Nanning (2004)
3. Gu, M.: Dictionary of Education, p. 200. Shanghai Education Press, Shanghai (1990)
4. Ji, L.: Physical Education Curriculum and Teaching Theory, pp. 21–24, 45. Guangxi Normal University Press, Guilin (2005)

Application of Genetic Algorithm in Enterprise Marketing

Junfeng Di[✉]

Department of Business Management, Laiwu Vocational and Technical College, No. 1 Shancai Street, Laiwu District, Jinan City 271100, Shandong Province, China

Abstract. Another application of genetic algorithm in forest product marketing is to summarize the distribution status of forest products and goods, point out that the distribution of forest products is a typical Vpr problem, and establish a mathematical model of the problem. By introducing the extended Liuhai crossing method, a new extended Liuhai crossing method with strong global search ability is constructed, and the original standard genetic algorithm is improved, The problem of optimizing the product distribution channel is presented and the strategy for solving this problem is based on the corresponding computational procedure, which allows a fast solution process of the above problem.

Keywords: Marketing · Genetic algorithm · Feature selection

1 Introduction of Genetic Algorithm

There are many ways to implement data mining information processing technology, including traditional computing technology and implementation.Generation of intelligent computing technology. In recent years, how to use computational intelligence technology to realize the key process of data mining has become a research hotspot of information processing technology. In this paper, the computational intelligence technology, especially the genetic theory, is introduced briefly.

Genetic algorithm is a new search and optimization technology, which simulates Darwin's theory of evolution, according to "the superior wins the inferior".With the help of selection, crossover, mutation and other operations, the optimal solution is gradually approached. It has implicit parallel mechanism and adaptability, so it is very suitable for multi-dimensional, nonlinear and multi peak problems [1]. Its main feature is that it uses a group search strategy, in which information is exchanged among individuals in a group, to represent complex phenomena using simple coding methods and replication mechanisms. It is not bound by restrictive assumptions about the search space, and does not require assumptions about continuity, the existence of derivatives, or one-to-one. Genetic algorithms are widely used in the fields of optimization, machine learning, and parallel data processing.

2 Basic Principle of Ant Colony Algorithm

2.1 Overview of Feature Selection

The task of feature selection is to select the m, (m < n) most effective features. The simplest method of feature selection is to select the most influential features for classification based on expert knowledge; the second method is to use mathematical filtering and comparison techniques to find the features with the most classification information. In order to accomplish the task of feature selection, two problems must be solved: one is the selection criteria that can be used to select feature groups to maximize separability. The other problem is to find a better algorithm that finds the best set of features in a short time.

2.2 Application of Feature Selection Based on Genetic Algorithm in Enterprise Marketing

It is mainly aimed at the feature selection of data rather than images, texts and rules. Its main purpose is to select some characteristic data that can represent all the information from a large amount of data, and use that characteristic data to build a solid database to accurately build future sales forecasting models, such as sales forecasting with time-series forecasting. In general, the number of samples is much larger than the number of traits. As you can see from the results of the analysis, the genetic algorithm uses a method that deals with multiple individuals in a population at the same time. This means that several solutions are evaluated simultaneously in the search space, which provides high global search performance and ease of parallelization [2]. In genetic algorithms, only the value of the fitness function is used to estimate genetic individuals, but the fitness function is not limited by continuous differentiability and has the advantage that it is fast and not subject to local extrema.

2.3 Preprocessing of Raw Data

Before using genetic algorithm for feature selection search, it is necessary to preprocess the original data, including eliminating outliers, normalizing the relative dynamic range, and dealing with missing data.

Laida criterion is the most commonly used criterion for judging and eliminating abnormal data. The basic idea of the Leyda criterion for assessing roughness is that it is based on a confidence probability given by a standard deviation limit of three times the measured value. If the error exceeds this limit, it is considered as roughness and not as random error. Measurements with large errors are called outliers and should be excluded from the measurement data.

If the total of the experimental data is normal, then

$$(p(|X - \mu|) > 3\sigma) \leq 0.003 \tag{1}$$

According to laida's criterion, the experimental data obeying normal distribution, if the residual error of a test value x_d:

$$V_d = x_d - \overline{x}(i = 1, \wedge, n, 1 \leq d \leq n) \tag{2}$$

Satisfy the following formula:

$$|V_d| > 3\sigma \tag{3}$$

Then x_d is rejected as an outlier. Where σ can be calculated according to Bessel formula:

$$\sigma = \sqrt{\sum_{i=1}^{n} (x_i - \bar{x})^2/(n-1)} \tag{4}$$

However, in computer programming, the following Bessel transformation formula is often used:

$$\sigma = \sqrt{\{\sum_{i=1}^{n} x_i^2 - (\sum_{i=1}^{n} x_i)^2/n\}(n-1)} \tag{5}$$

It should be noted that although it is widely used in scientific research, engineering and teaching to eliminate outliers by using the laida criterion, however, the scope of its application and how to eliminate outliers easily and quickly are rarely involved. Therefore, first of all, we need to make it clear that this criterion is applicable only when the number of measurements n is large, and at least $n > 10$ times. Otherwise, it is invalid to use this criterion. If $n \leq 10$, even if there are outliers with gross errors in the measurement column, they can not be judged and eliminated. Secondly, the raida criterion should be used repeatedly until no outliers are eliminated.

Because MATLAB is used as the programming tool in this paper, the mean function and STD function of Matlab can be used directly when taking the sample mean value α and sample mean square deviation α in the formula.

3 Application Design of Genetic Algorithm in Marketing

3.1 The General Design Idea of the Algorithm

The main goal and challenge of feature selection is to extract a small amount of data representative of the original characteristics of the data from a large set of sales data to provide reliable technical data for real-time performance and future forecasting of the entire forest products market [3]. In order to fully exploit the global exploratory power of genetic algorithms and their prominent role in the field of optimization, this work has been designed as follows:

Next, we divide the data into t subsegments and apply the genetic algorithm to each subsegment to find the maximum value. The initial population size of each subsegment is d/t. Finally, t attribute data are selected and one attribute is extracted for each subsegment.

Finally, we evaluate the accuracy of the feature data selection, i.e. the accuracy of the algorithm, by comparing the plots of the error function and curve fitting with the plots of the original data.

3.2 Selection of Fitness Function and Error Function

Selection of fitness function

$$F(x_i) = 2 - MAX + 2(MAX - 1)\frac{x_i - 1}{Nind - 1} \tag{6}$$

When applying the segmented genetic algorithm, the magnitude of the value of the error function is the criterion for selection. The individual with the smallest value of the error function is selected as the optimal solution, and this is the characteristic data selected in this section. Here, the value of the error is relatively small because the actual data requires an accuracy of 0.001, and the final error value is less than 0.00001, which is practically close to 0.

3.3 Coding Method and Generation of Initial Population

The data to be processed here has certain requirements: large data volumes, sales data, a wide range of values, dollar units, and even accuracy to three decimal places. Having presented and compared several coding methods described above, we can conclude that floating-point coding should be used. Floating point operations have significant advantages.

Initial populations have a great influence on genetic algorithms, and the generation of initial populations is one of the features of this section. Based on the advantages of floating-point coding, we decided to create loops in each section to further improve the computational efficiency of the genetic algorithm. Cycles and subscripts [3] allow us to select appropriate values of the original data as the initial generic model. Given the data accuracy requirements, the original data are successively decreased by 0.01 to form the lower limit of the original population in the first fielddr line, and the original data values are increased by 0.01 to form the upper limit of the original population in the second fielddr line, and then the initial fill function call statement is generated using floating-point code to have the Initial Fill created.

4 Conclusion

Combined with the specific example of product distribution route optimization problem, this chapter discusses the solution strategy of the real-time distribution route optimization problem. First, based on the mathematical description and ranking of the problem, the main research topics of this chapter are further defined and the mathematical model is constructed. Second, an extended bang crossover algorithm is proposed based on the standard genetic algorithm, as well as a corresponding improved genetic algorithm to eliminate the drawbacks inherent in the genetic algorithm, such as difficult choice of control parameters and fast convergence [4]. Finally, a computational program is developed to quickly solve the freight logistics routing optimization problem.

References

1. Yang, Z., Zhang, J., Peng, F.: Matlab Program Application and Problem Examples. Canghai Publishing House, Taichung (2000)
2. Liu, W.: Proficient in MATLAB and C/C++ Mixed Programming. Beijing University of Aeronautics and Astronautics Press, Beijing (2005)
3. Li, Q., Zhao, W.: Data Processing and Application of MATLAB. National Defense Industry Press, Beijing (2000)
4. David, R.A., et al.: Business and Economic Statistics. China Machine Press, Beijing

Molybdenum Disulfide Nanosheets for Efficient Hydrogen Evolution Reaction

Linxuan Chen(✉)

Ulink College of Shanghai, 559 South Laiting Road, Songjiang District, Shanghai 050023, China
chenlinxuan2020@yeah.net

Abstract. Replacing fossil fuels with renewable energy is of great importance to our daily life and build environmentally friendly society. Among all energy carriers, hydrogen is the most promising energy resource and attracts the most research attention. Compared to fossil fuel supplies and other energy carriers (CH4, C2H4), hydrogen reveals a variety of advantages, such as the clean final product (i.e. water), the various storable methods and many transportable ways.

Compared to bulk materials, nanomaterials with engineered effects and high area surfaces have offered more opportunities for energy related applications. Specially, two-dimensional MoS2 nanosheets which are graphene-like materials are more suitable for hydrogen evolution reaction.

Here, we synthesize MoS2 by a convenient hydrothermal method using KSCN and Na2MoO4 as sulfur and molybdenum precursors at 250 °C for 24 h. From the images of gate electron microscope and transmission electron microscope, it can be seen that the obtained MoS2 forms a nano probe or nano probe In 1m-coh electrolyte, the hydrogen activity on MoS2 nanoparticles was analyzed by three electrode system Considering that MoS2 film has ultra-high surface and unique electronic structure (Mo bit), MoS2 has higher characteristics and stability than MoS2, In particular, the number of excess particles on the MoS2 board is −50 mA/cm^2 390 mv. If oscillation occurs, mos2-510 mv.

The development of MoS2 nanosheets provides great insights and guidelines on the next-generation hydrogen evolution catalysts. Moreover, the improved hydrogen evolution activity will give a great boost to the improvement of electric vehicles for Tesla, BYD and so on. Resultantly, as in a new field, we are still facing some difficulties. As if we can develop the technology and skills more matured, we will succeed eventually.

Keywords: Renewable energy · Hydrogen evolution · MoS2 nanosheets · Electrocatalysts

1 Introduction

1.1 Renewable Energy

Renewable energy is an energy source that is continuously recycled and inexhaustible in nature, and is said to be relatively environmentally friendly compared to fossil fuels (1). Moreover, renewable energy is a primary energy in the process of energy development, which means that it is easy and direct to use (Fig. 1).

J. Macintyre et al. (Eds.): SPIoT 2021, LNDECT 98, pp. 962–966, 2022.
https://doi.org/10.1007/978-3-030-89511-2_135

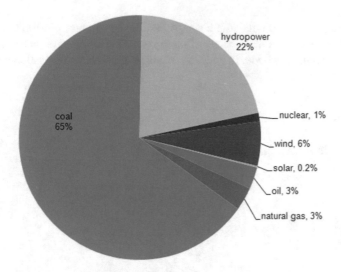

Fig. 1. Energy development for China *(1)*.

1.2 Hydrogen as an Energy Carrier

Hydrogen, as a kind of secondary energy,is a promising energy carrier. As we know that there are many advantages of hydrogen as an energy carrier. To start with, hydrogen has the smallest mass. Also, its thermal capacity is higher than most of the gas. *(40200 kJ/kg for liquid hydrogen, and 285.8 kJ/mol for gas hydrogen)*.

However, The trouble of making use of hydrogen still exists. For instance, the storage of hydrogen is not stable, It is easily exploded during the process. Moreover, the inefficiency of producing hydrogen is quite serious, because it will be costly if using available methods. Therefore, there is a need for us to find an efficient catalyst which can accelerate the speed of the reaction.

1.3 Approach for Chosen Synthesizing Hydrogen

One of the most advanced and cleanest methods of hydrogen production is splitting water with electricity. Globally, however, the industry is still small and slow to develop. It is still one of the most promising methods of producing hydrogen due to the simplicity of the production process, the fact that the only byproduct is water produced by the general reaction of hydrogen splitting, and the low energy consumption. Therefore, it is necessary to find an electrocatalyst that will increase its efficiency and speed, and it is worth investigating MoS2 and other catalysts.

1.4 Nanomaterials and Electrocatalysts for Hydrogen Evolution Reaction

In a broad sense, it is a general term for ultrafine particles in the nanoscale range in at least one three-dimensional space. Nanomaterials are materials whose structural units range in size from 1 nm to 100 nm. The main composition of molybdenite (MoS2) is a type of black solid powder with metallic luster. As for the advantages of MoS2 firstly, owing to its tiny structure, it is able to be equal to do some special operations like transferring medicine in the human's body. In addition, nanomaterial-made catalysts can accelerates the reaction faster.

2   Method

2.1   The Synthesis Method of MoS2 Nanosheets

See Fig. 2.

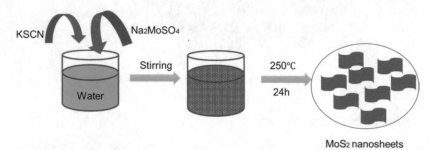

Fig. 2. Schematic drawing the synthesis of MoS2 nanosheets. KSCN and Na2MoSO4 are used as sulfur and Mo precursors, respectively. The reaction is carried out in the teflon-lined autoclave.

2.2   Equipment and Instruments

The morphology of MoS2 nanosheets was studied by scanning electron microscope (SEM); the SEM mechanism is shown in Fig. 3 and 4 below. We also used transmission electron microscopy (TEM) to observe the structure of MoS2 nanosheets and check whether prediction and reality match.

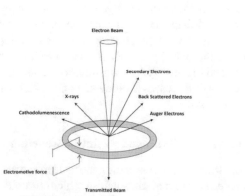

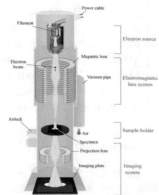

Fig. 3. The mechanism of SEM (5).   **Fig. 4.** The mechanism of TEM (6).

3 Result and Discussion

According to the record of the experiment, it reveals that MoS2 nanosheets indeed perform better compared to bulk MoS2.According to the LSV measurement, when the current density comes at -50 mA/cm^2,the potential of the bulk MoS2 is 510 mV, which shows the lowest efficient. The nanoscale MoS2 exhibits a potential of 390 mV at $-$ 50 mA/cm^2, which is 120 mV better than that of the bulk MoS2. This improvement reveals the advantages of nanomaterials. It is apparent that the MoS2 nanosheets with Pt has the best potential. It is a great proof of the nanosheet form of MoS2 indeed plays a better and more efficient role through the whole reaction as a catalyst for HER. More importantly, MoS2 nanosheets indeed give a great boost to the efficiency of the reaction, thus accelerating the speed of the reaction perpendicularly (Fig. 5).

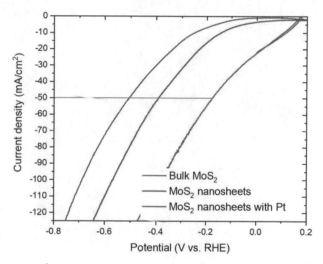

Fig. 5. Linear sweep voltametry measurements to determine the performance of MoS2 nanosheets

4 Conclusion and Outlook

In conclusion, it is shown that the performance of the MoS2 nanosheets are much better than bulk MoS2. As a result of all above, it is seen that MoS2 nanosheets indeed matches the need of a type of electrocatalyst. MoS2 nanosheets have the ability to be easily and safely applied into the reality. It can be used on the part of battery, which is able to accelerate the efficiency of water splitting using electricity, and there is going to be a better, more efficient, more tolerant, safer way to produce energy resources supporting the need of the reusable new energy vehicles, like Tesla, BYD, and so on as if its technique becomes matured.

References

1. https://www.e-education.psu.edu/ebf301/node/457
2. Sougou, et al.: https://baike.sogou.com/v141520556.htm
3. Sultan, S., et al.: Single atoms and clus-ters based nanomaterials for hydrogen evolution, oxygen evolution reactions, and full water splitting. Adv. Energy Mater. **9**, 1900624 (2019)
4. Chen, Y., Yang, K., Jiang, B., Li, J., Zeng, M., Lei, F.: Emerging two-dimensional nanomaterials for electrochemical hydrogen evolution. J. Mater. Chem. A **5**, 8187 (2017)
5. Koçak, A., et al.: Thin Film Preparation,Particle Size and Thickness Analysis
6. http://www.hk-phy.org/atomic_world/tem/tem02_e.html
7. Wang, H., Li, C., Zhang, J., et al.: Synthesis, properties, and optoelec-tronic applications of two-dimensional MoS2 and MoS2-based heterostructures. Chem. Soc. Rev. **47**, 6101 (2018)

Bayesian Decision for Nano Material Resolution Algorithm

Chengwen Zhang(⊠)

School of Electrical Engineering, Tongling University, Tongling 244000, Anhui, China

Abstract. With the rapid development of electronic industry information, notebook computers, mobile phones, tablet computers and other mobile electronic devices have almost covered every family, and become an indispensable part of people's daily life. However, with the rapid increase of the number of these wireless communication devices and high-frequency electronic devices, electromagnetic interference and electromagnetic pollution problems are becoming more and more serious. Therefore, around the design and development of new electromagnetic shielding materials, broadening the absorption frequency range of electromagnetic shielding materials, as well as lightweight research has become a research hotspot of scientists all over the world. Ni Zn ferrite has suitable permeability and dielectric constant, so it is a traditional microwave absorbing material. However, its high resistivity and narrow absorption bandwidth limit its application in many fields. In order to solve the above problems, a new two-dimensional layered transition metal carbides or carbonitrides (Menes) with high conductivity are introduced to improve the conductivity of ferrite materials and the impedance matching of composites.

Keywords: Keywords ZnO · Hydrothermal method · Coprecipitation method · Electromagnetic properties

1 Introduction

With the rapid development of modern science and technology, especially the harm of electromagnetic wave to the environment and the fourth generation of "stealth" fighters put into practice, electromagnetic compatibility, electromagnetic shielding and microwave absorption technology have attracted more and more attention of material researchers. Traditional microwave absorbing materials mainly include metal and ferrite materials, which have good mechanical properties and microwave absorption properties, However, it is difficult to be used as a single microwave absorbing material because of its high density and easy to corrode, It is widely used in electromagnetic shielding and microwave absorbing materials because of its conductivity, flexibility and machinability of conductive polymers, as well as the magnetic properties, mechanical strength and hardness of inorganic nanoparticles [1].

Among the conductive polymer materials, polyaniline (PANI) has been widely used in electromagnetic shielding materials, sensor materials, rechargeable batteries, metal

J. Macintyre et al. (Eds.): SPIoT 2021, LNDECT 98, pp. 967–971, 2022.
https://doi.org/10.1007/978-3-030-89511-2_136

anti-corrosion materials and microwave absorbing materials due to its simple synthesis, good environmental stability and unique electrochemical and physicochemical behaviors. Conductive polymer/ferrite nanocomposites not only have the advantages of flexibility and easy processing of organic polymers, but also have the mechanical strength and hardness of inorganic particles. At the same time, there may be synergistic effect between the two components, which has attracted extensive attention. Modern microwave absorbing materials are developing in the direction of dielectric and magnetic composite materials, because the composite of conductive polymer and ferrite can maintain the characteristics of two components, thus cutting the electromagnetic parameters of materials, broadening the microwave absorption band, reducing the material density, improving the absorption efficiency, improving the comprehensive performance, meeting the requirements of military stealth and microwave anechoic chamber, Therefore, the research of conductive polymer/ferrite composite nanomaterials has become a hot spot.

The preparation methods of core shell composite nanomaterials have been greatly enriched, including sol-gel method, ultrasonic chemistry method, layer by layer self-assembly method, microemulsion method and so on. The inverse microemulsion method (W/O type) is one of the effective and simple methods to prepare nanocomposites. In this method, nonpolar medium is used as continuous phase and water dissolved in reactants is used as dispersion phase to form isolated water droplets. These water nuclei are used as reaction sites, namely micro reactors, so the particles formed by the reaction are small and uniform, and the dispersion is good, and the reaction conditions are easy to control.

2   Test Methods and Raw Materials

2.1   Hydrothermal Method

A certain amount of functionalized $Ti_3C_2T_xMXenes$ was dissolved in water to prepare a solution of certain concentration. After ultrasonic dispersion for 30 min, $Ni(NO)_3 \cdot 6H_2O, Zn(NO)_3 \cdot 6H_2O$ and $FeSO_4 \cdot 6H_2O$ were prepared into a salt solution of certain concentration according to the stoichiometric ratio of $Ni^{2+} : Zn^{2+} : Fe^{3+}$ (0.5:0.5:2). The mixed solution of $Ti_3C_2T_xMXenes$ and salt solution was magnetically stirred for 60 min. 1 mol/L NaOH solution was used as precipitant to adjust pH $= 10.5 \pm 0.1$, After stirring for 10 min, the samples were transferred to a Teflon lined reactor.

2.2   Microstructure Analysis

The microstructure of the composite was completed by scanning electron microscope. For the materials with poor conductivity, it is necessary to spray gold before testing. For the solid powder materials with good conductivity, it can be directly coated on the conductive adhesive for micro morphology test. Scanning electron microscope (SEM) is a kind of scanning electron microscope (SEM) produced by Fei company in the United States. It is equipped with energy dispersive spectrometer (EDS). The current parameter is 10 μA and the voltage parameter is 20 kV.

The model of TEM is jm 2100, which is produced by Japan Electronics Co., Ltd. the morphology, structure and composition of the powder were analyzed under high pressure of 200 kV.

The specific surface area tester used in this experiment is ASAP2020, which has been heated to 200 °C for 360 min at the rate of 10 °C/min, and then degassed. After degassing, nitrogen adsorption and desorption tests were carried out at the relative pressure P/P_0 of 0.01–0.99, and the temperature of liquid nitrogen was about 77 K.

3 Preparation of Composite Materials by Coprecipitation Hydrothermal Method

Graphene, as a pioneer of new two-dimensional sheet materials, has proved that two-dimensional sheet materials have good application prospects in microwave absorbing materials [2]. Menes is similar to graphene with excellent microwave absorbing properties for its unique nanosheet structure and good thermal conductivity. Moreover, Menes has a natural "accordion like" structure, which is not easy to agglomerate. At the same time, oh and f groups on the surface are the most suitable ligands for transition metal Fe/Ni/Zn ions. If ferrite particles formed by Fe/Ni/Zn are used to modify or functionalize lamellar Menes as microwave absorbing materials, good results can be achieved.

Because the hydrothermal synthesis of NiZn ferrite is in the alkaline environment (pH = 10.5), the hydroxide precipitation generated by Fe^{2+} is unstable in the alkaline environment, the dissolved oxygen in water can easily oxidize Fe^2, and the oxidation of Fe^{2+} and the formation of nio5 ZnO can be realized at low temperature. Therefore, the crystallinity of ferrite synthesized with Fe^{2+} iron source is better than that of iron source, Another key factor of low temperature hydrothermal reaction is to select the appropriate iron source.

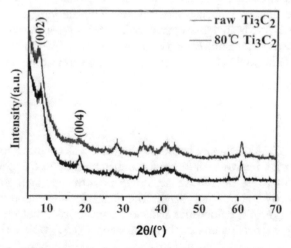

Fig. 1. XRD patterns before and after hydrothermal reaction for 4 h

It has been proved that the low-temperature coprecipitation hydrothermal reaction is feasible. The optimal temperature of hydrothermal synthesis is explored in the later part, so the experiment of ferrite synthesized at different temperatures is supplemented.

One is that we choose to synthesize NiZn ferrites from Fe^{3+} salts instead of traditional Fe^{2+} salts. Under alkaline conditions, the oxidation of Fe2 and the formation of $Ni_{0.5}Zn_{0.5}Fe_2O_4$ can be realized at low temperature. Because Fe^{2+} leaves are unstable in alkaline environment, dissolved oxygen in water can easily oxidize Fe^{2+}. The reaction process is as follows.

Hydrolysis of Fe^{2+}:

$$Fe^{2+} + OH^- \rightarrow FeOH^+ \tag{1}$$

Oxygen transfer:

$$O_2(g) \rightarrow O_2(aq) \tag{2}$$

The addition of isopropanol in the hydrothermal reaction can reduce the dielectric constant of the reaction system, and the solubility of salt also decreases with the decrease of the dielectric constant of the system [3]. Therefore, the decrease of the dielectric constant of the hydrothermal system can improve the growth rate of NiZn ferrite particles. In addition, the addition of isopropanol reduces the boiling point of water and increases the reaction pressure. However, the addition of excessive isopropanol will lead to the decrease of the polarity of the hydrothermal system, which is unfavorable to the growth of NiZn Ferrite crystalline phase particles. A very important factor is the addition of proper amount of isopropanol in water.

Due to multiple magnetic resonance, some different resonance peaks were observed in the test frequency range [4]. It should be noted that the existence of high initial permeability (μ) in the real part of the composite is beneficial to obtain large μ value and good impedance matching characteristics. The imaginary part (μ") of permeability decreases with the increase of frequency, which is a typical ferromagnetic behavior. In some frequency ranges, the imaginary part of permeability is negative, which is due to the electromagnetic energy radiation caused by the movement of the composite charge in the electromagnetic field, so that the imaginary part of the permeability in some frequency ranges is negative.

4   Conclusion

The addition of $Ti_3C_2T_xMXenes$ improves the electrical conductivity and dielectric loss of Ni Zn ferrite, improves the impedance matching of $Ti_3C_2T_x/Ni_{0.5}Zn_{0.5}Fe_2O_4$ composite, improves the microwave absorbing performance at high frequency and widens the effective absorption band. The possible absorption mechanisms of $Ti_3C_2T_x/Ni_{0.5}Zn_{0.5}Fe_2O_4$ composites include magnetic loss, dielectric loss, conductivity loss, multiple reflection and scattering between $Ti_3C_2T_xMXenes$ layers.

References

1. Liu, S., Liu, J., Dong, X.: Electromagnetic Wave Shielding and Absorbing Materials [M]. Chemical Industry Press, Beijing (2006)
2. Kang, Q.: New Microwave Absorbing Material. Science Press, Beijing (2006)

3. Li, G., Liu, X., Senlan, L.I., et al.: Hydrothermal synthesis of octahedral spinel ferrite. Chem. Res. Appl. **21**, 220–224 (2009)
4. Mao, W., Fang, K., Wu, Q., et al.: Conductive polyaniline carbonyl iron powder composite absorbing material. Acta Comp. Mater. **22**(1), 11–14 (2005)

EOR Based on Multi-sensor Fusion Technology

Shaoning Yuan[✉]

Geological brigade in No.7 Oil Production Plant of
Daqing Oilfield Company Ltd., Daqing 163517, China
yuanshaoning@petrochina.com.cn

Abstract. Since the founding of the people's Republic of China, China's Shantian exploration and development have been making continuous progress, especially in the past 40 years after the reform and opening up, China's oilfield development technology has developed rapidly. This study is mainly based on the related work of No.6 oil production plant of Zhongyuan oil Park Branch. Through the research, it mainly analyzes and discusses the relevant measures of EOR in the late stage of old oil development due to high water cut, hoping that the obtained results can cause everyone's attention and attention.

Keywords: High water cut oilfield · Recovery factor · Problems

1 Introduction

Oil is of great significance to China's economy, military, politics, diplomacy and many other aspects. At present, China is making greater efforts to develop oil fields, and the remaining oil in the reservoir is gradually showing a decentralized state. In the new situation of oilfield development, it is particularly important to adjust the relevant technologies and links of oilfield development to make it more suitable for the characteristics of oilfield development and ultimately improve the recovery factor of water cut oilfield.

2 Computer Data Structure and Calculation Method

2.1 Array

Array is a data type in aggregate state, which groups variable conditions of the same type in a certain order.

This data structure is mainly used in the application of various programming languages, and can be divided into integer, character, floating point, pointer and other types according to the type of data elements [1]. On the calculation condition, arglist parameter is used as the threshold list assigned to the array elements contained in Varian. If this parameter is not technically managed, an array with zero length will appear. For example, you can create an array by.

In terms of calculation conditions, arglist parameter is used as the threshold list assigned to array elements contained in Varian. If this parameter is not technically

J. Macintyre et al. (Eds.): SPIoT 2021, LNDECT 98, pp. 972–976, 2022.
https://doi.org/10.1007/978-3-030-89511-2_137

managed, an array with zero length will appear. For example, you can create an array by:

$$DimAA$$

$$AA(0) = 10 \tag{1}$$

$$AA(1) = 20 \tag{2}$$

At the same time, in the application of type array planning scheme, we can adjust its array creation method in the relatively complicated array content, and form the following mode to realize the optimization of the overall array.

2.2 Screening and Data Analysis

Usually, in a computer, a user's personalized request is usually composed of a series of subtasks with execution order constraints. The necessary condition for a task to be executed is that its predecessor task has been executed. If the predecessor task has not been executed, the current task can only wait. If a task has more than one antecedent task, the task can be executed only when all antecedent tasks have been completed. An entry node without a son node is called an exit node, as shown in Fig. 1. There are seven nodes in the graph. Each circle represents a node, which corresponds to a task. Each vertex contains two parts, which represents the second price task. The teaching words in the lower box represent the calculation cost [2]. Each arrow represents a source node with a directed edge. The corresponding task is the forward task of the corresponding task of the tail node. The number on the edge represents the communication cost of the two tasks connected by the edge.

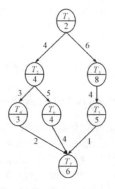

Fig. 1. Dag task diagram.

The first part of the system consists of an initial user request queue and three different types of user request queues. The initial queue is a first in first out (FFO) queue used to store the initial user requests of the system. When the user requests arrive at the system, they are inserted into the initial queue in order of their arrival. According to the obtained information related to user request, such as the request completion time, service quality, etc., the system analyzes these information quickly and accurately through big data analysis. According to the results of processing analysis, the importance level of user requests to users is determined. According to the importance level of user requests, the system allocates the user requests in the initial queue to the corresponding user request queue in turn. The most important user requests are dispatched to the queue tq1, the more important user requests are dispatched to the queue tq2, and the general user requests are dispatched to the queue tq3. The queue is used to store the corresponding three types of user requests with different importance levels, and assign the user requests to the corresponding scheduling queue in the second part.

3   Significance of Deep Oilfield Development

With the continuous development of oilfield exploitation, the exploitation speed of China's oilfields has been quite high, and the exploitation speed of the remaining oil fields (water bearing oilfields) has also reached more than 20%. In this context, if we only pay attention to the increase of production and ignore the improvement of recovery factor and recoverable reserves, we can only increase the speed of remaining oil production, which directly leads to the deterioration of oilfield production and the gradual reduction of oilfield energy, so this kind of production method is not scientific [3]. At the same time, the viscosity of China's onshore crude oil is very high, which is not conducive to exploitation to a certain extent, and its exploitation efficiency will not be too high even with the increase of water content. From here, we can also see that the amount of remaining oil in China is very large and has great exploitation value. To do a good job in the deep development of the oilfield, we need to constantly expand the water injection wave and volume, master the distribution of the remaining oil, and understand the characteristics of the remaining oil, so that we can continue to take the necessary measures to improve the production of the remaining oil on the premise of ensuring economic and efficient.

4   EOR Methods for Deep Development of High Water Cut Oilfield

4.1   Adjust the "Liquid" Structure of Oil Field Production

In the later stage of high water cut development in old oilfields, in order to enhance oil recovery, the hydrodynamic method of the former Soviet Union can be used to improve the direction of fluid flow and cyclic water injection, and fracturing can also be used to increase the liquid production of low water cut wells, and the wells can be sealed to reduce the liquid production of high water cut wells. Therefore, our country needs to speed up the research of these research and test methods. When carrying out cyclic water injection, we can analyze the different parts of water injection wells to carry out alternate water injection in batches. At the same time, we can apply a series of measures to improve the direction of liquid flow. We can also reduce the water injection volume first, and then adjust the liquid production. But the research of these methods need to continue scientific experiments and rigorous data simulation, to constantly sum up experience, in order to be able to more effectively improve the level. If the water content in the oilfield is constantly increasing, it is necessary to take water plugging, profile control and other ways to expand the volume of water injection sweep, so as to slow down the rising trend of water content. This way has obvious characteristics, which not only costs less, but also costs less, and can achieve significant results faster.

4.2   Optimize the Well Drilling Method

For oilfield drilling, it is necessary to fully consider the concentration and distribution of remaining oil, so as to determine the basic situation of drilling, which can be divided into two ways: non-uniform high efficiency well and uniform well distribution. If there are more layers in the well, the amount of remaining oil may be more. This method can effectively control the recoverable reserves. Moreover, this method is not only economic and accounting, but also has enough production capacity, which can be evenly infilled, and the uniform well pattern can be driven according to the polymer. When drilling a uniform well pattern, the remaining oil will be relatively small if the oil is evenly distributed to each well. And the productivity and economic profit of such single crystal can not meet the requirements of the standard better, and can not meet the standard. For the wells with high water content that may be drilled, it is not possible to adopt the method of uniform densification. It is necessary to add up the distribution map of sand body and small layer, from the overall point of view, and carry out the well drilling work of efficient adjustment wells from the places with abundant remaining oil, To a certain extent, this method can reduce the risk and improve the economic benefits. At the same time, although some oilfields do not seem to be able to develop.

4.3   Injection Production is Carried Out According to the Characteristics of Flow Unit

The so-called flow unit is the sand body flowing with each other in the reservoir. Each flow unit is different. Some flow units have no controlled corners. Some flow units have more water injection wells, but lack of necessary production wells. Some flow units have no water injection wells, and there are no production wells [4]. These problems are more serious. Therefore, it is necessary to ensure that the fine reservoir description can go deep into each sand body, and then divide the characteristics of the flow unit according to the situation and characteristics of these sand bodies. At the same time, the distribution of oil and water is analyzed, and the oil-water distribution map of flow unit is calculated. If necessary, the well may be drilled again. However, if the storage capacity of a flow unit is insufficient, the flow units of the well location need to be added up. When the reserves reach the standard, the well drilling work will be carried out.

5   Epilogue

In this paper, the author briefly analyzes the orientation of EOR in deep development of high water cut oilfield, and some subjective opinions are involved. At present, China has entered a new stage of high water cut and high recovery, so in order to improve the recovery of the oilfield, we must combine the characteristics of the deep development stage of the oilfield, and constantly take the necessary foundation and measures to improve its recovery. Only in this way can we better promote the recovery efficiency of the oilfield and promote the development of China's Oilfield work.

References

1. Jin, S., Wang, C.Q.: Research and application of nitrogen foam flooding technology for heterogeneous reservoirs in ultra-high water cut stage – comment on enhanced foam flooding technology for enhancing oil recovery. Xinjiang Geol. **37**(04), 585 (2019)
2. Song, Y.: Research and application of EOR technology in medium high permeability and high water cut oilfield. Tianjin, Dagang Oilfield Exploration and Development Research Institute of PetroChina, 13 December 2017
3. Fenglan, W.: research and application of key technologies for EOR in Daqing Changyuan ultra-high water cut oilfield. Heilongjiang Province, exploration and Development Research Institute of Daqing Oilfield Co., Ltd., 25 December 2014
4. Jielian, Y., Bin, H., Jiang, T., Zhi, G., Wang, L.: Research and application of EOR technology in high water cut oilfield. Inner Mongolia Petrochem. Ind. **04**, 91–93 (2006)

Application of ID3 Algorithm in Mental Health

Yanan Li[⊠]

Hubei College of Chinese Medicine, Hubei 434020, China

Abstract. Decision tree is a very popular data mining method. This paper introduces the theory of decision tree, analyzes the structure of decision tree, and discusses the idea of C5.0 algorithm and its advantages and disadvantages. At the same time, in order to deeply understand the main psychological symptoms and factors affecting college students' mental health, C5.0 algorithm is applied to college students' mental health evaluation data. According to the mining results, students' mental health problems can be more deeply understood, It is of practical significance to carry out mental health education for college students.

Keywords: Data mining · Decision tree · C5.0 algorithm · Mental health

1 Introduction

With the accelerating pace of life and the increasingly fierce social competition, college students are facing more and more pressure of study, life, emotion and employment. The resulting psychological problems are becoming increasingly prominent, which directly affect the healthy growth of students and the stability of campus. The mental health of college students has aroused the general concern of the whole society, The purpose of this study is to make up for the lack of traditional analysis and statistics of College Students' psychological problems, using data mining technology to study college students' psychological problems, so as to find the main psychological symptoms and factors that affect college students' psychological health, To provide a more scientific decision-making basis for school psychological counseling, to provide treatment programs for students with mental health problems, and to provide new methods of early prevention and intervention for college students' mental health, so as to make the school's mental health education more reasonable [1].

2 The Concept of Decision Tree

A decision tree is a tree structure similar to a flow chart. It is a kind of tree diagram used to represent a series of judgment processes for people to make a decision. This method is used to represent rules such as "what results will be obtained under what conditions". The decision tree is composed of decision nodes, branches and leaves [2]. The top node of the decision tree is called root node, Each branch is a new decision node or leaf node of the tree. Each decision node represents a problem or decision, usually

J. Macintyre et al. (Eds.): SPIoT 2021, LNDECT 98, pp. 977–981, 2022.
https://doi.org/10.1007/978-3-030-89511-2_138

corresponding to the attribute of the object to be classified. Each leaf node represents a possible classification result. In the process of traversing the decision tree from top to bottom, each node will encounter a test, Different test results on each node lead to different branches, and finally reach a leaf node.

3 Construction of Decision Tree

The core content of decision tree algorithm is how to construct a decision tree with high precision and small scale, which is generally divided into two steps: the first step is to generate the decision tree: the process of generating the decision tree from the data set (that is, the training sample set) divided during data preprocessing; the second step is to generate the decision tree from the data set (that is, the training sample set); The second step is decision tree pruning technology: the process of checking, correcting and modifying the generated decision tree. The main purpose is to prune those branches that affect the accuracy of pre balance. The data in the test data set divided by data preprocessing will play its existing value at this time, and verify the preliminary rules generated in the process of building decision tree from training set data, The smaller the size of the decision tree, the better, because it has predictive ability, the smaller the tree, the stronger, so we should try to build a small tree as much as possible.

3.1 Generation of Decision Tree

The generation process of decision tree is realized by the following steps:

(1) The further processing of training sample data is determined according to the actual needs and the characteristics of the processed data. The most important thing is to select the appropriate decision attribute, which can best reflect the particularity of the sample, and determine the value of each sample decision attribute.
(2) The current decision node of decision tree is selected based on the attribute that has the most classification and identification ability in the decision attribute set. Given the "index" in the training sample set, the best attribute is the attribute that has the most classification and identification ability [3].
(3) The training sample data set is divided into several subsets according to the different attribute values of the current decision nodes. If there are several attribute values, they form several subsets, because each value forms a subset.
(4) For each subset obtained in step 3, repeat steps 2 and 3 until the final subset is consistent. All tuples in the subset belong to the same category, the subset is obtained by traversing all decision attributes, and all remaining measurement attributes in the subset have exactly the same values, but the classification attributes are not the same, One of the three conditions is to stop further subset division according to these decision attributes.

3.2 Pruning of Decision Tree

The basic technique to overcome the noise in decision tree is to use pruning technique. Its purpose is to simplify the decision tree and make it easy to understand. Because the decision tree generation process adopts the top-down and divide and conquer strategy, the number of samples decreases with the increase of iteration depth. Although the time complexity of the algorithm is reduced, the overall distribution of samples is ignored, This is because in the deeper level of the sample partition algorithm, the statistical characteristics are concentrated in the subset of training samples. Due to the loss of general representativeness, it can not be used for the classification and prediction of new data, resulting in over matching phenomenon. Deleting the branches caused by noise data is the purpose of decision tree pruning, So as to avoid excessive matching of decision tree.

There are two common pruning methods for decision trees:

(1) Pre pruning: the most direct method of pre pruning is to specify the maximum depth of the growth of the decision tree in advance, which makes the decision tree unable to grow sufficiently. Another method is to test the tree nodes by using inspection technology to determine whether the corresponding branches of the decision tree are allowed to continue to grow, and a minimum allowable value can be specified in advance.

(2) Post pruning: Based on the sufficient growth of the decision tree, according to certain rules, the redundant leaf nodes or meaningless branches in the decision tree can be cut off. The feature of post pruning is pruning while checking. Its idea is: during the continuous pruning process of the decision tree, the data from the training sample set or the test sample set are usually used for checking, If there is a leaf node pruned, and the accuracy or other measures on the test set are still not reduced, then the leaf node is pruned.

4   Decision Tree Algorithm C5. 0

The earliest decision tree algorithm was put forward by hunter CLS (concept learning system). Later, Jr Quinlan put forward the famous ID3 algorithm in 1979, mainly for discrete attribute data, C4. 5 is an improved algorithm of ID3, which increases the discretization of continuous attributes, C5. 0 is C4. 5 classification algorithm applied to large data sets, mainly in the implementation efficiency and memory usage are improved.

4.1   C5.0 Algorithm Selection Criteria of Decision Tree Branches

C5.0 decision tree algorithm uses the information increment of attributes to determine the criteria of decision tree branches and find the best grouping variables and segmentation points. Let s be a sample set, and the target variable C has k classifications, which represents the number of samples belonging to class Ci, and s represents the number of samples of sample geometry S. then the information entropy of set s is defined as:

$$Info(S) = -\sum_{i=1}^{k} ((freq(C_i, S)/|S|) \times log_2(fre(C_i, S)/|S|)) \tag{1}$$

If an attribute variable t has n classifications, the conditional entropy of the attribute variable t is defined as:

$$Info = -\sum_{i=1}^{n} ((|T_i|/|T|) \times Info(T_i)) \tag{2}$$

The information gain brought by attribute variable t is as follows:

$$Gain(T) = Info(S) - Info(T) \tag{3}$$

4.2 Data Preprocessing

Data preprocessing is a very important part in the process of data mining. The data set processed by data mining usually not only has massive data, but also may have a large number of noise data, redundant data or incomplete data, such as students' missing or irregular filling in the test, which will lead to a large number of noise data in the database, Therefore, it is necessary to preprocess the data, which generally requires 70% of the workload in the mining process.

(1) Data extraction. Data extraction, also known as data sampling, can make the regularity and potential characteristics of data more obvious. In the test data, because the student number, name, answer of each question, test date and other attribute values are unique, mining these attributes has no significance. At the same time, the students tested are all grade 2019, and 97.8% of the students are Han nationality, which has no impact on the mining results, So delete these attributes.

(2) Data cleaning. Data cleaning includes missing value processing, abnormal data processing, noise data processing, duplicate data checking and data validation. The mental health assessment system for Chinese college students has dealt with some missing values of attributes, but the missing values of only child, student cadre, source, family structure and other attributes of students have not been dealt with, In this paper, we use the method of artificial filling, using most attribute values to fill the vacancy [4].

(3) According to the results of Chinese norm, the total score of SCL-90 of the test data is divided into four grades: 160 points or less is "Health (a)", 161 points or 200 points is "further examination (b)", 201 points or 250 points is "very obvious (c)", and 250 points or more is "more serious D" as the criteria of College Students' mental health; For "family monthly income" and other attributes that belong to continuous data, this paper divides them into three intervals according to the following: 2000 yuan is "low", 2000 yuan is "medium", 5000 yuan is "high".

5 Conclusion

This paper analyzes the structure of decision tree, discusses the idea of C5.0 algorithm and its advantages and disadvantages, and uses Clementine to construct C5.0 mining model of decision tree, and carries out data mining on College Students' mental health data. According to the mining results, this paper analyzes the main psychological symptoms and factors affecting college students' mental health, and gives some suggestions, It is very helpful to guide the relevant departments and personnel of mental health to make the correct counseling plan, assist decision-making, and pave the way for the development of students' physical and mental health.

References

1. Zhou, Y., Wang, S.: Medical Data Mining: A Case Study of SQL Server 2005. China Railway Publishing House, Beijing (2008)
2. Xiong, S., Nie, L.: Decision tree construction of student achievement analysis based on C5.0 algorithm. Sci. Tech. Inf. (8), 24–25 (2010)
3. Qi, W., Yan, J., Guo, L., Lu, R., Huang, S.: Application of association rules mining in college students' mental health evaluation system. J. Hunan Univ. Technol. **11**, 94–99 (2013)
4. Wu, X., Zhou, P., Peng, W.H.: Application of decision tree algorithm in college students' mental health evaluation. Comput. Appl. Softw. **201**(10), 240–244 (2015)

Author Index

J. Macintyre et al. (Eds.): SPIoT 2021, LNDECT 98, pp. 983–986, 2022.
https://doi.org/10.1007/978-3-030-89511-2